T0203939

ENAMINES
Synthesis, Structure, and Reactions

ENAMINES

Synthesis, Structure, and Reactions

SECOND EDITION,
REVISED AND EXPANDED

Edited by

A. Gilbert Cook

VALPARAISO UNIVERSITY
VALPARAISO, INDIANA

CRC Press
Taylor & Francis Group
Boca Raton London New York

CRC Press is an imprint of the
Taylor & Francis Group, an **informa** business

First published 1988 by Marcel Dekker, Inc.

Published 2019 by CRC Press
Taylor & Francis Group
6000 Broken Sound Parkway NW, Suite 300
Boca Raton, FL 33487-2742

© 1988 by Taylor & Francis Group, LLC
CRC Press is an imprint of Taylor & Francis Group, an Informa business

First issued in paperback 2019

No claim to original U.S. Government works

ISBN-13: 978-0-367-45139-4 (pbk)
ISBN-13: 978-0-8247-7764-7 (hbk)

This book contains information obtained from authentic and highly regarded sources. Reasonable efforts have been made to publish reliable data and information, but the author and publisher cannot assume responsibility for the validity of all materials or the consequences of their use. The authors and publishers have attempted to trace the copyright holders of all material reproduced in this publication and apologize to copyright holders if permission to publish in this form has not been obtained. If any copyright material has not been acknowledged please write and let us know so we may rectify in any future reprint.

Except as permitted under U.S. Copyright Law, no part of this book may be reprinted, reproduced, transmitted, or utilized in any form by any electronic, mechanical, or other means, now known or hereafter invented, including photocopying, microfilming, and recording, or in any information storage or retrieval system, without written permission from the publishers.

For permission to photocopy or use material electronically from this work, please access www.copyright.com (http://www.copyright.com/) or contact the Copyright Clearance Center, Inc. (CCC), 222 Rosewood Drive, Danvers, MA 01923, 978-750-8400. CCC is a not-for-profit organiza-tion that provides licenses and registration for a variety of users. For organizations that have been granted a photocopy license by the CCC, a separate system of payment has been arranged.

Trademark Notice: Product or corporate names may be trademarks or registered trademarks, and are used only for identification and explanation without intent to infringe.

Visit the Taylor & Francis Web site at
http://www.taylorandfrancis.com

and the CRC Press Web site at
http://www.crcpress.com

Library of Congress Cataloging-in-Publication Data

Enamines : synthesis, structure. and reactions.

   Includes bibliographies and index.
   1. Enamines. I. Cook. A. Gilbert
QD305.A7E53   1988   647'.04   87-22272
ISBN 0-8247-7764-6

Preface

The important renaissance of interest in enamine chemistry took place in 1954 with the first publication by Stork and co-workers on alkylation and acylation of enamines. Following that report, research in the field of enamine chemistry expanded dramatically. In the first edition of this book, the published reports of this research up through 1968 were reviewed and correlated. However, the amount of research in the area of enamines has continued to increase since 1968 at an accelerated rate.

There are several reviews that survey the more recent advances in enamine chemistry. But there seemed to be a need for a comprehensive review and correlation of the entire field of enamines. So this second edition of *Enamines: Synthesis, Structure, and Reactions* is presented as a comprehensive treatise on enamines. Six of the original eight chapters have been updated and rewritten by the editor, a new chapter on oxidation and reduction of enamines was added, and the final two chapters were updated and rewritten by the original authors.

The editor wishes again to extend his thanks to each of the contributors to this book who so willingly gave of their time and talents, and he wishes to extend his personal gratitude to Dr. Nelson J. Leonard for initially stimulating his interest in enamines as well as for his continuing interest over the years. The editor also wants to thank his wife, Nancy, for her patient support during the completion of this project.

"And these are but the outer fringe of His works, how faint the whisper we hear of Him! Who then can understand the thunder of His power?" Job 26:14.

A. Gilbert Cook

Contents

Preface iii
Contributors vii

1 Structure and Physical Properties of Enamines 1

A. Gilbert Cook

2 Methods and Mechanisms of Enamine Formation 103

Leroy W. Haynes and A. Gilbert Cook

3 Hydrolysis of Enamines 165

E. J. Stamhuis and A. Gilbert Cook

4 Electrophilic Substitutions and Additions to Enamines 181

G. H. Alt and A. Gilbert Cook

5 Oxidation and Reduction of Enamines 247

A. Gilbert Cook

6   Ternary Iminium Salts   275

    Joseph V. Paukstelis and A. Gilbert Cook

7   Cycloaddition Reactions of Enamines   347

    A. Gilbert Cook

8   Heterocyclic Enamines   441

    Otakar Červinka

9   Application of Enamines to Synthesis of Natural Products   531

    Gowrikumar Gadamasetti and Martin E. Kuehne

Index   701

Contributors

G. H. ALT Research Department, Agricultural Division, Monsanto Company, St. Louis, Missouri

OTAKAR ČERVINKA Department of Organic Chemistry, Prague Institute of Chemical Technology, Prague, Czechoslovakia

A. GILBERT COOK Department of Chemistry, Valparaiso University, Valparaiso, Indiana

GOWRIKUMAR GADAMASETTI Department of Chemistry, The University of Vermont, Burlington, Vermont

LEROY W. HAYNES Department of Chemistry, The College of Wooster, Wooster, Ohio

MARTIN E. KUEHNE Department of Chemistry, The University of Vermont, Burlington, Vermont

JOSEPH V. PAUKSTELIS Department of Chemistry, Kansas State University, Manhattan, Kansas

E. J. STAMHUIS Department of Chemical Engineering, The University of Groningen, Groningen, The Netherlands

1

Structure and Physical Properties of Enamines

A. GILBERT COOK *Valparaiso University, Valparaiso, Indiana*

I.	Introduction	1
II.	Electronic Structure	3
III.	Framework Structure	27
	A. Thermochemistry and Isomerization	27
	B. Enamines of Cyclic Ketones	34
	C. Enamines of Acyclic Ketones and Aldehydes	54
IV.	Physical Properties of Enamines	60
	A. Infrared Spectra	60
	B. Ultraviolet Spectra	66
	C. Nuclear Magnetic Resonance Spectra	66
	D. Mass Spectra	74
	E. Basicity	77
	References	83

I. INTRODUCTION

The term "enamine" was first introduced by Wittig and Blumenthal [1] as the nitrogen analog of the term "enol." Enamine chemistry was the subject of study by Mannich and Davidsen [2] in 1936, but it was the

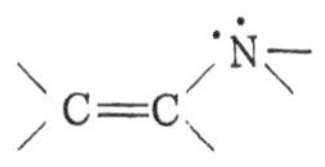

enamine 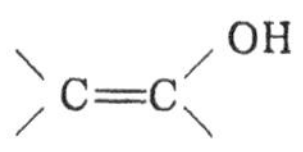   enol pioneering work of Stork and his co-workers [3,148] that strikingly brought it to the attention of the organic chemistry world.

The two most significant features of tertiary enamine reactions that make them so useful are: (a) the nucleophilic nature of the β-carbon atom that makes it susceptible to electrophilic attack (1); and (b) after addition of an electrophile to the β-carbon atom to form an iminium ion,

(1)   (2)

the resultant electrophilic character of the α-carbon atom making that carbon a good target for nucleophilic attack (2). β-Carbon protonation is very often the electrophilic addition involved in this second feature since the proton is a frequently used catalyst in organic reactions. The unique characteristics of enamines that enable them to have both these reaction features in the same molecule are well summarized by Jencks [4a] as he describes the importance of enamines in enzyme reactions. In speaking of the second feature of nucleophilic attack at the α-carbon of the iminium ion (2), he states:

> Although nitrogen itself is not strongly electronegative, it can act as an effective electron sink in such reactions by virtue of the fact that it is easily protonated and can form cationic unsaturated adducts easily. The ease of formation of these charged compounds more than makes up for the low electronegativity of nitrogen, so that nitrogen is generally more effective as an electron sink than oxygen, which is more electronegative but forms a cation at neutral pH with such great difficulty as to be relatively inactive as an electron acceptor by this mechanism.

Then he describes the first feature of electrophilic attack at the β-carbon atom (1), by saying:

> Conversely, the ease with which nitrogen can donate an electron pair permits it to act as a sort of low-energy carbanion; the enamine can easily donate electrons to form a new bond to a carbon atom (1), whereas the formation of a true carbanion is a much higher energy process.

Enamines are very sensitive to changes in the electronic environment caused by substituents being added to the system, and they are very sensitive to steric effects such as nonbonded interactions and stereoelectronic effects. These sensitivites are demonstrated in their reaction pathways being changed by relatively small electronic or steric modifications.

II. ELECTRONIC STRUCTURE

The electronic structure of an enamine can be most simply represented by the Lewis structure with contributing resonance forms indicated below by (3a) and (3b). For a shorthand notation of enamines which shows something of the electronic nature of enamines, this Lewis struc-

$$
\begin{array}{ccc}
\underset{\displaystyle \begin{array}{c}\diagup\\ C=C\\ \diagup\quad\diagdown\end{array}}{\overset{\displaystyle \diagdown\ddot{N}-}{}}
& \longleftrightarrow &
\underset{\displaystyle \begin{array}{c}(-)\ \diagup\\ :C-C\\ \diagup\quad\diagdown\end{array}}{\overset{\displaystyle \diagdown\ \overset{(+)}{N}-}{}}
\end{array}
$$

(a)   (b)

(3)

tural description serves very well. However, at a more sophisticated level this simple description of an enamine does not indicate the relative importance of (3b) as compared to (3a) for a given enamine molecule. It also tends to imply a geometric coplanarity of the alkene and amine portions of the system which is usually not present.

The chemical properties of an enamine are largely a function of the extent to which the lone-pair electrons on the nitrogen atom are delocalized into the π-system of the α,β-carbon-carbon double bond. Normally an amine has a pyramidal geometry with bond angles around the nitrogen of 109.5°. This implies sp^3-hybridization involving three of the hybrid orbitals forming σ-bonds with the three substituents, and the fourth sp^3-hybrid orbital being occupied by the lone-pair electrons. Normally, the two unsaturated carbons and their four substituent atoms in an alkene describe a plane. Each of the two alkene carbons is sp^2-hybridized, with the π-bond consisting of a molecular orbital described by orthogonal, overlapping p-orbitals. This π-molecular orbital is occupied by a pair of electrons.

In order to achieve *full* delocalization of the lone-pair of nitrogen electrons into the alkene π-system for the enamine ground state, the plane formed by the nitrogen and its two nonalkene substituents must be coplanar with the two carbons and four substituents of the alkene. Such coplanarity requires sp^2-hybridization for the nitrogen atom. The three sp^2-hybrid orbitals form σ-bonds with the three substituents, and the lone-pair electrons occupy the remaining p-orbital. This p-orbital is then orthogonal to the nuclear $\underset{\diagup}{\overset{\diagdown}{C}}=\underset{\diagdown}{\overset{\diagup N}{C}}$ system and parallel to the p-orbitals in the alkene π-system. This allows for maximum overlap of the parallel p-orbitals and delocalization of the nitrogen lone-pair electrons into the alkene π-system.

In an elegant piece of work, von Doering [8] and co-workers have shown experimentally that conjugative interaction in enamines does require the nitrogen lone-pair orbital and the alkene π-system orbitals to be parallel in order to maximize overlap. This was done by using the enamine, Δ^2-1-azabicyclo[3.2.2]nonene (4), and the allyl amine, Δ^3-1-azabicyclo[3.2.2]nonene (5), in which the nitrogens are forced to be pyramidal and the lone-pair electron orbitals are orthogonal to the π-alkene orbitals. The amount of strain in each isomer should be about equal, and any difference in entropy is negligible. It has been established that tertiary enamines (in which conjugative interaction is possible) are more thermodynamically stable than corresponding tertiary allyl amines by factor of from 5 to 6 kcal/mol [4b,4c,4d,37] (see Section III.A). Allowing compounds 4 and 5 to equilibrate in two different ways, the equilibrium constant was found to be one. So this

(4)

(5)

experimentally substantiates the theoretical conclusion that delocalization of nitrogen lone-pair electrons into the π-alkene system disappears when their respective orbitals are perpendicular to each other.

Ionization potentials as obtained from ultraviolet photoelectron spectroscopy [5] are a good source of experimental information about the energies of occupied molecular orbitals. The relation of these data to molecular orbital energies is shown by Koopman's theorem [6], which states that vertical ionization energies of a molecule are identically equal to the negative orbital energies determined by self-consistent field calculations. Generally, the photoelectron (PE) spectra of simple enamines show two broad distinct bands as the lowest energy bands in their spectra. This is indicative of relatively localized electrons. The lower-energy band is attributed to the lone-pair electrons on the nitrogen, and the higher-energy band is due to the alkene electrons. Often vibrational spacings are found in this second band corresponding to carbon-carbon double-bond stretching frequency in the radical cation [7]. For example, 1-(N,N-dimethylamino)cyclohexene (6a) gives a PE spectrum with a broad, low-energy band at 7.56 eV (attributable mainly to the nitrogen lone-pair) and the sharper higher-energy band at 9.7 eV (attributable mainly to the C=C orbital) [33]. The PE spec-

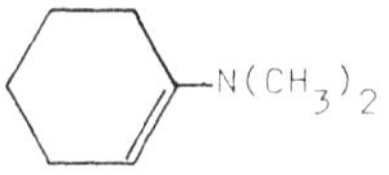

(6a) 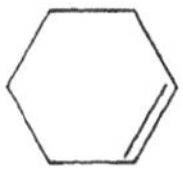 (6b) (6c) (7)

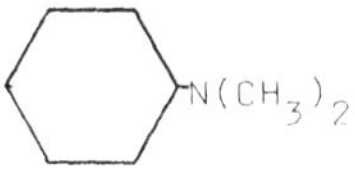

tra of the "component parts" of the molecule show 8.95 eV for cyclohexene (6b) [9] and 8.93 eV for dimethylamine (6c) [10]. The compound without the α,β-unsaturation, N,N-dimethyl-N-cyclohexylamine
(7), shows a PE spectrum band at 8.09 eV [33]. Thus, in being transformed into an enamine, the amine band undergoes a shift to lower
energies (8.21 or 8.09 eV to 7.56 eV) and the alkene band shifts to
higher energies (8.95 eV to 9.7 eV).

So from a knowledge of the general ordering of molecular orbitals
obtained from PE spectroscopy and semiempirical quantum mechanical
calculations using MNDO (modified neglect of diatomic overlap) [11],
a qualitative molecular orbital diagram showing the HOMO (highest
occupied molecular orbital) and the next lower in energy occupied
molecular orbital as well as the LUMO (lowest unoccupied molecular orbital) can be constructed. The molecule chosen to represent enamines
in this diagram is the simplest possible tertiary enamine, N,N-dimethyl-
aminoethene (8). Calculations for Figure 1 were made using ethylene π_u
and π_g fragments and an amine nitrogen p_z orbital, and forcing molecule
8 to be planar, but allowing every other independent parameter to vary to

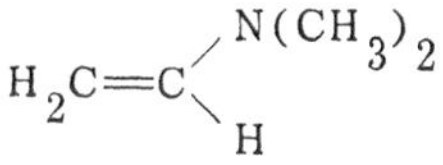

$$H_2C=C\begin{array}{c} N(CH_3)_2 \\ \\ H \end{array}$$

(8)

minimize the energy [12]. The orbital energy for the π_u-ethylene orbital is experimental [9], but all of the rest of the energies are from
MNDO calculations [12]. (The experimental IP for N,N-dimethylamine
was not used because an orthogonal p-orbital is desired for the planar
fragment, not a pyrimidal hybrid orbital.) Coefficients for the contributing atomic orbitals are shown next to each enamine orbital. It can
be seen that we have a typical interaction diagram involving an alkene
fragment with a π-donor system [12]. The lower alkene level is stabilized, and the higher amine nitrogen level (π-donor) is destabilized.
The destabilization of the upper level is greater than the stabilization
of the lower level, so the interaction is net destablizing. Since this
destabilizing interaction by a π-donor is maximal for planar geometry,
π-donors (in our case nitrogen) favor pyramidal over planar geometry
[13,14].

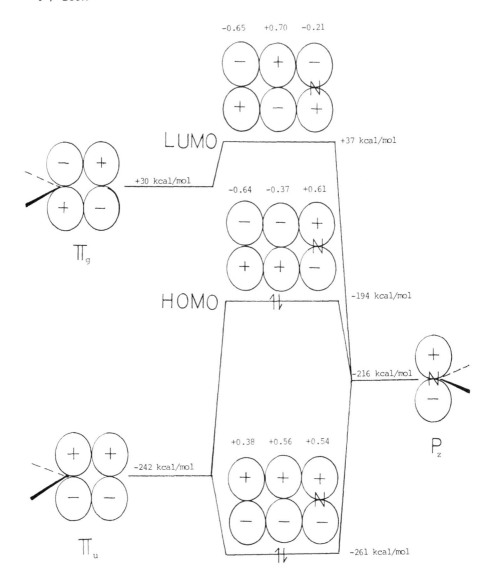

Figure 1 Interaction diagram construction of molecular orbitals for planar enamine.

Calculations on the simple primary enamine, vinylamine (9), have shown it to be nonplanar [15−18,44d]. An earlier microwave study of 9 led the investigators to the same conclusion [23]. This conclusion

$$CH_2 = CHNH_2$$

(9)

has been extended to tertiary enamines as well, based on the shape of bands in photoelectron spectra [7], on quantum chemical calculations [20−22], and on x-ray diffraction studies of crystalline enamines [24−26,47].

MNDO calculations on tertiary enamine (8) allowing all independent parameters to vary to minimize the energy also show a nonplanar enamine molecule [12,21]. This lowered the heat of formation of the molecule by 1.7 kcal/mol. The orbital energies of the LUMO and HOMO were lowered, and the energy of the next lower molecular orbital was raised [12]. The approximate resulting molecular orbitals (showing only major parallel p-orbital contributions) are shown in Figure 2 (the coefficients of contributing atomic p-orbitals are shown also).

The first IPs of amines and enamines can be correlated with the amount of s-character possessed by the nitrogen lone-pair orbital [27], namely, that as the first IP decreases, the amount of s-character in the lone-pair electron orbital also decreases. It has been observed that the values of first IPs of amines fall as one goes from secondary amines to the corresponding tertiary amines [28] (see Table 1). This decrease in IP takes place because replacing the hydrogen on the nitrogen with a more electropositive alkyl group causes the overall hybridization of the nitrogen atom to change in such a manner that the lone-pair electron orbital decreases in s-character. As a corollary to this, the σ-bond orbitals pointing toward the substituents simultaneously increase in s-character. This rehybridization is in accord with Walsh's rule [29] as restated by Bent [30] that increasing the electropositivity of the substituent on nitrogen decreases the s-character of the lone-pair [14]. It has also been found that the order of IP_1s for heterocyclic amines decreases in the order of increasing ring size, i.e., aziridine > azetidine > pyrrolidine > piperdine > hexamethylenimine [27]. This observation can also be explained on the basis of rehybridization of the nitrogen atom as the lone-pair orbital nitrogen decreases in s-character. The increase in ring size results in a decrease in s-character for the lone-pair orbital. We can now extent this use of IP_1 lowering to determine the decrease in s-character of the lone-pair nitrogen orbital to enamines. Increased amount of resonance interaction between the nitrogen lone-pair electrons (n) and the alkene π-system

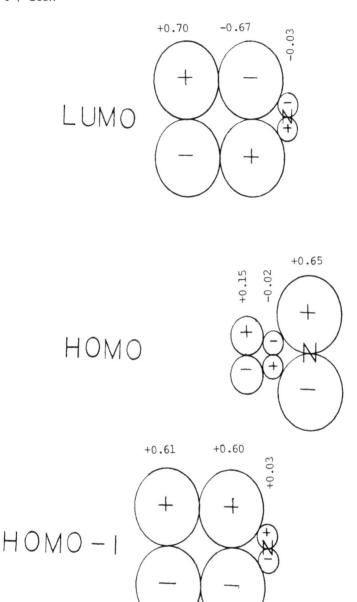

Figure 2 Molecular orbitals for nonplanar enamine 8.

Table 1 Vertical Ionization Potentials of Enamines, Amines, and Alkenes

Structure	R	IP_1	IP_2	Δ $(IP_2 - IP_1)$	Ref.
	cyclopentenyl	7.10	9.66	2.56	7
	cyclohexenyl	7.10	9.51	2.41	28
	$(CH_3)_2C=C-$	7.66	9.53	1.87	28
	$CH_3CH_2CH=CH-$	7.24	10.04	2.80	20
	$CH_3CH_2\overset{\mid}{C}=CHCH_3$	7.29	9.61	2.31	20
	H	8.77	—	—	27
	CH_3	8.29	—	—	27
	cyclopentyl	8.00	—	—	38
	cyclohexyl	7.96	—	—	28
	$(CH_3)_2CHCH_2-$	8.17	—	—	28
piperidinyl (N-R)	cyclopentenyl	7.4	9.55	2.15	7
	cyclohexenyl	7.42	9.31	1.89	28
	$(CH_3)_2C=CH-$	7.95	9.25	1.30	39
	$CH_3CH_2CH=CH-$	7.46	9.95	2.49	20
	$CH_3CH_2\overset{\mid}{C}=CHCH_3$	7.61	9.42	1.81	20

Table 1 (Continued)

Structure	R	IP$_1$	IP$_2$	Δ (IP$_2$ − IP$_1$)	Ref.
	H	8.64	—	—	27
	CH$_3$	8.29	—	—	27
	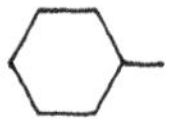 ⬡—	7.93	—	—	28
	(CH$_3$)$_2$CHCH$_2$—	8.16	—	—	28
	CH$_3$	7.57	10.40	2.83	37
	CH$_3$	7.49	10.15	2.66	32
	CH$_3$	8.24	—	—	32
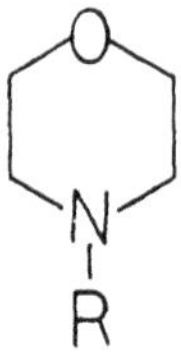	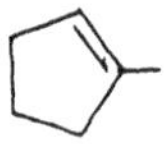	7.60	10.08	2.48	7
	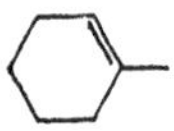	7.66	9.42	1.76	39
	(CH$_3$)$_2$C=CH—	8.20	9.41	1.21	40
	H	8.88	—	—	7
	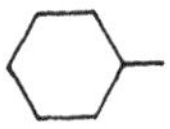 ⬡—	8.18	—	—	28
	(CH$_3$)$_2$CHCH$_2$—	8.46	—	—	28

Table 1 (Continued)

Structure	R	IP_1	IP_2	Δ $(IP_2 - IP_1)$	Ref.
	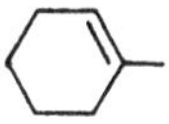 (cyclohexenyl)	7.46	9.76	2.30	20
(azetidine N–R)	$(CH_3)_2C=CH-$	7.56	9.97	2.41	20
	$CH_3CH_2CH=CH-$	7.62	10.17	2.55	20
	$CH_3CH_2\overset{\displaystyle |}{C}=CHCH_3$	7.48	9.79	2.31	20
	H	9.04	—	—	27
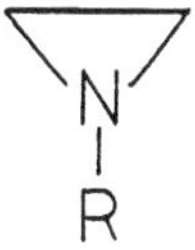 (aziridine N–R)	$CH_2=CH-$	8.75	11.05	2.3	43
	$(CH_3)_2C=CH-$	8.2	10.5	2.3	20
	$CH_3CH=CH-$	8.4	10.7	2.3	43
	$CH_3CH_2CH=CH-$	8.26	10.58	2.32	20
$(CH_3)_2N-R$	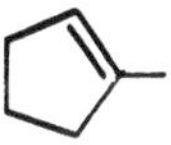 (cyclopentenyl)	7.46	9.76	2.30	33
	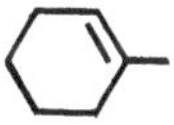 (cyclohexenyl)	7.56	9.70	2.14	33
$(CH_3)_2N-$	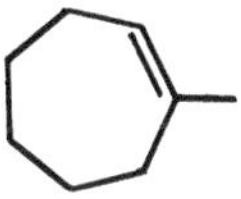 (cycloheptenyl)	7.57	9.50	1.93	33
	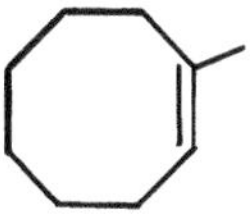 (cyclooctenyl)	7.50	9.55	2.05	33
	$(CH_3)_2C=CH-$	8.15	9.50	1.35	20
	$CH_3CH_2CH=CH-$	7.57	10.28	2.71	20

Table 1 (Continued)

Structure	R	IP_1	IP_2	$(IP_2 - IP_1)$	Ref.
	$CH_3CH_2\overset{\mid}{C}\!=\!CHCH_3$	7.61	9.69	2.08	20
	H	8.9	—	—	20
	CH_3-	7.82	—	—	9
	⬠-	8.09	—	—	33
	⬡-	8.34	—	—	33
$(CH_3CH_2)_2NR$	$(CH_3)_2C\!=\!CH-$	8.0	9.2	1.2	41
	CH_3CH_2-	7.50	—	—	9
$CH_2\!=\!CHR$	$H-$	10.51	—	—	5a
	CH_3-	10.01	—	—	42
⬡-R	$H-$	8.95	—	—	9

(π) results from decreased s-character for nitrogen orbital. So a lowering of IP_1 should mean greater n-π resonance interaction. This has been observed in many cases [28,32] (see Table 1). For example, N-cyclopentylpyrrolidine (10) has an IP_1 of 8.00 eV, whereas 1-(N-pyrrolidinyl)-cyclopentene (11) has in IP_1 of 7.10 eV, a decrease of 0.9 eV in going from a saturated amine to an enamine.

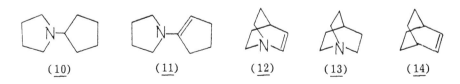

(10)　　　　(11)　　　　(12)　　　　(13)　　　　(14)

However, in using this method, one must take into account other effects that influence IP_1 values, such as simple inductive-field effects. The magnitude of the other sources has been studied using the enamine dehydroquinuolidine (12) and the corresponding saturated amine quinuclidine (13) [33]. Dehydroquinuclidine (12) is an α,β-unsaturated amine in which resonance interaction between the lone-pair nitrogen electrons and the alkene π-system is impossible because the potential interacting orbitals are orthogonal to each other. Many of the chemical and physical properties of 12 differ markedly from those of normal enamines [34]. Unsaturated amine 12 has a lower energy orbital for its lone-pair electrons than does its saturated counterpart 13 by 0.42 eV (8.44 eV for 12 versus 8.02 eV for 13). This stabilization is due to the inductive-field effect of the π-bond. Note that this inductive-field effect is in the opposite direction (saturated amine → enamine, IP_1 increase or n-orbital stabilization) from resonance interaction stabilization (saturated amine → enamine, IP_1 decrease or n-orbital destabilization).

As an interesting aside, the second ionization potential (IP_2) of enamine 12 (9.41 eV) is higher than its carbocyclic analog, bicyclo [2.2.2]oct-2-ene (14) (9.01 eV), by an amount similar to the IP_1 lowering discussed above. This means that the inductive-field effect of the nitrogen on the alkene system is similar in magnitude (about 0.4 eV) and direction (stabilizing the system) as the inductive-field effect of the alkene system on the nitrogen lone-pair orbital.

This method was also used to determine the amount of resonance interaction between an amine group and a benzene ring in a series of substituted anilines. Aniline was taken as the reference point for complete participation (IP 10.48 eV), and cyclohexylamine was used for the compound having no interaction (IP 8.53 eV) [35].

An extension of this method is the use of the bond separation between IP_1 and IP_2 for an enamine as a rough measure of the relative amount of interaction between an amino group and the double-bond unit. $\Delta IP(IP_2-IP_1)$ should be a maximum when the interaction is the greatest [20]. This argument is completely supported by quantum-chemical calculations [20,12] and is supported to a lesser degree by experimental observations (see Table 1). In contrast to the nonenamine-like character of dehydroquinuclidine (12), N-methyl-1,2,3,4-tetrahydropyridine (15) shows a very large ΔIP of 2.83 eV (7.57 and 10.40 eV), indicating probable strong n-π interaction [37]. This is reason-

(15a)

(15b)

able since compound 15's most stable conformer is the half-chair form
(15b) in which maximum overlap can take place with no rotation about
the carbon-nitrogen bond.

The rotational barrier about the C—N bond in enamines has been
determined theoretically [44d] and experimentally [36]. The barriers
seems to be on the order of 4—6 kcal/mol. Rotational barriers are
another good method for determining conjugative interaction in enamines.

Some other experimental methods that can be used to determine the
amount of n-π participation in an enamine molecule are ultraviolet spectroscopy, ^{1}H, ^{13}C, and ^{15}N nuclear magnetic resonance spectroscopy.
This topic will be discussed in later sections of this chapter that deal
specifically with those kinds of spectroscopy as applied to enamines.

Other experimental data that can be indicators of n-π interaction
are bond distances. As the s-character of a bond increases, the bond
becomes shorter [38]. In terms of enamine n-π interaction, as the
amount of interaction increases, the nitrogen atom σ-bonds increase in
s-character; that is, they approach sp^2-hybridization. Hence the
C—N bond distances shorten. Another way of describing this phenomenon in terms of resonance is to say that the relative importance of
resonance contributing form 3b becomes more important and form 3a
less important as n-π interaction increases. This means more double-
bond character to the C—N bond (and a shorter bond distance) and
less double-bond character to C=C (and a longer bond distance).
This parallel between bond distance and n-π interaction (using decreasing pyramidality at the nitrogen atom as a measure of n-π interaction)
was observed in the x-ray crystallographic studies of several crystalline enamines [25]. There is a pronounced shortening of the enamine
C—N bond distance (as n-π interaction increases) from about 1.42 to
about 1.38 Å. The C=C distance stayed almost constant, showing
only a slight lengthening with increased n-π interaction.

Quantum mechanical calculations of organic molecules such as enamines have reached a stage of sophistication at which they are very use-

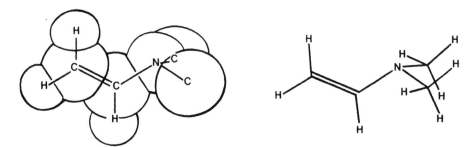

Figure 3 N,N-Dimethylaminoethene.

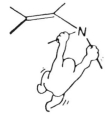

Figure 4 Distortion from planarity of an enamine. (From Ref. 19.)

ful in showing how certain chemical and physical properties arise, and they reliably predict some untested properties. Many calculations showing the extent of enamine n-π interaction along with other properties have been made [7,12,15−22,31,43−46]. The geometrical picture of ground state N,N-dimethylaminoethene (8) as calculated using the molecular mechanics semiempirical method [234] is shown in Figure 3.

There are two geometrical parameters that determine how much n-π interaction can take place. In other words, they describe how closely the unpaired electron orbital on nitrogen parallels the p-orbitals in the alkene π-system. These geometrical parameters are: (a) the pyramidality around the nitrogen atom (when the enamine geometry is planar, the nitrogen atom is sp^2-hybridization with an orthogonal p-orbital for long-pair electrons; when the enamine geometry is pyramidal, the nitrogen atom is sp^3-hybridization with a 109.5° sp^3-orbital for lone-pair electrons) (see Figure 5); and (b) the torsional twist around the C—N bond (see Figure 6). When the nitrogen atom is planar (completely nonpyramidal), the sum of the three bond angles is (3) (120°) = 360°. When the nitrogen atom is completely pyramidal, the sum of the three bond angles is (3) (109.5°) = 328.5°. So the percent pyramidality of a nitrogen atom is defined as $\dfrac{360 - \Sigma \text{ three N-bond angles}}{360 - 328.5}$ (100). This book on enamines deals primarily with simple tertiary hydrocarbon-substituted enamines. However, even though the chemical preparations and reactions of the non-hydrocarbon-substituted enamines is beyond the scope of this book, it is important to review the electronic changes brought about in enamines by the presence of these substituents. We will use structure 16 as our generalized template for a substituted

$$Y-CH=C{\overset{\displaystyle N-W}{\underset{\displaystyle X}{}}}^{R'}$$

(16)

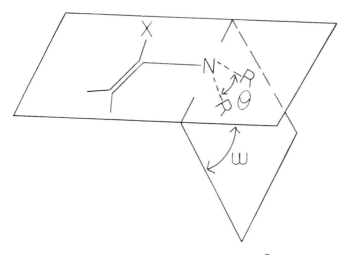

Figure 5 Enamine nitrogen pyramidality. $\ominus$ is $\angle$ RNR, $\sqcup\sqcup$ is angle between C-N bond extended and $\angle$ RNR bisector.

enamine. MNDO calculations [11] for variously (E)-substituted N,N-dimethylaminoethenes are shown in Table 2 [12]. All the parameters were allowed to be optimized in these molecules. Experimentally determined geometries for some crystalline enamines are shown in Table 3 [24,25,47].

Some general observations concerning the MNDO calculations of Table 2 follow. The simple enamine N,N-dimethylaminoethene ($\underline{8}$; X═Y═H) possesses a great deal of pyramidality (56%) and a great deal of torsional twist about the nitrogen atom (80°). In order to have maximum overlap betwene the lone-pair nitrogen orbital and the alkene π-system, pyramidality should be 0% and the torsional twist should be 0° (see Figures 5 and 6). However, calculations indicate that forcing the molecule into a planar configuration raises the energy by 5.6 kcal/mol. On the other hand, forcing the molecule into a planar configuration increases both the net positive charge at the α-carbon and the net negative charge at the β-carbon. This phenomenon can also be seen in the cases of other substituted enamines which are nearly planar, such such as compounds $\underline{22}-\underline{26}$. Starting with the Y substituents, compounds $\underline{25}$ and $\underline{26}$ are typical enaminones [51,108], $\underline{25}$ being a vinylogous amide and $\underline{26}$ being a vinylogous urethane. The calculations show nearly planar molecules for this system, which indicates a great deal of electron delocalization. This delocalization of electrons results in

larger-than-expected dipole moments [48]. Protonation takes place
primarily at the oxygen atom [49,50]. The properties of these compounds have been reviewed [51]. Enaminethiones [52,53] and enamine-
selenones [54] have also been studied. β-Cyanoenamines [55] (compound 22) and β-nitrosoenamines [56] (compound 24) have been observed.

β-Nitroenamines present a class of compounds that have been
called "push-pull" ethylenes [57] because of the presence of a strong
electron-donating group on one end of the alkene and a strong electron-

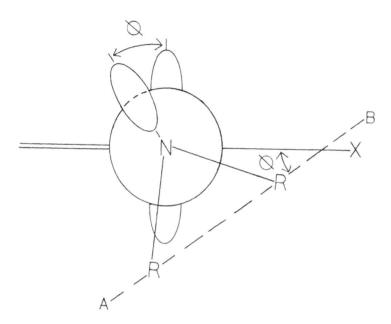

Figure 6 Torsional twist around an enamine carbon-nitrogen bond.
The points RNR represent a projection of these atoms on a plane orthogonal to the C—N bond. ϙ is the angle of torsional twist about the
C—N bond as described by the angle of intersection of line $\overline{AB}$ (which
intersects both R groups) and the NCX plane. The bisector of the
RNR obtuse angle is the lone-pair orbital geometry. So ϙ can also be
described as the torsional angle between the lone-pair orbital and the
lobe of the alkene p-orbital (orthogonal to the NCX plane) which gives
the smallest angle.

Table 2 MNDO Calculations on Some (E)-Substituted N,N-Dimethylaminoethenes[12] (All Independent Bond Distances, Bond Angles, and Dihedral Angles Allowed to Vary)

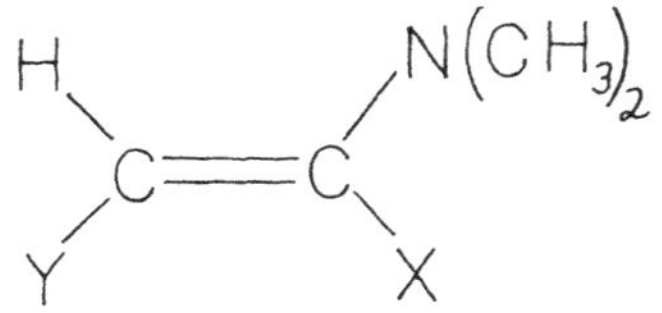

Cmpd. no.	X	Y	Pyramidal θ^a	Pyramidal (%)^b	Pyramidal $r_{=C-N}$ (Å)	Pyramidal $r_{C=C}$ (Å)	μ (Debye)	Charges α-C	Charges β-C	Charges $-N(CH_3)_2$	ΔH_f (kcal/mol)
(8)	H[c]	H	80°	57	1.44	1.34	0.78	0.01	−0.06	−0.09	20.2
(8[1])	H	H	0[d]	0[d]	1.42	1.36	1.11	0.09	−0.19	−0.06	25.8
(17)	H	CH_3	53	43	1.43	1.35	0.87	0.06	−0.14	−0.07	10.3
(18)	CH_3	H	89	45	1.44	1.36	0.75	0.06	−0.14	−0.09	13.5
(19)	CH_3	CH_3	87	28	1.44	1.36	0.55	0.04	−0.13	−0.10	6.6
(20)	H	F	22	31	1.42	1.36	2.67	0.04	+0.04	−0.04	−28.5
(21)	H	Cl	23	28	1.42	1.35	2.80	0.10	−0.09	−0.04	10.2
(22)	H	CN	5	9	1.40	1.36	4.80	0.17	−0.15	−0.01	47.0
(23)	H	NO_2	0	0	1.39	1.37	7.14	0.22	−0.23	+0.03	30.2

(24)	H	NO	2	0	1.40	1.37	4.69	0.20	−0.29	−0.01	23.1
(25)	H	—CCH$_3$ (=O)	1	9	1.41	1.35	5.89	0.18	−0.31	−0.05	563.3
(26)	H	—COCH$_3$ (=O)	14	8	1.40	1.37	5.29	0.17	−0.13	−0.00	103.7
(27)	(CH$_3$)$_2$N—	H	79	16	1.42	1.35	3.81	0.09	−0.02	−0.07	497.0
(28)	F	H	89	41	1.44	1.36	1.10	0.21	−0.08	−0.07	−29.9
(29)	Cl	H	89	39	1.41	1.34	1.54	0.10	−0.03	−0.02	8.9

a Absolute value taking smallest angle with respect to C=C—N plane (see Figure 6).

b Defined as $\left(\dfrac{360 - \Sigma \text{three N—bond angles}}{360 - 328.5}\right)$ (100).

c Also reported by Refs. 21 and 22.

d Molecule held rigid as far as being planar is concerned. Bond distances and N—CH$_3$ rotations allowed to vary.

Table 3 Geometry of Crystalline Enamines from X-ray Crystallography

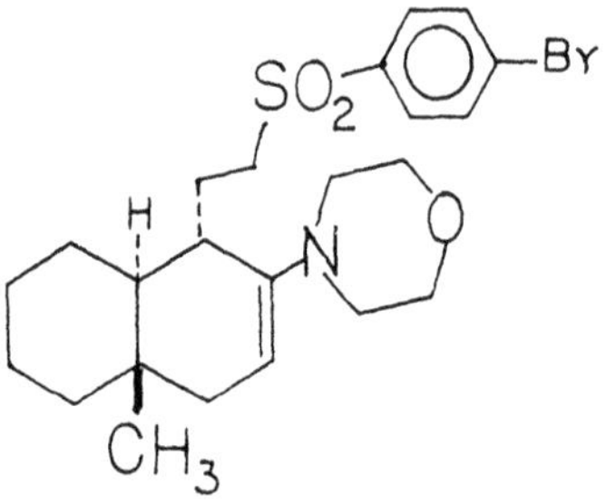

(30)

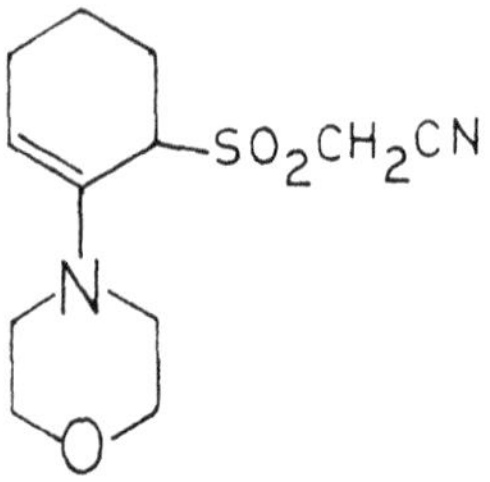

(31)

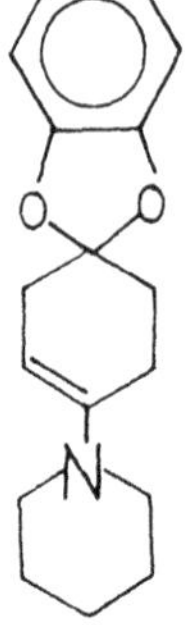

(32)

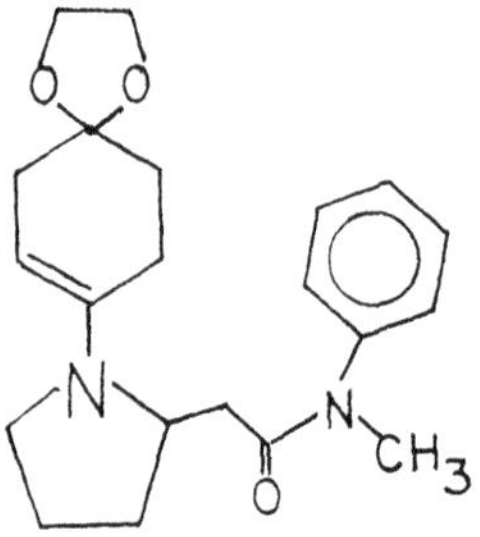

(33)

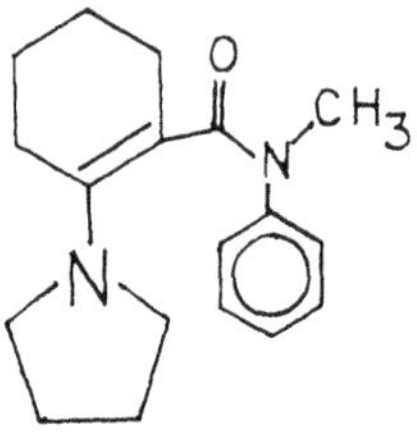

(34)

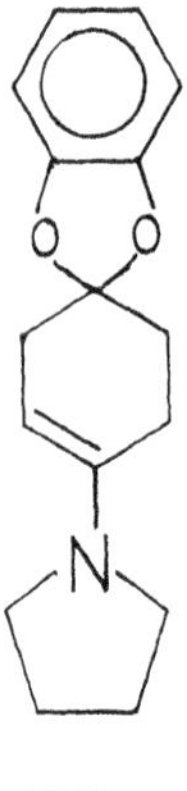

(35)

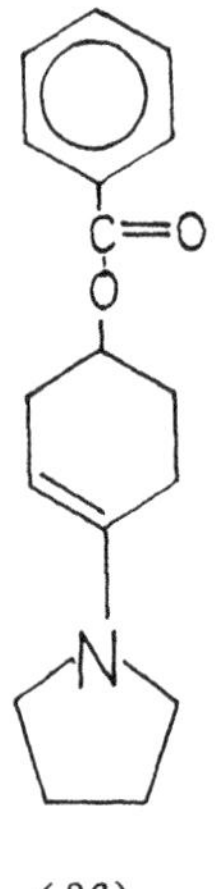

(36)

Table 3 Geometry of Crystalline Enamines from X-ray Crystallography

Compd. no.	ϑ^a	Pyramidal[b] (%)	$r_{=c-N}$ (Å)	$r_{c=c}$ (Å)	Ref.
(30)	46	70	1.42	1.35	24
(31)	36	60	1.426	1.334	47
(32)	26	47	1.410	1.342	25
(32)[c]	25	43	1.412	1.337	25
(33)	13	17	1.395	1.334	25
(33)[c]	11	7	1.385	1.348	25
(34)	4	15	1.393	1.349	25
(35)	5	3	1.380	1.366	25
(36)	3	0	1.381	(1.412)	25

[a]Absolute value taking smallest angle with respect to $>C=C<$ N plane (see Fig. 5).

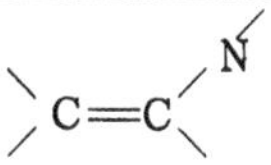

[b]Defined as $\left(\dfrac{360 - \Sigma \text{three N—bond angles}}{360 - 328.5}\right)$ (100).

[c]Two crystallographically independent molecules of 32 and 33.

withdrawing group at the other end. This means that resonance contributor 23a is very important in describing compound 23 [107]. Calculations show it to be completely planar with a very large dipole mo-

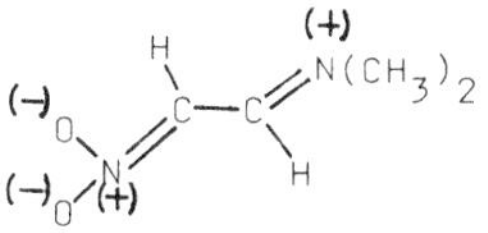

(23a)

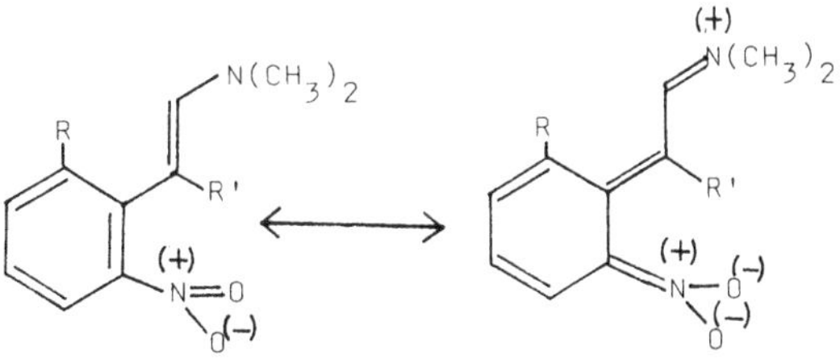

(30)

a: R=H, R'=H
b: R=NO$_2$, R'=H
c: R=H, R'=CH$_3$

ment (see Table 2). Enamine (30a) is deep red and is found experimentally to be a very stable molecule [116]. It undergoes acid hydrolysis only under very vigorous conditions. In enamines 30b and

30c such resonance interaction is sterically inhibited, and they are readily hydrolyzed to aldehydes.

The situation is very similar to that of p-nitroaniline (31) in which

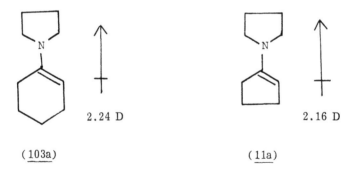

(31a)   (31b)

there is direct conjugate interaction between amino and nitro groups. This type of directly interacting groups on the benzene ring has resulted in aromatic molecules that deviate from behavior described by the normal Hammett substituent constants, σ. Consequently, new sets of modified substituent constants, σ⁻ and σ⁺, have to be used in the Hammett equation [58], indicating the importance of contributing resonance form 31a. Aniline, like enamine 8, has a lot of pyramidal character. Its angle of declination, ω (see Figure 5), is 39°21' [59] for a 72% pyramidal character. The amount of pyramidal character is greatly diminished by the presence of a p-nitro group [60], and the energy barrier for rotational twist about the Ar—N bond is increased significantly [61].

The dipole moments for the pyrrolidine enamines of cyclohexanone (103a) and cyclopentanone (11a) have measured at 25° in benzene solution, and they lie between 2.1 and 2.3 Debye [463,459].

2.24 D

2.16 D

(103a)   (11a)

The dipole moment for p-nitro-N,N-dimethyl aniline is 6.93 Debye, which indicates a 1.48-Debye interaction moment; i.e., the moment is that much greater than would be expected based on group moments alone with no interaction taken into account [62,63]. Applying this method to calculated and experimental dipole moments of β-nitroenamine 23 shows interaction moments of 3.18 D [12] and 2.90 D [463], respectively.

The infrared spectra of β-nitroenamines show unusually low values for symmetrical nitro vibration of $1250-1280$ cm^{-1} [457]. These values are $70-80$ cm^{-1} below the values for nitroalkenes. The high degree of electronic polarization present in these molecules is reflected in extremely weak Raman bands for all major functions groups [457].

The nitrogen-15-nmr spectrum of nitroenamine 23 shows a very large downfield shift of 66 ppm relative to the simple enamine 8. This is due to the strong deshielding effect of the nitro group on the enamine nitrogen [407]. The chemistry of β-nitroenamines has been reviewed [57,130]. β-Nitroenamines are less reactive toward electrophilic attack than are simple alkyl-substituted enamines [57,64]. It appears (from Table 2) that the β-carbon of β-nitroenamine 23 has a greater negative charge than the β-carbon of the simple enamines 8 or 17. Therefore, one would expect compound 23 to be more reactive toward electrophiles than simple enamines. However, this would only be true if such an electrophilic attack has an early transition state according to the Hammond postulate [65] or the Bell-Evans-Polyanyi principle [66]. It seems probable that the reaction between an electrophile and a β-nitroenamine has a late transition state, and if this is true, then the transition state would resemble the product. The cation product obtained from addition of an electrophile to the β-carbon of a β-nitroenamine would be expected to be a high-energy intermediate because the conjugate interaction between the enamine moiety and the nitro group, present in the enamine, has been destroyed in the iminium ion product. Calculations show that the energy difference between enamine and iminium ion product is about 24 kcal/mol greater in the case of β-nitroenamine 23 than in the case of simple enamine 8. So the activation energy for electrophilic attack of β-nitroenamine 23 is on the order of magnitude of 24 kcal/mol greater than the activation energy for electrophilic attack of simple enamine (8). This would explain the reluctance of β-nitroenamines to react with electrophiles as compared to simple enamines.

Some other β-substituted enamines have the following [67–71] substituents for Y: Y = R_2N-(1,2-enediamine) [87], ArSe- [72], $(CH_3)_3$

(+)

Si- [72,73], CH_3S- [74,75], $(CH_3)_2$S--enaminosulfonium salts seem to

(+) (−)   (+)

possess a high degree of ylid character, i.e., $(CH_3)_2$S—C—C=NR$_2$) [76–78,152], halogen- [74,79–84,87], p—MeAr—SO$_2$—(tosylates) [85, 86,88], $(EtO)_2$P(O)- (diethylphosphonoenamine) [89], R_2P(S)-(3-amino-2-phospholene sulfide) [158], and R_2C=CR— (extension of the conjugated alkene system) [87]. When Y is lithium (β-lithioenamines), the nucleophilicity of the enamine β-carbon is greatly increased to the extent that competitive N-alkylation vanishes [90,458]. Its reactions are stereospecific with retention of configuration [91]. Replacing X and Y [in (16)] with a π-bond results in ynamines (RC≡C—NR₂) [92, 93,167].

The X substituents (see 16) are (CH₃)₃Si-(α-trimethylsilyl) [94], R₂N- [64,87,95—97], CN- [26,98—100,252], RC(O)-(ketoenamine, chiefly amide in character) [101,102], and (EtO)₂P(O)-(α-diethylphosphonoenamine) [103—106,129]. α-Haloenamines (X = halide) are a particularly intriguing set of compounds [109—111]. They behave chemically as an equilibrium mixture of α-haloenamine (32) and keteniminium halide (33). As such, they represent synthons equivalent to both eno-

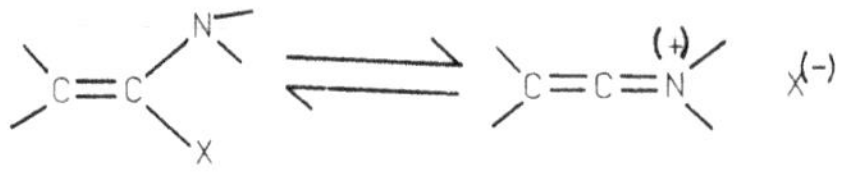

(32)   (33)

late and acylium ions derived from carboxylic acids. They show enhanced electrophilic character at the α-carbon and diminished nucleophilic character at the β-carbon. The extent of the enhancement or diminution can be regulated by which halogen is present. α-Iodoenamines display the most electrophilic character, and α-fluorenamines show the most nucleophilic character [112,113]. They have been well reviewed [114]. MNDO calculations on α-fluorenamine 28 and α-chloroenamine 29 are in good agreement with the experimental x-ray crystallographic studies of the two α-haloenamines 34a and 34b [26,114] (see Table 2) in several areas. First, the calculated C—Cl bond distance was lengthened from 1.75 to 1.78 Å in going from β-chloroenamine 21

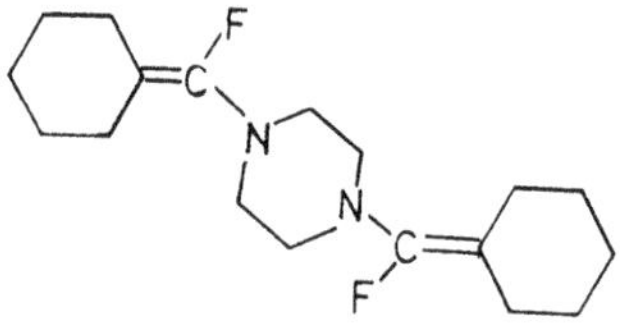

(34a) 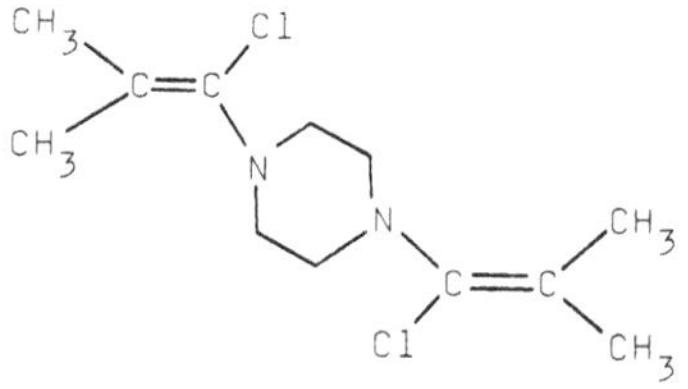   (34b)

to α-chloroenamine 29 [12]. A similar lengthening to 1.79 Å was observed in α-chloroenamine 34b. Second, the calculated torsional angle φ (see Figure 6) is 89° for both α-fluorenamine 28 and α-chloroenamine 29, which means that the lone-pair electron orbital on nitrogen is almost orthogonal with the alkene π-orbitals. The observed torsional angles φ for 34a and 34b are 73° and 83°, respectively. Finally, the calculated pyramidalities of 28 and 29 are 41% and 39%, respectively, comparing favorably (considering the difference in the amine moieties) with the pyramidalities of 34a and 34b as 66% and 65%, respectively.

α-Metallated enamines (X = metal) provide a good pairing with α-haloenamines for umpolung [115] (dipole reversal of reagent) at the α-carbon atom. α-Metallated enamines are good nucleophiles at C-1, and α-haloenamines are fine electrophiles at C-1 [117]. This class of compounds has been reviewed [87]. In this same general category of compound, γ-metallated enamines [(35)] have been synthesized and studied [118]. These may be considered homoenolate-type anions

$$(-)\ \underset{M^{(+)}}{C-C=C}\overset{N-}{\diagdown} \quad\longleftrightarrow\quad \overset{}{C=C-C}\underset{M^{(+)}}{(-)}\overset{N-}{\diagdown}$$

(35)

which have found great synthetic utility [119—124].

Substituents on the nitrogen other than alkyl or aryl also modified the nature of the enamine. When W = H (see 16) one has a secondary enamine or an imine. The equilibrium position of the tautomers (36)

$$\underset{(a)}{C=C}\overset{N-}{\diagdown}_H \quad\rightleftharpoons\quad \underset{(b)}{-C-C}\overset{N-}{\diagup}$$

(36)

depends on what other substituents are present [125,126], but for simple imines it lies almost entirely in form 36b. The chemistry of this group of compounds has been reviewed [127,147]. Other substituents on the nitrogen are: W = RC(O)- (enamide, thermally relatively un-reactive but photochemically very useful) [128], $(CH_3)_3Si-$ [131—136], NO_2-(N-nitrosoenamine) [137]. Metalloenamines (W = metal) have found many uses, undergoing many of the same reactions as simple enamines [127,138—143]. They are especially useful in asymmetrical syntheses [144]. Enamine N-oxides have also been synthesized [145].

The final factor that must be considered in discussing the electronic structure of tertiary enamines is the nature of the amine moiety. It was recognized by Stork in the very early work on enamines [148] that pyrrolidine enamines are more reactive toward electrophilic attack than piperidine enamines, which in turn are more reactive than morpholine enamines. The general order of reactivity of pyrrolidine > hexamethyl-

enimine > piperdine > morpholine has been observed by other workers
as well [151,169,170,224]. The magnitude of conjugative interaction
between amine and alkene in an enamine seems to parallel the reactivity
of the enamines toward electrophilic attack. This order is pyrrolidine >
dimethylamine > piperidine ~ azetidine > morpholine as shown by ionization potentials (see Table 1) [7,20] and [1]H- [20,50,149,150,153,224,
304,392] and ^{13}C-NMR [20,36,154−158] spectral analyses. In these
analyses it is assumed that increased shielding of the β-proton or β-
carbon, respectively, as shown by upfield chemical shifts, indicates
a greater amount of delocalization of nitrogen lone-pair electrons into
the alkene π-system.

In a study of a 3-amino-1-phospholene sulfide series (37) in which
the amino groups were pyrrolidine, piperidine, and morpholine, [31]P
NMR shifts were reported [158]. There was an upfield shift (greater
shielding) in the order morpholine, piperidine, pyrrolidine, reinforc-
ing the conclusion drawn from the other types of nmr spectroscopy.

(37)   (38)

The experimental data obtained from x-ray crystallography on seven
crystalline enamines (30−36, see Table 3) confirms this order of conjugative interaction also [24,25,47]. The two morpholine enamines
(30 and 31) show the greatest amount of torsional twist and pyramidality. The piperidine enamine (32) is next, and the pyrrolidine enamines
(33,34,35,36) show the smallest amount of these properties. In fact,
enamines 36 shows almost complete coplanarity between the amine system and the alkene system. A steady decrease in the alkene carbon-
nitrogen bond distances also follows this same order. In order to increase the amount of pyrrolidine nitrogen lone-pair electron participation in the alkene π-system, the isoxazolidine enamine 38 was synthesized and studied [159]. It was hoped that the lone-pair electrons on
the neighboring oxygen would cause repulsion of the nitrogen lone-pair
electrons, causing them to delocalize to a greater extent into the alkene
system. This was not observed, however, but rather a decrease in
conjugation was found due to the oxygen's inductive effect.

Use of azetidine enamines would seem to be a good compromise
between the amine being small enough to allow steric crowding in β-
dialkyl-substituted enamines, but not such a small ring that it cannot
have conjugative interaction with the alkene [20]. These kinds of
enamines have been synthesized [158], and some 3,3-dimethylazetidine
enamines (40) were studied [160,161]. The enamines formed much fas-

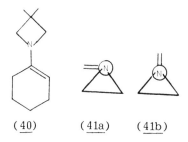

(<u>40</u>)   (<u>41a</u>)   (<u>41b</u>)

ter with this azetidine than with pyrrolidine, but the reaction properties were poorer than those of pyrrolidine, e.g., more N-alkylation
and less selectivity toward monoalkylation with the azetidine enamine
as compared to the pyrrolidine enamine.

In the three-membered ring, aziridine series, several acyclic
enamines were synthesized and studied [43]. The aziridino group has
a low π-donor capacity due to a structually and electronically enforced
high pyramidality at the N-atom with concomitant stabilization of the
lone electron pair. The simplest of these enamines, N-vinylaziridine
(<u>41</u>), exists as an equilibrium mixture of the *gauche* form (<u>41a</u>) and
the *trans*-bisected form (<u>41b</u>) with the former being the major component.

The observation, made on several enamines, for one of the bonds
emanating from the N-atom of an enamine to eclipse the C=C bond and
hence be syn-periplanar to the enamine C=C [20,25,162,299] was not
found in the MNDO calculations [12].

III. FRAMEWORK STRUCTURE

A. Thermochemistry and Isomerization

Equilibrium of allyl amines with enamines under basic conditions or in
presence of alkali metal has been extensively studied [37,44c,44d,163–
168,171–176,359,452,453]. With only a rare exception [177], the
equilibrium lies far to the side of the enamine [454]. For example N,N-
dimethylamino-2-propane (<u>42</u>) is isomerized with t-BuOK and DMSO
[163] first to *cis*-enamine <u>43</u> [164–166] and then to the thermodynamically more stable *trans*-enamine <u>44</u>. This type of isomerization was also
observed in the six-membered ring nitrogen-heterocyclic compounds
which already have a fixed *cis* configuration [4d,37]. For example, the

$$(CH_3)_2 \, N-CH_2CH{=}CH_2 \xrightarrow[\text{DMSO}]{\text{t-BuOK}} (CH_3)_2 \, N \diagup\!\!\!\diagdown \longrightarrow$$

$$CH_3 \qquad (CH_3)_2 \, N \diagdown\!\!\!\diagup CH_3$$

(<u>42</u>)   (<u>43</u>)   (<u>44</u>)

equilibrium between allylic amine 45 and enamine 46 lies far on the side
of the enamine with a ΔG of −4 kcal/mol. In a contest between phenyl

(45)   (46)   (47)   (48)

conjugative ability and that of the dimethylamineo group, the equilibrium
between 47 and 48 lies on the side of enamine 48 to the extent of ΔG =
−2.3 kcal/mol [178]. This isomerization has also been observed to take
place under thermal conditions [179]. Hine and associates have found
the dimethylamino substituent by far one of the best double-bond-stab-
ilizing substituents [178,180]. This is further born out by a comparison of the experimental [181] and MNDO-calculated [12] heats of formation of enamines and their corresponding allylamines found in Table 4.
The heats of hydrogenation are also listed in Table 4. The heats of
formation and hydrogenation both show the greater stability of the
enamine relative to the corresponding allylamine, even after correction
for differences of C=C substitution in the pairs of isomers [181]. Rearrangements of one enamine to an equilibrium mixture of isomeric enamines will take place under acid catalysis, but it will not take place
under neutral or basic conditions, as is discussed later [248,249].

Several [3,3] sigmatropic [183] rearrangements of enamines have
been reported [4b,184−191,193, 194]. For example, when 1-methyl-
2-(3-butenyl)-Δ²-tetrahydropyridine (49) was heated to 240−250° in a

(49)   (50a)   (50b)

sealed ampule for several hours, the product of 1,2-dimethyl-3-allyl-
Δ²-tetrahydropyridine (50b) [which is in tautomeric equilibrium with
its exocyclic form (50a)] was obtained in an almost quantitative yield
[4b]. A case of an enamine undergoing a Cope rearrangement is the
heating to 250° of enamine 51 (obtained from methyl vinyl ketone dimer
and pyrrolidine) resulting in the formation of 2-pyrrolidino-4-acetyl-
cyclohexene (52) [188,189]. Claisen [186], aza-Claisen [191], and

Table 4 Enthalpies of Formation and Hydrogenation of Some Enamines and Alkenes[a]

Compound	(R)	ΔH_f^0 (kcal/mol)	$\Delta H^0_{hydrogen}$ (kcal/mol)	Ref.
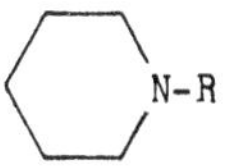 (N-R ring)	$CH_3CH{=}CH{-}$	−13.71	−23.65	181
	$CH_2{=}CHCH_2{-}$	−8.70	−28.66	181
	$CH_3CH_2CH_2{-}$	−37.36	—	181
	$CH_3CH_2CH{=}CH{-}$	−18.50	—	181
	$CH_3CH{=}CHCH_2{-}$	−13.16	−27.90	181
	$CH_3CH_2CH_2CH_2{-}$	−41.06	—	181
(cyclopentene)	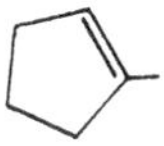	−11.70	−24.25	181
(cyclopentane)	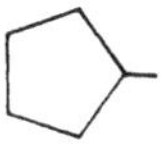	−35.95	—	181
(cyclohexene)	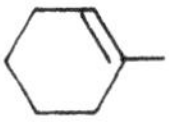	−23.86	−25.23	181
(cyclohexane)	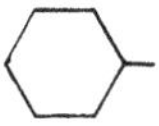	−49.09	—	181
$(CH_3)_2N{-}R$	$CH_3CH{=}CH{-}$	+10.26	—	12
	$CH_2{=}CHCH_2{-}$	+18.03	—	12
(cyclohexene)	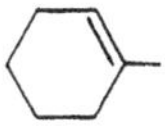	−1.28	−27.10	182,183
(cyclopentene)	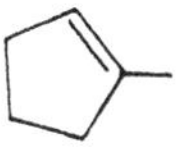	+7.73	−25.70	182,183

[a]All of the thermodynamic quantities listes here are experimental except those for 1-(N,N-dimethylamino)propene and 3-(N,N-dimethylamino)propene, which are MNDO calculations.

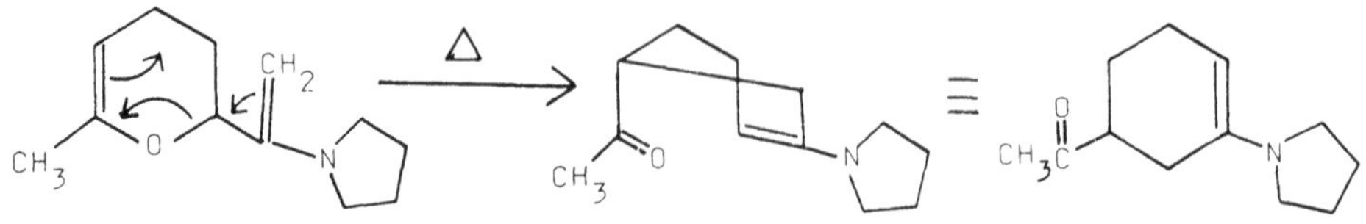

(51)   (52)

aza-Cope [185,223–228] rearrangements involving enamines have also been reported. A [1,3]-sigmatropic rearrangement of an enamine was observed when β-furfuryloxyenamine 53 was heated at 80° in an argon atmosphere to produce 2-pyrrolidino-3-(2-furyl)-2-methylpropanal (54).

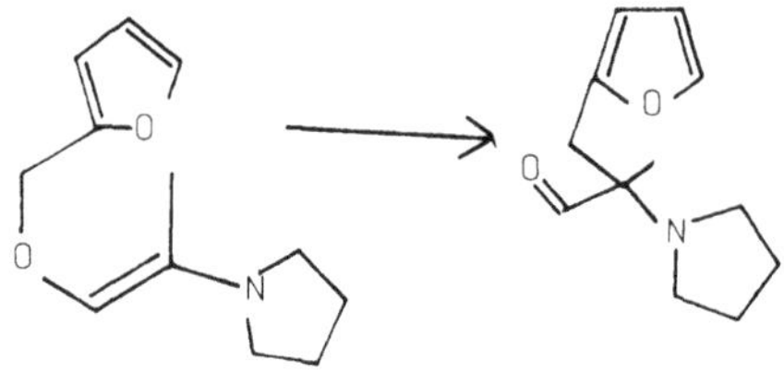

(53)   (54)

Nonconcerted rearrangements of N-alkylated enamines to their thermodynamically more stable C-alkylated enamine counterparts have been observed [148,223,224]. These arrangements have been shown to take place through a reversal of the N-alkylation followed by subsequent C-alkylation [225]. Methyl, ethyl, and benzyl groups are among the alkyl groups that have been observed undergoing these arrangements. N-Allyl-and N-propargylenamines probably rearrange through a [3,3]-sigmatropic mechanism (aza-Cope rearrangement), as mentioned above [185,223–228].

The cyclopropane ring, in some respects, resembles the double bond [196,201]. It is the "homoisomer" of ethylene. The prefix *homo* in nomenclature is used to indicate the addition of a skeletal atom to a well-known structure [197]. Winstein used the term homo in his concept of homoconjugation in which there is electron delocalization across intervening carbon atoms [198]. When a cyclopropane ring replaces the carbon-carbon double bond in an enamine, it can be considered a homoenamine. As such it should be tested to determine whether it undergoes enaminelike reactions with the nitrogen lone-pair electrons initiating the cyclopropyl ring opening. Cyclopropylamines have been found to be resistant to ring opening by acids or bases, and they do not readily react as homologous enamines with electrophiles [199]. However, heating of bicyclic [n.1.0]aminocyclopropanes such as 55 with aqueous alcohols gives the α-methylketone (56). The ring may also be opened by hydrides [200] and amines [339]. In comparison of the course of thermal rearrangement of isomeric chlorohomoenamines 57a and 57b, 57b gave products that seem to indicate enaminelike character since an electrocyclic ring opening pathway to the products obtained is sterically prohibited [202].

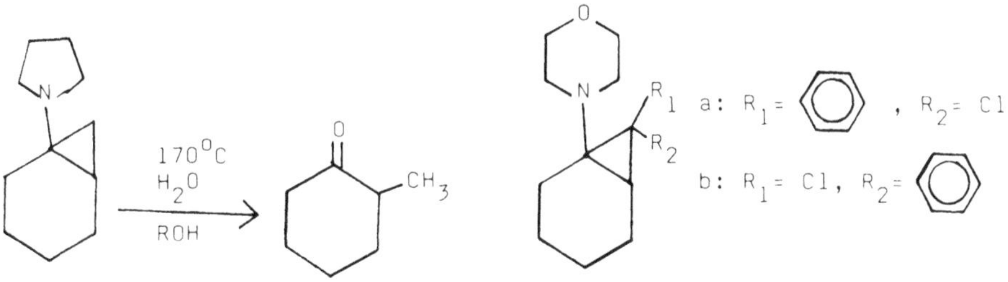

(55)   (56)   (57)

Cyclopropyl cyanoamine 58 rearranges under the mild conditions of AgBF$_4$/DME followed by LiBr in acetonitrile to give enammonium salt 61 via iminium salt 59 and enamine 60 [203]. This reaction has as an

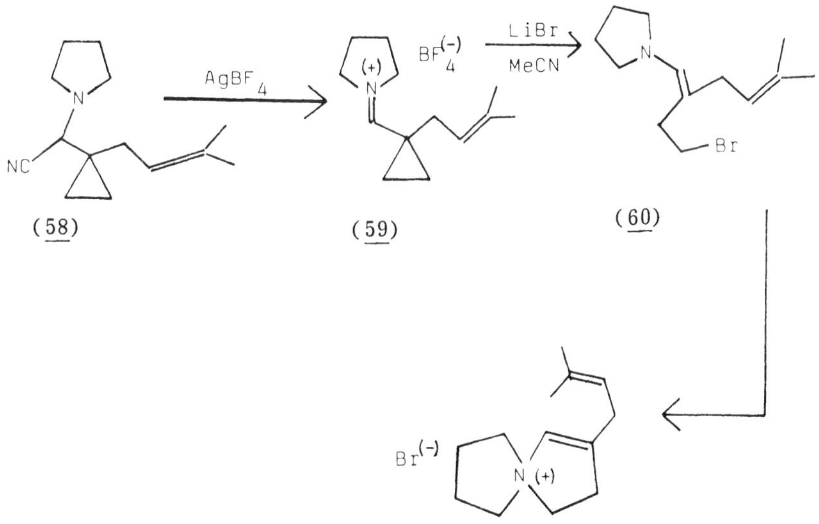

(58)   (59)   (60)

(61)

"alkene-instead-of-cyclopropyl" counterpart, the acid-catalyzed rearrangement of aminal 62 (corresponding to 58) (synthesized from N,N-dimethylaminoallene and dimethylamine) to enamine 64 (corresponding to 60) via iminium salt 63 (corresponding to 59) [204].

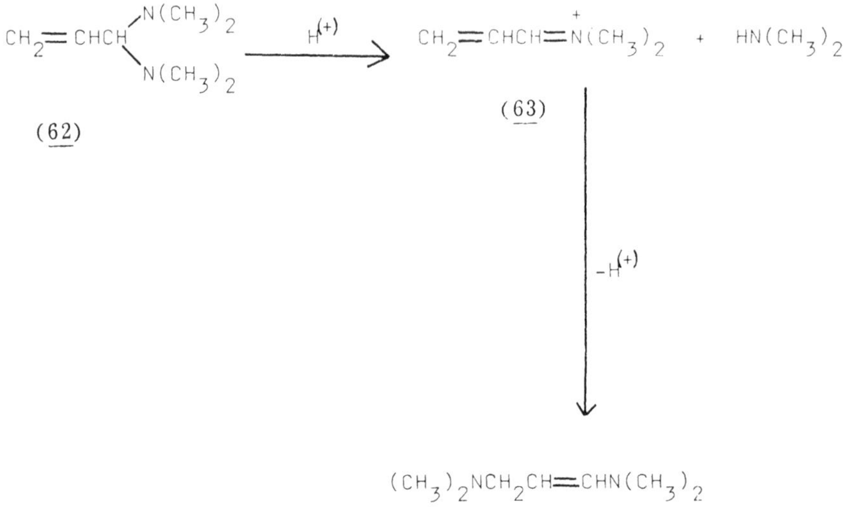

$$CH_2\!\!=\!\!CHCH\!\!\left\langle{}^{N(CH_3)_2}_{N(CH_3)_2}\right. \xrightarrow{H^{(+)}} CH_2\!\!=\!\!CHCH\!\!=\!\!\overset{+}{N}(CH_3)_2 \ + \ HN(CH_3)_2$$

(62) (63)

$$\downarrow {}_{-H^{(+)}}$$

$$(CH_3)_2NCH_2CH\!\!=\!\!CHN(CH_3)_2$$

(64)

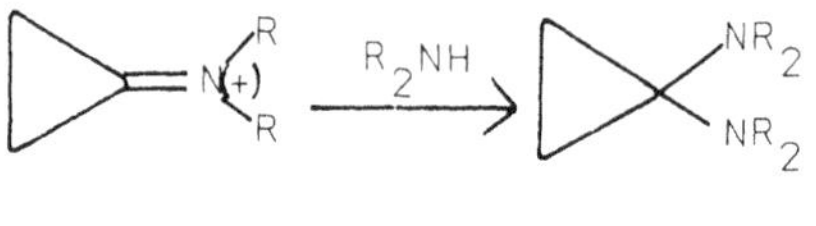

(65) (66)

Compounds in which the carbon-nitrogen iminium double bond are directly attached to the cyclopropyl group (such as compound 65) have been observed as intermediates in Favorskii-type reactions [205–209] and spectroscopically [210]. The final products of these reactions are aminals such as 66. Calculations using *ab inito* molecular orbital theory with STO-3G basis set have shown that the amino group greatly stabilizes the cyclopropyl cation [192].

The pyrrolidine aminal of bicyclo[3.1.0]hex-3-en-6-*endo*-carbaldehyde (67) can be isomerized to *exo*-aminal 68 under mild acid conditions

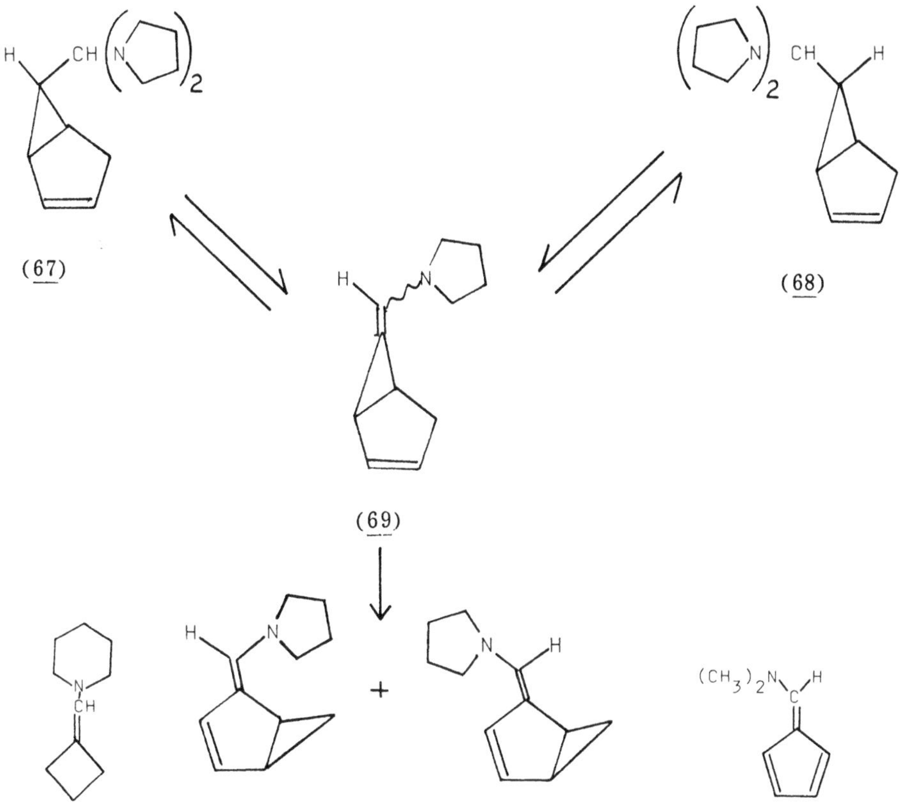

(67)

(68)

(69)

(73)   (70)   (71)   (72)

via homoallenic amine 69 [211]. At higher temperatures, a (1,3)-sig-
matropic rearrangement takes place, producing a 1:1 mixture of the
syn (70) and anti (71) pyrrolidino fulvenes [179,211–213]. Fulvenes
such as 6-(dimethylamineo)fulvene (72) are very stable [214,215].

We can now shift from enamines with the carbon-carbon double
bond attached directly to a three-membered ring to enamines with the
carbon-carbon double bond attached directly to a four-membered ring.
Attempts to isomerize cyclobutyl enamine 73 failed [216].

Transannular reactions [217] of enamines with the amine group on
one side of the medium-size ring and the carbon-carbon double bond on
the other side of the ring, positioned so that it can interact with the
amine nitrogen have been observed [218,219]. For example, protonation of compound 74 gives transannular bonded product 75. A transannular oxygen atom can interact with an enamine system to help stab-

CH₂ structure (74) → HClO₄ → structure (75) ClO₄⁻

(74)   (75)

ilize an intermediate such as an immonium ion by delocalization of the
positive charge [220]. Enamines have been found to undergo rearrangements with a sulfur atom acting as a neighboring group positive-charge
stabilizer across a six-membered ring [221]. Aryl group migration has
also been reported in aromatic enamines [220].

B. Enamines of Cyclic Ketones

Introduction

Enamines derived from the reaction of cyclic ketones and secondary
amines are the most commonly used and most useful of all the enamines
studied. Their value is enhanced by their potential as intermediates
in the synthesis of a large variety of natural products. The enamines
derived from five-, six-, and seven-membered-ring ketones have been
extensively studied, especially those of the six-membered ring. The
conformations of these ring systems are those of substituted cyclopentene, cyclohexene, and cycloheptene, respectively.

Cycloheptene can exist in boat, twist, and chair conformations,
with the chain conformation being the more stable form [229–234]. The
chair conformer (76) possesses C_s-symmetry and is 1.5 kcal/mol more
stable than the twist conformer (77), which possesses C_2-symmetry
[233]. The interconversion of the two chair conformers passes through
the boat form (78) with C_s-symmetry [234].

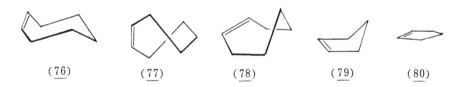

(76)   (77)   (78)   (79)   (80)

Cyclopentene exists principally in the envelope (puckered) form
(79) with C_s-symmetry [231,234]. The envelope conformers intercon-
vert by way of a C_{2v}-planar form (80) [234]. *Ab initio* calculations
have shown that envelope conformer 79 has a very low-energy barrier
to planarity [235].

The most stable conformers of cyclohexene are the half-chain forms (81) with C_2-symmetry [234,236–238,244]. These two conformers can be interconverted in a pseudorotationlike movement which passes through boat conformation transition state 82 with C_{2v}-symmetry and

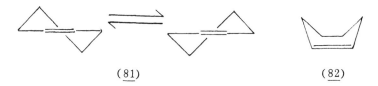

(81)　　　　　　　　　　　　　　　　(82)

an energy of 5.3 kcal/mol [234]. The half-chair is more stable than the half-boat by 2.7 kcal/mol [239]. Relative to cyclohexane, cyclo-hexene, is like a flattened chair, with only C-4 and C-5 having truly axial and equatorial bonds [240,241,244]. Bonds at C-3 and C-6 are somewhat differently oriented and are called pseudoaxial (a') and pseudoequatorial (e') (83). Due to the importance of the interac-

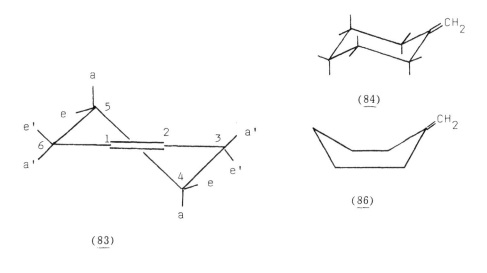

(84)

(86)

(83)

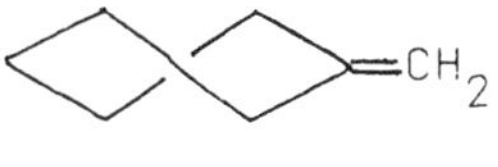

(85)

tion of nitrogen lone-pair electrons with the alkene π-system of cyclo-hexyl enamines, both in the free enamine itself and in the iminium ion intermediates resulting from electrophilic attack, the geometry of methylenecyclohexane is also very important. Its lowest energy con-former is the chair form (84), which is 4.4 kcal/mol more stable than the C_2-symmetry twist-boat form (85) [242], and even more stable than the C_s-symmetry boat form (86). The dihedral angle that the C-2 and C-6 equatorial bonds make with the methylene plane at C-1 in the stable chair conformer 84 is very small (~4°).

These geometries were the basis of two types of unique allylic strains in six-membered rings which were first proposed by Johnson and Malhotra [241,243,250]. The first type, involving a substituent

(87e)   (87a)

on the alkene-carbon of a cyclohexene ring (R) and a pseudoequatorial substituent on the adjacent alkane-carbon (R') (see 87e), is called $A^{(1,2)}$ strain (strain between groups on adjacent or 1- and 2-carbons of an allylic system). This strain results in the following rule: "If R and R' are moderately large, they will interfere with each other steric-ally to such an extent that that conformer having the axial substituent will be the favored form" [241], i.e., conformer 87a. The converse to this rule is that if R and R' are small in size then the conformer having the substituent equatorial will be the favored conformer, i.e., conformer 87e. So $A^{(1,2)}$ strain is most important when the groups at C-1 and C-6 are large, as well as when C-1 substituent conjugates with the double bond (for example, when the amine group in enamines conjugates with the alkene group). However, *this effect is not a very powerful effect* and so will not be a controlling effect unless all other major steric factors in the system cancel each other.

This $A^{(1,2)}$ strain will help determine which is the most abundant conformer at equilibrium, and it can be a factor in determining the direction from which a reagent attacks the alkene moiety. Reagent attack will be controlled by the stereoelectronic effect (conformational requirements of the groups involved in the reaction [244] which would demand as much continuous π-orbital overlap as possible in the transition state [245], and by steric effects (conformational require-ments of the nonreactive groups [244]).

The second type of strain involves steric interaction between a substituent on the exocyclic alkene-carbon (R) and the alkane-carbon adjacent to the ring alkene-carbon (R') as seen in 88e. This type of

(88e)   (88a)

strain is called $A^{(1,3)}$ strain (strain involving groups *syn* to one another on the C-1 and C-3 positions of an allylic system). The result of this strain is given in the following rule: "Where R and R' are moderate in size, they will interfere with each other drastically, in fact more so than if they were 1,3 diaxially related in a cyclohexane ring. Thus it could be expected that the conformational equilibrium should favor that form having the axial substituent" [24], i.e., form 88a. The converse is that if R and R' are small and do not interfere with each other, then conformer 88e should be favored. So $A^{(1,3)}$ strain will help to determine which is the most abundant conformer at equilibrium.

Both $A^{(1,2)}$ strain and $A^{(1,3)}$ strain are important effects as far as determining the product composition from electrophilic attack of enamines is concerned. Scheme 1 shows the electrophilic attack of 89 by an electrophile (X). Electrophilic attack takes place on 89 from one of two directions. If the electrophile approaches the double bond of 89 from the same side as the axial hydrogen on the adjacent carbon, then it is said to be adding by parallel attack [249] (also called equatorial attack or α-attack). If the electrophile attacks the double bond from the opposite side as the axial hydrogen on the adjacent carbon, then it is said to be adding by antiparallel attack [249] (also called axial or β-attack). Assuming that R and R' are of moderate to large size, $A^{(1,2)}$ strain would affect the reaction as follows: (1) Conformer 89a would be the more stable conformer because of $A^{(1,2)}$ strain in 89e; and if the reaction involved an early (reactantlike) transition state, then the lowest energy transition state would look like 89a, and either 90 or 91 would be the intermediate; (2) the stereoelectronic requirement is antiparallel attack [245,246], which presents no problem for the $A^{(1,2)}$ strain opposed conformer 89e, but with the $A^{(1,2)}$ strain favored conformer 89a antiparallel attack is sterically opposed by the axial R' group. So if an early transition state is involved, there are these effects which must be balanced against each other. $A^{(1,3)}$ strain would affect the reaction for a late (productlike) transition state. As far as ring conformations are concerned, products 90 and 93 (from anti-

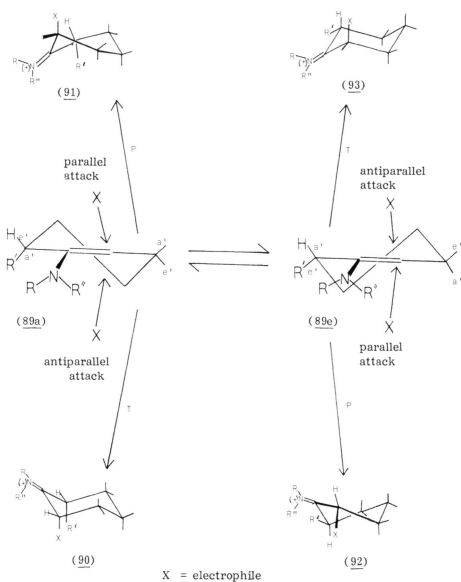

(91)

(93)

parallel
attack

antiparallel
attack

(89a)

(89e)

antiparallel
attack

parallel
attack

(90)

(92)

X = electrophile
P = parallel attack
T = antiparallel attack
a' = pseudoaxial
e' = pseudoequatorial

Scheme 1

parallel attack) would be favored because they are in the lower-energy chair conformations, whereas products 91 and 92 (from parallel attack) are in the higher-energy twist-boat conformations (about 4 kcal/mol energy difference). In the two chair conformations, 93 has $A^{(1,3)}$ strain between R' and R, and 90 possesses 1,3-axial interaction between R' and X, which 93 does not have. However, it has been reported that the energy of 1,3-diaxial methyl-methyl interaction is 3.7 kcal/mol [247], and $A^{(1,3)}$ methyl-methyl interactions are about 5−6 kcal/mol. Hence, $A^{(1,3)}$ interactions are usually greater than 1,3-diaxial interactions [248]. As far as the twist-boat conformations are concerned, 91 has neither $A^{(1,3)}$ strain between R' and R nor 1,3-axial interaction. On the other hand, 92 has $A^{(1,3)}$ strain, but does not have 1,3-axial interaction between R' and X. So the product formed is determined by where the balance lies in these effects. Stereoelectronic effects are not a factor in the further reaction of any of the products (90, 91, 92, or 93) since continuous overlap in the transition state is possible no matter from which side the reagent attacks.

In the case where R' is a hydrogen atom, then R and R' are not *both* of a moderate to large size. So conformers 89a and 89e are essentially identical. Then 89 would undergo an antiparallel attack by an electrophile because of the stereoelectronic factor and because (assuming a late transition state and a moderate to large electrophile, X) parallel attack would lead to $A^{(1,3)}$ strain between X and R''. This has been shown in the alkylation of the pyrrolidine enamine of 4-*t*-butyl-

(94)   (95)   (96)

cyclohexanone (94) which took place by antiparallel attack under stereoelectronic control to give iminium salt 95 [295]. The product of hydrolysis under nonequilibrating conditions led largely to the *trans*-2-alkyl-4-*t*-butylcyclohexanone (96).

The relative order of reactivity of enamines derived from a common secondary amine but differing cyclic ketones is the following: cyclopentanone > cycloheptanone and cyclooctanone > cyclohexanone [169, 170,224].

Enamines can be rearranged to an equilibrium mixture of isomeric enamines. But this isomerization does not take place thermally under basic conditions, nor does it take place thermally under neutral conditions [251,275].

Table 5 Isomer Distribution of Enamines of 2-Methylcyclohexanone

Amine	Trisubstituted isomer (%)	Tetrasubstituted isomer (%)	Ref.
Pyrrolidine	90	10	149
3,3-Dimethylazetidine	83	17	161
Dimethylamine	60	40	150
Morpholine	52	48	149
N-Phenylpiperazine	51	49	253,254
Piperidine	46	54	149
N-Methylpiperazine	45	55	253
Diethylamine	25	75	149
2,6-Dimethylpyrrolidine	≤10	≥90	150
N-Methylaniline	0	100	150

2-Substituted Ketones

The isomer distribution of enamines produced from 2-substituted cyclic ketones varies according to what amine moiety is present. Table 5 summarizes the isomer distribution of various enamines derived from 2-methylcyclohexanone as determined by nmr spectroscopy. It can be seen that the greater the conjugation of the lone-pair electrons of the amine nitrogen with the alkene π-system, the greater the amount of the trisubstituted isomer as compared to the tetrasubstituted isomer. So we find that pyrrolidine (which, as discussed in a previous section, possesses the greatest amount of n-π interaction) has the greatest percentage of trisubstituted isomer, whereas morpholine and piperidine have a much lower percentage. Support for this postulate is also found in the nmr spectra of these enamines, wherein the chemical shifts of the vinylic protons of the pyrrolidine enamines are at a higher field than those of the corresponding morpholine and piperidine enamines by 20–27 Hz. The N-methylaniline enamine of 2-methycyclohexanone (97) has been reported to consist exclusively of the tetrasubstituted

(97)

isomer [150]. Here the electron pair on the nitrogen atom can overlap predominantly with the phenyl group and not the enamine double bond, thus minimizing the steric interference between the C-2 methyl group and the substituents attached to the nitrogen.

(89t)

The tetrasubstituted isomer, 89t, can be destabilized by $A^{(1,3)}$ strain between the methyl group (R') and the α-methylene group on the nitrogen (R). This $A^{(1,3)}$ strain is present when these two groups are in the same plane, which would be the case when there is maximum overlap between the lone-pair nitrogen electrons and the alkene π-system. Hence, this $A^{(1,3)}$ strain causes the highly conjugative pyrrolidine enamine to exist principally as the more stable trisubstituted isomers (89a and 89e). However, for those enamines which are not highly conjugated (such as the morpholine or piperidine enamines), this $A^{(1,3)}$ steric strain can be relieved by rotation about the C—N bond and bending of the methylene-nitrogen bond toward a pyramidal geometry, thus separating the amine methylene group from the methyl group on the ring.

When the enamines of 2-phenylcyclohexanones are examined, different distributions between tri- and tetrasubstituted enamines are obtained which favor the trisubstituted enamines even more, as shown in Table 6. It is obvious that the large phenyl group causes greater $A^{(1,3)}$ strain in the tetrasubstituted isomer than could be made up for by conjugation of the carbon-carbon double bond with the phenyl ring. Kuehne [255] has reported that the pyrrolidine enamine of 2-phenyl-cyclohexanone failed to show any styrene-type absorption in the ultraviolet, which would have been exhibited by the tetrasubstituted isomer. The morpholine enamine of 2-propylcyclohexanone has been reported to be a 2:3 mixture of tri- and tetrasubstituted iosmers [256].

Table 6 Isomer Distribution of Enamines of 2-Phenylcyclohexanone

Amine	Trisubstituted isomer (%)	Tetrasubstituted isomer (%)	Ref.
Morpholine	93	7	253
Piperidine	91	9	253
N-Methylpiperazine	92	8	253
N-Phenylpiperazine	91-94	9-6	253

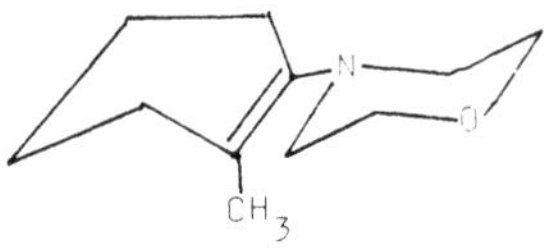

(98t)

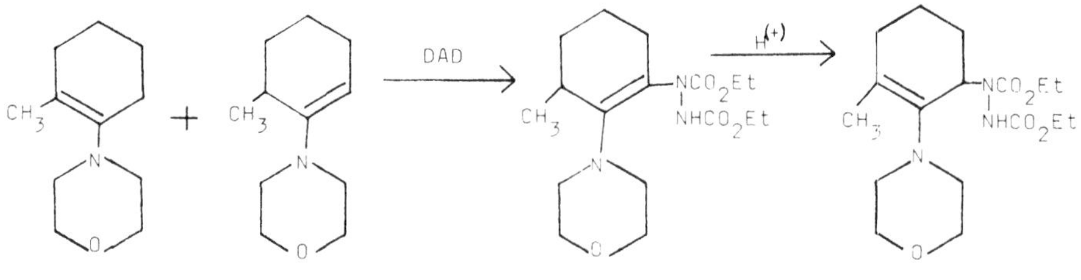

(98t)   (98r)   (99r)   (99t)

   This contrast between the isomeric composition of the 2-substituted cyclohexenyl enamines of highly conjugative pyrrolidine versus 2-substituted cyclohexenyl enamines of low conjugative morpholine is brought into even sharper relief when the 2-substituent is an alkoxy group [303]. The isomeric distribution of pyrrolidine enamines obtained from 2-methoxy, 2-ethoxy, and 2-isopropoxy cyclohexanones is more than 90% trisubstituted isomer and less than 10% tetrasubstituted isomer. In contrast, the isomeric distribution of the morpholine enamines obtained from these same cyclohexanones is about a 1:1 ratio between the two possible isomers.

The tetrasubstituted isomer of the morpholine enamine of 2-methyl-cyclohexanone (98t) should be expected to exhibit a lower degree of enamine-type reactivity toward electrophilic agents than the trisubstituted isomer (98r) because of the diminished electronic overlap. This was demonstrated to be the case when the treatment of the enamine with dilute acetic acid at room temperature resulted in the completely selective hydrolysis of the trisubstituted isomer within 5 min. The tetrasubstituted isomer was slow to react and was 96% hydrolyzed only after 22 hr [257].

The reaction between the isomeric mixtures of 2-substituted enamines and diethylazodicarboxylate (DAD) is selective in that only the trisubstituted enamine will react in a quantitative manner to form the 2,6-disubstituted enamine [258–260]. So this reaction in conjunction with nmr spectroscopy can be used for determination of the amount of trisubstituted isomer. For example, when mixtures such as morpholine enamines 98t and 98r were treated with diethylazodicarboxylate under mild conditions (5°C, 72 hr), the product obtained was 99r in a yield corresponding to the percentages of the trisubstituted enamines in the original mixture [261–265]. Enamine 99r was almost quantitatively converted to isomer 99t by equilibration with *p*-toluenesulfonic acid in refluxing benzene [265].

Diethylazodicarboxylate will attack the cyclohexanone enamine from either the parallel or antiparallel direction, depending on the steric effects in the late transition state [283].

An equalizing of the steric effects that favor the trisubstituted

(100)   (101)   (102)

enamine over the tetrasubstituted enamine was brought about in the cyclization of 100. As a result, the tetrasubstituted internal enamine (101) is favored over the trisubstituted internal enamine (102) in a 4:1 ratio [251], reflecting the more normal distribution between tetra- and trisubstituted alkenes.

The trisubstituted enamine can exist as either the quasiaxial (89a) or the quasiequatorial (89e) conformer. As was noted in the previous section, the quasiaxial conformer (89a) is more stable owing to the $A^{(1,2)}$ strain present in conformer 89e.

In 1954 Stork and co-workers [3a] reported that the alkylation of the pyrrolidine enamine of cyclohexanone (103) with methyl iodide followed by acid hydrolysis led to the monoalkylated ketone. It was

(103)   (104)   (105)

thus obvious that the enamine 105, derived by the loss of a proton from
the intermediate methylated imminium cation 104, failed to undergo any
further alkylation.

A rationale for the inert behavior of enamine 105 toward further
alkylation was put forward by Williamson [267], who argued that the
methyl group in 105 should assume an axial orientation if the overlap
between the electron pair on the nitrogen atom and the double bond is
to be maintained, since the alternate conformation the equatorial
methyl group is almost coplanar with the methylene group adjacent to
the nitrogen atom. The alkylation of the axial conformer of 105, being
subject to stereoelectronic control, would, therefore, involve a severe
1,3-diaxial alkyl-alkyl interaction. This would increase the energy of
the transition state. This explanation was a forerunner of the more
generalized $A^{(1,2)}$ strain in cyclohexenyl-type systems as proposed
by Johnson and Malhotra [241,243] and discussed in the previous section.

It was demonstrated by Johnson and Whitehead [268,275] that the
methyl group in the pyrrolidine enamine of 2-methylcyclohexanone
(105) is predominantly in the axial conformation. They found that care-
ful hydrolysis of the isomeric mixture of pyrrolidine enamine of 2-
methyl-4-t-butylcyclohexanone (106), (in which the equatorial 4-t-butyl
group essentially fixes the conformation) led to a 7:3 mixture of trans
and cis isomers of the ketone (107 and 108), indicating that the methyl
group in the enamine is largely in the axial orientation.

(106a,e)   (106t)   (107)   (108)

However, there is some doubt as to whether this ratio reflects exactly the original composition of pyrrolidine enamine 106 since some
equilibration between 106 and its iminium salt intermediate does take
place [275]. Hence, some scrambling is possible. But this equilibrium
would probably diminish the amount of trans isomer (107) present

rather than enhance it. So there should be a greater amount of *trans* isomer (quasiaxial methyl group) present in the isomeric mixture of pyrrolidine enamine 106 than the 7:3 ratio of ketones 107 and 108 shows. The nmr spectrum of 106 indicates 6% tetrasubstituted (106t), 19% tri-substituted quasiequatorial (106e), and 75% quasiaxial (106a). It was further shown that there is no stereoelectronic control of hydrolysis of the tetrasubstituted enamine (106t) as indicated by an approximate 50:50 mixture of *trans* (107) and *cis* (108) ketones being obtained upon hdyrolysis of pure 106t. A similar result was obtained upon hydrolysis of the N-methylaniline enamine of 2-methyl-4-*t*-butylcyclohexane (109) which exists almost exclusively as the tetrasubstituted isomer depicted [150]. So it seems as if the original composition of the trisubstituted enamine was a little less than a 4:1 quasiaxial (106a) to quasiequatorial (106e) mixture [275].

Johnson and co-workers [275] have further shown that the reduc-

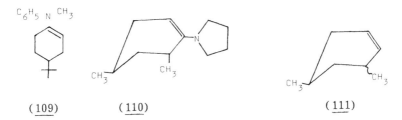

(109)   (110)   (111)

tive elimination of the pyrrolidine group from the pyrrolidine enamine of 2,4-dimethylcyclohexanone (110), which involved treating it with diborane [269,270], gave 3,5-dimethyl-Δ^2-cyclohexane (111) which on further reduction led to a 77:23 mixture of *trans*-1,3-dimethylcyclohex-ane (112) and *cis*-1,3-dimethylcyclohexane (113).

The proposed direct proof for the orientation of the methyl group in enamine 105 by Malhotra and Johnson [271] by hydrolyzing 105 with 50% deuterioacetic acid—deuterium oxide in diglyme solution has been shown not to be conclusive [272]. It has been shown, contrary to earlier reports [271,273,274], that there is not high axial stereoselec-tivity in the protonation and deuteriation of enamines due to an early transition state in this reaction [272]. This was one of the basic premises of the proposed proof.

The morpholine enamine of 2-methyl-4-*t*-butylcyclohexanone (114) exists at equilibrium as an isomeric mixture of 46% tetrasubstituted enamine (114t) and 54% trisubstituted enamines (114e and 114a) [275].

CH₃ structures

(112)   (113)

(114t)   (114e)   (114a)   (115)   (116)

The distribution between quasiequatorial (114e) and quasiaxial (114a) trisubstituted isomers is 50:50 (27% of each). This is consistent with the weak interacting ability of morpholine nitrogen lone-pair electrons with the rest of the enamine π-system, as discussed before. Since there is little interaction, the necessity of coplanarity of the morpholine nitrogen-methylene system and the cyclohexyl alkene system is gone. So the steric strains created by such coplanarity or near coplanarity disappear as the morpholine system is allowed to twist and bend. Hence, there is energetically little difference between the trisubstituted quasiequatorial or quasiaxial isomers (114e or 114a). Nor is there a very significant energy gap between the trisubstituted and the tetrasubstituted enamines (114e,a and 114t). Hydrolysis of 114 isomeric mixture produced an almost 50:50 mixture of *trans* (115) and *cis* (116) ketones, which is consistent with the nmr results. This can be explained as follows: It has been shown that, unlike the case of the pyrrolidine enamine of 2-methyl-4-*t*-butylcyclohexanone (106) discussed before, the morpholine enamine of this ketone does not equilibrate upon hydrolysis; therefore, there is no scrambling such as was observed with the pyrrolidine enamine. This difference between the hydrolysis of the pyrrolidine and morpholine enamines can be related to the relative stabilities of the *exo* and *endo* double bonds in 5- and 6-membered rings [276,277] and the relation of these relative stabilities to the iminium salts formed during the hydrolysis of pyrrolidine (5-membered ring) and morpholine (6-membered ring) enamine [275,278].

The early premise of the predominance of the quasiaxial methyl isomer in the isomeric mixture of enamines produced from 2,4-dimethylcyclohexanone was used in the total synthesis of the glutarimide antibiotic *dl*-dehydrocycloheximide both by Johnson and co-workers [268, 279,280] and by Schaeffer and Jain [281,282].

The presence of a substituent at the 2-position of a cyclohexanone greatly reduces the rate of its reaction with pyrrolidine to form an enamine by the usual azeotropic removal of water method. This is accentuated in the case of the α-tetralone (117) by the α-substituent being coplanar with the pyrrolidine nitrogen-methylene group in inter-

(117)

(118)

(119)

(120)

mediate 118 thus showing considerable steric interaction in that group and the aromatic perihydrogen [290−293]. Enamine 119 can be formed in good yield (87%) but only after an extended reaction time (88 hr) [294]. On the other hand, β-tetralone readily forms the pyrrolidine enamine. Substituting the smaller azetidine for pyrrolidine in the reaction with α-tetralone causes enamine 120 to form in a relatively short period of time [294]. This is because of the diminished steric interaction between the nitrogen-methylene group of the smaller ring amine and the aromatic perihydrogen. The UV spectrum of enamine 119 shows a definite steric inhibition to p-π overlap [294], but enamines of α-tetralone undergo electrophilic substitution at the β-carbon atom in good yields [12,294].

Enamines react with electrophilic olefins such as methyl vinyl ketone, acrolein, methyl acrylate, or acrylonitrile under mild conditions in a stereoselective manner [148]. With 2-alkylcyclohexanone enamines, the reaction usually results in 2,6-disubstituted products. The reason for the apparent avoidance of substitution of the tetrasubstituted isomer (89t) at the substituted alkene carbon is because of the $A^{(1,3)}$ strain introduced regardless of whether the attack is parallel or antiparallel. This is seen in Scheme 2, where tetrasubstituted enamine 89t can undergo parallel attack to give twist-chair intermediate 122, or it can undergo antiparallel attack to give chair intermediate 121. Because of the late transition state, the transition state resembles these intermediates with their $A^{(1,3)}$ strain. A lower-energy pathway is provided by electrophilic attack of the trisubstituted enamines (see Scheme 1).

However, there have been reports of 2,2-disubstituted products being obtained from the reaction of the enamines of 2-alkylcyclohexan-

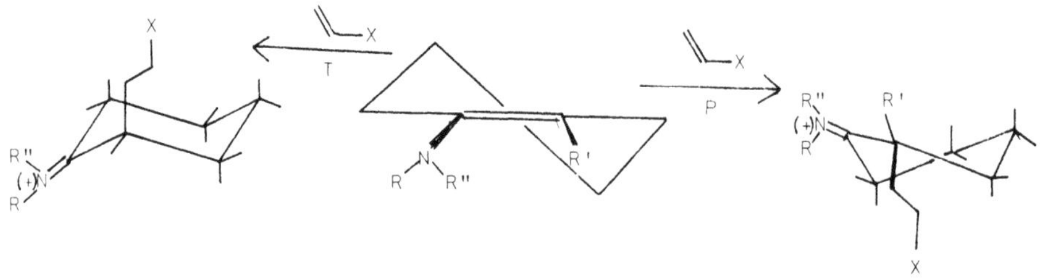

(121)   (89t)   (122)

```
P = parallel attack
T = antiparallel attack
```

Scheme 2

one with electrophilic olefins [194,284,285,287]. Sometimes there is more 2,2-disubstituted product than tetrasubstituted enamine isomer starting material as determined by the equilibrium ratio of tetrasubstituted enamine isomer to the trisubstituted isomer. Hickmott and Firrell [288] have explained this by suggesting that the more stable trisubstituted isomer reacts reversibly with the electrophilic olefin, whereas the less stable tetrasubstituted isomer reacts irreversibly to produce the 2,2-disubstituted product. Huffman and co-workers [289] have shown this explanation to be inadequate. They have postulated that the 2,2-disubstitution products come from conversion of the initially formed 2,6-disubstituted enamines (123) into their corresponding enolate anions. The enolate anions then react with a second molecule of the electrophilic olefin to produce the 2,2-disubstitution products (124).

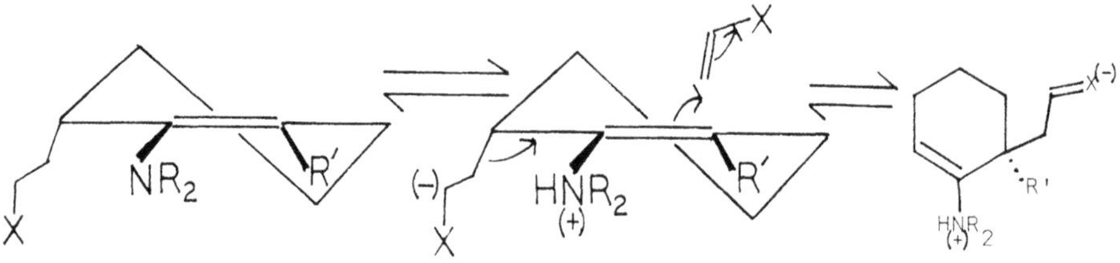

(123)   (124)

The introduction of a heteroatom at the 3-ring position of a cyclohexanone brings in a new factor influencing the location of the double bond of an enamine that is formed from the cyclic ketone. Enamines formed from 3-oxacyclohexanone (125a) [296,390], from 3-thiacyclohex-

a: X = O
b: X = S
c: X = NR'
d: X= SO₂

(125)   (126)   (127)   (128)   (129)

anone (125b) [296], or from 1-alkyl-3-piperidone [286,297], and morpholine or pyrrolidine produce the regioisomer with the double bond
toward the heteroatom (126) in greater than 80% yield. The ring hetero-
atoms with their nonbonding pairs of electrons can conjugatively interact
with the enamine system in isomer 126. A similar observation is
made concerning the pyrrolidine and morpholine enamines of 5,6-dihy-
dro-2H-thiopyran-3(4H)-one 1,1-dioxide (125d), which are produced
in greater than 90% yield in the isomeric form of 126d [298]. When a
2-methyl group is added onto cyclic β-ketosulfone 125d, namely 2-
methyl-5,6-dihydro-2H-thiopyran-3(4H)-one 1,1-dioxide (127), different enamine regioisomers are obtained when morpholine is the amine
moiety than when pyrrolidine is the amine group [86]. In the former
case, trisubstituted enamine 128 is formed as a single isomer, whereas
in the latter case, tetrasubstituted enamine 129 is produced as a single
isomer. Apparently steric effects are dominant in the morpholine
enamine (128), whereas electonic conjugative effects are more important
in the pyrrolidine enamine (129).

The isomeric mixtures of enamines resulting from the reaction of
2-alkylcyclopentanones with pyrrolidine [287,300] or N-alkylpiperazine
[254] consist principally of the trisubstituted form (130) as compared
to the tetrasubstituted form (131) in a 2 or 3:1 ratio. The enamines
produced from the five-membered sulfur heterocyclic ketone,3-oxotetra-

(130)   (131)   (132)   (133)

hydrothiophene (132), and secondary amines give a mixture of isomers
[266,301]. 1-Indanone (133) reacts quite readily with secondary
amines such as pyrrolidine to form the corresponding enamine [290,
302]. This is in contrast to the enamines of 1-tetralone (discussed
earlier), which are formed very slowly. In enamine formation from

1-indanone there is only minimal interaction between the pyrrolidine nitrogen-methylene group and the aromatic perihydrogen in the iminium ion intermediate. The resulting enamine also readily undergoes electrophilic alkylation [302].

3-Substituted Ketones

In the steroidal series, Heyl and Herr early reported in 1953 that the pyrrolidine [314] enamine of 5α-cholestan-3-one consisted of a mixture of Δ^2 and Δ^3 isomers (134). The isomeric composition of an equilibrium mixture of the pyrrolidine enamine of 3-methylcyclohexanone was first reported by Malhotra and co-workers [310] to be predominantly the Δ^6 isomer (see Table 7). The preponderance of the Δ^6 isomer in the mixture can be attributed to $A^{(1,2)}$ strain between the quasiequatorial methyl group and the vinylic hydrogen atom in the Δ^1 isomer (135).

(135)

(134)

The magnitude of the allylic strain is on the order of 0.6−0.78 kcal/mol [304,310]. The equilibrium isomeric composition of the enamines of the trans-decalin-2-one shows a ratio strongly favoring the Δ^2-isomer over the Δ^1-isomer (see Table 7). Only the Δ^2-isomers of the morpholine enamines from trans-decalin-2-one and 4-methyl-trans-decalin-2-one react with electrophiles such as β-nitrostyrene and phenyl vinyl ketone [308,316−318]. When conjugation of the enamine system with a 3-substituted aromatic ring is made possible, the Δ^1 isomer is the predominant isomer (see Table 7). 2,5-Disubstituted cyclohexyl enamines set up competition between the steric interactions with the 2-substituent and the 5-substituent. As can be seen in Table 7, the 2-substituent provides the greater strain, and hence the trisubstituted isomer is the major part of the isomeric mixture. Equilibrium between Δ^1 and Δ^6 isomers of 3-substituted enamines takes place very readily even at low temperatures [305,312,313]. This equilibrium is present in the enamines of 2-tetralone also [315] even though there is only a negligible amount of the Δ^6 isomer present [148,304,319,320]. This is shown by the reaction of electrophiles such as β-nitrostyrene

with the 2-tetralone enamines to give both C-1 and C-3 adduct even
under mild conditions [315]. β-Nitrostyrene reacts with these enamines
by an antiparallel attack unless steric hinderance blocks such an
approach [305,312].

The reaction of the 2,4,4-trimethylcyclopentanone with pyrrolidine
gives an isomeric mixture of enamines 136 and 137 in a 3:1 ratio and a

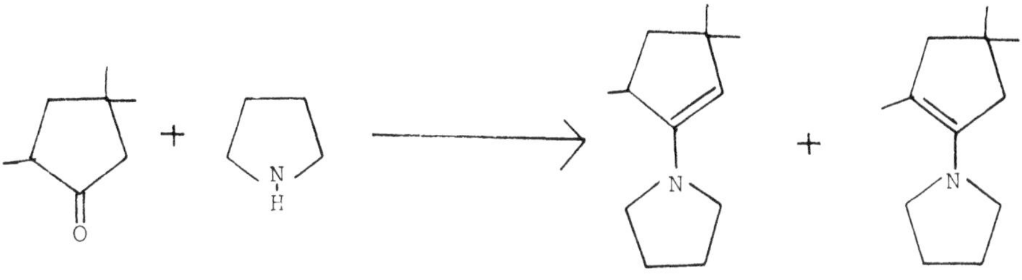

(136)   (137)

total yield greater than 90% [287]. The enamine of 2,2,4-trimethyl-
cyclopentanone could only be formed with difficulty.

Enamines derived from 2-indanone (138) are readily synthesized
[325,326] with the equilibrium lying far to the side of enamine product
[327]. For example, the pyrrolidine enamine of 2-indanone forms in

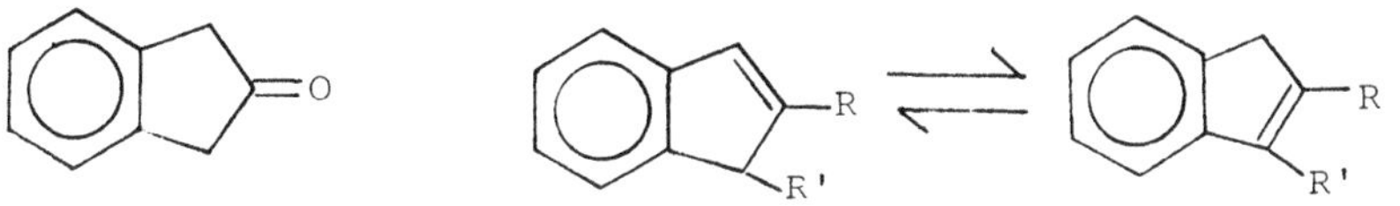

(138)   (139)   (140)

methanol at room temperature without the assistance of water being
removed. It could be heated with excess water at 80° for 2 hr, and
no 2-indanone could be detected by nmr [327]. A similar observation
has been reported with the morpholine derivative of 2-indanone [328].
That true equilibrium is taking place in the hot water is shown by the
trideuteration of the enamine when it is heated with D_2O [327]. As
would be expected, the enamines of 2-tetralone are less reactive toward
electrophilic reagents than the enamines of 1-tetralones [302] since the
iminum ion of the latter can be stabilized by resonance interaction with
the benzene ring whereas the iminium ion of the former cannot be stabilized in this manner. When a methyl substituent is present on the
5-membered ring of indene itself, the tetrasubstituted isomer (140,
R = H, R' = CH_3) is very strongly favored at equilibrium over the
trisubstituted isomer (139, R = H, R' = CH_3) (see Table 8) [323,324].

Table 7 Isomer Distribution of Enamines of 3-Substituted Cyclohexanones

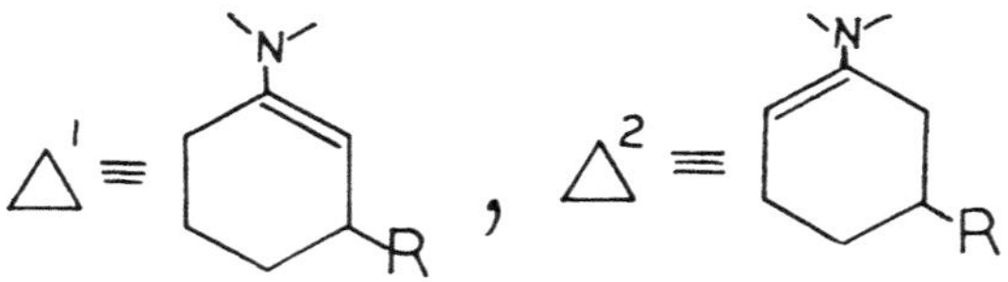

$\Delta^1 \equiv$ (structure), $\Delta^2 \equiv$ (structure), $\Delta^6 \equiv$ (structure)

Ketone	R		Amine				Ref.
			Pyrrolidine (% composition)		Morpholine (% composition)		
			Δ^1	Δ^6	Δ^1	Δ^6	
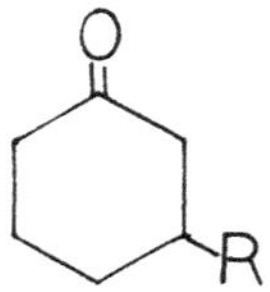 (3-substituted cyclohexanone)	CH_3		43	57	45	55	304,310,311
	$C(CH_3)_3$		45	55	—	—	305
	C_6H_5		100	0	100	0	304,305

Ketone	R	R'	Δ^1	Δ^6	Δ^1	Δ^6	Ref.
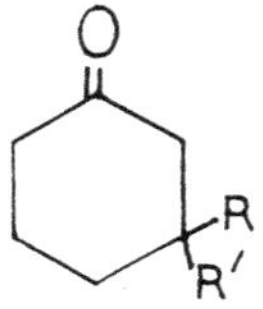 (3,3-disubstituted cyclohexanone)	CH_3	CH_3	20	80	25	75	306,304
	CH_3	C_2H_5	25	75	—	—	307

	R	R'	Trisub.	Tetrasub.	Trisub.	Tetrasub.	Ref.
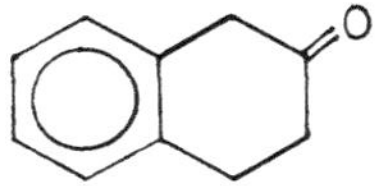			100	0	100	0	304
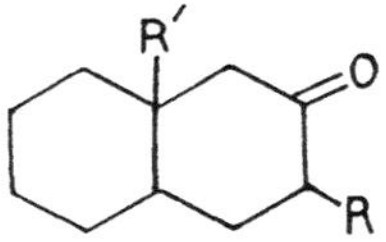	H	H	28	72	21	79	308,310
	H	CH$_3$	—	—	<10	>90	309
	OCH$_3$	H	90	10	46	54	303

	R	R'	Trisub.	Tetrasub.	Trisub.	Tetrasub.	Ref.
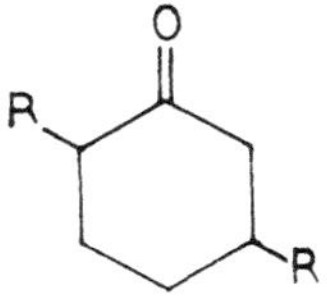	CH$_3$	CH$_3$	90	10	36	64	275,287
	CH(CH$_3$)$_2$	CH$_3$	90	10	—	—	275,287
	CH$_3$	CH(CH$_3$)$_2$	90	10	—	—	287
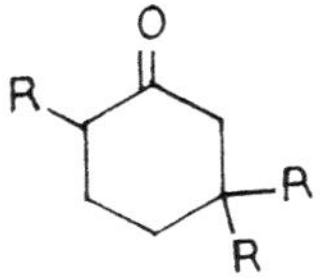			65	35	—	—	287

Table 8 Isomer Distribution of Enamines of Substituted 2-Indanones

R	R' = CH$_3$ ($\%$ composition)		R' = C$_6$H$_5$ ($\%$ composition)		Ref.
	Trisub.	Tetrasub.	Trisub.	Tetrasub.	
Pyrrolidino	47	53	90	10	321,302
Piperidino	61	39	68	32	321,302
Hexamethyleneimino	60	40	60	40	321,302
Morpholino	59	41	—	—	321
Dimethylamino	—	—	71	29	302
Hydrogen	<2	>98	—	—	323,324

But in the case of either methyl (R' = CH$_3$) or phenyl (R' = C$_6$H$_5$) substituted enamines, the equilibrium strongly shifts toward the trisubstituted isomer (139, R' = CH$_3$ or C$_6$H$_5$, R = amine group) (see Table 8). It has been shown by UV spectra that the phenyl substituent is twisted out of the plane of the enamine double bond and is not in conjugation with the enamine system [302]. The tetrasubstituted isomer has steric strain caused by interaction between the substituent and a potentially planar amine group, its planarity determining its degree of conjugation.

C. Enamines of Acyclic Ketones and Aldehydes

The synthesis of enamines from acyclic ketones using the usual method of heating with a secondary amine and removing the water produced has been reported [329,330,333]. However, this seemingly straightforward synthesis is complicated by the formation of cross-conjugated dienamines along with the monoenamines. For example, Madsen and Lawesson [331]

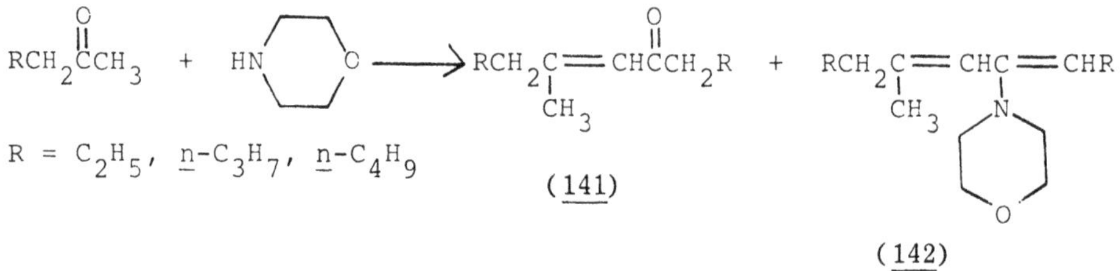

$$RCH_2\overset{\overset{\text{O}}{\|}}{C}CH_3 \ + \ HN\!\!\diagdown\!\!\diagup\!\!O \longrightarrow RCH_2C\!\!=\!\!CH\overset{\overset{\text{O}}{\|}}{C}CH_2R \ + \ RCH_2C\!\!=\!\!CHC\!\!=\!\!CHR$$

$$R = C_2H_5, \ \underline{n}\text{-}C_3H_7, \ \underline{n}\text{-}C_4H_9$$

(141)

(142)

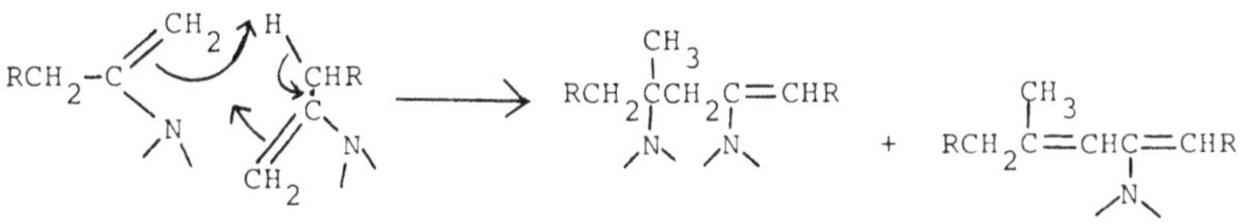

Scheme 3

reported that the treatment of *n*-alkyl methyl ketones with morpholine in the presence of *p*-toluenesulfonic acid for a short period of time results in the formation of a mixture of condensed ketone 141 and cross-conjugated enamine 142. Similar results were encountered by Bianchetti and co-workers [332], who found that the ketal derivatives of *n*-alkyl methyl ketones with morpholine led to cross-conjugated enamines. These authors have suggested the following probable scheme (see Scheme 3) for the dienamine formation. That the formation of the dienamine probably occurs by way of the monoenamine, as shown in Scheme 3, rather than via an unsaturated ketone condensation product from a simple ketone starting material has been shown as follows. Condensation of methyl *n*-propyl ketone with morpholine over a molecular sieve and in the absence of acid catalyst gave the corresponding mixture of monoenamine isomers. After storage for 12 months at 0–5°C, nmr signals due to the dienamine appeared and the nmr signals due to the monenamine became smaller [385].

Enamines formed from unsymmetrical acyclic ketones can exist as two different structural isomers. When the ketone from which the enamine is formed is an unsymmetrical methyl ketone, the choice of structural isomers is between a terminal double bond (143t) and a more fully substituted double bond (143s). The structural isomer distribu-

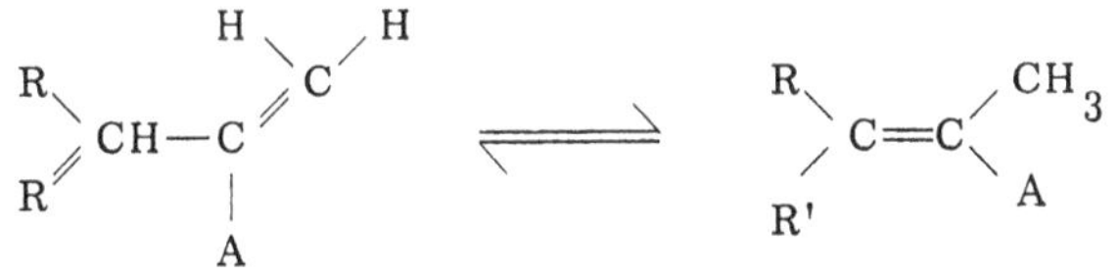

(143t)   (143s)

tions of the enamines derived from several unsymmetric methyl ketones and four different amines (dimethylamine, morpholine, N-methylaniline, and N-methyl-*o*-toluidine) are tabulated in Table 9.

Several determining factors are involved in the resultant isomeric distribution of these enamines. First, for simple alkenes with little steric strain, the more highly substituted alkene is thermodynamically more stable than the less substituted or terminal alkene [182]. So if this were the only determining factor, isomer 143s would be favored.

However, $A^{(1,3)}$ strain is also present. This strain is maximum in the more highly substituted isomer 143s, especially between the amine group A and substituent R'. The magnitude of the $A^{(1,3)}$ strain between these two groups will depend on the extent to which the amine group is conjugated with the alkene group. Maximum conjugation would mean an amine group coplanar with the alkene group and hence maximum strain. This can be seen by comparing enamines 143-1, 143-2, 143-3, and 143-4 in Table 9. It was noted in an earlier section that the relative conjugative interaction of amine groups in enamines decreases in the order dimethylamino > morpholino > N-methylanilino (the bottom position of the last amine is due to the delocalizing of the lone-pair nitrogen electrons into the aromatic ring rather than into the alkene). So enamine 143-1 has the most conjugative interaction and the greatest $A^{(1,3)}$ strain, and consequently it has the highest percentage of terminal isomer (143t) present of the four enamines being compared. Steric inhibition of resonance between the nitrogen and the aromatic ring caused by the ortho methyl group in 143-3 causes it to show a greater amount of terminal isomer than 143-4.

A similar ordering of isomeric distribution according to amino substituent is observed in grouping 143-5, -6, and -7, and 143-8, -9, and -10 in Table 9. In these groups it should be noted that $A^{(1,3)}$ strain between substituent R and the methyl group results in even greater percentages of terminal isomers. This is accentuated in enamine 143-11, where the t-butyl group maximizes this type of steric strain resulting in 100% terminal isomer being present.

On the other hand, the conjugative interaction between a phenyl ring and the enamine double bond that is present in the most highly substituted isomer (143s) of enamine 143-12 causes its isomeric distribution to favor isomer 143s 100%.

Strong resistance to a double bond exocyclic to a cyclopropane ring is illustrated once again in enamine 143-13's existing as the terminal isomer only. Cyclopentane rings favor exocyclic double bonds, but cyclohexane rings do not [276, 277]. These facts help to explain the isomeric distributions of enamines 143-14 and 143-15.

When enamines are produced by the titanium tetrachloride method of White and Weingarten [349], the kinetically controlled product is formed [338]. In the case of 143-1 and 143-2, the kinetically controlled products were the terminal enamine isomers (143t) [338]. It turns out that the less substituted (terminal) isomers are more reactive toward reagents such a phenylisocyanate or 4-nitrophenylazide [338]. The isomeric products that are formed do not correspond to the isomeric composition of the starting materials. This, however, is in accord with the Curtin-Hammett principle [350], which states that the proportion of the products in no way reflects the ground-state isomeric composition but depends only on the activation energies of the processes leading to the products

TABLE 9 Structural Isomer Distribution of Enamines of Acyclic Ketones

Cmpd. no. (143-n) n	R	R'	A	% Composition 143t	% Composition 143s	Ref.
1	CH_3	CH_3	Dimethylamino	50	50	334,335
2	CH_3	CH_3	Morpholino	30	70	336,337,338
3	CH_3	CH_3	N-Methyl-o-toluidino	30	70	335
4	CH_3	CH_3	N-Methylanilino	0	100	335
5	CH_3CH_2	H	Dimethylamino	55	45	335
6	CH_3CH_2	H	Morpholino	35	65	337
7	CH_3CH_2	H	N-Methylanilino	10	90	335
8	$(CH_3)_2CH$	H	Dimethylamino	65	35	335
9	$(CH_3)_2CH$	H	Morpholino	51	49	337,354
10	$(CH_3)_2CH$	H	N-Methylanilino	10	90	335
11	$(CH_3)_3C$	H	Morpholino	100	0	337
12	C_6H_5	H	Morpholino	0	100	337
13	$(CH_2)_2$		Morpholino	100	0	339
14	$(CH_2)_4$		Morpholino	9	91	337,354
15	$(CH_2)_5$		Morpholino	65	35	337,354

Unlike cyclopentanones or cyclohexanones, many acyclic ketones and aldehydes can also form E- (144-E) and Z-stereoisomers (144-Z) when they are allowed to react with secondary amines to form enamines. The stereoisomer distribution of several of these amines are shown in Table 10. The principal factor at work here is the $A^{(1,3)}$ strain. This strain is maximized for the interaction between the amine and R_1 (for 144-E) or R_2 (for 144-Z) when the conjugative interaction between the

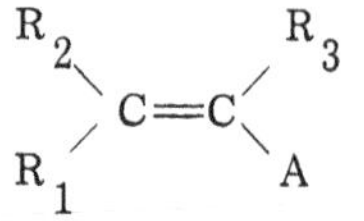

$$\begin{array}{cc} \underset{R_1}{\overset{R_2}{>}}C=C\underset{A}{\overset{R_3}{<}} & \underset{R_2}{\overset{R_1}{>}}C=C\underset{A}{\overset{R_3}{<}} \\ (144\text{-}E) & (144\text{-}Z) \end{array}$$

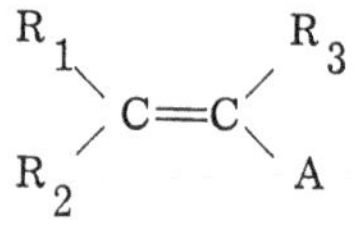

amine and the alkene is maximized, making the amine group coplanar. This is illustrated by comparing the distributions of enamines 144-1, 144-2, and 144-3. The diethylamino group undergoes the largest conjugative interaction of the three amines, so the most abundant isomer is the one with the smallest R_1 group, which is hydrogen in the E stereoisomer. Steric inhibition of resonance between the nitrogen and the aromatic ring by the ortho methyl group in enamine 144-2 causes that amine group to conjugate with the alkene almost to the extent of the diethylamino group. When that steric constraint is gone, however, as in enamine 144-3, the nitrogen delocalizes its electrons into the aromatic ring, greatly diminishing its conjugation with the alkene system. Hence, the strain between R_2 and R_3 becomes more important, and the Z-isomer (144-Z) is favored.

These same principles can be illustrated by comparing the isomer distributions of 144-4, -5, -6, -7, and -8 in Table 10. They all have a common amine, namely morpholine. The UV spectrum of 144-4 shows that the phenyl group is twisted out of the plane of the enamine double bond making the steric requirement of the phenyl less than that of the coplanar amine [348]. Finally, a comparison of the distributions of enamines 144-9 and 144-10 shows the pyrrolidine enamine (144-10) as having the much greater proportion of E-isomer as compared to Z-isomer [356] because of the greater conjugative interaction of pyrrolidine with the alkene group as compared to that of morpholine.

Relating the effect of various substituents on the 144-E ⇌ 144-Z equilibrium and making predictions about this equilibrium for untested compounds is facilitated by the linear free-energy equation shown below [346]. Notations for the E- and Z-isomers are obtained by ordering

$$\ln K = \ln \frac{[E]}{[Z]} = \rho_{c=c} [\lambda^d(R_1) - \lambda^d(R_2)][\lambda^d(R_3) - \lambda^d(R_4)]$$

Table 10 Stereoisomer Distribution of Enamines of Acyclic Ketones and Aldehydes

Cmpd. no. (144-n) n	R_1	R_2	R_3	A	% Composition		Ref.
					E	Z	
1	H	CH_3	C_2H_5	Diethylamino	86	14	340
2	H	CH_3	C_2H_5	N-Methyl-o-toluidino	83	17	341
3	H	CH_3	C_2H_5	N-Methylanilino	20	80	340,341
4	H	CH_3	C_6H_5	Morpholino	98	2	155
5	H	C_6H_5	C_6H_5	Morpholino	88	12	343,347
6	CH_3	C_6H_5	H	Morpholino	85	15	342,344,345
7	C_2H_5	C_6H_5	H	Morpholino	60	40	155
8	$(CH_3)_2CH$	C_6H_5	H	Morpholino	20	80	155
9	H	$(CH_3)_2CH$	H	Morpholino	72	28	356
10	H	$(CH_3)_2CH$	H	Pyrrolidino	97	3	356

the substituents R according to the priority sequence [451] with $R_2 > R_1$ and $R_4 > R_3$ in each pair. The $\lambda^d(R)$'s are substituent param-eters at double bonds, and $\rho_{C=C}$ is the sensitivity factor (0.95 for enamines).

IV. PHYSICAL PROPERTIES OF ENAMINES

A. Infrared Spectra

In the infrared spectra of enamines the double-bond stretching appears generally in the 1600–1680 cm^{-1} region (see Table 11). The intensity of these stretching bands is larger than that of the corresponding olefin stretching bands because of the presence of the more electronegative nitrogen atom instead of a hydrogen or a carbon atom. This results in a greater change in dipole moment upon stretching and hence a more intense band [342]. The intensity of this band is decreased as the symmetry about the carbon-carbon double bond is increased as with tetrasubstituted enamines, and it is increased with diminished symmetry about the stretching alkene bond such as in mono- or disubstituted enamines. The positions of the enamine C=C stretching absorption bands are very similar to those of the corresponding olefins [342]. For example, substitution of the hydrogens attached to the C=C bond by alkyl groups increases the stretching frequency, owing in part to out-of-phase interaction with the attached C—C bonds. So it is ob-served that tetrasubstituted enamines absorb at higher stretching fre-quencies than lesser substituted enamines (see Table 11). When the enamine double bond is exocyclic to a ring, the C=C stretching fre-quency increases as the methylene cycloalkane strain increases, such as when one moves from a methylene cyclohexane to a methylene cyclo-butane.

Delocalization of the C=C π-electrons causes a lowering of the C=C stretching frequency owing to a weakening of the C=C bond force constant. Conjugation with an aromatic ring system may actually lower the stretching frequency to below 1600 cm^{-1}. Similar conjugative effects are observed in dienamines [374], enaminoketones [373], and enamines derived from 5,6-dihydro-2H-thiopyran-3(4H)-one 1,1-dioxide [298] (see Table 11).

The particular amine moiety that is present in the enamine has a relatively small effect on the C=C stretching frequency of the enamine. The effect that is produced by changing the amine group present in the enamine can be generally related to the extent of conjugative inter-action between the nitrogen lone-pair electrons and the alkene π-system. When more conjugative interaction is taking place, the C=C stretching frequency is lowered.

Table 11 Infrared Double-Bond Stretching Frequencies of Some Enamines

Compound	R_1	R_2	R_3	ν_{max} Range (cm^{-1})	Ref.
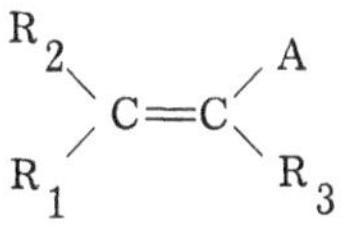	H	H	H	1622–1628	42,351
	Alk	H	H	1653–1657	43,351,352,353
	H	Alk	H	1646	43,351,353
	H	H	Alk	1615	322
	Alk	Alk	H	1660–1680	20,43,223,322,351,355
	Alk	H	Alk	1649	20
	H	H	Ar	1593–1610	322,357
	Ar,H		H	1630–1643	223,353
	Ar,Alk		H	1630–1640	223,355,358
	Ar	H	Ar	1617	343
	H	Ar	Ar	1611	343

| 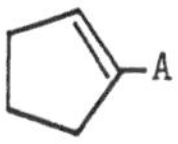 | | | | 1622–1635 | 216,253,322,351,360 |

	R_1	R_2			

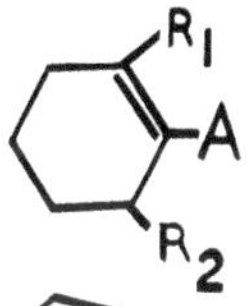	H	H		1630–1660	253,322,351,360,361
	H	Alk		1635–1645	253,275
	Alk	H		1666–1675	253,275

| 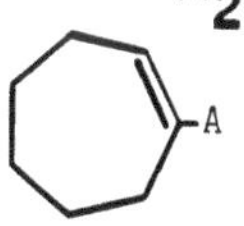 | | | | 1634–1670 | 253,357,361 |

| 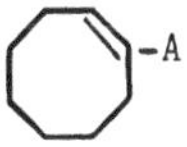 | | | | 1610–1642 | 253,357 |

| 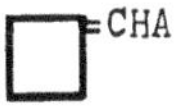 | | | | 1678 | 216 |

| 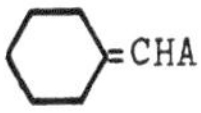 | | | | 1650 | 355 |

Table 11 (Continued)

Compound	R_1	R_2	R_3	ν_{max} Range (cm^{-1})	Ref.
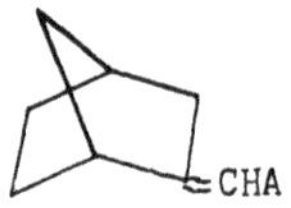				1664	362
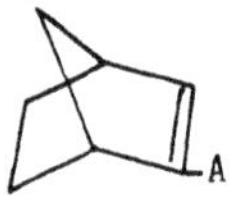				1600—1610	253,362,363,364,41
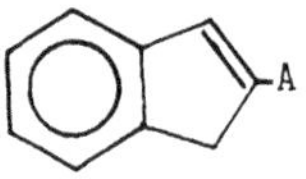				1567—1586	325
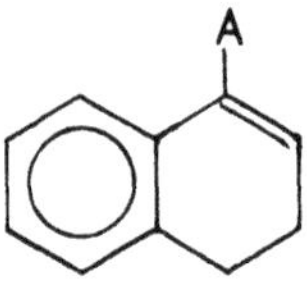				1615—1620	294
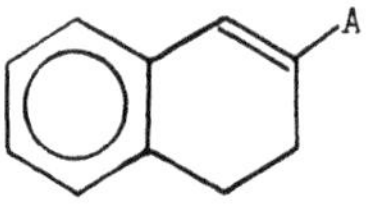				1610—1615	366
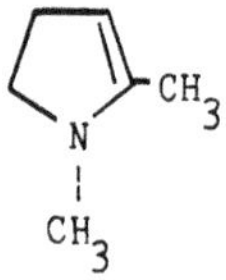				1632—1640	367,368
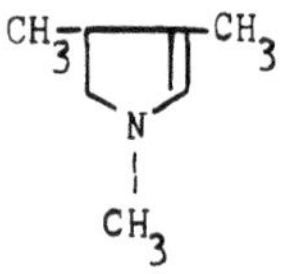				1648	368
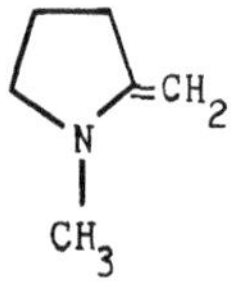				1677	367

Table 11 (Continued)

Compound	R_1	R_2	R_3	ν_{max} Range (cm^{-1})	Ref.
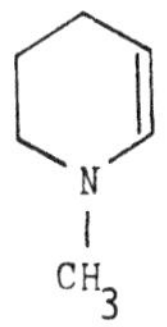 (N-methyl tetrahydropyridine)				1642	37
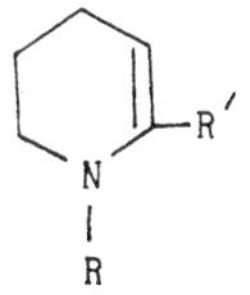 (N,R-tetrahydropyridine with R′)				1635–1650	369–372
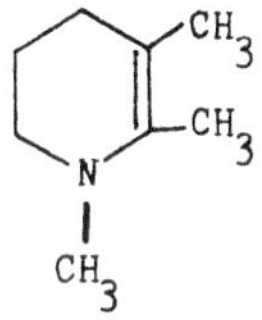 (N,CH₃-dimethyl tetrahydropyridine)				1657	371
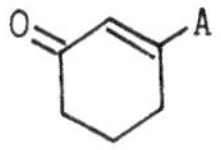 (cyclohexenone with A)				1546–1560	153,298
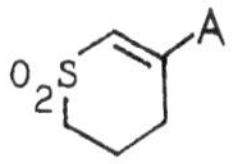 (sulfone ring with A)				1570	298
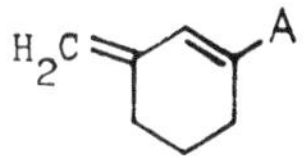 (methylene cyclohexene with A)				1620 (in phase) 1585 (out of phase)	374
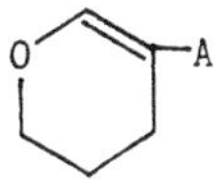 (dihydropyran with A)				1660	390

A = aliphatic amine group; R = aliphatic alkyl group (both cyclic and acyclic).

A hypsochromic shift (shift to higher frequencies) of about 20—50 cm^{-1} is observed in the C=C stretching region when enamines are converted to the corresponding iminum salts by the electrophilic addition of a proton at the β-carbon atom. This shift is accompanied by an enhancement in the intensity of the bond, as would be expected since a greater bond dipole moment is being produced by the protonation which would result in a greater dipole moment change during stretching of the iminium C=N+ bond (146). Leonard and co-workers [368 370,371,375,376] have used this absorption shift as a diagnostic tool for the determination of the position of the double bond with respect to the nitrogen atom. α-Haloenamines also show this shift when they

(147)   (145)   (146)

are protonated to form the iminium salts [377]. The magnitude of this shift can be very large, as in the case of the pyrrolidine enamine of 5,6-dihydro-2H-thiopyran-3(4H)-one 1,1-dioxide (129), which shows an upward shift of frequency (hypsochromic shift) of 95 cm^{-1} when its perchlorate salt is formed [298]. The greater magnitude of this shift is due to the alkene π-electrons moving away from conjugation with the sulfone group upon C-protonation. On the other hand, the pyrrolidine enamine of α-tetralone (119) has a very small upward frequency shift of 15 cm^{-1} when its iminium salt (118) is formed [12]. The alkene π-electrons of the enamine do not move out of conjugation with the aromatic ring in this case.

For enamines in which for steric reasons there is no conjugative interaction between the amine nonbonding electrons and the carbon-carbon double bond, only N-protonation takes place to form the enammonium ion (147). An example of this is found in neostrychnine [369]. In these cases, there is no appreciable difference in the C=C stretching region of the free enamine (145) and its enammonium salt (147). However, there have been other cases observed with enamines that can undergo either C-protonation or N-protonation to the imminium tion gave an isolated enammonium salt with a upward frequency shift (hypsochromic shift) of 21 cm^{-1}, and C-protonation to the imminium salt showed an upward frequency shift of 34 cm^{-1} [358]. With some trinitroarylenamines it has been reported that the upward shift of infrared enamine absorption bonds caused by protonation does not reliably indicate whether it was N-protonation or C-protonation that occurred [379—381].

Table 12   Ultraviolet Maxima of Some Enamines

	Compound			λ max (nm)	ϵ	Ref.
	A	R_1	R_2			
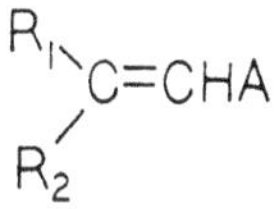 $\begin{array}{c}R_1\\ \diagdown\\ C{=}CHA\\ \diagup\\ R_2\end{array}$	Pyrrolidino	H	CH_3	234	4890	351
		CH_3	CH_3	235	6990	351
	Diethylamino	H	C_2H_5	232.5	8110	351
		CH_3	CH_3	233		41
	Piperidino	H	CH_3	226.5	5930	351
		C_2H_5	C_2H_5	227	4890	351
	Morpholino	H	C_5H_{11}	225	9960	351
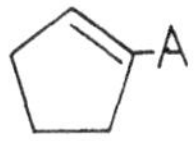 (cyclopentene)—A	Piperidino			222.5	8020	351
	Morpholino			220	6870	351
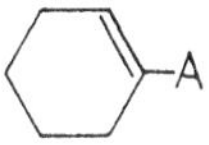 (cyclohexene)—A	Pyrrolidino			233		298
	Piperidino			224.5	8300	351
	Morpholino			222.5	7900	351
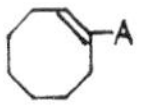 (cyclooctene)—A	Morpholino			228	8700	391
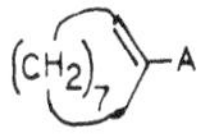 $(CH_2)_7$—A	Morpholino			223	9000	391
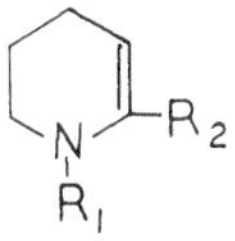 (piperidine ring, N-R_1, R_2)		C_2H_5	CH_3	231	5100	383
		CH_3	C_3H_7	228	2900	384
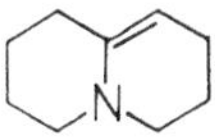 (quinolizidine)				228	5600	383
$CH_2{=}CHCH{=}CH{-}A$	Pyrrolidino			280	22,000	385
	Diethylamino			278	28,500	388
	Piperidino			273	34,200	388
	Morpholino			268	23,500	385,388
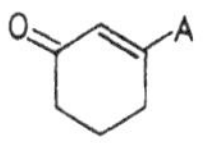 O=(cyclohexene)—A	Pyrrolidino			305	35,550	298
	Morpholino			298	30,900	298
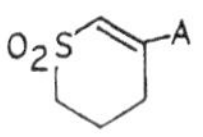 O_2S(thiopyran)—A	Pyrrolidino			251	28,200	298
	Morpholino			250	17,800	298

B. Ultraviolet Spectra

The introduction of an amine auxochrome onto the C=C chromophore of a alkene produces a bathochromic or red shift (shift to longer wavelengths) for the $\pi \to \pi^*$ electronic transition absorption maximum of significant magnitude. Most isolated alkenes show absorption maxima for this transition of about 190 ± 10 nm with an absorptivity of about 10,000 [378,382,387,389]. The enamines produced by addition of an amine group to a C=C chromophore show maxima at 230 ± 10 nm with about the same absorptivity as the original alkene (see Table 12). The results of some amine groups having greater conjugative interaction with the C=C π-system than others are observed in the ultraviolet spectra of enamines also. For example, the pyrrolidino enamines show consistently longer wavelength absorbance than the corresponding piperidino or morpholino enamines owing to the greater conjugative interaction of the lone-pair nitrogen electrons in the pyrrolidine group (see Table 12). Iminium salts formed from C-protonation of enamines exhibit maxima and absorptivities similar to those of their parent enamines [386]. However, enammonium ions produced by N-protonation of enamines show a shift to shorter wavelgnth for the maxima. The circular dichroisms of some iminium salts have been used in the assignment of absolute configurations [460].

Dienamines have ultraviolet absorption maxima that are even more strongly shifted to longer wavelengths when compared to simple enamines (see Table 12). Pyrrolidine dienamines absorb in the range 276—296 nm, piperidine dienamines in the range 268—280 nm, and morpholine dienamines in the range 263—272 nm [87]. So the stronger interacting pyrrolidine dienamines exhibit a larger bathochromic shift than the weaker interacting piperidine or morpholine enamines. An attempt has been made to extend Woodward's rules for predicting absorption maxima of conjugated diene systems to dienamine systems [385]. The pyrrolidine auxochrome was found to shift the parent diene absorption maximum by +63 nm, and morpholine shifts it by +50 nm. Formation of eniminium salts by δ-C-protonation of dienamines brings about either a bathochromic shift or no shift, whereas N-protonation to form dienammonium salts causes definite hypsochromic shifts [87,388]. Both cyclic enone and ensulfone systems show bathochromic shifts with amine auxochromes (see Table 12). Theoretical calculations of the electronic structures and spectral properties of enamines and dienamines have been made [44g].

C. Nuclear Magnetic Resonance Spectra

Nuclear magnetic resonance of hydrogen nuclei in enamines is a useful diagnostic tool for determining the degree of nitrogen electron pair-double bond interaction [149,150]. The β-proton of the enamine has greater shielding as the amount of n-π electron interaction increases. This is shown by an upfield chemical shift of that proton signal in the

Table 13 β-Vinyl H^1-NMR Chemical Shifts of Cyclic Enamines

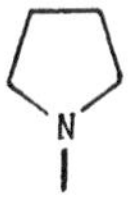

			(CH$_3$)$_2$N-				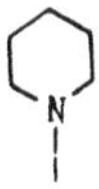 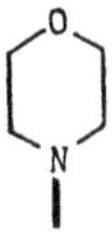 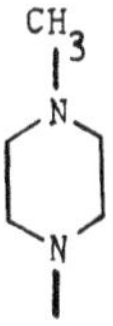 Ref.
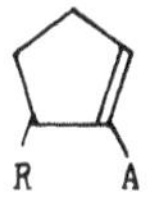 3.95	4.20	—	4.16	4.26	4.39	4.40	153,154,216, 253,322,392, 393
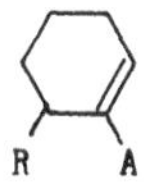 4.20	4.04	4.25	4.41	4.58	4.61	4.56	138,149,150, 153,154,161, 253,322,357, 360,392,394
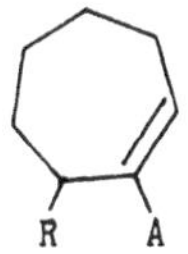 4.37	—	—	—	4.78		4.85	253,357,392, 395
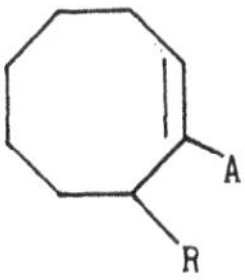 4.08	—	—	—	4.48	4.50	4.53	253,357,392
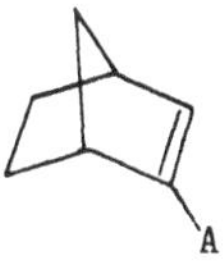 4.13	—	4.27	4.15	4.48	4.60	4.52	253,364,410

δ, ppm (variance of about ±0.05 for chemical shifts); A = amine; R= alkyl.

nmr spectrum. For example, in Table 13 it can be seen that the pyrrolidine enamine signals, which have the largest n-π electron interactions of all the enamines listed, are shifted the farthest upfield. On the other hand, morpholine and N-methylpyrazoline enamines, which show the smallest amount of nitrogen−double bond overlap, have β-vinyl proton signals which are shifted farthest downfield for a given cyclic system. However, all these cyclic enamine β-vinyl signals are shifted upfield relative to those of the corresponding cyclic alkenes. For example, cyclopentene and cyclohexene show olefinic proton signals at 5.60 and 5.59 ppm [397], respectively, whereas their corresponding enamines listed in Table 13 have β-vinyl proton signals at least 1 ppm upfield from those of the unsubstituted alkenes.

A similar shift of β-vinyl enamine protons is observed in acyclic enamines, as can be seen in compound numbers 12−14 and 18−20 in Table 14. Alkyl substitution in the α' position of cyclic enamines does not change the chemical shift of the β-protons appreciably. Some small variations in the chemical shifts of the enamines have been observed caused by whether the spectra are taken of neat samples or whether they are run in solution, and then depending on what solvents are used.

Chemical shifts of cyclic enamines depend on the ring size of the cycloalkene moiety also (see Table 13). Increasing the ring size from a five-membered ring through a 12-membered ring, the β-proton signal shifts downfield, reaching a maximum at the 7-membered ring, and then shifts upfield again as the ring size gets larger.

Extended conjugation of an enamine system with an aromatic ring β-substituent, such as in compound (148) [294], causes deshielding

A = morpholine

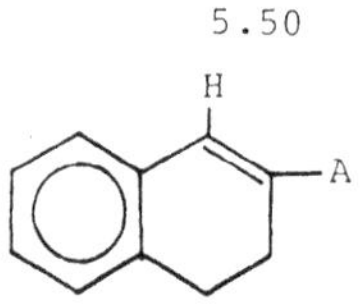

(148) 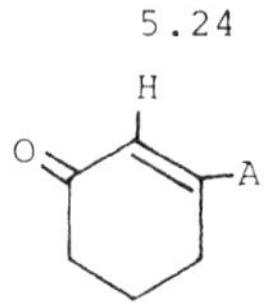   (149) 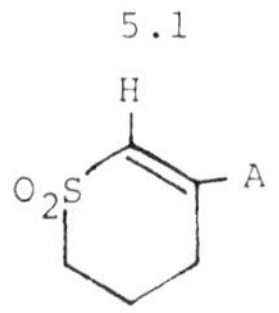   (150)

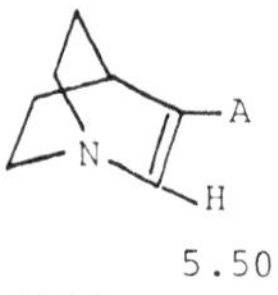

(151) 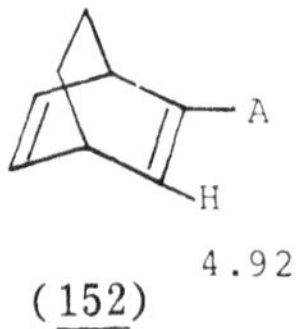   (152) 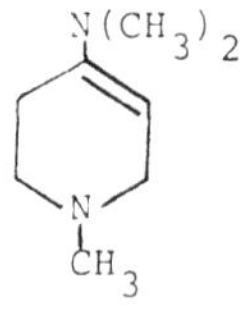   (153)

of the β-proton with the consequent downfield shift of the β-proton signal [294,302,365,366]. Extended conjugation of an enamine system with β-alkene substituents also brings about β-proton deshielding [398]. In a similar manner, conjugation with a carbonyl β-substituent such as enaminoketone 149 [153] or a sulfone β-substituent such as compound 150 [298] results in β-proton deshielding. Substitution of an electronegative halogen [81] or nitrogen at the β-position also results in β-proton deshielding. The latter is demonstrated by comparing β-proton chemical shifts of enamines 151 and 152 in which enamine 151 shows a downfield shift relative to enamine 152 [12,363]. Steric constraints exclude conjugation of the bridgehead nitrogen with the enamine in enamine 151 [405], so the effect of that nitrogen can only be an inductive-field effect. When the nitrogen is further removed, such as enamine 153, it has little or no effect on the β-proton chemical shift [138].

In acyclic enamines the structural isomer that has the β-vinyl proton *syn* to the amine moiety is more shielded and hence resonates at higher fields than the structural isomer that has the β-vinyl proton *anti* to the amine group [138,165,322,349,352,357,396]. Compare, for example, compounds 7 and 8 in Table 14. For those enamines that have both α- and β-vinyl hydrogens, the coupling is greater for the structural isomer with the hydrogens *trans* than with the hydrogens *cis* [165]. This is generally true for vicinal hydrogen-hydrogen coupling in all akenes [397].

The α-vinyl proton in an enamine is less shielded than the β-vinyl proton, as can be seen by the downfield chemical shifts of the former relative to chemical shifts of the latter in Table 14 (compare compounds 5 and 6, 9 and 12, 15 and 18, among other comparisons). It can be seen from Table 14 that the enamine α-vinyl protons become more deshielded (signals shift downfield) as one goes from the poorly interacting morpholine moiety to the more strongly interacting pyrrolidine moiety (see compounds 1−4, 9−11, and 15−17 in Table 14).

An empirical equation has been developed that is useful in predicting the chemical shifts of the protons of variously substituted ethylenes including enamines [401]. The chemical shifts are calculated starting with the equation

$$\delta = 5.28 + \sum Z$$

and substituting in the correct Z values from Table 15, including one of the secondary amine moiety values for the enamine.

Carbon-13 nmr of enamines show the α-vinyl carbon with signals in the range 134−169 ppm downfield from TMS and the β-vinyl carbon with signals in the range 90−134 ppm [37,40,154−157,402−404,461]. The α-alkene carbon signals are shifted downfield from the corresponding simple alkene owing to the deshielding effect of the attached nitrogen. The β-alkene carbon signals are shifted upfield relative to the corres-

Table 14 Vinyl ^{1}H-NMR Chemical Shifts of Acyclic Enamines

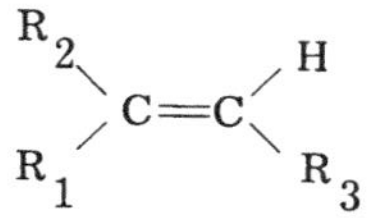

$$R_2 \diagdown \underset{R_1}{\overset{}{C}} = C \overset{H}{\underset{R_3}{\diagup}}$$

Cmpd. no.	R_1	R_2	R_3	Chemical shift (δ, ppm)	Ref.
1	CH_3	CH_3	Pyrrolidino	5.55	12
2	CH_3	CH_3	Dimethylamino	5.40	354, 400
3	CH_3	CH_3	N-Methylpyrazolino	5.32	253
4	CH_3	CH_3	Morpholino	5.26	322
5	Ar	CH_3	Morpholino	5.72	344
6	Ar	Morpholino	CH_3	4.61	396
7	Ar	Et_2N	CH_3	4.51	396

No.					
8	Et$_2$N	Ar	CH$_3$	5.07	396
9	CH$_3$	H	Pyrrolidino	5.72	165
10	CH$_3$	H	Piperidino	5.63	165
11	CH$_3$	H	Morpholino	5.50	165
12	Pyrrolidino	H	CH$_3$	4.04	165
13	Piperidino	H	CH$_3$	4.37	165
14	Morpholino	H	CH$_3$	4.75	165
15	H	CH$_3$	Pyrrolidino	6.07	165
16	H	CH$_3$	Piperidino	5.71	165
17	H	CH$_3$	Morpholino	5.82	165
18	H	Pyrrolidino	CH$_3$	4.01	165
19	H	Piperidino	CH$_3$	4.23	165
20	H	Morpholino	CH$_3$	4.34	165

Table 15   Z Values for Substituted Ethylenes [401]

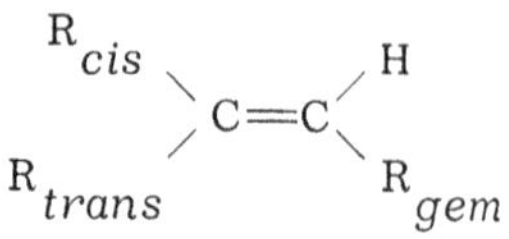

$$\begin{array}{c} R_{cis} \\ \diagdown \\ / \\ R_{trans} \end{array} C{=}C \begin{array}{c} H \\ / \\ \diagdown \\ R_{gem} \end{array}$$

| | Z (ppm) | | |
Substituent R^a	gem	cis	trans
—H	0	0	0
—Alkyl	0.44	−0.26	−0.29
—Alkyl ring	0.71	−0.33	−0.30
—CH_2O, —CH_2I	0.67	−0.02	−0.07
—CN	0.23	0.78	0.58
—C$\equiv$C (isol.)	0.98	−0.04	−0.21
—C$\equiv$C (conj.)	1.26	0.08	−0.01
—C$=$O (isol.)	1.10	1.13	0.81
—C$=$O (conj.)	1.06	1.01	0.95
—$C(O)NR_2$	1.37	0.93	0.35
—CO_2R (isol.)	0.84	1.15	0.56
—CO_2R (conj.)	0.68	1.02	0.33
—Aromatic	1.35	0.37	−0.10
—SO_2	1.58	1.15	0.95
—Cl	1.00	0.19	0.03
—Br	1.04	0.40	0.55
—NR_2 (R aliph.)	0.69	−1.19	−1.31
—NR_2 (R conj.)	2.30	−0.73	−0.81

aThe increments for "R conj." are used instead of those for "R isol." when the substituent R or the double bond in question is conjugated with additional substituents. The increments for "alkyl ring" are used when the substituent under consideration and the double bond form a ring.

ponding simple alkene signals because of the delocalization of the lone-pair nitrogen electrons into the alkene system and the consequent shielding effect. The magnitude of the α-effect by substituents de-shielding the alkene carbon to which they are attached varies widely. The contributions of the amine moieties to the chemical shift of the α-carbon are fairly constant at about 20−25 ppm downfield from the corresponding simple alkene. On the other hand, the contributions of the amine groups to the chemical shift of the β-carbon vary quite a bit, showing an upfield shift from the corresponding simple alkene due to n-π interaction shielding effects. The relative constancy of the α-carbon chemical shift and variability of the β-carbon chemical shift is illustrated with enamines 154 [37] and 155 [404]. Compound 156 possesses a nitrogen on the α-carbon that is sterically inhibited from having

(154)   (155)   (156)

interaction between its nitrogen lone-pair electrons and the alkene π-electrons. The chemical shift of its α-carbon is about the same as that of enamines 154 and 155 owing to a common electronegative nitrogen inductive-field effect. But the chemical shift of the β-carbon is much larger than that of similarly substituted enamine 155 because of the impossibility of n-π nitrogen-alkene interaction in compound 156.

The amine contribution to the chemical shift of the β-carbon of an enamine does reflect the size of the n-π interaction, between amine and alkene, but this is not always easy to observe because various steric shielding effects are superimposed on the electron delocalization effect. However, it has been observed that the enamines of cyclopentanone and cyclohexanone do have upfield chemical shifts for their β-carbons (relative to the corresponding simple cycloalkene) which correlate well with the known electron-donating ability of the attached amine gorup, namely pyrrolidine > diethylamine and azetidine > piperidine > morpholine. The chemical shift range is 93−100 ppm for the β-carbons of these cyclic enamines. The enamines of 3-keto-2-phospholene sulfide also show this correlation in their carbon-13 nmr spectra [107].

Carbon-13 nmr spectroscopy can also be used to distinguish between the E- and Z-configurations of acyclic enamines. The E-isomer β-carbons resonate at a higher field than those of the corresponding Z-isomer [155].

Table 16 ^{13}C NMR Increments for Chemical Shifts of Morpholine Enamines [406]

h g f e
$$\underset{k}{\overset{\text{h g f e}}{\diagdown}} C = C \underset{A}{\overset{\text{a b c d}}{\diagup}}$$

Substituents	Increments		Substituents	Increments	
	C_α	C_β		C_α	C_β
a	4.41	0.06	g	−1.35	−2.12
b	4.41	1.17	h	−0.93	−2.16
c	−0.36	2.05	k	−6.31	23.64
d	0.29	1.57	*y	0.53	−4.74
e	−5.96	13.40	*z	5.57	−7.92
f	−1.69	6.36	*A	147.15	84.12

[a]y is increment for a 6-membered ring; z is increment for a 5-membered ring; A is increment for morpholine substituent.

A table of empirical increments that can be used to calculate the chemical shifts of either the α-carbon or the β-carbon of morpholine enamines is shown in Table 16 [404,406]. The chemical shift is determined by adding together all of the appropriate increments.

The nitrogen-15 nmr spectra of enamines show chemical shifts of 299−309 ppm upfield of $D^{15}NO_3$ for simple cyclic enamines [36,408] and shifts upfield of 319−348 ppm for some acyclic enamines [36,462]. However, there does not seem to be a systematic correlation between ^{15}N chemical shifts and the ability of the amine group to have mesomeric interaction with the alkene. For ^{15}N chemical shifts, a decrease in electron density at a particular atom does not necessarily result in a downfield shift, as is generally the case for ^1H and ^{13}C chemical shifts [408,409]. Apparently with sp^2-hybridized nitrogens carrying a lone-pair, the second-order paramagnetic effect that is associated with the energy of the n → π* transition is an important influence on the shift and will usually dominate simple electron density effects [408].

D. Mass Spectra

One of the most common fragmentations of both acyclic and cyclic enamines is the homolytic cleavage of the bond to the double bond of the

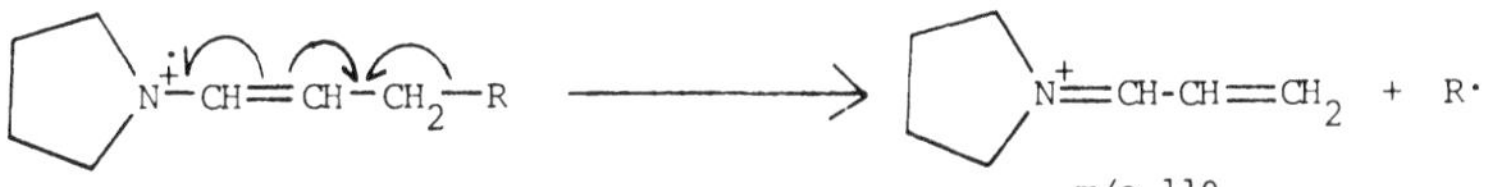

(157)   a: R = H
        b: R = alkyl

m/e 110

(158)

enamine. For example, enamine ion radical 157b, formed on impact with an electron beam, readily fragments into iminium ion 158 as the base peak and an alkyl radical [311]. This takes place more readily than loss of a hydrogen atom because elimination of a hydrogen atom is much less favorable than loss of an alkyl radical. Ion radical 157a cannot lose an alkyl radical in this manner, so it loses a hydrogen atom to form 158 only reluctantly, as shown by the fact that its parent peak is its base peak rather than 158.

This loss of an alkyl radical β to the double bond of the enamine can be used to identify whether there is branching on the enamine double-bond carbon furthest from the nitrogen [311]. For example, ion radical 159 will cleave to give iminium ion 160 as the base peak.

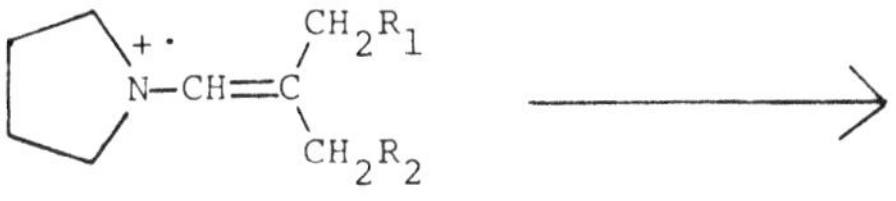

R_1 and/or R_2= alkyl

(159)

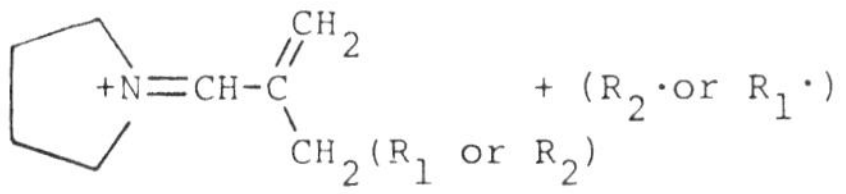

+ (R_2·or R_1·)

(160)

Ion 160 will have a higher m/e value than the ion obtained from the corresponding straight-chain enamine. If both R_1 and R_2 are alkyl or substituted alkyl groups, there will be competition between them as to which will be cleaved. The R group forming the more stable radical will predominate. For instance when R_1 = methyl and R_2 = $CH_2CO_2CH_3$, the latter cleavage predominates about 8 to 1 because it forms the more stable radical.

Cyclic enamines also show this loss of an alkyl radical β to an enamine's double bond. This is illustrated with the radical ion of the

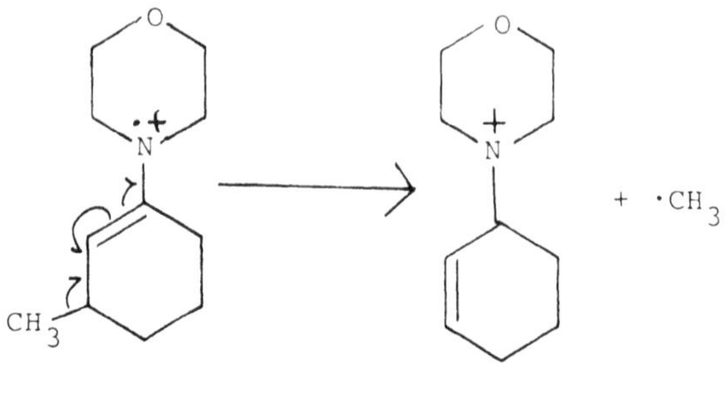

(161)   (162)

morpholine enamine of 3-methylcyclohexanone (161), which fragments
to iminium ion 162 and a methyl radical [311]. Another interesting example of this is with the tetrasubstituted isomers of the morpholine
enamine of 2-propylcyclohexanone (163).

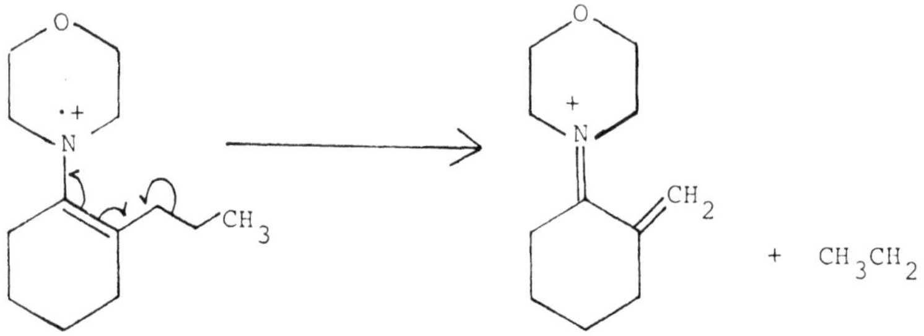

(163)

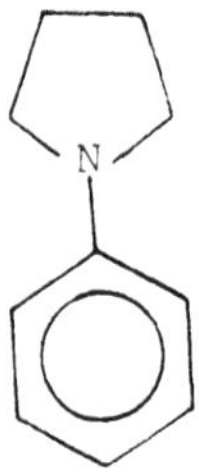

(164)

A significant process that takes place with enamines of cyclohexan-ones is aromatization [311,412]. The pyrrolidine enamine of cyclohex-anone gives, for instance, a significant peak in the mass spectrum for N-phenylpyrrolidine (164). This process is accentuated by an increased temperature of the injection port, and so it seems to be more dependent on temperature than on simple electron impact. Fragmentation is dom-inated by aromatization in hexacyclic dienamines [413].

Bicyclic enamines commonly show retro Diels-Alder fragmentations such as that shown by 165 [410].

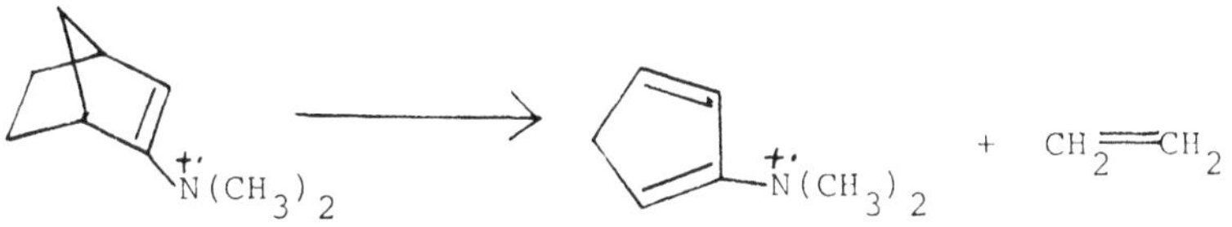

(165)

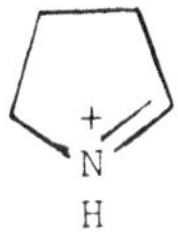

(166) 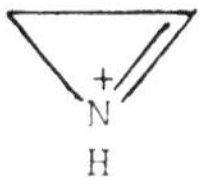   (167)

Fragmentation of the heterocyclic amine moiety often takes place. Pyrrolidine enamines will show peaks at m/e 70 corresponding to ion 166 [311], and aziridine enamines will have peaks at m/e 41 correspond-ing to ion 167 [43,361]. Greater fragmentation than this can also take place with heterocyclic amine groups [414].

E. Basicity

Comparing the basicities of enamines and their corresponding saturated amines has produced a variety of apparently contradictory answers in the literature over the past 45 years. The reason for the differing answers is because of variations in one or more of the following: (a) site of protonation; (b) medium in which protonation takes place; (c) structure of the enamine.

There are two possible sites for protonation in enamines, namely the nitrogen atom and the β-carbon atom [415–417]. In the saturated amine, protonation can only take place at the nitrogen atom. In solu-tion, N-protonation of the ambident enamine to form enammonium ion 168 is the kinetically favored product, whereas C-protonation to form

(168)   + H$^+$ (169)

iminium ion 169 is the thermodynamically favored product. The nature
of the protonating agent seems to be a factor as to where protonation
takes place in solution [418,428]. The hydronium ion, such as is
formed in 70% perchloric acid, leads initially at low temperatures to N-
protonation, whereas the softer carboxylic acids (where the proton is
situated on an uncharged oxygen) preferentially attack the softer
base site, namely the β-carbon atom. Existence of the N-protonated
enammonium salt was based on indirect evidence [148,169,419]. However, protonation of 2-methyl-1-(β-methylstyryl) piperidine (170) provided direct evidence of N-protonation by the formation of white,
crystalline N-protonated salt 171 [358,421—423]. This salt was synthesized by passing dry HCL gas into a benzene solution of 170 at below

(170)   (171)   (172)

0°C. Product 171 was isolated in a dry box. When enammonium salt
171 was warmed in methanol, it changed to iminium salt 172. This
reaction has been shown experimentally not to be an intramolecular
process [417]. Calculations indicate that no intramolecular proton shift
pathway exists, so the reaction probably takes place by way of a bimolecular and/or solvent-assisted process [17,44h].

Attempts have been made to correlate the first ionization potentials
of amines and enamines with their relative basicities in the gas phase
[27,28,434]. These correlation attempts have had limited success, but
there is no compelling theoretical reason why these correlations should
exist. Therefore, they may be simply fortuitous. A linear correlation
based on a strong theoretical foundation has been found between
basicities (expressed as proton affinities) and inner-shell binding
energies within a homologous series of amines [43]. In the gas phase,
the negative of the heat of reaction of

$$B + H^+ \longrightarrow BH^+$$

is defined as the proton affinity (PA) of B [411,424], that is,

$$PA = -[\Delta H_f(BH^+) - \Delta H_f(B) - \Delta H_f(H^+)]$$

This quantity represents the intrinsic basicity of B in the absence of solvent. Three important experimental techniques for determining proton affinities are ion cyclotron resonance [435−439,450], high-pressure mass spectrometry [440−443], and flowing afterglow [444]. Table 17 lists several amines and enamines with their proton affinities and solution pK_as.

It can be seen that in the case of dehydroquinuclidine (no. 2), where resonance interaction between the lone-pair nitrogen electrons and the alkene π-system is impossible, both the gas phase basicity (shown by the proton affinity) and the solution basicity decrease relative to the saturated quinuclidine (no. 1). Only N-protonation is involved, so a simple, unconjugated alkene attached to an amine nitrogen is electron withdrawing by the inductive/field effect and hence base weakening. However, in the cases of the rest of the enamines listed in Table 17 that can undergo n-π interaction, the PA of the enamine is greater than that of the corresponding saturated amine.

It has been shown that C-protonation occurs with enamines in the gas phase [430]. This means that the intrinsic C-protonation basicity of enamines is greater than the N-protonation of the corresponding saturated amines. This same general phenomenon is also observed in solution, as seen by comparing C-protonation pK_as of enamines with N-protonation pK_as of the corresponding saturated amines.

But when N-protonation of the enamine is what is being measured in solution, the pK_a of the enamine (and hence its basicity) is less than that of the corresponding saturated amine (e.g., compare nos. 12 and 13 or 14 and 15). This is due to the inductive/field effect of the alkene group as observed in nos. 1 and 2 also. The relative stabilities of the N-protonated enammonium ion and the C-protonated iminium ion in the gas phase have been calculated for vinylamine (173) using *ab-initio* calculations [17,44d,430,445] and for other enamines using semi-

$CH_2 = NH_2$

(173)

(174)

(175)

empirical MNDO calculations [12]. All the calculations show the C-protonated iminium ion to be more stable than the N-protonated enammonium ion by about 13−18 kcal/mol.

Table 17 Basicities of Some Amines and Enamines

Cmpd. no.		PA^a (kcal/mol)	pK$_a$ C-Protonation	pK$_a$ N-Protonation	Ref.
1	(bicyclic amine structure)	28.7	—	11.29	32,34, 426,429
2	(bicyclic amine structure)	25.9	—	9.82	32,34
3	(2,2-dimethyl-N-methylpiperidine) CH$_3$, CH$_3$, N-CH$_3$	27.4	—	10.3	32,432
4	(2,2-dimethyl-N-methyl enamine) CH$_3$, CH$_3$, N-CH$_3$	30.8	10.45	—	32,432
5	(2-methyl-N-methylpiperidine) CH$_3$, N-CH$_3$	—	—	10.26	427
6	(2-methyl-N-methyl enamine) CH$_3$, N-CH$_3$	—	11.43	—	427
7	(diethyl enamine) C$_2$H$_5$, N-CH$_3$, C$_2$H$_5$	—	9.47	—	446

Table 17 (Continued)

Cmpd. no.		PA[a] (kcal/mol)	pK$_a$		Ref.
			C-Protonation	N-Protonation	
8	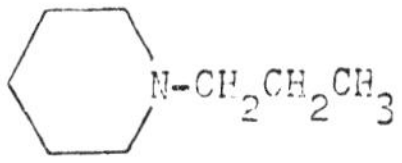 N-CH$_2$CH$_2$CH$_3$ (cyclohexyl)	—	—	10.23	427
9	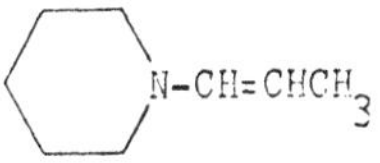 N-CH=CHCH$_3$ (cyclohexyl)	—	10.66	—	427
10	(CH$_3$)$_2$NCH$_2$CH$_2$CH$_3$	22	—	—	430
11	(CH$_3$)$_2$NCH=CHCH$_3$	24.3	—	—	430
12	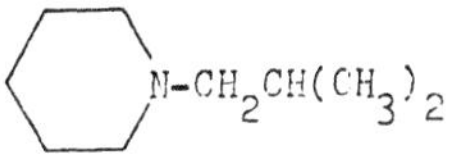 N-CH$_2$CH(CH$_3$)$_2$ (cyclohexyl)	—	—	10.44	431
13	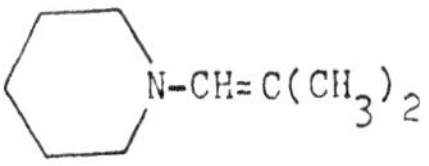 N-CH=C(CH$_3$)$_2$ (cyclohexyl)	—	—	8.35	431
14	(CH$_3$)$_2$NCH$_2$CH(CH$_3$)$_2$	23	—	9.91	430,455
15	(CH$_3$)$_2$N-CH=C(CH$_3$)$_2$	24.2	—	7.85	430,456
16	(CH$_3$)$_2$NCH(CH$_3$)CH$_2$CH$_3$	24	—	—	430
17	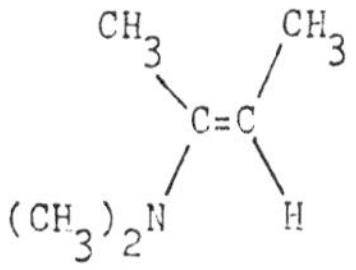 (CH$_3$)$_2$N-C(CH$_3$)=C(CH$_3$)(H)	32.6	—	—	430
18	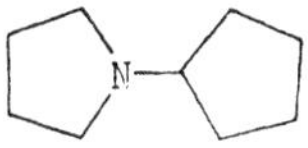 N-cyclopentyl (pyrrolidine)	29.7	—	—	32
19	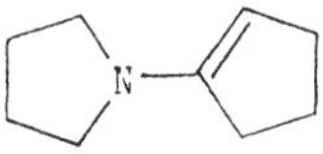 N-cyclopentenyl (pyrrolidine)	35.9	—	—	32

Table 17 (Continued)

Cmpd. no.	PA[a] (kcal/mol)	pK$_a$ C-Protonation	N-Protonation	Ref.
20	—	—	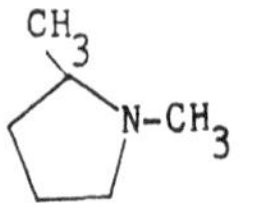 10.24	427
21	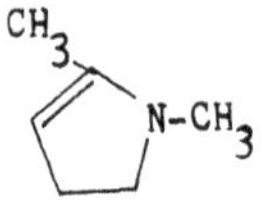 —	11.94	—	427
22	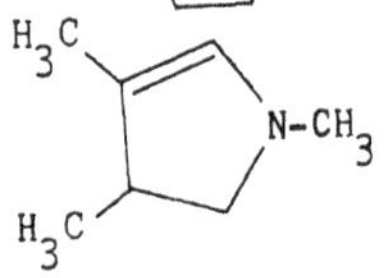 —	9.6	—	368

[a]Proton affinity reported relative to PA of NH$_3$ [433].

Solvent effects will attenuate basicity differences between amines and sometimes even reverse them [425,449]. For example, the relative order of base strengths of amines in water shows secondary amines to be generally stronger bases than tertiary amines [447]. However, the intrinsic basicity of these amines as shown by gas phase proton affinities indicates just the reverse [420,425,426]. Another example is the comparison of N-methylpiperidine (174) and 1-methyl-2,6-di-t-butyl-piperidine (175) in which amine 174 is 0.74 pK units more basic than amine 175 in water-ethanol solution, but 175 has a PA 8.9 kcal/mol greater than 174 [448]. This shows the much greater intrinsic basicity of amine 175 and, consequently, the very large solvent effect. This solvent effect is observed in comparing enamines with their corresponding saturated amines. For example, a comparison can be made in Table 17 between the relative stability of C-protonated enamine no. 4 and its free base as compared to corresponding N-protonated saturated amine no. 3 and its free base. In solution, the comparison shows an increased stability of the protonated form for enamine no. 4 of 0.2 kcal/mol (1.36 kcal/mol per pK$_a$ unit), whereas there is an increase in stability of the protonated form for enamine no. 4 of 3.4 kcal/mol in the gas phase. This type of leveling effect of solvent is quite general when relative gas- and solution-phase basicities are compared [430].

Hydrogen bonding is the most important contributor to differential solvation energies (energy difference between a species being in a gas state and its being in a solvent) [425]. This probably is the explanation for N-protonated enamines being found in solution but not in the gas phase. There should be much more hydrogen bonding to the N-protonated enammonium (and hence greater solvation) than to the C-protonated iminium ion. Therefore, the energy difference between the two ions is greatly diminished in solution, making production of the enammonium ion a viable protonation option.

As far as the structure of the enamine is concerned, α-alkyl substituents have a very pronounced base-strengthening effect on enamines both in the gas state (compare enamine nos. 15 and 17 in Table 17) and in solution (compare enamine nos. 4 and 6). In the gas phase this α-stabilizing effect is larger than in the saturated amines, but the N-substituent effects are smaller [430]. Therefore, the relative gas-phase basicities of amine-enamine pairs with no α-substituents in Table 17, such as nos. 10 and 11 or 14 and 15, are nearly the same. On the other hand, a comparison of the gas-phase basicities of enamines with α-substituents with that of their corresponding saturated amines (such as nos. 16 and 17 or 18 and 19) shows a large increase in enamine basicity. A similar effect has been noted in solution [399] with α-substituents increasing the C-protonation basicity relative to the corresponding saturated amine (nos. 5 and 6 or 20 and 21). Absence of α-substituents will result in very little change in basicity or possibly a decrease in basicity (see nos. 7 and 22).

REFERENCES

1. G. Wittig and H. Blumenthal, *Chem. Ber.*, *60*, 1085 (1927).
2. C. Mannich and H. Davidsen, *Chem. Ber.*, *69*, 2106 (1936).
3a. G. Stork, R. Terrell, and J. Szmuszkovicz, *J. Am. Chem. Soc.*, *76*, 2029 (1954).
3b. G. Stork and H. K. Landesman, *J. Am. Chem. Soc.*, *78*, 5128 (1956).
3c. G. Stork and H. K. Landesman, *J. Am. Chem. Soc.*, *78*, 5129 (1956).
4a. W. P. Jencks, *Catalysis in Chemistry and Enzymology*, McGraw-Hill, New York, 1969, pp. 111–112.
4b. D. Pocar, R. Stradi, and B. Gioia, *Gazz. Chim. Ital.*, *98*, 958 (1968).
4c. S. J. Martinez and J. A. Joule, *Tetrahedron*, *34*, 3027 (1978).
4d. R. M. Coates and E. F. Johnson, *J. Am. Chem. Soc.*, *93*, 4016 (1971).

5a. D. W. Turner, A. P. Baker, and C. R. Brundle, *Molecular Photoelectron Spectroscopy*, Interscience, New York, 1970.

5b. J. W. Rabalais, *Principles of Ultraviolet Photoelectron Spectroscopy*, Wiley, New York, 1977.

5c. H. Bock and B. G. Ramsey, *Angew. Chem. Int. Ed. Engl.*, *12*, 734 (1973).

6. T. Koopmans, *Physica*, *1*, 104 (1934).

7. L. M. Domelsmith and K. N. Houk, *Tetrahedron Lett.*, 1981 (1977).

8. W. von E. Doering, L. Birladeanu, D. W. Andrews, and M. Pognotta, *J. Am. Chem. Soc.*, *107*, 428 (1985).

9. L. L. Miller, G. D. Nordblom, and E. A. Mayeda, *J. Org. Chem.*, *37*, 916 (1972).

10. J. P. Maier and D. W. Turner, *J. Chem. Soc., Faraday Trans. 2*, *69*, 521 (1973).

11. M. J. S. Dewar and W. Thiel, *J. Am. Chem. Soc.*, *99*, 4899 (1977).

12. A. G. Cook, unpublished results.

13. T. A. Albright, J. K. Burdett, and W. H. Whangbo, *Orbital Interactions in Chemistry*, Wiley, New York, 1985.

14. C. C. Levin, *J. Am. Chem. Soc.*, *97*, 5649 (1975).

15. J. Teysseyre, J. Arriau, A. Dargelos, and J. Elguero, *J. Chim. Phys.*, *72*, 303 (1975).

16. J. Teysseyre, J. Arriau, A. Dargelos, J. Elguero, and A. R. Katritzky, *Bull. Soc. Chim. Belg.*, *85*, 39 (1976).

17. K. Muller and L. D. Brown, *Helv. Chim. Acta*, *61*, 1407 (1978).

18. R. Meyer, *Helv. Chim. Acta*, *61*, 1418 (1978).

19. K. Muller, *Chimia*, *34*, 310 (1980).

20. K. Muller, F. Previdoli, and H. Desilvestro, *Helv. Chim. Acta*, *64*, 2497 (1981).

21. C. Glidewell, *J. Mol. Struct.*, *89*, 349 (1982).

22. L. N. Koikov, P. B. Terent'ev, I. P. Gloriozov, and Y. G. Bundel, *Zh. Org. Khim.*, *20*, 917 (1984).

23. F. J. Lovas and F. O. Clark, *J. Chem. Phys.*, *62*, 1925 (1975).

24. M. Forchiassin, A. Risaliti, C. Russo, N. B. Pahor, and M. Calligaris, *J. Chem. Soc., Perkin Trans.*, *1*, 935 (1977).

25. K. L. Brown, L. Damm, J. D. Dunitz, A. Eschenmoser, R. Hobi, and C. Kratky, *Helv. Chim. Acta*, *61*, 3108 (1978).

26. P. M. VanMeerssche, G. Germain, J. P. Declercq and A. Colens, *Acta Cryst.*, *B35*, 907 (1979).

27. K. Yoshikawa, M. Hashimoto, and I. Morishima, *J. Am. Chem. Soc.*, *96*, 288 (1974).

28. F. P. Colonna, G. Distefano, S. Pignataro, G. Pitacco, and E. Valentin, *J. Chem. Soc., Faraday Trans.*, *2*, 1572 (1975).

29. A. D. Walsh, *Discuss Faraday Soc.*, *2*, 18 (1947).
30. H. A. Bent., *Chem. Rev.*, *61*, 275 (1961).
31. R. Meyer, *Chimia*, *31*, 55 (1977).
32. R. Houriet, J. Vogt, and E. Haselbach, *Chimia*, *34*, 277 (1980).
33. R. S. Brown, *Can. J. Chem.*, *54*, 1521 (1976).
34. C. Grob, A. Kaiser, and E. Renk, *Chem. and Ind. (London)* 598 (1957); C. Grob, A. Kaiser, and E. Renk, *Helv. Chim. Acta*, *40*, 2170 (1957).
35. S. A. Cowling and R. A. W. Johnstone, *J. Elect Spect. Rel. Phenom.*, *2*, 161 (1973).
36. W. Schwotzer and W. vonPhilipsborn, *Helv. Chim. Acta*, *60*, 1501 (1977).
37. P. Beeken and F. W. Fowler, *J. Org. Chem.*, *45*, 1336 (1980).
38. H. A. Bent, *J. Chem. Phys.*, *32*, 1259 (1960).
39. T. Itoh, K. Kaneda, I. Wananabe, S. Ideda, and S. Teranishi, *Chem. Lett.*, 227 (1976).
40. K. Kaneda, T. Itoh, N. Kii, K. Jitsukawa, and S. Teranishi, *J. Mol. Catalysis*, *15*, 349 (1982).
41. P. Dupuis, C. Sandorfy, and D. Vocelle, *Photochem. Photobiol.*, *39*, 391 (1984).
42. P. Masclet, D. Grosjean, G. Mouvier, and J. Dubois, *J. Elect. Spectr.*, *2*, 225 (1973).
43. K. Muller and F. Previdoli, *Helv. Chim. Acta*, *64*, 2508 (1981).
44a. H. Bock, G. Wagner, K. Wittel, J. Sauer, and D. Seebach, *Chem. Ber.*, *107*, 1869 (1974).
44b. F. Texier and J. Bourgois, *Bull. Soc. Chim. France*, 487 (1976).
44c. K. N. Houk, J. Sims, R. E. Duke, Jr., R. W. Strozier, and J. K. George, *J. Am. Chem. Soc.*, *95*, 7287 (1973).
44d. R. A. Eades, D. A. Weil, M. R. Ellenberger, W. E. Farneth, D. A. Dixon, and C. H. Douglass, Jr., *J. Am. Chem. Soc.*, *103*, 5372 (1981).
44e. A. F. Freimanis, Y. P. Stradyn, *Zh. Obshch. Chim.*, *39*, 631 (1969).
44f. K. Yamaguchi, *Int. J. Quantum Chem.*, *20*, 393 (1981).
44g. E. M. Evleth, *J. Am. Chem. Soc.*, *89*, 6445 (1967).
44h. J. Teysseyre, J. Arriau, A. Dargelos, and J. Elguero, *J. Chim. Phys, Phys.-Chim. Biol.*, *72*, 303 (1975).
44i. J. Dabrowski, K. Kamienska-Trela, and L. Kozerski, *Org. Mag. Resonance*, *6*, 43 (1974).
44j. J. C. Meslin, Y. T. N'Guessan, and H. Quiniou, *Tetrahedron*, *31*, 2679 (1975).
45. L. N. Koikov, P. B. Terent'ev, I. P. Gloriozov, and Y. G. Bundel, *Zh. Org. Khim.*, *20*, 1629 (1984).

46. N. Bodor and R. Pearlman, *J. Am. Chem. Soc.*, *100*, 4946 (1978).

47. M. P. Sammes, R. L. Harlow, and S. H. Simonsen, *J. Chem. Soc.*, *Perkin Trans.*, *2*, 1126 (1976).

48. H. Vanbrabant-Govaerts and P. Huyskens, *Bull. Soc. Chim. Belg.*, *90*, 767 (1981).

49. M. Azzaro, J. F. Gal, S. Geribaldi, B. Videau, and A. Loupy, *J. Chem. Soc., Perkin Trans.*, *2*, 57 (1983).

50. M. LeBlanc, G. Santini, J. Gallucci, and J. G. Riess, *Tetrahedron*, *33*, 1453 (1977).

51. J. V. Greenhill, *Chem. Soc. Rev.*, *6*, 277 (1977).

52. J. B. Rasmussen, R. Shabana, and S-O. Lawesson, *Tetrahedron*, *38*, 1705 (1982).

53. J. B. Rasmussen, R. Shabana, and S-O. Lawesson, *Tetrahedron*, *37*, 3693 (1981).

54. J. Liebscher and H. Hartmann, *Synthesis*, 521 (1976).

55. R. Helmers, *Angew. Chem. Int. Ed. Engl.*, *10*, 725 (1971).

56. Y. L. Chow and D. W. L. Chang, *Chem. Commun.*, 64 (1971).

57. S. Rajappa, *Tetrahedron*, *37*, 1453 (1981).

58a. H. H. Jaffe, *Chem. Rev.*, *53*, 191 (1953).

58b. R. W. Taft, Jr., *J. Am. Chem. Soc.*, *79*, 1045 (1957).

58c. H. C. Brown and Y. Okamoto, *J. Am. Chem. Soc.*, *80*, 4979 (1958).

59. D. G. Lister and J. K. Tyler, *Chem. Commun.*, 152 (1966).

60a. W. J. E. Parr and R. E. Wasylishen, *J. Mol. Struct.*, *38*, 272 (1977).

60b. K. N. Trueblood, E. Goldish and J. Donohue, *Acta Crystallogr.*, *14*, 1009 (1961).

60c. D. L. Hughes and J. Trotter, *J. Chem. Soc. (A)*, 2181 (1971).

61. G. Garbieu, R. Benassi, R. Grandi, U. M. Pagnoni, and F. Taddei, *J. Chem. Soc. Perkin Trans.*, *2*, 330 (1979).

62. O. Exner, *Dipoles Moments in Organic Chemistry*, Georg Thieme, 1975.

63. V. I. Minkin, O. A. Osipov, and Y. A. Zhdanov, *Dipole Moments in Organic Chemistry*, Plenum, New York, 1970.

64. S. Rajappa, S. Sreenivasan, B. G. Advani, R. H. Sommerville, and R. Hoffmann, *Ind. J. Chem.*, *15B*, 297 (1977).

65. G. Hammond, *J. Am. Chem. Soc.*, *77*, 334 (1955).

66. M. J. S. Dewar, *The Molecular Orbital Theory of Organic Chemistry*, McGraw-Hill, New York, 1969.

67. L. Duhamel, P. Duhamel, and G. Ple, *Bull. Soc. Chim. France*, 4423 (1968).

68. P. Duhamel, L. Duhamel and J-L. Klein, *Bull. Soc. Chim. France*, 2517 (1973).

69. L. Duhamel, P. Duhamel, and G. Ple, *Comp. Rend.*, *271*, 751 (1970).
70. P. Duhamel, L. Duhamel, and J-Y. Valnot, *Comp. Rend.*, *273*, 835 (1971).
71. P. Duhamel, L. Duhamel, and P. Siret, *Comp. Rend.*, *276*, 519 (1973).
72. L. Duhamel, J.-M. Poirier and N. Tedga, *J. Chem. Res. (S)*, 222 (1983).
73. S. Kuno and Y. Sato, *J. Organomet. Chem.*, *218*, 309 (1981).
74. P. Duhamel, L. Duhamel, and J. Chauvin, *Compt. Rend.*, *274*, 1233 (1972).
75. J. Barluenga, F. Aznar, and R. Liz, *Synthesis*, 304 (1984).
76. E. Vilsmaier and C. M. Klein, *Angew. Chem. Int. Ed. Engl.*, *18*, 800 (1979).
77. E. Vilsmaier, W. Troger, W. Sprugel, and K. Gagel, *Chem. Ber.*, *112*, 2997 (1979).
78. J. Firl, H. Braun, A. Amann, and R. Barnert, *Z. Naturforsch*, *35b*, 1406 (1980).
79. N. DeKimpe and N. Schamp, *Org. Prep. Proc.*, *13*, 241 (1981).
80. N. DeKimpe and N. Schamp, *Org. Prep. Proc.*, *15*, 71 (1983).
81. P. Granger, S. Chapelle, and J-M. Poirier, *Org. Mag. Resonance*, *14*, 69 (1980).
82. L. Duhamel, P. Duhamel, and J-M. Poirier, *Tetrahedron Lett.*, 4237 (1973).
83. L. Duhamel, P. Duhamel and J-M. Poirier, *Bull. Soc. Chim. France*, 221 (1972).
84. S. J. Huang and M. V. Lessard, *J. Am. Chem. Soc.*, *90*, 2432 (1968).
85. L. Marchetti and V. Passalacqua, *Ann. Chim.*, *57*, 1275 (1967).
86. S. Fatutta, G. Pitacco, and E. Valentin, *J. Chem. Soc.*, *Perkin Trans.*, *1*, 2735 (1983).
87. P. W. Hickmott, *Tetrahedron, 40*, 2989 (1984).
88. A. A. M. Houwen-Claassen, J. W. McFarland, B. H. M. Lammerink, L. Thijs, and B. Zwaenburg, *Synthesis*, 628 (1983).
89. M. S. Chattha and A. M. Aguiar, *J. Org. Chem.*, *38*, 820 (1973).
90. L. Duhamel and J-M. Poirier, *J. Am. Chem. Soc.*, *99*, 8356 (1977).
91. L. Duhamel and J-M. Poirier, *Bull. Soc. Chim. France*, 297 (1982).
92. J. Ficini, *Tetrahedron, 32*, 1449 (1976).
93. V. G. Granik, *Usp. Khim.*, *53*, 651 (1984).
94. J-P. Picard, A. Aziz-Elyusulfi, R. Calas, J. Dunogues, and N. Duffaut, *Organometallics, 3*, 1660 (1984).
95. H. Baganz and L. Domaschke, *Chem. Ber.*, *95*, 2095 (1962).

96. H. Weingarten and W. A. White, *J. Am. Chem. Soc.*, *88*, 850 (1966).

97. J. S. Hartman and E. C. Kelusky, *Can. J. Chem.*, *59*, 1284 (1981).

98. N. DeKimpe, R. Verhe, L. DeBuyck, and N. Schamp, *Chem. Ber.*, *116*, 3846 (1983).

99. J. Toye and L. Ghosez, *J. Am. Chem. Soc.*, *97*, 2276 (1975).

100. N. DeKimpe, R. Verhe, L. DeBuyck, H. Haskim, and N. Schamp, *Tetrahedron*, *32*, 3063 (1976).

101. C. H. Robinson, L. Milewich, and K. Huber, *J. Org. Chem.*, *36*, 211 (1971).

102. J. E. Baldwin, R. H. Fleming, and D. M. Simmons, *J. Org. Chem.*, *37*, 3963 (1972).

103. M. Fukuda, K. Kan, Y. Okamoto, and H. Sakurai, *Bull. Chem. Soc. Jpn.*, *48*, 2103 (1975).

104. H. Ahlbrecht and W. Farnung, *Synthesis*, 336 (1977).

105. H. Ahlbrecht and W. Farnung, *Chem. Ber.*, *117*, 1 (1984).

106. B. Costisella, I. Keitel, and H. Gross, *Tetrahedron*, *37*, 1227 (1981).

107. I. Allade, P. Dubois, P. Levillain, and C. Viel, *Bull. Soc. Chim. France*, 339 (1983).

108. C. B. Kanner and U. K. Pandit, *Tetrahedron*, *38*, 3597 (1982).

109. L. Ghosez, B. Haveaux, and H. G. Viehe, *Angew. Chem. Int. Ed. Engl.*, *8*, 454 (1969).

110. J. Marchand-Brynaert and L. Ghosez, *J. Am. Chem. Soc.*, *94*, 2869 (1972).

111. L. Ghosez, *Angew. Chem. Int. Ed. Engl.*, *11*, 852 (1972).

112. A. Colens, M. Demuylder, B. Techy, and L. Ghosez, *Nouveau J. Chim.*, *1*, 369 (1977).

113. A. Colens and L. Ghosez, *Nouveau J. Chim.*, *1*, 371 (1977).

114. L. Ghosez and J. Marchand-Bryaert, in *Iminuim Salts in Organic Chemistry* (H. Bohme and H. G. Vieke, eds.), in *Advances in Organic Chemistry* (E. C. Taylor, ed.), Vol. 9, Wiley-Interscience, New York, 1976.

115. D. Seebach, *Angew. Chem. Int. Ed. Engl.*, *18*, 329 (1979).

116. R. M. Acheson, G. N. Aldrich, M. C. K. Choi, J. O. Nwankwo, M. A. Ruscoe, and J. D. Wallis, *J. Chem. Res. Synop.*, 100 (1984).

117. C. Wiaux-Zamar, J-P. Dejonghe, L. Ghosez, J. F. Normat, and J. Villieras, *Angew. Chem. Int. Ed. Engl.*, *15*, 371 (1976).

118. H. Ahlbrecht, *Chimia*, *31*, 391 (1977).

119. H. W. Thompson and B. S. Huegi, *Chem. Commun.*, 636 (1973).

120. H. W. Thompson and B. S. Huegi, *J. Chem. Soc., Perkin Trans.*, *1*, 1603 (1976).

121. H. Ahlbrecht and G. Rauchschwalbe, *Synthesis*, 417 (1973).
122. H. Ahlbrecht and C. S. Sudheendranath, *Synthesis*, 717 (1982).
123. B. Costisella and H. Gross, *Tetrahedron*, *38*, 139 (1982).
124. B. Costisella, H. Gross, and H. Schick, *40*, 733 (1984).
125. A. deSavignac, M. Bon, and A. Lattes, *Bull. Soc. Chim. France*, 3167 (1972).
126. B. A. Shainyan and A. N. Mirskova, *Usp. Khim.*, *48*, 201 (1979).
127. P. H. Hickmott, *Tetrahedron*, *38*, 3363 (1982).
128. G. R. Lenz, *Synthesis*, 489 (1978).
129. G. Sturtz, *Bull. Chim. France*, 1345 (1967).
130. A. Krowczynski and L. Kozerski, *Synthesis*, 489 (1983).
131. A. G. Brook, C. Golino, and E. Matern, *Can. J. Chem.*, *56*, 2286 (1978).
132. H. Ahlbrecht and E.-O. Duber, *Synthesis*, 630 (1980).
133. W. Walter and H-W. Luke, *Angew. Chem. Int. Ed. Engl.*, *16*, 535 (1977).
134. A. M. Churakov, S. L. Ioffe, B. N. Khasapov, and V. A. Tartakovskii, *Izv. Akad. Nauk SSSR, Ser. Khim.*, 113 (1976).
135. R. J. P. Corriu, V. Huynk, J. J. E. Moreau, and M. Pataud-Sat, *Tetrahedron Lett.*, 3257 (1982).
136. M. Fourtinon, B. D. Jeso, and J-C. Pommier, *J. Organomet. Chem.*, *193*, 165 (1980).
137. R. Kupper and C. J. Michejda, *J. Org. Chem.*, *45*, 2919 (1980).
138. H. V. Hirsch, *Chem. Ber.*, *100*, 1289 (1967).
139. S. F. Martin, *Synthesis*, 633 (1979).
140. R. Knorr, A. Weiss, P. Low, and E. Rapple, *Chem. Ber.*, *113*, 2462 (1980).
141. B. D. Jeso and J-C. Pommier, *J. Organomet. Chem.*, *186*, C9 (1980).
142. B. D. Jeso and J-C. Pommier, *J. Organomet. Chem.*, *122*, C1 (1976).
143. J-M. Brocas, B. D. Jeso, and J-C. Pommier, *J. Organomet. Chem.*, *120*, 217 (1976).
144. D. E. Bergbreiter and M. Newcomb, in *Asymmetric Synthesis* (J. D. Morrison, ed.), Vol. 2, Academic Press, New York, 1983, p. 243.
145. J. S. Krouwer and J. P. Richmond, *J. Org. Chem.*, *43*, 2464 (1978).
146. V. P. Ivshin, V. F. Smirnov, and O. A. Yashukova, *Zh. Org. Khim.*, *19*, 1416 (1983).
147. S. Patai, editor, *The Chemistry of the Carbon Nitrogen Double Bond*, Interscience, New York, 1970.

148. G. Stork, A. Brizzolara, H. Landesman, J. Szmuszkovicz, and R. Terrell, *J. Am. Chem. Soc., 85,* 207 (1963).

149. W. D. Gurowitz and M. A. Joseph, *Tetrahedron Lett.,* 4433 (1965).

150. W. D. Gurowitz and M. A. Joseph, *J. Org. Chem., 32,* 3289 (1967).

151. M. E. Kuehne, *J. Am. Chem. Soc., 84,* 837 (1962).

152. T. Severin and I. Brautigam, *Chem. Ber., 112,* 3007 (1979).

153. E. J. Cone, R. H. Garner, and A. W. Hayes, *J. Org. Chem., 37,* 4436 (1972).

154. D. Tourwe, G. VanBinst, S. A. G. DeGraaf, and U. K. Pandit, *Org. Mag. Resonance, 7,* 433 (1975).

155. R. Stradi, P. Trimarco, and A. Vigevani, *J. Chem. Soc., Perkin Trans., 1,* 1 (1978).

156. M. G. Ahmed and P. W. Hickmott, *J. Chem. Soc., Perkin Trans., 2,* 838 (1977).

157. M. Ahmed, P. W. Hickmott, and R. D. Soelistyowati, *J. Chem. Soc., Perkin Trans., 2,* 372 (1978).

158. R. S. Jain, H. F. Lawson, and L. D. Quin, *J. Org. Chem., 43,* 108 (1978).

159. M. G. Ahmed, S. A. Ahmed, and P. W. Hickmott, *J. Chem. Soc., Perkin Trans., 1,* 2383 (1980).

160. T. Chen, H. Kato, and M. Ohta, *Bull. Chem. Soc. Jpn., 43,* 1913 (1970).

161. H. W. Thompson and J. Swistok, *J. Org. Chem., 46,* 4907 (1981).

162. E. L. Eliel, N. L. Allinger, S. J. Angyal, and G. A. Morrison, *Conformational Analysis,* Interscience, New York, 1965, pp. 19–21.

163. C. C. Price and W. H. Snyder, *Tetrahedron Lett.,* 69 (1962).

164. J. Sauer and H. Prahl, *Tetrahedron Lett.,* 2863 (1966).

165. J. Sauer and H. Prahl, *Chem. Ber., 102,* 1917 (1969).

166. M. Riviere and A Lattes, *Bull. Soc. Chim. France,* 4430 (1968).

167. G. Pitacco and E. Valentin, in *The Chemistry of Amino, Nitroso, Nitro Compounds and Their Derivatives,* (S. Patai, ed.), Part 1, Wiley, New York, 1982, p. 623.

168. A. H. Hubert, *J. Chem. Soc. (C),* 2048 (1968).

169. G. Opitz and A. Griesinger, *Ann., 665,* 101 (1963).

170. W. H. Daly, J. G. Underwood, and S. C. Kuo, *Tetrahedron Lett.,* 4375 (1971).

171. S. J. Martinez, L. Dalton, and J. A. Joule, *Tetrahedron, 40,* 3339 (1984).

172. M. Riviere and A. Lattes, *Bull. Soc. Chim. France,* 730 (1972).

173. G. T. Martirosyan, M. G. Indzhikyan, E. A. Grigoryan, and A. T. Babayan, *Arm. Khim. Zh., 20,* 275 (1967).

174. G. T. Martirosyan, E. A. Grigoryan, and A. T. Babayan, *Arm. Khim. Zh.*, *24*, 971 (1971).

175. A. T. Malkhasyan, L. V. Asratyan, and E. A. Grigoryan, *Arm. Khim. Zh.*, *30*, 239 (1977).

176. D. V. Grigoryan, A. Z. Gevorkyan, and A. T. Babayan, *Arm. Khim. Zh.*, *32*, 789 (1979).

177. W. R. Ashcroft, S. J. Martinez, and J. A. Joule, *Tetrahedron*, 37, 3005 (1981).

178. J. Hine, S.-M. Linden, A. Wang, and V. Thiagarajan, *J. Org. Chem.*, *45*, 2821 (1980).

179. M. K. Huber and A. S. Dreiding, *Helv. Chim. Acta*, 57, 748 (1974).

180. J. Hine and M. J. Skoglund, *J. Org. Chem.*, *47*, 4758 (1982).

181. M. Prochaza, V. Krestanova, M. Palecek, and K. Pecka, *Collect. Czech. Chem. Commun.*, *35*, 3813 (1970).

182. *API Tables*, Project 44, Carnegie Inst. Tech., Pittsburgh, 1953–61.

183. R. B. Turner and W. R. Meador, *J. Am. Chem. Soc.*, *79*, 4133 (1957).

184. R. B. Woodward and R. Hoffmann, *The Conservation of Orbital Symmetry*, Verlag Chemie, Weinheim, 1970.

185. P. Houdewind and U. K. Pandit, *Tetrahedron Lett.*, 2359 (1974).

186. K. A. Parker and R. W. Kosley, Jr., *Tetrahedron Lett.*, 341 (1976).

187. J. Oda, T. Igarashi, and Y. Inouye, *Bull. Inst. Chem. Res., Kyoto Univ.*, *54*, 180 (1976).

188. B. P. Mundy and W. G. Bornmann, *Synth. Commun.*, *8*, 227 (1978).

189. B. P. Mundy and W. G. Bornmann, *Tetrahedron Lett.*, 957 (1978).

190. T.-T. Wu, J. L. Moniot, and M. Shamma, *Tetrahedron Lett.*, 3419 (1978).

191. R. K. Hill and H. N. Khatri, *Tetrahedron Lett.*, 4337 (1978).

192. L. Radom, J. A. Pople, and P. V. R. Schleyer, *J. Am. Chem. Soc.*, *95*, 8193 (1973).

193. J. Barluenga, F. Aznar, R. Liz and M. Bayod, *Chem. Commun.*, 1427 (1984).

194. L. Birkofer and G. Daum, *Angew. Chem.*, *72*, 707 (1960).

195. L. Birkofer and G. Daum, *Chem. Ber.*, *95*, 183 (1962).

196. M. Yu. Lukima, *Russ. Chem. Revs.*, *31*, 419 (1962).

197. M. Orchin, F. Kaplan, R. S. Macomber, R. M. Wilson, and H. Zimmer, *The Vocabulary of Organic Chemistry*, Wiley, New York, 1980.

198. S. Winstein, *Quart. Rev. Chem. Soc.*, *23*, 141 (1969).

199. M. E. Kuehne and J. C. King, *J. Org. Chem.*, *38*, 304 (1973).

200. C. Kaiser, A. Burger, L. Zirngibl, C. S. Davis, and C. L. Zirkle, *J. Org. Chem.*, *27*, 768 (1962).

201. W. A. Bernett, *J. Chem. Ed.*, *44*, 17 (1967).
202. U. K. Pandit and S. A. G. deGraaf, *Chem. Commun.*, 659 (1972).
203. R. K. Boeckman, Jr., P. F. Jackson, and J. P. Sabatucci, *J. Am. Chem. Soc.*, *107*, 2191 (1985).
204. W. Klop, P. A. A. Klusener, and L. Brandsma, *Rec. Trav. Chim.*, *103*, 27 (1984).
205. J. Szmuszkovicz, E. Cerda, M. F. Grostic, and J. F. Ziersel, Jr., *Tetrahedron Lett.*, 3969 (1967).
206. J. Szmuszkovicz, D. J. Duchamp, E. Cerda, and C. G. Chidester, *Tetrahedron Lett.*, 1309 (1969).
207. H. H. Wasserman and M. S. Baird, *Tetrahedron Lett.*, 1729 (1970).
208. E. Jongejan, H. Steinberg, and T. J. deBoer, *Synth. Commun.*, 4, 11 (1974).
209. E. Jongejan, H. Steinberg, and T. J. deBoer, *Tetrahedron Lett.*, 397 (1976).
210. E. Jongejan, W. J. M. van Tilborg, C. H. V. Dusseau, H. Steinberg, and T. J. deBoer, *Tetrahedron Lett.*, 2359 (1972).
211. M. Rey and A. S. Dreiding, *Helv. Chim. Acta*, 57, 734 (1974).
212. A. G. Cook, S. B. Herscher, D. J. Schultz, and J. A. Burke, *J. Org. Chem.*, 35, 1550 (1970).
213. M. K. Huber, R. Martin, M. Rey, and A. S. Dreiding, *Helv. Chim. Acta*, *60*, 1781 (1977).
214. K. Hafner, K. H. Hafner, C. Konig, M. Kreuder, G. Ploss, G. Schulz, E. Sturm, and K. H. Vopel, *Agnew. Chem. Int. Ed. Engl.*, *2*, 123 (1963).
215.. K. Hafner, K. H. Vopel, G. Ploss, and C. Konig, in *Organic Synthesis*, (W. D. Emmons, ed.), Vol. 47, Wiley, New York, 1967, pp. 52–54.
216. K. L. Erickson, J. Markstein, and K. Kim, *J. Org. Chem.*, *36*, 1024 (1971).
217. N. J. Leonard, *Rec. Chem. Prog.*, *17*, 243 (1956).
218. F. L. Pyman, *J. Chem. Soc.*, 817 (1913).
219. R. A. Johnson, *J. Org. Chem.*, *37*, 312 (1972).
220. L. A. Paquette and R. W. Begland, *J. Am. Chem. Soc.*, *87*, 3784 (1965).
221. F. H. S. Deckers, W. N. Speckamp, and H. O. Huisman, *Chem. Commun.*, 1521 (1970).
222. D. Beck and K. Schenker, *Helv. Chim. Acta*, 54, 734 (1971).
223. E. Elkik and C. Francesch, *Bull. Soc. Chim. France*, 903 (1969).
224. M. E. Kuehne and T. Garbacik, *J. Org. Chem.*, 35, 1555 (1970).
225. W. J. M. van Tilborg, G. Dooyewaard, H. Steinberg, and T. J. deBoer, *Tetrahedron Lett.*, 1677 (1972).
226. K. C. Brannock and R. D. Burpitt, *J. Org. Chem.*, *26*, 3576 (1961).
227. G. Opitz, *Ann.*, *650*, 122 (1961).

228. E. Elkik, *Bull. Soc. Chim. France,* 972 (1960).
229. R. Pauncz and D. Ginsburg, *Tetrahedron, 9,* 40 (1960).
230. N. L. Allinger and W. Szkrybalo, *J. Org. Chem., 27,* 722 (1962).
231. N. L. Allinger and J. T. Sprague, *J. Am. Chem. Soc., 94,* 5734 (1972).
232. O. Ermer and S. Lifson, *J. Am. Chem. Soc., 95,* 412 (1973).
233. D. N. J. White and M. J. Bovill, *J. Chem. Soc., Perkin Trans.,* 2, 1610 (1977).
234. U. Burkert and N. L. Allinger, *Molecular Mechanics,* American Chemical Society, Washington, DC, 1982.
235. S. Saebo, F. Cordell, and J. Boggs, *Theochem., 13,* 221 (1983).
236. See Ref. 162, pp. 109–111.
237. F. R. Jensen and C. H. Bushweller, in *Advances in Alicyclic Chemistry* (H. Hart and Karabatsos, eds.), Vol. 3, Academic Press, New York, 1971.
238. L. N. Fergerson, *Alicyclic Chemistry,* Part 1, Franklin Publishing, 1973.
239. C. W. Beckett, N. K. Freeman, and K. S. Pitzer, *J. Am. Chem. Soc., 70,* 4227 (1948).
240. D. H. R. Barton, R. C. Cookson, W. Klyne and C. W. Shoppee, *Chem. Ind. (London),* 21 (1954).
241. F. Johnson, *Chem. Rev., 68,* 375 (1968).
242. N. L. Allinger, J. A. Hirsch, M. A. Miller, and I. J. Tyminski, *J. Am. Chem. Soc., 90,* 5773 (1968).
243. F. Johnson and S. K. Malhotra, *J. Am. Chem. Soc., 87,* 5492 (1965).
244. E. L. Eliel, *The Sterechemistry of Carbon Compounds,* McGraw-Hill, New York, 1962.
245. E. J. Corey and R. A. Sneen, *J. Am. Chem. Soc., 78,* 6269 (1956).
246. P. Deslongchamps, *Stereoelectornic Effects in Organic Chemistry,* Pergamon Press, New York, 1983.
247. N. L. Allinger and M. A. Miller, *J. Am. Chem. Soc., 83,* 2145 (1961).
248. P. W. Hickmott, P. J. Cox, and G. A. Sim, *J. Chem. Soc., Perkin Trans., 1,* 2544 (1974).
249. See Ref. 162, pp. 484–485.
250. S. K. Malhotra and F. Johnson, *J. Am. Chem. Soc., 87,* 5493 (1965).
251. S. Danishefsky and M. Feldman, *Tetrahedron Lett.,* 1131 (1965).
252. N. DeKimpe, R. Verhe, L. DeBuyck, and N. Schamp, *Synthesis,* 741 (1979).
253. H. Mazarquil and A. Lattes, *Bull. Soc. Chim. France,* 319 (1969).
254. H. Mazarquil and A. Lattes, *Tetrahedron Lett.,* 975 (1971).
255. M. E. Kuehne, *J. Am. Chem. Soc., 81,* 5400 (1959).

256. H. K. Jakobson, S-O. Lawesson, J. T. B. Marshall, G. Schroll, and D. H. Williams, *J. / Chem. Soc. B,* 940 (1966).
257. S. K. Malhotra, L. Duquette and F. Johnson, unpublished results.
258. A. Risaliti and L. Marchetti, *Ann. Chim. Rome, 55,* 635 (1965).
259. A. Risaliti, S. Fatutta, and M. Forchiassin, *Tetrahedron, 23,* 1451 (1967).
260. A. Risaliti and L. Marchetti, *Ann. Chim. Rome, 53,* 718 (1963).
261. A. Risaliti, L. Marchetti, and M. Forchiassin, *Ann. Chimica, 56,* 317 (1966).
262. A. Risaliti, M. Forchiassin, and E. Valentin, *Tetrahedron, 24,* 1889 (1968).
263. F. P. Colonna, S. Fatutta, A. Risaliti, and C. Russo, *J. Chem. Soc. (C),* 2377 (1970).
264. F. P. Colonna, M. Fochiassin, G. Pitacco, A. Risaliti, and E. Valentin, *Tetrahedron, 26,* 5289 (1970).
265. G. Pitacco, F. P. Colonna, C. Russo, and E. Valentin, *Gazz. Chim. Ital., 105,* 1137 (1975).
266. D. N. Reinhoudt, W. P. Trompenaars, and J. Geevers, *Synthesis,* 368 (1978).
267. W. R. N. Williamson, *Tetrahedron, 3,* 314 (1958).
268. F. Johnson and A. Whitehead, *Tetrahedron Lett.,* 3825 (1964).
269. J. W. Lewis and A. A. Pearce, *Tetrahedron Lett.,* 2039 (1964).
270. J. W. Lewis and A. A. Pearce, *J. Chem. Soc. (B),* 863 (1969).
271. S. K. Malhotra and F. Johnson, *Tetrahedron Lett.,* 4027 (1965).
272. P. W. Hickmott and K. N. Woodward, *Chem. Commun.,* 275 (1974).
273. H. O. House, B. A. Tefertiller, and H. D. Olmstead, *J. Org. Chem., 33,* 935 (1968).
274. J. P. Schaefer and D. S. Weinberg, *Tetrahedron Lett.,* 1801 (1965).
275. F. Johnson, L. G. Duquette, A. Whitehead, and L. C. Dorman, *Tetrahedron, 30,* 3241 (1974).
276. H. C. Brown, J. H. Brewster, and H. Schechter, *J. Am. Chem. Soc., 76,* 467 (1954).
277. H. C. Brown, *J. Org. Chem., 22,* 439 (1957).
278. W. Maas, M. J. Janssen, E. J. Stamhuis, and H. Wynberg, *J. Org. Chem., 32,* 1111 (1967).
279. F. Johnson, N. A. Starkovsky, A. C. Paton, and A. A. Carlson, *J. Am. Chem. Soc., 86,* 118 (1964).
280. F. Johnson, N. A. Starkovsky, A. C. Paton, and A. A. Carlson, *J. Am. Chem. Soc., 88,* 149 (1966).
281. H. J. Schaeffer and V. K. Jain, *J. Pharm. Sci., 52,* 509 (1963).
282. H. J. Schaeffer and V. K. Jain, *J. Org. Chem., 29,* 2595 (1964).

283. G. Pitacco, R. Toso, E. Valentin, and A. Risaliti, *Tetrahedron, 32*, 1757 (1976).
284. H. O. House and M. J. Schellenbaum, *J. Org. Chem., 28*, 34 (1963).
285. N. F. Firrell and P. W. Hickmott, *Chem. Commun.*, 544 (1969).
286. T. Masamune, H. Hayashi, M. Takasugi, and S. Fukuoka, *J. Org. Chem., 37*, 2343 (1972).
287. W. M. B. Konst, J. G. Witteveen, and H. Boelens, *Tetrahedron, 32*, 1415 (1976).
288. P. W. Hickmott and N. F. Firrell, *J. Chem. Soc., Perkin Trans., 1*, 340 (1978).
289. J. W. Huffman, C. D. Rowe, and F. J. Matthews, *J. Org. Chem., 47*, 1438 (1982).
290. R. F. Parcell and F. P. Hauck, *J. Org. Chem., 28*, 3468 (1963).
291. R. T. Parfitt, *J. Chem. Soc. (C)*, 140 (1967).
292. L. A. Paquette and M. Rosen, *J. Org. Chem., 33*, 2130 (1968).
293. L. H. Hellberg, R. J. Milligan, and R. N. Wilke, *J. Chem. Soc. (C)*, 35 (1970).
294. F. A. VanDerVlugt, J. W. Verhoeven, and U. K. Pandit, *Rec. Trav. Chim., 89*, 1258 (1970).
295. S. Karady, M. Lenfant, and R. E. Wolff, *Bull. Soc. Chim. France*, 2472 (1965).
296. J. A. Hirsch and X. L. Wang, *Synth. Commun., 12*, 333 (1982).
297. S. Danishefsky and R. Cavanaugh, *J. Org. Chem., 33*, 2959 (1968).
298. S. Fatutta, G. Pitacco, C. Russo, and E. Valentin, *J. Chem. Soc., Perkin Trans., 1*, 2045 (1982).
299. M. Hanack, *Conformation Theory*, Academic Press, New York, 1965.
300. U. K. Pandit and H. O. Huisman, *Tetrahedron Lett.*, 3901 (1967).
301. F. A. Buiter, J. H. S. Weiland, and H. Wynberg, *Rec. Trav. Chim., 83*, 1160 (1964).
302. A. L. Ham and P. R. Leeming, *J. Chem. Soc. (C)*, 2017 (1969).
303. M. Forchiassin, A. Risaliti, and C. Russo, *Gass. Chim. Ital., 109*, 33 (1979).
304. M. Charles, G. Descotes, J. C. Martin, and Y. Querou, *Bull. Soc. Chim. France*, 4159 (1968).
305. E. Valentin, G. Pitacco, F. P. Colonna, and A. Risaliti, *Tetrahedron, 30*, 2741 (1974).
306. J. Champagne, H. Favre, D. Vocelle, and I. Zbikowski, *Can. J. Chem., 42*, 212 (1963).
307. A. R. Greenaway and W. B. Whalley, *J. Chem. Soc., Perkin Trans., 1*, 1385 (1976).
308. M. Forchiassin, C. Russo, and A. Risaliti, *Gazz. Chim. Ital., 102*, 607 (1972).

309. M. E. Kuehne and T. J. Giacobbe, *J. Org. Chem.*, 33, 3359 (1968).
310. S. K. Malhotra, D. F. Moakley, and F. Johnson, *Chem. Commun.*, 448 (1967).
311. H. J. Jakobsen, S.-O. Lawesson, J. T. B. Marshall, G. Schroll, and D. H. Williams, *J. Chem. Soc. (B)*, 940 (1966).
312. F. P. Colonna, E. Valentin, G. Pitacco, and A. Risaliti, *Tetrahedron*, 29, 3011 (1973).
313. E. Valentin, G. Pitacco, and F. P. Colonna, *Tetrahedron Lett.*, 2837 (1972).
314. F. W. Heyl and M. E. Herr, *J. Am. Chem. Soc.*, 75, 1918 (1953).
315. G. Pitacco, F. P. Colonna, E. Valentin, and A. Risaliti, *J. Chem. Soc., Perkin Trans.*, 1, 1625 (1974).
316. M. Forchiassin, A. Risaliti, C. Russo, M. Calligaris, and G. Pitacco, *J. Chem. Soc., Perkin Trans.*, 1, 660 (1974).
317. M. Calligaris, M. Forchiassin, A. Risaliti, and C. Russo, *Gazz. Chim. Ital.*, 105, 689 (1975).
318. F. Fernandez, D. N. Kirk, and M. Scopes, *J. Chem. Soc., Perkin Trans.*, 1, 18 (1974).
319. N. A. Nelson, R. S. P. Hsi, J. M. Schuck, and L. D. Kahn, *J. Am. Chem. Soc.*, 82, 2573 (1960).
320. M. E. Kuehne, *J. Am. Chem. Soc.*, 83, 1492 (1961).
321. U. Edlund and G. Bergson, *Acta Chem. Scand.*, 25, 3625 (1971).
322. W. L. F. Armarego, *J. Chem. Soc. (C)*, 986 (1969).
323. G. Bergson and A.-M. Weidler, *Acta Chem. Scand.*, 17, 862 (1963).
324. F. Plinat and G. Bergson, *Arkiv Kemi*, 25, 109 (1965).
325. A. T. Blomquist and E. J. Moriconi, *J. Org. Chem.*, 26, 3761 (1961).
326. W. Schroth and G. W. Fischer, *Chem. Ber.*, 102, 575 (1969).
327. U. Edlund, *Acta Chem. Scand.*, 26, 2972 (1972).
328. P. K. Khandelwal, and B. C. Joshi, *Def. Sci. J.*, 21, 199 (1971).
329. M. E. Munk and Y. U. Kim, *J. Am. Chem. Soc.*, 86, 2213 (1964).
330. R. Jacquier, C. Petrus, and F. Petrus, *Bull. Soc. Chim. France*, 2845 (1966).
331. P. Madsen and S-O. Lawesson, *Rec. Trav. Chim.*, 85, 753 (1966).
332. G. Bianchetti, P. D. Croce, and D. Pocar, *Tetrahedron Lett.*, 2039 (1965).
333. D. Pocar, G. Bianchetti, and D. Croce, *Gazz. Chim. Ital.*, 95, 1220 (1965).
334. M. Colonna and L. Marchetti, *Gazz. Chim. Ital.*, 96, 1175 (1966).
335. D. Pocar, R. Stradi, and B. Gioia, *Gazz. Chim. Ital.*, 98, 958 (1968).
336. R. Carlson and C. Rappe, *Acta Chem. Scand. B*, 28, 1058 (1974).

337. R. Carlson, L. Nilsson, C Rappe, A. Babadjamian, and J. Metzger, *Acta Chem. Scand. B*, *32*, 85 (1978).
338. D. Pocar, R. Stradi, and G. Bianchetti, *Gazz. Chim. Ital.*, *100*, 1135 (1970).
339. D. Pocar, R. Stradi, and P. Trimarco, *Tetrahedron*, *31*, 2427 (1975).
340. G. Bianchetti, R. Stradi, and D. Pocar, *J. Chem. Soc.*, *Perkin Trans.*, *1*, 997 (1972).
341. R. Stradi and D. Pocar, *Chim. Ind. (Milan)*, *53*, 265 (1971).
342. N. B. Colthup, L. H. Daly, and S. E. Wiberley, *Introduction to Infrared and Raman Spectroscopy*, 2nd ed., Academic Press, New York, 1975.
343. M. E. Munk and Y. K. Kim, *J. Org. Chem.*, *30*, 3705 (1965).
344. L. Duhamel, P. Duhamel, S. Combrisson, and P. Siret, *Tetrahedron Lett.*, 3603 (1972).
345. L. Duhamel, P. Duhamel, and P. Siret, *Tetrahedron Lett.*, 3607 (1972).
346. R. Knorr, *Chem. Ber.*, *113*, 2441 (1980).
347. M. E. Munk and Y. K. Kim, *J. Am. Chem. Soc.*, *86*, 2213 (1964).
348. P. Y. Sollenberger and R. B. Martin, *J. Am. Chem. Soc.*, *92*, 4261 (1970).
349. W. A. White and H. Weingarten, *J. Org. Chem.*, *32*, 213 (1967).
350. Ref. 162, pp. 151–152.
351. G. Optiz, H. Hellmann, and H. W. Schubert, *Ann.*, *623*, 112 (1959).
352. R. Ishino and J. Kumanotani, *J. Org. Chem.*, *39*, 108 (1974).
353. R. Dulou, E. Elkik and A. Veillard, *Bull. Soc. Chim. France*, 967 (1960).
354. R. Carlson and L. Nilsson, *Acta Chem. Scand. B*, *31*, 732 (1977).
355. D. E. Heitmeier, J. T. Hortenstine, Jr., and A. P. Gray, *J. Org. Chem.*, *36*, 1449 (1971).
356. R. Carlson and A. Nilsson, *Acta Chem. Scand. B*, *38*, 49 (1984).
357. K. Taguchi and F. H. Westheimer, *J. Org. Chem.*, *36*, 1570 (1971).
358. H. Matsushita, Y. Tsujino, M. Noguchi, and S. Yoshikawa, *Chem. Lett.*, 1087 (1976).
359. M. Riviere and A. Lattes, *Bull. Soc. Chim. France*, 2539 (1967).
360. T. Chen, H. Kato, and M. Ohta, *Bull. Chem. Soc. Jpn.*, *39*, 1618 (1966).
361. S. C. Kuo and W. H. Daly, *J. Org. Chem.*, *35*, 1861 (1970).
362. A. G. Cook, W. C. Meyer, K. E. Ungrodt, and R. H. Mueller, *J. Org. Chem.*, *31*, 14 (1966).
363. A. G. Cook, W. M. Kosman, T. A. Hecht, and W. Koehn, *J. Org. Chem.*, *37*, 1565 (1972).

364. J. F. Stephen and E. Marcus, *J. Org. Chem.*, *34*, 2535 (1969).
365. J. W. Daly and B. Witkop, *J. Org. Chem.*, *27*, 4104 (1962).
366. U. K. Pandit, S. A. G. DeGraaf, C. T. Braams, and J. S. T. Raaphorst, *Rev. Trav. Chim.*, *91*, 799 (1972).
367. R. Lukes, V. Dekek and L. Novotñy, *Collect. Czech. Chem. Commun.*, *24*, 1117 (1959).
368. N. J. Leonard and A. G. Cook, *J. Am. Chem. Soc.*, *81*, 5627 (1959).
369. V. Prelog and O. Hafliger, *Helv. Chim. Acta*, *32*, 185 (1949).
370. N. J. Leonard and V. W. Gash, *J. Am. Chem. Soc.*, *76*, 2781 (1954).
371. N. J. Leonard and F. P. Hauck, *J. Am. Chem. Soc.*, *79*, 5279 (1957).
372. O. Cervinka, *Collect. Czech. Chem. Commun.*, *25*, 1174, 2675 (1960).
373. R. Jacquier, C. Petrus, F. Petrus and E. Valentin, *Bull. Soc. Chim. France*, 2629 (1969).
374. A. J. Birch, E. G. Hutchinson, and G. Subba Rao, *J. Chem. Soc. (C)*, 637 (1971).
375. N. J. Leonard, A. S. Hay, R. W. Fulmer, and V. W. Gash, *J. Am. Chem. Soc.*, *77*, 439 (1955).
376. N. J. Leonard, P. D. Thomas, and V. W. Gash, *J. Am. Chem. Soc.*, *77*, 1552 (1955).
377. P. Duhamel, L. Duhamel, and J.-M. Poirier, *Comp. Rend.*, *274*, 411 (1972).
378. E. S. Stern and T. C. J. Timmons, *Electronic Absorption Spectroscopy in Organic Chemistry*, St. Martin's Press, New York, 1971.
379. K-A. Kovar and M. Bojadiew, *Arch. Pharm.*, *315*, 883 (1982).
380. K-A. Kovar, F. Schielein, T G. Dekker, K. Albert, and E. Breitmaier, *Tetrahedron*, *35*, 2113 (1979).
381. K-A. Kovar and U. Schwiecker, *Arch. Pharm.*, *307*, 384 (1974).
382. A. I. Scott, *Interpretation of the Ultraviolet Spectra of Natural Products*, Pergamon Press, Oxford, 1964.
383. N. J. Leonard and D. M. Locke, *J. Am. Chem. Soc.*, *77*, 437 (1955).
384. H. Weitkamp and F. Korte, *Chem. Ber.*, *95*, 2896 (1962).
385. P. W. Hickmott, B. J. Hopkins, and C. T. Yoxall, *J. Chem. Soc. (B)*, 205 (1971).
386. G. Opitz, H. Hellmann, and H. W. Schubert, *Ann.*, *623*, 117 (1959).
387. H. H. Jaffe and M. Orchin, *Theory and Applications of Ultraviolet Spectroscopy*, Wiley, New York, 1962.
388. G. Opitz and W. Merz, *Ann.*, *652*, 139 (1962).
389. C. N. R. Rao, *Ultraviolet and Visible Spectroscopy*, 2nd. ed., Butterworths, London, 1967.

390. F. Eiden and K. T. Wanner, *Arch. Pharm.*, *317*, 958 (1984).
391. A. Kirrmann and C. Wakselman, *Bull. Soc. Chim. France*, 3766 (1967).
392. K. Nagarjan and S. Rajappa, *Tetrahedron Lett.*, 2293 (1969).
393. J. Weber and P. Faller, *Bull. Soc. Chim. France*, 783 (1975).
394. M. P. Strobel, L. Morin and D. Paquer, *Nouveau J. Chim.*, *4*, 603 (1980).
395. P. P. Lynch and P. H. Doyle, *Gazz. Chim. Ital.*, *98*, 645 (1968).
396. R. Stradi, D. Pocar, and C. Cassio, *J. Chem. Soc.*, *Perkin Trans.*, *1*, 2671 (1974).
397. H. Gunther, *NMR Spectroscopy*, Wiley, New York, 1980.
398. G-J. Martin, G. Lavielle, J-P. Dorie, G. Sturtz, and M-L. Martin, *Comp. Rend.*, *268*, 1004 (1969).
399. R. L. Hinman, *Tetrahedron*, *24*, 185 (1968).
400. L. Alias, P. Angibeaud, and R. Michelot, *Comp. Rend.*, *269*, 150 (1969).
401. C. Pascual, J. Meier, and W. Simon, *Helv. Chim. Acta*, *49*, 164 (1966).
402. U. Edlund, *Acta Chem. Scand.*, *27*, 4027 (1973).
403. K. Burger and F. Hein, *Ann.*, 853 (1982).
404. M. P. Strobel, L. Morin, D. Paquer, and C. C. Pham, *Nouveau J. Chim.*, *5*, 27 (1981).
405. G. VanBinst and D. Tourwe, *Org. Mag. Resonance*, *4*, 625 (1972).
406. D. Barillier, M. P. Strobel, L. Morin, and D. Paquer, *Tetrahedron*, *39*, 767 (1983).
407. L. Kozerski, K. Kamienska-Treba, and L. Kania, *Helv. Chim. Acta*, *66*, 2113 (1983).
408. P. W. Westerman and J. D. Roberts, *J. Org. Chem.*, *42*, 2249 (1977).
409. M. Azzaro, S. Geribaldi, and B. Videau, *Mag. Reson. Chem.*, *23*, 28 (1985).
410. K. G. R. Sundelin, R. A. Wiley, R. S. Givens, and D. R. Rademacher, *J. Med. Chem.*, *16*, 325 (1973).
411. J. Sherman, *Chem. Rev.*, *11*, 164 (1932).
412. P. P. Lynch, *Gazz. Chim. Ital.*, *99*, 787 (1969).
413. N. F. Firrell and P. W. Hickmott, *J. Chem. Soc. (C)*, 716 (1970).
414. J. S. Ballesteros and M. D. H. Hernandez, *Bol. Soc. Quim. Peru*, 6 (1980).
415. J. Elguero, R. Jacquier, and G. Tarrago, *Tetrahedron Lett.*, 4719 (1965).
416. J. Elguero, R. Jacquier, and G. Tarrago, *Tetrahedron Lett.*, 1112 (1966).
417. M. Liler, in *Advances in Physical Organic Chemistry* (V. Gold and D. Bethell, eds.), Vol. 11, Academic Press, New York, 1975, p. 267.

418. L. Alais, R. Michelot, and B. Tchoubar, *Comp. Rend., 273*, 261 (1971).
419. E. J. Stamhuis and W. Maas, *J. Org. Chem., 30*, 2156 (1965).
420. M. S. B. Munson, *J. Am. Chem. Soc., 87*, 2332 (1965).
421. H. Matsushita, Y. Tsujino, M. Noguchi and S. Yoshikawa, *Bull. Chem. Soc. Jpn., 50*, 1513 (1977).
422. H. Matsushita, Y. Tsujino, M. Noguchi, M. Saburi, and S. Yoshikawa, *Bull. Chem. Soc. Jpn., 51*, 201 (1978).
423. H. Matsushita, Y. Tsujino, M. Noguchi, M. Saburi, and S. Yoshikawa, *Bull. Chem. Soc. Jpn., 51*, 862 (1978).
424. E. M. Arnett, *Acc. Chem. Res., 6*, 404 (1973).
425. E. M. Arnett, *J. Chem. Ed., 62*, 385 (1985).
426. D. H. Aue, H. W. Webb, and M. T. Bowers, *J. Am. Chem. Soc., 98*, 318 (1976).
427. R. Adams and J. E. Mahan, *J. Am. Chem. Soc., 64*, 2588 (1942).
428. L. Nilsson, R. Carlson and C. Rappe, *Acta Chem. Scand. B, 30*, 271 (1976).
429. C. A. Grob, *Helv. Chim. Acta, 68*, 882 (1985).
430. M. R. Ellenberger, D. A. Dixon, and W. E. Farneth, *J. Am. Chem. Soc., 103*, 5377 (1981).
431. E. J. Stamhuis, W. Maas, and H. Wynberg, *J. Org. Chem., 30*, 2160 (1965).
432. E. M. Kosower and T. S. Sorensen, *J. Org. Chem., 27*, 3764 (1962).
433. R. A. Eades, K. Scanlon, M. R. Ellenberger, D. A. Dixon, and D. S. Marynick, *J. Phys. Chem., 84*, 2840 (1980).
434. D. W. Davis and J. W. Rabalais, *J. Am. Chem. Soc., 96*, 5305 (1974).
435. J. I. Braumann and L. K. Blair, *J. Am. Chem. Soc., 90*, 5636 (1968).
436. J. D. Baldeschwieler and S. S. Woodgate, *Acc. Chem. Res., 4*, 114 (1971).
437. M. T. Bowers, D. H. Aue, H. M. Webb, and R. T. McIver, Jr., *J. Am. Chem. Soc., 93*, 4314 (1971).
438. J. F. Wolf, R. H. Staley, I. Koppel, M. Taagepera, R. T. McIver, Jr., J. L. Beauchamp, and R. W. Taft, *J. Am. Chem. Soc., 99*, 5417 (1977).
439. D. H. Aue and M. T. Bowers, in *Gas Phase Ion Chemistry* (M. T. Bowers, ed.), Vol. 2, Academic Press, New York, 1979.
440. M. A. Haney and J. L. Franklin, *J. Phys. Chem., 73*, 4328 (1969).
441. J. L. Franklin and P. W. Harland, *Ann. Rev. Phys. Chem., 25*, 485 (1974).
442. F. Jordan, *J. Phys. Chem., 80*, 76 (1976).
443. R. Walder and J. L. Franklin, *Int. J. Mass Spect. Ion Physics, 36*, 85 (1980).

444. C. H. DePuy and V. M. Bierbaum, *Acc. Chem. Res.*, *14*, 146 (1981).
445. M. R. Ellenberger, R. A. Eades, M. W. Thomsen, W. E. Farneth, and D. A. Dixon, *J. Am. Chem. Soc.*, *101*, 715 (1979).
446. N. J. Leonard, K. Conrow, and R. R. Sauers, *J. Am. Chem. Soc.*, *80*, 5185 (1958).
447. H. K. Hall, Jr., *J. Phys. Chem.*, *60*, 63 (1956).
448. J. C. Day, *J. Am. Chem. Soc.*, *103*, 7355 (1981).
449. R. W. Taft, in *Progress in Physical Organic Chemistry* (R. W. Taft, ed.), Vol. 14, Interscience, New York, 1983, p. 247.
450. A. G. Marshall, *Acc. Chem. Res.*, *18*, 316 (1985).
451. R. S. Cahn, C. Ingold, and V. Prelog, *Angew. Chem. Int. Ed. Engl.*, *5*, 385 (1966).
452. A. Hattori, H. Hattori, and K. Tanabe, *J. Catalysis*, *65*, 245 (1980).
453. D. A. Evans, C. H. Mitch, R. C. Thomas, D. M. Zimmerman, and R. L. Robey, *J. Am. Chem. Soc.*, *102*, 5955 (1980).
454. H. Kumobayashi, S. Akutagawa, and S. Otsuka, *J. Am. Chem. Soc.*, *100*, 3949 (1978).
455. J. Hanson, *Svensk Kem. Tidskrift*, *67*, 256 (1955).
456. P. Y. Sollenberger and R. Martin, *J. Am. Chem. Soc.*, *92*, 4261 (1970).
457. D. L. Ostercamp and P. J. Taylor, *J. Chem. Soc., Perkin Trans.*, *2*, 1021 (1985).
458. G. Stork, C. S. Shiner, C.-W. Cheng, and R. L. Polt, *J. Am. Chem. Soc.*, *108*, 304 (1986).
459. R. Huisgen, H-U. Reissig, H. Huber, and S. Voss, *Tetrahedron Lett.*, 2987 (1979).
460. C. F. Wong and R. T. La Londe, *J. Org. Chem.*, *38*, 3225 (1973).
461. H. O. House, A. V. Prabhu, and W. V. Phillips, *J. Org. Chem.*, *41*, 1209 (1976).
462. W. von Philipsborn and R. Muller, *Angew. Chem. Int. Ed. Engl.*, *25*, 283 (1986).
463. J. A. Schlueter and A. G. Cook, unpublished results.

2

Methods and Mechanisms of Enamine Formation

LEROY W. HAYNES *The College of Wooster, Wooster, Ohio*

A. GILBERT COOK *Valparaiso University, Valparaiso, Indiana*

I.	Introduction	104
II.	Enamines from the Condensation of Aldehydes and Ketones with Secondary Amines	104
	A. General Aspects	104
	B. Mechanistic and Structural Considerations	110
	C. Secondary Reactions in Enamine Formation from Ketones or Aldehydes and Amines	115
III.	Enamines via Mercuric Acetate Oxidation of Tertiary Amines	118
	A. General Aspects	118
	B. Mechanistic Considerations	123
IV.	Enamines by Other Oxidative Processes	129
V.	Synthesis of Iminium Salts	130
VI.	Enamines by Reductive Processes	134
VII.	Enamines from Lactams and Organometallic Reagents	136
VIII.	Synthesis of Enamines Utilizing Various Compounds of Phosphorus, Mercury, Boron, Arsenic, Tin, Silicon, Germanium, and Titanium	137
IX.	Miscellaneous Preparations	145
	References	151

I. INTRODUCTION

The primary objectives of this chapter are to detail the methods by which enamines (α, β-unsaturated amines) [1] can be synthesized and the mechanisms of enamine formation. The enamines discussed are those in which the nitrogen is tertiary and, with the exception of a few selected examples, contain no other functional groups. The term "simple enamines" might be used to describe the majority of enamines noted in this chapter.

$$\begin{array}{c} \diagdown \\ \diagup \end{array} C{=}C{-}N \begin{array}{c} \diagup \\ \diagdown \end{array}$$
$$\mid$$

(<u>1</u>)

II. ENAMINES FROM THE CONDENSATION OF ALDEHYDES AND KETONES WITH SECONDARY AMINES

A. General Aspects

The most versatile method for preparing enamines involves the condensation of aldehydes and ketones with secondary amines [Eq. (1)]. Mannich and Davidsen [1] discovered that the reaction of secondary amines with aldehydes in the presence of potassium carbonate and at temperatures near 0°C gave enamines, whereas calcium oxide and elevated temperatures were required to cause a reaction between ketones and secondary amines, although usually in poor yield. The introduction by Herr and Heyl [2–5] of the removal of the water produced in the condensation by azeotropic distillation with benzene made possible the facile preparation of enamines from ketones and disubstituted aldehydes.

$$\begin{array}{c}\diagdown\\\diagup\end{array}\!NH + O{=}C\begin{array}{c}\diagdown\\\diagup\end{array} \rightarrow \begin{array}{c}\diagdown\\\diagup\end{array}\!N{-}C\begin{array}{c}\diagup\\\diagdown\end{array}\!\!\!\!\!\!\!\!\!\!\!\!\!\!\!\!\!C\!\!\diagup + H_2O \qquad\qquad (1)$$

This innovation was exploited by Stork and his co-workers [6–8] for a study of enamine formation from a variety of ketones and secondary amines.

A number of modifications of this general method have been published. The benzene may be replaced by toluene or xylene to give a reasonable rate of reaction [9,10]. An acid catalyst, p-toluenesulfonic acid [3,4,11], Dowex-50 [12], montmorillonite catalyst K10 [13], boron trifluoride etherate [14], or even acetic acid [15] can be employed for the normal condensations or when the uncatalyzed reaction is slow. As an alternative to removal of the water by means of a water separator, the water can be removed by passing the condensate through a drying agent such as calcium carbide [16], magnesium sulfate [17], calcium hydride [18], or a molecular sieve [14,19–26]. By replacing the normally employed potassium carbonate or calcium oxide [1] with granular calcium chloride, Blanchard [27] was able to synthesize the N,N-dimethyl- and N,N-diethylenamines of cyclopentanone and cyclohexanone in greater than 50% yields. The procedure simply involves stirring at room temperature a mixture of the ketone, the amine (in excess), and calcium chloride in ether. Barium oxide has also been used [28] to replace potassium carbonate or calcium oxide. Methanol, acetone, pyridine, or dimethylformamide may be used as solvents for the preparation of the pyrrolidinyl enamines of the Δ^4-3-ketosteroids [16]. Formation of an enamine may also be brought about by heating the ethyl ester of a tertiary amino acid and a ketone with KOH in 1-butanol [29].

The condensation of aldehydes with secondary amines has also received considerable attention since the original work of Mannich and Davidsen [1]. In their study to extend the scope of the Leuckart-Wallach reaction (the reductive alkylation of amines by aldehydes, usually formaldehyde, and ketones in the presence of formic acid), deBenneville and Macartney [30] synthesized a number of enamines derived from both aliphatic and aromatic aldehydes and several secondary amines. Herr and Heyl [2] first introduced their azeotropic procedure for the preparation of the piperidyl and morpholinyl enamines of steroidal aldehydes. Instead of using benzene or similar solvents to form water azeotropes, Benzing [31,32] used excess isobutyraldehyde in the preparation of enamines from this aldehyde. That enamines of aldehydes and dimethylamines can successfully be formed under quite different operating conditions is illustrated by two syntheses using xylene as the solvent. One preparation requires potassium hydroxide and a temperature of −15°C while acetaldehyde is added and then a temperature of 20°C for 20 hr [33]. The other [34] makes use of an autoclave containing the aldehyde, potassium carbonate, dimethylamine, and xylene, which is rocked for 4 hr at 100°C. N,N-Dimethylpropenylamine has been prepared by adding propionaldehyde to a mixture of anhydrous dimethylamine, ether, and Linde no 13X molecular sieve [35]. Enamines derived from aldehydes can be synthesized by an acid catalyzed transamination reaction involving a ketonic enamine and an aldehyde [36]. Another variation in the synthesis of these enamines is to use tris(dimethylamino) methane in place of the usual secondary amine [37].

There is a great diversity in the types of enamines that have been prepared by the condensation of aldehydes and ketones with secondary amines. In addition to the commonly used dialkylamines such as dimethyl-, diethyl-, and dipropylamine, highly hindered amines such as diisobutylamine have also been used [38]. The heterocyclic second-

ary amines that are routinely involved in enamine formation are pyro-
lidine, piperidine, hexamethylenimine, and morpholine. But other
heterocyclic amines that have been used are N-methyl- and N-phenyl-
piperazine (2) [39,40], azetidine (3) [41,42], and aziridine (4) [43].

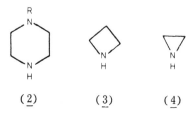

    (2)   (3)   (4)

    A wide variety of aldehydes and ketones have been used to prepare
enamines. Among the aliphatic aldehydes, α-fluoroaldehydes have been
allowed to react with heterocyclic secondary amines to produce β-
fluoroenamines (5), which hydrolyze less readily than corresponding

    (5)   (6)   (7)

simple enamines [44]. Among the carbocyclic aldehydes, cyclobutane-
carboxaldehyde (6, $n = 1$) [45], cyclohexanecarboxaldehyde (6, $n = 3$)
[46], and cyclooctanecarboxaldehyde (6, $n = 5$) [47] all form normal
enamines. However, cyclopropanecarboxaldehyde (6, $n = 0$) forms only
an aminal which is stable upon distillation [48]. Nitrogen heterocyclic
aldehydes (7) have also been converted into enamines [49]. The four-
and five-membered-ring aldehydes both give exocyclic enamine double
bonds when allowed to react with morpholine, but the six-membered-
ring aldehyde (7, $n = 3$) gives an 80:20 mixture of endocyclic and exocyclic enamines. Furanose aldehydes as found in aldo sugars will condense with secondary amines to form enamines [50].
    The most widely used carbocyclic ketones for the production of
enamines are cyclopentanones and cyclohexanones, but enamines of 7-
and 8-membered cyclic ketones [22,51] on up through 12-membered
cyclic ketones have also been made. Carbocyclic ketones with ester and
urethane functional groups present can also be used to produce enamines in good yields without interference by the ester or urethane groups
[326]. Heterocyclic ketones with the heteroatom β to the carbonyl
group (8) give a mixture of isomeric enamines (9 and 10) [52—54].

The regioisomer with the double bond toward the heteroatom (9) is overwhelmingly the major product. Similar observations were made with the thiopyranone 1,1-dioxide system [55] and with the phosphor-inanone sulfide system [56].

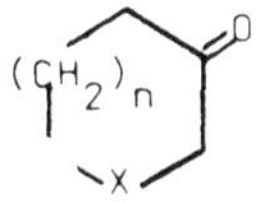

(8) X=S,O n=0,1   (9) 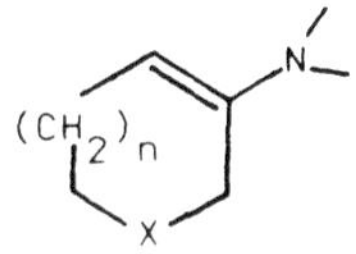   (10)

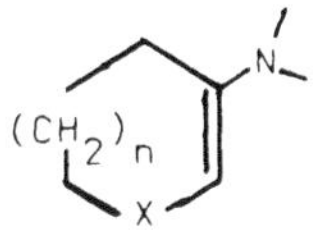

Enamines have been made from bicyclic aldehyde 11 [57–59] as well

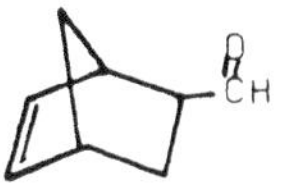

(11) 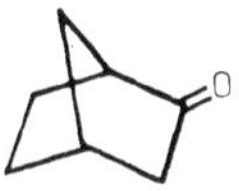   (12) 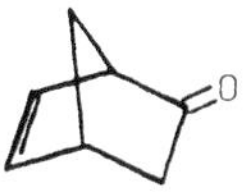   (13) 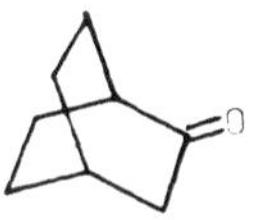   (14) 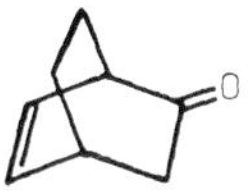   (15) 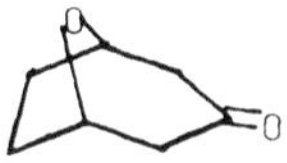   (16)

as from bicyclic ketones 12 [59,60], 13 [61], 14, and 15 [61]. In the cases of bicyclic ketones (13) and (15), special methods must be used since the usual acid-catalyzed reactions give primarily nonenamine products. A bicylcic ketone with an oxygen atom as one of the bridges (16) will also readily give an enamine (18).

Carbocyclic ketones with ketonic cycle fused to a benzene ring are precursors to an interesting set of enamines. α-Tetralone (17, n = 1) produces enamines very slowly under the usual conditions [62–66] but

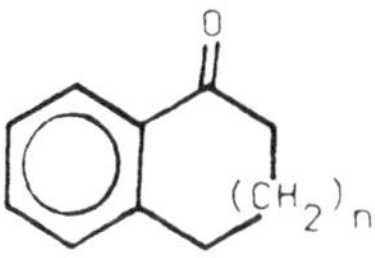

n=0,1,2   n=0,1,2   (19) 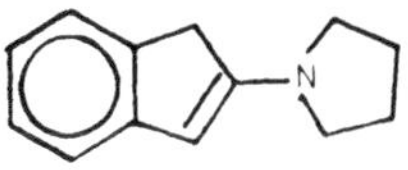   (20)
(17)   (18)

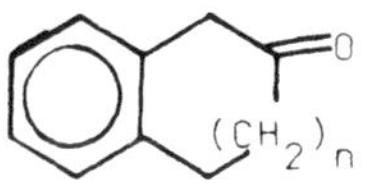

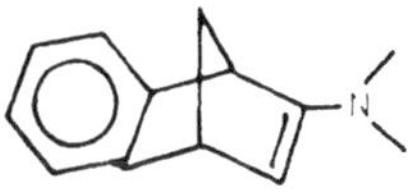

in excellent yields. β-Tetralone (18, n = 1), on the other hand, yields the corresponding enamines much more readily [9,63,67]. Bicyclic enamine 19 is readily formed from benzonorbornanone [59]. When a seven-membered ring is fused to the benzene ring, both the α- (17, n = 2) and the β-ketones (18, n = 2) easily give enamine products [65,68,69], the latter being potential precursors to colchicine. Both the α- (<u>17</u>, $n = 0$) and the β-ketones (<u>18</u>, $n = 0$) have been used to make enamines in the five-membered-ring series [62,70—74]. Enamines from 2-indanone form rapidly and in good yields at room temperature just by mixing the ketone and secondary amine together in a solvent such as methanol without any effort to remove water from the reaction mixture. The morpholine or pyrrolidine enamines (<u>20</u>) thus formed are so stable toward hydrolysis that they can be recovered unchanged from boiling water [72,74].

Osmotic pressure measurements and the NMR spectra of dilute solutions of the aminoaldehyde (<u>21</u>) indicated that the primary species in solution was the dimeric enamine (<u>22</u>) [75].

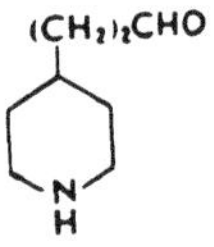

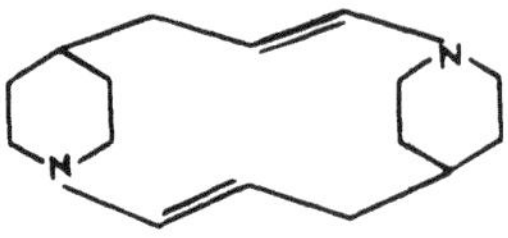

(<u>21</u>)   (<u>22</u>)

Intramolecular cyclizations take place in the production of exocyclic

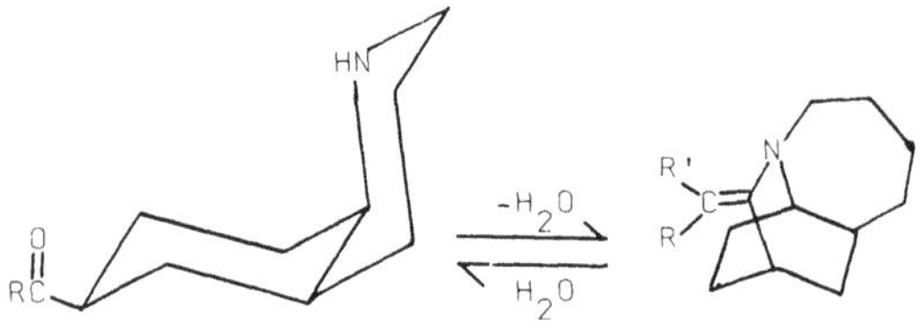

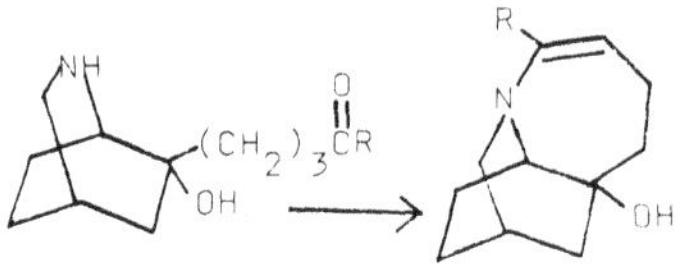

(<u>23</u>)   (<u>24</u>)

enamine <u>23</u> [76] and heterocyclic enamine <u>24</u> [77]. Other intramolecular cyclizations to enamines have also been observed [212]. Another intramolecular cyclization forms an isolatable, oxygen-bridged adamantane

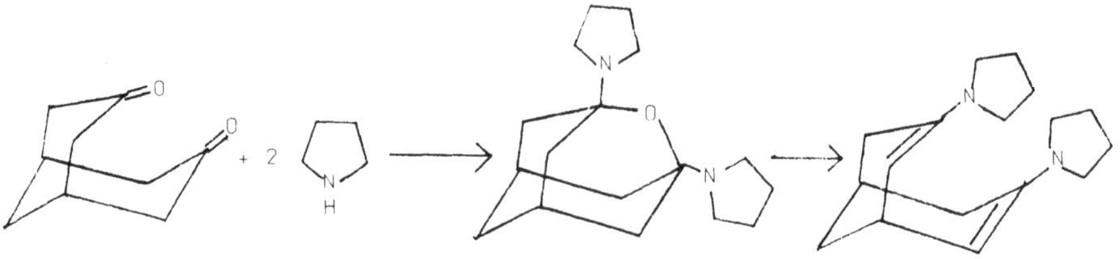

(<u>25</u>)   (<u>26</u>)

(2)

intermediate (<u>25</u>), which can then be dehydrated to dienamine <u>26</u> [see Eq. (2)] [78].

If a molecule contains both a ketonic and aldehydic carbonyl group, a secondary amine will react with the aldehydic carbonyl group to give a β-enamino ketone (28). This has been shown not only for 2-formylcyclohexanone (27) [79,80], but also in steroidal systems when the aldehyde and ketone groups are in five- or six-membered rings [81].

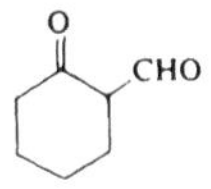

(27) 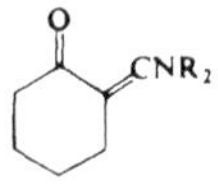   (28)

The rate of enamine formation is greatly reduced by increasing the amount of substitution alpha to the carbonyl group of the precursor ketone. This fact has been used as a basis for separation of a mixture of non-, mono- and dimethylated ketones (30,31,32) which were obtained by alkylation of enamine 29 followed by hydrolysis. The separation was accomplished by first allowing the mixture of ketones in refluxing benzene to react with gradually increasing amounts of morpholine until gas chromatographic analysis indicated that all of 30 had disappeared owing to formation of enamine 33. Then ketones 31 and 32 were removed and treated with increasing amounts of the more reactive amine pyrrolidine until glc showed all of 31 had disappeared as it formed enamine 34. Ketone 32 could then be distilled off and separation was complete [82] (see Scheme 1).

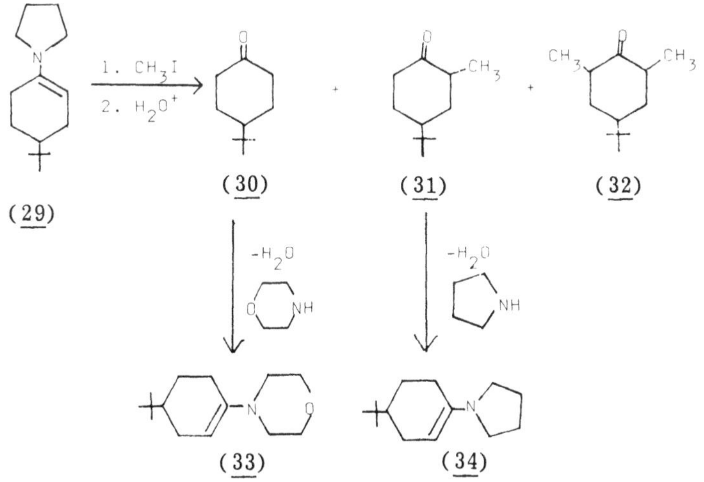

(29)   1. CH₃I   (30)   (31)   (32)
        2. H₂O⁺

                -H₂O   -H₂O

                (33)   (34)

Scheme 1

B. Mechanistic and Structural Considerations

The overall reaction pathway usually presented [1,9,19,36] for the preparation of an enamine from an aldehyde bearing an α-hydrogen and a secondary amine is given in Eqs. (3) and (4). Intermediate 35, which can be isolated in some cases [44,83–87], is called an aminal.

$$\overset{|}{\text{RCHCHO}} + 2\text{HNR}_1\text{R}_2 \xrightarrow{-\text{H}_2\text{O}} \overset{|}{\text{RCHCH}}(\text{NR}_1\text{R}_2)_2 \qquad (3)$$

$$(\underline{35})$$

$$(\underline{35}) \xrightarrow{\Delta} \text{RC}\overset{|}{=}\text{CHNR}_1\text{R}_2 + \text{HNR}_1\text{R}_2 \qquad (4)$$

Therefore, most investigators have used at least a twofold molar excess of amine to convert the aldehyde to the enamine in good yield.

That an animal is a necessary intermediate was first questioned by Herr and Heyl (2). They found that by using a slight excess of amine, the yield of the enamine from two of the steroidal aldehydes studied was 84%. Also, the β-fluoroenamines discussed earlier are formed in 60–.90% yield from equimolar amounts of the β-fluoroaldehyde and secondary cyclic amine [44]. However, neither of these studies was specifically designed to show whether or not aminals were intermediates.

Experiments designed to clarify the situation were carried out by Wittig and Mayer [88]. It was shown that changing the molar ratio of acyclic amine (diethylamine, di-n-butylamine, or diisobutylamine) to n-butyraldehyde from 1:1 to 2:1 did not affect the yield of enamine (53–64%, based on the aldehyde). Contrariwise, changing the ratio of cyclic amine (morpholine, piperidine, or pyrrolidine) to n-butyraldehyde from 1:1 to 2:1 boosted the yields from 52–57% to 80–85%. The authors interpret these data as indicating that the cyclic amines form aminals with n-butyraldehyde, but the open chain do not. Infrared evidence is stated as having shown that the aminal originates not from attack of excess amine on the enamine, which is stable under the conditions of the reaction, but from the N-hemiacetal (36). The conces-

$$\text{CH}_3\text{CH}_2\text{CH}_2\text{CH(OH)}(\text{NR}_2)$$

$$(\underline{36})$$

sion is made that longer reaction times, as used by Mannich and Davidsen (1), could produce an aminal from an enamine plus excess amine.

A study of the equilibrium between acetaldehyde and morpholine to yield hemiaminal 37 [Eq. (5)] showed the equilibrium lying far to the right with an equilibrium constant at 25°C of 20.5 and an ethalpy of reaction of −4.75 kcal/mol [89].

$$CH_3CHO \quad + \quad HN \underset{}{\overset{}{\bigcirc}} O \quad \rightleftharpoons \quad CH_3CH(OH)N \underset{}{\overset{}{\bigcirc}} O \tag{5}$$

$$(\underline{37})$$

It has been demonstrated that aminals and enamines are in equilibrium under certain conditions [83]. 1,1-Di(N-morpholino)ethane (38), when heated with excess diethylamine for 24 hr at 60°C and then treated with 4-nitro-phenylazide, gave a triazle (39) in 80% yield. The authors contend that for this to occur, the aminal (38) must be in

$$CH_3CH \left[N \bigcirc O \right]_2$$

$$(\underline{38})$$

$$(CH_3CH_2)_2N - \overset{N}{\underset{N}{\overset{\displaystyle N}{\bigtriangleup}}} - N \overset{}{\underset{}{\bigcirc}}$$

$$(\underline{39}) \quad NO_2$$

equilibrium with N-vinylmorpholine, which is eventually converted to N-vinyldiethylamine.

Additional evidence that a dynamic equilibrium exists between an enamine, N-hemiacetal, and aminal has been presented by Marchese [90]. It should be noted that no acid catalysts were used in the reactions of aldehydes and amines discussed thus far. The piperidino enamine of 2-ethylhexanal (0.125 mole), morpholine (0.375 mole), and p-toluenesulfonic acid (1.25×10^{-4} mole) diluted with benzene to 500 ml were refluxed for 5 hr. At the end of this time the enamine mixture was analyzed by vapor-phase chromatography, which revealed that exchange of the amino residue had occurred in a ratio of eight morpholine to one piperidine. Marchese proposed a scheme [Eqs. (6), (7), and (8)] to account for these results. Either the aminal (40) could

$$RCH{=}CHN \bigcirc \quad + \quad H^{(+)} \quad \rightleftharpoons \quad RCH_2\overset{(+)}{CHN} \bigcirc \tag{6}$$

$$RCH_2\overset{(+)}{CHN} \bigcirc \quad + \quad HN \bigcirc O \quad \rightleftharpoons \quad RCH_2CH \left[HN^{(+)} \bigcirc O \right] \left[N \bigcirc \right] \tag{7}$$

$$RCH_2CH \left[HN^{(+)} \bigcirc O \right] \left[N \bigcirc \right] \quad \rightleftharpoons \quad RCH_2CH \left[N \bigcirc O \right] \left[N \bigcirc \right] \quad + \quad H^{(+)} \tag{8}$$

$$(\underline{40})$$

break down upon distillation to give the mixture of enamines, or by
a series of similar equilibrium steps, the piperidine group could be
protonated and eventually lost as piperidine.

The evidence accumlated to data unfortunately is not conclusive.
The most accurate statement that probably can be made is that animals
are produced when aldehydes and secondary amines react, but the
aminals are not necessarily the direct precursor of the enamine.

The intermediacy of an aminal in the formation of enamines from
ketones and secondary amines is not usually proposed. The only direct
evidence for this is the infrared spectra of the reaction mixtures produced when dimethyl- or diethylamine was allowed to react with cyclohexanone or cyclopentanone in ether [27]. The spectra revealed the
presence of the enamine double bond (1640 cm^{-1}) prior to distillative
workup. General mechanisms for the noncatalyzed [9,91] and acid-
catalyzed reaction [90] have been offered. (See Scheme 2.) The only
kinetic data reported are in a Ph.D. thesis [90]. Integral-order kinetics were usually not obtained for the reaction of a number of ketones
with piperidine and a number of secondary amines with cyclohexanone.

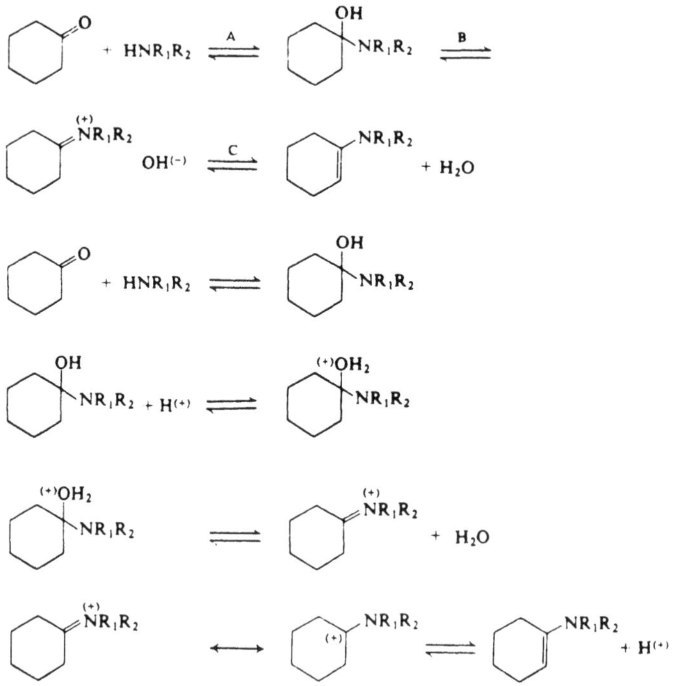

Scheme 2

A few of the combinations studied (cyclopentanone plus piperidine, pyrrolidine, and 4-methylpiperidine, and N-methylpiperazine plus cyclohexanone) gave reactions that were close to first-order in each reactant. Relative weights were based on the time at which a 50% yield of water was evolved. For the cyclohexanone-piperidine system the half-time ($t_{1/2}$) for the 3:1 ratio was 124 min and for the 1:3 ratio 121 min. It appears that an excess of ketone is just as effective as an excess of amine in driving the reaction to completion. The $t_{1/2}$ for the 1:1 ratio is 547 min. The conclusions drawn from the data in the thesis are similar to those of Stork and co-workers*:

> The rate is affected, not unexpectedly, by two factors: the basicity and steric environment of the secondary amino group and the nature and environment of the carbonyl group. Of the secondary amines used, pyrrolidine gives a higher reaction rate than the more weakly basic morpholine, while cyclic amines general- ly produce enamines faster than open-chain ones. This is of course what would be expected, but the fact that pyrrolidine reacts faster than piperidine may deserve comment. The basicity and steric en- vironment of the two bases are closely similar (pyrrolidine has $K = 1.3 \times 10^{-3}$, morpholine has $K = 2.44 \times 10^{-6}$, and piperidine has $K = 1.6 \times 10^{-3}$) and the differences in rate are probably to be ascribed to the different rates of the dehydration steps: The transition state with pyrrolidine involves making a trigonal carbon in a five-membered ring, and the faster rate of solvolysis of methylcyclopentyl chloride than that of the corresponding cyclo- hexyl compound (H. C. Brown, *J. Chem. Soc.*, 1956, 1248) cor- relates with the faster formation of an enamine from pyrrolidine than from piperidine. The effect of the ring size in the case of cyclic ketones is also notable: cyclopentanone reacts more rapidly, followed by cyclohexanone which is faster than the seven- and higher-membered ketones. If the rate of formation of enamines were solely a reflection of the rate of formation of the intermediate carbinolamines, cyclohexanone would form its enamine faster than cyclopentanone. If, on the other hand, the rate of dehydration of the carbinolamine were the controlling factor, then the seven- membered ring would be faster than the six. Since neither of these orders corresponds to the experimental one, the over-all rate is evidently not solely ascribable to any single one of the reversible steps A, B and C involved in the formation of the enamine.

---

*Reprinted from Ref. 9, p. 209, by courtesy of the American Chemical Society.

The morpholine aminal of acetone (41) has been synthesized using the TiCl$_4$ method, and it is found to be stable in the solid state. However, when it is melted or placed into solution, it decomposes into an

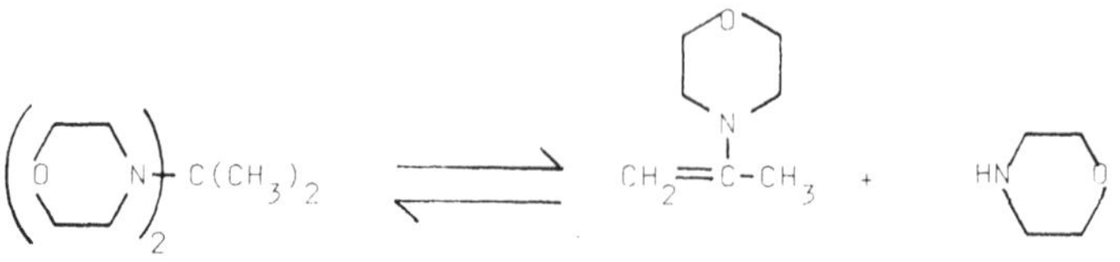

(9)

(41)

equilibrium mixture of aminal and enamines [Eq. (9)] [92]. Similar observations with aminals of ketones have been made by other workers [93]. Stable aminals and hemiaminals of ketones have been isolated for the cyclopropanone system [94–97]. For example, treatment of cyclo-

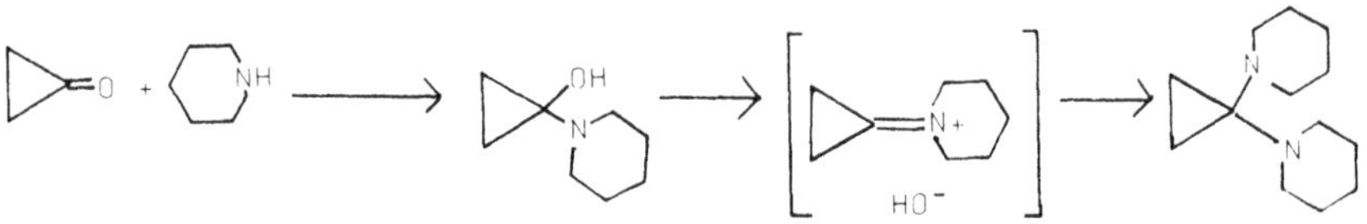

(42)   (43)   (44)

(10)

propanone with piperidine for 5 min in the cold followed by a rapid workup gave hemiaminal 42 as a solid product [97]. Further treatment of 42 with a second equivalent of piperidine at 25°C produced aminal 44 [Eq. (10)]. It was shown that iminium salt 43 probably is an intermediate in this latter reaction by allowing 42 to undergo a nucleophilic attack by an enamine to give a good yield of addition product. When hemiaminal 42 undergoes other similar nucleophilic attacks, the results are the same [98]. The intermediacy of a hemiaminal in the formation of dienamine 26 (discussed previously) from the reaction of a diketone and a secondary amine is indicated by the initial formation of dimerized hemiaminal condensation product 25 [78]. In the aldehyde family, the stable hemiaminal of quinuclidine 45 has been isolated [99], and treatment of iminium salt 46 with aqueous base gives dimeric hemiaminal 47 [100].

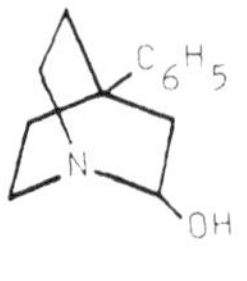

(45) 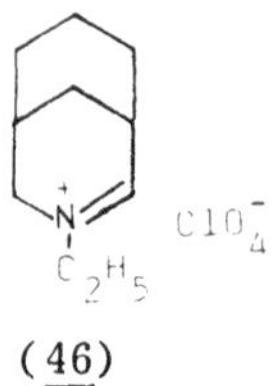   (46) 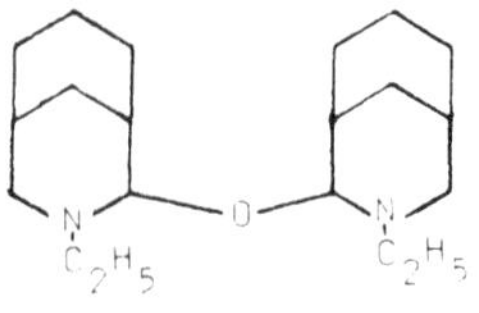   (47)

C. Secondary Reactions in Enamine Formation from Ketones or Aldehydes and Amines

Norcamphor (48) and hexamethylenimine (49) in xylene containing a catalytic amount of p-toluenesulfonic acid gave not only the enamine (50), but also the saturated amine (51) in about equal amounts [59]. Hexamethylenimine was believed to be the reducing agent since the

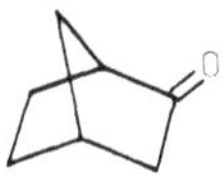

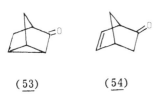

(48) 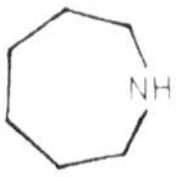   (49) 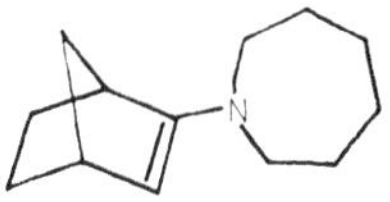   (50) 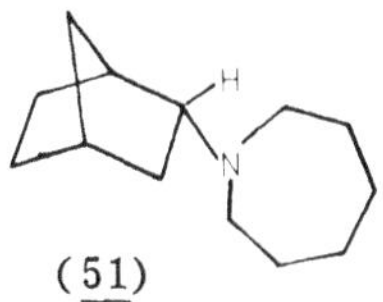   (51)

iminium perchlorate (52) could be converted to 51 in 60% yield with an excess of hexamethylenimine. This conclusion was challenged by Pat- more and Chafetz [101], who proposed that the saturated amine is

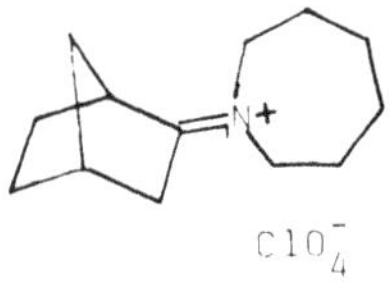

(52)

formed from the enamine by an initial protonation to give the iminium salt followed by an intermolecular hydride transfer from a second molecule of enamine.

However, Cook and Schulz [102] demosntrated conclusively that reduction of iminium salts by secondary amines is possible by isolating and characterizing the imine oxidation products, and by observing yields of greater than 50% for several other reductions. Allowing these reactions to take place in the absence of acid catalyst reduced the rates of reaction (refluxed for 14 days), but the enamine products were ob- tained in good yields with only small amounts of corresponding satura- ted amines [60].

The reactions of morpholine with tricyclo[2.2.1.0^{2,6}] heptan-3-one (53) and with 5-bicyclo[2.2.1]hepten-2-one (54) do not proceed in the normal manner [59,61,103]. The former reaction gave exo-5-morpho-

(53)   (54)

linobicyclo[2.2.1]heptan-2-one (55) in 64% yield by way of a homoeno-
late ion intermediate [103,104]. The reaction of unsaturated ketone

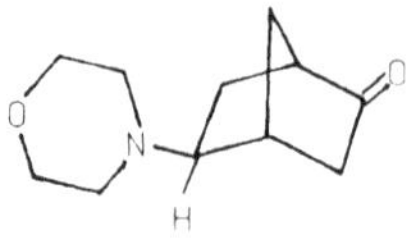

(55)

(56)

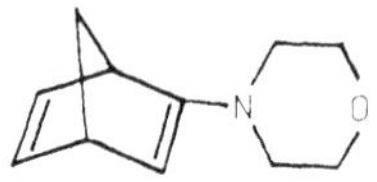

(57)

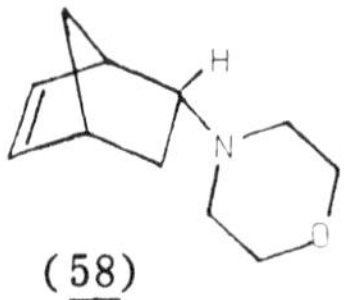

(58)

with morpholine in the presence of p-toluenesulfonic acid gave diamine
(56) in a 23% yield along with a small amount of enamine 57. In the
absence of an acid catalyst, 56 was produced in a diminished yield
along with a small amount of 57, but two new products appeared in
small amounts. These products are the reduced enamine 58 and ketoamine 55. The mechanism leading to the formation of diamine 56 would
involve production of carbocation 59 followed by homoconjugative
attack of this carbocation by another molecule of morpholine.

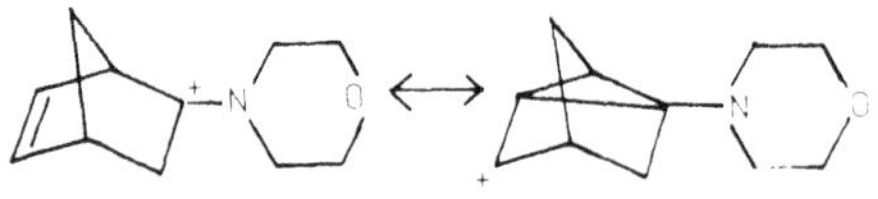

(59)

Rearrangements can also take place subsequent to the initial formation of an enamine. The acid-catalyzed reaction of acetophenone
with acyclic secondary amines results in the formation of the expected
enamine and a rearrangement product. The latter product arises from
the transfer of one of the amino N-alkyl groups to the enamine's β-
carbon to produce a ketimine [26].

An interesting preparation of substituted α-aminophenols has been
developed by Birkofer and Daum [105]. 2-Acylfurans (60) plus an
aliphatic secondary amine presumably condense to give the corresponding enamine (61) (not isolated), which undergoes thermal isomerization
to the o-aminophenol (62).

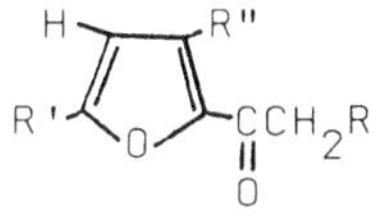

(60) 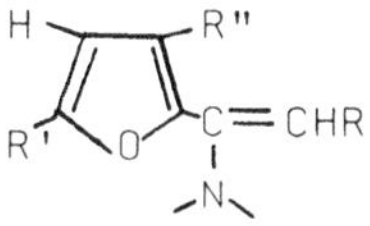   (61) 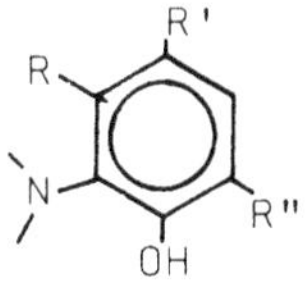   (62)

As was mentioned previously, cyclopropanecarboxaldehydes react with secondary amines to give very stable aminals which do not convert to enamines upon distillation [48]. Therefore, when bicyclo [3.1.0] hex-2-ene-6-*endo*-carboxaldehyde (63) is allowed to react with pyrrolidine, the expected *endo* aminal 64 is obtained [48,106—108]. However, when this aminal was heated to 80°C with a little acid catalyst, it isomerized to the exo form by way of enamine 65 [106—108]. When the

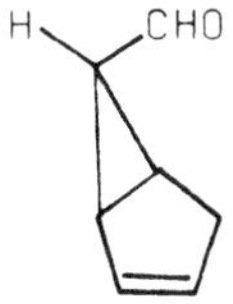

(63) 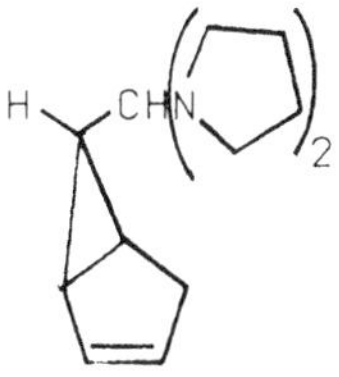   (64) 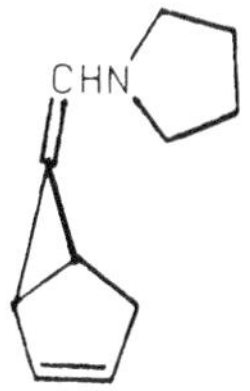   (65) 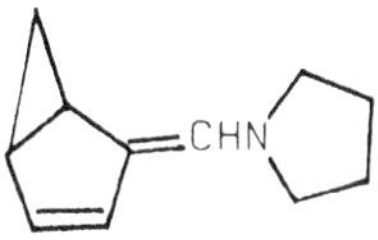   (66)

aminal is heated to 140°C, pyrrolidino-homofulvene 66 is obtained in a 1:1 mixture of *syn*- and *anti*-isomers via enamine 65 (see Chapter 1, Section III.A) [106—108].

The extra stability of an aromatic system sometimes causes enamine formation to be followed by an aromatization reaction. The acid-cata-

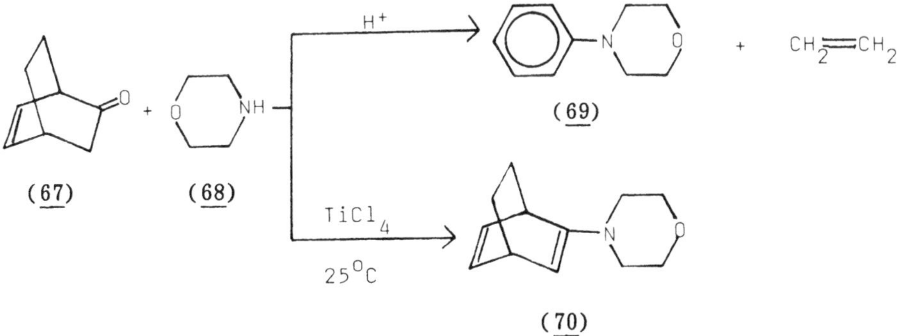

(67)   (68)   (69)   (70)

Scheme 3

lyzed reaction in refluxing xylene between bicyclo[2.2.2]oct-5-ene-2-one (67) and morpholine (68) in an attempt to make enamine 70 resulted instead in a symmetry-allowed (π2s + σ2s + σ2s) cycloreversion reaction to give N-phenylmorpholine (69) and ethylene as the products [61]. Enamine 70 can be synthesized at room temperature using titanium tetrachloride catalyst [61] (see Scheme 3). A retro-Diels-Alder reaction upon thermolysis of an enamine to form the aromatic molecule anthracene and an ynamine has been observed [322].

Oxidation of the initially formed enamine to an aromatic system was the result when the corresponding pyrrolidine enamines were attempted

(72)   (73)   (74)

(71)

to be synthesized from either diketone 71 [109] or pyrrolidone 73 [17]. In the former case, air oxidation was allowed to take place to give benzene compound 72, but in the latter case, oxidation to pyrrole 74 took place even in a nitrogen atmosphere.

III. ENAMINES VIA MERCURIC ACETATE OXIDATION OF TERTIARY AMINES

A. General Aspects

The oxidation of amines by mercuric acetate is an old reaction (110) which until recent years was employed primarily to modify alkaloid structures (111). A systemic study of the oxidizing action of mercuric acetate by Leonard and co-workers led to the development of a general method for the synthesis of enamines from cyclic tertiary amines. An observation made after a large number of compounds were oxidized, but which is worth noting at the onset, is that a tertiary hydrogen alpha to the nitrogen atom is removed preferentially to a secondary α-hydrogen.

The first compound studied [112] was quinolizidine (75), which can be readily converted to Δ$^{1(10)}$-dehydroquinolizidine (76) in 60% yield by the action of 4 moles of mercuric acetate in 5% aqueous acetic acid on 1 mole of the amine. Mercurous acetate precipitates as the reaction progresses at steam-bath temperatures. Extension of the reaction to more complicated systems gave similar results.

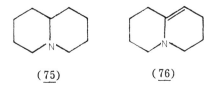

(75)   (76)

Sparteine (77) is oxidized to a mixture of isomers: $\Delta^{5,11}$-didehydrosparteine (78) and Δ^{5}-dehydrosparteine (79) [113]. The other two stereoisomers of sparteine, α-isosparteine (80) [114,115] and β-isosparteine (spartalupine) (81) [114,116] have been subjected to mercuric acetate oxidation, each giving $\Delta^{5,11}$-didehydrosparteine (78).

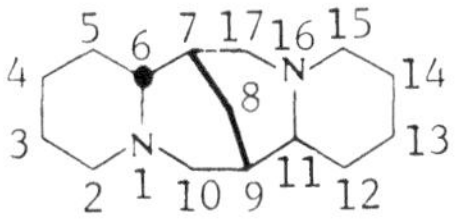

(77) 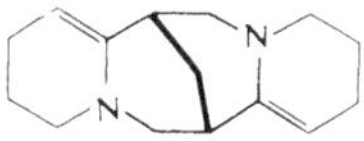   (78)

(79)

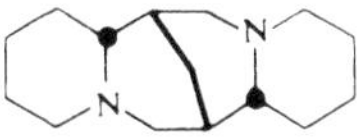

(80)    (81)

Bicyclic amines with nitrogen at the bridgehead and alpha to the bridgehead have been studied extensively. The methylquinolizidines gave the expected enamines [117], and either *cis*- or *trans*-1-methyldecahydroquinoline (82) gave a dehydrogenation-hydroxylation product (78) [118]. If the angular position is substituted, as in 84, oxidation takes place to give 85 [119]. The unsubstituted bicyclic amines,

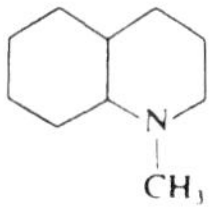

(82) 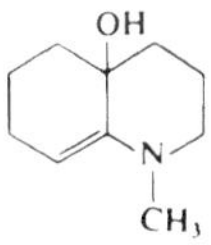   (83) 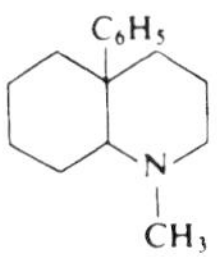   (84) 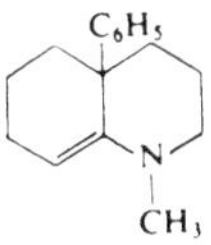   (85)

1-axabicyclo[4.3.0] nonane (86), R = H), 1-azabicyclo[5.3.0]decane, 1-azabicyclo[5.4.0]hendecane, and 1-azabicyclo-[5.5.0]dodecane, have

been converted to the corresponding enamine or isomeric enamines
[120]. In a later study [121], 1-azabicyclo[4.3.0]nonane (86, R = H),
several substituted 1-azabicyclo[4.3.0]nonanes, quinolizidine, and 1-
methylquinolizidine were dehydrogenated to the corresponding enamine
or isomeric enamines. The presence of an isomeric pair and the relative amounts of each were ascertained using NMR sepctroscopy. This
technique revealed that the mercuric acetate oxidation of 81 (R = H)
produces almost exclusively enamine 87 [121] rather than the mixture
of isomeric enamines, as previously believed [120]. Reinecke and
Kray [121] proposed several generalizations as a result of their study:

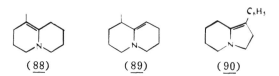

(86)   (87)

The double bond of indolizidine enamines (e.g., 87) prefers to be endo
to the six- and exo to the five-membered ring; substituents stabilize
the double bond (ratio of 88 to 89 is 2:1), and isomers whose double
bonds are conjugated are favored [enamine 90 is the apparent sole
product when 86 (R = C_6H_5) is dehydrogenated].

(88)   (89)   (90)

1,4-Diazobicyclo[2.2.2]octane (91) is recovered in 80% yield when
mercuric acetate oxidation is attempted at room temperature [122].
However, N-substituted derivatives of 2-azabicyclo[2.2.2]octane (92)
undergo oxidation readily to give immonium salts (93), dealkylation to
2-azabicyclo[2.2.2] octane (94), or lactams (95), depending on the
substituent group and the reaction temperature [123]. 2-Methyl- and

(91)   (92)   (93)   (94)   (95)

2-benzyl-2-azabicyclo[2.2.2] octane (92, R = R' = H and R = H,
R' = C_6H_5) yield the corresponding iminium salts (93), while 2-ethyl-
and 2-isopropyl-2-azabicyclo[2.2.2]octane (92, R = H, R' = CH_3 and
R = R' = CH_3) are dealkylated to 94. The amino alcohols (92, R = H,
R' = CH_2OH and R = H, R' = CH_2CH_2OH) when heated at 97°C gave

13.5% 95 (n = 2) and 45% 95 (n = 3), respectively. Such overoxidation has also been noted in pyrrolidino and piperidino amino alcohols
[122,124]. The iminium salts (93) cannot be converted to enamines
upon treatment with hydroxide since Bredt's rule would be violated.
The 3-alkyl-3-azabicyclo[3.3.1]nonanes (96) can also be dehydrogenated to the iminium salts, but once again, enamine formation is impossible [110,125].

(96)

The study of mercuric acetate oxidation of substituted piperidines
[126] established that the normal order of hydrogen removal from the
N-carbon is tertiary —C—H > secondary —C—H > primary —C—H,
an observation mentioned earlier in this section. The effect of substitution variations in the piperidine series can be summarized as
follows: 1-methyl-2,6-dialkyl and 1-methyl-2,2,6-trialkyl piperidines,
as model systems, are oxidized to the corresponding enamines; the
1,2-dialkyl and 1-methyl-2,5-dialkyl piperidines are oxidized preferentially at the tertiary α-carbon; the 1-methyl-2,3-dialkyl piper-
dines gave not only the enamines formed by oxidation at the tertiary
α-carbon but also hydroxylated enamines as found for 1-methylde-
cahydroquinoline (82) [118]; 1-methyl-2,2,6,6-tetraalkyl piperidines
and piperidine are resistant to oxidation by aqueous mercuric acetate;
and 1-methylpiperidine gave 1,1'-dimethyl-Δ^2-tetrahydroanabasine
(97) in 67% yield presumably by the dimerizaiton of the expected initial
oxidation product [126]. Both the enamine and the Δ^2-tetrahydroana-
basine were formed when 1,4,4-trimethylpiperidine was oxidized.

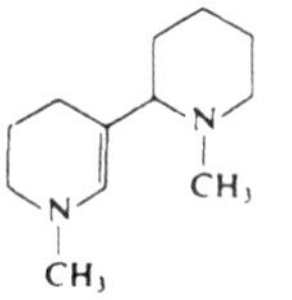

(97)

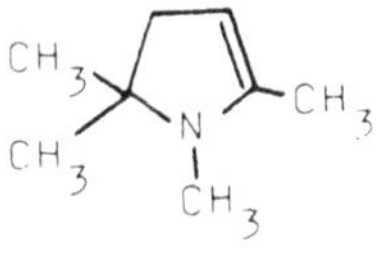

(98)

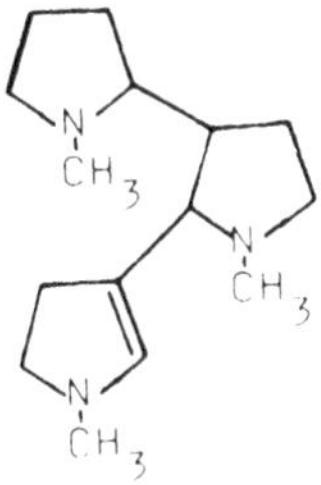

(99)

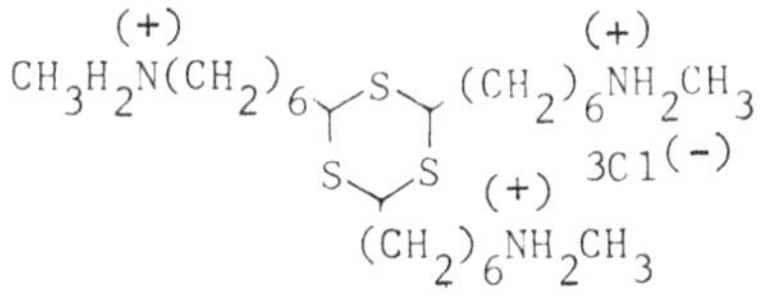

(100)

In the five-membered pyrrolidine ring series, in some cases such as 1,2,5,5-trimethylpyrrolidine only the monomeric enamine (98) was isolated (78% yield) [127]. In the case of the simplest member of the series, N-methylpyrrolidine, not only was the dimer isolated (analogous to 97 from oxidation of 1-methylpiperidine), but trimer 99 was also found.

Mercuric acetate oxidations of seven-, eight-, or nine-membered nitrogen-heterocyclic rings using the same experimental procedures as in the studies of piperidines and pyrrolidines resulted in crude products whose spectral properties indicated the presence of enamines, but enamines could not be isolated by distillation [127,128]. The cyclic iminium ions formed would be expected to be unstable relative to hydrolytic ring opening due to the ring strain present in medium-size rings. So when the oxidation experimental procedure was varied to include resaturation of the product solution with hydrogen sulfide followed by treatment with concentrated hydrochloric acid, 2,4,6-trithiane trihydrochlorides (100) were isolated [128]. This probably took place by conversion of the aminoaldehydes into aminothioaldehydes followed by trimerization.

Sometimes mercuric acetate oxidation goes beyond the formation of an enamine to give more highly oxidized products [126,127,129]. For example, in the oxidation of 1,3,3-trimethylpiperidine, 1,3,3-trimethyl-2-piperidone (101) was obtained as a minor product [126],

(101)

(102)

and the oxidation of 1,3,4-trimethylpyrrolidine produced pyrrole 102 as the "overoxidation" product [127].

Knabe has introduced mercuric acetate plus ethylenediaminetetraacetic acid (EDTA) as an oxidizing agent for tertiary amines [130]. The solvent employed is 1% aqueous acetic acid. Knabe's studies have centered on the oxidation of synthetic and naturally occurring 2-alkyl-1,2,3,4-tetrahydroisoquinolines [131--133], which bear various substituents. In this EDTA-complexed system, the mercuric acetate is

(103)

(104)

reduced to free mercury as it oxidizes the amine, whereas in the ordinary acetic acid system the mercuric acetate is reduced to mercurous acetate. Both two- and four-electron oxidations are observed to take place in this system. For example, mercuric ion complexed with EDTA will oxidize 103 to produce both 2-alkylquinolinium salts (104) and the 2-alkyl-3,4-dihydro- and 2-alkyl-1,4-dihydro-isoquinolinium salts (105 and 106, respectively) [131].

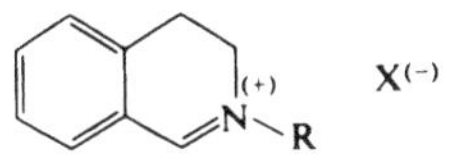

(105) 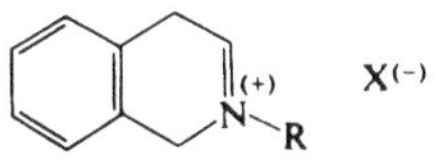   (106)

A regiospecific cyclization in the synthesis of yohimbine was effected by a four-electron oxidation of 107 followed by reduction of the iminium ion with sodium borohydride to give 108 [134]. This type of

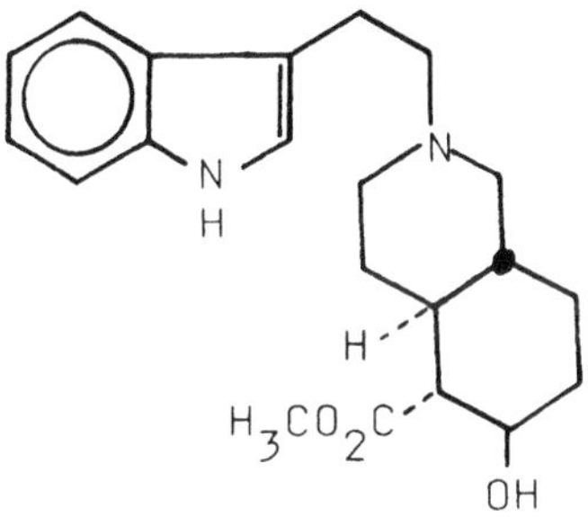

(107) 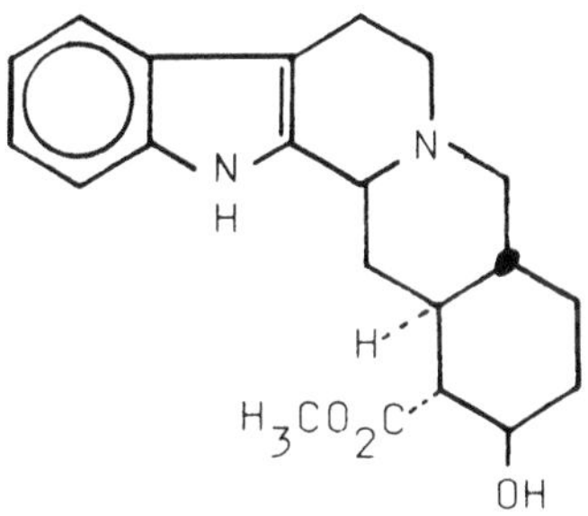   (108)

cyclization following mercuric acetate oxidation has been observed in other indole-type compounds also [135]. (This type of Bischler-Napieralski cyclization reaction will be further discussed in Section V.) In some N-aromatic cyclic amine systems the mercuric-EDTA system oxidizes the amines to amides and stops [136], but enamines were shown to be intermediates in these reactions.

B. Mechanistic Considerations

The mechanism proposed [112] for the mercuric acetate oxidation of tertiary amines involves the initial formation of a mercurated complex through the electron pair on nitrogen followed by a concerted removal of a proton from an α-carbon and cleavage of the mercury-nitrogen bond [Eq. (11)]. This four-center elimination implies that the removal of the α-hydrogen is the rate-determining step and also that a *trans*-coplanar relationship exists between the proton being removed and

the nitrogen-mercury complex. To determine whether the breaking
of an α carbon-hydrogen bond is the rate-determining step, the
oxidation of quinolizidine-10-d (109) was carried out [137] [see Eq.

$$+ \quad (Hg^0) + DOAc + OAc^{(-)} \tag{11}$$

(11)]. Quinolizidine was found to react faster than quinolizidine-10-d
by a factor of about 2.3. This factor was arrived at by spectral determination fo the amount of enamine and by weighing the mercurous
acetate precipitated after a given time interval. Importantly, the
quinolizidine-10-d recovered after partial oxidation was of unchanged
deuterium content.

These experiments verified that cleavage of the C—H bond is
occurring in the rate-limiting step, but proof of the necessity of a
trans relationship of the C—H to the nitrogen-mercury complex was
lacking. Indications that such a relation was necessary are found in
studies of the mercuric acetate oxidation of alkaloids, which will be
discussed subsequently.

The bicyclic amine 11-methyl-11-azabicyclo[5.3.1]hendecane
(110) provided a model system in which the hydrogens on the equivalent α-tertiary-carbon atoms cannot be *trans* to the nitrogen-mercury
bond in the mercurated complex and in which epimerization at these
α-carbons is impossible [138]. This bicyclic system is large enough
to accommodate a bond at the bridgehead. When the amine (110) was

oxidized, demethylation to the secondary amine (111) occurred since
one of the hydrogens of the N-methyl group, which has essentially
unrestricted rotation, not the hydrogens at C-1 and C-7, could be
aligned *trans* coplanar with the nitrogen-mercury complex. The lack
of any hydrogens that could be aligned *trans* coplanar with the nitrogen-mercury complex explains the previously mentioned failure of
1,4-diazabicyclo[2.2.2]octane (91) to undergo mercuric acetate oxidation [122].

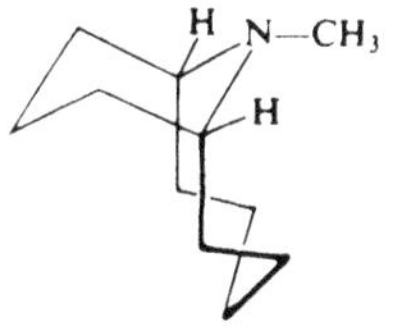

(110)

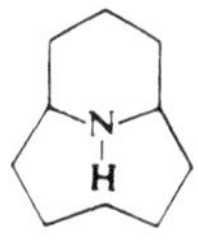

(111)

The mechanistic sequence as outlined for quinolizidine-10-*d* has metallic mercury as the reduced species. Mercurous acetate is the form in which the mercury eventually appears. It has been shown [137] that under the standard operating conditions, mercuric acetate will oxidize metallic mercury to the mercurous form. The proposal that a two-electron transfer occurs appears reasonable [112]. As noted previously, when mercuric ion complexed with EDTA is used, metallic mercury is formed [130]. Redox studies [139] of both oxidation systems have revealed the formation of mercurous ion as a common reaction intermediate.

The results observed in the oxidation of alkaloids that indicated something of the stereochemistry required for oxidation and prompted studies on model systems can now be interpreted more confidently. However, care must be used when basing steric differentiation on mercuric acetate oxidation studies since conditions must be employed that avoid epimerization at carbons alpha to the nitrogen.

The oxidation of sparteine (77) can give either dehydro- (79) or didehydro-sparteine (78). The hydrogen at C-6 is lost readily at room temperature to give dehydrosparteine [113], whereas refluxing is necessary to form didehydro-sparteine [140]. Thus, the hydrogen that is axial to two rings and to the electron pair on nitrogen is the one lost first. When these data were analyzed, it was assumed that all rings are in the chair form [138]. On the basis of a variety of observations, the preferred conformation of sparteine is now believed to be 112 with ring C in the boat form [141]. In β-isosparteine (81)

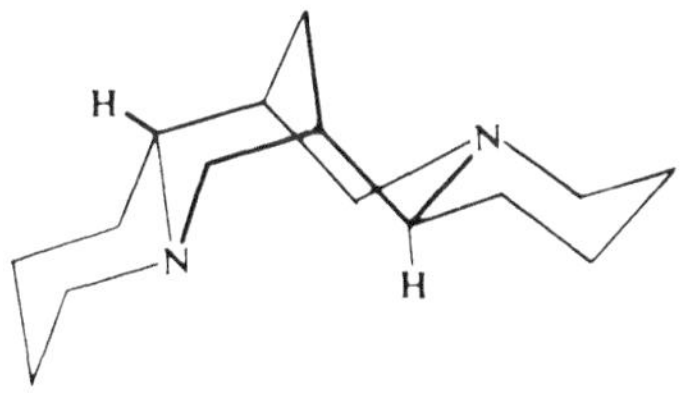

(112)

both tertiary protons at C-6 and C-11 are arranged is the *cis* position to the free electron pair on nitrogen as indicated by the NMR spectrum

[141]. Much earlier x-ray analysis showed that all rings in crystalline α-isosparteine (80) are present in stable chair conformations [142]. A comparative rate (extrapolated) at 65°C for α-isosparteine (5.0) and sparteine (1.0) has been calculated [143]. It has also been reported that β-isosparteine gave the dehydro derivative under mild conditions and the didehydro under more drastic conditions (times, temperatures not given) [116].

The indole alkaloids of the yohimbine-reserpine series exist in four configurations: normal (113), allo (114), pseudo (115), and epiallo (116). The results of the mercuric acetate oxidation of the indole

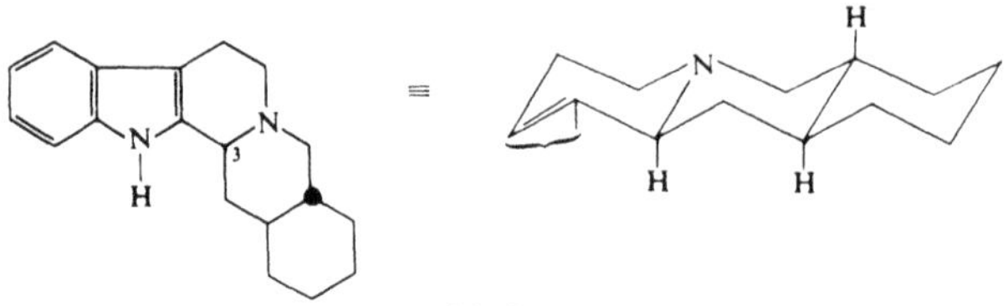

(113)

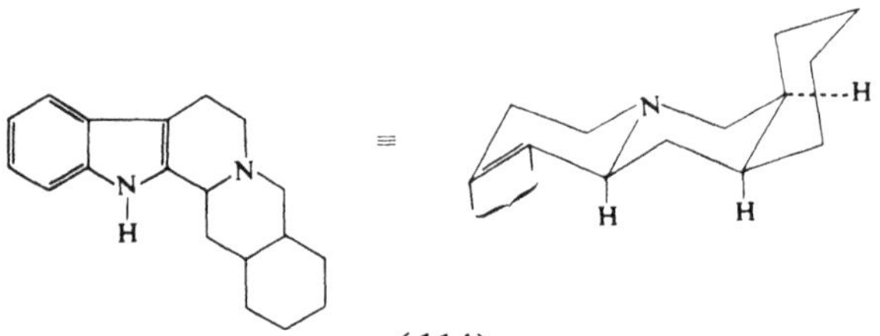

(114)

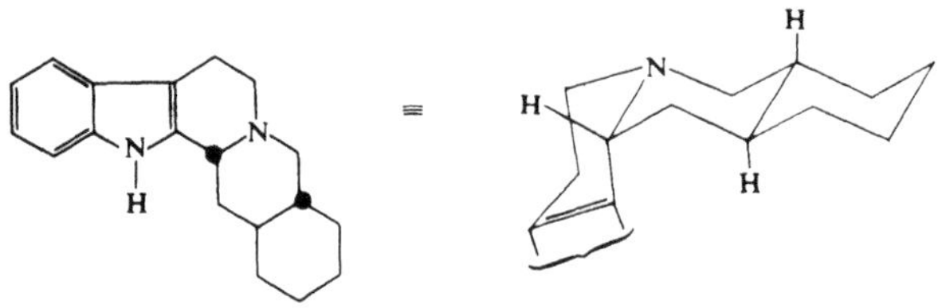

(115)

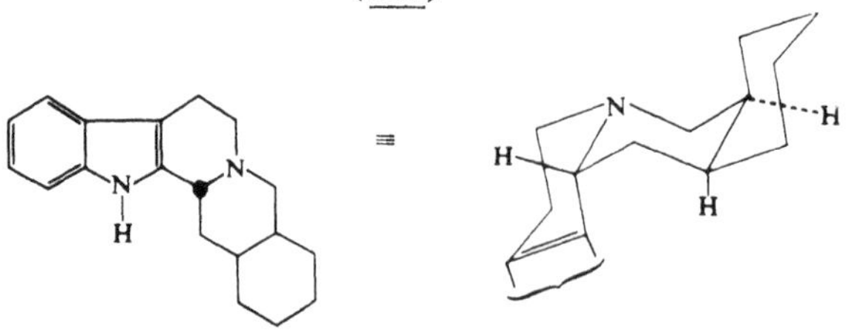

(116)

alkaloids are in general accord with the proposed *trans*-coplanar mechanism: namely, compounds with the normal (e.g., yohimbine) [144] and allo (allo- or α-yohimbine, isoreserpine) [145] are more readily oxidized at C-3 by mercuric acetate than those with the pseudo (pseudoyohombine) [144] and epiallo (epialloyohimbine, reserpine) [145] configurations.

Bohlmann and Arndt [146] have separated the possible stereoisomers of hexahydrojulolidine (117–119) and subjected them to mercuric acetate oxidation. The rates, which were followed by the precipitation of mercurous acetate, showed that isomer 117 reacted about five times faster than isomer 118, and isomer 119 reacted very slowly. The difference in rates between 117 and 118, both of which have tertiary α-hydrogens *trans* to the nitrogen electron pair, was explained by pointing out the greater relief of nonclassical strain occurs in the oxidation of 117 as compared to 118. Isomer 119 has no tertiary α-hydrogens *trans* to the nitrogen electron pair except when it is in an unfavorable boat conformation.

(117)   (118)   (119)

A kinetic study of the mercuric acetate oxidation of 1-alkyl-3,4-dimethyl-piperidines (120) and 3-alkyl-3-azabicyclo[3.3.1]nonanes (121) was made to evaluate the effect of the N-alkyl group on the rate of oxidation and to contrast these two ring systems [125]. The maximum factor in the piperidine system was 2.15 (1.00 for 120, R = t-butyl, and 2.15 for 120, R = n-propyl) and in the bicyclic system 6.00 (1.00 for 121, R = methyl, and 6.00 for 121, R = n-butyl). Thus, the

(120)   (121)

size of the alkyl group had little or no effect in the piperidine series and a moderate effect in the bicyclic series, but not in an order based on the usual steric effects of the groups. The order within a series did not depend on the relative basicities of the amines. The oxidation followed reasonable second-order kinetics. As illustrated by the rate constants for 1,3,5-trimethylpiperidine (1.92×10^{-4}, 70°C) and 3-methyl-3-azabicyclo[3.3.1]nonane (0.64×10^{-3}, 50°C), the bicyclic amines reacted many times more rapidly than the monocyclic amines. A possible explanation, similar to that offered to rationalize differences in rates between the hexahydrojulolidine isomers 117 and 118 [146], is the greater relief of nonbonded interactions in the amine-mercuric acetate complexes of the bicyclic amines when product is formed.

Conformational effects appear to be important in determining which tertiary α-hydrogen is removed. For example, 4-phenyl-, 4-*p*-nitrophenyl-, and 4-*p*-methoxyphenylquinolizidine (122) all are oxidized to the corresponding $\Delta^{5(10)}$-iminium salts (123) and not to the conjugated Δ^4-iminium salts (124) [147]. The authors judged that steric hindrance was responsible or that the conformation of the 4-substituted quinolizidines did not contain ideal chairs, so that the *trans* hydrogen on C-4 was not coplanar with the unshared pair of electrons

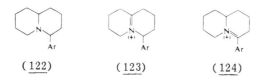

(122)   (123)   (124)

on the nitrogen atom. Even 4-methylquinolizidine is oxidized to the $\Delta^{5(10)}$-iminium salt [117]. In both these studies it was not stated whether a pure epimer or a mixture of epimers was oxidized. A later report [141] did give the NMR spectra of both epimers of 4-phenyl-quinolizidine (125, *trans*; 126, *cis*). The *trans* epimer was shown to yield the $\Delta^{5(10)}$-iminium salt (123, Ar = C_6H_5).

(125)

(126)

The substituted 1,2,3,4-tetrahydroisoquinolines studied by Knabe and Roloff [148] also produce some puzzling results difficult to rationalize when oxidized with mercuric EDTA. Compounds 127 and 128 are not attacked to any great extent, whereas several higher alkyl derivatives (129) R = ethyl, *t*-butyl, *n*-butyl; R',R" = —OCH$_3$ or —OCH$_2$O—) are oxidized.

It was found, during a study of synthetic routes leading to quinine analogs, that oxidation of a specific bicyclic pyrazoline derivative with mercuric acetate gives an enaminelike pyrazole [149].

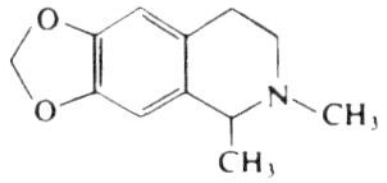

(127)

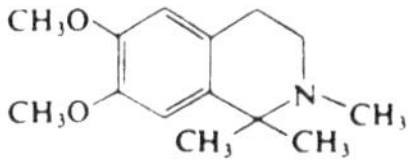

(128)

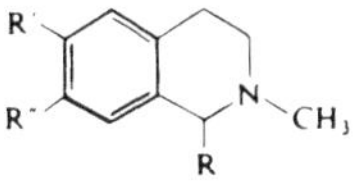

(129)

IV. ENAMINES BY OTHER OXIDATIVE PROCESSES

Although mercuric acetate is by far the most commonly used oxidizing agent to transform tertiary amines into enamines, it is not the only possible oxidizing agent that can be used. However, many of the oxidizing agents that have been used either have not been extensively studied and/or give relatively poor yields. Some of the metal salts that have been observed to oxidize tertiary amines to enamines are copper (II) chloride [150], copper (I) chloride with oxygen [151], palladium (II) chloride [152], manganese dioxide [153], and permanganate ion [154]. Iodine oxidizes trialkylamines to enamines [321]. Iodine pentafluoride [155,156] and chlorine dioxide [157] have also been found to oxidize tertiary amines to enamines. When a pentane solution of tributylamine was treated with ozone at $-78°C$, a 34% yield of the corresponding enamine (130) was obtained [158].

$$(CH_3CH_2CH_2CH_2)_2NCH=CHCH_2CH_3$$

(130)

Some of organic reagents that have been used to oxidize tertiary amines to enamines are N-bromosuccinimide [159], benzoylperoxide [160], quinones [161,162,210], and diethyl azodiformate [163,164]. For example, N-cyclohexylpyrrolidine produced enamine 131 when

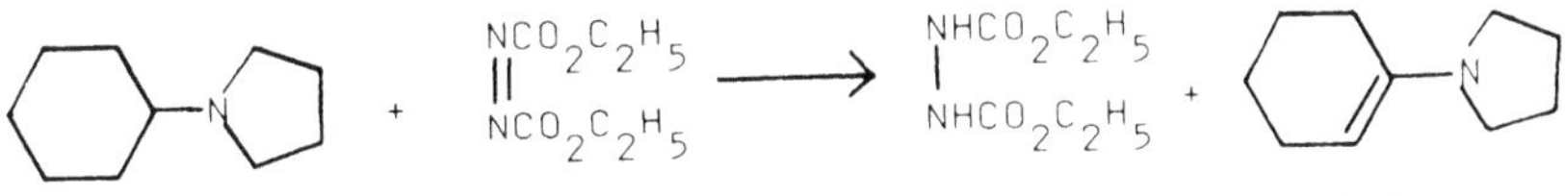

(131)

(12)

allowed to react with diethylazodiformate [163] [Eq. (12)]. The photo-reduction of aromatic ketones with tertiary amines also causes the formation of enamines [165,166].

The formation of iminium salts by treatment of tertiary amines with trityl ions or by anodic oxidation of tertiary amines is discussed in Section V.

V. SYNTHESIS OF IMINIUM SALTS

Section III described the preparation of enamines by mercuric acetate oxidation of tertiary amines. The initial product in these oxidations is the ternary iminium salt, which is converted to the enamine or mixture of enamines by reaction with base. Thus, iminium salts synthesized by methods other than the oxidation of tertiary amines or the protonation of enamines are potential enamine sources.

The alkylation of imines by an alkyl halide to give an iminium salt will be illustrated by selected reactions. A more complete survey is available [167]. Decker and Becker [168] prepared a number of iminium salts (132, for example) by mixing methyl iodide and aromatic imines in benzene. 2,5-Dimethyl-2-pyrroline (133) has been alkylated and the product (134) decomposed by potassium hydroxide to yield

$$C_6H_5CH_2CH_2N{=}CHC_6H_5 + CH_3I \rightarrow \underset{(+)}{C_6H_5CH_2CH_2\overset{\overset{\displaystyle CH_3}{|}}{N}{=}CHC_6H_5} \ I^{(-)}$$

(132)

1,2,5-trimethyl-Δ^2-pyrroline (135) [169]. It is possible that the endocyclic enamine (135) was contaminated with the exocyclic enamine

(133)   (134)   (135)   (136)

(136) since 1,2-dimethyl-Δ^2-pyrroline has been shown to be in equilibrium with its exocyclic isomer [170]. The bicyclic exocyclic enamine (139) has been prepared by alkylation of the imine (137) followed by treatment of the iminium salt (138) with base [171]. This route gave a purer product in better yield than the mercuric acetate oxidation of the tertiary amine (140).

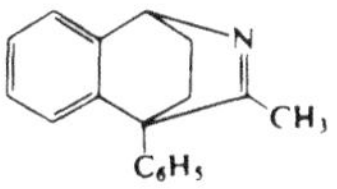

(137) 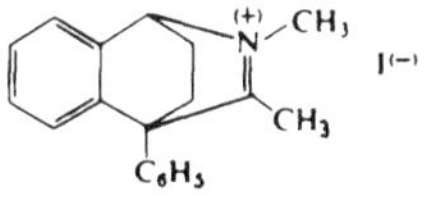   (138) 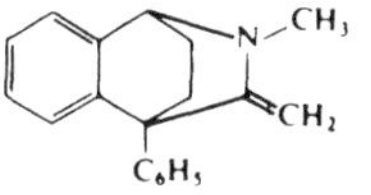   (139) 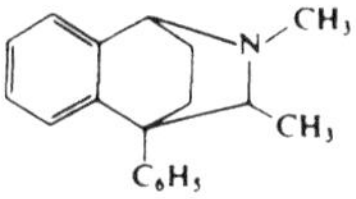   (140)

A variation on the alkylation of imines that yields enamines directly has been developed [172]. The imine is converted to the ambident ion by sodium amide or sodium hydride. An alkylating agent $[(CH_3CH_2)_3OBF_4, (CH_3CH_2)_2SO_4,$ or $CH_3CH_2I]$ is then added to the cooled solution of the imine anion. For example, N-ethylcyclohexanone imine (141) can be converted to 1-diethylaminocyclohexene (142) in 53% yield using sodium amide as the base and $(CH_3CH_2)_3OBF_4$ as the alkylating agent.

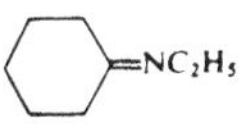

(141) 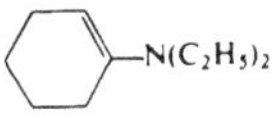   (142)

An intramolecular alkylation of imines to produce endocyclic enamines via iminium salt has been developed by Evans [211] [Eq. (13)]. The imine carbanion was produced from the imine by a $-30°C$

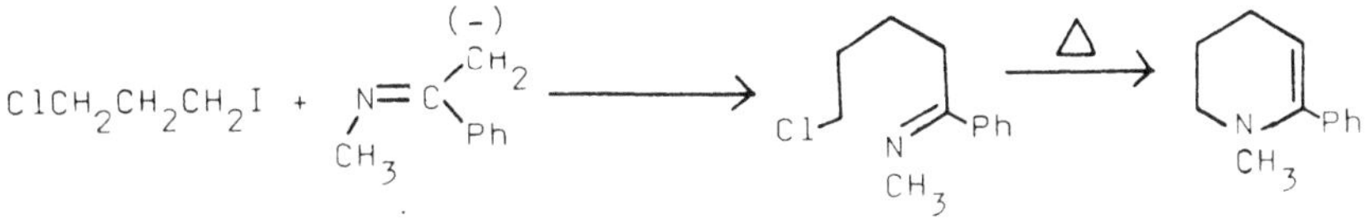

(13)

solution of lithium diisopropylamide in 1:1 THF-hexane. The enamine product was isolated in an 85% yield.

Ternary iminium complex salts can be prepared by direct combination of an aldehyde or ketone with a secondary amine complex salt [173]. An adaptation of this procedure employing the perchlorate salts of secondary amines provides a simple method for preparation of the readily crystallized and nonhydroscopic ternary iminium perchlorates [174] [Eq. (14)]. This method has been extended to the synthesis

(14)

of bridgehead bicyclic iminium salts by intramolecular cyclization of cyclic ketones possessing alkyl, secondary amine salt side chains

[175]. The bridgehead double bonds formed are in accord with Wiseman's revision of Bredt's rule [176].

Ternary iminium salts have been prepared by cleavage of a covalent C—Y bond in a system. Either silver nitrate or silver iodide

$$\overset{\diagdown}{\underset{\diagup}{N}}-\overset{|}{\underset{|}{C}}-Y$$

can be added to a solution of appropriately substituted aliphatic amino nitriles (143) in absolute ethanol to give the corresponding iminium salts (144) in yields ranging from 20 to 60% [177]. The following is an

$$R_2C(NR_2)CN \qquad \overset{(+)}{R_2C=NR_2} \quad X^-$$

(143)   (144)

example in which the direct addition of an aldehyde to the secondary amine salt did not produce the desired iminium salt, but the more indirect route involving an α-nitrile group did achieve the wanted product. Pyrrolidinium fluoroborate was allowed to react with cyclopropylaldehyde 145 at 40°C in a medium of KCN, MgSO$_4$, and THF to produce 146. Nitrile 146, upon treatment with silver borofluorate, gave iminium salt 147 [178] [Eq. (15)].

(15)

Decarbonylation of α-tertiary amino acids in phosphorus oxychloride results in iminium salt products [179,180,205]. This reaction is regiospecific and gives high yields of iminium salts. Oxidative decarboxylation of N,N-dialkyl α-amino acids with sodium hypochlorite to produce enamines will be discussed in Section IX.

Dimethyl(methylene)ammonium iodide (149) can be synthesized by heating (iodomethyl)trimethylammonium iodide (148). In tetrahydrothiophene dioxide at about 150°C for 10–15 min [181]. An S$_N$2 substitution by the iodide ion nucleophile and concomitant iodide ion elimination by the second iodine atom appears to be the reaction mechanism. Iminium salt 149 was readily obtained in 96% yield by treatment of aminal 150 with trimethylsilyl iodide [182] [Eq. (16)]. This latter method can also be used to produce alkyl-substituted salts as well.

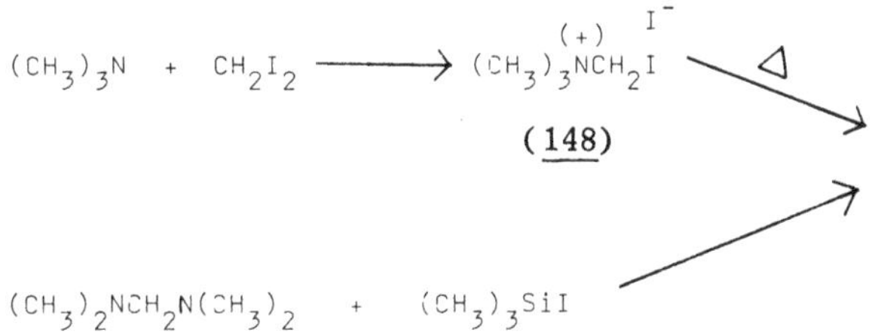

$$(CH_3)_3N \;+\; CH_2I_2 \longrightarrow (CH_3)_3\overset{(+)}{N}CH_2I \;\; \overset{I^-}{\triangle}$$

$$(\underline{148})$$

$$\longrightarrow (CH_3)_2\overset{(+)}{N}{=}CH_2 \quad (16)$$

$$(\underline{149})$$

$$(CH_3)_2NCH_2N(CH_3)_2 \quad + \quad (CH_3)_3SiI$$

$$(\underline{150})$$

Electrophilic substitution of a tertiary amide on an aromatic ring system results in formation of an iminium salt via the Bischler-Napieralski reaction. For example, in the total synthesis of ajmalicine, the cyclization of lactam 151 to iminium salt 152 was carried out using phosphorus oxychloride [183]. Similar cyclizations have been carried out by others [184].

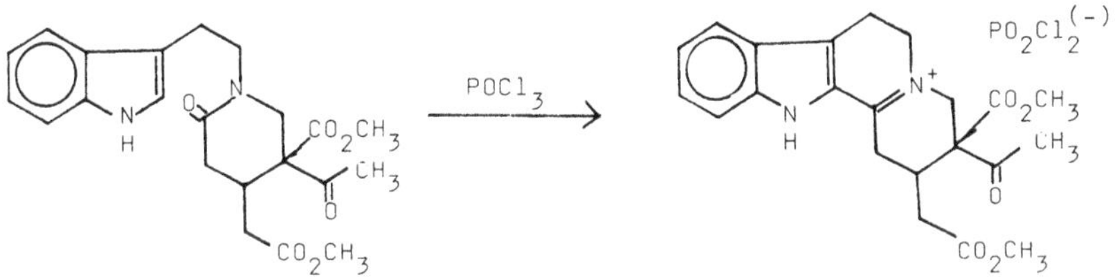

$$(\underline{151}) \qquad\qquad\qquad (\underline{152})$$

Cycloaddition of keteniminium ions to olefins gives cyclobutyliminium salts [185,186]. An example of an intramolecular cycloaddition is shown in Scheme 4 [187].

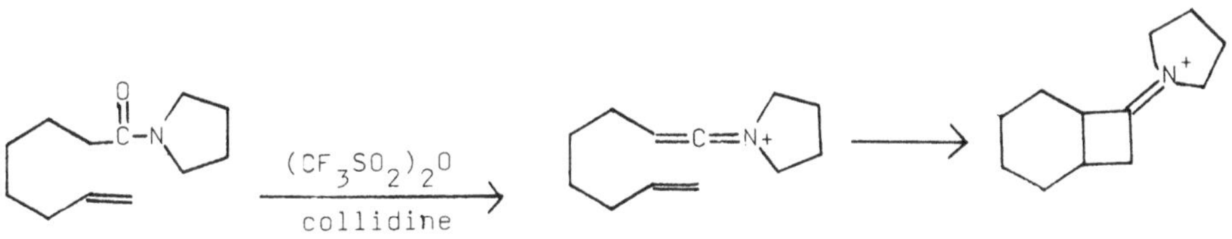

Scheme 4

Using the Polonovski reaction [188,189] in a modified form [190, 191], various iminium salts have been made including dimethyl(methylene)ammonium salt 149 (see [Eq. (15)]) [192]. The modified reaction involves mild treatment of the tertiary amine N-oxide with trifluoroacetic acid in methylene chloride to give the immonium trifluoroacetate. This reaction has been used in the total synthesis of some indole alkaloids [193].

Trityl salts can be used to oxidize tertiary amines to iminium salts by means of a hydride transfer mechanism [194,195]. Anodic oxidation of tertiary amines also, apparently, gives iminium ion intermediates [196–202].

VI. ENAMINES BY REDUCTIVE PROCESSES

The preparation of enamines by reduction of aromatic heterocyclic bases and their quaternary salts is not always the most useful approach [203]. But is has been used successfully in the reduction of isoquinolinium salts to conjugated enamines with either sodium borohydride [204] or lithium aluminum hydride [205] as the reducing agent. The lithium aluminum hydride reduction of N-acyl enamines has been used with both fruitful and unsuccessful results. A series of 3-N-acetyl-Δ^2-cholestenes (153) has been prepared by desulfurization of the appropriate thiazolidine (154) [206,207]. Lithium aluminum hydride reduction of the N-acyl enamine (153, R = CH_3CH_2) gave an

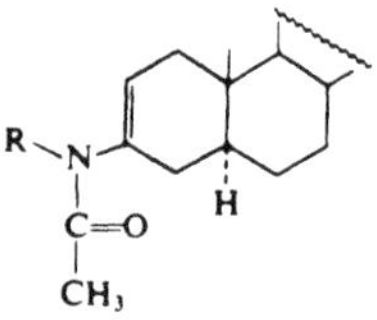

(153) 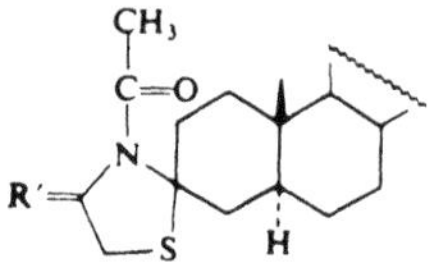   (154) 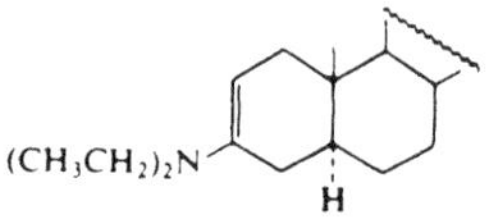   (155)

unstable enamine (155) which decomposed readily to 3-cholestanone. The steroidal N-acetyl enamines (156 and 157, R = $C_6H_5CH_2$) can be

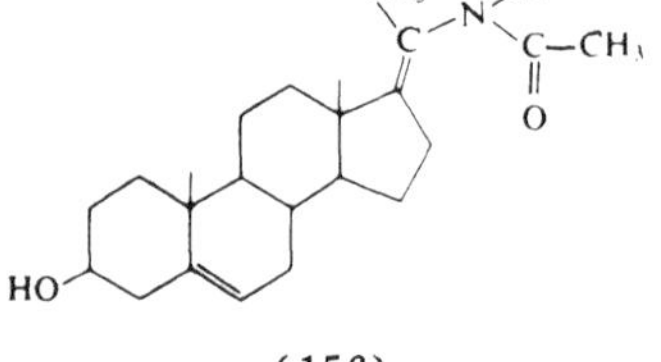

(156) 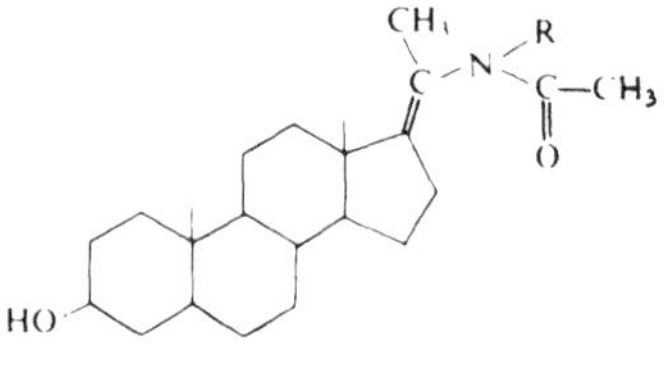   (157)

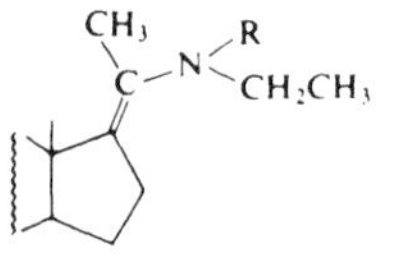

(158) 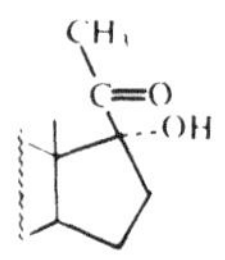   (159)

reduced by lithium aluminum hydride in tetrahydrofuran to the corresponding enamines (158, R = $C_6H_5CH_2$) in 90 and 68% yield, respectively [208]. Attempts to reduce the enamine (156, R = CH_3) led
to the formation of the impure enamine (158, R = CH_3), which decomposed to the hydroxy ketone (159).

The simpler enamide, 1-styryl-2-pyrrolidone (160), is reduced by
lithium aluminum hydride in refluxing ether to 1-styrylpyrrolidine
(161) in 52% yield [209].

(160)   (161)

The reduction of lactams with diisobutylaluminum hydride (DIBAL)
has proven to be a valuable method for synthesizing cyclic enamines
[213–215]. N-Benzyl-3-ethyl-2-piperidone (162) is readily reduced to
enamine 163 [213]. However, it has been reported that for some

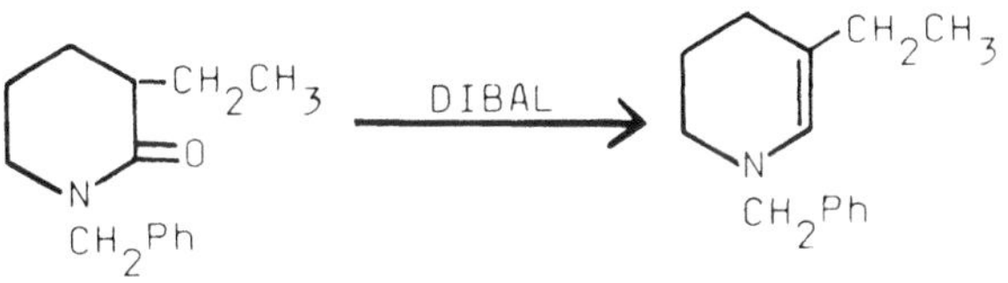

(162)   (163)

lactams, lithium aluminum diethoxyhydride is a better reducing agent
for producing enamines than DIBAL [216].

The lithium-n-propylamine reducing system has been found capable
of reducing julolidine (164) to Δ^5-tetrahydrojulolidine (165, 66% yield)
and 1-methyl-1,2,3,4-tetrahydroquinoline to a mixture of enamines
(87% yield), 1-methyl-Δ^8-octahydroquinoline (166) and 1-methyl-Δ^9-
octahydroquinoline (167) [217]. This route to enamines of bicyclic and

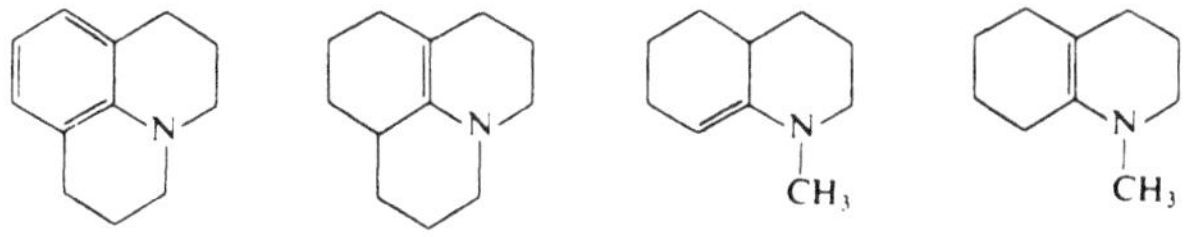

(164)   (165)   (166)   (167)

tricyclic systems avoids hydroxylation, which occurs during mecuric
acetate oxidation of certain bicyclic and tricyclic tertiary amines (118,

146; see Section III.A). Birch reductions of N,N-dimethylanilines normally give conjugated cyclohexadienamines [218,219].

VII. ENAMINES FROM LACTAMS AND ORGANOMETALLIC REAGENTS

This method of preparation has been developed primarily by Lukes [203,220]. N-Methyl lactams (168) with five- and six-membered rings plus Grignard reagents yield the 1-methyl-2-alkyl pyrrolines (169, n =1) and 1-methyl-2-alkyl piperideines (169, n =2), respectively, plus 2,2-dialkylated bases (170) as by-products [203]. For example, 1,3-dimethyl-2-piperidone (171), with a threefold excess of ethylmag-

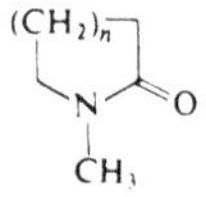

(168) 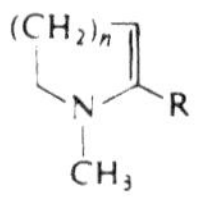   (169) 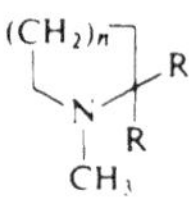   (170)

nesium iodide, yielded 40% of 2,2-diethyl-1,3-dimethylpiperidine (172) and 32% of 1,3-dimethyl-2-ethyl-Δ^2-tetrahydropyridine (173) [126].

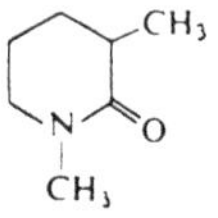

(171) 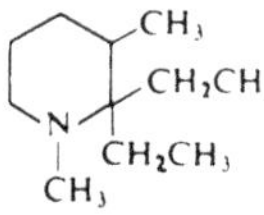   (172) 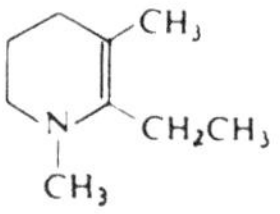   (173)

If the position alpha to the lactam carbonyl is disubstituted, exocyclic enamines (174, R = H or CH$_3$) are produced [221].

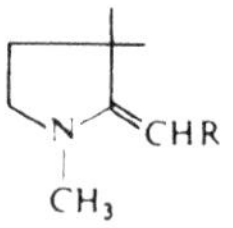

(174)

This method provides a route to certain medium-ring-sized enamines [222,223] not obtainable by other methods. 1-Methyl-2-phenyl-1-azacyclo-hept-2-ene (175) can be prepared by the reaction of N-methylcaprolactam with phenyl magnesium bromide [222], an

enamine that cannot be prepared by mercuric acetate oxidation [222]
(see also Section III.A).

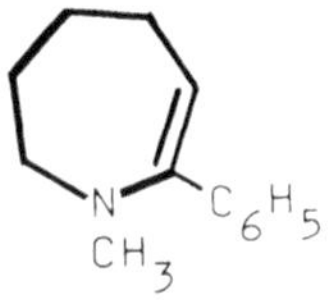

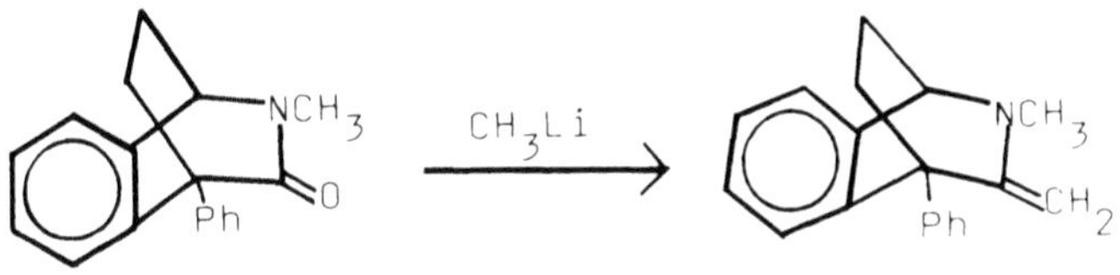

(175)   (176)   (177)

These are cases in which alkyllithium reagents are superior to
Grignard reagents in reacting with lactams. Treatment of bicyclic
lactam 176 with methyl magnesium iodide gave only a low yield of
enamine 177 in a very sluggish reaction, whereas the reaction of
lactam 176 with methyllithium produces enamine 177 quantitatively in a
vigorous reaction [224].

Acyclic enamines can be prepared by allowing N,N-dialkylformam-
ides to react with Grignard reagents [225]. Several sterically hindered "aldehyde-type" enamines can be synthesized in this manner
which cannot readily be made by direct condensation of an aldehyde
with a hindered secondary amine. Use of alkyllithium reagents with
this reaction gave poorer yields than use of Grignard reagents.

VIII. SYNTHESIS OF ENAMINES UTILIZING VARIOUS COMPOUNDS OF PHOSPHORUS, MERCURY, BORON, ARSENIC, TIN, SILICON, GERMANIUM, AND TITANIUM

Speziale and his co-workers have carried out comprehensive studies of
the reactions of phosphorus compounds. It has been shown [226] that
the reaction of N,N-dialkyl-α-trichloroacetamides (178) with phosphites
and phosphines give trichlorovinylamines (179). In general, the trialkyl-phosphines gave somewhat higher yields (60–83%) and purer
products than the phsophorous esters. An additional advantage in

$$Cl_3CCO(NR_2) + R'_3P \rightarrow Cl_2C{=}CCl(NR_2) + R'_3PO$$

(178)   (179)

using the trialkyl-phosphines is that the reaction can be carried out
at room temperature rather than at 150°C, as is necessary with the
phosphorous esters.

In a later paper, Speziale and Smith [227] investigated the reaction of trivalent phosphorus compounds with N-monosubstituted α-trichloroacetamides and α-trichloroacetamide. The products were imidoyl chlorides (180) and dichloroacetonitrile (181), respectively. The intermediacy of enamines (182) was assumed. For the monosubstituted amides the enamine (182, R = C_6H_5 or C_2H_5) can tautomerize

$$Cl_2CHC(=NR)Cl \qquad Cl_2CHCN$$

$$(180) \qquad\qquad (181)$$

to the more stable imidoyl chloride (180). The unsubstituted amide would give an enamine (182, R = H) that could also tautomerize to an

$$Cl_2C=CCl(NHR)$$

$$(182)$$

imidoyl chloride (180, R = H), which in turn would yield dichloroacetonitrile.

In this paper Speziale and Smith [227] described experiments that led them to modify the mechanism proposed earlier [226] for the reaction of trivalent phosphorus compounds with haloamides. The first step is considered to be attack of the trivalent phosphorus compound on a chlorine atom of the halo amide (183) to produce a resonance-stabilized enolate ion (184). This is reasonable since under conditions where the trichloroamide (183, X = Cl) and N,N-diethyl-2,2-dichlorophenyl-

$$
\begin{array}{c}
X \quad\; O \\
| \quad\;\; \| \\
Cl-C-C-N(C_2H_5)_2 + R_3P \rightarrow \\
| \\
Cl
\end{array}
$$

$$(183)$$

$$
\begin{array}{c}
(+) \\
R_3PCl +
\end{array}
\left[
\begin{array}{c}
O \\
\overset{(-)}{\|} \\
Cl-\overset{\_}{C}-C-N(C_2H_5)_2 \\
| \\
X
\end{array}
\leftrightarrow
\begin{array}{c}
O(-) \\
| \\
Cl-C=C-N(C_2H_5)_2 \\
| \\
X
\end{array}
\right]
$$

$$(184)$$

acetamide (183, X = C_6H_5) react readily and in high yield, the fluoroamide (183, X = F), the dichloropropionamide (183, X = CH_3), and the dichloroacetamide (183, X = H) react poorly. These results support the contention that in the first step a negative charge is formed which is stabilized by the ability of a chlorine atom (184, X = Cl) and a phenyl group (184, X = C_6H_5) to delocalize electrons through resonance. The next step involves the formation of an intermediate ion pair (185). That the chloride ion is ion-paired is indicated by the fact that no

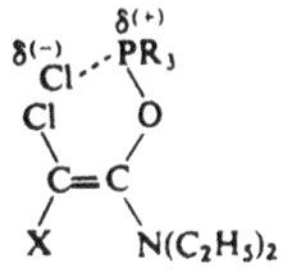

$$\underset{X}{\overset{\delta(-)}{Cl}}\overset{\overset{\delta(+)}{PR_3}}{\underset{\underset{N(C_2H_5)_2}{}}{\overset{O}{\underset{}{\big|}}}}C=C$$

(185)

$$Cl_2C = CBr[N(C_2H_5)_2]$$

(186)

N,N-diethyl-1-bromo-2,2-dichlorovinyl amine (186) could be detected by vapor-phase chromatography when the reaction of N,N-diethyl-2, 2,2-trichloroacetamide (183, X = Cl) and tributylphosphine was carried out in chloroform solution in the presence of a molar equivalent of tetra-propylammonium bromide.

Intermediate 185 could then collapse to 187 which upon loss of the trialkyl- or triarylphosphine oxide would give the enamine (188). The conversion of 185 to 188 is probably best viewed as a concerted process.

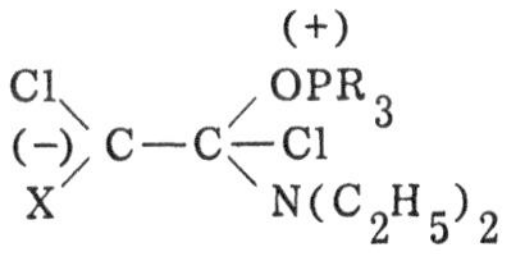

$$\underset{X}{\overset{Cl}{\underset{}{}}}\overset{(-)}{C}-\overset{(+)}{\underset{N(C_2H_5)_2}{\overset{OPR_3}{\underset{}{C}}}}Cl$$

(187)

$$ClXC{=}CCl[N(C_2H_5)_2]$$

(188)

It has been reported by Burgada and co-workers [228–230] that highly enolized ketones form enamines when they are treated with tris[dimethylamino]phosphine. Only condensation products are formed when slightly enolized ketones are treated with this reagent. Formation of enamines by allowing cyclic ketones to react with hexa-methylphosphoric triamide (HMPT) has also been reported [231].

The use of the Wittig reaction (reaction of a phosphorus ylid or phosphorane with an aldehyde or ketone) has had limited use in the synthesis of enamines. The transannular enamine 189 has been made

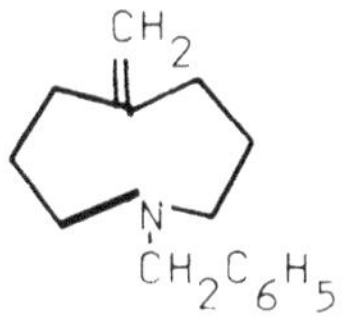

(189)

from the corresponding keto-amide using this method [262]. However, modified forms of this method using phosphoryl-stabilized anions [263, 264] have been increasingly important in the production of enamines. The Wadsworth-Emmons modification involves the use of a phosphonate ester in place of a triphenyl alkyl phosphonium salt [265]. The synthesis of the α-aminoalkylphosphonate ester (190) starting material

$$(R'CH{=}NR_2)Cl^{(-)} + (EtO)_3P \rightarrow (EtO_2P(O)CH(NR_2)R' \qquad (17)$$

$$(190)$$

necessary for making enamine 191 is readily carried out by Arbuzov reaction of an iminium salt with triethyl phosphite [Eq. (17)]. Phos-

$$(EtO)_2P(O)CH(NR_2)R' \xrightarrow[\text{(2)R''R'''CO}]{\text{(1)NaH}} R_2N{-}C(R'){=}C(R'')(R'' \, ')$$

$$(190) \qquad\qquad\qquad (191)$$

phonate ester 190 then forms an ylid with strong base, which subsequently reacts with an aldehyde or ketone to form enamine 191 [266–273]. The Horner modification of the Wittig reaction uses diphenyl alkyl phosphine oxide in place of the triphenyl allyl phosphonium salt [274,275]. So treatment of (morpholinomethyl)diphenylphosphine oxide (192) with n-butyllithium at 0°C afforded a colored, stable anion which then is allowed to react with a ketone or an aldehyde. Then the

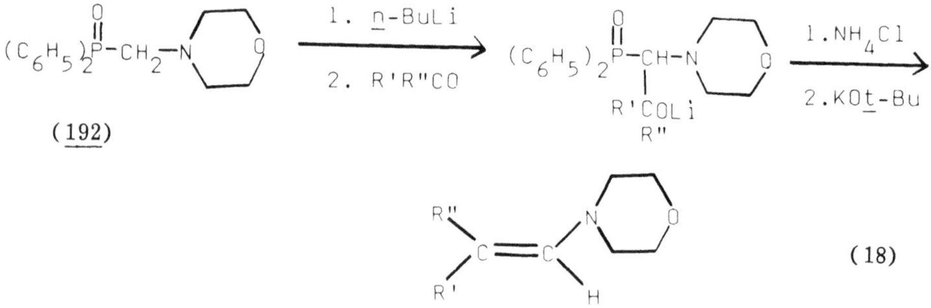

$$(18)$$

final treatment is hydrolysis to the alcohol followed by elimination of diphenylphosphinate to give the enamine product [Eq. (18)] [276–280]. This modification of the Wittig reaction for the synthesis of enamines appears quite often to be superior to the other modification in that the starting reagents are often crystalline, their reactivity and yields are often higher, the diphenylphosphinic acid by product is very water soluble and so easily removed, and a greater range of possible aldehydes or ketones can be used with them [279].

A synthetic method using dimethyl (diazomethyl) phosphonate [(193)] with a base, an amine, and a ketone or aldehyde at $-78°C$

$$R'R''CO + (CH_3O)_2P(O)CHN_2 \xrightarrow[\text{KO-}t\text{-Bu}]{R_2NH} (R')(R'')C{=}CHNR_2 \quad (19)$$

$$(193)$$

seems like a versatile and relatively simple method of making enamines [Eq. (19)] [281].

Seyferth et al. [232] have also synthesized N,N-diethyltrichloro-vinylamine (179, R = C_2H_5) from the reaction of triethylamine and phenyl(trichloromethyl)mercury. The best yield was 23%, obtained when a benzene solution of the amine (45 mM) was added to a refluxing solution of phenyl(trichloromethyl)mercury (10 mM) in benzene.

$$C_6H_5HgCCl_3 + (C_2H_5)_3N \rightarrow (C_2H_5)_2NCCl{=}CCl_2 + C_6H_5HgCl$$

$$(179)$$

Nelson and Pelter [233] have shown that a mixture of tris(pyrro-lidinyl)borane (1.1 mole), a ketone (1 mole), pyrrolidine 1.4 mole), and a catalytic amount of *p*-toluenesulfonic acid in refluxing benzene for about 30 min gave the corresponding pyrrolidine enamine (194) in

$$(194)$$

70–85% yield. The formation of the enamine is slow if free base is absent, or if there is no acid catalyst. No mechanism for the reaction was proposed, although it is probably similar to that given by Nelson and Pelter [233] for the conversion of carboxylic acids to amides using trisdialkylaminoboranes. Bicyclic enamines were prepared using the bicyclic ketone, tris(dimethylamino)borane, dimethylamine, and potassium carbonate in an autoclave heated to 95–105°C [234].

Both van Hirsch [235] and Weingarten and White [84] have reported the amination of aldehydes and ketones by tris(dimethylamino)-arsine (195) to yield the corresponding gem diamine or enamine. Von Hirsch's yields ranged from 67 to 87%, and Weingarten and White's yields were about 10% higher (when direct comparisons could be made). Weingarten and White's use of toluene or diethyl ether as a diluent

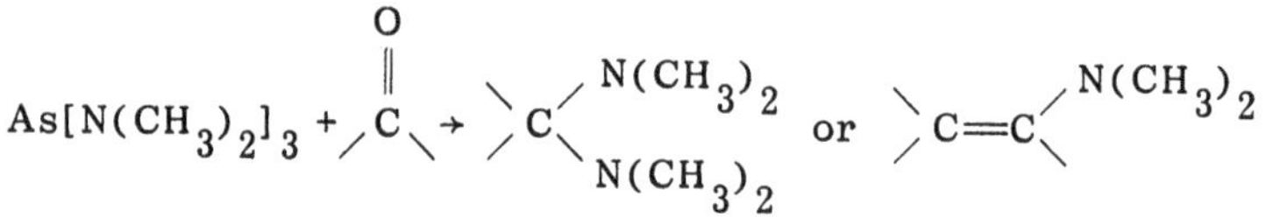

$$As[N(CH_3)_2]_3 + \underset{O}{\overset{\parallel}{C}} \rightarrow C \underset{N(CH_3)_2}{\overset{N(CH_3)_2}{<}} \quad or \quad C = C \overset{N(CH_3)_2}{<}$$

(195)

might be the reason. This method is quite useful since it offers a way by which the dimethylamine group can be introduced without using dimethylamine itself. Von Hirsch [235] extended the method to the preparation of piperidino- and pyrrolidino-enamines in 80% yield using tripiperidinosarsine and tripyrrolidinoarsine.

Secondary amine complexes of tin, silicon, germanium, and titanium [M(NR$_2$)$_4$·4HCl, where M = Sn, Si, Ge, or Ti] have been successfully used with 3-pentanone to yield the corresponding enamines in 25–32% yields [236].

The tin tetraamine complex of dimethylamine, when allowed to react with bicyclo[2.2.1]heptan-2-one (norcamphor, 48, see Section II.C), produced enamine 196 in 58% yield [237]. This compares favorably with

(48)   (196)

the 62% yield of enamine 196 obtained by treating ketone 48 with tris-(dimethylamino)borane [234] (see above). Aminostannanes react with aldehydes and ketones to produce enamines [238–240]. However, a competing reaction is the formation of an enoxytin compound (see Scheme 5).

Some moderate success has been achieved in making enamines by using the trimethylsilyl derivative of various secondary amines with ketones and aldehydes [241,242]. The reaction goes best with di-methylamine as the amine moiety. It can be carried out at room temperature with an acid catalyst or at elevated temperatures without an acid catalyst [242].

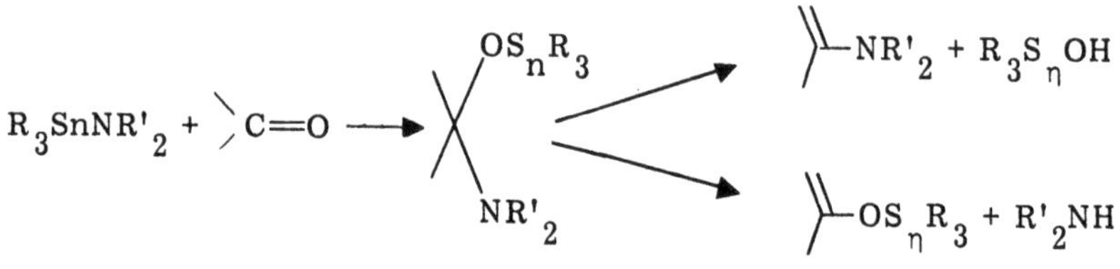

$$R_3SnNR'_2 + \underset{}{\overset{}{>}}C{=}O \longrightarrow X \begin{matrix} OS_nR_3 \\ \\ NR'_2 \end{matrix} \begin{matrix} \nearrow & -NR'_2 + R_3S_nOH \\ \\ \searrow & -OS_nR_3 + R'_2NH \end{matrix}$$

Scheme 5

One of the most general and most powerful techniques for synthesizing enamines involves the use of titanium tetrachloride. This method was first reported by Weingarten and White [84] as they used tetrakis(dimethylamino)titanium. With this compound it was possible to prepare N,N-dimethyl(1-isopropyl-2-methylpropenyl)amine (<u>197</u>) from diisopropyl ketone. If benzaldehyde, formaldehyde, or acetaldehyde is used, the corresponding gem diamine or aminal (<u>194</u>) is formed.

$$(CH_3)_2CHCCH(CH_3)_2 \overset{O}{\parallel} + Ti[N(CH_3)_2]_4 \longrightarrow (CH_3)_2CHC \overset{\diagup C(CH_3)_2}{\diagdown N(CH_3)_2}$$

(<u>197</u>)

Since tetrakis(dialkylamino)titanium compounds must be synthesized, White and Weingarten [243] sought a more versatile synthetic pathway. They found that a stoichiometric mixture of titanium tetrachloride, secondary amine, and aldehyde or ketone produced enamines directly and rapidly. The yields ranged from 55% for the mixture of enamines formed from morpholine and methylisopropyl ketone to 94% for the enamine formed from dimethylamine and methyl *t*-butyl ketone. The hindered ketone 2,5-dimethylcyclopentanone could be converted to an enamine, but the more hindered ketone 2,6-di-*t*-butylcyclohexanone was inert. However, this method is ideally suited for synthesis of enamines from sterically hindered ketones. For example the synthesis of the morpholine enamine of methyl *t*-butyl ketone by the use of molecular sieves to remove water under optimal conditions produced less than 0.5% yield after 150-hr reaction time. Using the TiCl₄ technique, yields of 65—70% were obtained after 3 hr, and 80—87% yields were produced after 10--12 hr [21].

Subsequent to the original publications by White and Weingarten in which the TiCl₄ method was discussed, this technique has been frequently used to synthesize a variety of enamines [66,244—247,320]. Using this method, 1-cyclopropyl-1-morpholino-2-cyclopentylidenemethane (<u>198</u>) can be formed in a 71% yield free from the cyclopropylidene tautomer [104] [Eq. (20)]. This technique can also be used to

(20)

(<u>198</u>)

make enamines from ketones possessing various functional groups such as nitriles, esters, amides, or sulfides [248].

A comparison of the molecular sieves and titanium tetrachloride techniques was carried out in which a multivariate method of optimization was used for each technique [21]. The molecular sieve method was generally slower and restricted to unhindered ketones. A further study of optimizing the reaction conditions for the TiCl$_4$ technique was carried out using multivariate strategies [249]. It was found that the optimal conditions consist of the addition of the ketone to a preformed complex between the secondary amine and titanium tetrachloride. The ratio of secondary amine to titanium tetrachloride for various amine-ketone reactions ran from 6:1 to 10:1 [249]. A direct comparison of this optimized modified procedure and the original procedure used by White and Weingarten [243] was made using the morpholine enamine of isobutyrophenone. The original procedure produced a 62% yield after several hours, whereas the optimized modified procedure gave an 87% yield after 15 min [250]! This modified procedure can be used to produce enamines from aldehydes, cyclic and acyclic ketones, or aryl alkyl ketones.

It has been observed that too high an amine-to-titanium-tetra-chloride ratio decreases the yield of enamine, as does a ratio that is too low [249]. A reaction mechanism has been proposed (Scheme 6) in which either a titanium-coordinated immonium ion or a titanium-

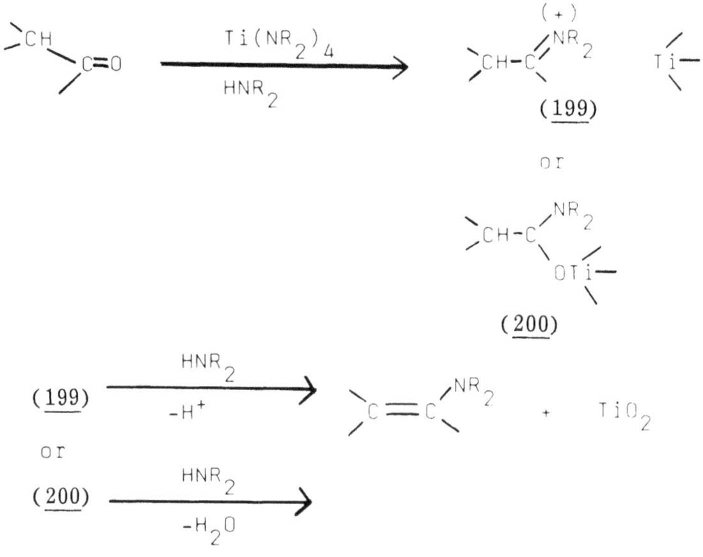

Scheme 6

coordinated carbinolamine (psuedo-base, 200) is formed in the first step, and base-catalyzed deprotonation or dehydration takes place in the second step [249].

IX. MISCELLANEOUS PREPARATIONS

1-(1-Cyclopenten-1-yl)piperidine (201) reacted with N-methyl-3-bromopropylamine hydrobromide to yield 74% of a mixture of enamines (202) [62]. The proposed mechanism involved an amine exchange to give the enamine 203, which underwent internal alkylation.

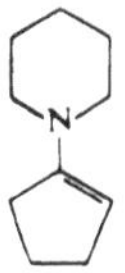

(201) 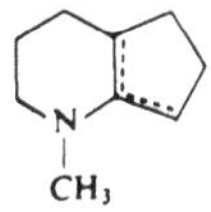   (202) 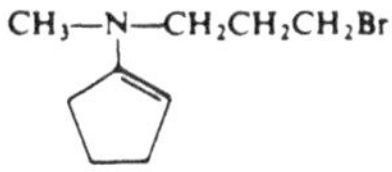   (203)

Displacement of vinyl fluoride of chloride by secondary amines has given some unusual enamines, as illustrated for the preparation of 1,1-difluoro-2-piperidino-3-phenyl-2-cyclobutene (204) [251], 1,1-difluoro-2,4-dipiperidino-3-phenyl-2-cyclobutene (205) [252], and 2-phenyl-3-(1'-aziridinyl)-2-cyclohexenone (206) [253]. Similar reactions have been observed with acyclic, perfluorinated alkenes [254].

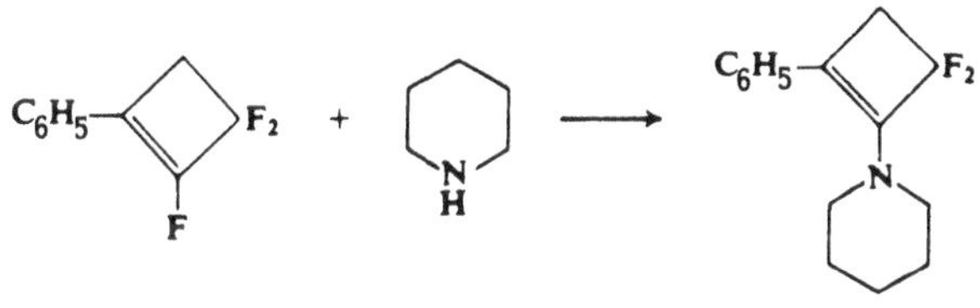

(204)

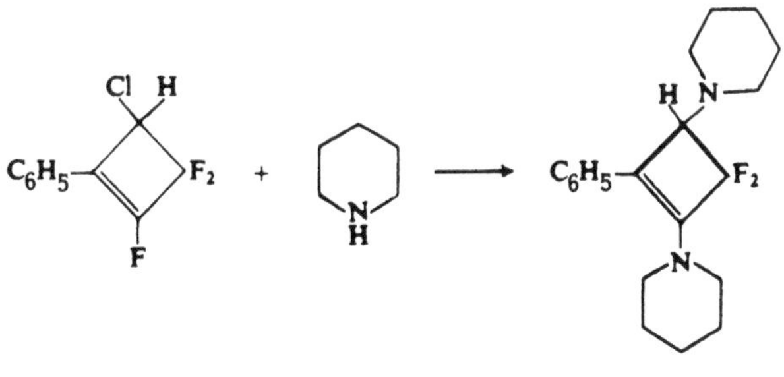

(205)

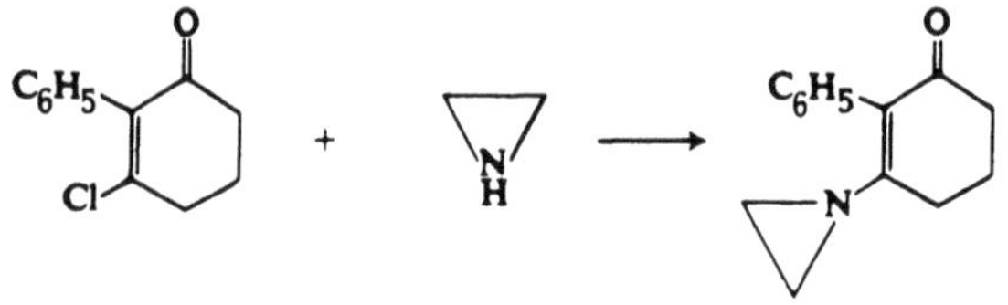

(206)

N-(2-Bromoallyl)-ethylamine with sodium amide in liquid ammonia gave N-ethylallenimine (207) [255−257]. In a similar vein intramo-

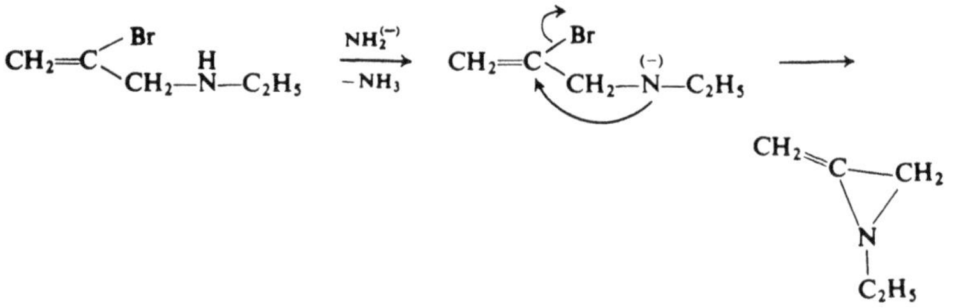

(207)

lecular alkylation of imines by alkylhalides (method of Evans [221,258]; see Section V) and by cyclopropyl or cyclobutyl groups [259−261] (Scheme 7) produces heterocyclic enamines with endocyclic double bonds.

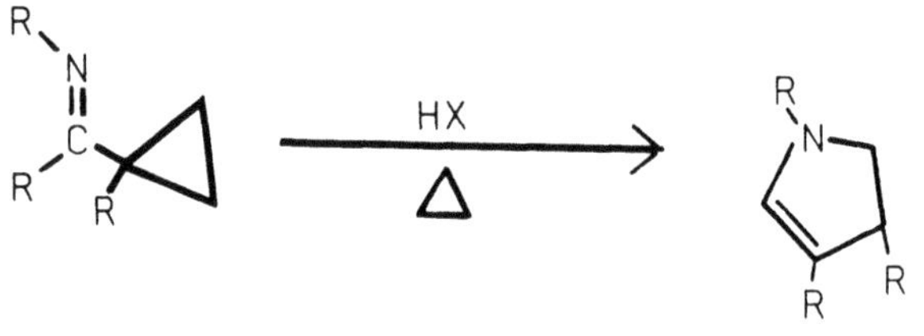

Scheme 7

Sterically hindered enamines can be synthesized from β-chloro-enamine 208 by treating it with a sterically crowded alkyllithium com-

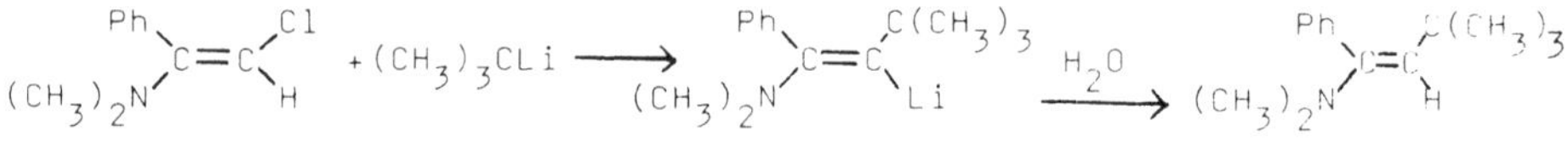

(208)   (209)

pound such as *t*-butyllithium at −70°C and then adding methyl iodide
to give enamine 209 in 75% yield [289].

N,N-Dimethylformamide diethyl acetal was first reported by Meerwein and co-workers [282], and it was used by them to react with
cyclopentadiene to form 6-(dimethylamino) fulvene (210). It will also

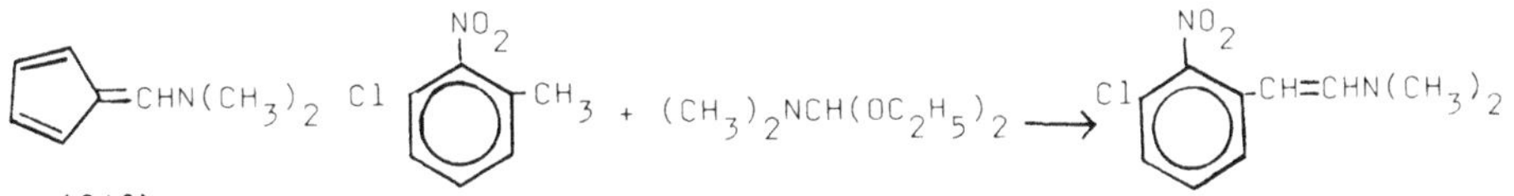

(210)   (211)   (212)

react with suὐstituted toluenes in good yield to form conjugated
enamines [283−288,323−325]. For example, 3-chloro-2-nitrotoluene
(211) when refluxed for 24 hr with N,N-dimethylformamide diethylacet-
al in dimethylformamide gave 40% yield of enamine 212 [284]. The
Vilsmeier-Haack reaction [316], which involves the reaction of dimethylformamide and $POCl_3$ with active aromatic systems or active methylene
systems, has been shown to first produce an enamine intermediate
[317,318].

A "one-pot" synthesis which avoids aldehydes as precursors, replacing them by substituted acetic acids, has been reported [290].

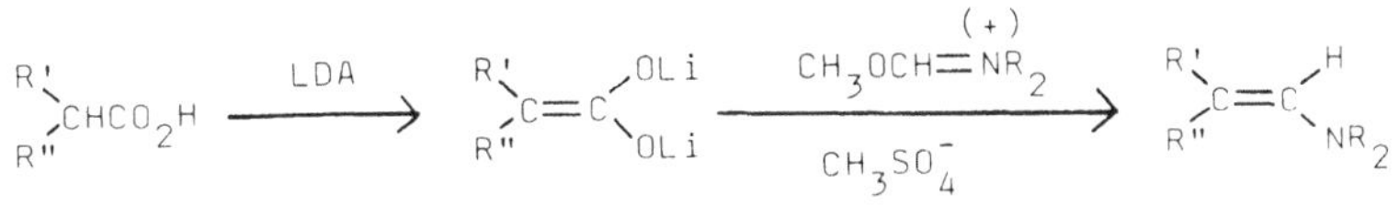

Scheme 8

It involves the double deprotonation of an "enolizable" carboxylic acid
with lithium diisopropylamide (LDA) followed by treatment with a
methoxymethaniminium methyl sulfate (obtained from dimethyl sulfate
and the appropriately substituted formamide) (Scheme 8). Not only
does this method avoid the synthesis of an aldehyde to make the
enamine, but it may be used to obtain the aldehyde itself by hydrolysis
of the enamine.

Deyrup and Kuta [291] reported the production of the 1,3-dipolar
azomethine ylide 214 generated by deprotonation of di-*tert*-butylketene-
N-methyl-N-ethyliminium fluorosulfonate (213) with sodium bis(trimethylsilyl) amide. This azomethine ylide was then trapped with norbornene to give enamine (215). C-Protonation of this enamine with
HCl produced iminium salt (216), which is very stable toward attack
by methyllithium, $NaBH_4$, NaI, and aqueous NaOH.

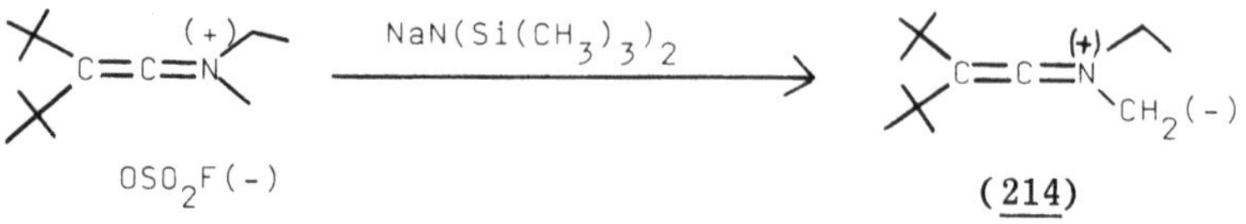

(213) → (214)

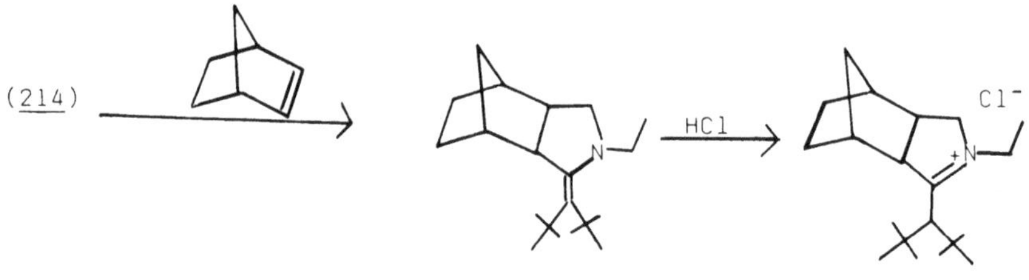

(215)   (216)

The use of the nitrimino group in place of the carbonyl group in the synthesis of enamines has been reported [292]. So the treatment of camphor nitrimine (217) with piperidine in acetonitrile solvent (with

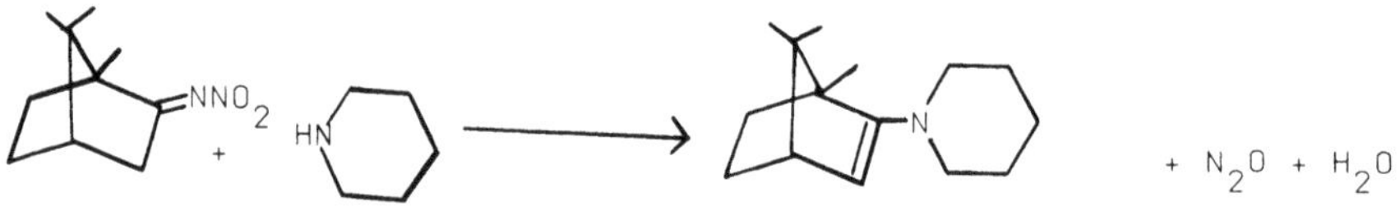

(217)   (218) $+ N_2O + H_2O$

molecular sieves present) gave bicyclic enamine 218 in an 87% yield. Enamines can be made in this way that cannot be made by the direct technique of secondary amine plus ketone.

Heating cyclic amides possessing propanoic or butanoic acid side chains with soda lime yield cyclic enamines with a new five- or six-membered-ring formation vis cyclodehydration and decarboxylation [293–295].

The preparation of enamines by heating secondary amines and ketals was originated by Hoch [296] and has been extended by Bianchetti and co-workers [297–299]:

$$C_6H_5C(OCH_2CH_3)_2CH_3 + CH_3NHC_6H_5 \longrightarrow$$

1-Piperidino-2-nitroethene (220) and 1-morpholine-2-nitroethene (221) were the final products when a slight excess of the appropriate secondary amine was caused to react with ethoxymethylenemalonate (219) in the presence of nitromethane [300]. The reaction sequence proposed was:

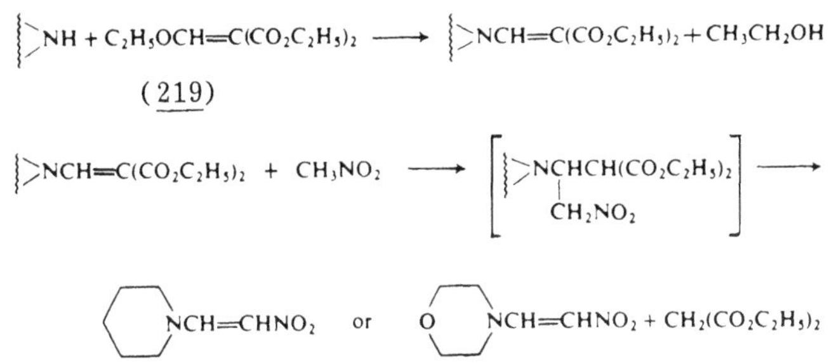

$$\text{\textbar}\text{NH} + C_2H_5OCH{=}C(CO_2C_2H_5)_2 \longrightarrow \text{\textbar}\text{NCH}{=}C(CO_2C_2H_5)_2 + CH_3CH_2OH$$

(219)

$$\text{\textbar}\text{NCH}{=}C(CO_2C_2H_5)_2 + CH_3NO_2 \longrightarrow \left[\text{\textbar}\text{NCHCH}(CO_2C_2H_5)_2 \atop \qquad \qquad \overset{|}{C}H_2NO_2 \right] \longrightarrow$$

(220)   (221)

This method was extended to the preparation of aminonitropropenes, but only piperidine and morpholine of the several secondary amines studied were found effective.

The dehydrobromination and dequaternization of 1,1,3-trimethyl-2-bromomethylpyrrolidinium bromide (222) has been accomplished by dry distillation from potassium acetate [301]. Since the product was isolated as the perchlorate salt, no conclusion can be drawn as to whether the original reaction mixture contained the exocyclic enamine (223) or the endocyclic enamine (224) or a mixture of both. Dehydro-

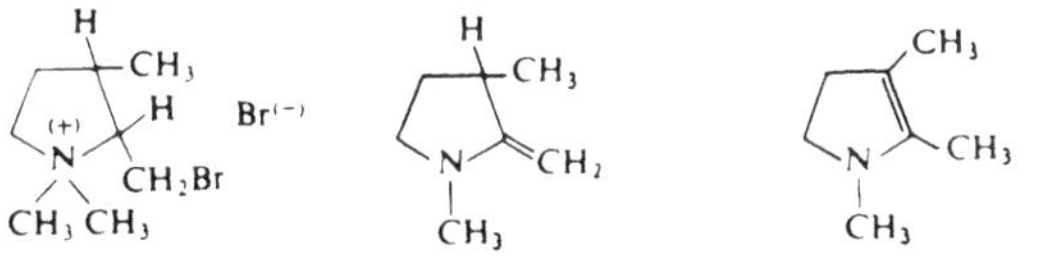

(222)   (223)   (224)

halogenation of β-halo tertiary amines [302,303], dehydration of β-hydroxy tertiary amines [304], dehydrocyanation of α-aminonitriles [305], and dehydroxysilanation of β-hydroxysilylamines [306] are further methods of making enamines.

The addition of secondary amines to 1-cyanoallenes (225) results in the formation of enamines in 80–90% yield [307]. Addition can occur at the 1,2 or 2,3 double bonds so that a mixture of isomeric enamines (226 and 227) is formed. The ratio of products is influenced by the alkyl substituents on the cyanoallenes and the structure of the secondary amine.

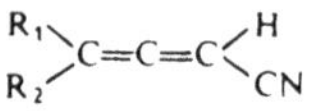

$$\underset{R_2}{\overset{R_1}{>}}C=C=C\overset{H}{\underset{CN}{<}} \qquad \underset{R_2}{\overset{R_1}{>}}C=C\overset{CH_2CN}{\underset{N-}{<}} \qquad \overset{R_1R_2CH}{\underset{-N}{|}}\overset{|}{C}=C\overset{H}{\underset{CN}{<}}$$

$$(225) \qquad\qquad (226) \qquad\qquad (227)$$

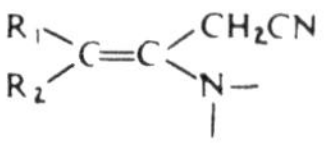

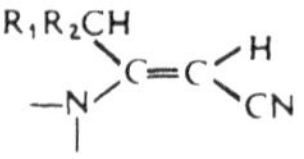

Aminomercuration of terminal alkynes and aziridine or an aromatic amine and mercury(II) acetate followed by alkaline sodium borohydride demercuration is a useful method of synthesizing enamines [308,309]. Thallium(III) acetate may be used in palce of mercury(II) acetate [309]. It has since been shown that only *catalytic amounts* of thallium(III) acetate [310] or mercury(II) chloride [311] are necessary to cause addition of secondary aromatic amines to terminal alkynes.

Perfluoroalkylenamines (229) can be obtained by treating a 1-hydrylperfluoroalk-1-yne (228) with a secondary amine at room tempera-

$$R_F CF_2 C{\equiv}CH + R_2NH \rightarrow R_F CF_2 CH{=}CHNR_2$$

$$(228) \qquad\qquad\qquad (229)$$

ature or lower [312,313]. The more basic the secondary amine, the faster the reaction proceeds.

The reaction of a propargyl alcohol with an amide acetal can give an enamine product [314]. Treatment of 3-methyl-4-penten-1-yn-3-ol (230) with dimethylacetamide diethyl acetal gives enamine 231 following the pathway of Scheme 9.

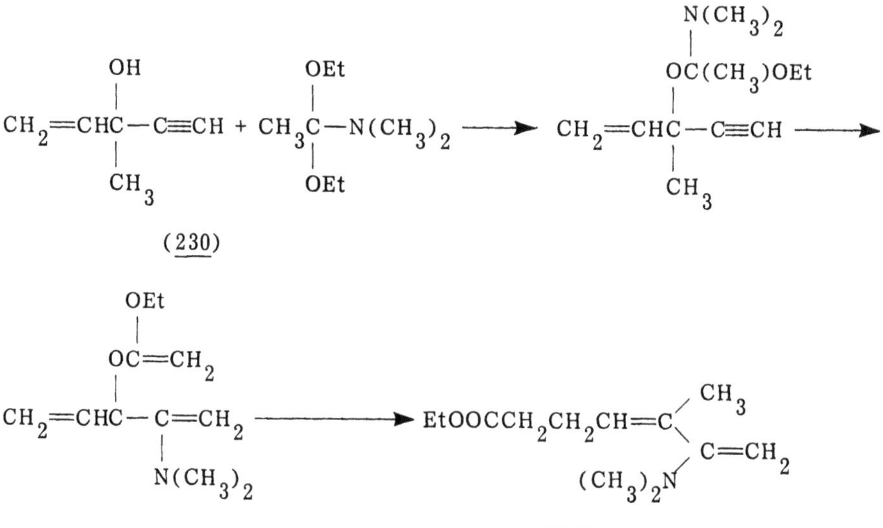

Scheme 9

The first step is alcohol exchange with the acetamide acetal followed by addition of the dimethylamino group from the amide acetal moiety to the alkyne group. Then a Claisen rearrangement takes place to give enamines 231.

Van Tamelen and co-workers [315] have reported a useful and specific synthetic method for production of enamines by the oxidative decarboxylation of N,N-dialkyl α-amino acids with sodium hypochlorite.

The neutral or base-catalyzed isomerization of allylamines into enamines is discussed in Chapter 1 (Section III.A).

Transamination reactions between enamines and secondary amines have been commonly observed, and it has been systematically studied [319].

REFERENCES

1. C. Mannich and H. Davidsen, *Ber.*, *69*, 2106 (1936).
2. M. E. Herr and F. W. Heyl, *J. Am. Chem. Soc.*, *74*, 3627 (1952).
3. F. W. Heyl and M. E. Herr, *J. Am. Chem. Soc.*, *75*, 1918 (1953).
4. M. E. Herr and F. W. Heyl, *J. Am. Chem. Soc.*, *75*, 5927 (1953).
5. F. W. Heyl and M. E. Herr, *J. Am. Chem. Soc.*, *77*, 488 (1955).
6. G. Stork, R. Terrell, and J. Szmuszkovicz, *J. Am. Chem. Soc.*, *76*, 2029 (1954).
7. G. Stork and H. K. Landesman, *J. Am. Chem. Soc.*, *78*, 5128 (1956).
8. G. Stork and H. K. Landesman, *J. Am. Chem. Soc.*, *78*, 5129 (1956).
9. G. Stork, A. Brizzolara, H. Landesman, J. Szmuszkovicz, and R. Terrell, *J. Am. Chem. Soc.*, *85*, 207 (1963).
10. R. Fusco, G. Bianchetti, and S. Rossi, *Gazz. Chim. Ital.*, *91*, 825 (1961).
11. J. D. Roberts, Editor, *Org. Syn.*, *41*, 65 (1961).
12. A. Mondon, *Ber.*, *92*, 1461 (1959).
13. S. Hünig, E. Benzing, and E. Lücke, *Ber.*, *90*, 2833 (1957).
14. M. F. Semmelhack, Editor, *Org. Syn.*, *62*, 191 (1984).
15. R. Jacquier, C. Petrus, and F. Petrus, *Bull. Soc. Chim. France*, 2845 (1966).
16. J. L. Johnson, M. E. Herr, J. C. Babcock, A. E. Fonken, J. E. Stafford, and F. W. Heyl, *J. Am. Chem. Soc.*, *78*, 430 (1956).
17. P. A. Zoretic, F. Barcelos and B. Branchaud, *Org. Prep. Proced. Int.*, *8*, 211 (1976).
18. M. F. Ansell, J. S. Mason, M. P. L. Caton, *J. Chem. Soc., Perkin Trans.*, *I*, 1061 (1984).
19. A. A. Brizzolara, Ph.D. thesis, Columbia University, New York, 1960.
20. C. Djerassi and B. Tursch, *J. Org. Chem.*, *27*, 1041 (1962).

21. R. Carlson, R. Phan-Tan-Luu, D. Mathieu, F. S. Ahouande, A. Babadjamian, J. Metzger, *Acta Chem. Scand.*, *B 32*, 355 (1978).
22. K. Taguchi and F. H. Westheimer, *J. Org. Chem.*, *36*, 1570 (1971).
23. T. Takayuki, N. Kawai, *Jpn. Pat. 79,117,466* (1979). (C.A., *92*, 128710y (1980)).
24. D. P. Roelofsen and H. Van Bekkum, *Rec. Trav. Chim.*, *91*, 605 (1972).
25. B. P. Mundy and W. G. Bornmann, *Tetrahedron Lett.*, 957 (1978).
26. P. Wittig and R. Mayer, *Z. Chem.*, *7*, 306 (1967).
27. E. P. Blanchard, Jr., *J. Org. Chem.*, *28*, 1397 (1963).
28. R. Dulou, E. Elkik, and A. Veillard, *Bull. Soc. Chim. France*, 967 (1960).
29. M. Takeda, H. Inoue, M. Konda, S. Saito and H. Kugita, *J. Org. Chem.*, *37*, 2677 (1972).
30. P. L. deBenneville and J. H. Macartney, *J. Am. Chem. Soc.*, *72*, 3073 (1950).
31. E. Benzing, *Angew. Chem.*, *71*, 521 (1959).
32. E. Benzing (to Monsanto Chemical Co.), *U.S. Pat. 3,074,940* (1963).
33. J. R. Geigy, *Brit. Pat. 832,078* (1960).
34. K. C. Brannock and R. D. Burpitt, *J. Org. Chem.*, *26*, 3576 (1961).
35. K. Brannock, A. Bell, R. D. Burpitt, and C. A. Kelly, *J. Org. Chem.*, *29*, 801 (1964).
36. J. M. Beaton (to Upjohn Co.) *U.S. Pat. 4,257,949* (1981) (C.A., *95*, 741 (1981)).
37. H. G. Hauthal and D. Schied, *Z. Chem.*, *9*, 62 (1969).
38. K. U. Acholonu and D. K. Wedegaertner, *Tetrahedron Lett.*, 3253 (1974).
39. H. Mazarguil and A. Lattes, *Bull. Soc. Chim. France*, 319 (1969).
40. H. Mazarguil and A. Lattes, *Bull. Soc. Chim. France*, 112 (1971).
41. T-Y. Chen, H. Kato, and M. Ohta, *Bull. Chem. Soc. Jpn.*, *39*, 1618 (1966).
42. H. W. Thompson and J. Swistok, *J. Org. Chem.*, *46*, 4907 (1981).
43. K. Muller and F. Previdoli, *Helv. Chim. Acta*, *64*, 2508 (1981)
44. E. Elkik and H. Assadi-Far, *Compt. Rend.*, *C 263*, 945 (1966).
45. K. L. Erickson, J. Markstein, and K. Kim, *J. Org. Chem.*, *36*, 1024 (1971).
46. A. G. Cook, Ph.D. Thesis, Univ. of Illinois, Urbana, 1959.
47. R. V. Stevens, Editor, *Org. Syn.*, *61*, 129 (1983).

48. A. G. Cook, S. B. Herscher, D. J. Schultz and J. A. Burke, *J. Org. Chem.*, *35*, 1550 (1970).
49. P. Duhamel, L. Duhamel and P. Siret, *Comp. Rend.*, *276*, 519 (1973).
50. J. M. J. Tronchet, B. Baehler and A. Bonenfant, *Helv. Chim. Acta, 59*, 941 (1976).
51. J. A. Hirsch and F. J. Cross, *J. Org. Chem.*, *36*, 995 (1971).
52. F. A. Buiter, J. H. Sperna Weiland, and H. Wynberg, *Rec. Trav. Chim.*, *83*, 1160 (1964).
53. J. A. Hirsch and X. L. Wang, *Syn. Commun.*, *12*, 33 (1982).
54. F. Eiden and K. T. Wanner, *Arch. Pharm.*, *317*, 958 (1984).
55. S. Fatutta, G. Pitacco and E. Valentin, *J. Chem. Soc.*, *Perkin Trans.*, *I*, 2735 (1983).
56. J. B. Rampal, K. D. Berlin and N. Satyamurthy, *Phosphorus Sulfur*, *13*, 179 (1982).
57. L. A. Paquette, *J. Org. Chem.*, *29*, 2851 (1964).
58. F. Kasper, *Z. Chem.*, *5*, 153 (1965).
59. A. G. Cook, W. C. Meyer, K. E. Ungrodt, and R. Mueller, *J. Org. Chem.*, *31*, 14 (1966).
60. J. F. Stephen and E. Marcus, *J. Org. Chem.*, *34*, 2535 (1969).
61. A. G. Cook, W. M. Kosman, T. A. Hecht and W. A. Koehn, *J. Org. Chem.*, *37*, 1565 (1972).
62. R. F. Parcell and F. P. Hauck, Jr., *J. Org. Chem.*, *28*, 3468 (1963).
63. R. T. Parfitt, *J. Chem. Soc. (C)*, 140 (1967).
64. L. A. Paquette and M. Rosen, *J. Org. Chem.*, *33*, 2130 (1968).
65. L. H. Hellberg, R. J. Milligan and R. N. Wilke, *J. Chem. Soc. (C)*, 35 (1970).
66. F. A. Van der Vlugt, J. W. Verhoeven and U. K. Pandit, *Rec. Trav. Chim.*, *89*, 1258 (1970).
67. U. K. Pandit, K. deJonge and H. O. Huisman, *Rec. Trav. Chim.*, *88*, 149 (1969).
68. T. A. Crabb and K. Schofield, *J. Chem. Soc.*, 4276 (1958).
69. T. A. Crabb and K. Schofield, *J. Chem. Soc.*, 643 (1960).
70. A. L. Ham and P. R. Leeming, *J. Chem. Soc. (C)*, 2017 (1969).
71. A. T. Blomquist and E. J. Moriconi, *J. Org. Chem.*, *26*, 3762 (1961).
72. P. K. Khandelwal and B. C. Joshi, *Def. Sci. J.*, *21*, 199 (1971).
73. U. Edlund and G. Bergson, *Acta Chem. Scand.*, *25*, 3625 (1971).
74. U. Edlund, *Acta Chem. Scand.*, *26*, 2972 (1972).
75. W. L. Meyer and R. G. Olsen, *Can. J. Chem.*, *45*, 1459 (1967).
76. D. Thon and W. Schneider, *Chem. Ber.*, *109*, 2743 (1976).
77. D. Thon and W. Schneider, *Liebigs Ann.*, 2094 (1976).
78. H. Stetter and K. Komorowski, *Chem. Ber.*, *104*, 75 (1971).
79. L. P. Vinogradova, G. A. Kogan, and S. I. Zavialov, *Izv. Acad. Nauk SSSR, Otd. Kim. Nauk*, 1954 (1964); *Bull. Acad. Sci. USSR, Div. Chem. Sci. (English Transl.)*, 979 (1964).

80. R. Jacquier and G. Maury, *Bull. Soc. Chim. France*, 320 (1967).
81. R. O. Clinton, A. J. Manson, F. W. Stonner, R. L. Clarke, K. F. Jennings, and P. E. Shaw, *J. Org. Chem.*, *27*, 1148 (1962).
82. P. W. Hickmott, *Tetrahedron, 38*, 1975 (1982).
83. P. Ferruti, D. Pocar, and G. Bianchetti, *Gazz. Chim. Ital.*, *97*, 109 (1967).
84. H. Weingarten and W. A. White, *J. Org. Chem.*, *31*, 4041 (1966).
85. F. Danusso, P. Ferruti, and G. Peruzzo, *Atti Accad. Naz. Lincei, 39*, 498 (1965).
86. P. Ferruti, A. Segre, and A. Fere, *J. Chem. Soc. (C)*, 2721 (1968).
87. L. Duhamel, P. Duhamel, and P. Siret, *Tetrahedron Lett.*, 3607 (1972).
88. P. Wittig and R. Mayer, *Z. Chem.*, *7*, 57 (1967).
89. B. Gaux and P. L. Henaff, *Comp. Rend.*, *271*, 1093 (1970).
90. J. S. Marchese, Ph.D. thesis, University of Maryland, College Park, 1964.
91. J. Szmuszkovicz, in *Advances in Organic Chemistry: Methods and Results*, Vol. 4 (R. A. Raphael, E. C. Taylor, and H. Wynberg, eds.), Wiley-Interscience, New York, 1963, p. 10.
92. G. Gianchetti, D. Pocar, and R. Stradi, *Gazz. Chim. Ital.*, *100*, 726 (1970).
93. R. Carlson, L. Nilsson, C. Rappe, A. Babadjamian, and J. Metzger, *Acta Chem. Scand.*, *B32*, 85 (1978).
94. W. J. M. Van Tilborg, S. E. Schaafsma, H. Steinberg, and T. J. DeBoer, *Rec. Trav. Chim.*, *86*, 417 (1967).
95. N. J. Turro and W. B. Hammond, *Tetrahedron Lett.*, 3085 (1967).
96. J. Szmuszkovicz, D. J. Duchamp, E. Cerda, and C. G. Chidester, *Tetrahedron Lett.*, 1309 (1969).
97. H. H. Wasserman and M. S. Baird, *Tetrahedron Lett.*, 1729 (1970).
98. W. J. M. VanTilborg, G. Dooyewaard, H. Steinberg, and T. J. deBoer, *Tetrahedron Lett.*, 1677 (1972).
99. H. Bochow and W. Schneider, *Chem. Ber.*, *108*, 3475 (1975).
100. W. Schneider and H. Gotz, *Ann.*, *653*, 85 (1962).
101. E. L. Patmore and H. Chafetz, *J. Org. Chem.*, *32*, 1254 (1967).
102. A. G. Cook and C. R. Schulz, *J. Org. Chem.*, *32*, 473 (1967).
103. A. G. Cook and W. M. Kosman, *Tetrahedron Lett.*, 5847 (1966).
104. D. Pocar, R. Stradi, and P. Trimarco, *Tetrahedron, 31*, 2427 (1975).
105. L. Birkofer and G. Daum, *Ber.*, *95*, 183 (1962).
106. M. Rey and A. S. Dreiding, *Helv. Chim. Acta*, *57*, 734 (1974).
107. M. K. Huber and A. S. Dreiding, *Helv. Chim. Acta*, *57*, 748 (1974).

108. M. K. Huber, R. Martin, M. Rey, and A. S. Dreiding, *Helv. Chim. Acta, 60,* 1781 (1977).
109. N. J. Leonard and R. R. Sauers, *J. Org. Chem., 21,* 1187 (1956).
110. J. Tafel, *Ber., 24,* 1619 (1892).
111. See footnote 5, in Ref. 112.
112. N. J. Leonard, A. S. Hay, R. W. Fulmer, and V. W. Gash, *J. Am. Chem. Soc., 77,* 439 (1955).
113. N. J. Leonard, P. D. Thomas, and V. W. Gash, *J. Am. Chem. Soc., 77,* 1552 (1955).
114. L. Marion and N. J. Leonard, *Can. J. Chem., 29,* 355 (1951).
115. D. Kettelhack, M. Rink, and K. Winterfield, *Arch. Pharm., 287,* 1 (1954).
116. M. Carmack, B. Douglas, E. W. Martin, and H. Suss, *J. Am. Chem. Soc., 77,* 4435 (1955).
117. N. J Leonard, R. W. Fulmer, and A. S. Hay, *J. Am. Chem. Soc., 78,* 3457 (1956).
118. N. J. Leonard, L. A. Miller, and P. D. Thomas, *J. Am. Chem. Soc., 78,* 3463 (1956).
119. C. F. Koelsch and D. L. Ostercamp, *J. Org. Chem., 26,* 1104 (1961).
120. N. J. Leonard, W. J. Middleton, P. D. Thomas, and D. Choudhury, *J. Org. Chem., 21,* 344 (1956).
121. M. G. Reinecke and L. R. Kray, *J. Org. Chem., 31,* 4215 (1966).
122. Y-L. Chang, Ph.D. thesis, University of Illinois, Urbana, 1961.
123. W. Schneider and R. Dillmann, *Arch. Pharm., 298,* 43 (1965).
124. N. J. Leonard and W. K. Musker, *J. Am. Chem. Soc., 82,* 5148 (1960).
125. L. W. Haynes, Ph.D. thesis, University of Illinois, Urbana, 1961.
126. N. J. Leonard and F. P. Hauck, Jr., *J. Am. Chem. Soc., 79,* 5279 (1957).
127. N. J. Leonard and A. G. Cook, *J. Am. Chem. Soc., 81,* 5627 (1959).
128. N. J. Leonard and W. K. Musker, *J. Am. Chem. Soc., 81,* 5631 (1959).
129. N. J. Leonard, K. Conrow, and R. R. Sauers, *J. Am. Chem. Soc., 80,* 5185 (1958).
130. J. Knabe, *Arch. Pharm., 292,* 416 (1959).
131. J. Knabe, H. Roloff, and U. R. Shukla, *Arch. Pharm., 298,* 879 (1965) and earlier papers.
132. J. Knabe and H. P. Herbort, *Arch. Pharm., 300,* 774 (1967).
133. J. Knabe and H. P. Herbort, *Ann., 710,* 133 (1967).
134. G. Stork and R. N. Guthikonda, *J. Am. Chem. Soc., 94,* 5109 (1972).

135. J. Gutzwiller, G. Pizzolato, and M. Uskokovic, *J. Am. Chem. Soc.*, *93*, 5907 (1971).
136. H. Mohrle and J. Gerloff, *Arch. Pharm.*, *311*, 672 (1978).
137. N. J. Leonard and R. R. Sauers, *J. Am. Chem. Soc.*, 79, 6210 (1957).
138. N. J. Leonard and D. F. Morrow, *J. Am. Chem. Soc.*, *80*, 371 (1958).
139. H. Moehrle, H. Rohrer, and W. Altenschmidt, *Arch. Pharm.*, *298*, 814 (1965); ibid., *298*, 350 (1965).
140. N. J. Leonard and R. E. Beyler, *J. Am. Chem. Soc.*, *72*, 1316 (1950).
141. F. Bohlmann, D. Schumann, and C. Arndt, *Tetrahedron Lett.*, 2705 (1965).
142. M. Przybylska and W. H. Barnes, *Acta Cryst.*, *6*, 377 (1953).
143. F. Bohlmann, W. Weise, D. Rahtz, and C. Arndt, *Ber.*, *91*, 2176 (1958).
144. F. L. Weisenborn and P. A. Diassi, *J. Am. Chem. Soc.*, *78*, 2022 (1956).
145. E. Wenkert and D. K. Roychaudhuri, *J. Org. Chem.*, *21*, 1315 (1956) and *J. Am. Chem. Soc.*, *80*, 1613 (1958).
146. F. Bohlmann and C. Arndt, *Ber.*, *91*, 2167 (1958).
147. F. Bohlmann and P. Strehlke, *Tetrahedron Lett.*, 167 (1965).
148. J. Knabe and H. Roloff, *Ber.*, *97*, 3452 (1964).
149. D. R. Bender and D. L. Coffen, *J. Org. Chem.*, *33*, 2504 (1968).
150. J. F. Weiss, G. Tollin, and J. T. Yoke, III, *Inorg. Chem.*, *3*, 1344 (1964).
151. R. A. Jerussi and M. R. McCormick, *Chem. Commun.*, 639 (1969).
152. R. McCrindle, G. Ferguson, G. J. Arsenault, A. J. McAlees, and D. K. Stephenson, *J. Chem. Res. (S)*, 360 (1984).
153. H. B. Henbest and M. J. W. Stratford, *J. Chem. Soc. (C)*, 995 (1966).
154. H. Shechter and S. S. Rawalay, *J. Am. Chem. Soc.*, *86*, 1706 (1964).
155. G. A. Olah and J. Welch, *Synthesis*, 419 (1971).
156. G. A. Olah, *Acc. Chem. Res.*, *13*, 330 (1980).
157. D. H. Rosenblatt, A. J. Hayes, Jr., B. L. Harrison, R. A. Streaty, and K. A. Moore, *J. Org. Chem.*, *28*, 2790 (1963).
158. H. B. Henbest and M. J. W. Stratford, *J. Chem. Soc.*, 711 (1964).
159. S. Dunstan and H. B. Henbest, *J. Chem. Soc.*, 4905 (1957).
160. D. Buckley, S. Dunstan, and H. B. Henbest, *J. Chem. Soc.*, 4901 (1957).
161. D. Buckley, S. Dunstan, and H. B. Henbest, *J. Chem. Soc.*, 4880 (1957).

162. D. Buckely, H. B. Henbest, and P. Slade, *J. Chem. Soc.*, 4891 (1957).
163. M. Colonna and L. Marchetti, *Gazz. Chim. Ital.*, *99*, 14 (1969).
164. L. Marchetti, *J. Chem. Soc.*, *Perkin Trans.* 2, 1977 (1977).
165. S. G. Cohen and H. M. Chao, *J. Am. Chem. Soc.*, *90*, 165 (1968).
166. S. G. Cohen, N. Stein, and H. M. Chao, *J. Am. Chem. Soc.*, *90*, 521 (1968).
167. J. Goerdeler in *Methoden der Organishchen Chemie*, Vol. XI/2 (Houben-Weyd), Thieme, Stuttgart, 1958, pp. 616—618.
168. H. Decker and P. Becker, *Ann.*, *395*, 328, 362 (1913).
169. G. G. Evans, *J. Am. Chem. Soc.*, *73*, 5230 (1951).
170. O. Cervinka, *Collection Czech. Chem. Commun.*, *25*, 1183 (1960).
171. G. N. Walker and D. Alkalay, *J. Org. Chem.*, *32*, 2213 (1967).
172. G. J. Heiszwolf and H. Kloosterziel, *Chem. Commun.*, 767 (1966).
173. See footnote 5, 9-12, in Ref. 174.
174. N. J. Leonard and J. V. Paukstelis, *J. Org. Chem.*, *28*, 3021 (1963).
175. H. Newman and T. L. Fields, *Tetrahedron*, *28*, 4051 (1972).
176. G. L Buchanan, *Chem. Soc. Rev.*, *3*, 41 (1974).
177. H. G. Reiber and T. D. Stewart, *J. Am. Chem. Soc.*, *62*, 3026 (1940).
178. R. K. Boeckman, Jr., P .F. Jackson, and J. P. Sabatucci, *J. Am. Chem. Soc.*, *107*, 2191 (1985).
179. R. T. Dean, H. C. Padgett, and H. Rapoport, *J. Am. Chem. Soc.*, *98*, 7448 (1976).
180. I. G. Csendes, Y. Y. Lee, H. C. Padgett, and H. Rapoport, *J. Org. Chem.*, *44*, 4173 (1979).
181. J. Schreiber, H. Maag, N. Hashimoto, and A. Eschemoser, *Angew. Chem. Int. Ed. Engl.*, *10*, 330 (1971).
182. T. A. Bryson, G. H. Bonitz, C. J. Reichel, and R. E. Dardis, *J. Org. Chem.*, *45*, 524 (1980).
183. E. E. vanTamelen, C. Placeway, G. P. Schiemenz, and I. G. Wright, *J. Am. Chem. Soc.*, *91*, 7359 (1969).
184. A. Buzas, J.-P. Jacquet, and G. Lavielle, *J. Org. Chem.*, *45*, 32 (1980).
185. A. Sidani, J. Marchand-Brynaert, and L. Ghosez, *Angew Chem. Int. Ed. Engl.*, *13*, 267 (1974).
186. L. Ghosez and J. Marchand-Brynaert in *Iminium Salts in Organic Chemistry* (H. Bohme and H. G. Viehe, eds.), Part 1, Wiley, New York, 1976.
187. I. Marko, B. Ronsmans, A.-M. Hesbain-Frisque, S. Dumas, and L. Ghosez, *J. Am. Chem. Soc.*, *107*, 2192 (1985).
188. M. Polonovski and M. Polonovski, *Compt. Rend.*, *184*, 331 (1927).

189. M. Polonovski and M. Polonovski, *Bull. Soc. Chim. France, 41*, 1190 (1927).

190. A. Cave, C. Kan-Fan, P. Potier, and J. Le Men, *Tetrahedron, 23*, 4681 (1967).

191. A. Cave and R. Michelot, *Compt. Rend., 265*, 669 (1967).

192. A. Ahond, A. Cave, C. Kan-Fan, H-P. Husson, J. deRostolan, and P. Potier, *J. Am. Chem. Soc., 90*, 5622 (1968).

193. H-P. Husson, L. Chevolot, Y. Langlois, C. Thal, and P. Potier, *Chem. Commun.,* 930 (1972).

194. R. Damico and C. D. Broaddus, *J. Org. Chem., 31*, 1607 (1966).

195. H. Volz and H-H. Kiltz, *Ann., 752*, 86 (1971).

196. P. J. Smith and C. K. Mann, *J. Org. Chem., 34*, 1821 (1969).

197. M. Masui, H. Sayo, and Y. Tsuda, *J. Chem. Soc. (B),* 973 (1968).

198. L. C. Portis, V. V. Bhat, and C. K. Mann, *J. Org. Chem., 35*, 2175 (1970).

199. M. Masui and H. Sayo, *J. Chem. Soc. (B),* 1593 (1971).

200. S. D. Ross, *Tetrahedron Lett.,* 1237 (1973).

201. J. E. Barry, M. Finkelstein, E. A. Mayeda, and S. D. Ross, *J. Org. Chem., 39*, 2695 (1974).

202. L. C. Portis, J. T. Klug, and C. K. Mann, *J. Org. Chem., 39*, 3488 (1974).

203. K. Blaha and O. Cervinka, *Adv. Heterocyclic Chem., 6*, 170 and 172 (1966).

204. G. Thuillier, B. Marcot, J. Craunes, and P. Rumpf, *Bull. Soc. Chim. France,* 4770 (1967).

205. R. T. Dean and H. Rapoport, *J. Org. Chem., 43*, 2115 (1978).

206. C. Djerassi, N. Crossley, and M. A. Kielczewski, *J. Org. Chem., 27*, 1112 (1962).

207. N. S. Crossley, C. Djerassi, and M. A. Kielczewski, *J. Chem. Soc.,* 6253 (1965).

208. W. Fritsch, J. Schmidt-Thome, H. Ruschig, and W. Haede, *Ber., 96*, 68 (1963).

209. H. Boehme and G. Berg, *Ber., 99*, 2127 (1966).

210. R. Foster, *Rec. Trav. Chim., 83*, 711 (1964).

211. D. A. Evans, *J. Am. Chem. Soc., 92*, 7593 (1970); D. A. Evans and L. A. Domeier, *Organic Synthesis* (R. E. Ireland, ed.), Vol. 54, Wiley, New York, 1974, p. 93.

212. F. E. Ziegler and P. A. Zoretic, *Tetrahedron Lett.,* 2639 (1968).

213. R. V. Stevens, R. K. Mehra, and R. L. Zimmerman, *Chem. Commun.,* 877 (1969).

214. F. Bohlmann, H.-J. Muller, and D. Schumann, *Chem. Ber., 106*, 3026 (1973).

215. A. G. Schultz, R. D. Lucci, J. J. Napier, H. Kinoshita, R. Ramanathan, P. Shannon, and Y. K. Lee, *J. Org. Chem., 50*, 217 (1985).

216. R. D. Glass and H. Rapoport, *J. Org. Chem.*, *44*, 1324 (1979).

217. N. J. Leonard, C. K.Steinhardt, and C. Lee, *J. Org. Chem.*, *27*, 4027 (1962).

218. A. J. Birch, E. G. Hutchinson, and G. S. Rao, *Chem. Commun.*, 657 (1970).

219. A. J. Birch, E. G. Hutchinson, and G. S. Rao, *J. Chem. Soc.* (*C*)., 637 (1971).

220. See footnotes 19-21, in Ref. 126.

221. R. Lukes, V. Dedek, and L. Novotny, *Collection Czech. Chem. Commun.*, *24*, 1117 (1959).

222. O. Cervinka and L. Hub, *Tetrahedron Lett.*, 463 (1964).

223. O. Cervinka and L. Hub, *Collection Czech. Chem. Commun.*, *30*, 3111 (1965).

224. G. N. Walker and D. Alkalay, *J. Org. Chem.*, *36*, 491 (1971).

225. C. Hansson and B. Wickberg, *J. Org. Chem.*, *38*, 3074 (1973).

226. A. J. Speziale and R. C. Freeman, *J. Am. Chem. Soc.*, *82*, 903 (1960).

227. A. J. Speziale and L. R. Smith, *J. Am. Chem. Soc.*, *84*, 1868 (1962).

228. R. Burgada, *Bull. Soc. Chim. France*, 3548 (1967).

229. R. Burgada and H. Normant, *Comp. Rend.*, *267*, 1854 (1968).

230. S. R. Burgada and J. Roussel, *Bull. Soc. Chim. France*, 192 (1970).

231. R. S. Monson, D. N. Priest, and J. C. Ullrey, *Tetrahedron Lett.*, 929 (1972).

232. D. Seyferth, M. E. Gordon, and R. Damrauer, *J. Org. Chem.*, *32*, 469 (1967).

233. P. Nelson and A. Pelter, *J. Chem. Soc.*, 5142 (1965).

234. K. G. R. Sundelin, R. A. Wiley, R. S. Givens, and D. R. Rademacher, *J. Med. Chem.*, *16*, 235 (1973).

235. H. von Hirsch, *Ber.*, *100*, 1289 (1967).

236. G. E. Manoussakis and J. A. Tossidis, *Prakt. Panelleniou Chem. Synedrion, 4th*, 86 (1972); *C.A.*, *85*, 93724g (1976).

237. D. W. Boerth and F. A. Van Catledge, *J. Org. Chem.*, *40*, 3319 (1975).

238. J.-C. Pommier and A. Roubineau, *J. Organomet. Chem.*, *17*, P25 (1969).

239. J.-C. Pommier and A. Roubineau, *J. Organomet. Chem.*, *50*, 101 (1973).

240. J.-M. Brocas and J.-C. Pommier, *J. Organomet. Chem.*, *92*, C7 (1975).

241. T. G. Selin, *U.S. 3,621,060* (1971); *C.A.*, *76*, 45258a (1972).

242. R. Comi, R. W. Franck, M. Reitano, and S. M. Weinreb, *Tetrahedron Lett.*, 3107 (1973).

243. W. A. White and H. Weingarten, *J. Org. Chem.*, *32*, 213 (1967).

244. S. C. Kuo and W. H. Daly, *J. Org. Chem.*, *35*, 1861 (1970).

245. D. Pocar, R. Stradi, and G. Bianchetti, *Gazz. Chim. Ital.*, *100*, 1135 (1970).

246. E. Jongejan, H. Steinberg, and T. J. deBoer, *Syn. Commun.*, *4*, 11 (1974).

247. T. Morinaga, Y. Nakazawa, K. Arimoto, K. Takahashi, Y. Arai, *Ger. Offen. 2,546,192* (1976); *C.A.*, *85*, 33036Z (1976).

248. A. Nilsson and R. Carlson, *Acta Chem. Scand.*, *B38*, 523 (1984).

249. R. Carlson, A. Nilsson, and M. Stromqvist, *Acta Chem. Scand.*, *B37*, 7 (1983).

250. R. Carlson and A. Nilsson, *Acta Chem. Scand.*, *B38*, 49 (1984).

251. Y. Kitahara, M. C. Caserio, F. Scardiglia, and J. D. Roberts, *J. Am. Chem. Soc.*, *82*, 3106 (1960).

252. E. F. Jenny and J. Druey, *J. Am. Chem. Soc.*, *82*, 3111 (1960).

253. H. W. Whitlock, JR., and G. L. Smith, *Tetrahedron Lett.*, 1389 (1965).

254. D. C. England and J. C. Piecara, *J. Fluorine Chem.*, *17*, 265 (1981).

255. M. G. Ettlinger and F. Kennedy, *Chem. Ind. (London)*, 166 (1956).

256. A. T. Bottini and J. D. Roberts, *J. Am. Chem. Soc.*, *79*, 1462 (1957).

257. A. T. Bottini and V. Dev, *J. Org. Chem.*, *27*, 968 (1962).

258. L. Duhamel and J. M. Poirier, *Tetrahedron Lett.*, 2437 (1976).

259. R. V. Stevens and L. E. DuPree, Jr., *Chem. Commun.*, 1585 (1970).

260. R. V. Stevens, *Acc. Chem. Res.*, *10*, 193 (1977).

261. C. P. Forbes, G. L. Wenteler, and A. Wiechers, *Tetrahedron*, *34*, 487 (1978).

262. R. A. Johnson, *J. Org. Chem.*, *37*, 312 (1972).

263. J. Boutagy and R. Thomas, *Chem. Rev.*, *74*, 87 (1974).

264. W. S. Wadsworth, Jr., in *Organic Reactions* (W. G. Dauben, ed.), Vol. 25, Wiley, New York, 1977.

265. W. S. Wadsworth, Jr., and W. D. Emmons, *J. Am. Chem. Soc.*, *83*, 1733 (1961).

266. H. Zimmer and J. P. Bercz, *Ann.*, *686*, 107 (1965).

267. H. Gross and E. Hoft, *Angew. Chem. Int. Ed. Engl.*, *6*, 353 (1967).

268. H. Gross and W. Buerger, *J. Prakt. Chem.*, *311*, 395 (1969).

269. H. Bohme, M. Haake, and G. Auterhoff, *Arch. Pharm.*, *305*, 88 (1972).

270. S. Sato, *Nippon Kagaku Kaishi*, 1780 (1975).

271. S. F. Martin and R. Gompper, *J. Org. Chem.*, *39*, 2814 (1974).

272. S. F. Martin, *J. Org. Chem.*, *41*, 3337 (1976).

273. S. F. Martin, T. S. Chou, and C. W. Payne, *J. Org. Chem.*, *42*, 2520 (1977).

274. L. Horner, H. Hoffmann, and H. G. Wippel, *Chem. Ber.*, *91*, 61 and 64 (1958).

275. L. Horner, H. Hoffmann, H. G. Wippel, and G. Klahre, *Chem. Ber.*, *92*, 2499 (1958).
276. N. L. J. M. Broekhof, F. L. Jonkers, and A. van der Gen, *Tetrahedron Lett.*, 2433 (1979).
277. N. L. J. M. Broekhof, F. L. Jonkers, and A. van der Gen, *Tetrahedron Lett.*, 2671 (1980).
278. B. H. Bakker, D. S. T. A-Lim, and A. van der Gen, *Tetrahedron Lett.*, 4259 (1984).
279. N. L. J. M. Broekhof and A. van der Gen, *Rec. Trav. Chim.*, *103*, 305 (1984).
280. N. L. J. M. Broekhof, P. van Elburg, D. J. Hoff, and A. van der Gen, *Rec. Trav. Chim.*, *103*, 317 (1984).
281. J. C. Gilbert and U. Weerasooriya, *J. Org. Chem.*, *48*, 448 (1983).
282. H. Meerwein, P. Borner, O. Fuchs, H. J. Sasse, H. Schrodt, and J. Spille, *Chem. Ber.*, *89*, 2060 (1956); H. Meerwein, W. Florian, N. Schon, and G. Stopp, *Ann.*, *641*, 1 (1961).
283. R. F. Abdulla and R. S. Brinkmeyer, *Tetrahedron*, *35*, 1675 (1979).
284. E. E. Garcia, L. E. Benjamin, and R. I. Fryer, *J. Heterocycl. Chem.*, *11*, 275 (1974).
285. J. J. Baldwin, K. Mensler, and G. S. Ponticello, *J. Org. Chem.*, *43*, 4879 (1978).
286. U. Hengartner, A. D. Batcho, J. F. Blount, W. Leimgruber, M. E. Larscheid, and J. W. Scott, *J. Org. Chem.*, *44*, 3748 (1979).
287. R. M. Acheson, G. N. Aldrich, M. C. K. Choi, J. O. Nwankwo, M. A. Ruscoe, and J. D. Wallis, *J. Chem. Res. Synop.*, 100 (1984).
288. A. P. Kozikowski, M. N. Greco, and J. P. Springer, *J. Am. Chem. Soc.*, *104*, 7622 (1982).
289. L. Duhamel and J.-M. Poirier, *J. Org. Chem.*, *44*, 3585 (1979).
290. R. Knorr, P. Low, and P. Hassel, *Synthesis*, 785 (1983).
291. J. A. Deyrup and G. S. Kuta, *J. Org. Chem.*, *43*, 501 (1978).
292. F. Bondavalli, P. Schenone, and A. Ranise, *Synthesis*, 830 (1979).
293. I. Murakoshi, A. Kubo, J. Saito, and J. Haginiwa, *Yakugaku Zasshi*, *88*, 900 (1968).
294. I. Murakoshi, K. Takada, and J. Haginiwa, *Yakugaku Zasshi*, *89*, 1661 (1969).
295. S. Miyano, T. Somehara, M. Nakao, and K. Sumoto, *Synthesis*, 701 (1978).
296. J. Hoch, *Compt. Rend.*, *200*, 938 (1935).
297. R. Fusco, G. Bianchetti, and D. Pocar, *Gazz. Chim. Ital.*, *91*, 849 (1961).
298. G. Bianchetti, P. Dalia Croce, and D. Pocar, *Gazz. Chim. Ital.*, *94*, 606 (1964).

299. R. Fusco, G. Bianchetti, D. Pocar, and R. Ugo, *Gazz. Chim. Ital.*, *92*, 1040 (1962).

300. C. D. Hurd and L. T. Sherwood, Jr., *J. Org. Chem.*, *13*, 471 (1948).

301. R. Lukes and J. Pliml, *Collection Czech. Chem. Commun.*, *26*, 471 (1961).

302. P. L.-F. Chang and D. C. Dittmer, *J. Org. Chem.*, *34*, 2791 (1969).

303. J. S. Krouwer and J. P. Richmond, *J. Org. Chem.*, *43*, 2464 (1978).

304. S. A. Fine and R. L. Stern, *J. Org. Chem.*, *35*, 1857 (1970).

305. H. Ahlbrecht, W. Raab, and C. Vonderheid, *Synthesis*, 127 (1979).

306. P. F. Hudrlik, A. M. Hudrlik, and A. K. Kulkarni, *Tetrahedron Lett.*, 139 (1985).

307. P. M. Greaves and S. R. Landor, *Chem. Commun.*, 322 (1966).

308. P. F. Hudrlik and A. M. Hudrlik, *J. Org. Chem.*, *38*, 4254 (1973).

309. J. Barluenga and F. Aznar, *Synthesis*, 704 (1975).

310. J. Barluenga and F. Aznar, *Synthesis*, 195 (1977).

311. J. Barluenga, F. Aznar, R. Liz, and R. Rodes, *J. Chem. Soc. Perkin 1*, 2732 (1980).

312. M. LeBlanc, G. Santini, and J. G. Riess, *Tetrahedron Lett.*, 4151 (1975).

313. M. LeBlanc, G. Santini, J. Gallucci, and J. G. Riess, *Tetrahedron*, *33*, 1435 (1977).

314. K. A. Parker, J. J. Petraitis, R. W. Kosley, Jr., and S. L. Buchwald, *J. Org. Chem.*, *47*, 389 (1982).

315. E. E. van Tamelen, V. B. Haarstad, and R. L. Orvis, *Tetrahedron*, *24*, 687 (1968).

316. A. Vilsmeier and A. Haack, *Chem. Ber.*, *60*, 119 (1927).

317. M. A. Kira and A. Bruckner-Wilhelms, *Acta Chim. Acad. Sci. Hung.*, *56*, 47 (1968).

318. M. A. Kira, A. Bruckner-Wilhelms, F. Ruff, and J. Borsy, *Acta Chim. Acad. Sci. Hung.*, *56*, 189 (1968).

319. P. D. Croce, P. Ferruti, and R. Stradi, *Gazz. Chim. Ital.*, *97*, 589 (1967).

320. S. S. P. Chou and C. W. Chu, *J. Chin. Chem. Soc. (Taipei)*, *31*, 351 (1984).

321. D. H. Wadsworth, M. R. Detty, B. J. Murray, C. H. Weidner, and N. G. Haley, *J. Org. Chem.*, *49*, 2676 (1984).

322. Z. Jabry, M-C. Lasne, and J-L. Ripoll, *J. Chem. Res.(S)*, 188 (1986).

323. A. Batcho and W. Leimgruber, *U.S. Pat. 3,732,245* (1973).

324. H. Maehr, J. Smallheer, J. F. Blunt, and L. J. Todaro, *J. Org. Chem., 46*, 5019 (1981).
325. H. Maehr and J. M. Smallheer, *J. Org. Chem., 49*, 1549 (1984).
326. R. J. Friary, J. M. Gilligan, R. P. Szajewski, J. J. Falci, and R. W. Franck, *J. Org. Chem., 38*, 3487 (1973).

3
Hydrolysis of Enamines

E. J. STAMHUIS   *The University of Groningen, Groningen,*
The Netherlands

A. GILBERT COOK   *Valparaiso University, Valparaiso, Indiana*

I.   Introduction   165

II. Kinetics and Mechanism of the Hydrolysis of Simple
     Tertiary Enamines   166
     A. Hydrolysis in Alkaline and Neutral Solution   166
     B. Hydrolysis in Weakly Acidic Solution   171
     C. Hydrolysis in Strongly Acidic Solution   176

III. Structure and Reactivity   177

     References   179

I. INTRODUCTION

Hydrolysis of simple enamines appears to be very easy, and decomposition to the corresponding carbonyl compound and the secondary amine can be achieved readily for most enamines by adding water to them. Basicity as well as resonance may be considered important factors which, among other effects, will determine the rate of proton addition from water. Not less important is the question of where the proton will add, on nitrogen or on the β-carbon atom. It is well known that carbon alkylation of enamines is mainly restricted to strongly electrophilic halides [1]. The use of weakly electrophilic halides, such as primary alkyl halides, leads to the very likely irreversible formation of quaternary ammonium salts, in which the double bond is unreactive for further electrophilic attack, thus preventing the desired carbon alkylation.

Knowledge of the mechanism enables one to obtain more insight into the various factors that determine the extent of reaction along both pathways. In this chapter special attention will be given to the kinetics and mechanisms of the hydrolysis of simple enamines (only tertiary enamines will be considered).

II. KINETICS AND MECHANISM OF THE HYDROLY-SIS OF SIMPLE TERTIARY ENAMINES

A. Hydrolysis in Alkaline and Neutral Solution

Experimental evidence, obtained in protonation [2,3], acylation [1,4], and alkylation [1,4,5-7] reactions, always indicates a concurrence between electrophilic attack on the nitrogen atom and the β-carbon atom in the enamine. Concerning the nucleophilic reactivity of the β-carbon atom in enamines, Opitz and Griesinger [8] observed, in a study of salt formation, the following series of reactivities of the amine and carbonyl components: pyrrolidine and hexamethylene imine ≫ piperidine > morpholine > ethylbutylamine; cyclopentanone ≫ cycloheptanone; cyclooctanone > cyclohexanone; monosubstituted acetaldehyde > disubstituted acetaldehyde.

Important for the hydrolysis is the observation [8] that protonation of enamines with hydrogen chloride does not immediately lead to immonium salts, but in most, if not all, cases first to the formation of the corresponding enammonium ions, which afterward rearrange more or less rapidly to the more stable immonium ions [Eq. (1)]:

$$\left[-\underset{|}{C}=\underset{|}{C}-\overset{+}{\underset{H}{N}}\langle \right] Cl^- \longrightarrow \left[H-\underset{|}{C}-\underset{|}{C}=\overset{+}{N}\langle \right] Cl^- \qquad (1)$$

These results have led to the conclusion [9] that the formation of enammonium salts is kinetically controlled, whereas the protonation on the β-carbon atom is subject to thermodynamic control (see Chapter 1, Section IV. E).

The same behavior has been observed in the attack of electrophiles on the ambident enolate anions, of which many reactions are closely related to those of enamines Eq. (2):

$$\left. \begin{array}{c} -\underset{|}{C}=\underset{|}{C}-O^- \\ \updownarrow \\ -\bar{\underset{|}{C}}-\underset{|}{C}=O \end{array} \right\} + E^+ \left\langle \begin{array}{c} -\underset{|}{C}=\underset{|}{C}-OE \\ \\ E-\underset{|}{C}-\underset{|}{C}=O \end{array} \right. \qquad (2)$$

The heats of reaction for O-alkylation and C-alkylation of enolate anions clearly show that the latter reactions lead to the thermodynamically more stable products [10].

Hydrolysis of an enamine yields a carbonyl compound and a secondary amine. Only a few rate constants are mentioned in the literature. The rate of hydrolysis of 1-(β-styryl)piperidine and 1-(1-hexenyl)piperidine have been determined in 95% ethanol at 20°C [11]. The values for the first-order rate constants are 4×10^{-5} sec^{-1} and approximately 10^{-3} sec^{-1}, respectively. Apart from steric effects, the difference in rate may be interpreted in terms of resonance stabilization by the phenyl group on the vinyl amine structure, thus lowering the nucleophilic reactivity of the β-carbon atom of that enamine.

The kinetics of the hydrolysis of 4-(2-methylpropenyl)morpholine, 1-(2-methylpropenyl)piperidine, and 1-(2-methylpropenyl)pyrrolidine have been investigated [12,13]. Results obtained from rate measurements of 4-(2-methylpropenyl)morpholine [1] in dilute phosphate buffers are shown in Figure 1.

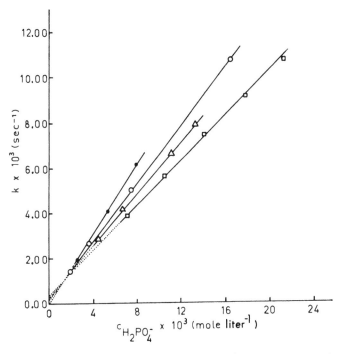

Figure 1 First-order rate constants for the hydrolysis of 4-(2-methylpropenyl)morpholine in aqueous phosphate buffers at 25°C as a function of the concentration of $H_2PO_4^-$. pH values: ● 7.30; ○ 6.30; △ 6.00; □ 5.79. (Reprinted with permission from Stamhuis and Maas [13]. Copyright by the American Chemical Society.)

The slope of the straight lines in Figure 1 is pH-dependent. This has been explained on the ground of an equilibrium between the free enamine and the nitrogen-protonated species. This acid-base equilibrium is built up very rapidly [Eq. (3)] and causes a decrease in concentration of the reactive enamine molecules immediately after the enamine is dissolved in the buffer solution. Only the fraction $K/(K + a_{H_3O^+})$ of the total amount is present as free enamine molecules.

$$(CH_3)_2C=CH-\overset{+}{\underset{H}{N}}\diagdown\bigcirc O + H_2O \; \overset{K}{\rightleftharpoons} \; (CH_3)_2C=CH-N\diagdown\bigcirc O + H_3O^+ \qquad (3)$$

$$(\underline{1})$$

The double bond of the nitrogen-protonated species is stable with respect to electrophilic attack under the reaction circumstances, since the free electron pair on nitrogen is no longer available for interaction with the π-electrons of the double bond. General acid catalysis has been clearly demonstrated for the hydrolysis of these enamines in the pH range 4.10—9.40. The general rate equation for the first-order rate constants, k, is given by Eq. (4):

$$k = \frac{K}{K + a_{H_3O^+}} (k_{H_3O^+} \, a_{H_3O^+} + k_{HA} \, c_{HA} + k_{H_2O} \, c_{H_2O}) \qquad (4)$$

in which K is the dissociation constant of the enammonium ions in Eq. (3), and $k_{H_3O^+}$, k_{HA}, and k_{H_2O} the second-order rate constants for the catalyzing agents H_3O^+, the acid component of the buffer HA, and water, respectively, but in the case of enamine 1 the contribution of the water-catalyzed reaction is negligible. Tables 1, 2, and 3 give the values of the second-order rate constants and K values at different temperatures.

Table 1 Second-Order Rate Constants and Dissociation Constants of 4-(2-Methylpropenyl)Morpholine

Temp. (°C)	$k_{H_2O^+}$ (liter mole^{-1} sec^{-1})	$k_{H_2PO_4^-}$ (liter mole^{-1} sec^{-1})	k_{HOAc} (liter mole^{-1} sec^{-1})	K (mole liter^{-1})
25.00	3.1×10^2	0.76	1.8	3.4×10^{-6}
39.60	9.5×10^2	—	5.4	5.8×10^{-6}
50.65	19.3×10^2	—	11.6	8.1×10^{-6}

Source: Ref. 13.

Table 2 Second-Order Rate Constants and Dissociation Constants of
1-(2-Methylpropenyl)Piperidine

Temp. (°C)	$k_{H_3O^+}$ (liter mole^{-1} sec^{-1})	k_{H_2O} (liter mole^{-1} sec^{-1})	$k_{H_3BO_3}$ (liter mole^{-1} sec^{-1})	K (mole liter^{-1})
25.00	0.14×10^6	0.97×10^{-5}	1.1×10^{-2}	4.4×10^{-9}
39.60	0.63×10^6	4.4×10^{-5}	4.0×10^{-2}	4.3×10^{-9}
50.24	1.8×10^6	11.3×10^{-5}	6.0×10^{-2}	3.3×10^{-9}

Source: Ref. 13.

For this type of reaction, the value of the solvent deuterium isotope effect is often a conclusive argument for the proposed mechanism
[14]. Rate measurements of 1 in acetic acid–acetate buffers in light
and heavy water resulted in an isotope effect $k_{H_3O^+}/k_{D_3O^+}$ of 2.5

and k_{HOAc}/k_{DOAc} of 9. A rate-determining proton transfer to the
β-carbon atom of the enamine has been proposed and accounts for the
experimental results [14,15] [Eq. (5)].

$$\tag{5}$$

$R_1R_2N{-} = $ morpholino-
  piperidino-
  pyrrolidino-

The protonated intermediate in Eq. (5) is very reactive and could not
be observed spectroscopically under the reaction circumstances. Fast
hydration to isobutyraldehyde and the secondary amine occurred [13].
This mechanism is exactly analogous to that of the hydrolysis of
enolate anions [16], as is to be expected.

   The kinetic behavior of the hydrolysis of the investigated enamines
permits also a mechanism in which a nucleophilic attack on the α-carbon
atom of the nitrogen-protonated enamine is the rate-determining step.
However, replacement of chloride ions by perchlorate ions, of which
the nucleophilicity is much smaller, has no influence on the rate, as
one would expect for this mechanism. Moreover, since not acetic acid,

Table 3 Second-Order Rate Constants and Dissociation Constants of
1-(2-Methylpropenyl)Pyrrolidine

Temp. (°C)	$k_{H_3O^+}$ (liter mole^{-1} sec^{-1})	k_{H_2O} (liter mole^{-1} sec^{-1})	$k_{H_3BO_3}$ (liter mole^{-1} sec^{-1})	K (mole liter^{-1})
0.00	0.69×10^6	1.2×10^{-5}	2.6×10^{-2}	1.7×10^{-9}
15.00	3.1×10^6	6.6×10^{-5}	12×10^{-2}	1.4×10^{-9}
25.00	8.4×10^6	17×10^{-5}	20×10^{-2}	1.4×10^{-9}

Source: Ref. 13.

but acetate ions are now the catalytic species in solutions of light as
well as heavy water, an explanation for the large isotope effect
k_{HOAc}/k_{DOAc} = 9 can hardly be given. This makes a rate-determining nucleophilic attack on the α-carbon of enammonium ions therefore
very unlikely.

It is noteworthy that the kinetics indirectly provided the evalua-
tion of the basicities of these enamines [Eq. (4)]. The pK_a values
for 4-(2-methylpropenyl)morpholine, 1-(2-methylpropenyl)piperidine,
and 1-(2-methylpropenyl)pyrrolidine are 5.47, 9.35, and 8.84, respectively [17]. Since the protonation of the β-carbon atom does not
possess the character of a real equilibrium at pH 7 and up [for compound 1 even at pH 1 and up] the basicity must be fully ascribed to
the equilibrium between enamine and the corresponding nitrogen-protonated conjugate acid.

The hydrolysis of the morpholine, dimethylamine, piperidine, and
pyrrolidine enamines of propionphenone (2) has also been studied [18].
The hydrolysis of all the enamines (pyrrolidine uncertain) of propio-

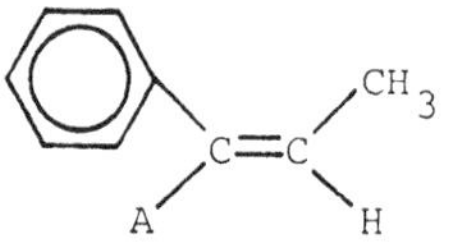

A = morpholino-

dimethylamino-

piperidino-

pyrrolidino-

(2)

X = H-, CH$_3$-, Cl-, NO$_2$-

phenone [(2)] shows general acid catalysis at pH greater than 10, but
the rate is independent of the hydrogen ion concentration. The rate-
determining step is the C-protonation of the free enamine. At pH
10.38 the Hammett ρ constant is-1.29, indicating that the rate is decreased by electron withdrawal from the reaction center (see Fig. 2).
The morpholine enamine hydrolysis reaction shows general acid
catalysis down to pH 5, whereas the other enamines of propiophenone
do not show general acid catalysis below pH 10.

B. Hydrolysis in Weakly Acidic Solution

From Eq. (4) and the date of Tables 1, 2, and 3 it can easily be calculated that the rate of hydrolysis of these enamines should rapidly
reach a maximum value in weakly acidic solutions at decreasing pH,
assuming that no buffer is used. The rate should then be constant
and pH-independent. For enamine 1, Eq. (4) reduces to $k = Kk_{H_3O^+}$,

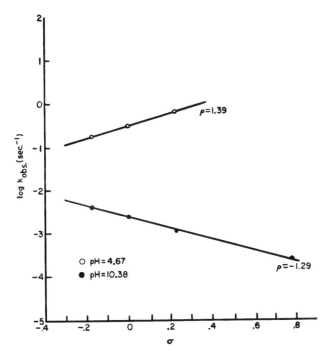

Figure 2 Logarithms of the observed first-order rate constants for
the hydrolysis of *para*-substituted morpholine enamines of propiophenone as a function of the Hammett σ constant. (Reprinted with permission from Sollenberger and Martin [18]. Copyright by the American Chemical Society.)

and this rate must be the same as that at the intersection of the straight lines in Figure 1. This appeared to be true for the observed rate at pH 2 [13].

However, at lower pH a sharp decrease in rate has been observed for 1-(2-methylpropenyl)pyrrolidine (3), as shown by Figure 3, which indicates that Eq. (4) is no longer valid at lower pH values. This sudden decrease in rate cannot be explained by the mechanism proposed in Section II.A. Kinetic measurements on compound 3 in aqueous acetate buffers and in dilute solutions of perchloric acid have clearly demonstrated that in weakly acidic media the hydrolysis is subject to general base catalysis (Figure 4). These results have been explained by assuming a change in rate-determining step from general acid-catalyzed formation of immonium ions to general base-catalyzed hydration of these ions to the amino alcohol [Eq. (6)]. The observed first-order rate

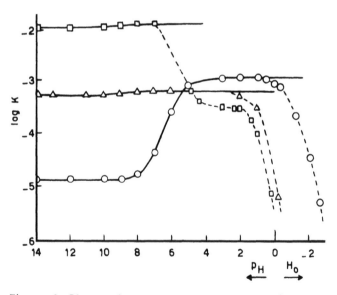

Figure 3 Observed versus calculated rates of hydrolysis of enamines at 24.8°C. Drawn lines: calculated log k values. Observed values: ○ for 4-(2-methylpropenyl)morpholine; △ for 4-(2-methylpropenyl)piperidine; □ for 1-(2-methylpropenyl)pyrrolidine. Values are corrected for buffer contributions. (Reprinted with permission from Maas, Janssen, Stamhuis, and Wynberg [19]. Copyright by the American Chemical Society.)

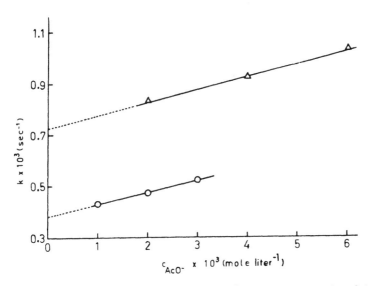

$$(CH_3)_2C{=}CH{-}\overset{+}{\underset{H}{N}}\diagdown \quad + \ H_2O \quad \underset{}{\overset{K}{\rightleftharpoons}} \quad (CH_3)_2C{=}CH{-}N\diagdown \quad + \ H_3O^+ \quad \textbf{(a, fast)}$$

$$(CH_3)_2C{=}CH{-}N\diagdown \quad + \ H_3O^+ \quad \underset{k_{-H_2O^+}}{\overset{k_{H_2O^+}}{\rightleftharpoons}} \quad (CH_3)_2CH{-}CH{=}\overset{+}{N}\diagdown \quad + \ H_2O \quad \textbf{(b, fast)} \qquad (6)$$

$$(CH_3)_2CH{-}CH{=}\overset{+}{N}\diagdown \quad + \ H_2O + B \quad \overset{k_B}{\longrightarrow} \quad (CH_3)_2CH{-}\overset{\overset{\displaystyle OH}{|}}{C}H{-}N\diagdown \quad + \ HB^+ \quad \textbf{(c, slow)}$$

constant of the overall reaction is

$$k = \frac{Kk_{H_3O} \cdot a_{H_3O} \cdot}{Kk_{H_3O^+} a_{H_3O^+} + Kk_{-H_3O^+} + k_{-H_3O^+} a_{H_3O^+}} \cdot \sum_i k_{B_i} c_{B_i} \qquad (7)$$

In the pH range of interest, $Kk - H_3O^+ << k - H_3O^{+a}H_3O^+$ (see Table 3). From observations of Optiz and Griesinger in the rearrangement of tertiary enammonium salts to immonium salts [8] and from NMR spectra of enamine 1 in perchloric acid solution [19], it may be concluded that as soon as the reactions of Eqs. (6a) and (6b) have

Figure 4 First-order rate constants for the hydrolysis of 1-(2-methyl-propenyl)pyrrolidine in acetate buffers (24.8°C). pH values: ○, 4.41; △, 4.49. (Reprinted with permission from Maas, Janssen, Stamhuis, and Wynberg [19]. Copyright by the American Chemical Society.)

reached equilibrium, the concentration of enammonium ions may be ne-
glected with regard to the concentration of immonium ions. Therefore, Eq. (7) reduces to

$$k = \sum_i k_{Bi} c_{Bi} \tag{8}$$

At pH values lower than 4, the concentration of hydroxide ions becomes too small to give an appreciable contribution to the overall rate,
and only water acts as the catalyzing base (Figure 3). In this pH
range the rate is, therefore, pH-independent, as is predicted from
the data of Table 4.

Direct attack of acetate ions on the α-carbon atom of the immonium
ions in the acetate buffer solutions is unlikely, but the catalyzing action involves the removal of a proton from a water molecule in its attack on the immonium ion.

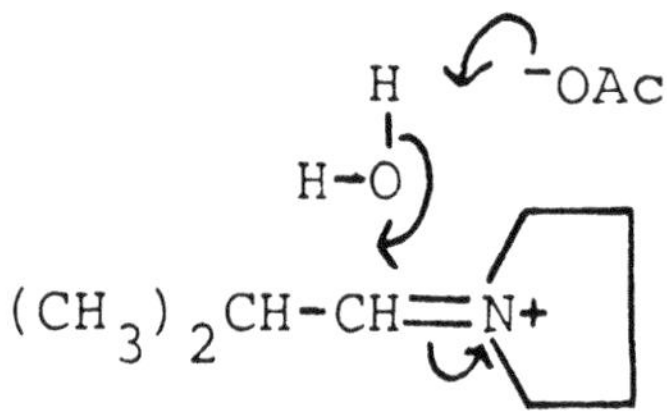

A similar concerted mechanism accounts for the water-catalyzed reaction, which becomes predominant at pH values lower than 4. The
transition in rate-determining step has not been observed for the
other two enamines.

The attack of the iminium ions of propiophenone enamines by
water is the rate-determining step over a wide pH range [18]. The
hydrolysis of the morpholine enamine changes to this rate-determining
step at about pH 5 as one decreases pH, whereas the other enamines
change the limiting step at about pH 10 (see Figure 5).

The next change in the rate-limiting step takes place at pH 1.
The reason for this wide pH range for this rate-determining step as
compared to the enamines of isobutyraldehyde is because of the greater stability of the propiophenone iminium ions. This greater stability
is due to the delocalization of the positive charge into the aromatic
ring (see Eq. (9)).

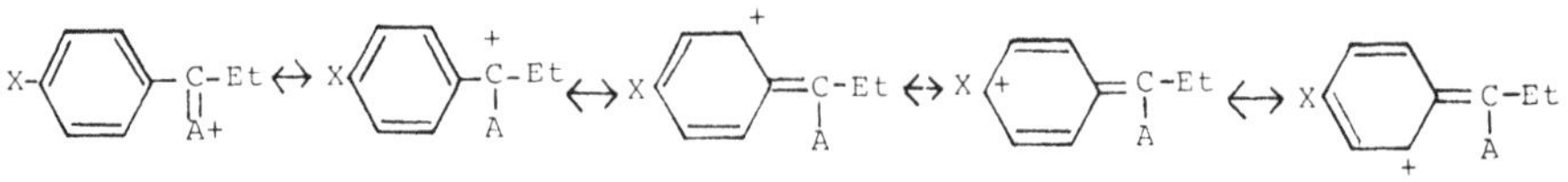

$$\tag{9}$$

Table 4 Second-Order Rate Constants for the Hydrolysis of Immonium Ions, Derived from 1-(2-Methylpropenyl)Pyrrolidine at 24.8°C

k_{OH^-} (liter mole^{-1} sec^{-1})	k_{OAc^-} (liter mole^{-1} sec^{-1})	k_{H_2O} (liter mole^{-1} sec^{-1})
4×10^5	4×10^2	5.5×10^{-6}

Source: Ref. 19.

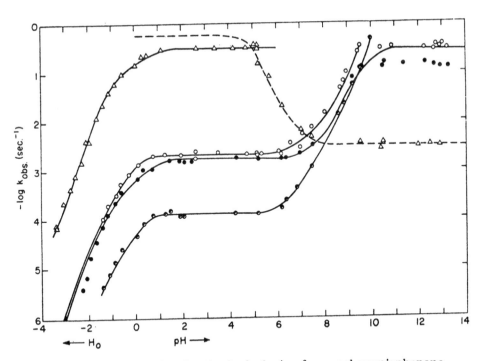

Figure 5 pH-rate profile for the hydrolysis of several propiophenone enamines at 25°C: △, morpholine; ○, dimethylamine; ●, piperidine; ◉, pyrrolidine. (Reprinted with permission from Sollenberger and Martin [18]. Copyright by the American Chemical Society.)

The Hammett ρ function at pH 4.67 is 1.39 (see Figure 2), showing that the rate is increased by electron withdrawal from the reaction center.

It is interesting to note that the hydrolysis of certain Schiff bases in weakly acidic solutions shows a similar mechanism [20]. N-protonated-substituted benzylidine-t-butylamines react with hydroxide ions to amino alcohols in the rate-determining step, and at lower pH the rate is almost entirely determined by attack of water on the protonated Schiff bases as a consequence of the rapidly decreasing concentration of hydroxide ions.

C. Hydrolysis in Strongly Acidic Solution

The rate of hydrolysis of 4-(2-methylpropenyl)morpholine (1), 1-(2-methylpropenyl)piperidine, and 1-(2-methylpropenyl)pyrrolidine (3) decreases sharply around pH 0 (Figure 3), as does the rate of hydrolysis of 1-(4-morpholinol)-1-phenyl-2-methylethene (2), 1-piperidino-1-phenyl-2-methylethene, 1-(N,N-dimethylamino)-1-phenyl-2-methylethene, and 1-pyrrolidino-1-phenyl-2-methylethene (see Figure 5). Such a decrease in rate is not to be expected on the basis of Eq. (4), which predicts a constant and pH-independent rate at lower pH values.

Obviously, Eq. (4) is no longer valid for any of the seven enamines in strongly acidic solution. Since the kinetics of the hydrolysis of enamine (3) and the propiophenone enamines show a rate-determining formation of the intermediate amino alcohol, the formation of which becomes pH-independent around pH 2, it is evident that a further decrease in rate at still lower pH indicates another change in the rate-determining step. A slow decomposition of the amino alcohol to isobutyraldehyde or propiophenone and secondary amine has been proposed, as this accounts for the sharp decrease in rate at pH 0 [19]. The amino alcohol, to a large extent nitrogen-protonated in acidic media, is in rapid equilibrium with the uncharged derivative and the dipolar structure [Eq. (10)].

$$-\overset{\underset{\displaystyle |}{}}{\underset{\underset{\displaystyle |}{}}{C}}-\overset{\overset{\displaystyle OH}{\underset{\displaystyle |}{}}}{\underset{\underset{\displaystyle |}{}}{C}}-N< \ + \ H^+ \ \underset{}{\overset{K_1}{\rightleftharpoons}} \ -\overset{\underset{\displaystyle |}{}}{\underset{\underset{\displaystyle |}{}}{C}}-\overset{\overset{\displaystyle OH}{\underset{\displaystyle |}{}}}{\underset{\underset{\displaystyle |}{}}{C}}-\overset{+}{\underset{\underset{\displaystyle H}{}}{N}}< \ \overset{K_2}{\rightleftharpoons} \ -\overset{\underset{\displaystyle |}{}}{\underset{\underset{\displaystyle |}{}}{C}}-\overset{\overset{\displaystyle \bar{O}}{\underset{\displaystyle |}{}}}{\underset{\underset{\displaystyle |}{}}{C}}-\overset{+}{\underset{\underset{\displaystyle H}{}}{N}}< \ + \ H^+ \qquad (10)$$

It has been concluded from an estimate of K_1 and K_2 that the uncharged amino alcohol, as well as the dipolar structure, is present in small concentrations [21]. The decomposition is strongly retarded as the pH is lowered, and this phenomenon has been explained by assuming the switterions to be the active intermediates [Eq. (11)].

$$-\overset{\underset{\displaystyle |}{}}{\underset{\underset{\displaystyle |}{}}{C}}-\overset{\overset{\displaystyle \overset{\displaystyle O^-}{\frown}}{}}{\underset{\underset{\displaystyle |}{}}{C}}\overset{+}{\underset{\underset{\displaystyle H}{}}{N}}< \ \longrightarrow \ -\overset{\underset{\displaystyle |}{}}{\underset{\underset{\displaystyle |}{}}{C}}-\overset{\underset{\displaystyle |}{}}{C}{=}O \ + \ HN< \quad \text{(slow)} \qquad (11)$$

The Hammett ρ function levels off in these strong acid solutions, showing a constant value for ρ of 2.20 at $H_0 = -2.28$ and -3.32. This means that the rate is increased by electron-withdrawing groups.

A strong argument in favor of the proposed mechanism [19] over other possible reaction pathways is that the enhanced driving force, present in the ammonium alcoholate ion, is necessary to expel the strongly basic amine. The kinetics of the hydrolysis of Schiff bases, derived from strongly basic amines, if carried out under acidic conditions, have been explained to occur also via similar dipolar intermediates [20]. The slow decomposition of the Zwitterion is an uncatalyzed reaction. This is supported by the observation that isobutyraldehyde reacts rapidly with strongly basic secondary amines to produce amino alcohols without the aid of acid catalysts [22].

The proposed mechanism indicates that all the foregoing reactions in the sequence are now more or less rapidly established equilibria, and the complete reaction scheme can be written as shown in Scheme 1.

The concentrations of the different intermediates are determined by the equilibrium constants. The observation of immonium ions [Eq. (5)] in strongly acidic solutions by ultraviolet and NMR spectroscopy also indicates that these equilibria really exist [19,23]. The equilibria in aqueous solutions are of synthetic interest and explain the convenient method for the preparation of 2-deuterated ketones and aldehydes by hydrolysis of enamines in heavy water [24].

It has been noticed that the reverse reaction of Eq. (5) is a particular type of the Hofmann elimination reaction [23] via either an E2 or an E1cB mechanism. An E2 mechanism seems to be more obvious for this reaction than an e1cB mechanism, however.

Finally, it should be noted that at alkaline and neutral pH, a concerted mechanism, involving β-carbon protonation of the enamine and a simultaneous addition of a water molecule, leading to the amino alcohol, can be rejected, since the immonium ion appeared to be a real and detectable intermediate.

III. STRUCTURE AND REACTIVITY

The hydrolysis of enamines proceeds via a number of separate reactions, which may be considered as equilibria. Which reaction in the sequence becomes rate-determining depends on the pH of the solution as well as on the structure of the intermediates. Under alkaline and neutral conditions, the order of reactivity of the enamines discussed in the foregoing sections is: pyrrolidino- > dimethylamino- > piperidino- > morpholino-. One would expect the rate of protonation in the first step to be strongly dependent on the basicity of the enamine. This is true for the morpholino, dimethylamino, and piperidino enamines, but the much higher protonation rate of the pyrrolidino en-

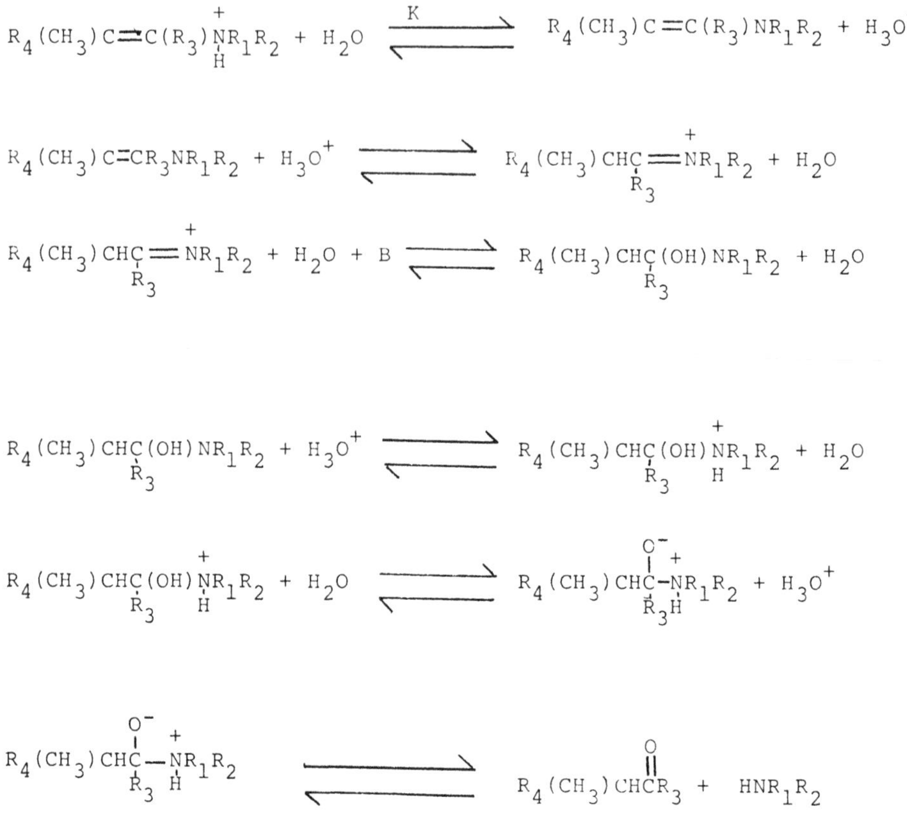

$$R_4(CH_3)C \overset{+}{\underset{H}{\rightleftharpoons}} C(R_3)\overset{+}{N}R_1R_2 + H_2O \quad \xrightarrow{\quad K \quad} \quad R_4(CH_3)C = C(R_3)NR_1R_2 + H_3O^+$$

$$R_4(CH_3)C = CR_3NR_1R_2 + H_3O^+ \quad \rightleftharpoons \quad R_4(CH_3)CH\underset{R_3}{\overset{+}{C}} = NR_1R_2 + H_2O$$

$$R_4(CH_3)CH\underset{R_3}{\overset{+}{C}} = NR_1R_2 + H_2O + B \quad \rightleftharpoons \quad R_4(CH_3)CH\underset{R_3}{C}(OH)NR_1R_2 + H_2O$$

$$R_4(CH_3)CH\underset{R_3}{C}(OH)NR_1R_2 + H_3O^+ \quad \rightleftharpoons \quad R_4(CH_3)CH\underset{R_3}{C}(OH)\overset{+}{\underset{H}{N}}R_1R_2 + H_2O$$

$$R_4(CH_3)CH\underset{R_3}{C}(OH)\overset{+}{\underset{H}{N}}R_1R_2 + H_2O \quad \rightleftharpoons \quad R_4(CH_3)CH\underset{R_3}{\overset{\overset{O^-}{|}}{C}}-\overset{+}{\underset{H}{N}}R_1R_2 + H_3O^+$$

$$R_4(CH_3)CH\underset{R_3}{\overset{\overset{O^-}{|}}{C}}-\overset{+}{\underset{H}{N}}R_1R_2 \quad \rightleftharpoons \quad R_4(CH_3)CH\overset{\overset{O}{\|}}{C}R_3 + HNR_1R_2$$

HNR_1R_2 = morpholine
piperidine
pyrrolidine
dimethylamine

Scheme 1

amine in comparison with the piperidino compound, of which the basicities are closely similar, cannot be explained in this way. The tendency
of the five-membered ring of the amine part in the molecule to form an
energetically favorable exocyclic double bond accounts for the much
higher reactivity of this enamine than the corresponding piperidino
enamine and has been observed in a large number of electrophilic addition reactions [1].

Notwithstanding the expected and also observed high reactivity
of the intermediate isobutyraldehyde immonium ions, the stabilization
of the exocyclic double bond in the pyrrolidino derivative evidently

prevents rapid nucleophilic attack of water, and the hydration of this ion to the amino alcohol becomes a slow, general base-catalyzed process in weakly acidic solutions [Eq. (6)]. All the propiophenone iminium ions are stabilized by charge delocalization in the aromatic ring. So they all show this slow water hydration of iminium ion step over a wide pH range.

The difference in reaction rates of the amino alcohols to isobutyraldehyde or propiophenone and the secondary amine in strong acidic solutions is determined by the reactivity as well as the concentration of the intermediate zwitterions [Figure 3, Eq. (11)]. Since several of the equilibrium constants of the foregoing reactions are unknown, an estimate of the relative concentrations of these dipolar species is difficult. As far as the reactivity is concerned, the rate of decomposition is expected to be higher, as the basicity of the secondary amines is lower, since the necessary driving force to expel the amine will increase with increasing basicity of the secondary amine.

ACKNOWLEDGMENT

Tables 1–3 were originally published in the *Journal of Organic Chemistry*. The American Chemical Society has kindly granted permission for reproduction.

REFERENCES

1. G. Stork, A. Brizzolara, H. Landesman, J. Szmuszkovicz, and R. Terrell, *J. Am. Chem. Soc.*, *85*, 207 (1963).
2. N. J. Leonard and V. W. Gash, *J. Am. Chem. Soc.*, *76*, 2781 (1954).
3. G. Optiz, H. Hellmann, and H. W. Schubert, *Ann.*, *623*, 117 (1959).
4. G. Stork, R. Terrell, and J. Szmuszkovicz, *J. Am. Chem. Soc.*, *76*, 2029 (1954).
5. G. Optiz and H. Mildenberger, *Ann.*, *649*, 26 (1961).
6. G. Optiz, H. Hellmann, H. Mildenberger, and H. Suhr, *Ann.*, *649*, 36 (1961).
7. G. Optiz, H. Mildenberger, and H. Suhr, *Ann.*, *649*, 47 (1961).
8. G. Optiz and A. Griesinger, *Ann.*, *665*, 101 (1963).
9. K. Brodersen, G. Optiz, D. Breitinger, and D. Menzel, *Ber.*, *97*, 1155 (1964).
10. C. D. Gutsche, *The Chemistry of Carbonyl Compounds*, Prentice-Hall, Englewood Cliffs, NJ, 1967, pp. 34, 35, 84.
11. R. Dulou, E. Elkik, and A. Veillard, *Bull. Soc. Chim. France*, *1960*, 967.

12. E. J. Stamhuis and W. Maas, *Rec. Trav. Chim.*, *82*, 1155 (1963).

13. E. J. Stamhuis and W. Maas, *J. Org. Chem.*, *30*, 2156 (1965).

14. F. A. Long and J. Bigeleisen, *Trans. Faraday Soc.*, *55*, 2077 (1959).

15. C. A. Bunton and V. J. Shiner, Jr., *J. Am. Chem. Soc.*, *83*, 42, 3207, 3214 (1961).

16. T. Riley and F. A. Long, *J. Am. Chem. Soc.*, *84*, 522 (1962).

17. E. J. Stamhuis, W. Maas, and H. Wynberg, *J. Org. Chem.*, *30*, 2160 (1965).

18. P. Y. Sollenberger and R. B. Martin, *J. Am. Chem. Soc.*, *92*, 4261 (1970).

19. W. Maas, M. J. Janssen, E. J. Stamhuis, and H. Wynberg, *J. Org. Chem.*, *32*, 1111 (1967).

20. E. H. Cordes and W. P. Jencks, *J. Am. Chem. Soc.*, *85*, 2843 (1963).

21. W. Maas, Ph.D. thesis, University of Groningen, 1966, pp. 79, 80.

22. J. Hine and J. Mulders, *J. Org. Chem.*, *32*, 2200 (1967).

23. J. Elguero, R. Jacquier, and G. Tarrago, *Tetrahedron Lett.*, 4719 (1965).

24. J. P. Schaefer and D. S. Weinberg, *Tetrahedron Lett.*, 1801 (1965).

4
Electrophilic Substitutions and Additions to Enamines

G. H. ALT *Monsanto Company, St. Louis, Missouri*

A. GILBERT COOK *Valparaiso University, Valparaiso, Indiana*

I.   Introduction   182

II. Protonation   182

III. Alkylation   182

   A. Reaction with Alkyl Halides   183
   B. Reaction with Electrophilic Olefins   189
   C. Reaction with Acetylenes   194
   D. Arylation   195
   E. Reaction with Aldehydes and Ketones   199
   F. Reaction with Iminium Ions   202

IV. Acylation   204

   A. Reaction with Carboxylic Acid Chlorides,
       Anhydrides, and Ketenes   204
   B. Reaction with Sulfonyl and Sulfenyl Derivatives   215
   C. Reaction with Isocyanates and Isothiocyanates   218

V.   Reaction with Other Electrophiles   220

   A. Halogenation   220
   B. Reaction with Cyanogen Halides   223
   C. Reaction with Aromatic Diazonium Salts   224
   D. Reaction with Aromatic Azides   226
   E. Reaction with Diethyl Azodicarboxylate and
       Activated α-Diazo Compounds   227
   F. Reaction with Diborane and Aluminum Hydrides   228

G. Metal-Complex-Catalyzed Reactions 229

H. Miscellaneous 231

References 232

I. INTRODUCTION

One of the two most important chemical characteristics of enamines is the nucleophilic nature of the β-carbon atom, which makes this atom very susceptible to *electrophilic attack* [1]. (The second important chemical characteristic is the electrophilic character of the α-carbon atom in an iminium ion, making that atom a good target for nucleophilic attack; see Chapter 6). An incoming electrophile has two sites for possible attack: the nitrogen atom and the β-carbon atom. The topic of electrophilic substitution of enamines has been reviewed recently [2—4].

II. PROTONATION

The most elementary electrophile for attacking an enamine is the proton. The proton can add to the nitrogen to give an enammonium ion

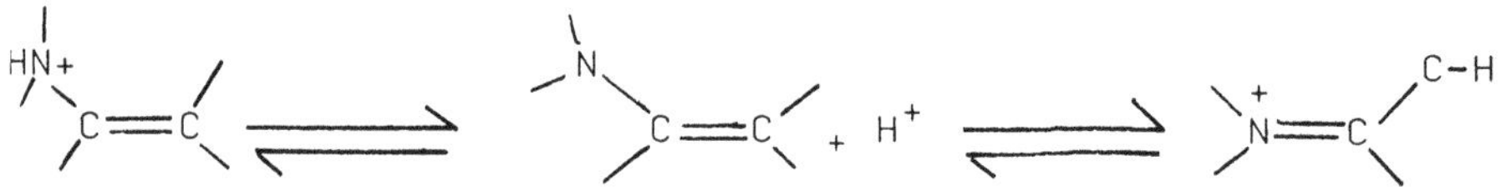

(1) (2)

(1), which is the kinetically favored product. Or the proton can add to the β-carbon atom, resulting in the formation of an iminium ion (2), which is the thermodynamically most stable product. The conditions under which each of these species are formed, along with full discussions of this reaction, are given elsewhere in this book (Chapter 1, Section IV.E and Chapter 6, Section II.B.1); so this reaction will not be further discussed here.

III. ALKYLATION

Alkylation of enamines can take place on carbon or on nitrogen. The theoretical considerations and reaction conditions that determine whether C- or N-alkylation takes place have been studied extensively [1, 5—16]. These studies have shown that the facility with which alkylation takes place depends on the ease of formation of a trigonal atom in the transition state, and on the nature of the enamine, the alkylating agent, and the solvent.

A. Reaction of Alkyl Halides

The principal synthetic problems encountered in using alkylation of enamines with simple unactivated alkyl halides as a vehicle for α-alkylation of ketones are: (a) the chief reaction is usually N-alkylation rather than C-alkylation; (b) mixtures of unalkylated, monoalkylated, and dialkylated products are obtained along with N-alkylated products; (c) aldol condensation products can be obtained with aldehyde enamines. These alkylation reactions have been reviewed [17].

Aldehyde and Acyclic Ketone Enamines

Because of self-condensation under the conditions of the alkylation reaction, enamines derived from acetaldehyde or monosubstituted acetaldehydes cannot usually be alkylated with simple unactivated alkyl halides [4]. Enamines derived from aldehydes disubstituted on the β-carbon, such as those derived from isobutyraldehyde, are alkylated on nitrogen by alkyl halides [4]. One method that has been used to achieve C-alkylation with aldehyde enamines using simple unactivated alkyl halides is use of a bulky secondary amine to make the enamine [18,19].

Activated alkyl halides provide easier access to C-alkylated products (as opposed to N-alkylated products), but the increased activity also results in greater amounts of dialkylation products. So it has been reported that allyl [1,6,20−24], benzyl [20−22], and propargyl [25] halides, along with α-halocarbonyl compounds [26], will C-alkylate enamines. The mechanism for the allyl halide addition involves, after initial N-alkylation, a [3,3]-sigmatropic rearrangement (a 3-Aza-

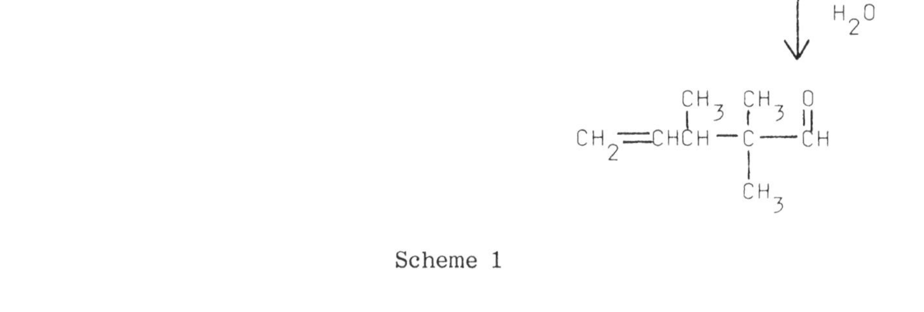

Scheme 1

Cope rearrangement [29] sometimes called an Aza-Claisen rearrangement), such as is illustrated in Scheme 1 [13,20,22,24]. Hydrolysis of the rearrangement product gave 2,2,3-trimethyl-4-pentenal. Support for this mechanism was provided by the alkylation of 2-N-dimethyl-N-(2-propenyl)-1-propene (3) by methyl tosylate, which on hydrolysis gave 2,2-dimethyl-4-pentenal (4). Similarly, alkylation [25] of

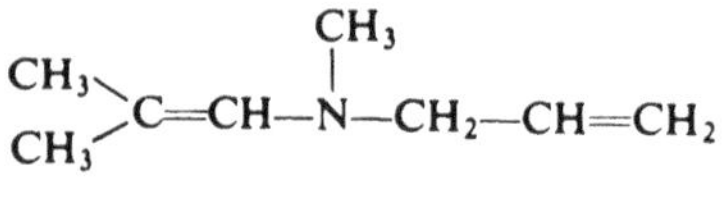

(3) 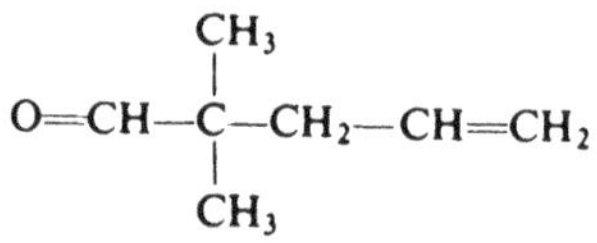   (4)

1-N-pyrrolidino-2-methyl-1-propene (5) with propargyl bromide gave initial N-alkylation to (6) with subsequent rearrangement to the allene (7). N-Alkylation of enamine 8 with benzyl bromide also occurs, and

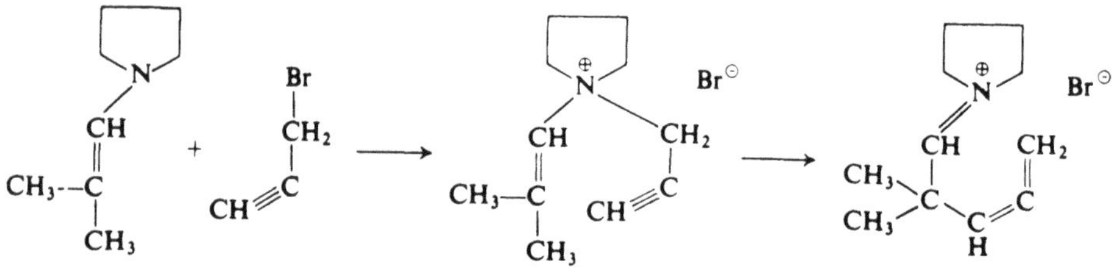

(5)   (6)   (7)

further heating of the reaction mixture leads to the C-alkylated product (9), probably by an intermolecular mechanism [13,20]. Support

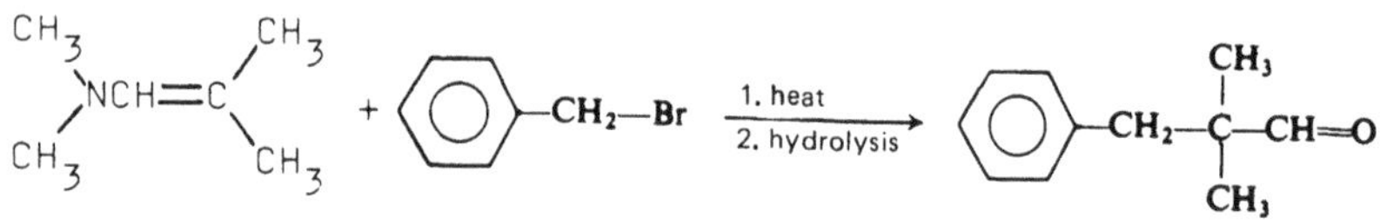

(8)   (9)

for initial N-alkylation is provided by the reaction of N-isobutenyl-N-methylbenzylamine (10) with methyl iodide, which gave α,α-dimethylhydrocinnamaldehyde (11) on hydrolysis [20]. When N-alkylation is

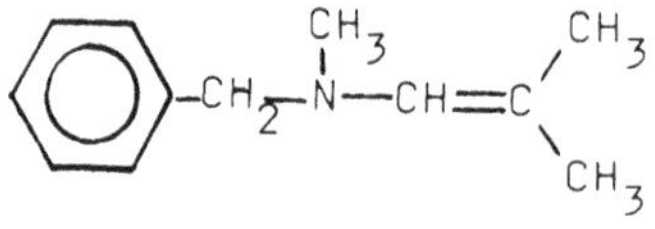

(10) 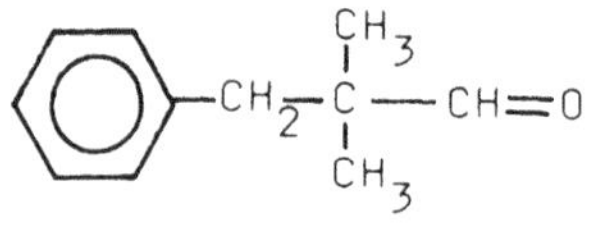   (11)

not possible for steric reasons, C-alkylation appears to occur directly [25]. Solvents of high dielectric also favor C-alkylation [6,9]. Thus 1-N-pyrrolidino-2-methyl-1-propene (5) with allyl bromide in ether gave only 20% of C-alkylated product, whereas in acetonitrile over 50% of this product is obtained.

Alkylation of aldehyde enamines using benzyl α-bromoacetate (12) gave a benzylated product (which means that α-bromoacetate is the leaving group) rather than the benzyl acetate alkylation product (with

$$BrCH_2CO_2CH_2C_6H_5$$

(12)

a bromide leaving group) [27].

A series of heterocyclic and carbocyclic cations were used to alkylate some aldehyde enamines as well as an acyclic ketone enamine [28]. These cations were trityl, indolizinium, dithiolium, pyrylium, thiapyrylium, selenoapyrylium, and tellurapyrylium cations.

Alkylation of some acyclic ketone enamines with α-halocarbonyl compounds has also been reported [30,31].

Cyclic Ketone Enamines

Simple unactivated alkyl halides give poor to fair yields of C-alkylated products with cyclic ketone enamines [1,14–16,32–39,55]. The exceptions to this rule are β-tetralone derivatives [1]. For example, β-tetralone 13 will produce, via its pyrrolidine enamine, alkylated

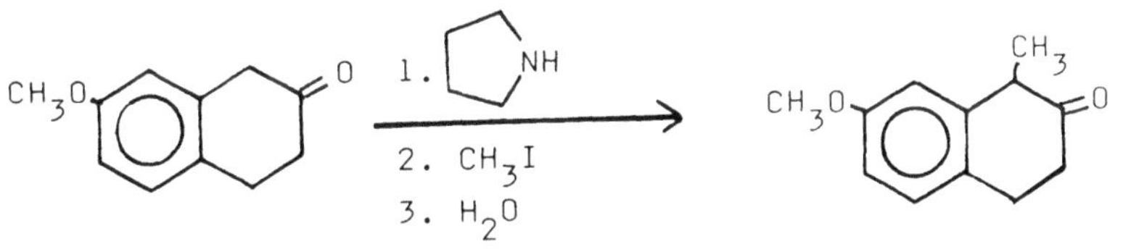

(13)   (14)

ketone 14 in an 80% yield [32]. In cyclic enamines, the size of both the olefin and the amine rings is an important factor in determining the ratios of C-alkylation to N-alkylation [9,14]. Pyrrolidine enamines of cyclic ketones give some of the best yields of C-alkylation [1], whereas morpholine enamines give poorer yields [1,38] or no C-alkylation product at all [37]. As was true with aldehyde enamines, enamines derived from sterically hindered enamines improved the yields of C-alkylations [16].

With enamines of cyclic ketones, direct C-alkylation occurs with allyl and propargyl as well as alkyl halides. The reaction is again sensitive to the polarity of the solvent [9]. The pyrrolidine enamine of cyclohexanone, on reaction with ethyl iodide in dioxane, gave 25% of 2-ethylcyclohexanone on hydrolysis, whereas in chloroform the yield was increased to 32%. 1-Pyrrolidino-1-cycloheptene gave approximately equal quantities of the C- and N-alkylated products in dioxane, and 1-pyrrolidino-1-cyclooctene and 1-pyrrolidino-1-cyclononene afforded N-alkylated products exclusively under similar conditions [9].

It has been concluded, with regard to the mechanism of C-alkylation, that it does not take place by intramolecular N-to-C transfer of the alkyl group, at least in the case of dienamines [15]. Rather, the N-alkylation is equilibrated with the free enamine and the alkyl halide reactants, and the enamine undergoes C-alkylation with the alkyl halide by a slower reaction [15].

There is a greater amount of C-alkylation when activated halides are used. Some of the activated halides used are allyl [1,40,41,55, 70,71], benzyl [42−44,70], and propargyl [45,46] halides, along with α-halocarbonyl compounds [47−50] and α-haloesters [51−54,70]. In the case of alkylation with benzyl halides, changing from dioxane to acetonitrile solvent greatly improved the yield [44]. The mechanism for the allyl halide C-alkylation of cyclic ketone enamines is similar to that for aldehyde enamines in that N-alkylation takes place first, followed by an Aza-Cope rearrangement [40].

Asymmetrical induction has been reported for the alkylation of the (+)-*trans*-2,5-dimethylpyrrolidine enamine of cyclohexanone with alkyl halides and allyl bromides [36]. These reactions took place with good optical purities and with very little dialkylation product. Asymmetrical induction has also been observed with the alkylation of the L-proline ester enamines of cyclohexanone with allyl bromide, ethyl bromoacetate, and benzyl bromide [70,71]. Allyl bromide gave the best results, and a bulky ester moiety on L-proline such as a *t*-butyl group brought about the best optical yields [70,71]. The solvent tetrahydrofuran produced the best optical yields.

One of the advantages of the enamine alkylation reaction over direct alkylation of the ketone under the influence of strong base is that the major product is the monoalkylated derivative [1,9]. When dialkylation is observed, it occurs at the least substituted carbon, in contrast to alkylation with base, where the α-disubstituted product is formed. Dialkylation becomes the predominant reaction when a strong organic base is added and an excess of alkyl halide is used [9]. Thus, 1-N-pyrrolidino-1-cyclohexene (15), on treatment with 2 moles of allyl bromide in the presence of ethyl dicyclohexylamine (a strong organic base that is not alkylated under the reaction conditions) gave a 95% yield of 2,6-diallylcyclohexanone (16). The course of the reac-

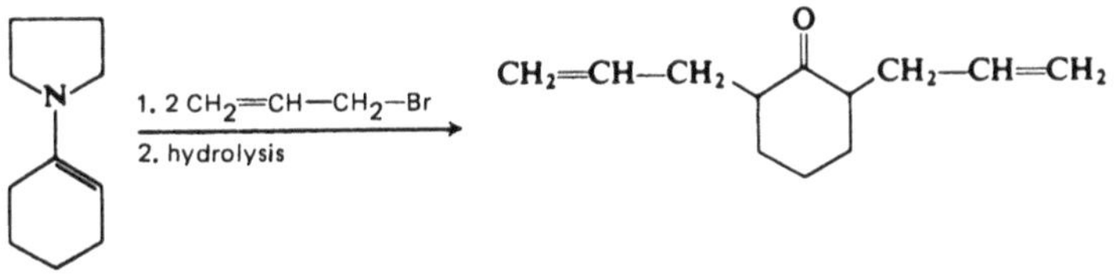

(15)   (16)

tion appears to involve dehydrohalogenation of the intermediate iminium salt (17) to the new enamine (18), which then undergoes further
alkylation. Evidence that alkylation in this case is directly on carbon

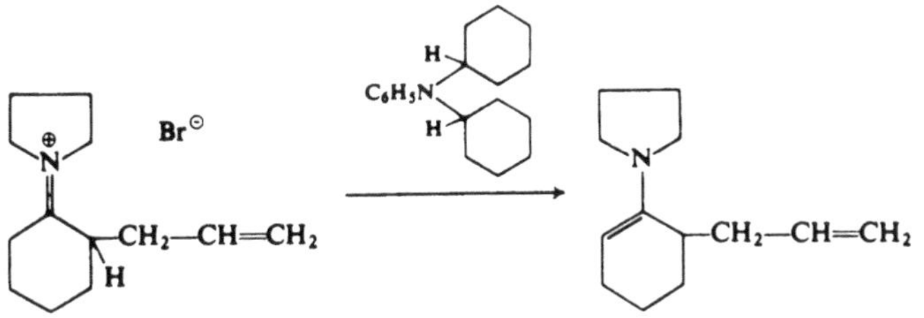

(17)   (18)

[9] rather than on nitrogen [25], followed by a rearrangement to carbon as in the case of the aldehyde enamines, is provided by the alkylation of 1-N-pyrrolidino-1-cyclopentene (19). Alkylation of 19 with
crotyl bromide followed by hydrolysis gave more than 80% of the 2-
crotyl cyclopentanone (20) and less than 20% of the 2-(α-methyl allyl)
cyclopentanone (21).

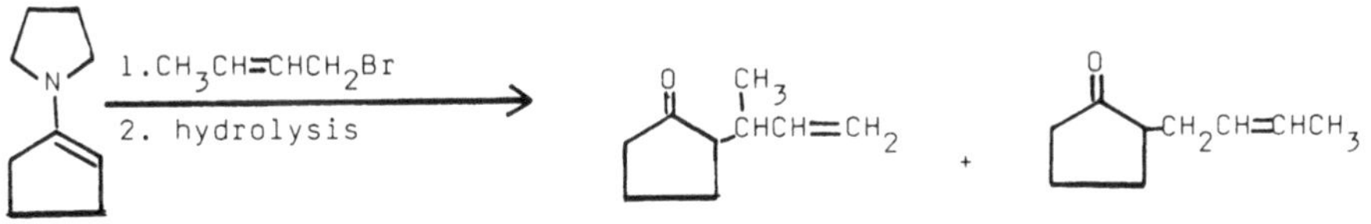

(19)   (20)   (21)

Other types of activated alkyl groups that have been used to
alkylate cyclic ketone enamines include 3,5-dimethyl-4-chloromethyl-
isoxazole (22a) [56], trichloro-s-triazine (22b) [57], 10-methylacri-
dinium salts [58], and 7-(acylamino)-5-phenylfurazano-[3,4-d] pyrim-

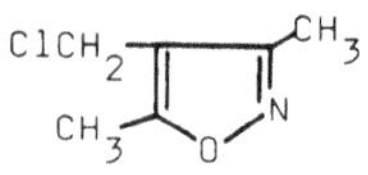

(22a) 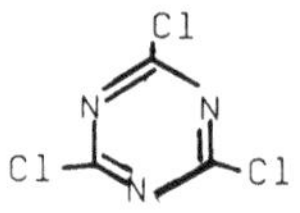   (22b) 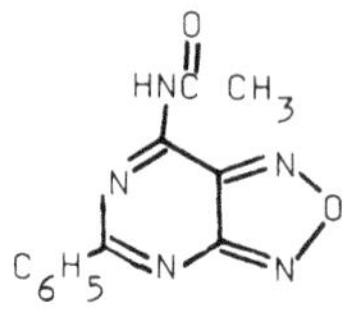   (22c)

idine (22c) [59]. In the last case, reaction takes place at the 7-position of 22c, and this reaction is a reversible alkylation, an unusual reaction in that respect [59]. Mannich bases are another source of alkylation reagents for enamines [60–62], including both α-phenyl and α-alkylthiomethylation agents [63].

Heterocyclic Enamines

The first example of an enamine alkylation reaction was the conversion of 1,3,3-trimethyl-2-methyleneindoline, the so-called Fisher's base, which was shown to give 1,3,3-trimethyl-2-isopropylideneindoline with methyl iodide rather than a quaternary salt [64,65]. More recently, other heterocyclic enamines have been shown to undergo similar reactions. Thus 1-methyl-2-ethylidenepyrrolidine gave the iminium salt (23) on alkylation with methyl iodide [66]. The latter then loses hydrogen iodide to give 1-methyl-2-isopropyl-2-pyrroline (24). In the benzothiazole series, there was some confusion in the

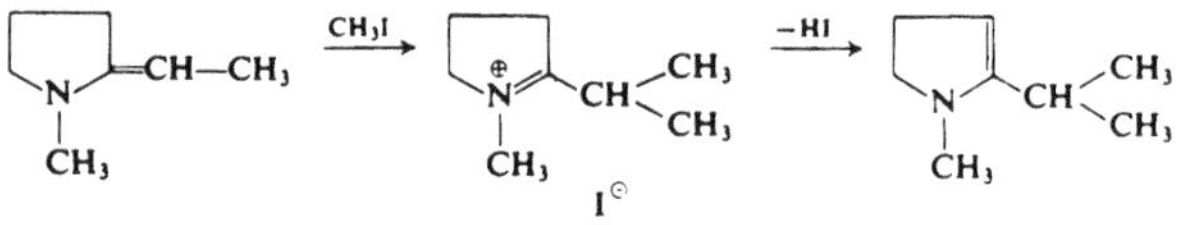

(23)   (24)

literature as the simplest enamine (25) exists as the dimer (26) [67]. The dimer undergoes alkylation on sulfur [68] rather than at the "enamine" carbon of 26. However, it has been shown [69] that monomer-

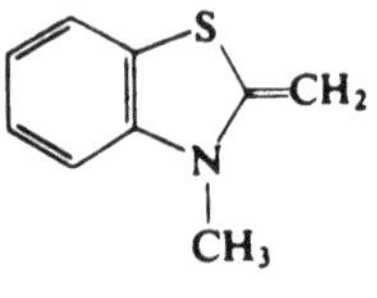

(25) 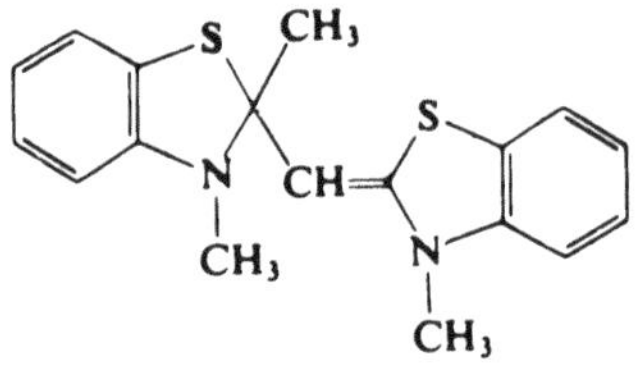   (26)

ic enamines such as 27 react normally to give the benzothiazoline salt (28) on alkylation with alkyl and benzyl halides.

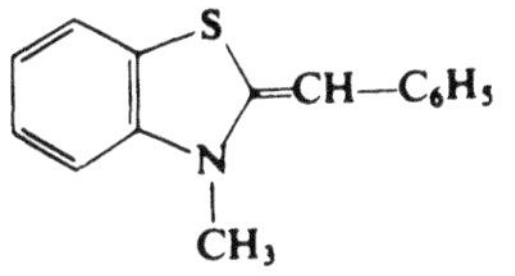

(27) 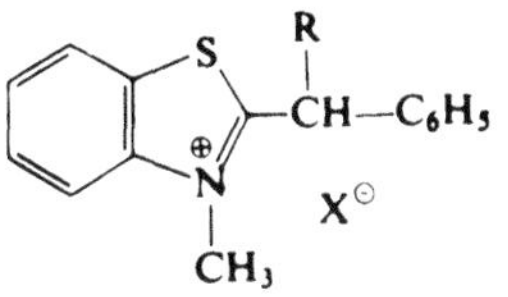   (28)

A more complete discussion of electrophilic additions to hetero-cyclic enamines is given in Chapter 8.

B. Reaction with Electrophilic Olefins

Enamines are readily alkylated by olefins activated by electron with-drawing substituents. N-Alkylation by one of these olefins is revers-ible, whereas C-alkylation usually is not, so that a good yield of mono-alkylated product is the rule. A few examples of reversible C-alkyl-ation have been reported [72,73,134].

The range of reactions that can take place between enamines and electrophilic olefins includes [2+2] cycloadditions, [4+2] cycloaddi-tions, and Michael additions to give linearly substituted adducts. The cycloaddition reactions are thoroughly discussed in Chapter 7. So in this chapter we will be primarily concerned with the Michael addition reactions. The Michael reaction is a two-step reaction that proceeds through a dipolar zwitterion intermediate (29) (see Chapter 7).

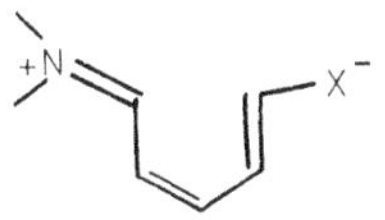

(29)

The stereochemical course of the alkylation of cyclic enamines by electrophilic olefins is determined by a number of steric and stereo-electronic effects. These effects have been thoroughly discussed in Chapter 1 (Section III.B.), so that discussion will not be repeated here. However, it should be noted that electrophilic olefins generally attack enamines of cycloalkanones in a stereospecific antiparallel man-ner except when steric hindrance inhibits this approach [120,121, 136].

The series of electrophilic olefins that typically undergo Michael addition to enamines are α,β-unsaturated ketones [1,72,74−81,100, 135], α,β-unsaturated nitriles [1,12,70,71,74−89,107,108,388], α,β-unsaturated esters, diesters, or amides [1,70,71,83,84,87,90−106], α,β-unsaturated sulfones [107−115,136,137], and nitroolefins [100, 110,116−134,138]. The typical reaction for the pyrrolidine enamine of cyclohexanone with one of these electrophilic olefins is shown in Scheme 2.

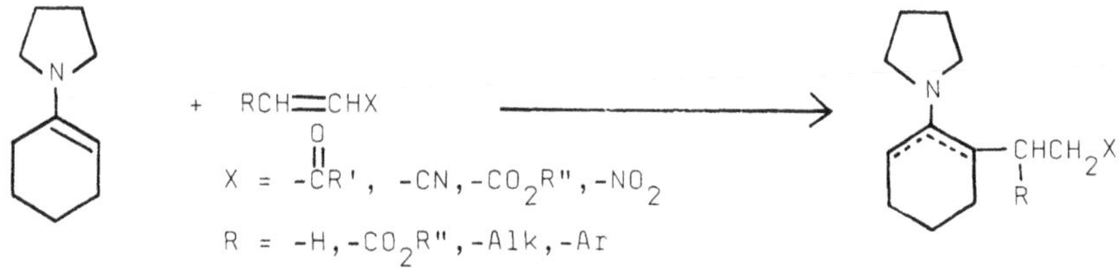

$$X = -\overset{\overset{\text{O}}{\|}}{\text{C}}R', \quad -CN, -CO_2R'', -NO_2$$

$$R = -H, -CO_2R'', -Alk, -Ar$$

Scheme 2

Monosubstituted products are usually obtained, but 2,6-dialkylated products can be obtained by varying the reaction conditions and the reagent molar ratios. The usual positioning of the two alkyl groups is 2,6, but occasionally 2,2-disubstitution products are obtained. The reason for this regiochemistry is discussed in Chapter 1 (Section III.B.).

An olefin that possesses both a keto and a nitrile group, β-cyanovinylphenylketone (30), adds to the pyrrolidine enamine of cyclohex-

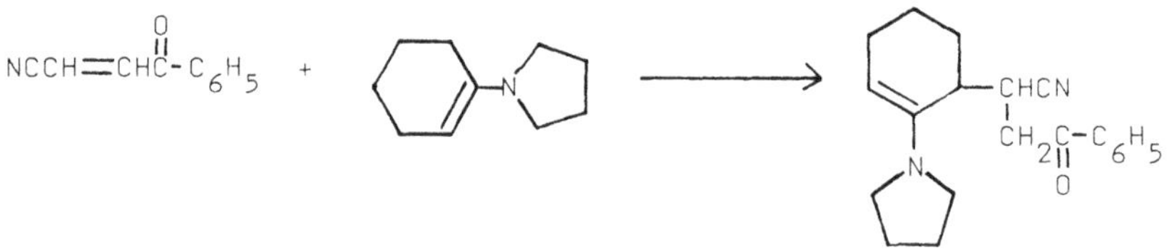

(30)   (15)   (31)

anone to give addition product 31 with the phenylketo group as the controlling substituent [74]. The morpholine enamine of cyclohexanone will not react with 30 at room temperature [74]. In competition between an ester group and a nitro group using nitroester olefin 32

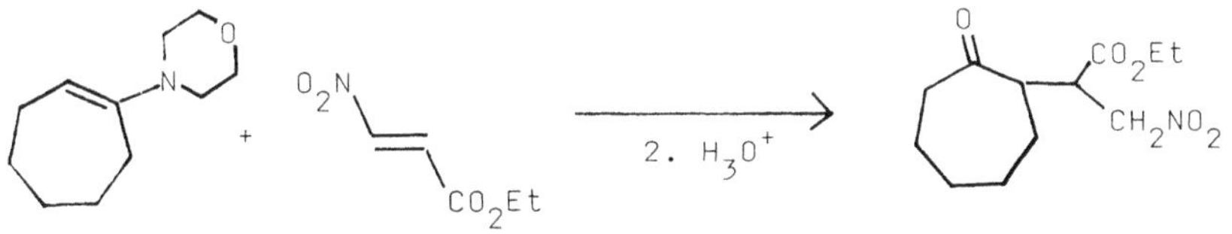

(32)   (33)

with the morpholine enamine of cycloheptanone, the nitro group seems
to have the stronger electrophilic influence in producing 33 [98].
    The reaction of the pyrrolidine enamine of cyclohexanones (R=H,

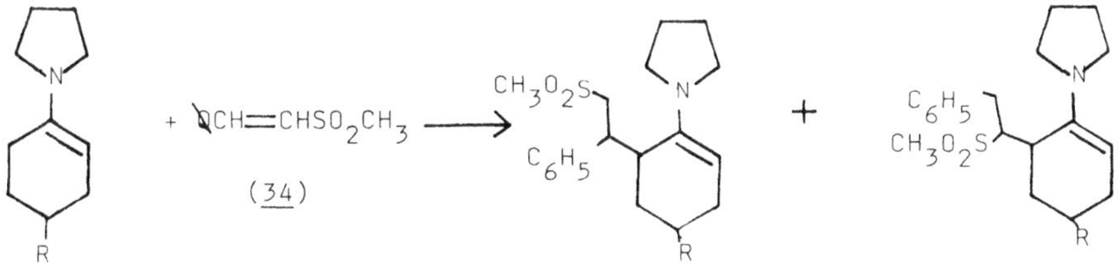

(34)   (35a)   (35b)

CH$_3$, t—Bu) with Z- and E-methyl-β-styryl sulphone (34) in acetonitrile solvent can produce the β-adduct (35a) and/or the α-adduct
(35b). The direction of attack with the E-sulfone (and to a lesser
extent the Z-sulfone) is determined by the size of the substituent on
the 4-position [115,118,136,137]. The larger the substituent, the
greater the amount of β-adduct (35a) formed [115]. In ethanol solvent both Z- and E-sulfone give β-adduct (35a) almost exclusively
[115].
    Treatment of the morpholine enamine of 5-methyl-*trans*-decalin-
2-one (36a) with p-bromophenyl vinyl sulfone gives an equimolar mixture of 3- and 1-substituted products (37a and 37b) [114], whereas
β-nitrostyrene reacts with the morpholine enamine of *trans*-decalin-
2-one (36b) to give exclusively adduct 38 from antiparallel attack at
C-3 of the enamine [122]. With the pyrrolidine enamine of 2-tetralone,
β-nitrostyrene also gives only the C-3 adduct [123], but with the
morpholine enamine it gives a 4:1 ratio of C-1 to C-3 [123].

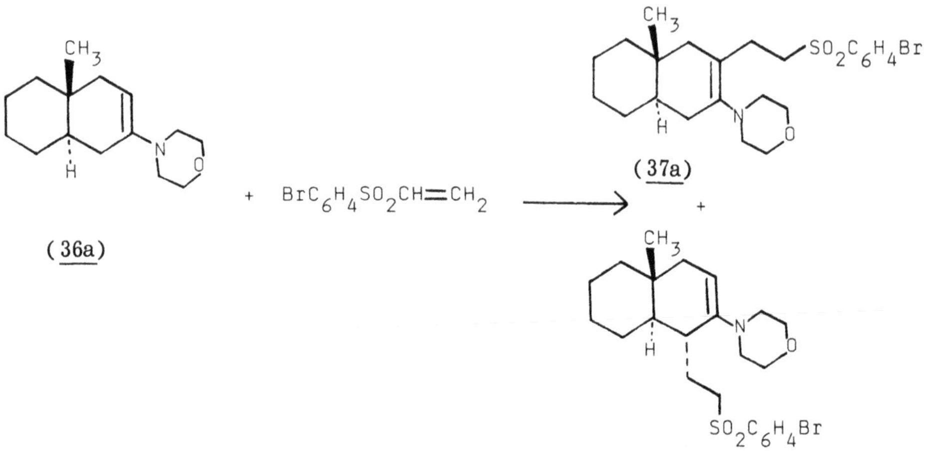

(36a)   + $BrC_6H_4SO_2CH{=}CH_2$ ⟶

(37a)

+

(37b)

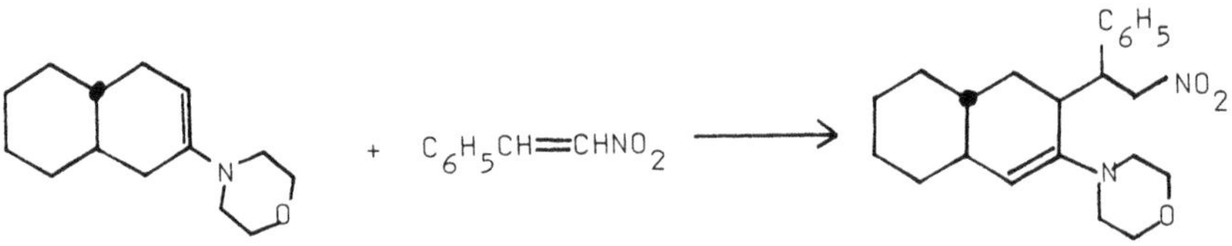

(36b)   + $C_6H_5CH{=}CHNO_2$ ⟶

(38)

β-Nitrostyrene reacts with the morpholine enamine of cyclohexanone (39) to give adduct 40a in a quantitative yield [116].

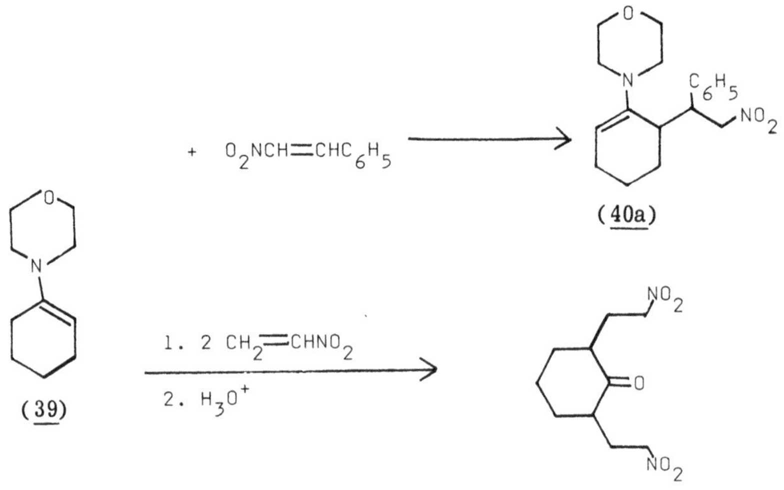

+ $O_2NCH{=}CHC_6H_5$ ⟶

(40a)

(39)

1. 2 $CH_2{=}CHNO_2$

2. H_3O^+

(40b)

Nitroalkylation by β-nitrostyrene is highly stereoselective [119, 120]. This is due to its sensitivity to stereoelectronic and steric requirements. It attacks by an antiparallel mechanism unless steric hindrance is present in the parent enamine. This high selectivity, however, is not general for all nitroolefins. α-Nitrostyrene is also less stereoselective toward enamines than β-nitrostyrene [131]. 1-Nitropropene, for example, attacks the same enamines in both a parallel and antiparallel manner [121]. 2,6-Disubstitution takes place when the morpholine enamine of cyclohexanone (39) is allowed to react with 2 moles of nitroethylene to form adduct 40b in a 60% yield [129].

The reactions between nitroolefins and enamines take place not only in good chemical yields, but also in excellent diastereomeric [139] yields of greater than 90% [130,132−134]. A topological rule for carbon-carbon-bond-forming processes between prochiral centers [140] in enamines and nitroolefins (as well as other systems) has been proposed [130] and shown to be very useful [132−134].

The alkylation of enamines with 2- and 4-vinyl pyridines such as 41 has been reported [141].

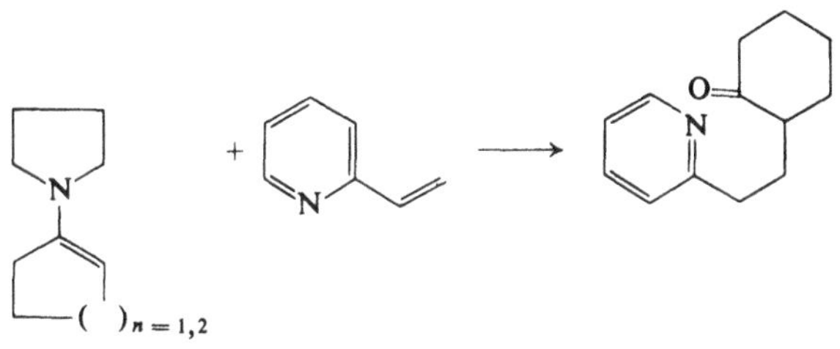

(41)

When β-propiolactone is allowed to react with an enamine with heating, a ketoamide is produced [196−198]. For example, when enamine 39 is treated with β-propiolactone, ketoamide 42 is formed in

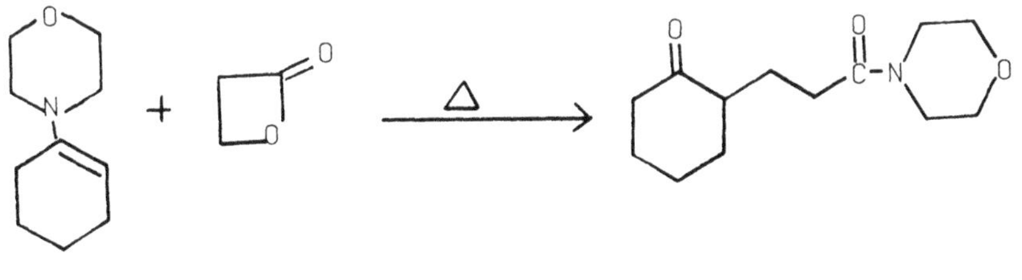

(39)   (42)

the absence of water in an 83% yield [196].

C. Reaction with Acetylenes

Nonactivated terminal acetylenes have been added to enamines derived from aldehydes. A long reaction time or catalysis by copper(I) chloride is necessary. Thus the dimethylamino enamine of isobutyraldehyde (43) formed adduct 44 on heating with phenylacetylene [142].

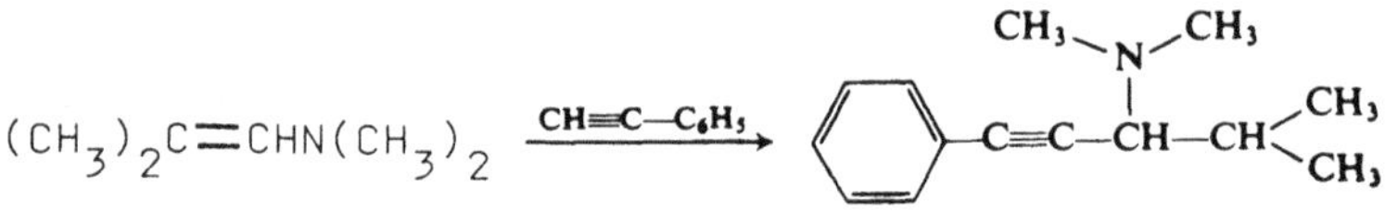

$(CH_3)_2C = CHN(CH_3)_2$ $\xrightarrow{CH \equiv C-C_6H_5}$

(43)   (44)

A more conventional cycloaddition occurs with activated acetylenes. However, the intermediate cyclobutene adducts undergo rearrangement to give insertion of two carbon atoms into the enamine chain [143,144]. Thus the enamine (43) reacted with methyl propiolate to give the dienamino ester (45), presumably via the cycloaddition product [143]. Dimethyl acetylenedicarboxylate reacts similarly

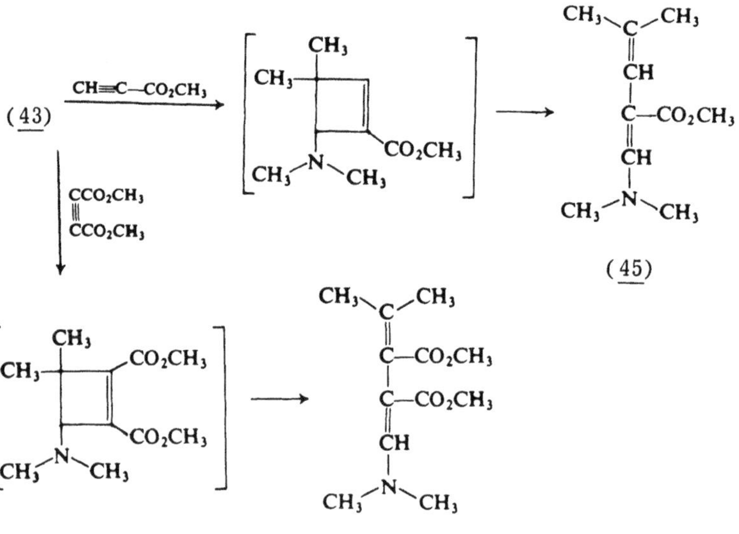

(43) $\xrightarrow{CH \equiv C-CO_2CH_3}$

(45)

(46)

to give 46. Again, the cycloaddition is presumed to be the initial step [144].

Dimethyl acetylenedicarboxylate (DMAD) as it adds to enamines is very solvent and temperature dependent as to whether cycloaddition reactions take place or whether a simple Michael addition reaction occurs [145].

The dehydroaporphine, dehydronuciferine (47), when treated with

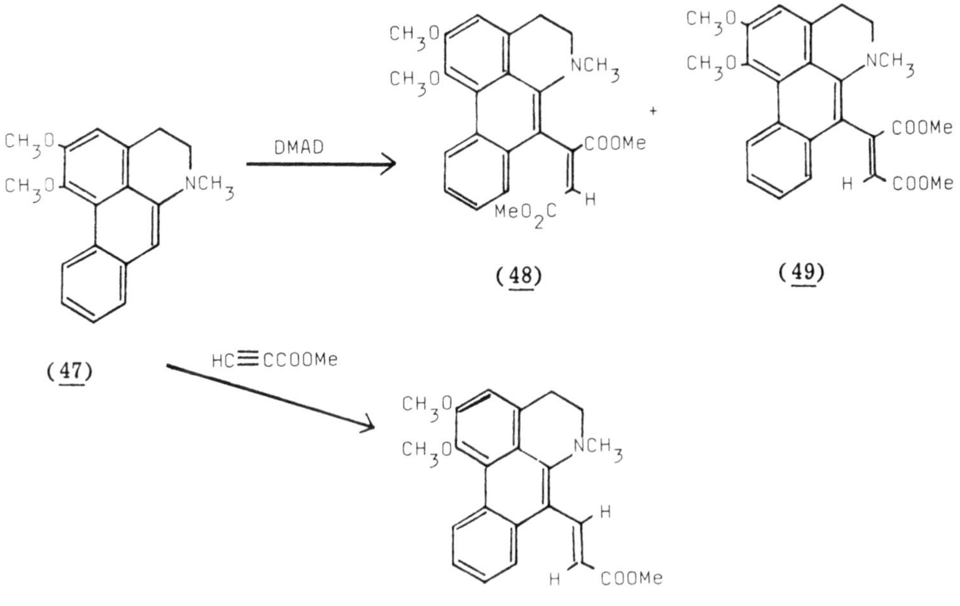

(48)

(49)

(47)

(50)

DMAD produces a mixture of stereoisomeric Michael addition products 48 and 49 in 52 and 40% yields, respectively [146]. The proton transfer in the zwitterion intermediate was shown to be intramolecular even though the reaction took place more readily in a methanolic solvent mixture than in an aprotic solvent [146]. The reaction of 47 with methyl propiolate gave a single crystalline adduct 50 [146]. Perfluoro-2-butyne readily forms Michael reaction adducts with enamines [147].

D. Arylation

Aryl halides with a halogen activated by electron-withdrawing groups react with pyrrolidine enamines of cyclic ketones [148–151,166] to give the α-arylated ketones after hydrolysis. Enamine 15 with 2,4-dinitrochlorobenzene gave an excellent yield of 2-(2,4-dinitrophenyl) cyclohexanone. The reaction was shown to go via the enamines 52

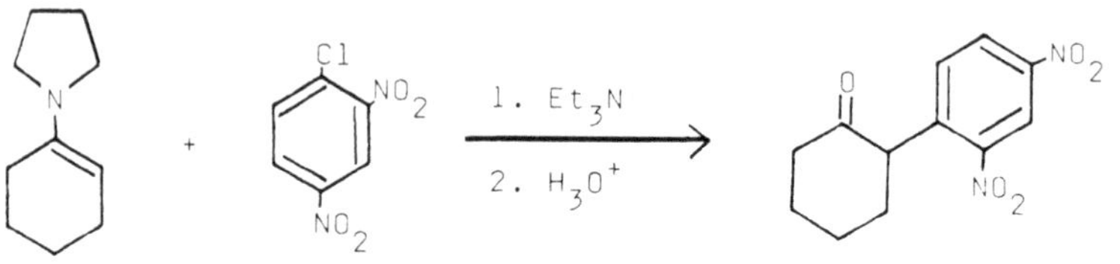

(15)   (51)

and 53 as intermediates, as acylation of the reaction mixture gave both possible acylated ketones on hydrolysis. Reductive cyclization of 51 gave the aminotetrahydrocarbazole [54]. The pyrrolidine en-amine of 2-methylcyclohexanone is arylated in the 6 position.

Good yields are obtained from the pyrrolidine enamines, poor yields from the piperidine enamines, and the morpholine enamines fail to react. Other activated halides, e.g., 2-chloro-5-nitropyridine,

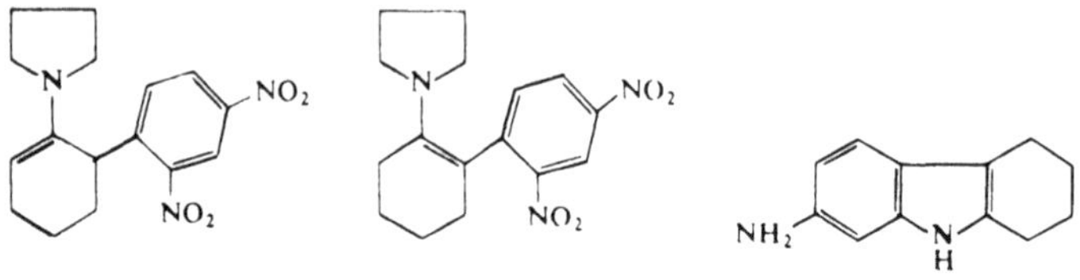

(52)   (53)   (54)

4-chloro-3-nitropyridine, and 2-chloro-4,5-dicarbethoxypyrimidine, have been shown to react equally well [148]. Similar arylations of heterocyclic enamines, e.g., 1,3,3-trimethyl-2-methylene indoline and 2-benzylidene-3-methylbenzo-thiazoline, have been reported. [69,152].

Enamine 15 did not undergo C-arylation with p-nitrochlorobenzene under these conditions, and at higher temperatures N-arylation and subsequent cleavage with formation of N-(4-nitrophenyl) pyrrolidine takes place [148].

Low yields of C-arylated ketones have also been obtained by re-action of pyrrolidine enamines with diaryl iodonium salts [148].

Heteroarylation of enamines with 1,3,4-thiadiazolium has been reported [153].

Treatment of 1-(N-morpholino)cyclopentene with methoxyphenyl-lead triacetate (55) in chloroform solvent at room temperature caused

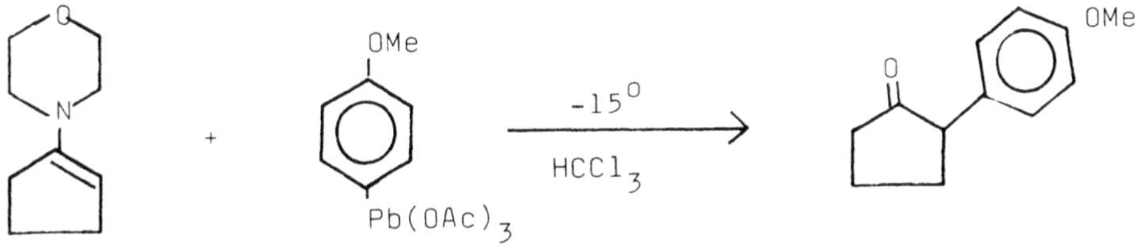

(55)   (56)

a very exothermic reaction to take place, producing, after hydrolysis, arylation product 56 in an 82% yield [154].

In the reactions of benzyne with enamines, arylated enamines or aminobenzocyclobutenes can be obtained, depending on reaction conditions and the structure of the enamine (see Scheme 3) [148]. Thus, the presence of a proton source such as a secondary amine will favor the enamine product through capture of the zwitterionic intermediate, whereas in the absence of protons one sees increased collapse of the intermediate to a benzocyclobutene [148,155]. There seem to be indications that there is an intramolecular transfer of the α-proton of the immonium salt [156].

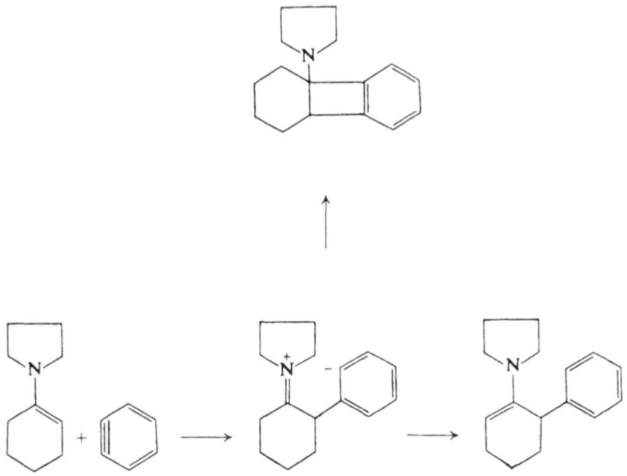

(15)   Scheme 3

The presence of an α'-substituent, found in the pyrrolidine enamine of 2-methylcyclohexanone, blocks the possibility of an intramolecular proton transfer in the zwitterionic intermediate, and thus only the benzocyclobutane is formed in this reaction (see Scheme 4). This result provided the first direct support for the requirement of axial

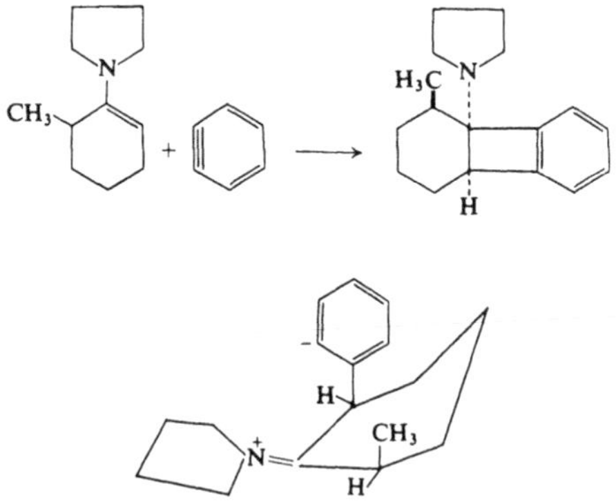

Scheme 4

α-attack and axial orientation of an α'-substituent in the transition state of some electrophilic attacks on a cyclohexenamine, with formation of a carbon-to-carbon bond (see Chapter 1, Section III.B) [156].

Arylation of enamines with p-benzoquinones takes a somewhat different course [148,157−165,167,168]. Enamine 57 reacts exothermally with p-benzoquinone in benzene solution to give 2-(dimethylamino)-2,3-dihydro-3,3-dimethyl-5-benzofuranol (58). The reaction of enamines with quinone dibenzenesulfonimide proceeds similarly [148, 161−164]. The product from enamine 15 is the tetrahydrocarbazole derivative (59).

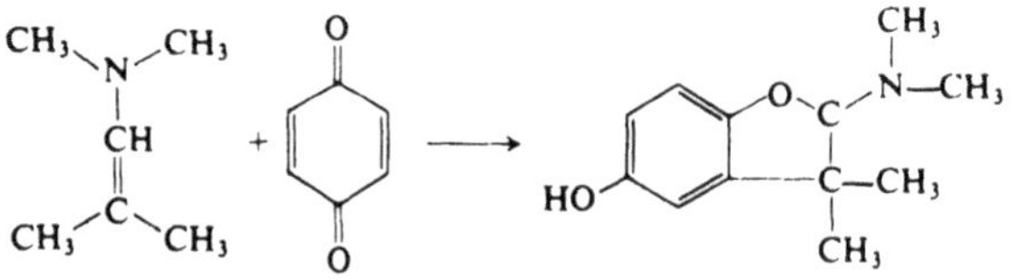

(57)   (58)

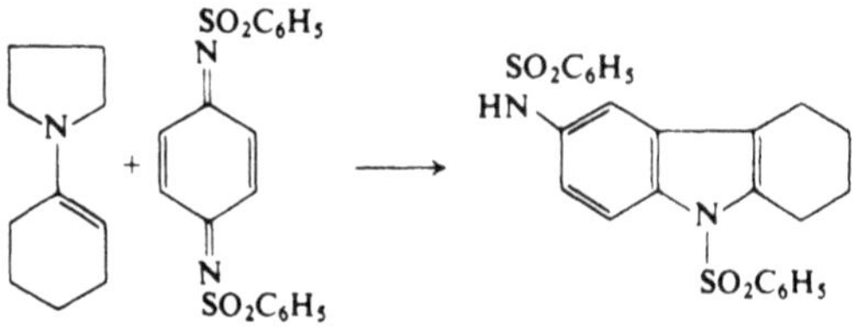

(15)   (59)

E. Reaction with Aldehydes and Ketones

The reaction of the dimethylamine or piperidine enamine of isobutyraldehyde with paraformaldehyde gives the substituted aminopivalaldehydes [169], which have also been obtained by the Mannich reaction of isobutyraldehyde, formaldehyde, and the corresponding amine [169]. Thus, enamine 57 gave β-dimethylaminopivalaldehyde [60]. A mechanistic study using formaldehyde-d_2 showed that this is a simple variation of the Mannich reaction, as a part of enamine 57 is hydrolyzed

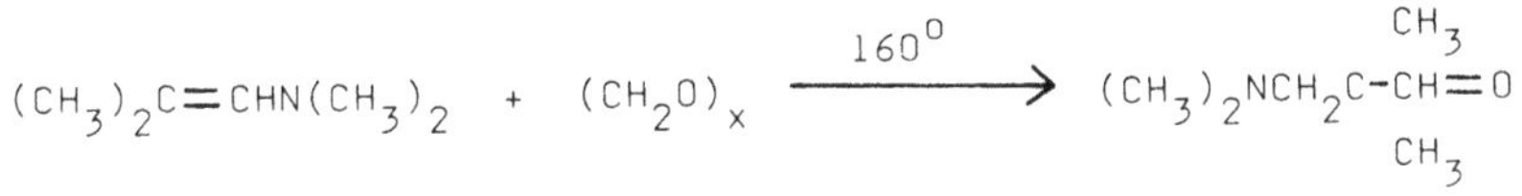

$(CH_3)_2C{=}CHN(CH_3)_2$
+
$(CH_2O)_x$
160°
$(CH_3)_2NCH_2C\text{-}CH{=}O$
CH_3
CH_3

(57) (60)

into its components by the residual water in the paraformaldehyde [157].

When chloral was used as the aldehyde, two equivalents reacted with one equivalent of the enamine (61), regardless of the ratio of reactants or order of addition, to give 2,6-bis(trichloromethyl)-5,5-dimethyl-4-morpholino-*m*-dioxane (62) in 83% yield [157]. Hydrolysis of 62 with hydrochloric acid at room temperature gave the hemiacetal (63), but when heated with acid, the aldol product (64) was formed.

2 Cl_3CCHO
HCl
25°
HCl
heat

(61) (62) (63) (64)

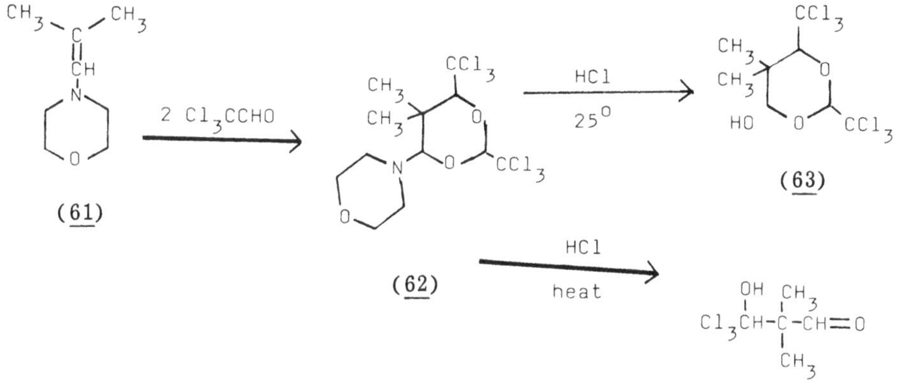

Enamines of cyclic ketones react stereoselectively with chloral to give good yields of only one of the possible 1:1 adduct isomers [170, 171], for example adduct 65.

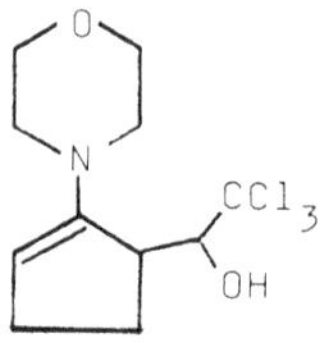

(65)

Acetaldehyde when treated with morpholine condenses with another molecule of acetaldehyde to form 3-morpholinobutanal, probably
via the enamine [172]. The reaction between enamine 66 and ester

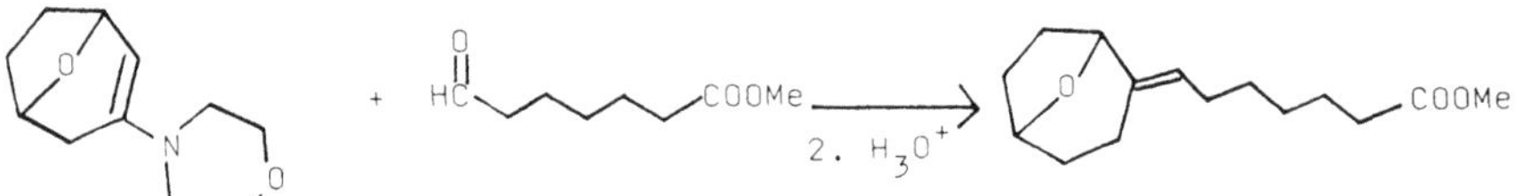

(67)   (68)

(66)

aldehyde 67 gives condensation product 68 [38]. Other condensations
of this type between aldehydes and enamines have been reported
[173–175]. The first step in these reactions appears to be the formation of a dipolar species such as 69, which can undergo proton transfer and elimination of water to give 70, which on hydrolysis gives the
2-alkylidene cyclanone.

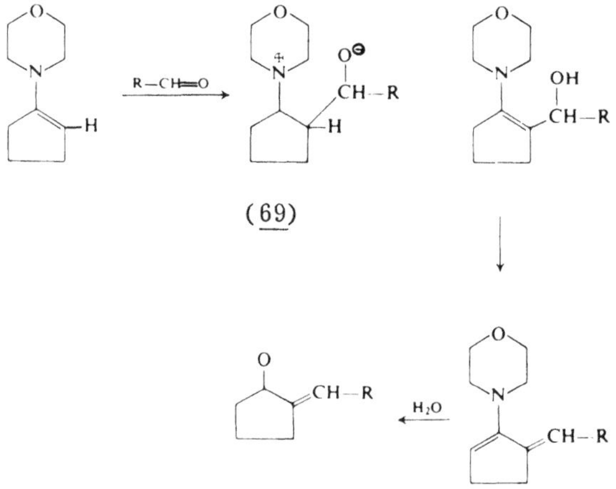

(69)

(70)

The intermediacy of dipolar species such as 69 has been demonstrated by reaction of enamines with 2-hydroxy-1-aldehydes of the
aromatic series [176]. Enamine 39 reacts in benzene solution at room
temperature with 2-hydroxy-1-naphthaldehyde (71) to give the crystalline adduct (72) in 91% yield. Oxidation with chromium trioxide-
pyridine of 72 gave 73 with β-elimination of the morpholine moiety.
Palladium on charcoal dehydrogenation of 73 gave the known 1,2-ben-
zoxanthone [176]. The condensation of benzaldhyde with enamines

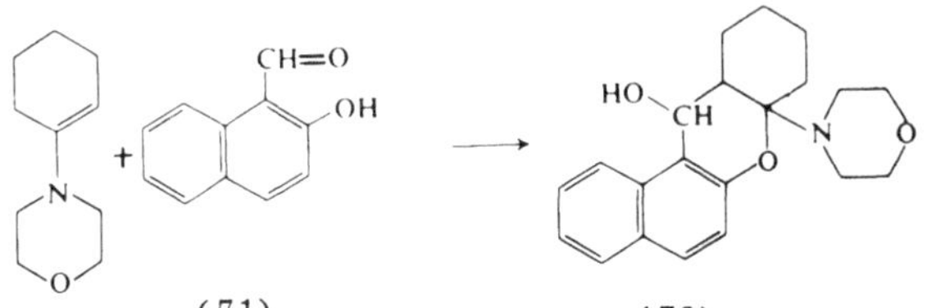

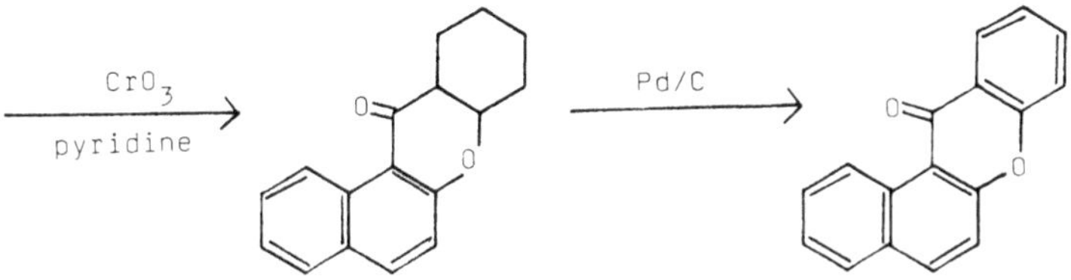

is also involved in the formation of 3,5-dibenzylpyridine from piperidine and benzaldehyde [177,178].

Imines (Schiff bases) have been reported to add to enamines [73,
181]. For example, treatment of 1-(N-morpholino)-4-*t*-butylcyclo-
hexene with benzylideneaniline (74) in absolute methanol produced

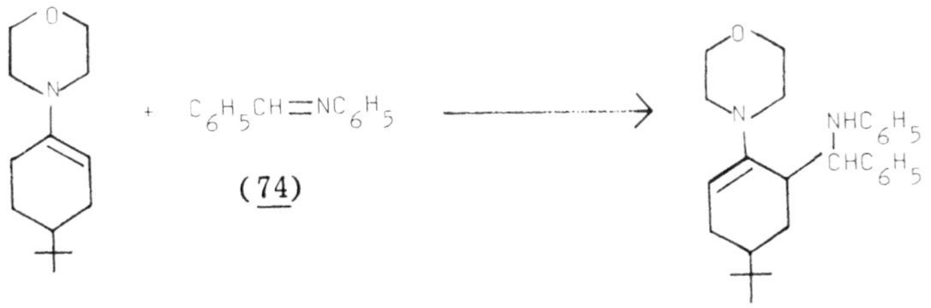

trans adduct 75 in about a 30% yield [181].

Ketones will also condense with enamines [179]. The long-known catalyses of some ketone condensation reactions by secondary amines can be postulated to have their basis in the reactions of enamine intermediates with ketones (see Scheme 5). The unsuitability of methyl ketones for azeotropic enamine formation is based on this phenomenon. Some studies in cyclization reactions have added further support to this concept [180].

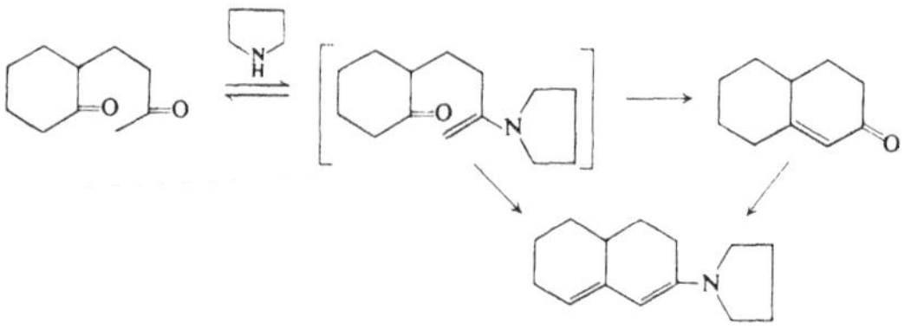

Scheme 5

F. Reaction with Iminium Ions

Ternary iminium ions can act as electrophiles and attack the β-carbon of enamines (see Chapter 2, Section III.A and Chapter 6, Section III. D) [182–188]. This is illustrated by the reaction between enamine 61 and iminium salt 76 to give adduct 77. Hydrolysis of this adduct

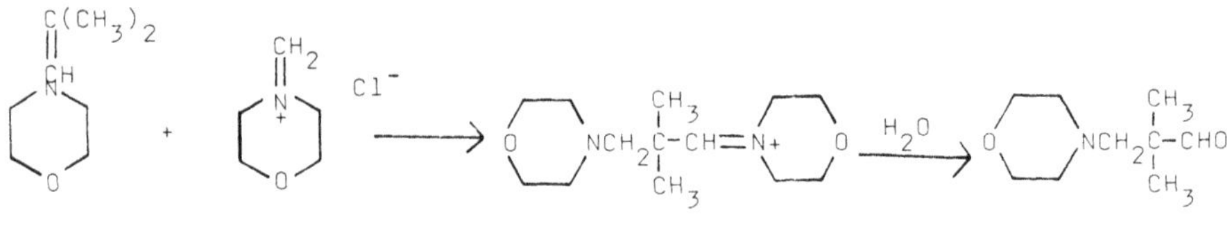

(61)   (76)   (77)   (77a)

produced aminoaldehyde 77a in an 87% yield [188].

In a similar manner, enamines undergo Mannich reactions with formaldehyde and dimethylamine in acidic media to form an amine

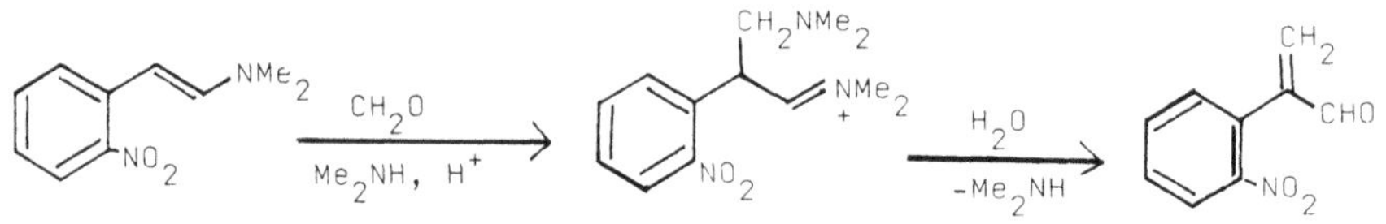

Scheme 6

iminium ion intermediate (see, for example, Scheme 6) [189,190]. Enamines can be alkylated with the Vilsmeier-Haack reaction [191], the final result after hydrolysis being net formation of the enamine [190,192,193,259]. The Vilsmeier reagent is the iminium ion des-

$$\text{cribed by } (\overset{Cl}{\underset{}{H\overset{+}{C}}} = \overset{}{NMe_2} \leftrightarrow \overset{Cl}{\underset{}{H\overset{}{C}}} - \overset{}{\underset{+}{NMe_2}})$$ [259]. The final product can

be changed into a thioaldehyde in this reaction by allowing the Vilsmeier salt to react in situ with NaHS (see Scheme 7) [194]. Enamines

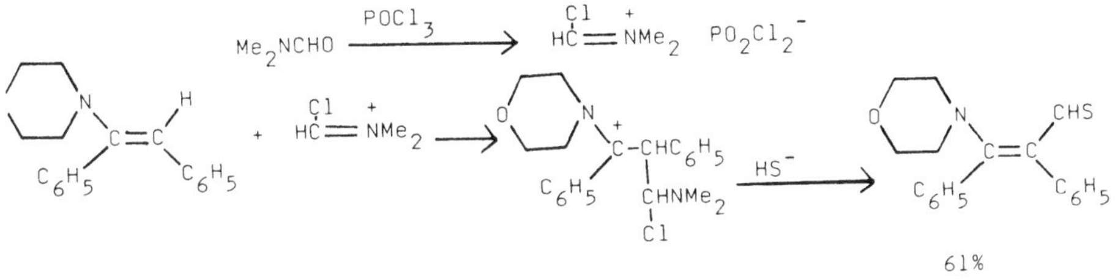

Scheme 7

will also react with an iminium salt similar to the Vilsmeier salt, namely N-(dichloromethylene)-N,N-dimethylammonium chloride (78), to give a similar addition product [195].

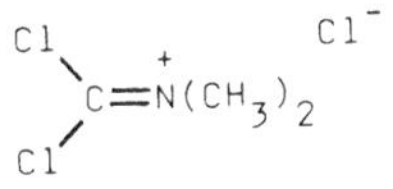

(78)

Nitrilium salts $(RC \equiv \overset{+}{N}R' \leftrightarrow R\overset{+}{C} = NR')$ can be used to alkylate enamines such as 39 to produce adducts such as 79 in good yields [199].

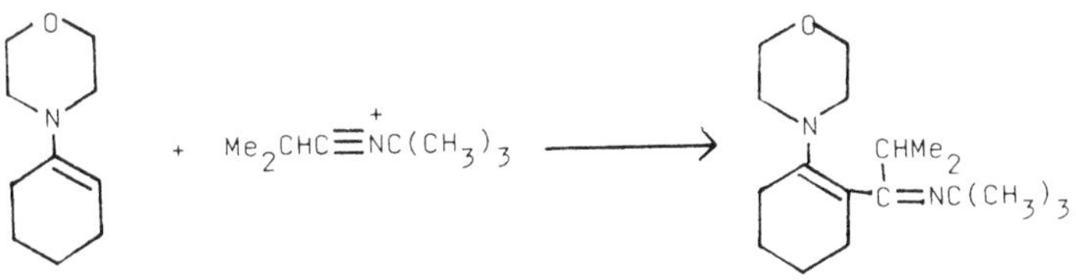

(39)   (79)

IV. ACYLATION

Acylation of enamines can take place on carbon or on nitrogen. In contrast to the N-alkylated products, the N-acylated products either are not stable or are acylating agents, so that C-acylation is the normal mode of reaction. The enamines are stronger bases than the acylated enamines, so that in acylation with acid chlorides half an equivalent of the enamine is lost to the acylation reaction by salt formation. Loss of enamine in this manner can be avoided by the addition of an organic base such as triethylamine [200]. It has also been shown that the less reactive morpholine enamines give better yields of acylated products [200]. The acylation of enamines with acid anhydrides or acid chlorides having no α-hydrogen atom appears to be a straightforward reaction at the enamine carbon. The reaction of acid chlorides having an α-hydrogen atom often involves a cycloaddition of the ketene formed by dehydrohalogenation of the acid chloride by the basic enamine. The initial product in these cases is a cyclobutanone derivative which may sometimes be isolated or which subsequently undergoes rearrangement to the normal C-acylated product.

A. Reaction with Carboxylic Acid Chlorides, Anhydrides, and Ketenes

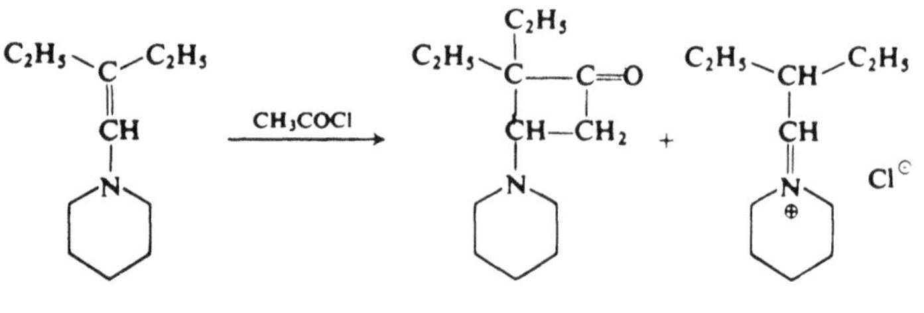

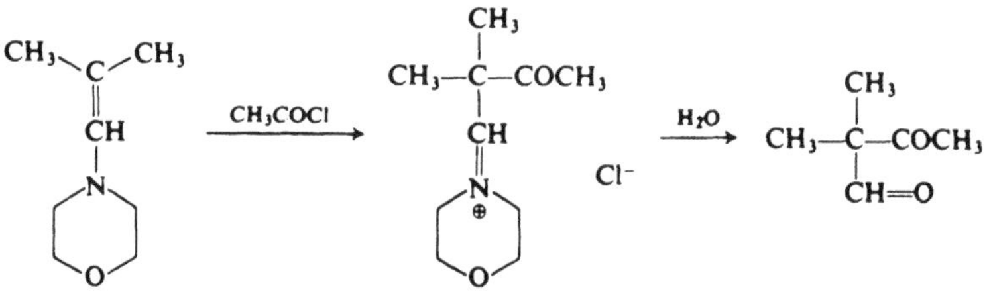

The reaction of the enamine (80) with acetyl chloride was reported [200] to afford no acyl derivative but the aminocyclobutanone (81) and the hydrochloride of the enamine. The acylation of aldehyde enamines was reinvestigated [201] and shown to proceed normally when the enamine is added to the acid chloride. The morpholine enamine of isobutyraldehyde (61), on being added to an ether solution of acetyl chloride, afforded the iminium salt (82), from which the ketoaldehyde (83) was obtained in 66% yield by hydrolysis [201]. This acylation might still be assumed to proceed via the aminocyclobutanone with subsequent rearrangement, but it was shown [201] that the hydrochloride salt of the aminocyclobutanone (84), prepared by an alternate method, was not rearranged under the reaction conditions. The intermediacy of 84 therefore seems unlikely. Under

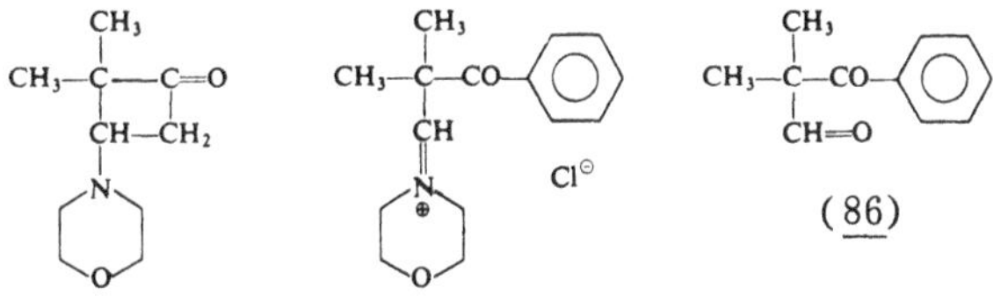

(84)   (85)   (86)

similar reaction conditions in dioxane, the enamine (61) was acylated by benzol chloride to give the iminium salt (85) and the ketoaldehyde (86) in 86 and 72% yields, respectively. This reaction was not affected by the addition of base, since the salt (85) was isolated even in the presence of triethylamine. When the enamine (87) was acylated with benzoyl chloride in the presence of triethylamine, triethylamine hydrochloride was precipitated instead of the iminium salt (88). Since hydrolysis of the reaction mixture gave the ketoaldehyde (90), it appears reasonable that the salt (88) lost the elements of HCl in this case to give the enamino ketone 1-N-morpholino-2-benzoyl-1-butene (89).

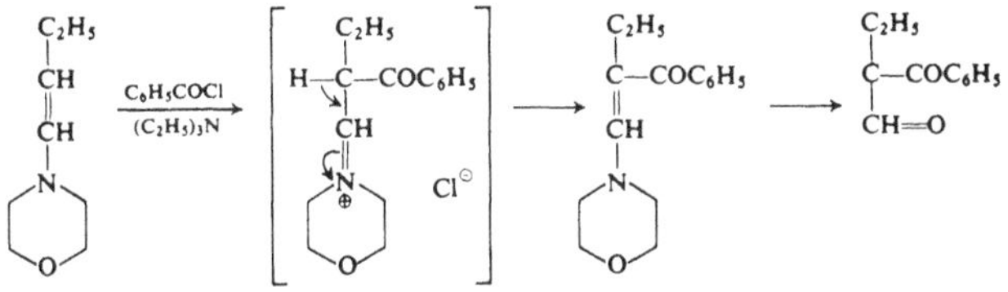

(87)   (88)   (89)   (90)

Other studies of C acylation of β,β-disubstituted aldehyde enamines have been reported [202]. The reaction of aldehyde enamines with ketenes has been well investigated [206–211] and shown to give cyclobutanone derivatives. A complete discussion of the cycloaddition

of ketene to all types of enamines is given in Chapter 7, Section III.A.
Most enamine acylations have been carried out with triethylamine as
an auxiliary base to prevent salt formation and consequent removal
of an equivalent amount of enamine from the reaction medium. If an
α-hydrogen is present in the acyl halide, acylations by this method
generally proceed through initial generation of ketene, which acts as
the acylating agent. When excess morpholine enamine is used as the
auxilary base, ketene is not formed and direct acylation by the acid
chloride is observed [222,223]. N-Methylpiperazine, with a built-in
extra basic amino group, has been used in this reaction, but with
limited success [246].

The stability of the cyclobutanone derivative is dependent on the
structures of the enamine and the ketene. Thus reaction of the enamine 8 with dimethyl ketene gave the thermally stable 3-dimethylami-
no-2,2,4,4-tetramethylcyclobutanone (91). Reaction of 8 with ketene,
on the other hand, gives the 3-dimethylamino-2,2-dimethylcyclobuta-
none (92), which on heating rearranges to the enamino ketone (93),
which is not the product that would be expected from acetylation of
8. An alternate mode of ring opening is shown by the amino cyclobutanone (94) derived from reaction of the piperidine enamine of

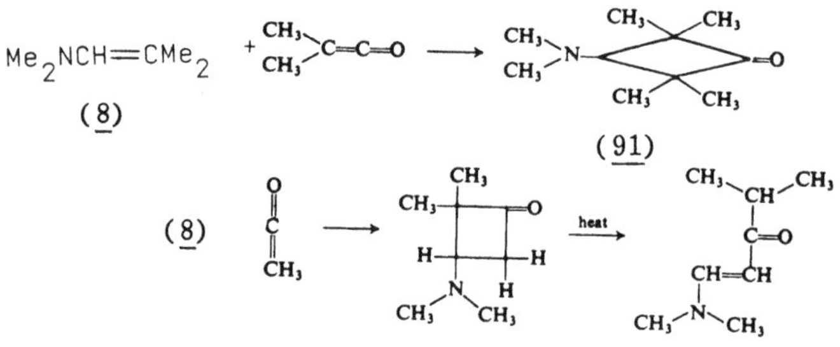

butyraldehyde and dimethyl ketene. In this case, the enamino ketone
(95) formed is the product that would be expected from acylation of
the enamine with isobutyryl chloride.

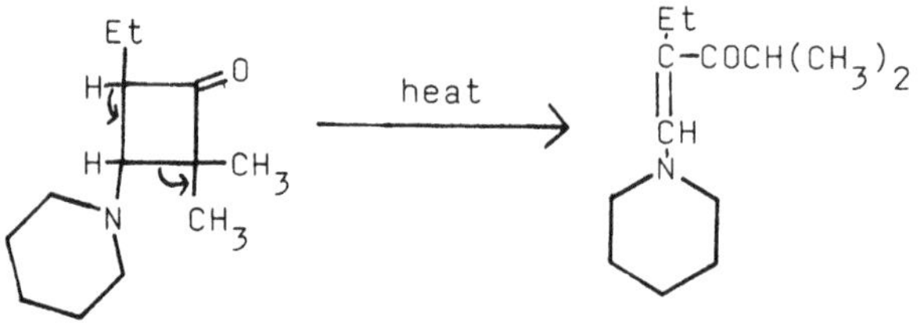

Unsymmetrical acyclic ketone enamines are acylated on the more highly substituted β-carbon if the substituent is an aromatic ring [203], but in the cases of nonaromatic substituents, the acylation takes place at the least substituted β-carbon, often in a quantitative yield [204,205]. If the most substituted isomer is present, it acts only as hydrogen chloride scavenger [205]. An example of this regiospecific reaction is treatment of unsymmetrical acyclic enamine 96

$$(CH_3)_2CHCH_2C{=\!\!=}CH_2 \quad + \quad C_6H_5\overset{O}{\overset{\|}{C}}Cl \quad \longrightarrow \quad (CH_3)_2CHCH_2C{=\!\!=}CH{-}\overset{O}{\overset{\|}{C}}C_6H_5$$

(96)   (97)

with benzoyl chloride to give acylated enamine 97 in a quantitative yield [204]. Morpholine enamines are preferable for acylations because these enamines rarely give rise to self-condensation products, whereas dimethylamine, diethylamine, and pyrrolidine enamines undergo self-condensation reactions [205]. The enamines derived from cyclic ketones are readily acylated with acyl and aroyl halides [1,200, 203,212—221]. The morpholine enamines give the best yields. Enamine 39 undergoes acylation to give, after acid hydrolysis of the intermediate enamino ketone (98), the 1,3-diketone (99) in high yield.

$$\underset{(39)}{\text{(structure)}} \xrightarrow[(C_2H_5)_3N]{RCOCl} \underset{(98)}{\text{(structure)}{-}COR} \longrightarrow \underset{(99)}{\text{(structure)}{-}COR}$$

(39)   (98)   (99)

The reported synthesis of several natural products has involved the acylation of enamines (see Chapter 9). The syntheses of antineoplastic cycloheximide and related compounds [228—230], steroids [231], optically active lupinine [232], and both iboga and epiiboga alkaloids [233] are examples of this.

The acylation of enamines has been used with long-chain acid chlorides [224]. The preparation and elongation of fatty acids [215, 225,226] and substituted aliphatic acids [227] have been carried out in this way. This is because cleavage of 99 with strong sodium hydroxide gives ketoacid 100, which is readily reduced by the Wolff-Kishner method to the saturated acid. A similar sequence of reactions can be carried out starting with the cyclopentanone enamine, and this method allows lengthening the chain of a carboxylic acid by five or six carbon atoms [212,213].

(99) $\xrightarrow{\text{NaOH}}$ HOOC-$(CH_2)_5$-COR $\xrightarrow[\text{NaOH}]{\text{NH}_2\text{NH}_2}$ HOOC-$(CH_2)_6$-R

(100)

The dicarboxylic acid chlorides from sebacic and azelaic acid react with 2 moles of enamine to give the tetraketone 101, which on base cleavage and reduction gives a dicarboxylic acid with chain length increased by 12 carbon atoms. In this way, tuberculostearic acid has been prepared [216,217].

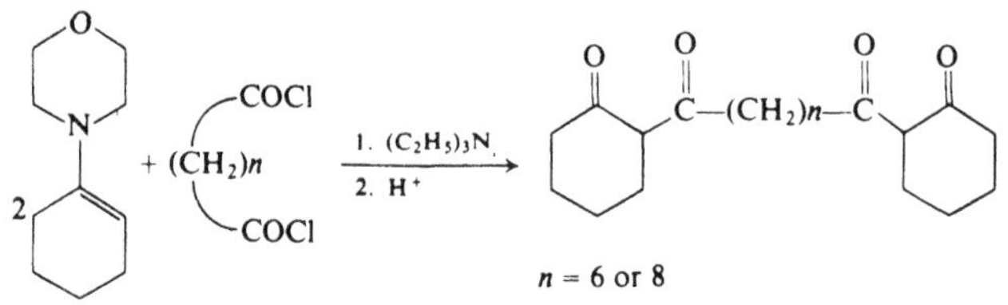

$n = 6$ or 8

(39)   (101)

Dicarboxylic acid dichlorides with less than seven carbon atoms do not always react to give tetraketones similar to 101, but instead undergo an intramolecular acylation [200] to give on hydrolysis the vinylogous acid anhydride (102), e.g., from succinyl chloride and the enamine 39. An example of a dicarboxylic acid dihalide which,

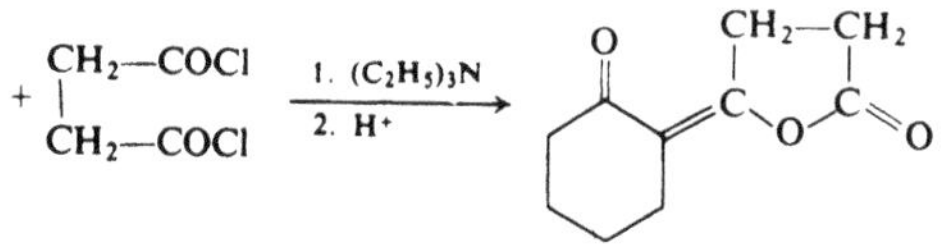

(102)

when allowed to react with an enamine, does lead to a tetraketone has been reported in oxalyl bromide [234].

In the acylation of enamines derived from 3-substituted cyclohexanones, 6-acylated products were favored over 2-acylated products [235].

Anhydrides have also been employed [1,256]. The mixed anhydride of acetic and formic acid reacts with the enamine 39 to give a 50% yield of 2-hydroxymethylene cyclohexanone (103). Acylation of

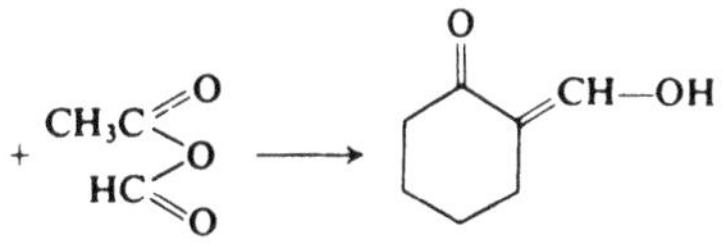

(103)

N,N-dimethylamino, piperidino, or morpholino enamines of isobutyr-aldehyde with trichloroacetic anhydride in tetrahydrofuran solvent at room temperature gave α-trichloromethyl-β-trichloroacetyl adducts of the enamine [255]. This reaction proceeds through initial β-acet-ylation by the anhydride to give the acetylated iminium trichloroace-tate. This is followed by decarboxylation resulting in α-trichloro-methylation. The yields range from 47 to 62%.

   Trifluoroacetic anhydride when allowed to react with enamines in tetrahydrofuran at room temperature gave β-acylation products [257]. Subsequent treatment of these products by increased temperature or with silica gel or alumina caused formation of 1,3-oxazines.

   Although esters do not usually react with enamines and can, in fact, be substituents in the azeotropic preparation of enamines, they can be used in acylation reactions when these involve intramolecular cyclizations. Such reactions have been observed even at room tem-perature when they lead to the formation of five- and six-membered vinylogous lactams (see Scheme 8) [236]. This type of reaction has also been used in the synthesis of azasteroids and alkaloids [237–239].

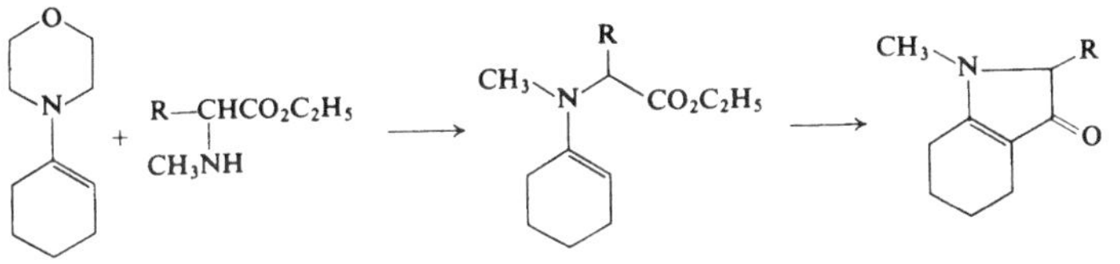

Scheme 8

Acylation of enamines of cyclic ketones with acid chlorides having an α-hydrogen in the presence of triethylamine usually proceeds via the ketene and subsequent cycloaddition [240,241]. The intermediate cyclobutanone is then opened to give the enamino ketone which is hy-drolyzed to the 2-acyl cyclohexanone. In the case of enamines of larger cyclic ketones, the alternate mode of the cyclobutanone opening predominates, with the formation of ring-expanded 1,3-diketones upon hydrolysis [222,242,243]. A complete discussion of cycloaddition re-actions of ketene with enamines is found in Chapter 7, Section III.A.

Acylation of the enamine (39) with α,β-unsaturated acid chlorides has been shown [244] to give bicyclo[3.3.1]nonan-2,9-diones. Acryloyl-chloride on reaction with the enamine [39] and subsequent hydrolysis gave bicyclo[3.3.1]nonan-2,9-dione (104). Mechanistic studies indicate reversible N-acylation takes place first followed by C-alkylation of the β-carbon of the enamine by the β-carbon of the acyl halide [2, 244,245]. This C-alkylation reaction takes place stereoselectively from the axial side of the enamine double bond in a cyclohexyl system (see Chapter 1, Section III.B and Chapter 7, Section III.A) [2,245].

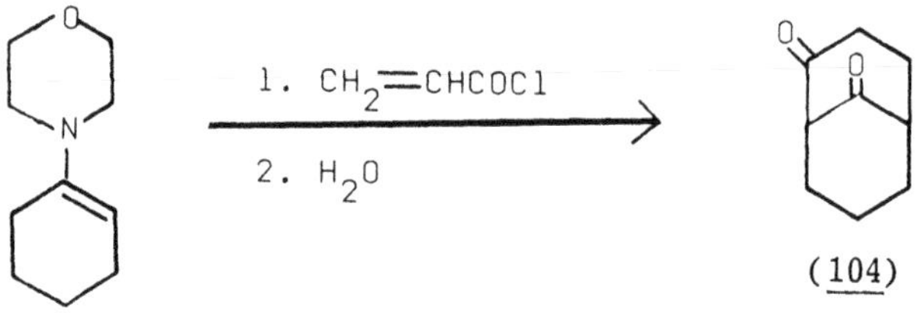

$$1. \ CH_2\!\!=\!\!CHCOCl$$
$$2. \ H_2O$$

(104)

(39)

Acylation of the heterocyclic enamines, e.g., 1,3,3-trimethyl-2-methyl-eneindoline (105) and 2-benzylidene-3-methylbenzothiazoline (27), takes place normally at the methylene carbon [69,152], with both acyl and aroyl chlorides in the presence of base.

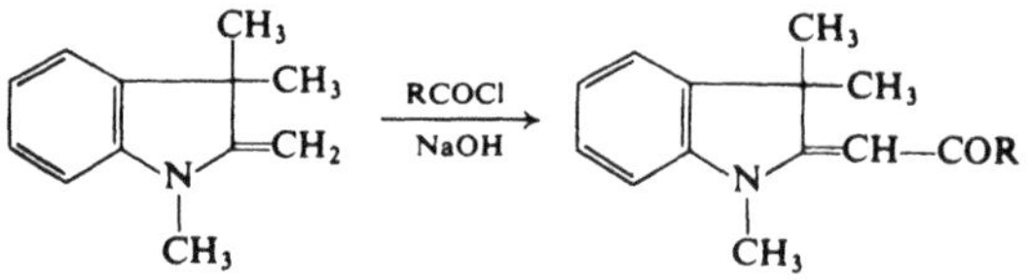

RCOCl, NaOH

(105)

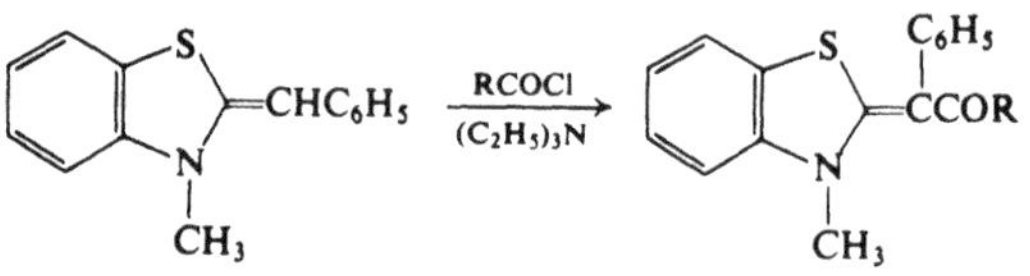

RCOCl, $(C_2H_5)_3N$

(27)

The acylation of the enamino ketone products obtained by initial acylation of an enamine can take place on oxygen or on carbon. Although reaction at nitrogen is a possibility, the N-acylated products are themselves acylating agents, and further reaction normally takes place. The first reported acylation of enamino ketones [200] was that of 106, prepared by acylation of enamine 39, which was shown to have undergone O-acylation because on mild hydrolysis the enol ester (107) could be isolated. A similar reaction took place with other aliphatic acid chlorides [213] and with dibasic acid chlorides (e.g., with succinyl chloride to give 102 above).

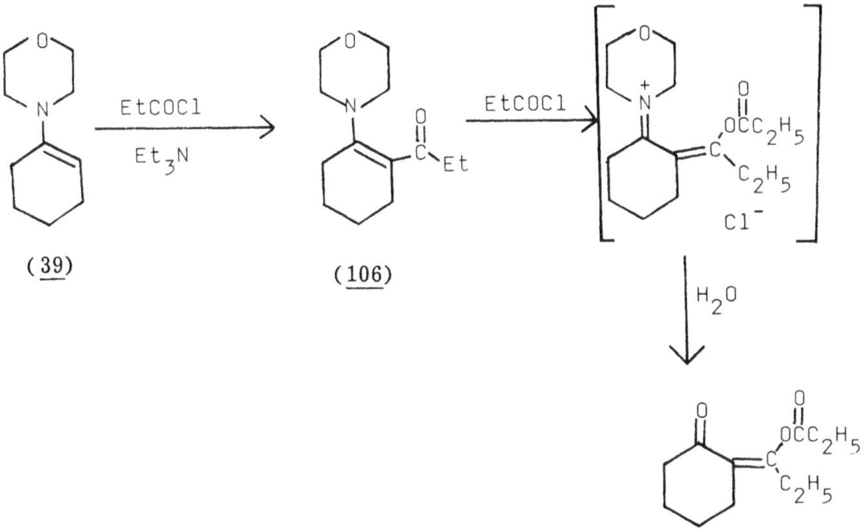

(39)   (106)

(107)

Acylation of an enaminoketone with a second mole of aromatic acid chlorides was believed to occur on carbon. The dibenzoylation of the enamine (39) with benzoyl chloride in the presence of triethylamine has, however, been shown to give a mixture of three products [248, 250]. The major components are the *cis* and *trans* isomers of the O-acylated enamino ketone (108a and b), and the minor isomer is the 2,6-diacylated enamine (109). It would thus appear that O-acylation

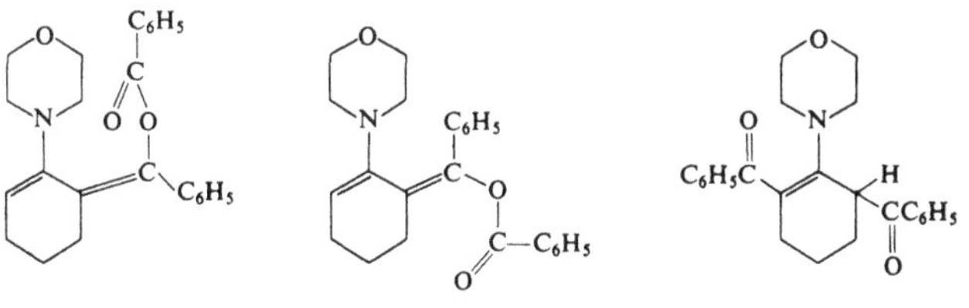

(108a)   (108b)   (109)

is the normal course of the acylation of enamino ketones. Surprising-
ly, the enamino ketones 110 and 111 undergo reaction with acid chlorides not having an α-hydrogen (e.g., benzoyl and pivalyl chlorides)
to give the products of C-acylation [112]. This result has been ra-

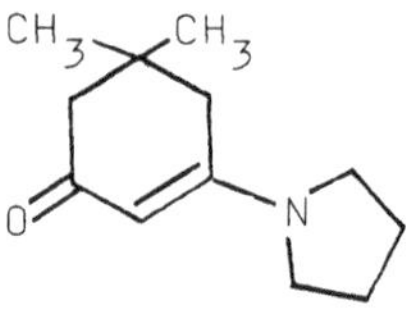

(110) 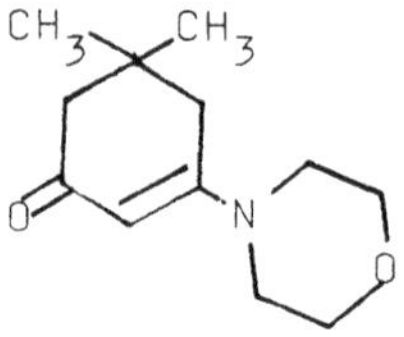   (111) 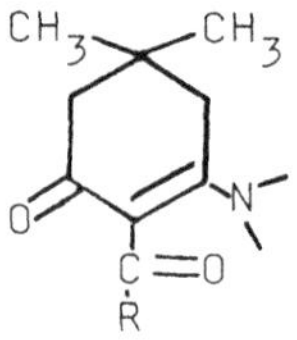   (112)

tionalized by consideration of the stability of the intermediate iminium
salts [251]. O-Acylation would give 113, whereas C-acylation would
give 114. The latter can undergo loss of a proton to give the product, whereas 113 cannot, but can revert to reactants, so that in this
case initial O-acylation may occur, but the reaction is reversible and

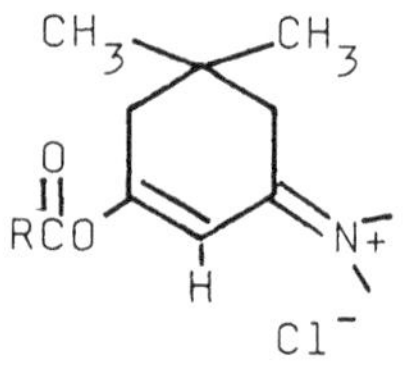

(113) 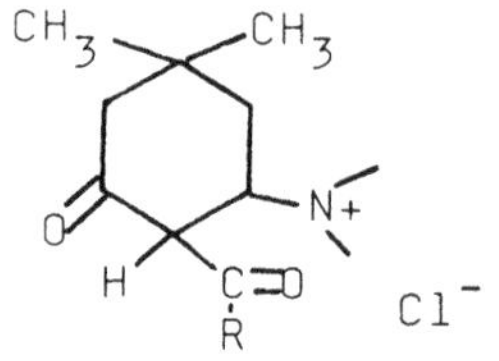   (114)

does not lead to products. The O-acylated salt can be isolated [252].
Thus benzoylation of heterocyclic enamino ketone 115 gave O-benzoyl-
ated salt 116. Further evidence for O-acylation is provided by reaction of enamino ketone 110 with trichloroacetyl chloride [69] to give

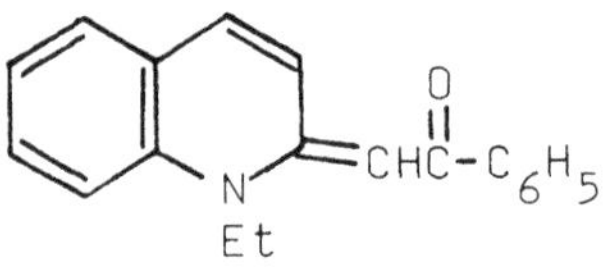

(115) 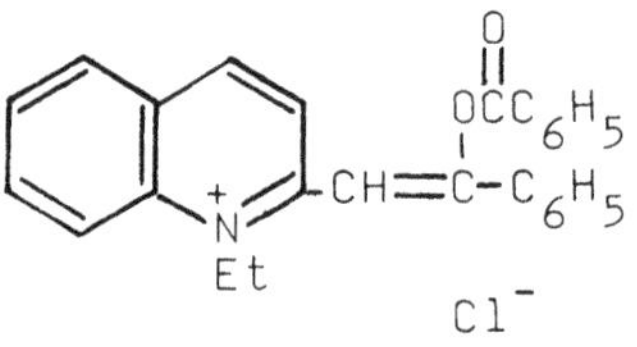   (116)

1-(3-chloro-5,5-dimethyl-2-cyclohexen-1-ylidene)pyrrolidinium chloride (117). The latter must have been formed by a reaction sequence involving initial O-acylation and subsequent addition of chloride ion to the cation with expulsion of trichloroacetate (i.e., the better leaving group). Salt 117 is also formed by reaction of enamino ketone 110 with phosphorus pentachloride [69].

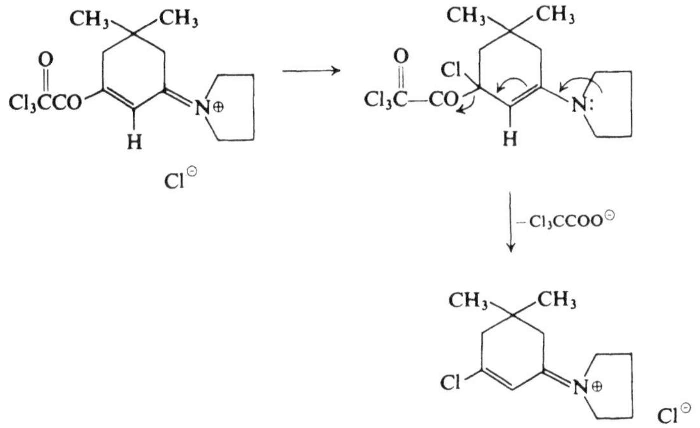

(117)

It is of interest to note that 2,4,6-trinitrochlorobenzene reacts similarly with 110 to give the cation of 117 isolated as the perchlorate. In the reaction of enamino ketone 111 with trichloroacetyl chloride [253,254], the chloroiminium cation undergoes reaction with trichloromethyl anion (formed by decarboxylation of the expelled trichloroacetate) to give the trichloromethyl derivative (118).

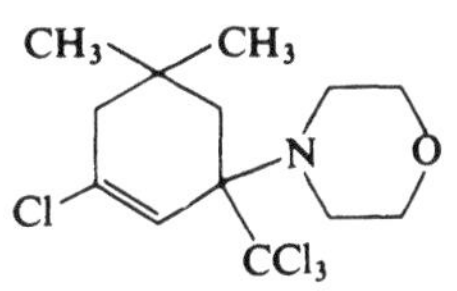

(118)

The enamino ketone (110) reacts with acetyl chloride [251] to give the C-acylated product (112), whereas the corresponding reaction with the enamino ketone (111) gives only the hydrochloride of 111. As neither 110 nor 111 undergoes reaction with ketene, this suggests that 110, the enamino ketone derived from pyrrolidine, reacts as an enamine toward acetyl chloride, but 111, derived from the weaker base morpholine, does not. It is still a strong enough base to remove the elements of hydrochloric acid from acetyl chloride. Enamino ketone 119 and enamino ester 120, on the other hand, react with 2 moles of ketene in the manner expected to give the α-pyrone derivatives 120 and 122, respectively [206].

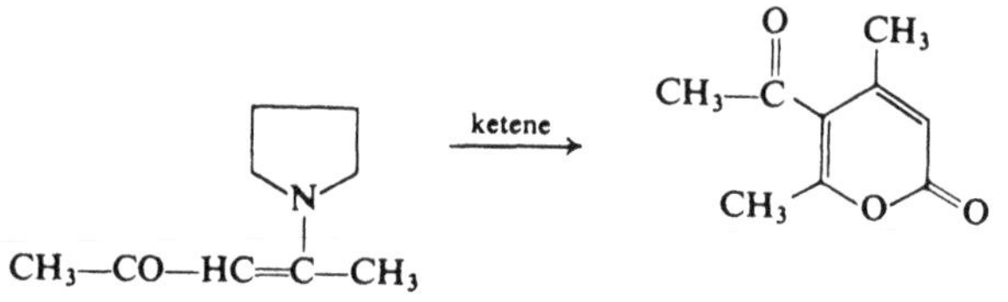

CH$_3$—CO—HC=C—CH$_3$

(119)   ketene   (120)

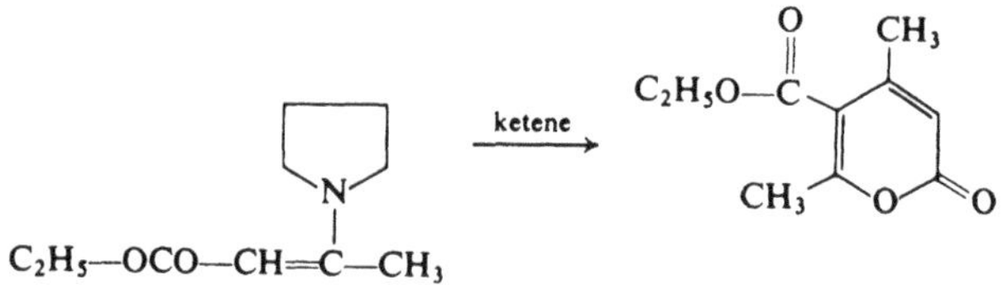

C$_2$H$_5$—OCO—CH=C—CH$_3$

(121)   ketene   (122)

Hexafluoropropene oxide (123) reacts readily with morpholine

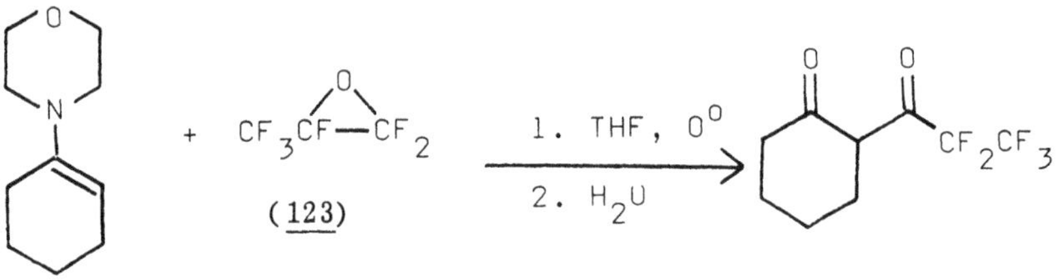

+   CF$_3$CF—CF$_2$ (123)   1. THF, 0°   2. H$_2$O   CF$_2$CF$_3$

(39)   (124)

enamines (such as 39) to give, after hydrolysis, fluorinated diketones (such as 124) [258]. The initial step of the reaction is the isomerization of hexafluoropropene oxide (123) into pentafluoropropionyl fluoride by the basic enamine. This is followed by β-carbon acylation by the acyl fluoride.

    Treatment of aldehyde or cyclic ketone enamine with 1 mole of phosgene gives, after hydrolysis, amidealdehyde 126 with β,β-disubstituted acyclic enamine 125 [260,261]. With cyclic enamine 39, triketone 127 or one of its derivatives is obtained [260,261].

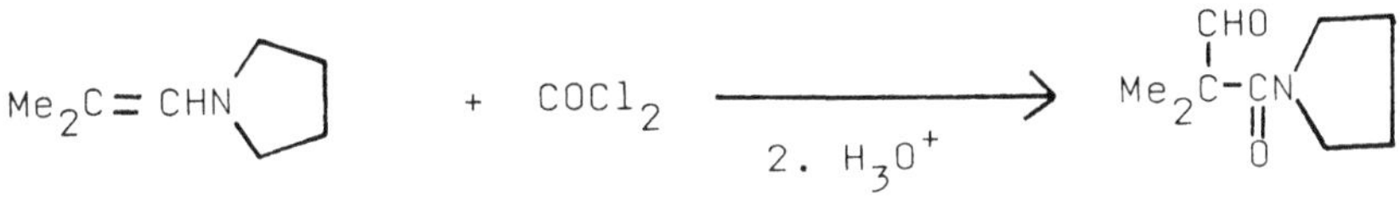

$Me_2C = CHN$ (pyrrolidine ring) + $COCl_2$ $\xrightarrow{\hspace{1cm}}$ Me_2C-CN (pyrrolidine ring) with CHO and O

2. H_3O^+

(125)   (126)

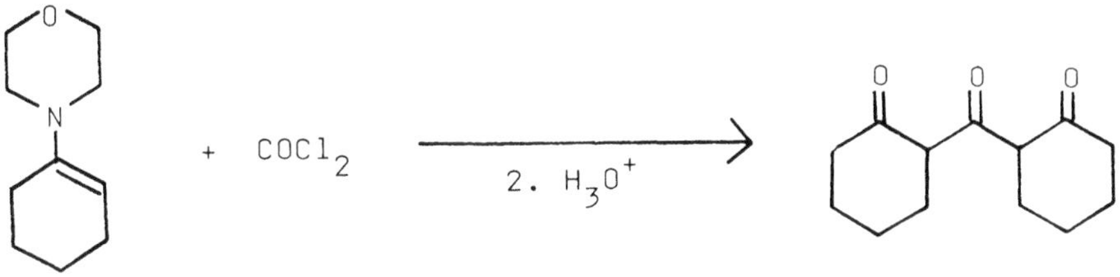

(morpholine enamine of cyclohexanone) + $COCl_2$ $\xrightarrow[\text{2. } H_3O^+]{\hspace{1cm}}$ (127 structure)

(39)   (127)

Enamines have been reported to react with diethyl pyrocarbonate to give aldehydo or keto esters [262].

When either cyclic ketone enamines or acyclic aldehyde enamines are treated with trialkyl orthoformates, β-aldehydo- or β-ketoacetals

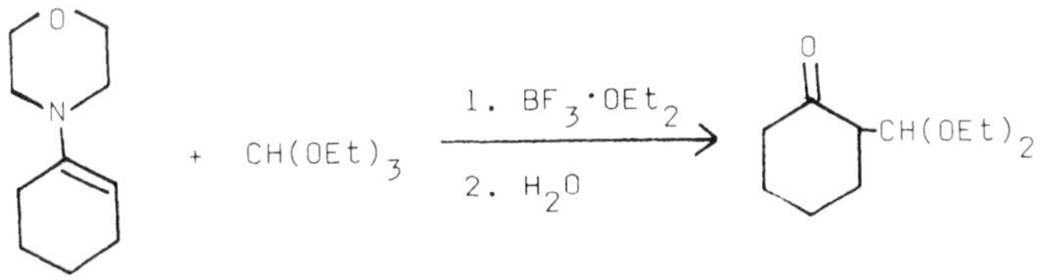

(morpholine enamine of cyclohexanone) + $CH(OEt)_3$ $\xrightarrow[\text{2. } H_2O]{\text{1. } BF_3 \cdot OEt_2}$ (128 structure) $-CH(OEt)_2$

(39)   (128)

are formed [263,264]. For example, enamine 39 produces β-ketoacetal 128 in an 85% yield at 0°C [263,264]. Acetals react with enamines to give β-alkoxyketones or β-alkoxyaldehydes [264].

B. Reaction with Sulfonyl and Sulfenyl Derivatives

Alkyl sulfonyl chlorides, having an α-hydrogen atom, react with enamines derived from aldehydes and cyclic ketones in the presence of triethylamine to give cyclic sulfones. A complete discussion of these cycloaddition reactions is found in Chapter 7, Section IV.B.

Acyclic sulfones have been isolated from these reactions [110,111, 127,265–274]. For example, reaction of enamine 15 with phenylmeth-ane sulfonyl chloride [266] gave benzyl-2-oxocyclohexyl sulfone. The

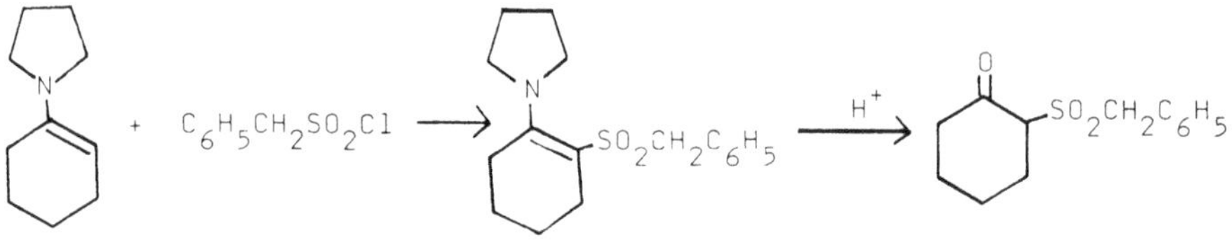

(15)   (129)

formation of acyclic sulfones is favored by increasing substitution at
the α-carbon of the sulfonyl chloride and also of the enamine [277,
278].

The aromatic sulfonyl chlorides which have no α-hydrogen and
thus cannot form sulfenes give acylic sulfones. Thus, 1-piperidino-
propene on reaction with benzene sulfonyl chloride [277] gave 2-ben-
zenesulfonyl-1-piperidinopropene (130). Similarly, the enamine (15)
reacts with p-toluenesulfonyl chloride to give the 2-p-toluenesulfonyl-
cyclohexanone (131) on hydrolysis [278]. Other examples of this have
been reported [275].

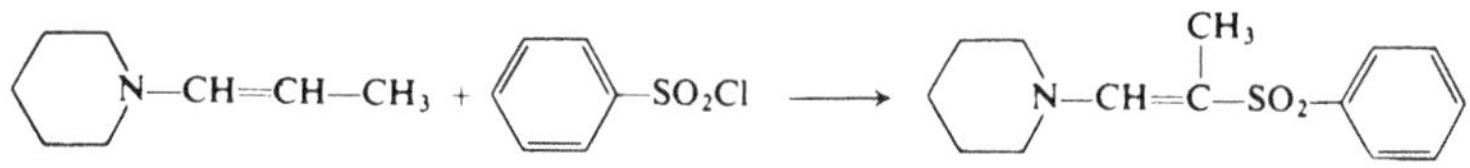

(130)

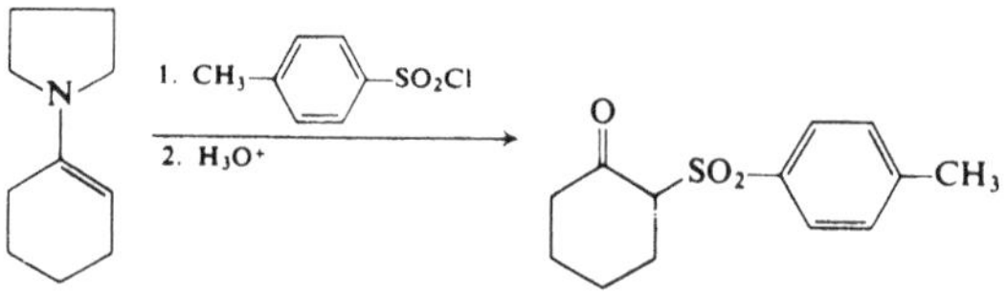

(15)   (131)

Treatment of cyclic enamines with p-toluenethiosulfonate (132)
and triethylamine in acetonitrile solvent gave, after hydrolysis, an

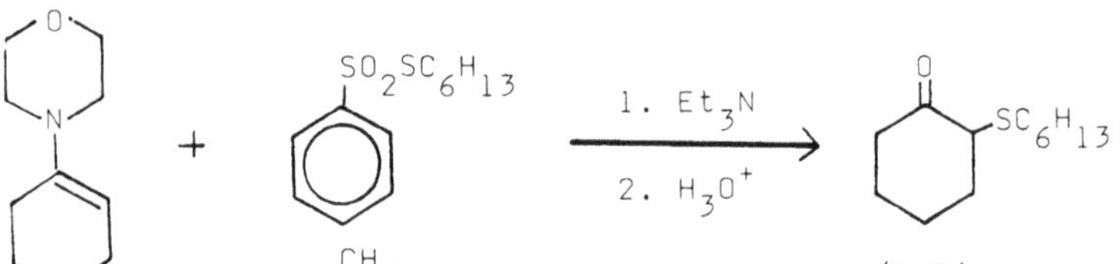

(39)   (132)   (133)

α-alkylthiolated ketone [279]. For example, enamine 39 forms ketone
133 in a 70% yield when allowed to react with 132 [279].

The reaction of enamines with p-toluenesulfinic acid and an activating agent such as phenyl phosphorodichloridate resulted in the
formation (after hydrolysis) of β-keto sulfoxides [280].

1-(N-Morpholino)cyclohexene (39), which possesses a β-hydrogen,
reacts with a mixture of diphenyldiazomethane and sulfur dioxide to

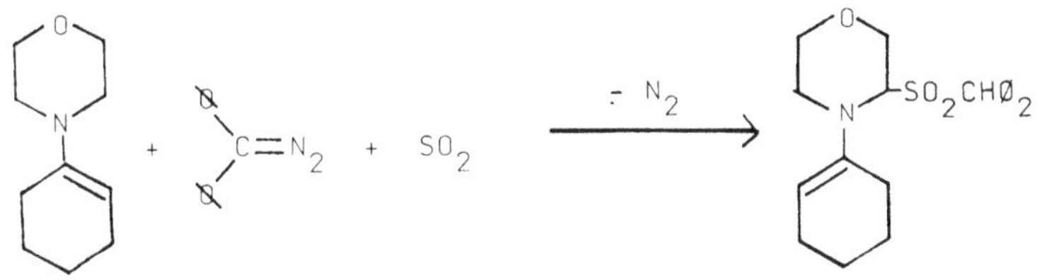

(39)   (134)

give adduct 134 [281]. Sulfonyl imides (RN = SO$_2$), which are isoelectronic with sulfur trioxide and sulfenes, attack enamines to produce sulfonamides [282].

The pyrrolidine enamine of cyclohexanone (15) has been shown
to react with o-, m-, and p-nitrobenzenesulfenyl chlorides [278]. A
mixture of the 2-mono- and 2,6-bis(o-, m-, and p-nitrophenylsulfenyl)
cyclohexanones is obtained on hydrolysis. Only the monosubstituted
derivative (135) [155] is obtained from 6-methyl-1-pyrrolidino cyclohexene and o-nitrobenzenesulfenyl chloride. An interesting variant of this

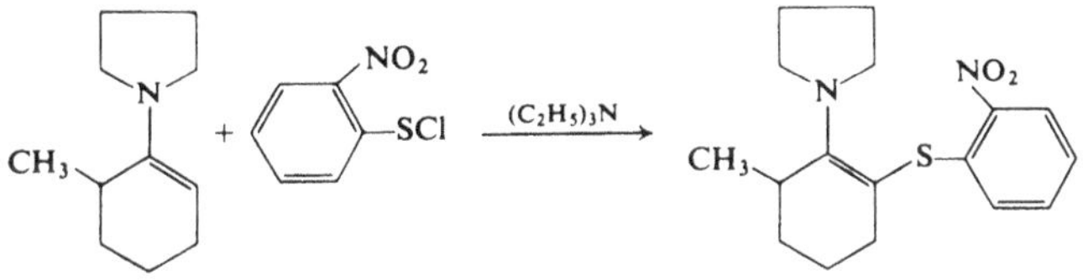

(135)

reaction is the formation of 2-thiaadamantane-4,8-dione by hydrolysis
of the reaction product of the bispyrrolidine enamine of bicyclo[3.3.1]
nonane-2,6-dione with sulfur dichloride [106].

A convenient method for preparing α-sulfenylated ketones is by
treating enamines with N-phenylthiophthalimide followed by acid hydrolysis [283]. The reaction of enamines with N-sulfonylisonitrile
dichloride or S-methyl-N-sulfonylthiocarbimidayl chloride leads to an
N-sulfonylketenimine [286,287]. A Michael-type addition reaction occurs when enamines are treated with (phenylsulfonyl)propadiene
[390].

C. Reaction with Isocyanates and Isothiocyanates

The reaction of isocyanates with enamines disubstituted at the β-carbon gives β-amino-β-lactams [284,285]. Thus the enamine (43) reacted exothermally with phenylisocyanate to give (20) dimethyl-1-phenyl-4-dimethylamino-2-acetidinone (136), which was converted by acid hydrolysis to 2-formyl-2-methyl propionanilide (137). For a

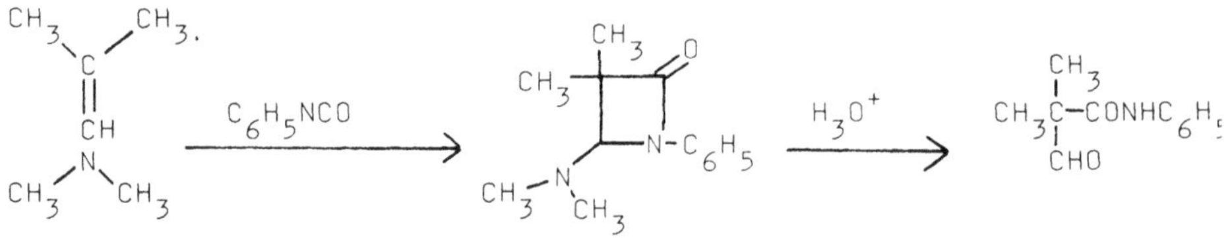

(43)   (136)   (137)

comprehensive discussion of the cycloaddition reactions of isocyanates and isothiocyanates with enamines, see Chapter 7, Section IV.C. Enamines of cyclic ketones do not form cycloaddition products, but give the mono- or dicarboxanilides [288–292]. Thus enamine 39 on reaction with one equivalent of phenyl isocyanate gave 138. Treatment of 39 with two equivalents, or 138 with one equivalent, of phenyl isocyanate gave the 2,6-disubstituted product 139. Mild acid hydrolysis of 138 and 139 produced the corresponding cyclohexanone(2-mono- and 2,5-di)carboxanilides [288]. Proof that the second mole of phenyl

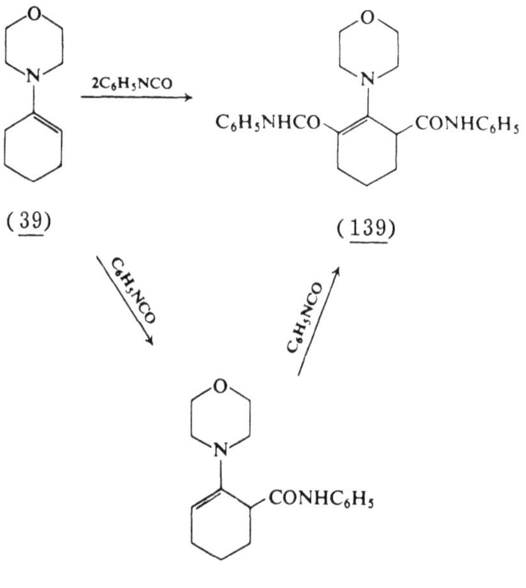

(39)   (139)

(138)

isocyanate did not react at the nitrogen of 138 was provided by the reaction of the enamine from 2-methylcyclohexanone, which gave only the monocarboxanilide on reaction with excess phenyl isocyanate.

Phenyl isocyanate, when allowed to react with the isomeric enamines of 2-substituted cycloalkanones (152a and b) at 5°C, reacts

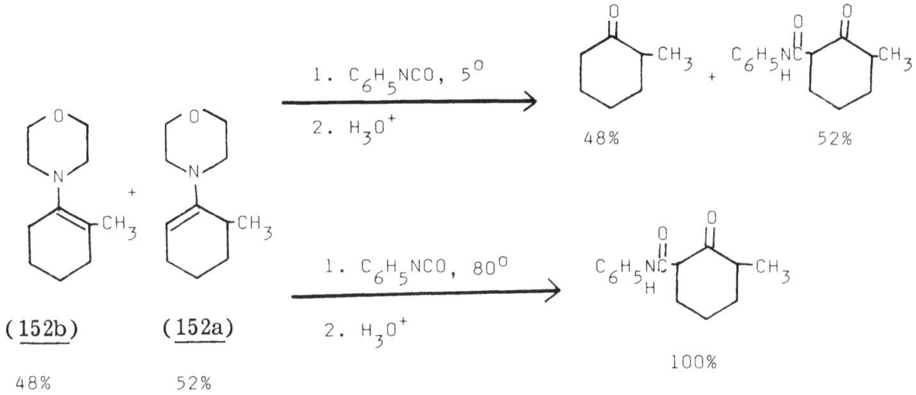

Scheme 9

quantitatively with the less substituted enamine (152a), but it does not react with the tetrasubstituted enamine (152b) (see Scheme 9) [290]. At the higher temperature of 80°C where equilibration between the isomeric enamine can take place, essentially all of the enamine reacts with phenyl isocyanate to form the product adduct.

Enamines of acyclic ketones when treated with phenyl isocyanate produce carboxanilides also [273,295].

The difunctionality of the enamines of cyclic ketones toward phenyl isocyanate provides the ideal situation for a potential polymer. The properties of the polyamides produced by addition of various diisocyanates to the enamines of cyclic ketones have been reported [294].

The reactions of the enamines of cyclic ketones with alkyl isocyanates [296], α-naphthyl isocyanate [297], acyl isocyanates [296], phenyl isothiocyanates [293,296], and acyl isothiocyanates [296,298] have also been reported. The products are the corresponding carboxamides. The products from the isothiocyanates have been utilized as intermediates in the preparation of various heterocyclic compounds [299].

The rate of addition of isocyanates to enamines increases with decreasing basicity of the isocyanate nitrogen and with increasing basicity of the enamine nitrogen [296].

An enamine in conjunction with an isocyanide and methanol has been used to help esterify an N-protected amino acid under very mild conditions [300].

V. REACTION WITH OTHER ELECTROPHILES

A. Halogenation

Halogenation of a vinyl quaternary ammonium salt such as neurine (140) involves addition of a whole mole of halogen across the carbon-

$$CH_2{=}CHN^+(CH_3)_3 \quad HO^-$$

(140)

carbon double bond [312]. But halogenation of enamines is formally analogous to protonation with salt formation. Thus steroidal enamine 141 undergoes bromination [301] to give β-bromo iminium bromide 142, which is readily hydrolyzed to β-bromo aldehyde 143.

This method of bromination has been employed in the selective bromination [302] of ketone 144. While direct bromination results in bromination not only in the position alpha to the ketone but also in the aromatic ring, bromination of enamine 145 and subsequent hydrolysis gave only monobrominated product 146.

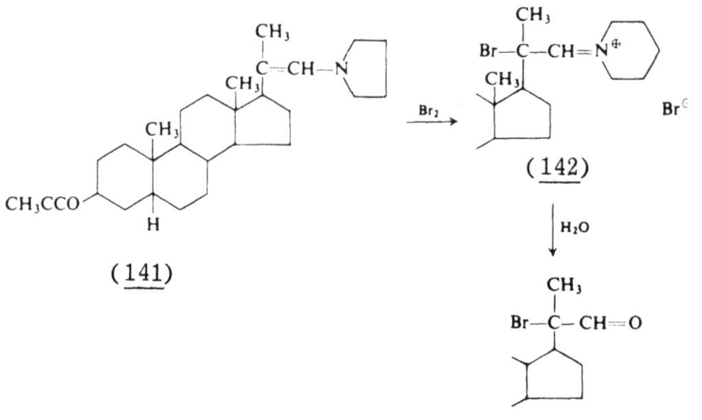

(141)

(142)

(143)

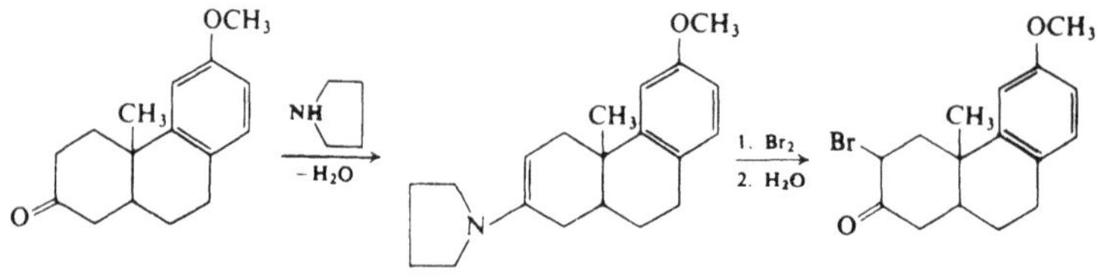

(144)

(145)

(146)

This bromination reaction has also been applied to the syntheses of a number of α-bromoaldehydes [303] and 2-α-bromocholestanone [304] along with other systems [305–310,336]

Chlorination of enamines has also been reported [69,311]. Thus trichlorovinylamine 147 has been chlorinated to 148, which was not isolated but treated with hydrogen sulfide to give the thioamide (149)

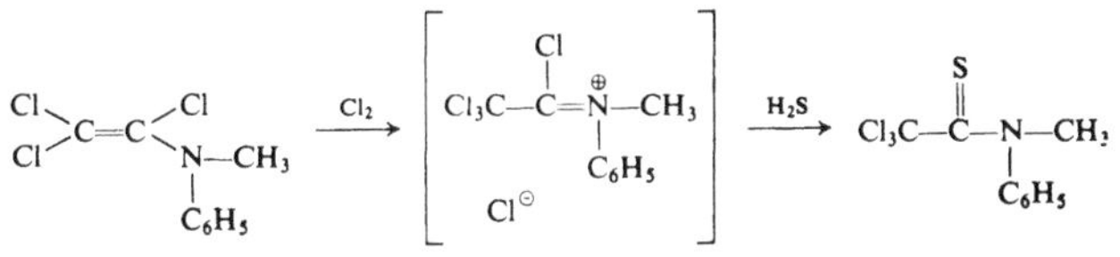

(147)   (148)   (149)

[311]. A stable chloroiminium chloride (151) has been isolated from the chlorination of the heterocyclic enamine (150) [69].

(150)   (151)

Halogenation of unsymmetrical enamines obtained from ketones is regioselective, with the least substituted enamine providing the kinetically controlled product by its faster reaction with the halogen [313–317]. For example, enamine 152 exists as both the tri-(152a) tetrasubstituted (152b) isomers, with the trisubstituted being the predominant species present in a 52-to-48 ratio of 152a and 152b (see Chapter 1, Section III.B). Addition of 0.52 mole of bromine gave the product

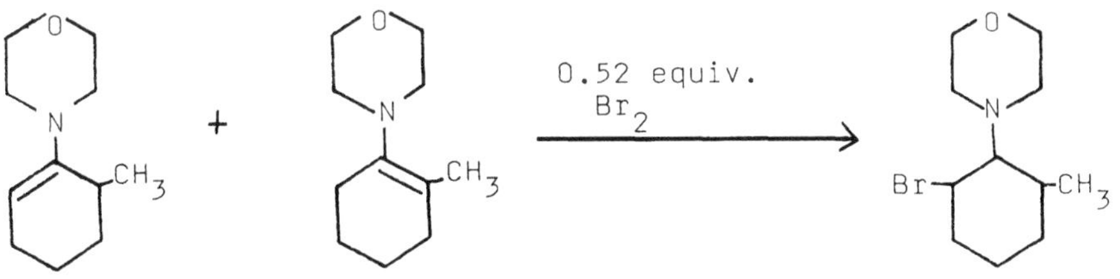

(152a)   (152b)   (153)

from bromine addition to the trisubstituted enamine (153) as a solid [318]. The other isomer can be isolated from the filtrate. This same feat can be accomplished by C-protonation of the enamine isomeric mixture followed by regioselective deprotonation of the iminium ion's least substituted carbon (see Chapter 6, Section IV). The pure least substituted enamine isomer obtained in this manner can then be brominated in the usual manner [316].

Chlorination of enamines with hexachloroacetone is also regiose-lective, with the least substituted enamine isomer being the major tar-get for chlorine substitution [318,319].

Sometimes halogenation of enamines will be followed by a rear-rangement brought about by treatment of the brominated enamine with

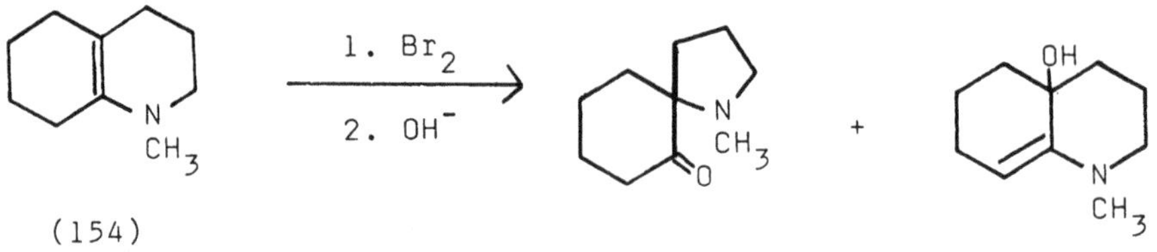

(154)

(154)   (155)   (156)

base [320,321]. This is illustrated with heterocyclic enamine 154 giving products 155 and 156 [321].

Stable β-chloro and bromo enamines have been obtained by re-action of enamines with the corresponding N-halosuccinimides [322–324]. Other halogenation reagents with the halogen attached directly to a heteroatom are N-chloro-p-toluenesulfonamide [325] and dialkyl-chloramines [326]. The latter reagent shows rearrangement products as well as normal addition products, probably via a cyclic dialkyl-aziridinium ion intermediate. Fluorination of enamine 39 with 1-fluoro-2-pyridone (157) gives product 158 under mild conditions [327].

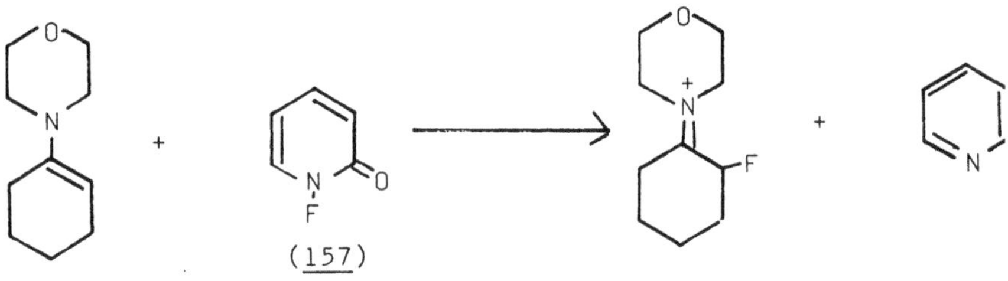

(39)   (157)   (158)

The onium ion, bromodimethylsulfonium bromide (159), can be readily prepared as a crystalline solid by allowing bromine to react with dimethyl sulfide. This reagent reacts with enamine 39 at room

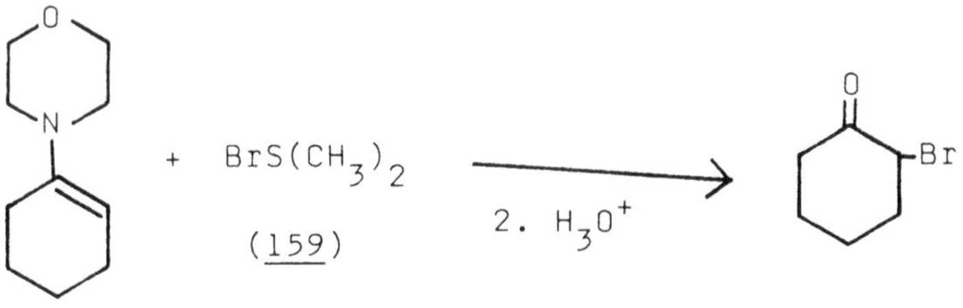

| (39) | (159) | (160) |

temperature to give <u>160</u> in an 80% yield [329]. The reaction of dimeth-yl (succinimido)sulfonium chloride with enamines has been used to synthesize chlorinated enamines [324,329].

Enamines from steroidal ketones have been fluorinated by means of perchloryl fluoride [330–334] to give the α-fluorinated ketones. Fluorination of an enamine with CF_3OF gave good yields of fluoroke-tones [335].

B. Reaction with Cyanogen Halides

The pyrrolidine enamines of the cyclic ketones cyclopentanone through cyclononanone have been reacted with cyanogen chloride to give high yields of the corresponding α-cyanoketones on hydrolysis [338]. Thus enamine <u>15</u> on reaction with one equivalent of cyanogen chloride in the presence of one equivalent of triethylamine in dioxane gave a 60% yield of 2-cyanocyclohexanone (<u>161</u>) on hydrolysis. The corre-sponding piperidine and morpholine enamines are less satisfactory in this reaction and gave yields of only 19 and 6% of <u>161</u>, respectively. The pyrrolidine enamine of 2-methyl-cyclohexanone and 2-phenylcyclo-hexanone gave 2-cyano-6-methylcyclohexanone and 2-cyano-6-phenyl-cyclohexanone in 66 and 77% yields, respectively. Only in the 2-methylcyclohexanone case was a small amount of 2-cyano-2-methyl-cyclohexanone observed [338].

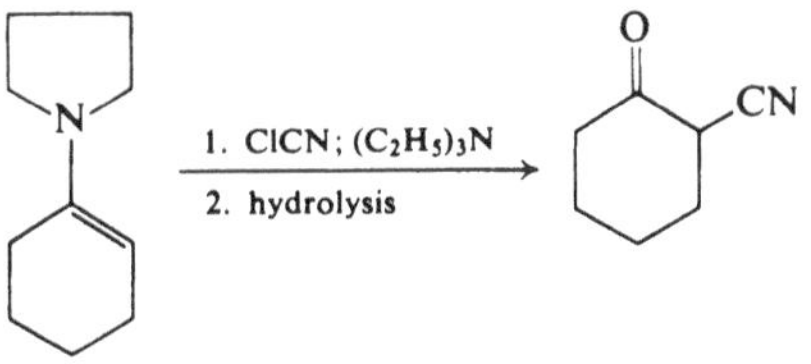

| (<u>15</u>) | (<u>161</u>) |

The pyrrolidine enamine of 2-tetralone (<u>162</u>) was converted to 1-cyano-2-tetralone, which exists almost entirely in the enolic form

(163), by reaction with cyanogen chloride [339]. Reaction of 162 with cyanogen bromide gave N-naphthylpyrrolidine (165), presumably via unstable bromoenamine 164. The latter observation is in accord with the mode of reaction of the heterocyclic enamine (105) with cyanogen bromide, which resulted in the formation of 166 [340]. These

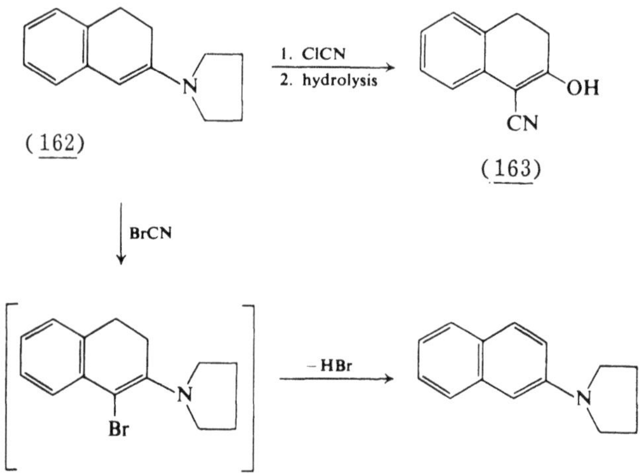

observations are also in agreement with the opposite polarization of cyanogen bromide and chloride [341,342]. Other examples of the use

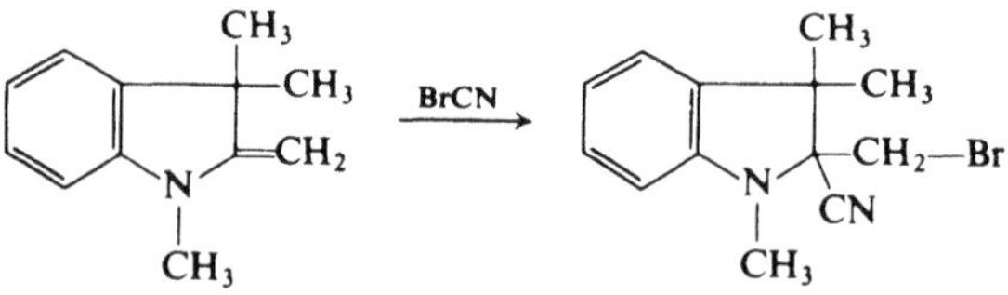

of cyanogen bromide in reactions with enamines have been reported [336,337].

C. Reaction with Aromatic Diazonium Salts

Aldehyde enamines react with aromatic diazonium salts in two ways, depending on the degree of substitution at the enamine carbon [305, 343]. Thus the piperidine enamine of butyraldehyde (167) reacted

with p-nitrophenyl-diazonium chloride to give the p-nitrophenylhydrazone of the α-keto aldehyde (168). Enamine 169 from isobutyral-

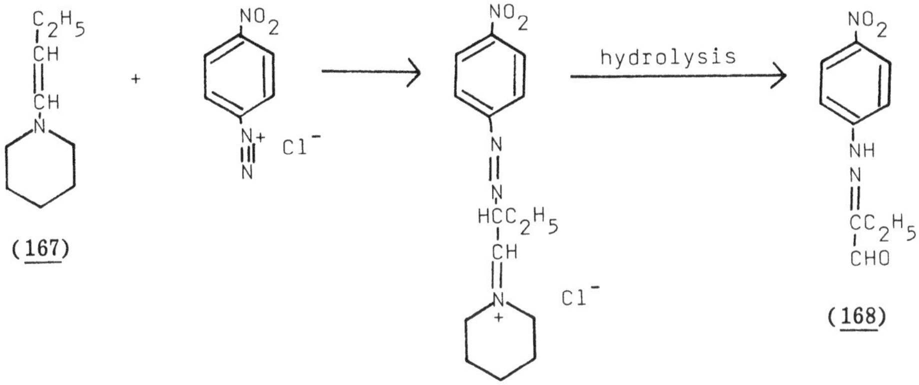

(167)   hydrolysis   (168)

dehyde on treatment with p-nitrophenyl-diazonium chloride, on the
other hand, gave the p-nitrophenylhydrazone of acetone (170) and
presumably N-formyl piperidine, although the latter was not isolated.

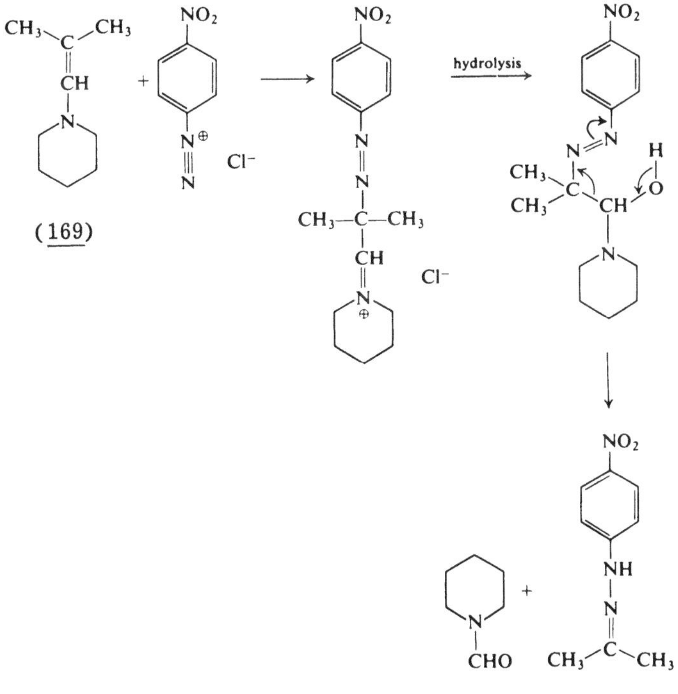

(169)   hydrolysis

(170)

Enamines of cyclic ketones react similarly [148,344]. Thus enamine 15 gave a good yield of the monophenylhydrazone of 1,2-cyclo-
hexanedione (171) on reaction with phenyldiazonium fluoborate and
subsequent hydrolysis [68]. These products have been cyclized to
the corresponding indoles [344]. In contrast the heterocyclic enamine,

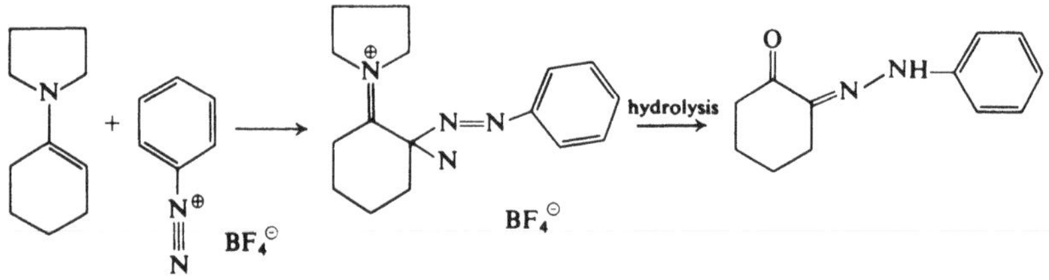

(15)   (171)

1,3,3-trimethyl-2-methyleneindoline (105) gave an azo compound [345,
346].

The iminium hydrazone derivatives obtained from the reaction of
aryldiazonium salts with β-carboxylic ester enamines can be thermally
cyclized to cinnoline-3-esters [347].

D. Reaction with Aromatic Azides

Enamines from cyclic ketones give derivatives of triazole. A complete
discussion of this type of cycloaddition reaction is given in Chapter
7, Section IV.C. Tosyl azide reacts differently to give sulfonamide
derivatives [348]. The morpholine enamine from dibenzylketone (172),
for instance, reacted with tosylazide to give 173 and phenyldiazo-
methane (174), which was trapped with acetic acid giving benzyl acetate [348].

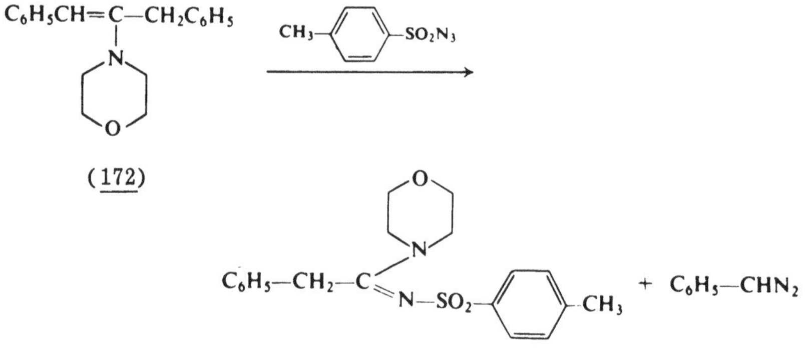

(172)

(173)   (174)

E. Reaction with Diethyl Azodicarboxylate and Activated α-Diazo Compounds

Diethyl azodicarboxylate (DAD) is an electrophile that is highly reactive toward enamines to form adducts such as 175 [349–351].

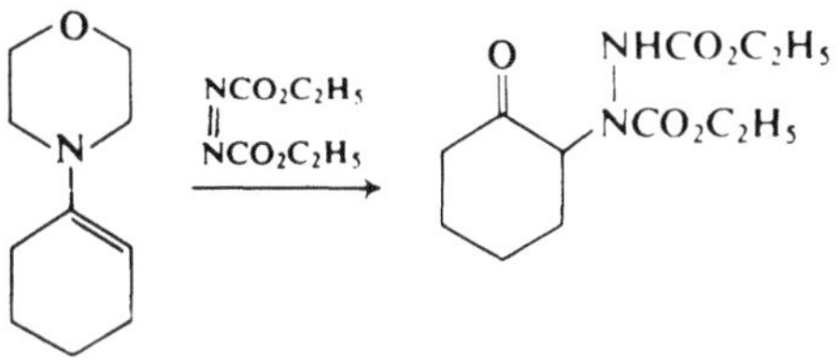

(39)   (175)

Diethyl azodicarboxylate provides a good chemical means of determining the isomeric composition of enamines of unsymmetrically substituted cyclic ketones (see Scheme 10) [273, 290, 351–355]. The stereo-

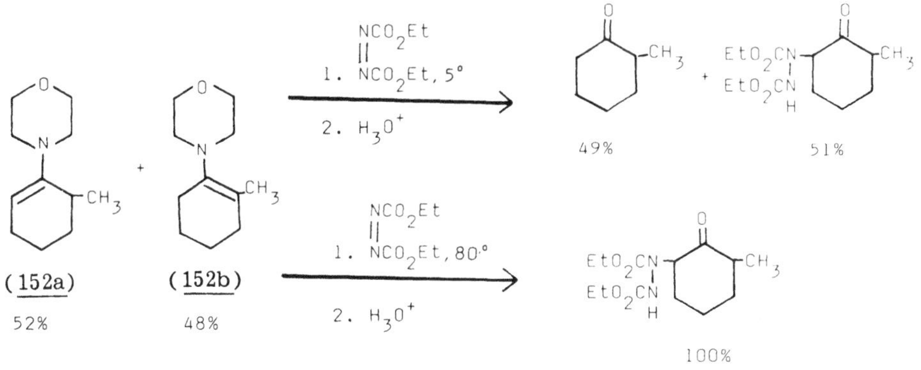

Scheme 10

chemistry of the addition of this ester to cyclic enamines has been intensively investigated and the results reported [290, 352–355].

Taking a mixture of isomers of the morpholine enamine of *trans*-2-decanone (176) and treating it with DAD results in the DAD mole-

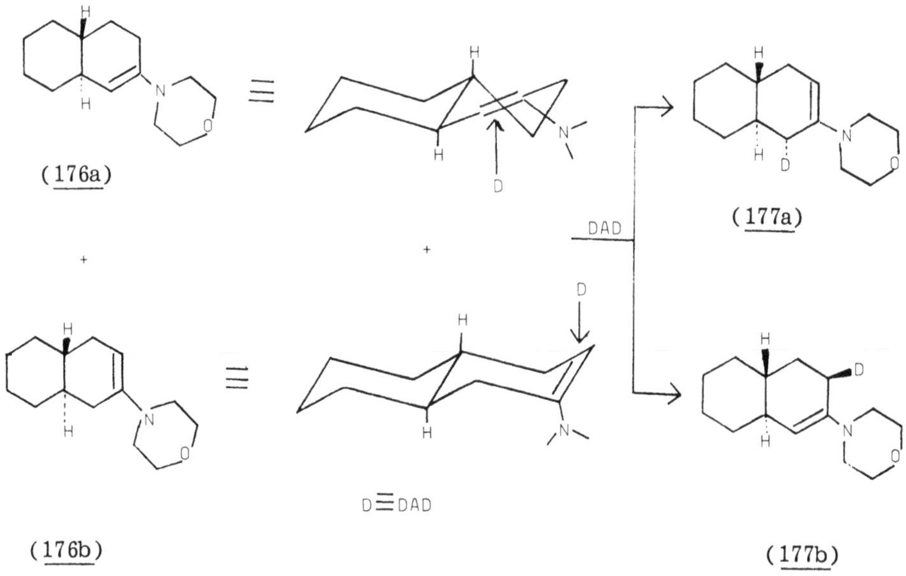

(176a)

+

(176b)

DAD

D≡DAD

(177a)

(177b)

cule attacking from the least hindered side of the enamine [352]. In the case of 176a, the attack is parallel, giving 177a because of the C-6, C-8, and C-10 axial hydrogens which would interfere with antiparallel attack. No such interference is present in 176b, so the DAD attack is antiparallel to give 177b [352] (see Chapter 1, Section III.B).

Diazo compounds activated by an α-carbonyl group [356], an α-phosphoryl group [357], or α-sulfonyl groups [358] react with enamines. In the cases in which cycloaddition does not take place, attachment is made to the β-carbon of the enamine by the terminal, electrophilic diazo nitrogen.

F. Reaction with Diborane and Aluminum Hydrides

Diborane and certain aluminum hydrides such as mixed hydride reagents lithium aluminum hydride and aluminum chloride are electrophilic in nature and can add directly to enamines. The reaction involves addition of the alumino cation and a hydride ion to the α- and β-positions, respectively, of the enamine. This adduct can then have either the aluminum functional group replaced by a hydrogen during hydrolysis to form the saturated amine [359] or the aluminium and amine functional groups eliminated to form the corresponding unsaturated compound [360,361]. The net effect of the first pathway is hydrogenation of the enamine, whereas that of the second pathway is hydrogenolysis of the enamine. Of the three mixed hydride reagents, $AlCl_2H$, $AlClH_2$, and AlH_3, the proportion of the olefin was greatest with AlH_3 and least with $AlCl_2H$ [360]. Hydrogenolysis of enamines also takes place when they are treated with diisobutyl aluminum hydride [362].

Diborane adds to enamines, with the hydride ion going to the α-position and the borane group to the β-position, except when steric conditions around the β-carbon prohibit it [360]. Treatment of this intermediate with refluxing acetic acid produces the saturated amine [363] or the aminoborinic acid [364]. When the intermediate is refluxed with acetic or propionic acids in diglyme, the corresponding alkenes are obtained in good yields through the hydrogenolysis reaction (see Scheme 11) [364,365]. Hydroboration of an enamine followed by oxidation of the resulting α-aminoborane gives excellent yields of the corresponding α-aminoalcohols [366−369]. Under some conditions hydroboration brings about reduction of the enamine [368,369]. Hydroboration of enamines with one equivalent of borane methyl sulfide in THF gives 2-(dialkylamino) organoboranes. Treating these with methanol gives the corresponding dimethyl boronate [370].

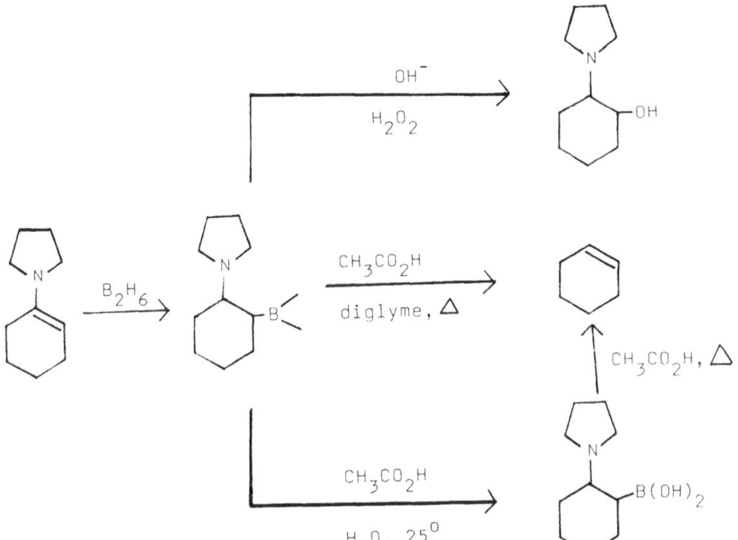

Scheme 11

G. Metal-Complex-Catalyzed Reactions

Iron carbonyl complexes have been used to activate electrophiles in their addition reactions to enamines [371,372]. For example, (1,2,3, 4,5-pentahapto-2-methoxycyclohexadienyl)-(tricarbonyl) (iron fluor-

oborate) (178) reacts with 1-(N-pyrrolidino)cyclohexene (15) to produce, after hydrolysis, an isomer mixture of adduct 179 in a 69%

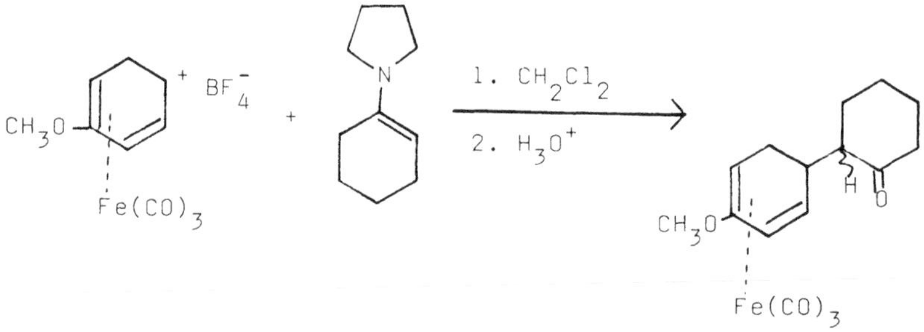

(178)   (15)   (179)

yield [371]. It has been shown that enamines are intermediates in
the formation of Simon-Awe complexes from disodium pentakis(cyano-
C)nitrosylferrate (II) [373].

Iodobenzene alkylates the β-carbon of an enamine in the presence
of nickel carbonyl to give, after hydrolysis, a β-diketone in good
yields [374]. The reaction between pentacarbonyltungsten complex
180 and enamine 39 produced, after hydrolysis, adduct 181 in a 90%
yield [375].

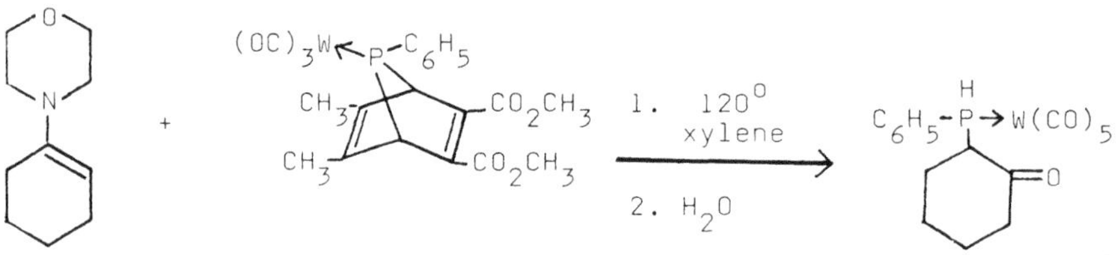

(39)   (180)   (181)

Allyl acetate or allyl phenoxide will add to enamines in an allyla-
tion reaction in the presence of palladium complex catalysts such as
that formed from palladium acetate and triphenylphosphine [376,377].
Butadiene reacts with a cycloalkanone enamine under these conditions
to produce, after hydrolysis, a 2-(2,7-octadienyl)cycloalkanone as
the principal product [377]. Using palladium chloride as a catalyst,
allene adds to 1-(N-pyrrolidino)cyclohexene (15) resulting in product
182 [378].

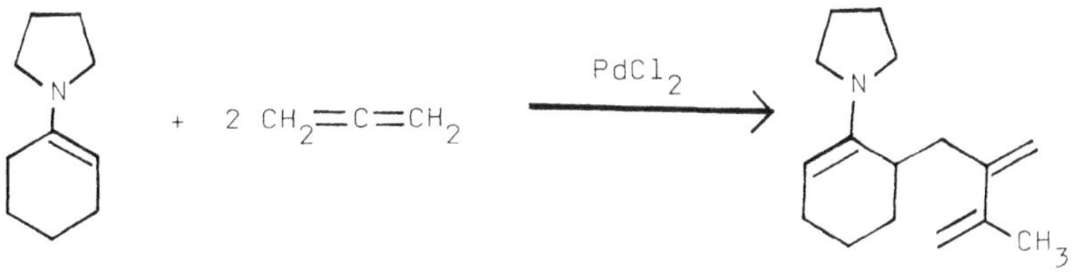

(15)   (182)

H. Miscellaneous

The transition metal complex [PtCl$_2$(C$_2$H$_4$)]$_2$ (Zeise's dimer) reacts with enamines to form brightly colored air-stable product crystals [379]. Group IV and V halides form β-substituted organometallic immonium salts as products [380].

Treatment of enamines with phenylselenenyl chloride [381] or phenyl selenocycanate [382] results in 1,2-addition across the carbon-carbon double bond, with the phenylselenenyl group attacking the β-carbon and the chloride or cyanide attacking the α-carbon. For example, the reaction of enamine 61 with phenyl selenocycanate gives

(61)   (183)

adduct 183. There are indications that this reaction is both regiospecific and stereospecific [382].

Nitrosobenzene reacts with enamine 184 to form hydroxylamine 185

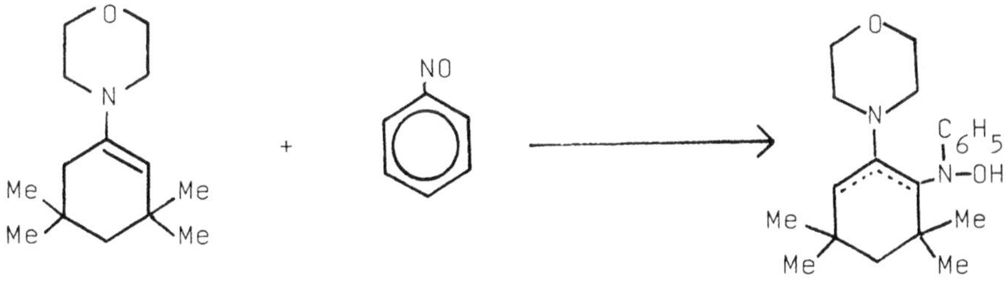

(184)   (185)

[383,389]. Enamines react with isonitrosomalonitrile tosylate to form
4-amino-2-aza-1,3-dienes through substitution of the tosylate group
[384]. Trimethylene or ethylene dithiotosylate attacks the β-carbon
of an enamine with the loss of a p-toluenesulfinate ion [385]. A second, internal displacement then gives the synthetically useful dithiane
or dithiolane. Reactions between 6-nitro-2-quinoxalone and enamines
result in alkylation of the enamine β-carbon by the 3-position of the
heterocycle [386]. It has been reported that 2-methylthio-3-methyl-
benzthiazolium salts attack enamine electrophilically with the loss of
methylsulfide [387].

REFERENCES

1. G. Stork, A. Brizzolara, H. Landesman, J. Szmuszkovicz, and
   R. Terrell, *J. Am. Chem. Soc.*, *85*, 207 (1963).
2. P. W. Hickmott, *Tetrahedron*, *38*, 1975 (1982).
3. G. Pitacco and E. Valentin in *The Chemistry of Amino, Nitroso,
   and Nitro Compounds and Their Derivatives* (S. Patai, ed.), Part
   1, Wiley, New York, 1982, p. 623.
4. V. G. Granik, *Russ. Chem. Rev.*, *53*, 383 (1984).
5. A. T. Blomquist and E. J. Moriconi, *J. Org. Chem.*, *26*, 3761
   (1961).
6. E. Elkik, *Bull. Soc. Chim. Fr.*, 972 (1960).
7. G. Opitz and H. Mildenberger, *Angew. Chem.*, *72*, 169 (1960).
8. G. Opitz and H. Mildenberger, *Liebigs Ann. Chem.*, *649*, 26
   (1961).
9. G. Opitz, H. Mildenberger, and H. Suhr, *Liebigs Ann. Chem.*,
   649, 47 (1961).
10. J. J. Panousse, *C. R. Seances Acad. Sci.*, *233*, 260, 1200 (1951).
11. S. Karady, M. Lenfant, and R. E. Wolff, *Bull. Soc. Chim. Fr.*,
    2479 (1965).
12. W. R. Williamson, *Tetrahedron*, *3*, 314 (1958).
13. E. Elkik and C. Francesch, *Bull. Soc. Chim. Fr.*, 903 (1969).
14. M. E. Kuehne and T. Garbacik, *J. Org. Chem.*, *35*, 1555 (1970).
15. U. K. Pandit, W. A. Zwart, and P. Houdewind, *Tetrahedron
    Lett.*, 1997 (1972).
16. T. J. Curphey, J. C. Hung, and C. C. C. Chu, *J. Org. Chem.*,
    40, 607 (1975).
17. J. K. Whitesell and M. A. Whitesell, *Synthesis*, 517 (1983).
18. T. J. Curphey and J. C. Hung, *J. Chem. Soc. Chem. Commun.*,
    510 (1967).
19. T-L. Ho and C. M. Wong, *Synth. Commun.*, *4*, 147 (1974).
20. K. C. Brannock and R. D. Burpitt, *J. Org. Chem.*, *26*, 3576
    (1961).

21. G. Opitz, H. Hellmann, M. Mildenberger, and H. Suhr, *Liebigs Ann. Chem.*, *649*, 36 (1961).
22. A. Kirrmann and E. Elkik, *C. R. Seances Acad. Sci.*, *267*, 623 (1968).
23. J. Oda, T. Igarashi, and Y. Inouye, *Bull. Inst. Chem. Res, Kyoto Univ.*, *54*, 180 (1976).
24. P. M. McCurry, Jr. and R. K. Singh, *Tetrahedron Lett.*, 3325 (1973).
25. G. Opitz, *Liebigs Ann. Chem.*, *650*, 122 (1961).
26. K. U. Acholonu and D. K. Wedegaertner, *Tetrahedron lett.*, 3253 (1974).
27. G. Kalaus, P. Gyory, L. Szabo, and C. Szantay, *J. Org. Chem.*, *43*, 5017 (1978).
28. D. H. Wadsworth, M. R. Detty, B. J. Murray, C. H. Weidner, and N. F. Haley, *J. Org. Chem.*, *49*, 2676 (1984).
29. H. Heimgartner, H-J. Hansen, and H. Schmid in *Iminium Salts in Organic Chemistry* (H. Bohm and H. G. Viehe, eds.), in *Advances in Organic Chemistry* (E. C. Taylor, ed.), Vol. 9, Part 2, Wiley-Interscience, New York, 1979, pp. 658–700.
30. L. Nilsson and C. Rappe, *Acta Chem. B, 30*, 1000 (1976).
31. M. Julia, S. Julia, and C. Jeanmart, *C. R. Seances Acad. Sci.*, *251*, 249 (1960).
32. J. G. Murphy, J. H. Ager, and E. L. May, *J. Org. Chem.*, *25*, 1386 (1960).
33. S. Saito and E. L. May, *J. Org. Chem.*, *26*, 4536 (1961).
34. G. Stork and J. W. Schulenberg, *J. Org. Chem.*, *84*, 284 (1962).
35. A. R. Greenaway and W. B. Whalley, *J. Chem. Soc.*, *Perkin Trans.*, *I*, 1385 (1976).
36. J. K. Whitesell and S. W. Felman, *J. Org. Chem.*, *42*, 1663 (1977).
37. S. Fatutta, G. Pitacco, C. Russo, and E. Valentin, *J. Chem. Soc.*, *Perkin Trans.*, *I*, 2045 (1982).
38. M. F. Ansell, J. S. Mason, and M. P. L. Caton, *J. Chem. Soc.*, *Perkin Trans.*, *I*, 1061 (1984).
39. A. G. Cook, W. M. Kosman, T. A. Hecht, and W. Koehn, *J. Org. Chem.*, *37*, 1565 (1972).
40. P. Houdewind and U. K. Pandit, *Tetrahedron Lett.*, 2359 (1974).
41. H. Bieraugel and U. K. Pandit, *Recl. Trav. Chim. Pays-Bas*, *98*, 296 (1979).
42. R. A. Benkeser, R. F. Lambert, P. W. Ryan, and D. G. Stoffey, *J. Am. Chem. Soc.*, *80*, 6573 (1958).
43. P. Schenone and G. Minardi, *Gazz. Chim. Ital.*, *100*, 945 (1970).
44. B. L. Jensen and D. P. Michaud, *Synthesis*, 848 (1977).
45. G. F. Hennion and F. X. Quinn, *J. Org. Chem.*, *35*, 3054 (1970).
46. A. Doutheau and J. Gore, *Tetrahedron*, *32*, 2705 (1976).

47. H. E. Baumgarten, P. L. Creger, and C. E. Villars, *J. Am. Chem. Soc.*, *80*, 6609 (1958).
48. K. Sisido, S. Kurozumi, and K. Utimoto, *J. Org. Chem.*, *34*, 2661 (1969).
49. W. Reid and F. Batz, *Liebigs Ann. Chem.*, *755*, 32 (1972).
50. M. S. Manhas, J. W. Brown, and U. K. Pandit, *Heterocycles*, *3*, 117 (1975).
51. D. M. Locke and S. W. Pelletier, *J. Am. Chem. Soc.*, *81*, 2246 (1959).
52. E. D. Bergmann and E. Hoffmann, *J. Org. Chem.*, *26*, 3555 (1961).
53. R. B. Miller and E. S. Behare, *J. Am. Chem. Soc.*, *96*, 8102 (1974).
54. S. Carlsson, A. A. El-Barbary, and S-O. Lawesson, *Bull. Soc. Chim. Belg.*, *89*, 643 (1980).
55. M. Barthelemy, J-P. Montheard, and Y. Bessiere-Chretien, *Bull. Soc. Chim. Fr.*, 2725 (1969).
56. G. Stork, M. Ohashi, H. Kamachi, and H. Kakisawa, *J. Org. Chem.*, *36*, 2784 (1971).
57. Y. Bessiere-Chretien and H. Serne, *Bull. Soc. Chim. Fr.*, 2039 (1973).
58. O. N. Chupakhin, V. N. Charushin, and E. O. Sidorov, *Khim. Geterotsikl. Soedin*, 666 (1979).
59. L. E. Crane, G. P. Beardsley, and Y. Maki, *J. Org. Chem.*, *45*, 3827 (1980).
60. T. I. Bieber and M. T. Dorsett, *J. Org. Chem.*, *29*, 2028 (1964).
61. V. I. Mikhailov, V. D. Sholle, E. S. Kagan, and E. G. Rozantsev, *Izv. Akad. Nauk SSSR, Ser. Khim.*, 1639 (1976).
62. F. LeGoffic, A. Ahond, and A. Goyuette, *Fr. Demande* 2,152,374 (1973); *CA*, *79*, 78773e (1973).
63. K. Suzuki and M. Sekiya, *Synthesis*, 297 (1981).
64. C. Zatti and A. Ferrantine, *Chem. Ber.*, *23*, 2302 (1890).
65. G. Plancher, *Chem. Ber.*, *31*, 1488 (1898).
66. R. Lukes and V. Dedek, *Chem. Listy*, *51*, 2059 (1957), *Collect. Czech. Chem. Commun.*, *23*, 2046 (1958).
67. J. Metzger, H. Larive, E. Vincent, and R. Dennilauler, *J. Chem. Phys*, *60*, 944 (1963).
68. F. S. Babichev, *J. Gen. Chem. USSR (Engl. Transl.)*, *20*, 1904 (1950).
69. G. H. Alt, *J. Org. Chem.*, *33*, 2858 (1968).
70. K. Hiroi, K. Achiwa, and S. Yamada, *Chem. Pharm. Bull.*, *20*, 246 (1972).
71. K. Hiroi and S. Yamada, *Chem. Pharm Bull.*, *21*, 47 (1973).
72. D. C. Cook and A. Lawson, *J. Chem. Soc., Perkin Trans. I*, 1112 (1974).

73. C. Berti, L. Greci, and L. Marchetti, *Gazz. Chim. Ital.*, *105*, 993 (1975).
74. M. Colonna and L. Marchetti, *Gazz. Chim. Ital.*, *96*, 1175 (1966).
75. L. H. Hellberg and M. F. Stough, III, *Acta Chem. Scand.*, *21*, 1368 (1967).
76. L. H. Hellberg, R. J. Milligan, and R. N. Wilke, *J. Chem. Soc.* (*C*), 35 (1970).
77. F. P. Colonna, S. Fatutta, A. Risaliti, and C. Russo, *J. Chem. Soc.* (*C*), 2377 (1970).
78. F. Weisbuch and G. Dana, *Tetrahedron*, *30*,2873 (1974).
79. L. Anandan and G. K. Trivedi, *Indian J. Chem.*, *16B*, 428 (1978).
80. Y. K. Singh and R. B. Rao, *Chem. Letters*, 653 (1979).
81. J. R. L. Smith, R. O. C. Norman, M. E. Rose, and A. C. W. Curran, *J. Chem. Soc.*, *Perkin Trans.*, *I*, 2863 (1979).
82. H. Christol, C. Montginoul, and F. Plenat, *C. R. Seances Acad. Sci.*, *265*, 836 (1967).
83. N. F. Firrell and P. W. Hickmott, *J. Chem. Soc.*, *Chem. Commun.*, 544 (1969).
84. H. Mazarguil and A. Lattes, *Bull. Soc. Chim. Fr.*, 3874 (1972).
85. F. Plenat, C. Montginoul, and H. Christol, *Bull. Soc. Chim. Fr.*, 691 (1973).
86. H. Kolind-Andersen and S.-O. Lawesson, *Bull. Soc. Chim. Belg.*, *86*, 543 (1977).
87. E. G. Rozantsev, M. Dagonneau, E. S. Kagan, V. I. Mikhailov, and V. D. Sholle, *J. Chem. Res.* (*S*), 260 (1979); *J. Chem. Res.* (*M*), 2901 (1979).
88. A. G. Angoh and D. L. J. Clive, *J. Chem. Soc.*, *Chem. Commun.*, 941 (1985).
89. R. G. Glushkov, N. I. Koretskaya, A. I. Ermakov, G. Y. Shvarts, and M. D. Mashkovshii, *Khim.-Farm. Zh.*, *9*, 6 (1975); *CA*, *84*, 180439r (1976).
90. S. Danishefsky, G. Rovnyak, and R. Cavanaugh, *J. Chem. Soc.*, *Chem. Commun.*, 636 (1969).
91. K. Igarasyi, J. Oda, Y. Inouye, and M. Ohno, *Agr. Biol. Chem.*, *34*, 811 (1970).
92. E. A. Lissi and J. C. Scaiano, *J. Chem. Soc.*, *Chem. Commun.* 457 (1971).
93. H. Stetter and K. Komorowski, *Chem. Ber.*, *104*, 75 (1971).
94. S. A. Vartanyan and E. A. Abgaryan, *Arm. Khim. Zh.*, *25*, 609 (1972).
95. S. Danishefsky and G. Rovnyak, *J. Org. Chem.*, *39*, 2924 (1974).
96. A. G. Cook, Ph.D. thesis, University of Illinois, Urbana, 1959.

97. H. Sliwa and G. Condonnier, *J. Heterocycl. Chem.*, *12*, 809 (1975).
98. J. W. Patterson and J. E. McMurry, *J. Chem. Soc., Chem. Commun.*, 488 (1971).
99. P. W. Hickmott and N. F. Firrell, *J. Chem. Soc., Perkin Trans.*, *I*, 340 (1978).
100. S. J. Blarer and D. Seebach, *Chem. Ber.*, *116*, 2250 (1983).
101. K. C. Brannock, R. D. Burpitt, V. W. Goodlett, and J. G. Thweatt, *J. Org. Chem.*, *29*, 813 (1964).
102. O. Cervinka, *Coll. Czech. Chem. Commun.*, 25, 1174 (1960).
103. O. Cervinka, *Coll. Czech. Chem. Commun.*, 25, 1183 (1960).
104. O. Cervinka, *Coll. Czech. Chem. Commun.*, 25, 2675 (1960).
105. D. Becker and H. J. E. Loewenthal, *J. Chem. Soc.*, 1338 (1965).
106. U. K. Pandit and H. O. Huisman, *Tetrahedron Lett.*, 3901 (1967).
107. C. D. Gutsche and D. M. Bailey, *J. Org. Chem.*, *28*, 607 (1963).
108. H. L. Lochte and A. G. Pittman, *J. Org. Chem.*, *25*, 1462 (1960).
109. A. Risaliti, S. Fatutta, and M. Forchiassim, *Tetrahedron*, *23*, 1451 (1967).
110. M. Forchiassin, E. Valentin, A. Risaliti, and S. Fatutta, *Tetrahedron Lett.*, 1821 (1966).
111. A. Risaliti, S. Fatutta, M. Forchiassin, and C. Russo, *Ricerca Sci.*, *38*, 827 (1968).
112. S. Fatutta and A. Risaliti, *J. Chem. Soc., Perkin Trans.*, *I*, 2387 (1974).
113. M. Calligaris, M. Forchiassin, A. Risaliti, and C. Russo, *Gazz. Chim. Ital.*, *105*, 689 (1975).
114. M. Forchiassin, A. Risaliti, C. Russo, N. B. Pahor, and M. Calligaris, *J. Chem. Soc., Perkin Trans.*, *I*, 935 (1977).
115. F. Benedetti, S. Fabrissin, and A. Risaliti, *Tetrahedron*, *40*, 977 (1984).
116. M. E. Kuehne and L. Foley, *J. Org. Chem.*, *30*, 4280 (1965).
117. A. Risaliti and M. Forchiassin, *Tetrahedron*, *22*, 6331 (1966).
118. A. Risaliti, M. Forchiassin, and E. Valentin, *Tetrahedron*, *24*, 1889 (1968).
119. E. Valentin, G. Pitacco, and F. P. Colonna, *Tetrahedron Lett.*, 2837 (1972).
120. F. P. Colonna, E. Valentin, G. Pitacco, and A. Risaliti, *Tetrahedron*, *29*, 3011 (1973).
121. E. Valentin, G. Pitacco, F. P. Colonna, and A. Risaliti, *Tetrahedron*, *30*, 2741 (1974).
122. M. Forchiassin, A. Risaliti, C. Russo, M. Calligaris, and G. Pitacco, *J. Chem. Soc., Perkin Trans.*, *I*, 660 (1974).

123. G. Pitacco, F. P. Colonna, E. Valentin, and A. Risaliti, *J. Chem. Soc.*, 1625 (1974).
124. G. Pitacco, A. Risaliti, M. L. Trevisan, and E. Valentin, *Tetrahedron, 33*, 3145 (1977).
125. M. C. Moorjani and G. K. Trivedi, *Indian J. Chem.*, *16B*, 405 (1978).
126. G. Pitacco and E. Valentin, *Tetrahedron Lett.*, 2339 (1978).
127. F. Benedetti, G. Pitacco, and E. Valentin, *Tetrahedron, 35*, 2293 (1979).
128. H. C. Mutreja and D. N. Reinhoudt, *Recl. Trav. Chim. Pays-Bas, 99*, 241 (1980).
129. D. Ranganathan, C. B. Rao, S. Ranganathan, A. K. Mehrotra, and R. Iyengar, *J. Org. Chem.*, 45, 1185 (1980).
130. D. Seebach and J. Golinski, *Helv. Chim. Acta, 64*, 1413 (1981).
131. P. Bradamante, G. Pitacco, A. Risaliti, and E. Valentin, *Tetrahedron Lett.*, *23*, 2683 (1982).
132. S. J. Blarer, W. B. Schweizer, and D. Seebach, *Helv. Chim. Acta, 65*, 1637 (1982).
133. S. J. Blarer and D. Seebach, *Chem. Ber.*, *116*, 3086 (1983).
134. D. Seebach, A. K. Beck, J. Golinski, J. N. Hay and T. Laube, *Helv. Chim. Acta, 68*, 162 (1985).
135. H. Molines and C. Wakselman, *J. Chem. Soc., Perkin Trans.*, *I*, 1114 (1980).
136. S. Fabrissin, S. Fatutta, N. Malusa, and A. Risaliti, *J. Chem. Soc., Perkin Trans.*, *I*, 686 (1980).
137. S. Fabrissin, S. Fatutta, and A. Risaliti, *J. Chem. Soc., Perkin Trans.*, *I*, 109 (1981).
138. A. Risaliti, L. Marchetti, and M. Forchiassin, *Ann. Chim. Rome, 56*, 317 (1966).
139. V. Prelog and G. Helmchen, *Angew. Chem. Int. Ed. Engl.*, *21*, 567 (1982).
140. K. R. Hanson, *J. Am. Chem. Soc.*, *88*, 2731 (1966).
141. G. Singerman and S. Danishefsky, *Tetrahedron Lett.*, 2249 (1964).
142. K. C. Brannock, R. D. Burpitt, V. W. Goodlett, and J. G. Thweatt, *J. Org. Chem.*, *28*, 1462 (1963).
143. K. C. Brannock, R. D. Burpitt, V. W. Goodlett, and J. G. Thweatt, *J. Org. Chem.*, *29*, 818 (1964).
144. K. C. Brannock, R. D. Burpitt, V. W. Goodlett, and J. G. Thweatt, *J. Org. Chem.*, *28*, 1464 (1963).
145. W. Verboom, G. W. Visser, W. P. Trompenaars, D. N. Reinhoudt, S. Harkema, and G. J. vanHummel, *Tetrahedron, 20*, 3525 (1981).
146. M. F. Menachery, J. M. Saa, and M. P. Cava, *J. Org. Chem.*, *46*, 2584 (1981).

147. J. C. Blazejewski and D. Cantacuzene, *Tetrahedron Lett.*, 4241 (1973).
148. M. E. Kuehne, *J. Am. Chem. Soc.*, *84*, 837 (1962).
149. K-A. Kovar and U. Schwiecker, *Arch. Pharm. (Weinheim)*, *307*, 390 (1974).
150. K-A. Kovar, F. Schielein, T. G. Dekker, K. Albert, and E. Breitmaier, *Tetrahedron*, *35*, 2113 (1979).
151. M. Coenen, *Liebigs Ann. Chem.*, *633*, 78 (1960).
152. M. Coenen, *Angew. Chem.*, *61*, 11 (1949).
153. G. Scherowsky and H. Mattoubi, *Liebigs Ann. Chem.*, 98 (1978).
154. G. L. May and J. T. Pinhey, *Aust. J. Chem.*, *35*, 1859 (1982).
155. D. Lednicer, U.S. 3,862,232 (1975); *CA*, *82*, 139854t (1975).
156. M. E. Kuehne and C. E. Bayha, unpublished results, 1962.
157. K. C. Brannock, R. D. Burpitt, H. E. Davis, H. S. Pridgen, and J. G. Thweatt, *J. Org. Chem.*, *29*, 2579 (1964).
158. K. Ley and R. Nast, *Angew. Chem.*, *79*, 150 (1967).
159. H. B. Henbest and P. Slade, *J. Chem. Soc.*, 1555 (1960).
160. G. Domschke, *Z. Chem.*, *4*, 29 (1964).
161. G. Domschke, *Chem. Ber.*, *98*, 930 (1965).
162. G. Domschke, *Chem. Ber.*, *98*, 2920 (1965).
163. G. Domschke, *Chem. Ber.*, *99*, 934 (1966).
164. G. Domschke, *Chem. Ber.*, *99*, 939 (1966).
165. V. I. Shvedov and A. N. Grinev, *Zh. Org. Khim.*, *1*, 1125 (1965).
166. C. Wakselman and J. C. Blazejewski, *J. Chem. Soc., Chem. Commun.*, 341 (1977).
167. G. Domschke and H. Oelmann, *J. Prakt. Chem.*, *311*, 800 (1969).
168. J. Valderrama and J. C. Vega, *An. Quim*, *73*, 1212 (1977).
169. C. Mannich, *Chem. Ber.*, *65*, 378 (1932).
170. A. Takeda, S. Tsuboi, F. Sakai, and M. Tanabe, *J. Org. Chem.*, *39*, 3098 (1974).
171. C. Nolde and S.-O. Lawesson, *Bull. Soc. Chim. Belg.*, *86*, 313 (1977).
172. B. Gaux and P. LeHenaff, *C. R. Seances Acad. Sci., Ser. C*, 271 (1970).
173. J. W. Lewis, P. L. Meyers, and M. J. Readhead, *J. Chem. Soc. (C)*, 771 (1970).
174. R. E. Harmon, H. N. Subbarao, and S. K. Gupta, *Synth. Commun.*, 117 (1971).
175. L. Bifkofer, S. M. Kim, and H. D. Engels, *Chem. Ber.*, *95*, 1495 (1962).
176. L. A. Paquette, *Tetrahedron Lett.*, 1291 (1965).
177. W. D. Burrows and E. P. Burrows, *J. Org. Chem.*, *28*, 1180 (1963).

178. E. P. Burrows, R. F. Hutton, and W. D. Burrows, *J. Org. Chem.*, *27*, 316 (1962).
179. A. G. Schultz and Y. K. Yee, *J. Org. Chem.*, *41*, 561 (1976).
180. T. A. Spencer and K. K. Schmiegel, *Chem. Ind. (London)*, 1765 (1963).
181. S. Tomoda, Y. Takeuchi, and Y. Nomura, *Tetrahedron Lett.*, 3549 (1969).
182. N. J. Leonard and F. P. Hauck, Jr., *J. Am. Chem. Soc.*, *79*, 5279 (1957).
183. N. J. Leonard and A. G. Cook, *J. Am. Chem. Soc.*, *81*, 5627 (1959).
184. H. H. Wasserman and M. S. Baird, *Tetrahedron Lett.*, 1729 (1970).
185. D. Beke, C. Szantay, and M. Barczai-Beke, *Liebigs Ann. Chem.*, *636*, 150 (1960).
186. R. F. Parcell, *J. Am. Chem. Soc.*, *81*, 2596 (1959).
187. N. J. Leonard and W. J. Musliner, *J. Org. Chem.*, *31*, 639 (1966).
188. H. Bohme, K. Osmers, and P. Wagner, *Tetrahedron Lett.*, 2785 (1972).
189. A. D. Batcho and W. Leimgruber, German Offen. 2, 057,840 (1971); *CA*, *75*, 63605 (1971).
190. U. Hengartner, A. D. Batcho. J. F. Blount, W. Leimgruber, M. E. Larscheid, and J. W. Scott, *J. Org. Chem.*, *44*, 3748 (1979).
191. A. Vilsmeier and A. Haack, *Chem. Ber.*, *60*, 119 (1927).
192. W. Ziegenbein, *Angew. Chem. Int. Ed. Engl.*, *4*, 358 (1965).
193. Z. Arnold, *Experientia*, *15*, 415 (1959).
194. M. Muraoka and T. Yamamoto, *J. Chem. Soc., Chem. Commun.*, 1299 (1985).
195. H. G. Viehe, T. vanVyve, and Z. Janousek, *Angew. Chem. Int. Ed. Engl.*, *11*, 916 (1972).
196. G. Schroll, P. Klemmensen, and S-O. Lawesson, *Acta Chem. Scand.*, *18*, 2201 (1964).
197. G. Schroll, P. Klemmensen, and S-O. Lawesson, *Tetrahedron Lett.*, 2869 (1965).
198. G. Schroll, P. Klemmensen, and S-O. Lawesson, *Arkiv. Kemi*, *26*, 317 (1966).
199. D. Baudoux and R. Fuks, *Bull. Soc. Chim. Belg.*, *93*, 1009 (1984).
200. S. Hunig, E. Benzing, and E. Lucke, *Chem. Ber.*, *90*, 2833 (1957).
201. T. Inukai and R. Yoshizawa, *J. Org. Chem.*, *32*, 404 (1967).
202. S-R. Kuhlmey, H. Adolph, K. Rieth, and G. Opitz, *Liebigs Ann. Chem.*, 617 (1979).
203. P. Rosenmund, D. Sauer, and W. Trommer, *Chem. Ber.*, *103*, 496 (1970).

204. L. Nilsson, *Acta Chem. Scand. B*, *33*, 203 (1979).
205. L. Nilsson, *Acta Chem. Scand. B*, *33*, 710 (1979).
206. G. A. Berchtold, G. R. Harvey, and G. E. Wilson, *J. Org. Chem.*, *26*, 4776 (1961).
207. G. A. Berchtold, G. R. Harvey, and G. E. Wilson, *J. Org. Chem.*, *30*, 2642 (1965).
208. G. Opitz and F. Zimmermann, *Liebigs Ann. Chem.*, *662*, 178 (1963).
209. R. H. Hasek and J. C. Martin, *J. Org. Chem.*, *26*, 4775 (1961).
210. G. Opitz, H. Adolph, M. Kleemann, and F. Zimmermann, *Angew. Chem.*, *73*, 654 (1961).
211. G. Opitz, M. Kleemann, and F. Zimmermann, *Angew. Chem.*, *74*, 32 (1962).
212. S. Hünig, E. Lucke, and E. Benzing, *Chem. Ber.*, *91*, 129 (1958).
213. S. Hünig and E. Lucke, *Chem. Ber.*, *92*, 652 (1959).
214. S. Hünig and W. Lendle, *Chem. Ber.*, *96*, 909, 913 (1960).
215. S. Hünig and W. Eckart, *Chem. Ber.*, *95*, 2493 (1962).
216. S. Hunig and M. Salzwedel, *Angew. Chem.*, *71*, 339 (1959).
217. S. Hünig and M. Salzwedel, *Chem. Ber.*, *99*, 823 (1966).
218. R. Jacquier and G. Maury, *Bull. Soc. Chim. France*, 320 (1967).
219. T. Smuszkovicz and L. L. Skaletzky, *J. Org. Chem.*, *32*, 3300 (1967).
220. G. Opitz and E. Tempel, *Leibigs Ann. Chem.*, *699*, 74 (1966).
221. R. Jacquier, C. Petrus, F. Petrus, and M. Valentin, *Bull. Soc. Chim. France*, 2629 (1969).
222. H. J. Buysch and S. Hünig, *Angew. Chem.*, *78*, 145 (1966).
223. S. Hünig and H. Hock, *Tetrahedron Lett.*, 5215 (1966).
224. W. W. Christie, F. J. Gunstone, and H. G. Prentice, *J. Chem. Soc.* 5768 (1963).
225. K. Sisido and M. Kawanisi, *J. Org. Chem.*, *27*, 3723 (1962).
226. M. S. R. Nair, H. H. Mathur, and S. C. Bhattacharyya, *Tetrahedron*, *19*, 905 (1963).
227. S. Yurugi, M. Numata, and T. Fushimi, *Yakugaku Zasshi*, *80*, 1170 (1960).
228. F. Johnson, N. A. Starkovsky, A. C. Paton, and A. A. Carlson, *J. Am. Chem. Soc.*, *88*, 149 (1966).
229. F. Johnson, U.S. Pat. 3,153,651 (1965); *CA*, *62*, 1580b (1965).
230. F. Jonnson, U.S. Pat. 3,657 (1965); *CA*, *62*, 1669g (1965).
231. G. I. Fujimoto and R. W. Ledun, *J. Org. Chem.*, *29*, 2059 (1964).
232. S. I. Goldberg and I. Ragade, *J. Org. Chem.*, *32*, 1046 (1967).
233. P. Rosenmund, W. H. Haase, J. Bauer, and R. Frische, *Chem. Ber.*, *108*, 1871 (1975).

234. M. Mühlstadt and J. Reimer, *Z Chem.*, *4*, 70 (1964).
235. G. Descotes and Y. Querou, *C. R. Seances Acad. Sci.*, *263*, 1231 (1966).
236. M. E. Kuehne and L. Foley, unpublished results, 1964.
237. M. E. Kuehne and C. Bayha, *Tetrahedron Lett.*, 1311 (1966).
238. W. Sobotka, W. N. Beverung, G. G. Munoz, J. C. Sircav, and A. I. Meyers, *J. Org. Chem.*, *30*, 3667 (1965).
239. J. Bohlmann and O. Schmidt, *Chem. Ber.*, *97*, 1354 (1964).
240. G. Opitz and M. Kleemann, *Liebigs Ann. Chem.*, *665*, 759 (1966).
241. S. Hünig and H. Hoch, *Chem. Ber.*, *105*, 2216 (1972).
242. A. Kirrmann and C. Wakselman, *Liebigs Ann. Chem.*, *665*, 115, (1963).
243. S. Hünig, H. J. Buysch, H. Hoch, and W. Lendle, *Chem. Ber.*, *100*, 3996 (1967).
244. P. W. Hickmott and J. R. Hargreaves, *Tetrahedron*, *23*, 3151 (1967).
245. P. W. Hickmott, P. J. Cox, and G. Sim, *J. Chem. Soc., Perkin Trans., I*, 2544 (1974).
246. L. Li and R. Su, *Gaodeng Xueriao Huaxue Xuebao*, 5, 366 (1984); *CA*, *101*, 191855u (1984).
247. R. D. Campbell and J. A. Jung, *J. Org. Chem.*, *30*, 3711 (1965).
248. R. Helmers, *Acta Chem. Scand.*, *19*, 2139 (1965).
249. R. Helmers, *Tetrahedron Lett.*, 1905 (1966).
250. L. Li and C. Chang, *Gaodeng Xueriao Huaxue Xuebao*, 5, 829 (1984); *CA*, *102*, 166677a (1985).
251. G. H. Alt, *J. Org. Chem.*, *29*, 798 (1964).
252. G. H. Alt, *J. Org. Chem.*, *31*, 2384 (1966).
253. G. H. Alt and A. J. Speziale, *J. Org. Chem.*, *31*, 1340 (1966).
254. G. H. Alt and A. J. Speziale, *J. Org. Chem.*, *31*, 2073 (1966).
255. T. Morimoto and M. Sekiya, *Chem. Pharm Bull.*, *26*, 1586 (1978).
256. J. Weber and P. Faller, *C. R. Hebd. Seances Acad. Sci., Ser. C*, *281*, 389 (1975).
257. W. Verboom and D. N. Reinhoudt, *J. Org. Chem.*, *47*, 3339 (1982).
258. T. Ishihara, T. Seki, and T. Ando, *Bull. Chem. Soc. Jpn.*, *55*, 3345 (1982).
259. C. Jutz in *Iminium Salts in Organic Chemistry* (H. Bohm and H. G. Viehe, eds.), in *Advance in Organic Chemistry* (E. C. Taylor, ed.), Vol. 9, Part 1, Wiley-Interscience, New York, 1976, pp. 225–342.
260. A. Halleux and H. G. Viehe, *J. Chem. Soc. (C)*, 881 (1970).
261. I. Belsky, *Tetrahedron*, *28*, 771 (1972).
262. O. E. Krivoshchekova and A. A. Shamshurin, *Zh. Org. Khim.*, 7, 474 (1971).

263. O. Takazawa and T. Mukaiyama, *Chem. Letters*, 1307 (1982).
264. O. Takazawa, K. Kogami, and K. Hayashi, *Bull. Chem. Soc. Jpn.*, *57*, 1876 (1984).
265. R. Fusco, S. Rossi, and S. Maiorana, *Chem. Ind. (Milan)*, *44*, 873 (1962).
266. J. J. Looker, *J. Org. Chem.*, *31*, 2973 (1967).
267. J. Elguero, R. Jacquier, and G. Tarrago, *Bull. Soc. Chim. Fr.*, 1149 (1968).
268. S. Bradamante, S. Maiorana, and G. Pagani, *J. Chem. Soc. Perkin I*, 282 (1972).
269. O. Tsuge, S. Iwanami, and S. Hagio, *Bull. Chem. Soc. Jpn.*, *45*, 237 (1972).
270. M. Furukawa, S. Tsuiji, Y. Kojima, and S. Hayashi, *Chem. Pharm Bull.*, *21* 1965 (1973).
271. A. Etienne and B. Desmazieres, *J. Chem. Res. (S)*, 484 (1978); *J. Chem. Res. (M)*, 5501 (1978).
272. R. A. Ferri, G. Pitacco, and E. Valentin, *Tetrahedron*, *34*, 2537 (1978).
273. P. Bradamante, M. Forchiassin, G. Pitacco, C. Russo, and E. Valentin, *J. Heterocycl. Chem.*, *19*, 985 (1982).
274. A. Bender, D. Guenther, L. Willms, and R. Wingen, *Ger. Offen.* DE 3,323,511 (1985); *CA*, 103, 7742x (1985).
275. R. T. LaLonde, A. I.-M. Tsai, and C. Wong, *J. Org. Chem.*, *41*, 2514 (1976).
276a. G. Opitz and K. Rieth, *Tetrahedron Lett.*, 3977 (1965).
276b. G. Opitz, *Angew. Chem.*, *79*, 161 (1967).
277. G. Opitz and H. Adolph, *Angew. Chem.*, *74*, 77 (1962).
278. M. Kuehne, *J. Org. Chem.*, *28*, 2124 (1963).
279. D. Scholz, *Monatsh. Chem.*, *115*, 655 (1984).
280. Y. Noguchi, K. Kurogi, M. Sekioka, and M. Furukawa, *Bull. Chem. Soc. Jpn.*, *56*, 349 (1983).
281. T. Tanabe and T. Nagai, *Bull. Chem. Soc. Jpn.*, *50*, 1179 (1977).
282. T. Nagai, T. Shingaki, M. Inagaki, and T. Ohshima, *Bull. Chem. Soc. Jpn.*, *52*, 1102 (1979).
283. T. Kumamoto, S. Kobayaski, and T. Mukaiyama, *Bull. Chem. Soc. Jpn.*, *45*, 866 (1972).
284. M. Perelman and S. A. Mizsak, *J. Am. Chem. Soc.*, *84*, 4988 (1962).
285. G. Opitz and J. Koch, *Angew. Chem.*, *75*, 167 (1963).
286. R. Neidlein and U. Askani, *Arch. Pharm. (Weinheim)*, *310*, 820 (1977).
287. R. Neidlein and U. Askani, *Synthesis*, 48 (1975).
288. G. Berchtold, *J. Org. Chem.*, *26*, 3043 (1961).
289. R. Fusco, G. Bianchetti, and S. Rossi, *Gazz. Chim Ital.*, *91*, 825 (1961).

290. E. P. Colonna, M. Forchiassin, G. Pitacco, A. Risaliti, and E. Valentin, *Tetrahedron*, 26, 5289 (1970).
291. S. C. Kuo and W. H. Daly, *J. Org. Chem.*, 35, 1861 (1970).
292. A.-B. A. G. Ghattas, K. A. Jorgensen, and S-O. Lawesson, *Acta Chem. Scand. Ser. B*, 36, 505 (1982).
293. G. Griss and H. Machleidt, *Liebigs Ann. Chem.*, 738, 60 (1970).
294. W. H. Daly and W. Kern, *Makromol. Chem.*, 108, 1 (1967).
295. D. Pocar, R. Stradi, and B. Gioia, *Gazz. Chim. Ital.*, 98, 958 (1968).
296. S. Hünig, K. Hübner, and E. Benzing, *Chem. Ber.*, 95, 926 (1962).
297. S. D. Sharma, P. K. Gupta, and A. L. Gauba, *Indian J. Chem.*, 16B, 424 (1978).
298. H. Singh and R. K. Mehta, *Indian J. Chem.*, 15B, 786 (1977).
299. S. Hünig and K. Hübner, *Chem. Ber.*, 95, 937 (1962).
300. G. Skorna and I. Ugi, *Chem. Ber.*, 112, 776 (1979).
301. R. L. Pedersen, J. L. Johnson, R. P. Holysz, and A. C. Ott, *J. Am. Chem. Soc.*, 79, 1115 (1957).
302. M. E. Kuehne, *J. Am. Chem. Soc.*, 83, 1492 (1961).
303. R. Tiollais, H. Bouget, J. Huet, and A. LePennec, *Bull. Soc. Chim. Fr.*, 1205 (1964).
304. M. E. Kuehne and T. J. Giacobbe, *J. Org. Chem.*, 33, 3359 (1968).
305. J. R. Geigy, Brit. 832,078 (1960); *CA*, 57, 20877f (1960).
306. P. Duhamel, L. Duhamel, C. Collet, and A. Haider, *C. R. Seances Acad. Sci.*, 273, 1461 (1971).
307. H. Ahlbrecht and M. T. Reiner, *Tetrahedron Lett.*, 4901 (1971).
308. M. Takeda, H. Inoue, M. Konda, S. Sato, and H. Kugita, *J. Org. Chem.*, 37, 2677 (1972).
309. L. Duhamel, P. Duhamel, and J-M. Poirier, *Tetrahedron Lett.*, 4237 (1973).
310. M. Gobbini, P. Giacconi, and R. Stradi, *Synthesis*, 940 (1983).
311. A. J. Speziale and L. R. Smith, *J. Org. Chem.*, 28, 3492 (1963).
312. R. R. Renshaw and J. C. Ware, *J. Am. Chem. Soc.*, 47, 2989 (1925).
313. I. J. Borowitz, E. W. R. Casper, R. K. Crouch, and K. C. Yee, *J. Org. Chem.*, 37, 3873 (1972).
314. R. Carlson and C. Rappe, *Acta Chem. Scand., Ser. B*, 28, 1058 (1972).
315. R. Carlson and C. Rappe, *Acta Chem. Scand., Ser. B*, 31, 485 (1977).
316. R. Carlson, *Acta Chem. Scand., Ser. B*, 32, 646 (1978).
317. L. Duhamel and J-C. Plaquevent, *Bull. Soc. Chim. Fr.*, II-239 (1982).

318. F. M. Laskovics and E. M. Schulman, *Tetrahedron Lett.*, 759 (1977).

319. F. M. Laskovics and E. M. Schulman, *J. Am. Chem. Soc.*, 99, 6672 (1977).

320. L. Duhamel and J-M. Poirier, *Tetrahedron Lett.*, 2437 (1976).

321. L. Duhamel and J-M. Poirier, *J. Org. Chem.*, 44, 3576 (1979).

322. S. J. Huang and M. V. Lessard, *J. Am. Chem. Soc.*, 90, 2432 (1968).

323. J.-J. Riehl and F. Jung, *C. R. Seances Acad. Sci.*, 270, 2009 (1970).

324. E. Vilsmaier, W. Sprugel, and K. Gagel, *Tetrahedron Lett.*, 2475 (1974).

325. I. Dyong and Q. Lam-Chi, *Angew. Chem. Int. Ed. Engl.*, 18, 933 (1979).

326. T. Wada, J. Oda, and Y. Inouye, *Agr. Biol. Chem.*, 36, 799 (1972).

327. S. T. Purrinton and W. A. Jones, *J. Fluorine Chem.*, 26, 43 (1984).

328. G. A. Olah, Y. D. Vankar, and M. Arvanaghi, *Tetrahedron Lett.*, 3653 (1979).

329. E. Vilsmaier, W. Tröger, W. Sprügel, and K. Gagel, *Chem. Ber.*, 112, 2997 (1979).

330. R. B. Gabbard and E. V. Jensen, *J. Org. Chem.*, 23, 1406 (1958).

331. J. Warnant, A. Guillemette, and B. Goffinet, Ger. Pat. 1,159,434; *CA*, 61, 1921h (1964).

332. B. J. Magerlein and F. Kagan, U.S. Pat. 3,232,960; *CA*, 64, 1187 (1966).

333. S. Nakanishi, R. L. Morgan, and E. V. Jensen, *Chem. Ind. (London)*, 1136 (1960).

334. R. Joly and J. Warnant, *Bull. Soc. Chim. Fr.*, 569 (1961).

335. D. H. R. Barton, L. S. Godinho, R. H. Hesse, and M. M. Pechet, *J. Chem. Soc., Chem. Commun.*, 804 (1968).

336. N. DeKimpe, R. Verhe, L. DeBuyck, and N. Schamp, *Chem. Ber.*, 116, 3846 (1983).

337. H. Ahlbrecht and D. Liesching, *Synthesis*, 495 (1977).

338. M. E. Kuehne, *J. Am. Chem. Soc.*, 81, 5400 (1959).

339. R. T. Parfitt, *J. Chem. Soc. (C)*, 140 (1967).

340. O. Mumm, H. Hinz, and J. Diedericksen, *Chem. Ber.*, 72, 2107 (1939).

341. F. Fairbrother, *J. Chem. Soc.*, 180 (1950).

342. G. Lord and A. A. Wolf, *J. Chem. Soc.*, 2546 (1954).

343. J. W. Crary, O. R. Quayle, and C. T. Lester, *J. Am. Chem. Soc.*, 78, 5584 (1956).

344. V. I. Shedov, L. B. Altukhova, and A. N. Grinev, *J. Org. Chem. USSR*, 2, 1608 (1966).

345. W. König and J. Müller, *Chem. Ber.*, *57*, 144 (1924).
346. W. König, *Chem. Ber.*, *57*, 891 (1924).
347. C. B. Kanner and U. K. Pandit, *Tetrahedron*, *37*, 3513 (1981).
348. D. Pocar, G. Bianchetti, and P. DallaCroce, *Gazz. Chim. Ital.*, *95*, 1220 (1965).
349. A. Risaliti and L. Marchetti, *Ann. Chim. (Rome)*, *55*, 635 (1965).
350. A. Risaliti, S. Fatutta, and M. Forchiassin, *Tetrahedron*, *23*, 1451 (1967).
351. M. Colonna and L. Marchetti, *Gazz. Chim. Ital.*, *99*, 14 (1969).
352. M. Forchiassin, C. Russo, and A. Risaliti, *Gazz. Chim. Ital.*, *102*, 607 (1972).
353. S. Fatutta, A. Risaliti, C. Russo, and E. Valentin, *Gazz. Chim. Ital.*, *102*, 1008 (1972).
354. G. Pitacco, F. P. Colonna, C. Russo, and E. Valentin, *Gazz. Chim. Ital.*, *105*, 1137 (1975).
355. G. Pitacco, R. Toso, E. Valentin, and A. Risaliti, *Tetrahedron*, *32*, 1757 (1976).
356. R. Huisgen, H.-U. Reissig, H. Huber, and S. Voss, *Tetrahedron Lett.*, 2987 (1979).
357. W. Welter, M. Regitz, and H. Heydt, *Chem. Ber.*, *111*, 2290 (1978).
358. U. Schollkopf, E. Wiskott, and K. Riedel, *Liebigs Ann. Chem.*, 387 (1975).
359. J. Sansoulet and Z. Welvart, *Bull. Soc. Chim. Fr.*, 77 (1963).
360. J. M. Coulter, J. W. Lewis, and P. P. Lynch, *Tetrahedron*, *24*, 4489 (1968).
361. J. W. Lewis and P. P. Lynch, *Proc. Chem. Soc.*, 19 (1963).
362. L. I. Zakharkin and L. A. Savina, *Izv. Akad. Nauk SSSR, Ser. Khim.*, 1695 (1964).
363. J. A. Marshall and W. S. Johnson, *J. Org. Chem.*, *28*, 421 (1963).
364. J. W. Lewis and A. A. Pearce, *Tetrahedron Lett.*, 2039 (1964).
365. J. W. Lewis and A. A. Pearce, *J. Chem. Soc. (B)*, 863 (1969).
366. J. J. Barieux and J. Gore, *Bull. Soc. Chim. Fr.*, 1649 (1971).
367. F. Bondavalli, P. Schenone, and A. Ranise, *J. Chem. Res. (S)*, 257 (1980); *J. Chem. Res. (M)*, 3256 (1980).
368. J. J. Barieux and J. Gore, *Tetrahedron*, *28*, 1537 (1972).
369. J. J. Barieux and J. Gore, *Tetrahedron*, *28*, 1555 (1972).
370. C. T. Goralski, B. Singaram, and H. C. Brown, Proceedings of 192nd ACS National Meeting, Sept. 7–12, 1986, Anaheim, CA.
371. R. E. Ireland, G. G. Brown, Jr., R. H. Stanford, Jr., and T. C. McKenzie, *J. Org. Chem.*, *39*, 51 (1974).
372. A. Rosan and M. Rosenblum, *J. Org. Chem.*, *40*, 3621 (1975).
373. W. Wiegrebe and M. Vilbig, *Z. Naturforsch. B*, *37*, 490 (1982).
374. Y. Seki, S. Murai, M. Ryang, and N. Sonada, *J. Chem. Soc., Chem. Commun.*, 528 (1975).

375. A. Marinetti and F. Mathey, *Organometallics*, *3*, 1492 (1984).
376. H. Onoue, I. Moritani, and S.-I Murahashi, *Tetrahedron Lett.*, 121 (1973).
377. J. Tsuji, *Bull. Chem. Soc. Jpn.*, *46*, 1896 (1973).
378. D. R. Coulson, *J. Org. Chem.*, *38*, 1483 (1973).
379. A. J. Kunin and D. B. Brown, *J. Organomet. Chem.*, *212*, C 27 (1981).
380. H. Weingarten and J. S. Wagaer, *Syn. Inorg. Metal.-Org. Chem.*, *1*, 123 (1971).
381. D. R. Williams and K. Nishitani, *Tetrahedron Lett.*, *21*, 4417 (1980).
382. S. Tomoda, Y. Takeuchi, and Y. Nomura, *Tetrahedron Lett.*, *23*, 1361 (1982).
383. J. W. Lewis, P. L. Myers, and J. A. Ormerod, *J. Chem. Soc., Perkin Trans.*, *I*, 2521 (1972).
384. J. P. Schoeni and J. P. Fleury, *Tetrahedron*, *31*, 671 (1975).
385. R. B. Woodward, I. J. Pachter, and M. L. Scheinbaum, *J. Org. Chem.*, *36*, 1137 (1971).
386. O. N. Chupakhin, V. N. Charushin, and Y. V. Shnurov, *Zh. Org. Khim.*, *16*, 1064 (1980).
387. E. Fanghanel, K. Behrmann, and K. Siwik, *Z. Chem.*, *21*, 355 (1981).
388. F. W. Vierhapper and E. L. Eliel, *J. Org. Chem.*, *40*, 2734 (1975).
389. R. A. Abramovitch, S. R. Challand, and Y. Yamada, *J. Org. Chem.*, *40*, 1541 (1975).
390. K. Hayakawa, M. Takewaki, I. Fujimoto, and K. Kanematsu, *J. Org. Chem.*, *51*, 5100 (1986).

5

Oxidation and Reduction of Enamines

A. GILBERT COOK *Valparaiso University, Valparaiso, Indiana*

I.   Introduction   247

II.   Electrochemical Oxidation and Reduction   248

III. Disproportionation   250

IV.   Chemical Oxidation   251

      A. By Gaining Oxygen   251
      B. By Losing Hydrogen   261

V.   Chemical Reduction   262

      References   268

I. INTRODUCTION

Enamines are unusually susceptible to oxidation, and their immonium salts are readily reduced. The relatively narrow range of reactions that will be defined as oxidation reactions in this chapter fall into one of the following three categories: (a) a reaction in which an electron is completely lost from the enamine; (b) a reaction in which an oxygen (or a closely related proxy such as a highly oxidized nitrogen) either replaces a hydrogen in the enamine or adds to the enamine; and (c) a reaction in which hydrogens are lost. Reduction reactions will be considered the converse reactions, namely: the complete gain of an electron by the enamine, replacement of nitrogen by hydrogen, or addition of hydrogen to the enamine.

II. ELECTROCHEMICAL OXIDATION
AND REDUCTION

Electrolytic oxidations of enamines in solvents such as acetonitrile
normally undergo a one-electron oxidation to a cation radical. These
oxidations take place at relatively low oxidation potentials [1]. For
example 1-(N,N-dimethylamino)cyclohexene (1), 2,5-dimethyl-1-(N,N-

(1)   (2)   (3)

dimethylamino)cyclohexene (2), and 1-phenyl-1-(N,N-dimethylamino)
ethene (3) have oxidation potentials (relative to saturated calomel elec-
trode) of 0.42, 0.38, and 0.70 volt, respectively [1]. Comparison of
these oxidation potentials with those of trimethylamine (1.07 volts),
aniline (0.95 volt), and 1,4-cyclohexadiene (1.85 volts) demonstrates
their relatively low values [2]. It has also been observed that enamines derived from cyclopentanone are more easily oxidized than the
corresponding enamines from cyclohexanone. Furthermore, pyrrolidine enamines are more readily oxidized than those of the corresponding piperidine or morpholine enamines [3]. This order of oxidation
potentials generally corresponds to the gas-phase first-ionization potentials of enamines (see Chapter 1, Section II) in which the pyrrolidine enamines are seen to have lower first-ionization potentials than
corresponding piperidine and morpholine enamines. A similar correlation of oxidation potentials and gas-phase ionization potentials for a
large variety of organic compounds has been observed and reported
[4], and the use of oxidation potentials as electrochemical reactivity
indices has been reviewed [5].

The products of electrochemical oxidation have been isolated.
Electrooxidation of (E)-1-(4-morpholino)-1,2-diphenylethene (4) pro-

(4)   (5)   (6)   (7)

duces desoxybenzoin (5) along with oxidation products benzoin (6)
and benzil (7) [6].

Electrooxidation has also been used synthetically for the preparation of various organic compounds. For example electrolysis of 1-(4-morpholino)cyclopentene (8) in methanol with sodium methoxide as supporting electrolyte gives a mixture of methoxyenamines 9 and 10 in a

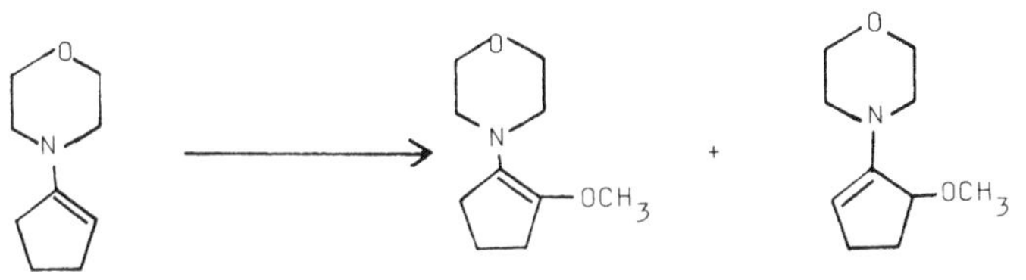

(8)   (9)   (10)

76% yield [7]. When methanolic solutions of some pyrrolidine, morpholine, or piperidine enamines are electrooxidized in the presence of methyl acetoacetate, acetylacetone, or dimethyl malonate anions, substitution of the anion takes place at the β-position of the enamine in good yields [3]. This is shown by the electrooxidation of 1-(4-mor-

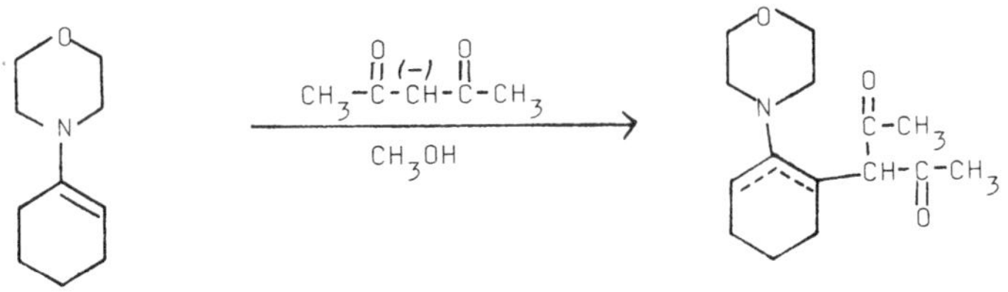

(11)   (12)

pholino)cyclohexene (11) in the presence of acetylacetone anion to produce β-substituted enamine 12 in a 60% yield [3].

The use of a mediatory system in electrochemical oxidations makes it possible to achieve oxidations at lower potentials and under milder conditions than by direct anodic oxidation of the substrate [9]. With potassium iodide as the mediator of an aqueous solution of an enamine, β-ketoamines are produced upon electrooxidation [10]. An example of this is the electrochemical oxidation of the pyrrolidine enamine of citronellal (13) in the presence of KI to give a 48% yield of ketoamine (14).

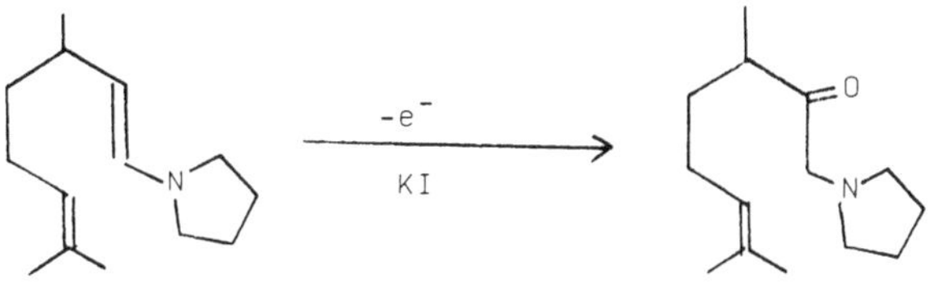

(13)   (14)

Electrochemical reduction of simple tertiary enamines appears to occur only with the C-protonated iminium ion form [1,11,12]. The electrochemical reduction of an iminium salt leads to an initial formation of a free radical [1,13]. The odd electron is probably localized at the α-carbon atom, and these radicals can then form a dimeric product [13,14].

III. DISPROPORTIONATION

A disproportionation reaction refers to a reaction in which the substrate is both oxidized and reduced. This type of reaction occurs with enamines primarily when the end product of oxidation is an aromatic ring. Refluxing a solution of 1-(4-morpholino)cyclohexene (11) in dioxane for 15 min in the presence of 10% Pd on charcoal gave N-

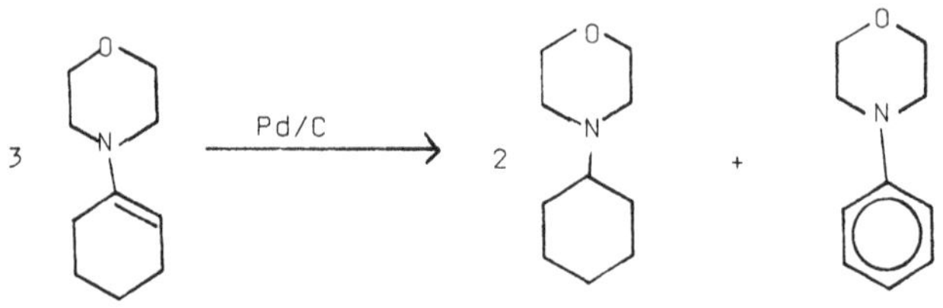

(11)　　　　　　　　　(15)　　　　　　　　(16)

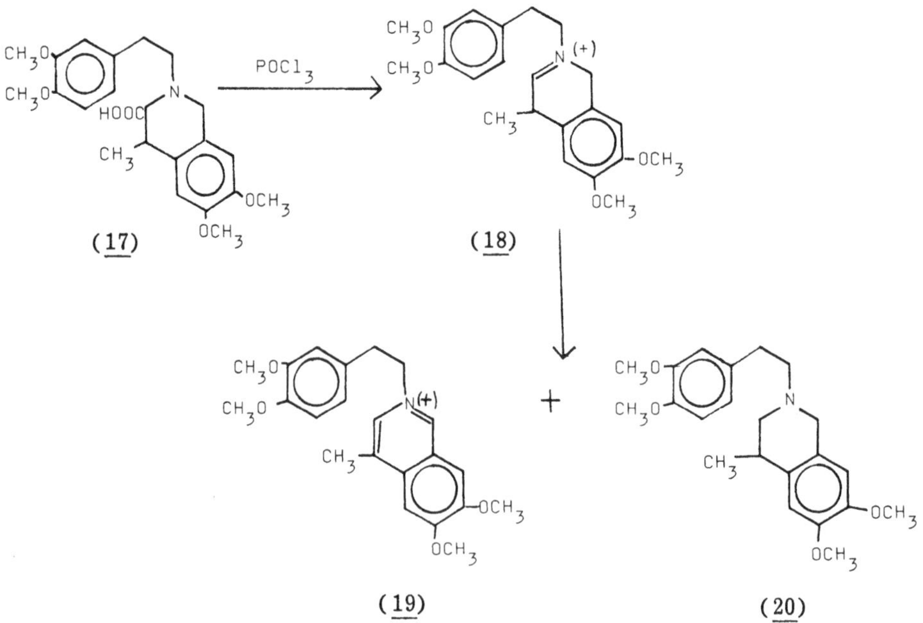

(17)　　　　　　　　　　　(18)

(19)　　　　　　　　　　　(20)

cyclohexylmorpholine (15) and N-phenylmorpholine (16) in a 2:1 ratio
quantitative yield [15]. Similar results were obtained when the amine
moiety on the cyclohexene ring was pyrrolidino, 4-methylpiperazino,
N,N-diethylamino, or N-methylanilino. Decarbonylation of amino acid
17 (an intermediate in the synthesis of some berbine derivatives) with
POCl₃ led to disproportionation side products 19 and 20 via intermediate 18 as well as the berbine cyclization product. A possible disproportionation reaction involving 1-(N,N-dimethylamino)cyclohexene
in hexamethylphosphoric triamide has been reported, but the exact
nature of the reaction has not been conclusively determined [17].

IV. CHEMICAL OXIDATION

A. By Gaining Oxygen

Oxidation of an enamine brought about by an oxygen being gained
can result in products in which one of the following has occurred:
one hydrogen has been replaced; two hydrogens have been replaced;
an addition to a carbon-carbon double bond has taken place with or
without carbon-carbon single-bond cleavage.

  Hydroxylation of an enamine by addition of vicinal hydroxyl
groups to the carbon-carbon double bond of an enamine has been carried out using osmium tetroxide [8].

  The formation of α-acetoxyketones by oxidation of enamines with
thallic acetate has been studied in detail [18–20] and found to be of
preparative value (80% yields) particularly in five- and six-membered-
ring ketone derivatives (see Eq. [1]). Lead tetraacetate causes a

$$\text{(1)}$$

similar reaction to take place when it is allowed to react with an enamine [19,21], but α-aminoketones are often the major product [19].
This reaction proceeds through intermediate 21 [19]. Sometimes a
Favorski-type rearrangement takes place, such as when 1-(N-pyrrolidino)cyclohexene (22) is allowed to react at room temperature for 30
hr with lead tetraacetate, boron trifluoride etherate, and ethanol in
benzene solvent to produce ethyl cyclopentylcarboxylate (23) [22].

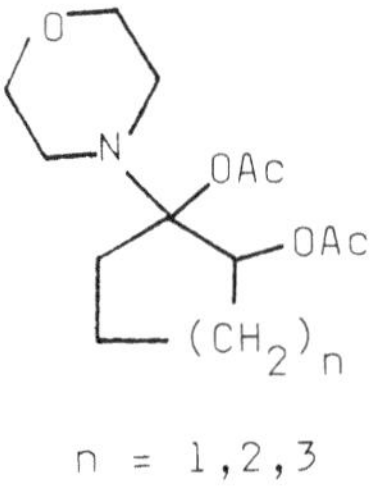

(21) 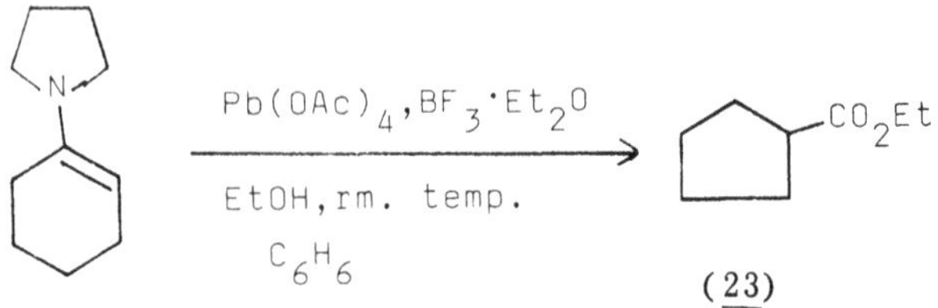   (22)   (23)

n = 1,2,3

During the oxidation of tertiary amines with mercuric acetate to
form enamines, sometimes further oxidation of the enamine by mercuric acetate takes place ([50]; see also Chapter 2, Section III.A). Ar-
yllead triacetates react with enamines to give a mixture of arylated
products and acetoxylated products; the exact composition of the mixture depends on the enamine used [23]. Reactions of benzoylperoxide
with morpholinocyclohexene and morpholinocyclopentene give α-ben-
zoyloxyketones [24,25].

α-Ketoenamines can be obtained by allowing an enamine to reaction with some aromatic nitro compounds. For example when 1-(N-

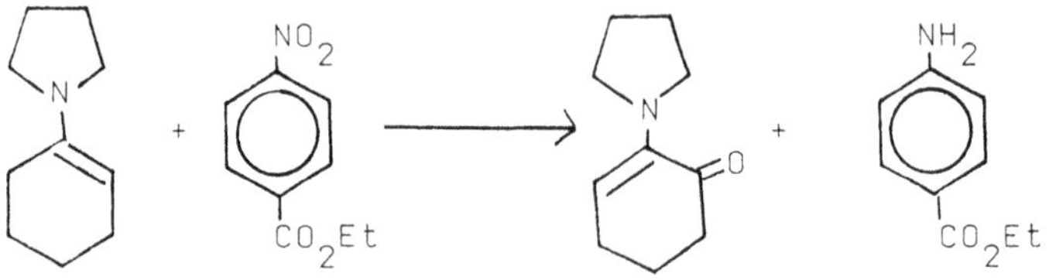

(22)   (24)

pyrrolidino)-cyclohexene (22) is treated with ethyl p-nitrobenzoate
(24), an α-ketoenamine results along with ethyl p-aminobenzoate [26].
Reactions of enamines with selenium dioxide give low yields of α-ketoenamines [27].

An α-oximino enamine is formed by the reaction of nitrosyl chloride and triethylamine with morpholinecyclohexene (11) [28,167]. α-

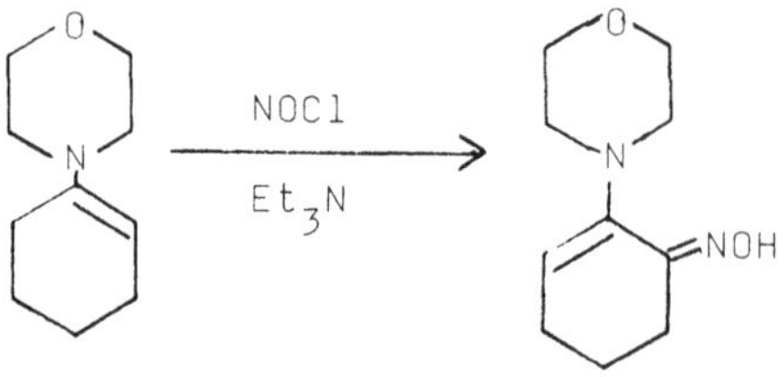

(11)

oximino enamines can also be synthesized by treating enamines with alkyl nitrite [29,30], nitrous acid [31,32], or nitric oxide/oxygen [33].

Sulfur-substituted enamines can be made by treating enamines with dichlorosulfane [34], dichlorodisulfane [34], or *bis*(N,N-dimethylthiocarbamoyl)disulfide [35]. Treatment of 1-(N-pyrrolidino)cyclohexene (22) with trimethylene dithiotosylate followed by acid hydrolysis gives a 45% yield of 2,2-(trimethylenedithio)cyclohexanone (25) [36,37].

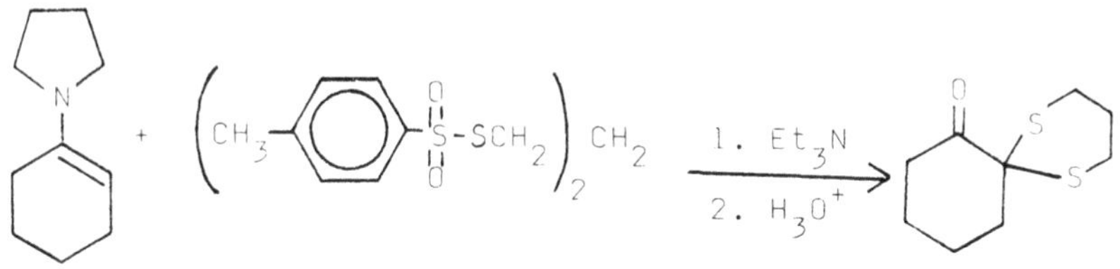

(22)   (25)

Potassium permanganate oxidation of an enamine can either replace a β-hydrogen with a hydroxyl group [38,39] or it can oxidatively cleave the carbon-carbon double bond [39].

Ozonolysis of an enamine normally gives oxidative cleavage of the carbon-carbon double bond. For example progesterone (27) is formed from ozonolysis of 22-(N-piperidino)-bisnor-4,20(22)-choladien-3-one (26) [40]. Likewise, iminium salts show oxidative cleavage on ozonol-

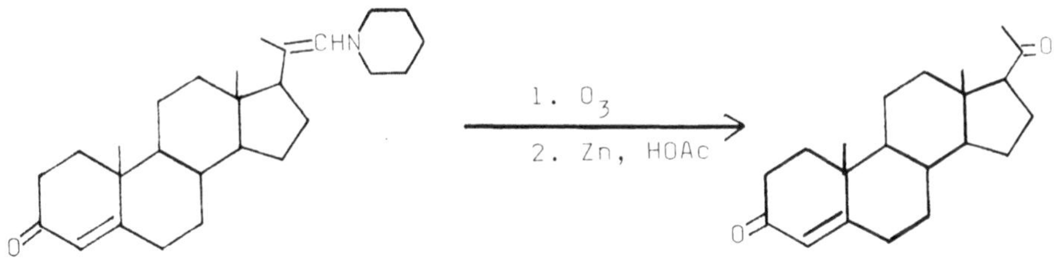

(26)   (27)

ysis (see Chapter 6) [41,42]. However, with certain enamines ozonolysis produces only replacement of a β-hydrogen with a hydroxyl group or a mixture of this product along with oxidative double-bond cleavage. For example, 1-(t-butyl)-1-(4-morpholino)ethene (28) when ozonized gives a mixture of 40% substitution product 29 and 20% oxi-

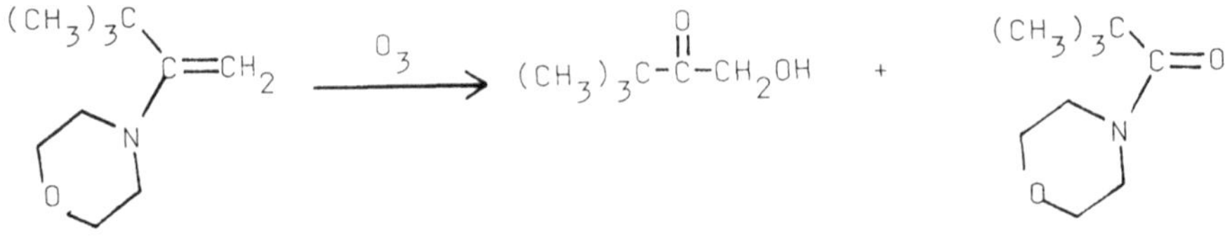

$(CH_3)_3C$ ―C=CH₂ (with morpholino N attached) (28) → O_3 → $(CH_3)_3C$-C(=O)-C-CH₂OH (29) + $(CH_3)_3C$-C=O (morpholino) (30)

dative cleavage product 30 [43,44]. In a similar manner, 2,6-dimethyl-1-(4-morpholino)cyclohexene (31) gives 35% of the substitution product 32 on ozonolysis [44].

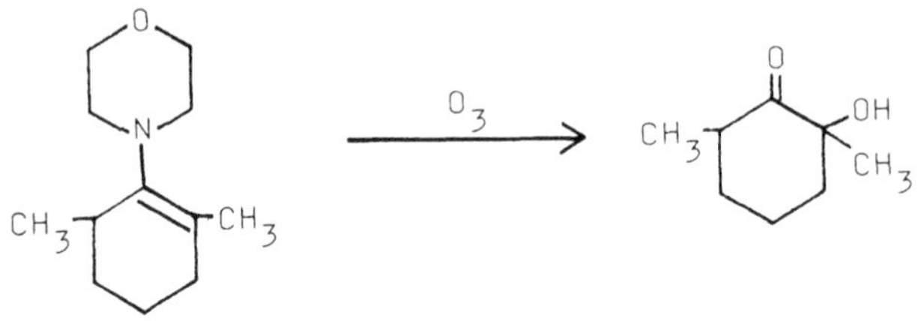

(31) → O_3 → (32)

Hydrogen peroxide at room temperature and for a short reaction period can add to enamines such as 8 to give stable peroxides such as 33 (74% yield) [45,46]. However, over an extended period, hydrogen

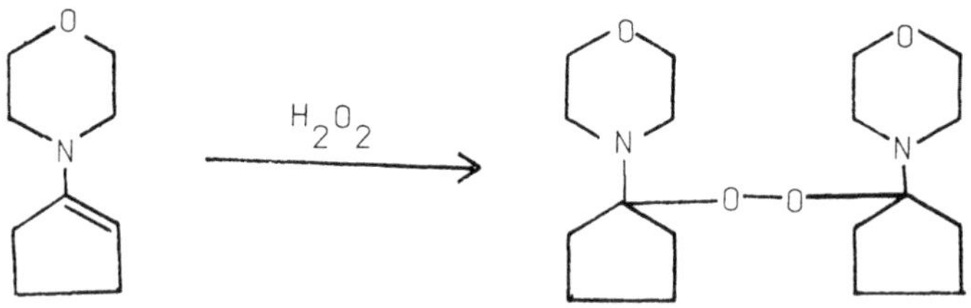

(8) → H_2O_2 → (33)

peroxide will oxidize enamines. For example, steroidal enamine 34, on treatment with 30% hydrogen peroxide, produces oxidative cleavage product 35 in a 53% yield [47]. Peracids can react similarly to give

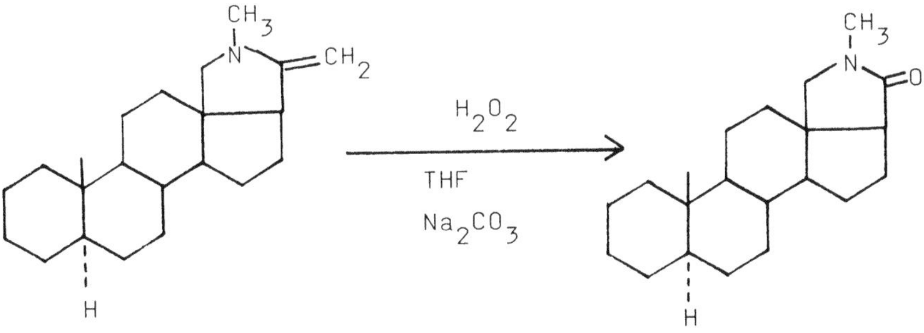

(34)   (35)

products oxidized at the α-carbon atom [47,48]. Oxidation of 6-acet-yl-1-(4-morpholino)cyclohexene (36) with hydrogen peroxide gives cyclopentanecarboxmopholide (37) [49].

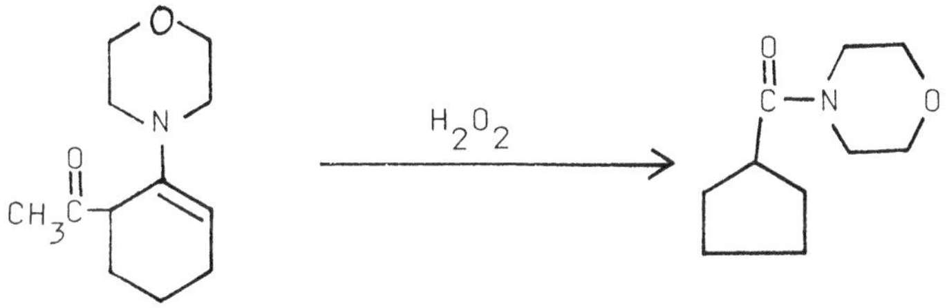

(36)   (37)

Oxidative cleavage of the carbon-carbon double bond in an ena-mine may also be accomplished by using sodium dichromate [51–53], nitric acid [54], periodate [32,55], or ruthenium tetroxide [56]. For example, treatment of the piperidine enamine of methyl 3β-formyl-26-trityloxycyclolandan-29-oate (38) with ruthenuim tetroxide produces 39 in a 71% yield [56].

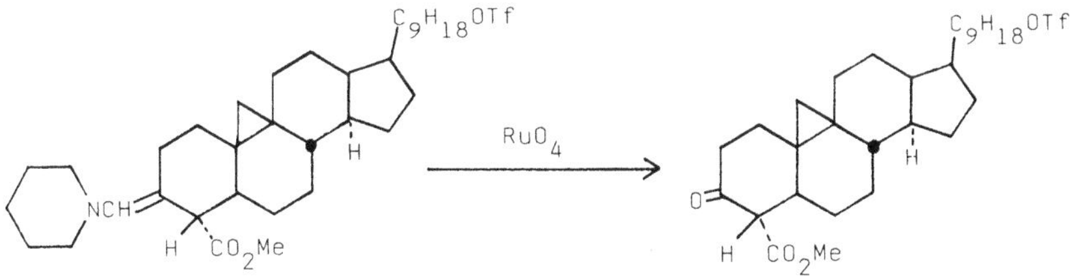

(38)   (39)

Oxygen itself can oxidize enamines producing identifiable products. For example, the uncatalyzed treatment of 1-(N,N-di-*n*-butyl-amino)butene (40) with oxygen at room temperature gives N,N-di-*n*-

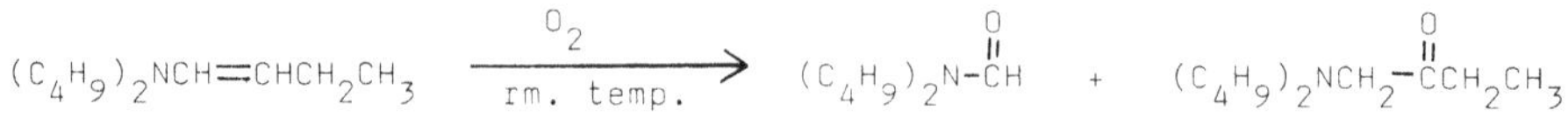

$$(C_4H_9)_2NCH\!=\!CHCH_2CH_3 \quad \xrightarrow[\text{rm. temp.}]{O_2} \quad (C_4H_9)_2N\!-\!\overset{\displaystyle O}{\overset{\displaystyle \|}{C}}H \quad + \quad (C_4H_9)_2NCH_2\!-\!\overset{\displaystyle O}{\overset{\displaystyle \|}{C}}CH_2CH_3$$

(40)   (41)   (42)

butylformamide (41) and 1-(N,N-di-*n*-butylamino)-2-butane (42). The former came about by oxidative cleavage of the carbon-carbon double bond and was obtained in a 19% yield, whereas the latter was obtained in a 34% yield [57]. 1-(N-Pyrrolidino)-cyclohexene (22) is oxidized by this treatment to α-pyrrolidinocyclohexanone (43) and 6-oxo-1-

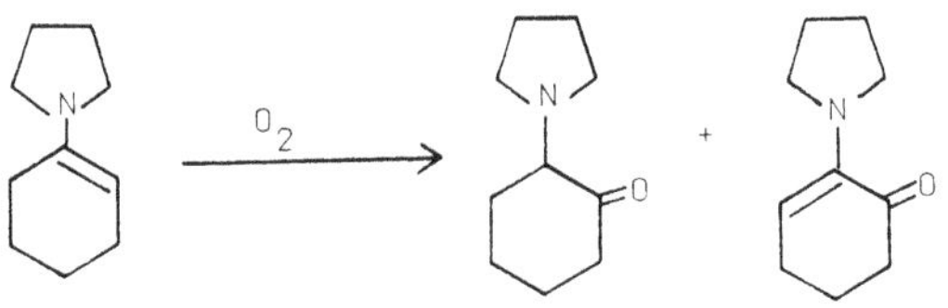

(22)   (43)   (44)

pyrrolidinocyclohexene (44) [57]. On the other hand, 1-(4-morpholino)cyclohexene (11) resists oxidation by uncatalyzed oxygen at room

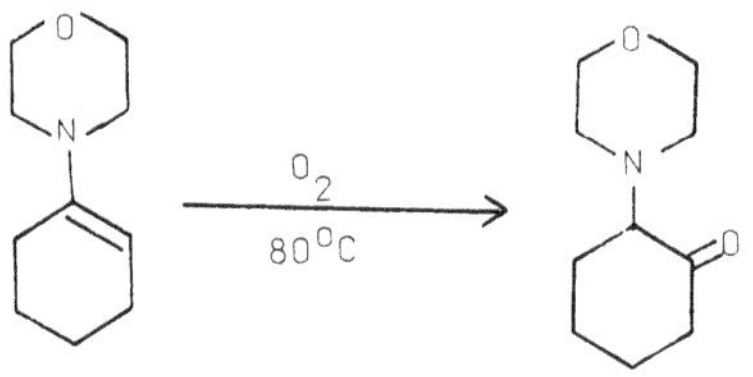

(11)   (45)

temperature, but at 80°C it reacts to form α-morpholinocyclohexanone (45) and some gum [57]. Other enamines have also been reported to be stable to oxygen [58].

Uncatalyzed oxygen oxidation of enamines has also been observed in indole series enamines [59]. Some enamines derived from α,β-unsaturated ketones are oxidized by oxygen [60]. For example, the pyrrolidine enamine of 10-methyl-$\Delta^{1(9)}$-octal-2-one (46) reacts with oxygen at room temperature to produce, after acid hydrolysis, 10-methyl-$\Delta^{1(9)}$-octalin-2,8-dione (47) [60].

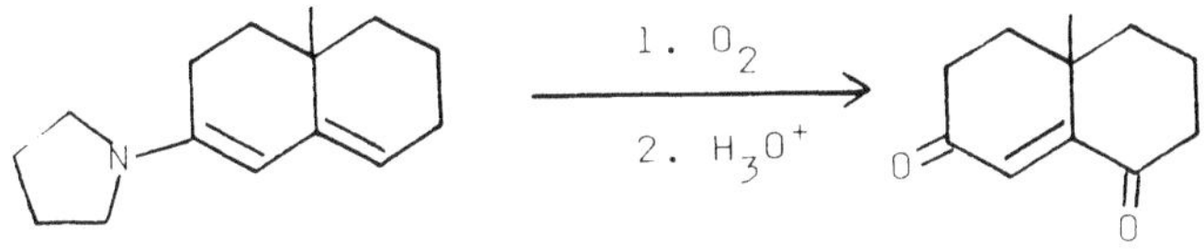

(46)   (47)

The use of copper salt catalysts with the oxygenation of enamines has been found to greatly facilitate the reaction [60–66]. In general, enamines with no β-vinylic hydrogen, such as enamine 48a, undergo

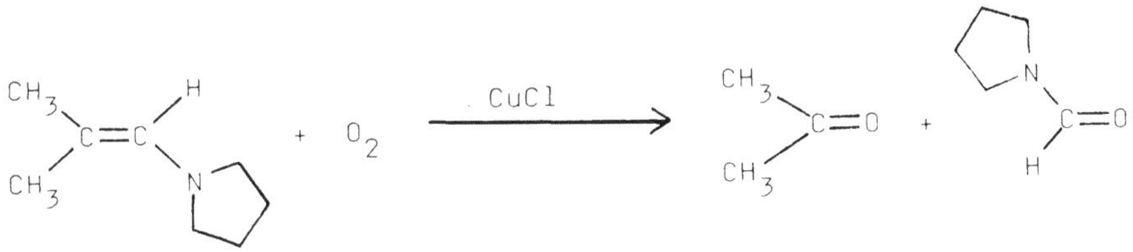

(48a)   (49)

oxidative cleavage in high yields [67,68], in this case to produce acetone and N-formyl pyrrolidine (49) in an 80% yield [68]. Enamines with a β-vinylic hydrogen, such as enamine 11, give four major prod-

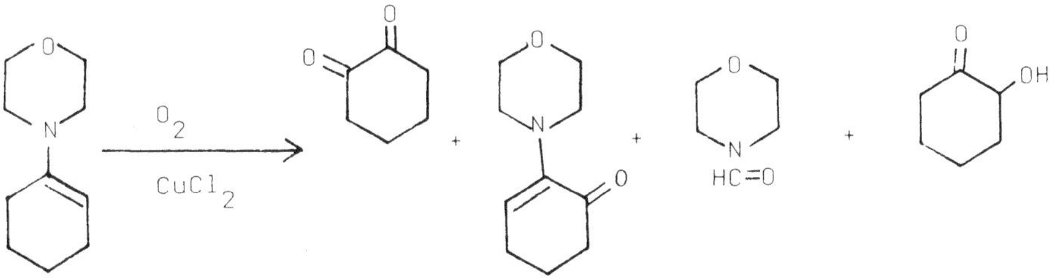

(11)   (50a)   (50b)   (50c)   (50d)

ucts in modest to low yields [67]. For enamine, 11; these are 1,2-cyclohexanone (50a; 30% yield), 1-(4-morpholino)-6-oxo-1-cyclohexene (50b; 12% yield), N-formylmorpholine (50c, 10% yield), and 2-hydroxy-cyclohexanone (50d; 7% yield) [68]. These are all noncleavage products.

The amine moiety of an enamine is determinative as far as the relative rates of copper-salt-catalyzed oxidation reactions are concerned. There is a fairly good correlation between these relative rates and the first-ionization potentials in a given series of enamines [67,68] (see Chapter 1, Section II). Other transition metal salts besides those of copper can be used. The activity of transition metals decreases in the following order: Cu>Fe>Co>>Ni. Copper halides are the salts with the highest activity. Either Cu(I) or Cu(II) can be used as catalyst, but apparently the actual catalyzing species is Cu(II) [68].

Photooxygenation of enamines gives products similar to those obtained by copper-salt-catalyzed oxidation of enamines [57,69—77]. For example, 1-piperidino-2-methylpropene (48b), when photolyzed in the presence of oxygen and a sensitizer, produces acetone and

(48b)

N-formylpiperidine in an essentially quantitative yield [69]. When 1-(4-morpholino)cyclohexene (11) is photooxygenated in methanol, it produces diketone 50a as the major product (75% yield) [75].

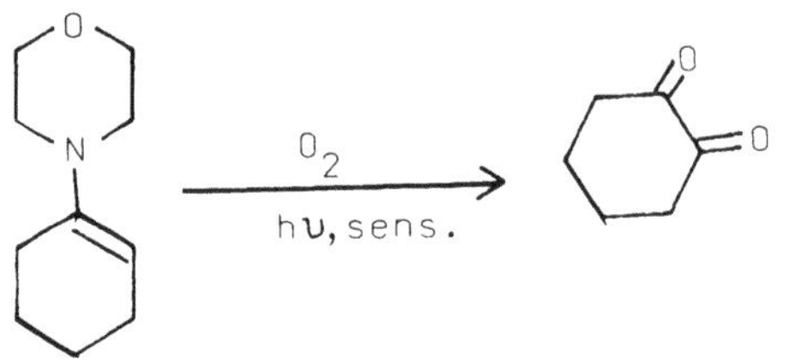

(11)   (50a)

The mechanism of these photooxygenation reactions of enamines has been extensively studied. It has been shown that singlet-state oxygen ($^1O_2$) is the reactant species involved [78—80]. When 2,6-di-t-butylphenol, a good free-radical inhibitor, was added to some of the reaction mixtures, there was no effect on the product yield [69], strongly suggesting a non-free-radical mechanism. A 1,2-dioxetane

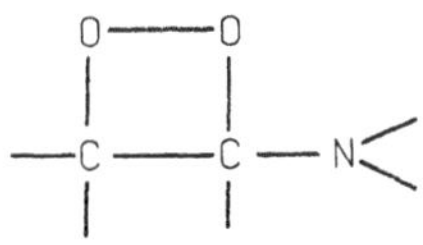

(51)

intermediate (51) was isolated and identified at -78°C as a crystalline
solid [75,78]. The dioxetane was stable at -78°C for several days,
but on warming to room temperature, violent decomposition to oxidative cleavage products took place.

The singlet-state oxygen ($^1\Delta g$) is produced by energy transfer
from an excited sensitizer to ground-state oxygen. This singlet-state
oxygen is 22.5 kcal/mol higher in energy than its ground triplet state
($^3\Sigma g^-$) [81,82]. The two electrons are paired in this excited singlet
state, and the properties of this species are those of a very reactive
olefin. A common reaction of singlet oxygen with monoalkenes having
an allylic hydrogen is the symmetry-allowed [83] concerted "ene"
reaction [81,82] [Eq. (2)]. However, production of a dioxetane (51)

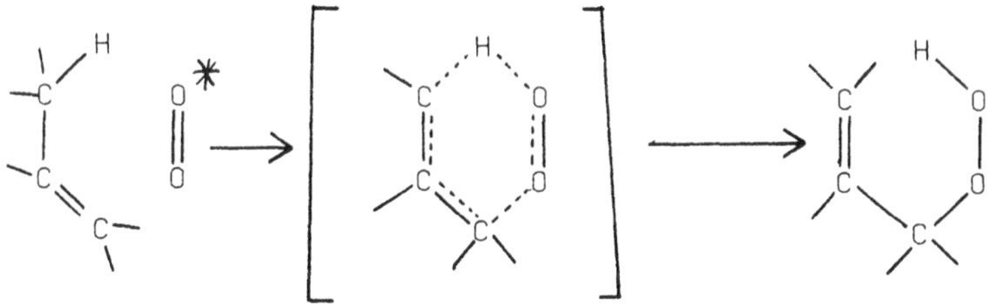

(2)

from an enamine by a concerted reaction would be a symmetry-forbid-
den ($\pi 2s + \pi 2s$) process. So it appears that the 1,2-dioxetane is
formed by at least a two-step process. The three proposed candidates
for the initially formed adduct between the enamine and singlet oxygen are a 1,4-diradical (52) [84], a 1,4-zwitterion (53) [85–87], and

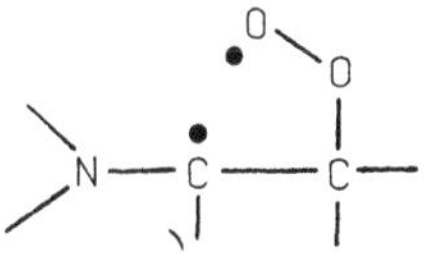

(52)

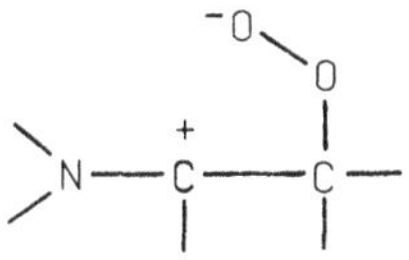

(53)

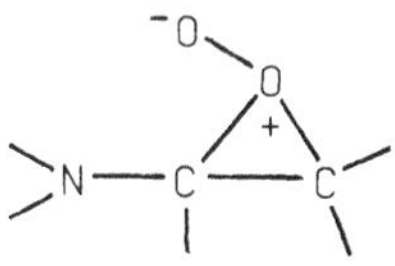

(54)

a peroxirane (perepoxide) (54) [88]. Theoretical studies using both *ab initio* MO calculations [87] and semiempirical calculations [85,87] seem to indicate that the 1,4-zwitterion (53) is the energetically favored intermediate for electron-rich enamines. By electron-rich enamines is meant enamines whose amine moiety strongly interacts with the alkene moiety in electron donation. The extent of this interactive electron donation parallels the first-ionization potentials of the enamines ([87]; see also Chapter 1, Section II). On the other hand, the 1,4-diradical pathway is energetically preferred for substituted ethylenes [87]. Those enamines which are not so electron-rich (some of those with morpholine moieties, for example) are indeterminate as to whether they proceed by the 1,4-zwitterion or the 1,4-diradical pathway. Protic solvents such as methanol are very effective in stabilizing the 1,4-zwitterion by hydrogen bonding [87].

It has been further determined, by a series of rate studies involving variously substituted aryl enamines and singlet oxygen, that the rate-determining step in the mechanism is a charge transfer step [79,80]. So the overall reaction mechanism is as described in Scheme

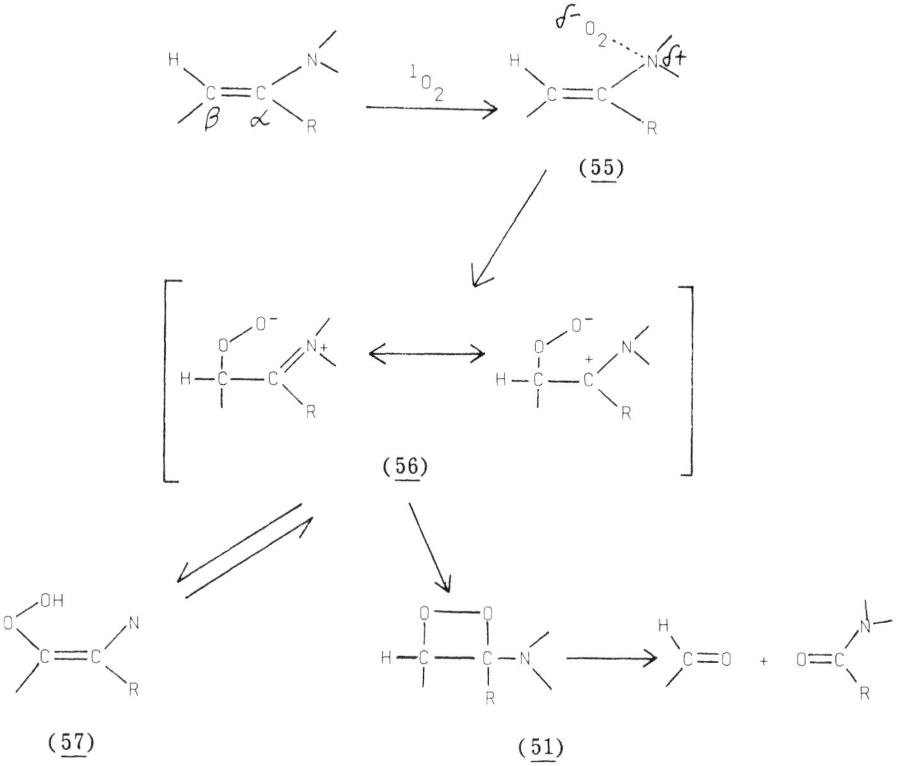

Scheme 1

1. The charge transfer complex (55) can rearrange to the zwitterion (56), which can then either go to the hydroperoxide (57) through protolysis involving a β-proton or transform into the 1,2-dioxetane (51). The thermal retrocycloaddition or cleavage of dioxetane 51 via a concerted, suprafacial process to two carboxyl-containing fragments is symmetry-allowed as long as one of the fragments is in an excited state. Emission of light has been observed in the decomposition of such a dioxetane, indicating that one of the fragments was probably in an excited state [77].

The formation of products other than oxidative cleavage of the carbon-carbon double bond (that is, formation of 1,2-diketones, α-hydroxyketones, α-ketoamine, and free secondary amine) is generally observed only when a β-hydrogen is present, such as in 1-(4-morpholino)cyclohexene (11) [75]. This can be accounted for by the fact that hydroperoxide 57 can only form in enamines with a β-hydrogen. This hydroperoxide could then go on to form the products mentioned. It has been reported that an α-phenyl group on the enamine (i.e., R = C$_6$H$_5$) also results in formation of these other products [74]. An alternative pathway for the production of these compounds is shown via the dioxetane [74].

B. By Losing Hydrogen

The most common driving force for the oxidative loss of hydrogen atom(s) in an enamine is aromatization of the reactant enamine. One reagent used to carry out this aromatization is oxygen [89–91]. For example, the pyrrolidine enamine of 2-oxo-1,2,3,4-tetrahydro-carbazole (58) can be air-oxidized to produce 2-pyrrolidinocarbazole (59) [91]. This type of aromatization can also be accomplished by the use

(58)   (59)

of trimethylene dithiotosylate [36] or palladium on charcoal in the presence of a hydrogen acceptor [92,93]. This is demonstrated by the reaction of 1-(4-morpholino)cyclohexene (11) with 10% Pd/C and styrene to produce N-phenylmorpholine (16) in a 70% yield [92] (see Section III of this chapter for an analogous disproportionation reaction). It was found that styrene is a better hydrogen acceptor than stilbene in this system [92].

(60)   (11)   (16)

Hydride ions can be abstracted from enamines by trityl ions to give the corresponding eniminium salt [94]. For example, enamine 11 can be oxidized to N-(2-cyclohexen-1-yliden)morpholinium tetrafluoroborate (60) by trityl tetrafluoroborate in a 91% yield.

N,N-Dimethylamino enamines can be oxidized to nitriles using hydroxylamine-O-sulfonic acid by oxidative loss of methyl groups (probably via the iminium salt) [166].

V. CHEMICAL REDUCTION

The chemical reduction of enamines by hydride depends on the prior generation of an iminium salt [95–97]. One of the two major methods for hydride reduction of iminium salts involves the use of metal hydrides such as potassium borohydride [98,99], sodium borohydride [97,100–102], lithium aluminum hydride [102–104], sodium cyanoborohydride [105,106], zinc-modified sodium cyanoborohydride [107], lithium tri-sec-butylborohydride (L-Selectride) [97], sodium bis(2-methoxyethoxy)aluminium hydride (Red-al) [97], diborane [108], t-butylamine borane [97], or carboxylhydridoferrates [109]. These reductions are more extensively discussed in Chapter 6 of this book. The second major method for hydride reduction of iminium salts is using formic acid [110–112]. This method will also be more completely discussed in Chapter 6.

Iminium salts can be reduced by using zinc and acetic acid [113–115].

NADPH or NADH (reduced forms of coenzyme nicotinamide adenine dinucleotide) reduce iminium ions in some biological systems, such as the reduction of cathenamine (20,21-didehydroajmalicine) (61) to indole alkaloid ajmalicine (62) in a 68% yield [116,120]. The Hantzsch ester (2,6-dimethyl-3,5-dicarboethoxy-1,4-dihydropyridine) (64) and 1-benzyl-1,4-dihydronicotinamide (63) have been used as simple models of NADH for the reduction of iminium salts [117–119]. It was shown by deuterium labeling that the reducing hydride ions come from the 4-position of the Hantzsch ester or the dihydronicotinamide [119]. Using these compounds to reduce iminium salts derived from

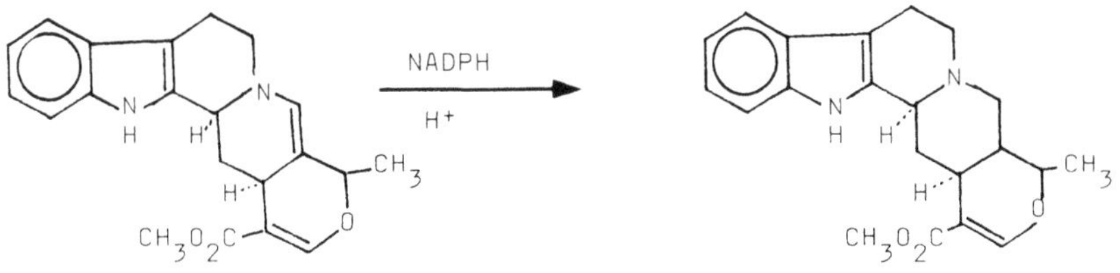

(61)

(62)

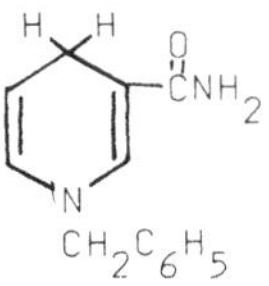

(63) 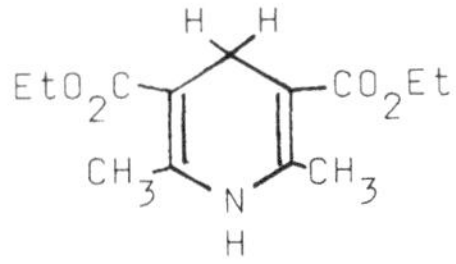   (64) 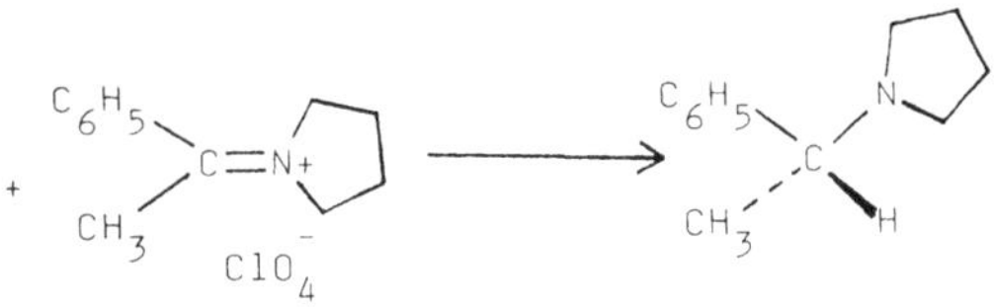   (65)   (66)

Δ^4-3-ketosteroids resulted in the hydrogen being transferred exclusively to the less hindered diastereotopic face of the substrate iminium group [117,118]. Reduction of 1-phenylethylidenepyrrolidium perchlorate (65) with the Hantzsch ester (64) resulted in the asymmetrical synthesis of (R)-N-(-phenylethyl)pyrrolidine (66) [119].

Phosphorous acid is an excellent reducing agent for iminium salts [121]. For example, enamine 11 is reduced to N-cyclohexylmorpholine

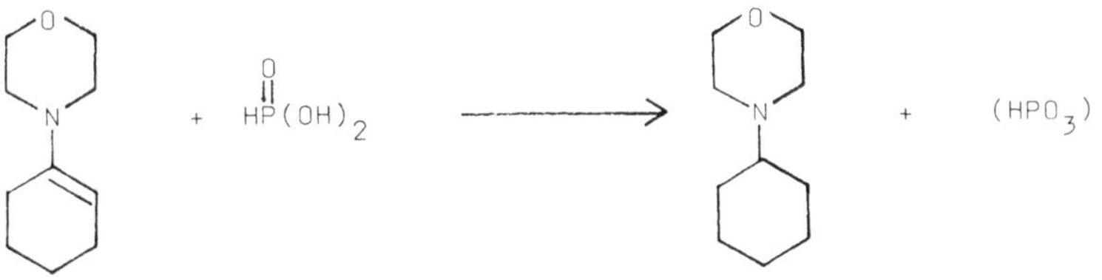

(11)   (15)

(15) by heating it at 90–100°C with phosphorous acid for 30 min to give a 91% yield. Phosphorous acid is slightly less stereoselective than formic acid. When the reductions were carried out in the presence of water, the yields were reduced. Reductions of iminium salts using 30% hypophosphorous acid give much lower yields [122].

The Sommelet reaction involves the reaction of hexamethylenetetramine (67) (iminium ion masked as a formaldehyde aminal) and halide to form the an α-arylalkyl aryl aldehyde or ketone [123–126]. The

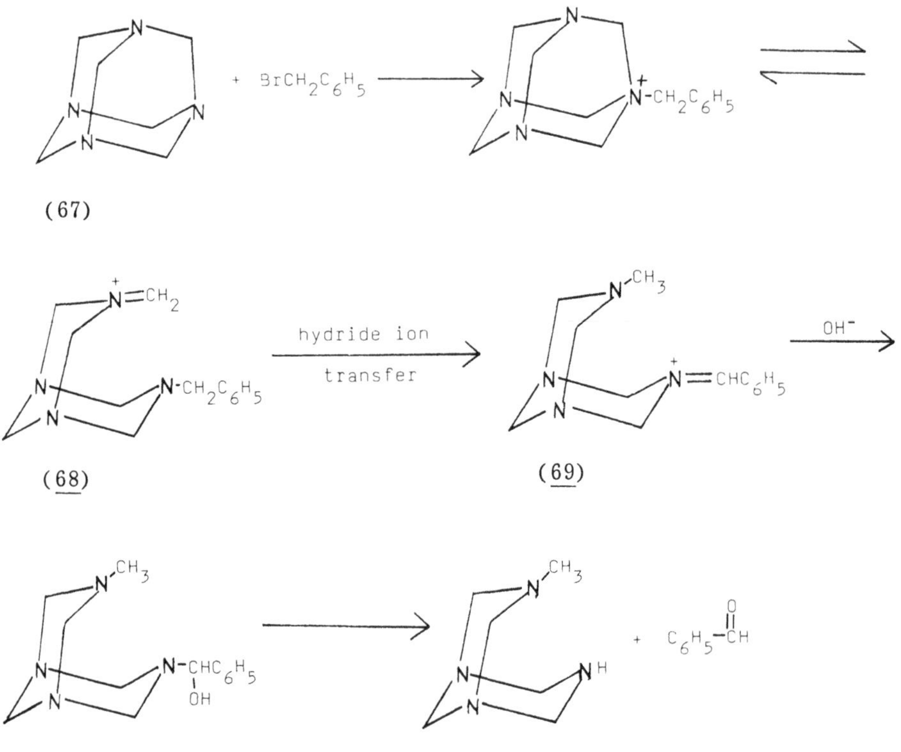

(67)

hydride ion

transfer

OH⁻

(68)   (69)

Scheme 2

mechanism for the reaction is shown in Scheme 2 [126—128]. Iminium ion 68 is reduced by an internal hydride transfer from the benzyl carbon to give iminium ion 69 because of the greater stability of the latter ion having its positive charge delocalized into the aromatic ring [129].

It has been observed that iminium salts can be reduced by sec-ondary amines [130—135], but these salts are not reduced by tertiary amines [131] except in very special cases ([136], Sommelet reaction discussed above). For example, treatment of N-2-bicyclo[2.2.1]hep-tylidenehexameth yleniminium perchlorate with excess hexamethyleni-mine produced exclusively the *endo* isomer of 2-N-hexamethylenimino-bicyclo[2.2.1]heptane (71) in a 60% yield [131]. The suggestion by Patmore and Chafetz [133] that the iminium ion is reduced by another enamine molecule rather than a secondary amine can be rejected on the basis of: (a) actual yields of saturated amines are often greater than 50% (the maximum possible yield if enamines are doing the reduc-

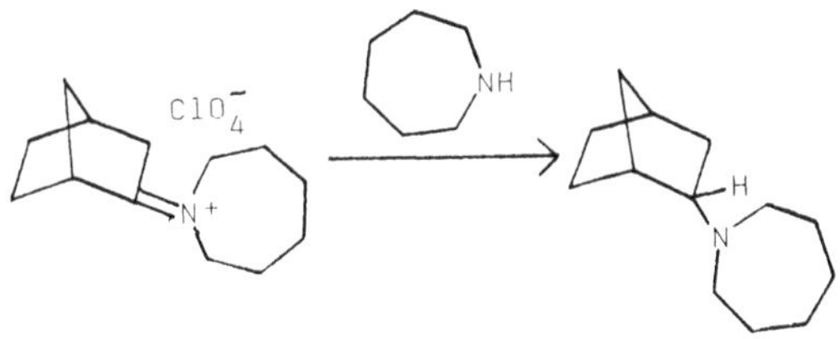

(70)   (71)

ing on a 1:1 basis); (b) reduction of iminium salts by fellow enamines would require an energetically unfavorable removal of a hydride ion from a bridgehead position in the case of bicyclic enamines; (c) the oxidation products expected if secondary amines are the reducing agents were isolated and characterized for the reactions shown by

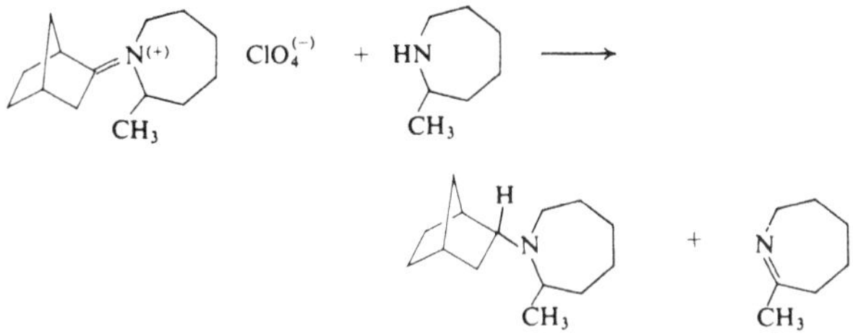

(3)

(72)

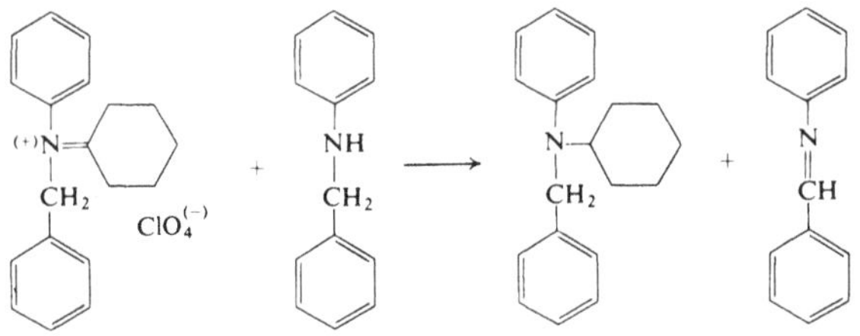

(4)

(73)

Eqs. (3) and (4), namely imines 72 and 73 [131]. Morpholine is poorer as a reducing agent than some of the other secondary amines, such as hexamethylenimine or pyrrolidine [131]. The relative abilities of various amines to transfer hydrogens to alkenes using catalysts such as Pd/C and RhCl(PPh$_3$)$_3$ as transfer agents has been studied [137,138]. These studies tend to support the observations made above; namely, pyrrolidine is good at transferring hydrogen but morpholine is rather poor, and secondary amines are better than tertiary amines.

Some iminium salts are reduced to saturated amines when treated with triphenylphosphine followed by basification [139]. Hydrogenolysis of the amine moiety has been observed to take place when the Grignard adduct to an iminium acetate is pyrolyzed [140].

Boranes (B_2H_6 derivatives) and alanes (AlH_3 derivatives) are electrophilic complex hydrides which can add directly to enamines. The reaction involves the addition of the alumino or borano group to the β-position of the enamine, and the hydride ion to the α-position. The addition of alanes leads to organoaluminum intermediates which can be oxidized to aminoalcohols, hydrolyzed to tertiary amines, or form olefins by elimination of the tertiary amine group. The alanes that have been added to enamines and show these reactions are aluminum hydrogen dichloride (LiAlH$_4$ and AlCl$_3$) [141—143], dialkyl aluminum hydrides [144], and trialkylaluminium [145].

Addition of diborane to an enamine can, after a subsequent step, lead to one of or a mixture of the following products: an α-hydroxylamine [146—152], a saturated amine (net hydrogenation) [148,149, 152—154], and/or an olefin (net hydrogenolysis) [155—158] (see Scheme 3).

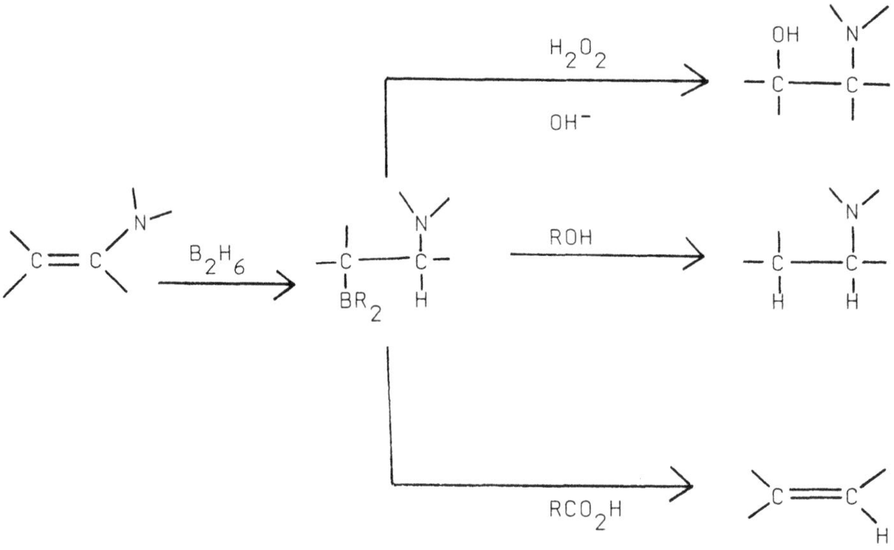

Scheme 3

The use of 9-borabicyclo[3.3.1]nonane(9-BBN) to react with an enamine in place of diborane sometimes improves the yield of the desired product [152].

Hydroboration of enamines can be used as one step in a sequence to bring about a carbonyl transposition [149–152,158]. One method involves acylation of an enamine and the Baeyer-Villiger reaction [158] (see Scheme 4). Another method uses pyrolysis of an amine

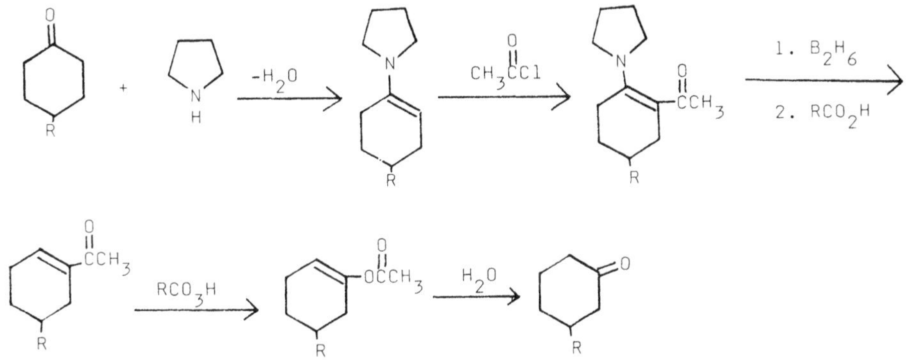

Scheme 4

oxide (Cope reaction) [149–152] (see Scheme 5).

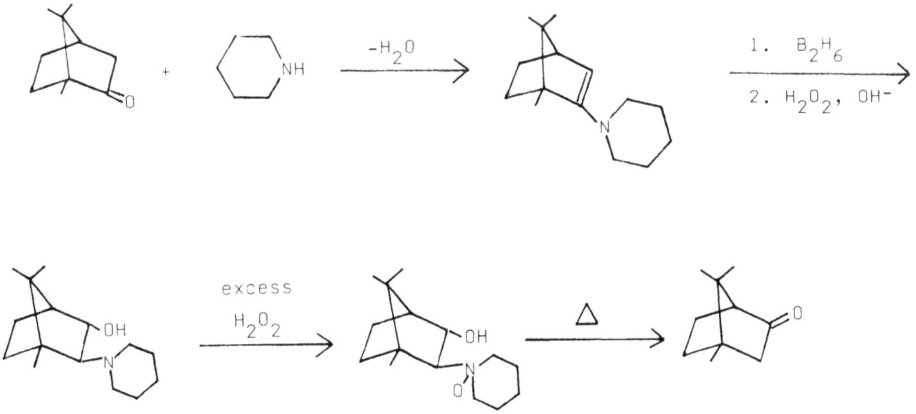

Scheme 5

Another method of reducing enamines is mercuration with mercuric acetate at low temperature in an aprotic solvent followed by reductive demercuration with sodium borohydride [159].

Enamines can be reduced via catalytic hydrogenation using catalysts such as PtO_2 [160−162], Pd/C [151], and Lindlar catalyst [163]. Catalytic hydrogenation has been shown to give different stereoisomers depending on what catalysts and solvents are used [164].

REFERENCES

1. J. M. Fritsch, H. Weingarten, and J. D. Wilson, *J. Am. Chem. Soc.*, *92*, 4038 (1970).
2. C. K. Mann and K. K. Barnes, *Electrochemical Reactions in Nonaqueous Systems*, Marcel Dekker, New York, 1970.
3. T. Chiba, M. Okimoto, H. Nagai, and Y. Takata, *J. Org. Chem.*, *44*, 3519 (1979).
4. L. L. Miller, G. D. Nordblom, and E. A. Mayeda, *J. Org. Chem.*, *37*, 916 (1972).
5. L. Eberson and K. Nyberg, in *Advances in Physical Organic Chemistry* (V. Gold and D. Bethell, eds.), Vol. 12, Academic Press, New York, 1976.
6. S. J. Huang and E. T. Hsu, *Tetrahedron Lett.*, 1385 (1971).
7. T. Shono, Y. Matsumura, H. Hamaguchi, T. Imanishi, and K. Yoshida, *Bull. Chem. Soc. Jpn.*, *51*, 2179 (1978).
8. J. P. Kutney and F. Bylsma, *J. Am. Chem. Soc.*, *92*, 6090 (1970).
9. R. Dietz and H. Lund, in *Organic Electrochemistry* (M. M. Baizer, ed.), Marcel Dekker, New York, 1973.
10. T. Shono, Y. Matsumura, J. Hayashi, M. Usui, S-I. Yamane, and K. Inoue, *Acta Chem. Scand, Ser B*, *37*, 491 (1983).
11. M. K. Polievkto, A. B. Grigor'ev, V. G. Granik, and R. G. Glushkov, *Zh. Obshch. Khim*, *43*, 1151 (1973).
12. A. B. Grigor'ev, V. G. Granik, and M. K. Polievktov, *Zh. Obshch. Khim*, *46*, 404 (1976).
13. C. P. Andrieux and J. M. Saveant, *Bull. Soc. Chim. France*, 4671 (1968).
14. A. A. Pozdeeva, V. A. Chernova, V. P. Yur'ev, and G. A. Tolstikov, *Zh. Obshch. Khim.*, *43*, 664 (1973).
15. G. Bianchetti, D. Pocar, and A. Marchesini, *Rend. 1st. Lombardo Sci. Lettere A*, *99*, 223 (1965); *CA*, *65*, 13581d (1966).
16. R. T. Dean and H. Rapoport, *J. Org. Chem.*, *43*, 4183 (1978).
17. R. S. Monson, D. N. Priest, and J. C. Ullrey, *Tetrahedron Lett.*, 929 (1972).

18. M. E. Kuehne and T. J. Giacobbe, *J. Org. Chem.*, *33*, 3359 (1968).
19. F. Corbani, B. Rindone, and C. Scolastico, *Tetrahedron*, *29*, 3253 (1973).
20. M. Montury and J. Gore, *Tetrahedron Lett.*, 219 (1977).
21. F. Corbani, B. Rindone, and C. Scolastico, *Tetrahedron*, *31*, 455 (1975).
22. Z. Cekovic, J. Bosnjak, and M. Cvelkovic, *Tetrahedron Lett.*, 2675 (1980).
23. G. L. May and J. T. Pinhey, *Aust. J. Chem.*, *35*, 1859 (1982).
24. R. L. Augustine, *J. Org. Chem.*, *28*, 581 (1962).
25. S. O. Lawesson, H. J. Jakobsen, and E. H. Larsen, *Acta. Chem. Scand.*, *17*, 1188 (1963).
26. S. Danishefsky and R. Cavanaugh, *Chem. and Ind. (London)*, 2171 (1967).
27. M. E. Kuehne and E. Underwood, unpublished results, 1967.
28. H. Metzger, *Tetrahedron Lett.*, 203 (1964).
29. G. Drefahl, G. Heublein, and G. Tetzloff, *J. Prakt. Chem.*, *311*, 162 (1969).
30. R. Sudo and M. Takahashi, *Jap. Pat.*, 7009531 (1970); *CA*, *73*, 14296 (1970).
31. J. R. Mahajan, G. A. L. Ferreira, H. C. Araujo, and B. J. Nunes, *Tetrahedron Lett.*, 3025 (1974).
32. J. R. Mahajan, B. J. Nunes, H. C. Araujo, and G. A. L. Ferreira, *J. Chem. Res. (S)*, 284 (1979).
33. R. H. Fischer and H. M. Weitz, *Synthesis*, 791 (1975).
34. M. Muhlstadt, P. Schneider, and D. Martinetz, *J. Prakt. Chem.*, *315*, 935 (1973).
35. E. Fanghanel, *J. Prakt. Chem.*, *317*, 123 (1975).
36. R. B. Woodward, I. J. Pachter, and M. L. Scheinbaum, *J. Org. Chem.*, *36*, 1137 (1971).
37. R. B. Woodward, I. J. Pachter, and M. L. Scheinbaum, in *Organic Syntheses* (R. E. Ireland, ed.), Vol. 54, Wiley, New York, 1974, p. 39.
38. J. Iwasa and S. Naruto, *Yakugaku Zasshi*, *86*, 534 (1966).
39. S. Naruto, H. Nishimura, and H. Kaneko, *Tetrahedron Lett.*, 2127 (1972).
40. M. E. Herr and F. W. Heyl, *J. Am. Chem. Soc.*, *74*, 3627 (1952).
41. G. Opitz and W. Merz, *Ann.*, *652*, 139 (1962).
42. G. Opitz and A. Griesinger, *Ann.*, *665*, 101 (1963).
43. M. P. Strobel, L. Morin, and D. Paquer, *Tetrahedron Lett.*, 523 (1980).
44. M. P. Strobel, L. Morin, and D. Paquer, *Nouv. J. Chim.*, *4*, 603 (1980).

45. A. Rieche, E. Schmitz, and E. Beyer, *Ber.*, *92*, 1206 (1959).
46. A. Rieche, E. Schmitz, and E. Beyer, *Ber.*, *92*, 1212 (1959).
47. P. Milliet, A. Picot, and X. Lusinchi, *Tetrahedron*, *37*, 4201 (1981).
48. A. Picot, P. Milliet, and X. Lusinchi, *Tetrahedron Lett.*, 1577 (1976).
49. L. P. Vinogradova and S. I. Zav'yalov, *Izv. Akad. Nauk SSSR, Ser. Klim.*, 1795 (1966).
50. N. J. Leonard, L. A. Miller, and P. D. Thomas, *J. Am. Chem. Soc.*, *78*, 3463 (1956).
51. D. A. Shepherd, R. A. Donia, J. A. Campbell, B. A. Johnson, R. P. Holysz, G. Slomp, J. E. Stafford, R. L. Pederson, and A. C. Ott, *J. Am. Chem. Soc.*, *77*, 1212 (1955).
52. G. Slomp, Y. F. Shealy, J. L. Johnson, R. A. Donia, B. A. Johnson, R. P. Holysz, R. L. Pederson, A. O. Jensen, and A. C. Ott, *J. Am. Chem. Soc.*, *77*, 1216 (1955).
53. G. K. Trivedi, P. S. Kalsi, and K. Chadravarti, *Tetrahedron*, *20*, 2631 (1964).
54. Y. Ogata, Y. Sawaki, and Y. Kuriyama, *Tetrahedron*, *24*, 3425 (1968).
55. D. H. Rammler, *Biochemistry*, *10*, 4699 (1971).
56. M. C. Desai, H. P. S. Chawla, and S. Dev, *Tetrahedron*, *38*, 379 (1982).
57. R. A. Jerussi, *J. Org. Chem.*, *34*, 3648 (1969).
58. J. E. Huber, *Tetrahedron Lett.*, 3271 (1968).
59. G. Massiot, F. S. Oliveira, and J. Levy, *Tetrahedron Lett.*, 177 (1982).
60. S. K. Malhotra, J. J. Hostynek, and A. F. Lundin, *J. Am. Chem. Soc.*, *90*, 6565 (1968).
61. V. Van Rheenen, *Chem. Commun.*, 314 (1969).
62. V. Van Rheenen, U.S. Patent 3,661,942 (1972); *CA*, *77*, 102041w (1972).
63. J. R. L. Smith and Z. A. Malik, *J. Chem. Soc.* (*B*), 920 (1970).
64. T-L. Ho, *Synth. Commun.*, *4*, 135 (1974).
65. E. Balogh-Hergovich and G. Speier, *React. Kinet. Catal. Lett.*, *3*, 139 (1975).
66. J. R. Bull and A. Tuinman, *S. Afr. J. Chem.*, *32*, 17 (1979).
67. T. Itoh, K. Kaneda, I. Watanabe, S. Ikeda, and S. Teranishi, *Chem. Lett.*, 227 (1976).
68. K. Kaneda, T. Itoh, N. Kii, K. Jitsukawa, and S. Teranishi, *J. Mol. Catal.*, 349 (1982).
69. C. S. Foote and J. W-P. Lin, *Tetrahedron Lett.*, 3267 (1968).
70. C. S. Foote, *Acc. Chem. Res.*, *1*, 104 (1968).
71. I. Saito, M. Imuta, Y. Yakahashi, S. Malsugo, and T. Matsuura, *J. Am. Chem. Soc.*, *99*, 2005 (1977).
72. K. Pfoertner and K. Bernauer, *Helv. Chim. Acta*, *51*, 1787 (1968).

73. K. Miyano, Y. Ohfune, S. Azuma, and T. Matsumoto, *Tetrahedron Lett.*, 1545 (1974).
74. W. Ando, T. Saiki, and T. Migita, *J. Am. Chem. Soc.*, 97, 5028 (1975).
75. H. H. Wasserman and S. Terao, *Tetrahedron Lett.*, 1735 (1975).
76. A. Murai, C. Sato, H. Sasamori, and T. Masamune, *Bull. Chem. Soc. Jpn.*, 49, 499 (1976).
77. F. McCapra, Y. C. Chang, and A. Burford, *J. Chem. Soc. Chem. Commun.*, 608 (1976).
78. C. S. Foote, A. A. Dzakpasu, and J. W-P. Lin, *Tetrahedron Lett.*, 1247 (1975).
79. N. H. Martin and C. W. Jefford, *Tetrahedron Lett.*, 22, 3949 (1981).
80. N. H. Martin and C. W. Jefford, *Helv. Chim. Acta*, 65, 762 (1982).
81. R. W. Denny and A. Nickon, *Organic Reactions*, 20, 133 (1973).
82. D. R. Kearns, *Chem. Rev.*, 71, 395 (1971).
83. R. B. Woodward and R. Hoffman, *The Conservation of Orbital Symmetry*, Verlag Chemie, Weinheim, 1970.
84. L. B. Harding and W. A. Goddard, III, *J. Am. Chem. Soc.*, 99, 4520 (1977).
85. M. J. S. Dewar and W. Thiel, *J. Am. Chem. Soc.*, 97, 3978 (1975).
86. K. Yamaguchi and T. Fueno, *Tetrahedron Lett.*, 3433 (1979).
87. K. Yamaguchi, *Int. J. Quantum Chem.*, 20, 393 (1981).
88. P. D. Bartlett, *Chem. Soc. Rev.*, 5, 149 (1976).
89. N. J. Leonard and R. R. Sauers, *J. Org. Chem.*, 21, 1187 (1956).
90. P. A. Zoretic, F. Barcelos, and B. Branchaud, *Org. Prep. Proced. Int.*, 8, 211 (1976).
91. Y. Kumar, A. K. Saxena, P. C. Jain, and N. Anand, *Indian J. Chem., Sect. B*, 19B, 996 (1980).
92. G. Van Binst, R. Baert, M. Biesemans, C. Mortelmans, and R. Salsmans, *Bull. Soc. Chim. Belg.*, 85, 1 (1976).
93. S. L. Keely, Jr., A. J. Martinez, and F. C. Tahk, *Tetrahedron*, 26, 4729 (1970).
94. M. T. Reetz, W. Stephan, and W. F. Maier, *Synth. Commun.*, 10, 867 (1980).
95. N. J. Leonard, A. S. Hay, R. W. Fulmer, and V. W. Gash, *J. Am. Chem. Soc.*, 77, 439 (1955).
96. N. J. Leonard and F. P. Hauck, Jr., *J. Am. Chem. Soc.*, 79, 5279 (1957).
97. R. O. Hutchins, W-Y. Su, R. Sivakumar, F. Cistone, and Y. P. Stercho, *J. Org. Chem.*, 48, 3412 (1983).
98. J. J. Panouse, *Compt. Rend.*, 233, 260 (1951).
99. J. J. Panouse, *Compt. Rend.*, 233, 1200 (1951).
100. K. Schenker, *Angew. Chem.*, 72, 638 (1960).

101. M. Ferles, Collect. Czech. Chem. Commun., 23, 479 (1958).

102. J. W. Daly and B. Witkop, J. Org. Chem., 27, 4104 (1962).

103. M. Ferles, Collect. Czech. Chem. Commun., 24, 2221 (1959).

104. D. Cabaret, G. Chauviere, and Z. Welvart, Tetrahedron Lett., 4109 (1966).

105. R. F. Borch, M. D. Bernstein, and H. D. Durst, J. Am. Chem. Soc., 93, 2897 (1971).

106. R. F. Borch in Organic Syntheses (H. O. House, ed.), Vol. 52, John Wiley, New York, 1972, p. 124.

107. S. Kim, C. H. Oh, J. S. Ko, K. H. Ahn, and Y. J. Kim, J. Org. Chem., 50, 1927 (1985).

108. O. Cervinka and L. Hub, Tetrahedron Lett., 463 (1964).

109. T. Mitsudo, Y. Watanabe, M. Tanaka, S. Alsuta, K. Yamamoto, and Y. Takegami, Bull. Chem. Soc. Jpn., 48, 1506 (1975).

110. R. Lukes, Collect. Czech. Chem. Commun., 10, 56 (1938).

111. R. Lukes and J. Jizba, Chem. Listy, 47, 1366 (1953).

112. N. J. Leonard and R. R. Sauers, J. Am. Chem. Soc., 79, 6210 (1957).

113. F. L. Weisenborn and P. A. Diassi, J. Am. Chem. Soc., 78, 2022 (1956).

114. A. Buzas, J-P. Jacquet, and G. Lavelle, J. Org. Chem., 45, 32 (1980).

115. G. Rossey, A. Wick, and E. Wenkert, J. Org. Chem., 47, 4745 (1982).

116. J. Stoeckigt, H. P. Husson, C. Kan-Fan, and M. H. Zenk, J. Chem. Soc., Chem. Commun., 164 (1977).

117. U. K. Pandit, F. R. Mas Cabre, R. A. Gase, and M. J. de Nie-Sarink, J. Chem. Soc., Chem. Commun., 627 (1974).

118. U. K. Pandit, R. A. Gase, F. R. Mas Cabre, and M. J. de Nie-Sarink, J. Chem. Soc., Chem. Commun., 211 (1975).

119. N. Baba, K. Nishiyama, J. Oda, and Y. Inouye, Agric. Biol. Chem., 40, 1441 (1976).

120. P. Heinstein, J. Stoeckigt, and M. H. Zenk, Tetrahedron Lett., 141 (1980).

121. D. Redmore, J. Org. Chem., 43, 992 (1978).

122. A. G. Cook, unpublished results.

123. M. Sommelet, C. R. Acad. Sci. Paris, 157, 852 (1913).

124. M. Sommelet, Bull. Soc. Chim. Fr., 13, 1085 (1913).

125. S. J. Angyal, Org. React., 8, 197 (1954).

126. N. Blazevic, D. Kolbah, B. Belin, V. Sunjic, and F. Kajfez, Synthesis, 161 (1979).

127. P. Le Henaff, C. R. Acad. Sci. Paris, 253, 2706 (1961).

128. P. Le Henaff, Ann. Chim. (Paris), 7, 367 (1962).

129. G. E. Stokker and E. M. Schultz, Synth. Commun., 12, 847 (1982).

130. A. G. Cook, W. C. Meyer, K. E. Ungrodt, and R. Mueller, J. Org. Chem., 31, 14 (1966).

131. A. G. Cook and C. R. Schulz, *J. Org. Chem.*, *32*, 473 (1967).
132. C. Kaiser, A. Burger, L. Zirngibl, C. S. Davis, and C. L. Zirkle, *J. Org. Chem.*, *27*, 768 (1962).
133. E. L. Patmore and H. Chafetz, *J. Org. Chem.*, *32*, 1254 (1967).
134. J. F. Stephan and E. Marcus, *J. Org. Chem.*, *34*, 2535 (1969).
135. L. D. Quin and R. C. Stocks, *J. Org. Chem.*, *39*, 686 (1974).
136. C. W. Thornber, *J. Chem. Soc.*, *Chem. Commun.*, 238 (1973).
137. T. Nishiguchi, K. Tachi, and K. Fukuzumi, *J. Org. Chem.*, *40*, 237 (1975).
138. T. Nishiguchi, H. Imai, Y. Hirose, and K. Fukuzumi, *J. Catal.*, *41*, 249 (1976).
139. G. Opitz, A. Griesinger, and H. W. Schubert, *Ann.*, *665*, 91 (1963).
140. P. P. Lynch and P. H. Doyle, *Gazz. Chim. Ital.*, *98*, 645 (1968).
141. J. Sansoulet and Z. Welvart, *Bull. Soc. Chim. France*, 77 (1962).
142. J. W. Lewis and P. P. Lynch, *Proc. Chem. Soc.*, 19 (1963).
143. J. M. Coulter, J. W. Lewis, and P. P. Lynch, *Tetrahedron*, *24*, 4489 (1968).
144. L. I. Zakharkin and L. A. Savina, *Izv. Akad. Nauk SSSR, Ser. Khim.*, 1695 (1964).
145. A. Alberola and F. J. L. Lopez, *An. Quim.*, *73*, 893 (1977).
146. I. J. Borowitz and G. L. Williams, *J. Org. Chem.*, *32*, 4157 (1967).
147. J-J. Barieux and J. Gore, *Bull. Soc. Chim. Fr.*, 3978 (1971).
148. J-J. Barieux and J. Gore, *Tetrahedron*, *28*, 1537 (1972).
149. J-J. Barieux and J. Gore, *Tetrahedron*, *28*, 1555 (1972).
150. F. Bondavalli, P. Schenone, A. Ranise, and S. Lanteri, *J. Chem. Soc.*, *Perkin Trans.*, *1*, 2626 (1980).
151. F. Bondavalli, P. Schenone, and A. Ranise, *J. Chem. Res. (S)*, 257 (1980); *J. Chem. Res. (M)*, 3256 (1980).
152. Y. K. Yee and A. G. Schultz, *J. Org. Chem.*, *44*, 719 (1979).
153. J. A. Marshall and W. S. Johnson, *J. Org. Chem.*, *28*, 421 (1963).
154. T. Kudo and A. Nose, *Yakugaku Zasshi*, *94*, 1475 (1974).
155. J. W. Lewis and A. A. Pearce, *Tetrahedron Lett.*, 2039 (1964).
156. J. W. Lewis and A. A. Pearce, *J. Chem. Soc. (B)*, 864 (1969).
157. R. W. Hyde, *Dissertation Abstr.*, *B27*, 1090 (1966).
158. M. Montury and J. Gore, *Tetrahedron*, *33*, 2819 (1977).
159. R. D. Bach and D. K. Mitra, *J. Chem. Soc.*, *Chem. Commun.*, 1433 (1971).
160. C. Mannich and H. Davidsen, *Chem. Ber.*, *69*, 2106 (1936).
161. F. Korte and A. K. Bocz, *Chem. Ber.*, *99*, 1918 (1966).
162. H. Mazarguil and A. Lattes, *C. R. Acad. Sci. Paris*, *267*, 724 (1968).
163. H. Mazarguil and A. Lattes, *Bull. Soc. Chim. France*, 119 (1971).

164. D. C. Horwell and G. H. Timms, *Synth. Commun.*, *9*, 223 (1979).
165. D. C. Snyder, *J. Organomet. Chem.*, *301*, 137 (1986).
166. H. Biere and R. Russe, *Tetrahedron Lett.*, 1361 (1979).
167. P. Bravo and C. Ticozzi, *Gazz. Chim. Ital.*, *105*, 91 (1975).

6
Ternary Iminium Salts

JOSEPH V. PAUKSTELIS *Kansas State University, Manhattan, Kansas*

A. GILBERT COOK *Valparaiso University, Valparaiso, Indiana*

I.	Introduction	276
II.	Structure, Preparation, and Detection of Iminium Salts	276
	A. Structure of Iminium Salts	276
	B. Preparation of Iminium Salts	277
	C. Detection of Iminium Salts	288
III.	Addition of Nucleophiles to Iminium Salts	297
	A. Addition of Organometallic Reagents	297
	B. Addition of Hydride	302
	C. Addition of Diazoalkanes	313
	D. Addition of Other Nucleophiles	317
IV.	Deprotonation of Iminium Salts	328
V.	Cycloadditions of Iminium Salts	331
VI.	Photochemical Reactions of Iminium Salts	332
VII.	Rearrangements of Iminium Salts	333
	References	334

I. INTRODUCTION

This chapter will be surveying the structure, preparation, and reactions (primarily nucleophilic additions) of iminium salts. The field of iminium salts has been extensively reviewed in a two-volume work (1976 and 1979) edited by Bohme and Viehe [1]. This chapter will restrict its consideration to simple ternary iminium salts and will not cover the variously substituted iminium salts also dealt with in the aforementioned two volume work.

II. STRUCTURE, PREPARATION, AND DETECTION OF IMINIUM SALTS

A. Structure of Iminium Salts

The Lewis structure of ternary iminium ions involves two contributing forms shown by 1a and 1b.

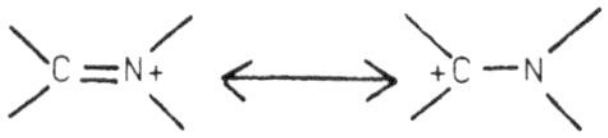

(1a)   (1b)

Quantum mechanical calculations on the simplest possible ternary iminium ion, the N,N-dimethylmethyleniminium ion (2), showed it to be about 21 kcal/mol *less* stable than the isomeric iminium ion 3. This

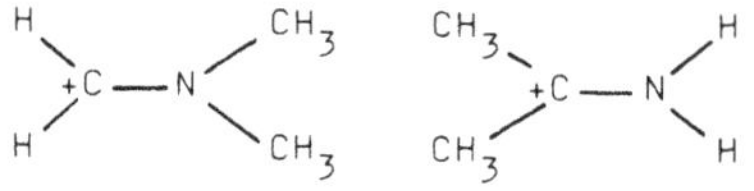

(2)   (3)

result was obtained both by *ab initio* STO-3G calculations [2,3] and by semiempirical MNDO calculations [4,5]. The optimum geometry required for these ions in order to minimize their energy is planar, that is, the two olefin hydrogen atoms, the three carbon atoms, and the nitrogen atom all lying in one plane. The positive charge for the ions is calculated to be predominantly localized at the carbon atoms with a charge of +0.35 for ion 2 and a charge greater than +0.4 for ion 3 [2]. These same calculations show an increase in Mulliken population for nitrogen in going from the less stable ion 2 to the more stable ion 3. The total nitrogen electron population is 7.16 for 2 and 7.32 for 3, or a net charge on the nitrogen atom of -0.16 for 2 and -0.3 for 3. This increase in the total nitrogen electron population for iminium ion 3 over iminium ion 2 appears to be the source of the increased stability

of 3 as compared to 2 [2]. In a similar vein, the N,N-dimethylisopro-
pylideniminium ion (4) was found by MNDO calculations to be about
16 kcal/mol *less* stable than isomeric iminium ion 5 [4].

CH₃ structure — (4), CH₃CH₂ structure — (5), CH= structure — (6)

(4)   (5)   (6)

MNDO calculations have shown the C-protonated form of enamine
2-(N,N-dimethylamino)propene, namely the N,N-dimethylisopropyliden-
iminium ion (4), to be about 18 kcal/mol more stable than the N-protonated form (6) (see Chapter 1, Section IV.E) [4].

Single crystal x-ray diffraction studies have been carried out on
N,N-dimethylisopropylideniminium perchlorate (4a) [6]. The body-
centered tetragonal crystal showed a planar iminium ion with C = N +

$$CH_3 \quad 125° \quad CH_3$$
$$117° \quad 1.51 \text{ Å}$$
$$117° \quad 1.30 \text{ Å} \quad ClO_4^-$$
$$117° - N \quad 1.51 \text{ Å}$$
$$CH_3 \quad + \quad CH_3$$

(4a)

distance of 1.30 Å and C—CH₃ and N—CH₃ distances of 1.51Å. The
< C–C–C and < H₃C–N–CH₃ are both 125.4°, and the < H₃C–N–C
are both 117.3°.

B. Preparation of Iminium Salts

Protonation of Enamines

Iminium salts are readily available from C-protonation of the corresponding enamines (see Chapter 1, Section IV.E). Experimentally,
the procedure is very simple: The enamine, dissolved in ether or
some other solvent, is treated with anhydrous hydrogen halide, 70%
perchloric acid, or trifluoroacetic acid. The iminium salt usually sep-
arates as a solid and is then collected. Protonation at low temperatures provides evidence that N-protonation occurs first, followed by
rearrangement to a C-protonated iminium salt. The indirect evidence
for N-protonation of a variety of enamines at -70°C comes from reaction of ozone, diazomethane, or lithium aluminum hydride (LAH) with
N-protonated salts [7]. Freshly prepared N-isobutylidenepiperidinium

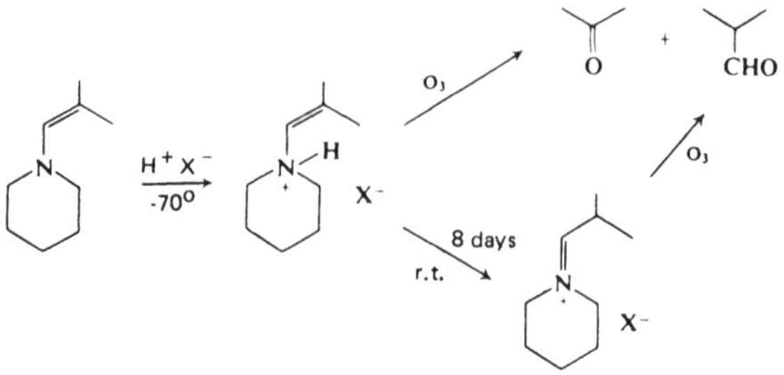

Scheme 1

hexachlorostannate gives a mixture of isobutyraldehyde and acetone after ozonolysis and reductive isolation. If the salt is allowed to stand for 8 days prior to ozonolysis, only isobutyraldehyde is obtained (see Scheme 1) [7]. In a similar manner N-(2-ethyl)butylidenepiperidinium hexachlorostannate gives mostly diethyl ketone and only a little of 2-ethylbutanal when a fresh solution is ozonized. If it is allowed to stand for a period of time, the only product obtained by ozonolysis is 2-ethylbutanal [7].

The determination of position of protonation by reaction with diazomethane was performed as follows: The enamine was treated at -70°C with ethereal hydrogen chloride, and the suspension of precipitated salt was treated with diazomethane and allowed to warm slowly to -40°C, at which temperature nitrogen was liberated. The reaction with LAH was carried out similarly except that an ether solution of LAH was added in place of diazomethane. The results from reaction of diazomethane and LAH [7] are summarized in Table 1.

The close agreement of the three methods supports the contention that protonation at low temperatures first occurs at nitrogen and is followed by a proton shift to give the iminium salt [7]. Other indirect evidence for initial N-protonation has also been found [8,9].

Direct evidence for initial N-protonation was provided by the isolation and identification of enammonium salt 8 from low-temperature

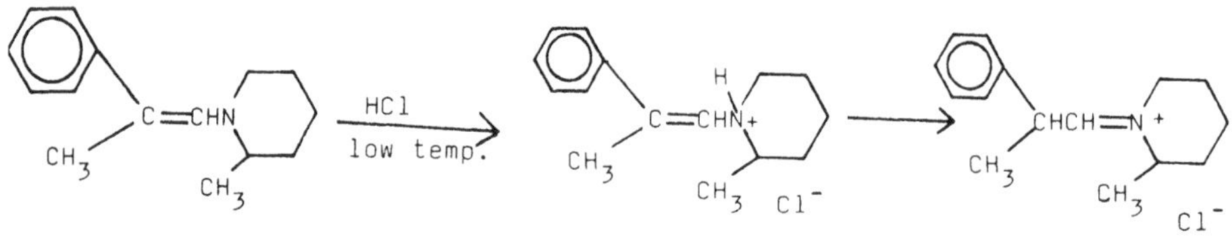

(7)   (8)   (9)

protonation of 2-methyl-1-(β-methylstyrl)-piperidine (7) with anhydrous hydrogen chloride [10-13]. When this white, crystalline N-

Table 1 Position of Enamine Protonation

Compound	Time at room temperature	Diazomethane % protonation		LiAlH$_4$ % protonation	
		At C	At N	At C	At N
1-N-Morpholylbutene	0	16	74	10	80
	60 min	93	0	83	0
1-N-Pyrrolidylisobutene	0	2	80	5	83
	48 hr	90	0	91	0
1-N-Morpholylisobutene	0	0	82	2	84
	48 hr	83	5	92	0
1-N-Pyrrolidylcyclohexene	0	93	2	91	0
	24 hr	95	0	—	—
1-N-Piperidylcyclohexene	0	12	81	25	68
	24 hr	89	8	87	7
1-N-Morpholylcyclohexene	0	11	86	18	76
	19 hr	81	10	93	2

protonated salt 8 is allowed to stand in a dry solvent at room temperature, it readily changes to the corresponding iminium salt 9. The relative stability of enammonium salt 8 is probably due to the stabilization of the enamine carbon-carbon double bond by conjugation with the benzene ring.

In solution, the N-protonated enammonium ion is the kinetically favored product, whereas the C-protonated iminium ion is the thermodynamically favored product. In the gas phase only C-protonation of enamines occurs [14].

The commonly used protonating agents are hydrogen halides [10, 15–19,28], perchloric acid [20–27,29,30], and trifluoroacetic acid [19,31,32]. Use of hydrogen halides and perchloric acid leads initially at low temperatures to N-protonation, whereas addition of trifluoroacetic acid leads initially to C-protonation [19,31].

Some of the variety of ternary iminium salts that were produced by protonation of the enamine are illustrated as follows: the endocyclic heterocyclic iminium salt, $\Delta^{5(10)}$-dehydroquinolizidinium perchlorate (10) [21], exocyclic iminium salt 11 [25], acyclic iminium salt 12 [25], and the unusually stable aromatic iminium salt 13 [26,27].

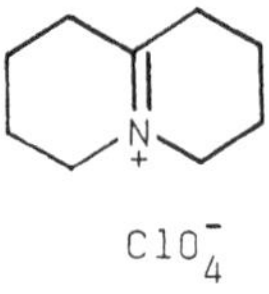

(10)

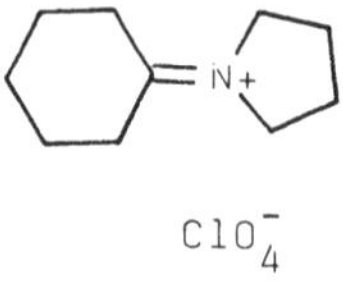

(11)

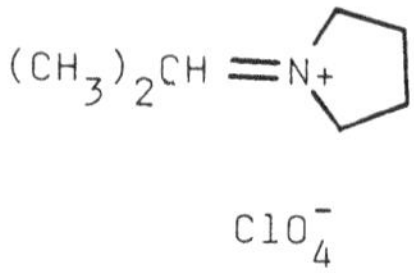

(12)

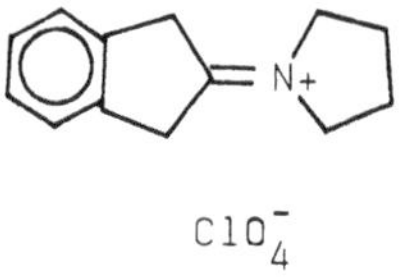

(13)

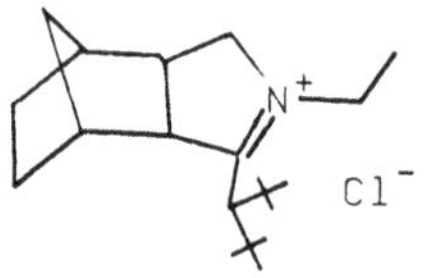

(14)

Iminium salt 14 is inert toward deprotonation and/or hydrolysis by aqueous sodium hydroxide, addition and/or deprotonation by methyl-lithium, reduction by sodium borohydride, or dealkylation by sodium iodide [28]. The *tert*-butyl groups provide steric inhibition toward any of these reactions taking place.

Protonation can take place more rapidly with one isomeric form of an enamine than with the other. For example, treatment of a one-to-one isomeric mixture of enamines 15 and 16 with a limited amount of dry hydrogen chloride gas showed protonation taking place pref-

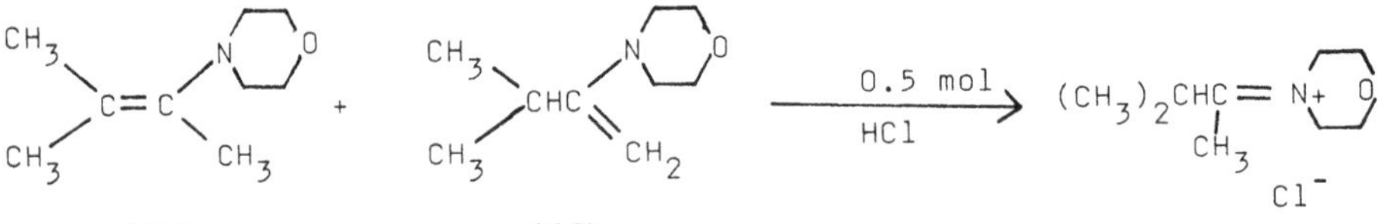

(15) + (16)

0.5 mol / HCl

(17) + (15)

erentially with isomer 16 to produce iminium salt 17, with enamine 15 remaining unprotonated [19]. This same regioselective protonation of enamines has been observed when ion-exchange resins are used as protonating agents also [333].

Enamine protonation can also take place stereospecifically. For example, protonation of bicyclic enamine 18 with perchloric acid in ether produced the *trans*-fused iminium perchlorate 19 as the kinet-

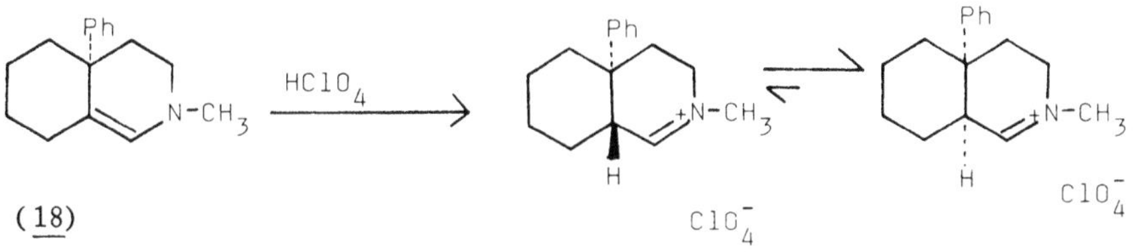

(18)   (19)   (20)

ically favored product. This salt then isomerized upon standing a few days to the thermodynamically more stable *cis*-fused perchlorate 20 [29].

Addition of other electrophilic agents to the β-carbon of an enamine also produces iminium salts (see Chapter 4).

Alkylation of Imines

The alkylation of aldimines and ketimines as a method for obtaining iminium salts is useful primarily for the preparation of iminium salts not accessible by any of the new methods. The preparation of 21 and 22 illustrates the conversion of ketimines to iminium salts [33,34].

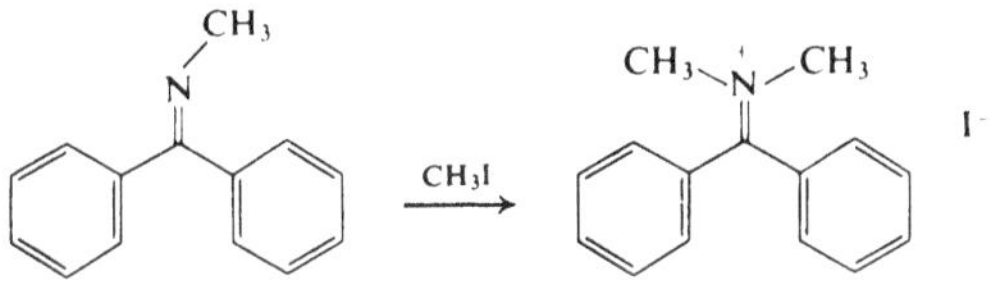

(21)

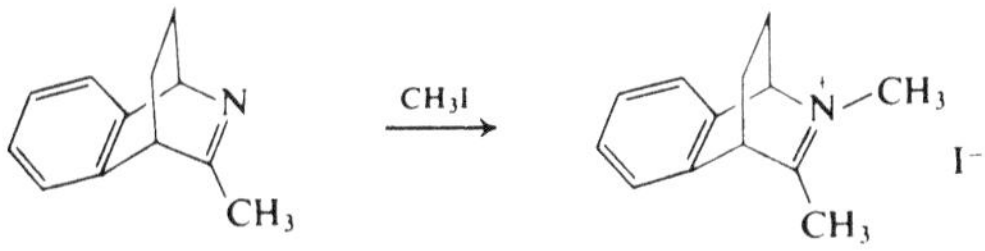

(22)

Elimination Reactions

The heterolytic cleavage of a C—Y bond in

$$N-\overset{|}{\underset{|}{C}}-Y$$

can result in the formation of an iminium salt [35,36]. The most common Y substituents are amino-, alkoxy-, alkylthio-, cyano-, and carboxylic acid groups. An extensive review covering elimination reactions of this type from the literature through 1975 has been published [37]. The acid-catalyzed cleavage of α-aminoethers (23) or aminals (24) to form iminium salts proceeds by protonation and elim-

NCH$_2$—O$\sim$R $\xrightarrow{\text{HX}}$ NCH$_2$—$\overset{H}{\underset{+}{O}}\sim$R $\quad$ $\longrightarrow$ $\quad$ $\overset{\backslash}{_+}$N=CH$_2$ $\quad$ + $\quad$ CH$_3$OH

$\qquad\qquad\qquad\qquad\qquad\qquad$ X$^-$ $\qquad\qquad\qquad\qquad$ X$^-$

(23)

$\overset{\backslash}{/}$N-CH$_2$-N$\overset{/}{\backslash}$ $\xrightarrow{\text{HX}}$ $\overset{\backslash}{/}$N-CH$_2$-$\overset{|}{\underset{|}{N}}\overset{+}{}$ $\longrightarrow$ $\overset{\backslash}{/}$N=CH$_2$ $\quad$ + $\quad$ H-N$\overset{/}{\backslash}$

$\qquad\qquad\qquad\qquad\qquad\qquad\qquad$ H $\qquad\qquad\qquad$ X$^-$

(24)

ination of an alcohol or amine [35,36]. For example, dimethyl(methylene)ammonium trifluoroacetate (26a) (one of a family of Mannich reagents, so called because of their utility in Mannich reactions) is produced when *bis*(dimethylamino)methane (25) is treated with trifluoroacetic acid [38,39]. The N,N-dimethylcyclopropaniminium ion (28) can be formed from the corresponding aminal (27) by methylating the

(CH$_3$)$_2$NCH$_2$N(CH$_3$)$_2$ $\qquad\xrightarrow{\text{CF}_3\text{CO}_2\text{H}}\qquad$ (CH$_3$)$_2$$\overset{+}{\text{N}}$=CH$_2$

$\qquad\qquad$ (25) $\qquad\qquad\qquad\qquad\qquad\qquad\qquad\qquad$ CF$_3$CO$_2^-$

$\qquad\qquad\qquad\qquad\qquad\qquad\qquad\qquad\qquad\qquad\qquad\qquad$ (26a)

aminal with methyl fluorosulfonate at -78°C and allowing it to warm

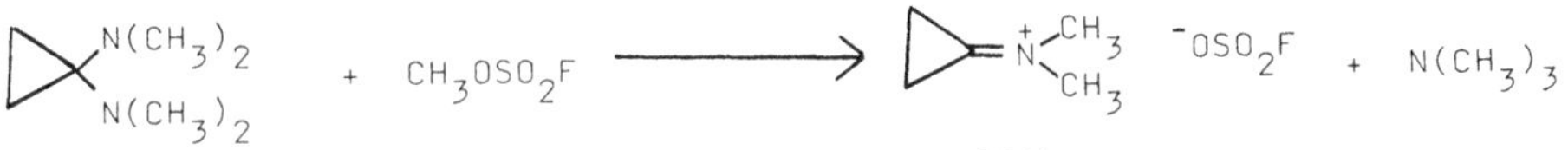

(27) + CH₃OSO₂F → (28) + N(CH₃)₃

(28)

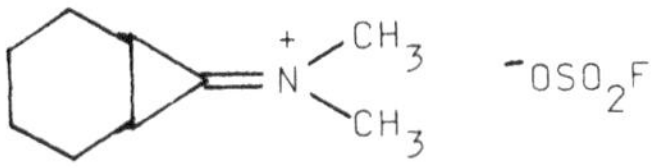

(29)

to room temperature [40a]. Bicyclic iminium salt 29 can be synthesized in a similar manner [40b]. However, when the 2,2-dimethyl
derivative of aminal 27 was treated with methyl fluorosulfonate, only
polymeric material was obtained. But treatment of this aminal with
dimethylchloronium ion in SO_2 (from SbF_5 and excess CH_3Cl) at -78°C
produces the corresponding 2,2-dimethyl derivative of iminium salt
28 [40c].

A convenient method of converting aminals to iminium salts in
good yields involves allowing the aminal to react with an acyl halide
[41–44]. A demonstration of this reaction is the treatment of aminal
25 with acetyl chloride to give Mannich reagent 26b as a crystalline
solid [43].

$(CH_3)_2NCH_2N(CH_3)_2$ + $CH_3\overset{O}{\overset{\|}{C}}-Cl$ → $(CH_3)_2\overset{+}{N}{=}CH_2$ Cl^- + $(CH_3)_2N\overset{O}{\overset{\|}{C}}-CH_3$

(25)

(26b)

$(CH_3)_2NCH_2N(CH_3)_2$ + $(CH_3)_3SiI$ → $(CH_3)_2\overset{+}{N}{=}CH_2$ I^-

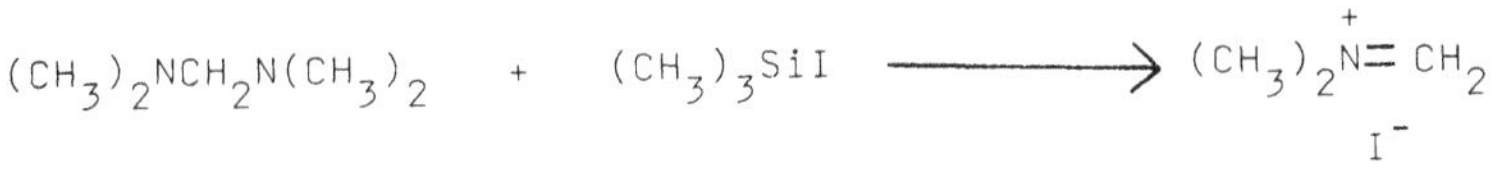

(25)

(26c)

Excellent yields of iminium salts have been obtained from cleavage
of tetraalkyl aminals using trimethylsilyl iodide [45]. The "Eschenmoser" [46] Mannich salt 26c can be formed in this manner in a 96%
yield [45].

Iminium salts can also be prepared from α-aminonitriles and silver nitrate [47]. An example of this type of reaction is found in the

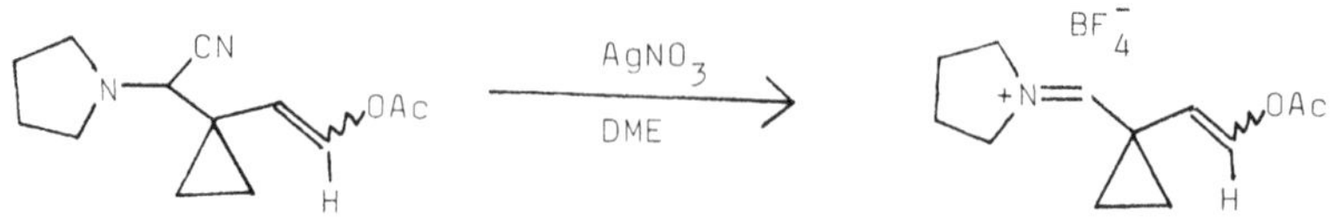

(30)   (31)

reaction of α-aminonitrile 30 with silver nitrate to give iminium salt 31 [48].

A high-yield, single-step, regiospecific technique for preparing iminium salts is by the decarbonylation of α-tertiary amino acids in phosphorus oxychloride [49–51]. The N-β-phenylethyl-Δ^1-pyrrolidinium ion (32) is produced in a 93% yield from N-β-phenylethylproline [49].

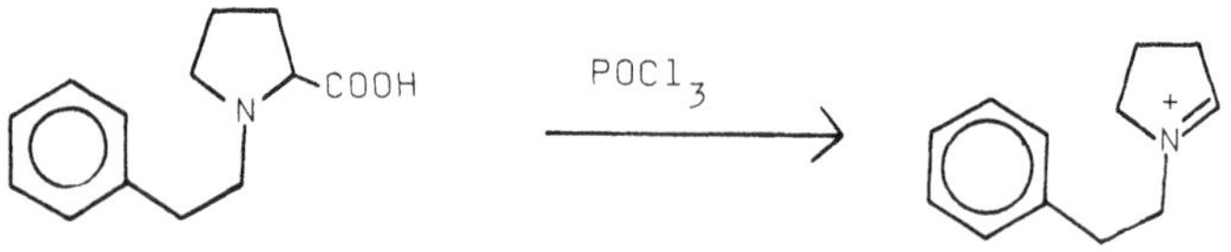

(32)

The cleavage of a carbon-hydrogen bond (hydride abstraction) on the α-carbon of a tertiary amine to produce an iminium ion has been reported using aryl diazonium fluoroborates as the abstracting agents [52], or using trityl (triphenylmethyl) carbocation as the abstracting agent [53,56-58]. Hydride abstraction from 7-(N,N-diethylamino)cycloheptatriene (33) using the tropenylium ion gives the non-

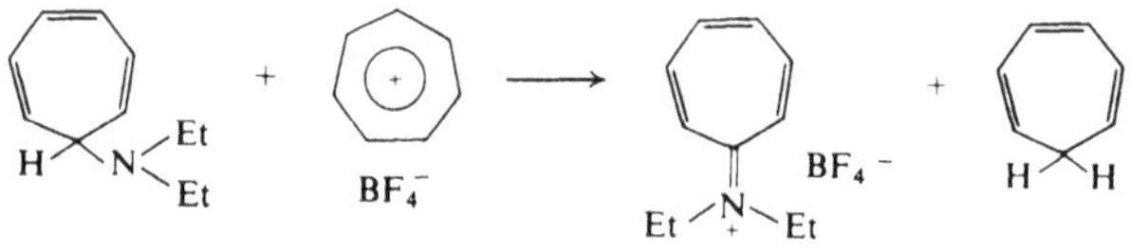

(33)   (34)

benzenoid aromatic ion, tropenylideniminium salt 34 [54,55]. This unique class of compounds had been prepared once before by another route, but not examined in detail [59].

Elimination of a group or atom from a quaternary nitrogen atom and also from a β-carbon atom, respectively, of a tertiary amine derivative is the basis for some important synthetic methods for iminium salts (see Scheme 2). An example of type A in Scheme 2 is the modi-

Scheme 2

fication of the Polonovski reaction [60] as developed by Potier and coworkers [61-69,318-321]. This reaction involves treatment of a tertiary amine oxide with trifluoroacetic anhydride (X = CF_3CO_2; Y = H; Nu = $CF_3CO_2^-$). This method has been used to synthesize Mannich base 26a [62]. The reaction of 1-methyl-3-ethyl-3-piperi-

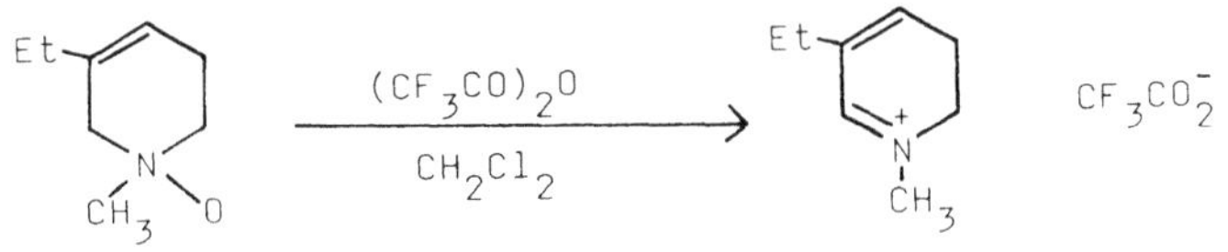

(35)   (36)

deine N-oxide (35) with trifluoroacetic anhydride in methylene chloride leads to the regiospecific formation of iminium salt 36 [38].

Reaction type B in Scheme 2 is the category in which another important type of iminium salt synthesis belongs, namely the thermal fragmentation of halomethylammonium halides (X = CH_3; Y = halogen; Nu = halide ion) [46,70-77]. The "Eschenmoser" Mannich salt 26c was originally produced by this method (see Scheme 3) [46].

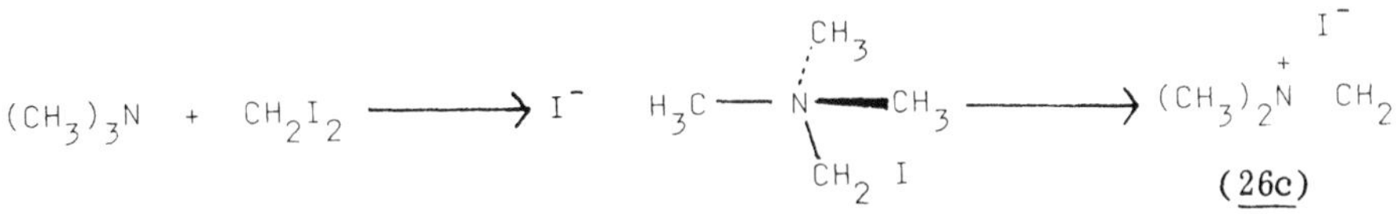

Scheme 3

Condensation Reactions

The initial investigation of the reaction of aldehydes and ketones with complex secondary amine salts was that of Lamchen and co-workers [78]. A few salts had been observed before by Zincke and Wurker [79], but the reaction was not examined in detail. Lamchen and co-workers prepared a number of compounds that were presumed to be iminium salts. The amine salts were halostannates, halobismuthates, haloantimonates, and hexahaloplatinates. Among the reported products were N-ethylidenepiperidinium (37) and N-cinnamilidenetetrahydroisoquinoline (38) salts.

(37)   (38)

An adaptation of this procedure employing perchlorate and fluoroborate salts has been reported by Leonard and Paukstelis [80]. The general reaction is between an aldehyde or ketone and a secondary amine perchlorate or fluoroborate to give an iminium salt [81–83]. So, for example, N,N-dimethylisopropylideniminium perchlorate (4) is formed from acetone and dimethylammonium perchlorate in a 92%

(4)

yield [80]. Tetraphenylborate secondary amine salts have also been used [84]. However, salts with simple anions such as chloride, bromide, nitrate, and sulfate which were investigated were far inferior to perchlorate or fluoroborate salts in the preparation of iminium salts [80,85]. Orthoformate esters have been added in some cases to help remove the water formed [86]. There are differences as to the relative reactivity of the secondary amine perchlorates, with pyrrolidine perchlorate being the most reactive and piperidine perchlorate being much less reactive [87].

Intramolecular formation of iminium salts using this condensation method between ketones and the salts of secondary amines has led to the formation of some interesting bicyclic iminium salts, such as 39a, and some intriguing bicyclic enamines, such as 39b [88].

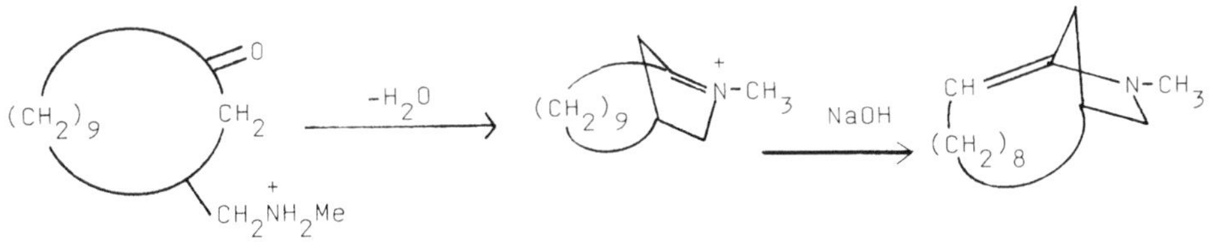

(39a)   (39b)

The indirect condensation of diphenylcyclopropanone and a secondary amine has produced the nonbenzenoid aromatic cyclopropenyl carbocation **40** [89,90]. This salt is so stable that it can be recrystallized undecomposed from hot water. The nonbenzenoid aromatic sub-

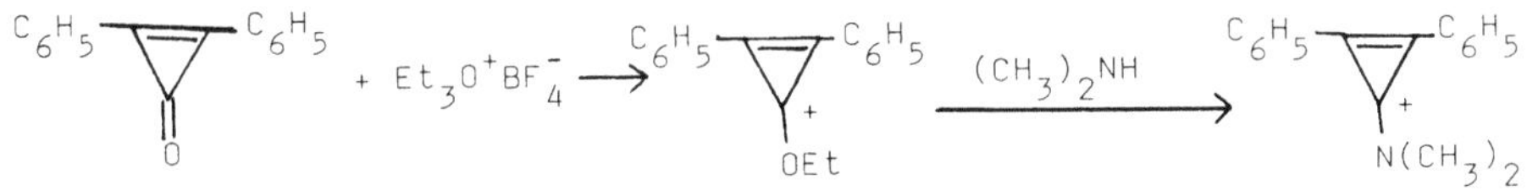

(40)

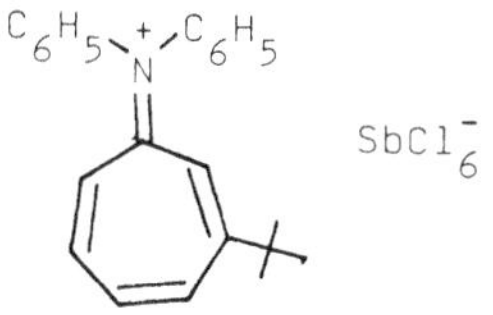

(41)

stituted tropenylideniminium salt **41** was synthesized indirectly from the corresponding tropone and secondary amine [91].

Iminium salts can also be produced when aromatic compounds or compounds with active methyl groups are treated with amides under Vilsmeier-Haack [92,93] reaction conditions [94,95]. This is illustrated by the double cyclization reaction of amide **42a** to iminium salt

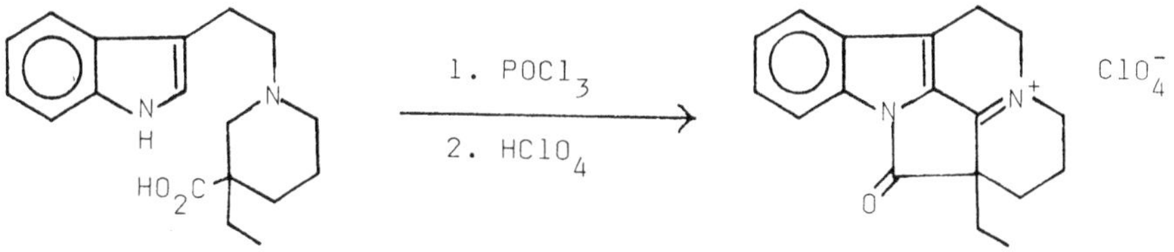

(42a)   (42b)

42b, a precursor in the synthesis of (+)-(E)-norvincamone [95]. A similar reaction produces 6-(dimethylamino)fulvenes (43), which shows a small amount of cyclopentadienyl anion nonbenzenoid aromatic char-

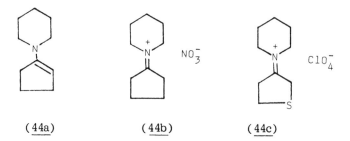

(43)

acter with a corresponding small amount of iminium ion character [96–99]. Fulvenes themselves will produce iminium salts under Vilsmeier-Haack reaction conditions [96,98,99].

C. Detection of Iminium Salts

The presence of iminium salts can be detected by chemical means or by spectroscopic methods. The chemical means of detecting iminium salts are reactions with nucleophiles and are the subject of this review. The spectroscopic methods are more useful for rapid identification. A review of iminium salt spectra has been written [110].

The ultraviolet spectra of simple ternary iminium salts absorb in the 220-235-nm region [17], which is an absorbing region similar to that of the parent enamine [103] (see Chapter 1, Section IV.B). For

(44a)	(44b)	(44c)

example, 1-(N-piperidino)cyclopentene (44a) shows a maximum at 222.5 nm (ε 8020), whereas its iminium salt (44b) has a maximum at 222.5 nm (ε 4140), essentially no shift in going to the iminium salt. Both small bathochromic and small hypsochromic shifts are observed from iminium salt formation, as well as no wavelength shift at all. Dienamines and diiminium salts show bathochromic shifts relative to simple enamines and simple iminium salts [18]. Relative to each other, there is no simple correlation between the free dienamine and its diiminium salt. Therefore, ultraviolet spectroscopy is not a good diagnostic tool for iminium salt formation. However, α-thioiminium perchlorate 44c gives a weak bond in acetonitrile solvent in the 275-295-nm region owing to charge-transfer interaction between the iminium ion group and the sulfide group [111].

The first correlation for the determination of enamines and iminium salts was that of Leonard and co-workers [20], who prepared a series of enamines and the corresponding iminium salts and compared the infrared spectra. There was observed a shift of 20–50 cm^{-1} toward higher frequencies whenever an enamine was converted to its iminium salt.

For enamines in which, for steric reasons, there is no conjugative interaction between the amine nonbonding electrons and the carbon-carbon double bond, only N-protonation takes place to form the enammonium ion (e.g., neostrychnine [112]). In these cases, there is no appreciable difference in the C = C stretching region of the free enamine and its enammonium salt. However, there have been other cases observed with enamines which can undergo either C-protonation or N-protonation in which N-protonation gave an isolated enammonium salt with an upward frequency shift of 21 cm^{-1}, and C-protonation to the iminium salt showed an upward frequency shift of 34 cm^{-1} [10]. With some trinitroarylenamines, it has been reported that the upward shift of infrared enamine absorption bands caused by protonation does not reliably indicate whether it was N-protonation or C-protonation that occurred [113–115].

The observed infrared maxima for some enamines and their corresponding iminium salts along with some other iminium salts are given in Table 2. The enamines of cyclic ketones have maxima in the range 1640–1625 cm^{-1}, depending on the ring size. The pyrrolidinium salts of the cyclic ketones have maxima of 1705, 1665, 1655, and 1649 cm^{-1} for cyclopentylidene-, cyclohexylidene-, cycloheptylidene-, and cyclooctylidne-pyrrolidinium perchlorates, respectively. The dimethyliminium salts derived from a cyclobutanone and a cyclopropanone show maxima at 1730 and 1810 cm^{-1}, respectively.

For enamines derived from a given cyclic or bicyclic ketone, the secondary amine used to form the enamine is the determining factor as to the magnitude of the hypsochromic (higher-frequency) shift when the corresponding iminium salt is formed. The general order of secondary amine groups with decreasing magnitude of frequency shift is dimethylamine > pyrrolidino > piperidino > morpholino > hexamethylenimino. This is most strikingly illustrated by the series of bicyclo[2.2.1]heptenyl enamines, which have almost identically the same low-frequency enamine C = C stretching (about 1600 cm^{-1}) owing to the strained ring system, but there are very large shifts to higher frequencies (one as large as 110 cm^{-1}) when the iminium salts are formed.

TABLE 2 Infrared Frequencies of Some Enamines and Iminium Salts

Structure	A	Enamine film, ν_{max}, cm^{-1}	Iminium Ion nujol, ν_{max}, cm^{-1}	Δ	Ref.
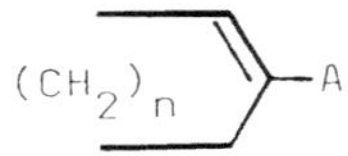	2-Pyrrolidino	1630	1705	75	100
	2-Piperidino	1634	1698	64	101,102
	2-Hexamethylenimino	1630	1672	42	81
	3-Pyrrolidino	1638	1665	27	25
	3-Piperidino	1637	1644	7	17,103
	3-Morpholino	1641	1640	-1	80,103
	4-Pyrrolidino	1629	1655	26	100
	5-Pyrrolidino	1625	1649	24	100
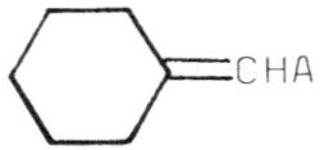	Pyrrolidino	1662	1693	31	25
$(CH_3)_2C=CHA$	Pyrrolidino	1674	1697	23	25
	Dimethylamino	1680	1705	25	4
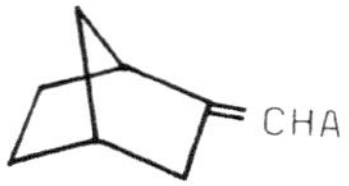	Pyrrolidino	1664	1695	31	81

Structure	Substituent				
	Pyrrolidino	1660	1695	35	4
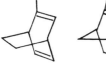	Morpholino	1620	1650	30	104
	Hexamethyleneamino	1600	1680	80	81, 105
	Morpholino	1600	1690	90	81, 105
	Piperidino	1600	1700	100	81, 105
	Pyrrolidino	1600	1705	105	81, 105
	Dimethylamino	1610	1720	110	81, 106
	Hexamethylenimino	—	1695	—	81
	Morpholino	—	1720	—	4
	Dimethylamino	—	1723	—	81
	Pyrrolidino	—	1727	—	81
	Dimethylamino	—	1740	—	81

TABLE 2 continued

Structure	A	Enamine film, ν_{max}, cm^{-1}	Iminium Ion nujol, ν_{max}, cm^{-1}	Δ	Ref.
	Dimethylamino	—	1730	—	107
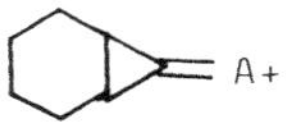	Dimethylamino	—	1810	—	40b
$CH_2 = A+$	Piperidino	—	1666	—	17
	Diisopropylamino	—	1670	—	108
	9-ABN	—	1680	—	109
	Dimethylamino	—	1682	—	46

R_1	R_2	R_3	R_4	R_5				
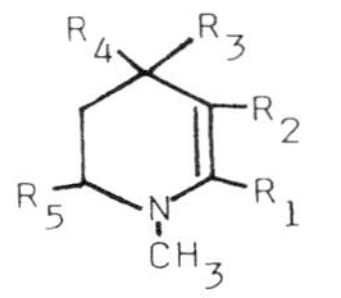 CH_3	H	H	H	H	1649	1690	41	23
CH_3	CH_3	H	H	H	1657	1682	25	23
H	CH_3	H	H	H	1673	1698	25	23
H	H	CH_3	CH_3	H	1645	1707	62	23
CH_3	H	H	H	CH_3	1650	1679	29	23
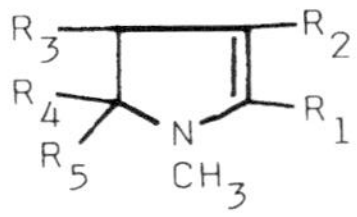 CH_3	H	H	H	H	1640	1701	61	24
H	CH_3	CH_3	H	H	1648	1693	45	24
CH_3	H	H	CH_3	CH_3	1635	1677	42	24
					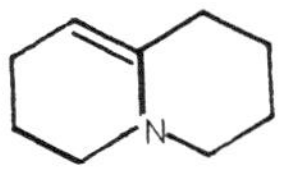 1652	1696	44	21

TABLE 3 ^{1}H NMR Correlations for Some Iminium Salts [100]

Structure	R_1	R_2	Chemical shifts (δ, ppm)			
			$^+N-CH_2$	$^+N=C-H$	$^+N=C-CH_3$	$^+N=C-CH_2-$
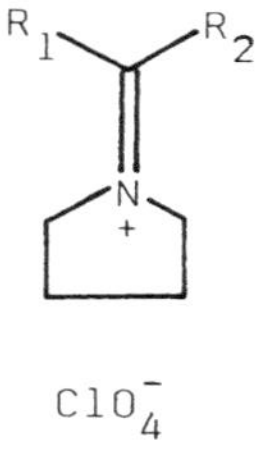	CH_3	CH_3	3.97 m	—	2.52 (J = 1.4)	—
	CH_3	Et	4.00 m	—	2.53 (J = 1.3)	2.77 (J = 8.0)
	Et	Et	4.50 m	—	—	2.78 (J = 8.0)
	$-(CH_2)_5-$		3.97 m	—	—	2.78 m
	Ph	CH_3	4.27 m	—	2.79	—
			3.90 m			
	Ph	H	4.20 m	8.99 (J = 2.1)	—	—
	PhCH=CH	CH_3	4.05 m	—	2.57	—

H	$(CH_3)_2CH$	4.26 m	8.34 (J = 9.0)	—	—
		4.05 m	(J = 2.0)		
H	$(Et)_2CH$	4.34 m	8.42 (J = 10.0)	—	—
		4.13 m	(J = 1.9)		
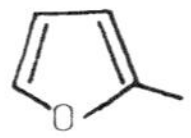 [furan]	H	4.25	8.78	—	—
PhCH=CH	H	4.08	8.68 (J = 10.0)	—	—
			(J = 1.8)		
$(CH_3)_3C$	H	4.18	8.27 (J = 2.1)	—	—
CH_3	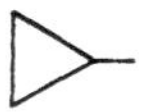 [cyclopropyl]	4.00	—	2.0	—

TABLE 4 ^{1}H NMR Correlations for Some Bicyclic Iminium Salts [116]

Structure	R_1	R_2	R_3	R_4	$+N-CH_2$ (5-ring)	(δ, ppm) (6-ring)	$+N=C\ CH_2$ (5-ring)	(δ, ppm) (6-ring)
(6-ring bicyclic, ClO_4^-)	H	—	—	—	—		—	2.79
	CH$_3$	—	—	—	—		—	2.75
(5,6-ring bicyclic, ClO_4^-)	H	H	H	H	4.15	3.70	3.20	2.74
	CH$_3$	H	H	H	4.21	3.71	3.18	2.90
	H	H	CH$_3$	H	4.12	3.71	3.3	—
	H	H	CH$_3$	CH$_3$	4.30	3.94	—	2.9
	CH$_3$	CH$_3$	H	H	4.23	3.76	3.20	2.72
	CH$_3$	H	CH$_3$	H	4.12	3.75	3.4	2.69
	H	H	Et	H	4.14	3.76	3.3	
	H	H	H	Ph	4.46	3.99	—	

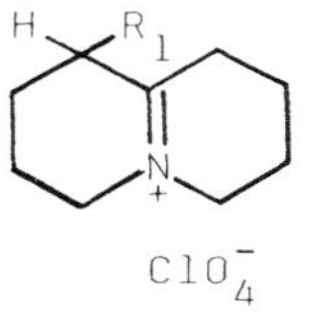

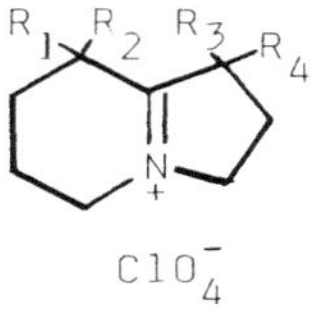

A compilation of proton magnetic resonance signals for various acyclic and cyclic iminium salts is given in Tables 3 and 4 for comparison purposes and for determination of trends. It should be noted that the simplest possible ternary iminium salt with an iodide anion (26c) has its signal for $\overset{+}{N}-CH_3$ at $\delta = 3.67$ and for $\overset{+}{N} = CH_2$ at 8.18 ppm [46]. The same simple iminium ion with a trifluoracetate anion (26a) has signals at $\delta = 3.87$ and 7.95 ppm, respectively [62]. The simplest symmetrically substituted iminium salt (4a) has its signal for $\overset{+}{N}-CH_3$ at $\delta = 7.54$ ppm and for $\overset{+}{N}-C-CH_3$ at $\delta = 2.47$ ppm, with a coupling constant between the two of 2.1 Hz [80]. In ring systems, the signals for methylene hydrogens adjacent to $C = N+$ in five-membered rings occur about 0.3-0.4 ppm downfield from the signals for corresponding methylene hydrogens in six-membered rings.

Carbon-13 nuclear magnetic resonance chemical shifts for some ternary iminium salts are shown in Table 5. This type of spectroscopy is a good indicating method as to whether the C-protonated iminium ion or the N-protonated enammonium ion is present. The chemical shift of the $C = N+$ carbon appears in the 170-195-ppm region [19,32, 117]. The very stable iminium salt 14 shows a chemical shift of 195.2 ppm for the α-carbon atom [28]. But the α-carbon in the enammonium ion has a chemical shift of only between 100 and 130 ppm [115]. Therefore, this is a more reliable technique for detecting iminium salt formation than the infrared spectroscopic shift method.

(45)

The ^{15}N NMR spectrum of iminium salt 45 showed a chemical shift of -218.12 ppm with respect to a CH_3NO_2 reference [118].

III. ADDITION OF NUCLEOPHILES TO IMINIUM SALTS

A. Addition of Organometallic Reagents

Organometallic reagents react with iminium salts to give C-alkylated products. The reactions can be divided into two categories: the reactions of pyridinium, quinolinium, and isoquinolinium salts; and the reactions of simple iminium salts. Most of the observations have been made with Grignard reagents, but from the examples available, it appears that lithium reagents react in the same way.

TABLE 5 ^{13}C NMR Chemical Shifts of Some Iminium Salts

| Structure | R$_1$ | Anion | δ, ppm relative to TMS | | | Ref. |
			C-1	C-2	C-2'	
(ring structure)	H	C 10$_4^-$	189.75	34.0	53.95	115
	CH$_3$	C 10$_4^-$	189.4	33.4	54.7	115
	(CH$_3$)$_3$C	C 10$_4^-$	190.0	34.5	54.7	115
(structure)	—	Cl$^-$	167.6	—	47.6	117
	—	CF$_3$CO$_2^-$	170.0	—	50.5	117
(structure)	—	C 10$_4^-$	186.2	—	—	56
(structure)	—	BF$_4^-$	180.8	33.3	63.9	109
					53.9	

A review of the literature prior to 1953 on reactions of pyridinium, quinolinium, and isoquinolinium salts is available [119]. The reactions will be described here only briefly. The initial observation was that of Freund et al. [120-125], who found that treatment of various derivatives of hydrastinine (46) with Grignard reagents yielded addition products, such as 47.

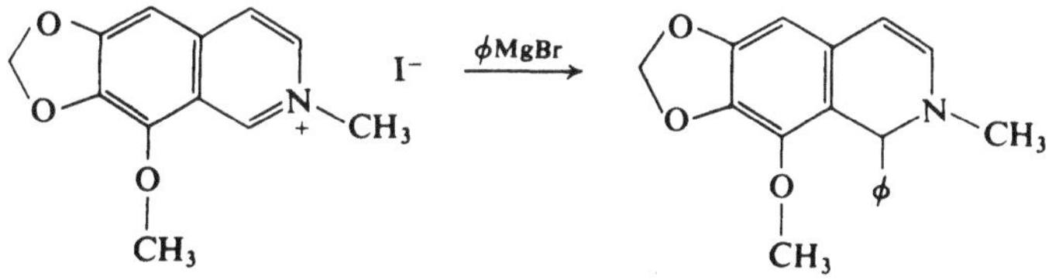

(48)

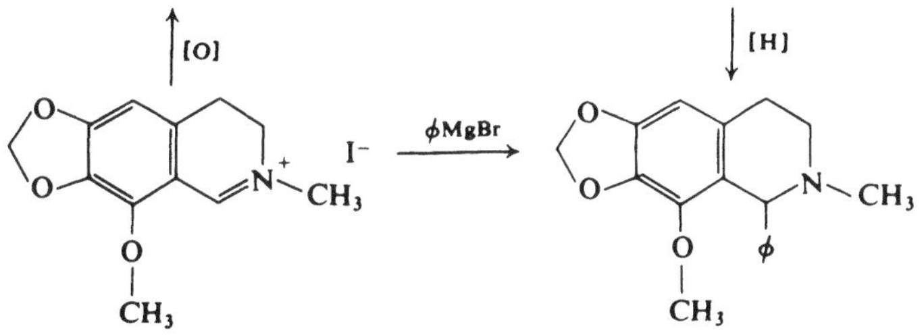

(46)

(47)

Oxidation of 46 to 48 followed by reaction with phenylmagnesium bromide and reduction gave 47, indicating that the isoquinolinium salt (48) reacted with the Grignard reagent at the same position. Another example of addition to quinolinium salts is that of Craig [126], who found that treatment of 2,4-dimethylquinolinium methiodide (49) with methyl magnesium iodide gave a C-alkylated product whose structure was shown to be 50.

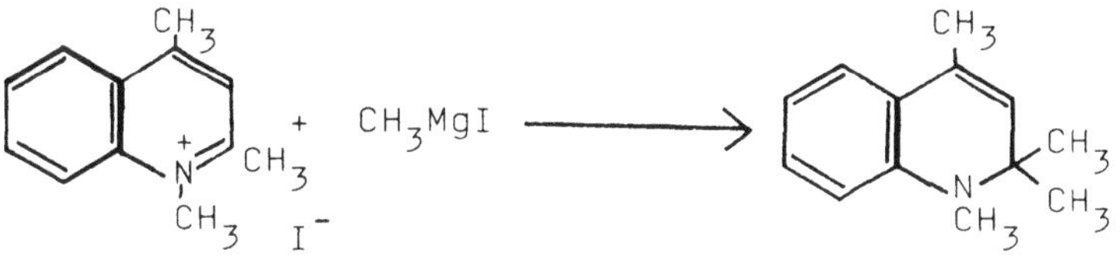

(49)

(50)

Since the initial discovery, there have been several investigations that have examined the Grignard reaction with quinolinium and isoquinolinium salts [127,128]. In general, quinolinium salts (51) give products from addition at the 2 position (52), and isoquinolinium salts (53) give products from addition at the 1 position (54).

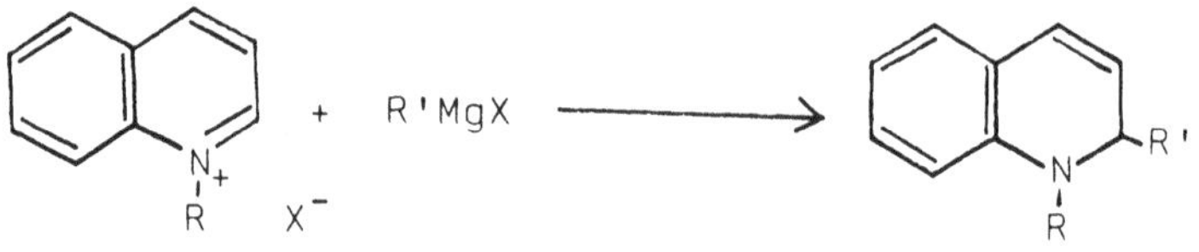

(51)   (52)

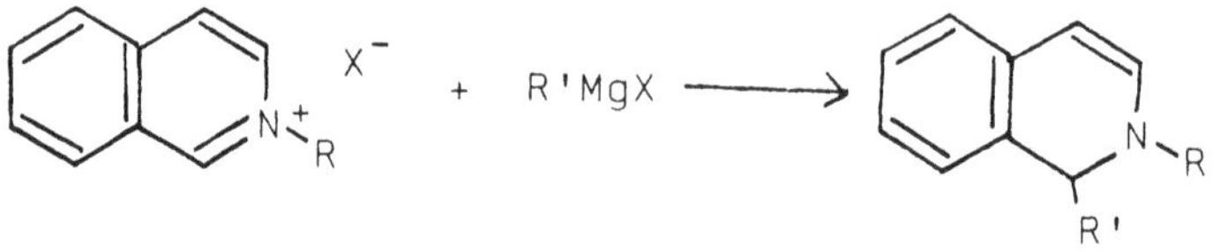

(53)   (54)

The reaction of quinolinium and isoquinolinium salts with dialkyl cadmium has been observed to be slow and occurs in relatively poor yield. The structures of the products are the same as those obtained from reaction with Grignard reagents, and the yields have ranged from 0 to 20% [129]. Leading references to other observed reactions of quinolinium and isoquinolinium salts with Grignard reagents can be found in the above-cited review [119].

Simple iminium salts react with organometallic reagents in an analogous way to yield addition products [7,21−23,33,56,57,130−142]. The results of many such addition reactions are listed in Table 6. The yields of addition compounds are good, generally in the range of 60−80%. A typical example is the reaction of $\Delta^{5(10)}$-dehydroquinolizidinium perchlorate (10) with benzyl magnesium chloride to give 10-benzylquinolizidine (55) in 68% yield [131].

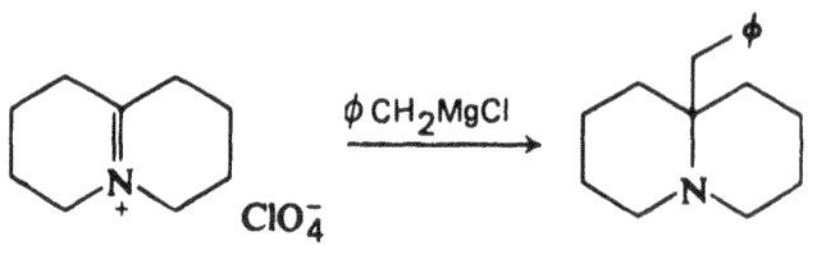

(10)   (55)

The addition of a Grignard reagent to an acetate iminium salt of a cyclic ketone produces a C-alkylated complex. Pyrolysis of this Grignard complex gives mixtures of alkylcycloalkenes and alkylidenecycloalkanes [328].

There are several reports that alkylated pyridine N-oxides (56) react with Grignard reagents to give 2-alkylated pyridines (57) [143, 144].

TABLE 6 Reaction of Grignard Reagent with Iminium Salts

Compound		Grignard reagent	Ref.
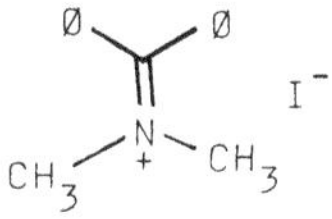		ϕCH_2MgCl	141
		CH_3MgI	142
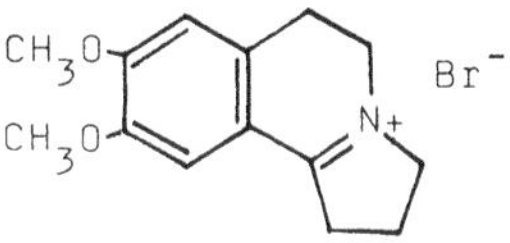		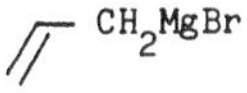 CH_2MgBr	133
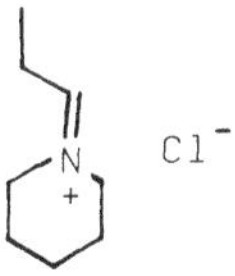		CH_3MgI	7
		ϕCH_2MgCl (82)	
	All combinations of R, R' = CH$_3$, Et, n-Pr, n-Bu	CH_3MgI	134
		EtMgBr	134
		PrMgBr	
		BuMgBr	
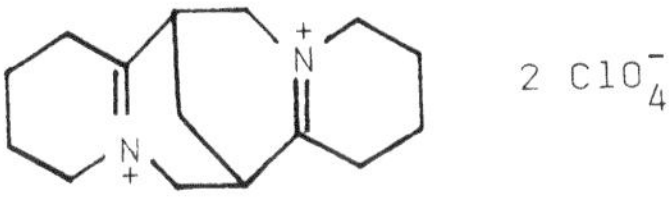	2 ClO$_4^-$	MeMgI	22
		EtMgBr	
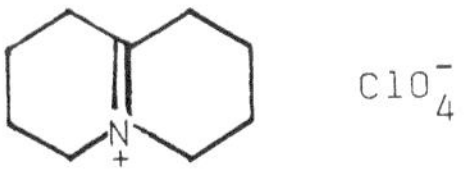	ClO$_4^-$	MeMgI	21
		EtMgI (71)	
		ϕCH_2MgCl (68)	
	ClO$_4^-$	MeMgI	131
	ClO$_4^-$	R—⬡—MgBr	140
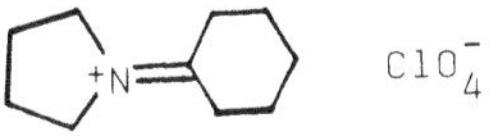	ClO$_4^-$	R = CH$_3$, CH$_3$O	140

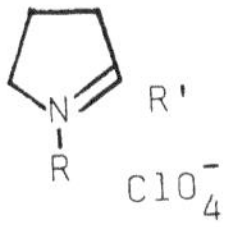

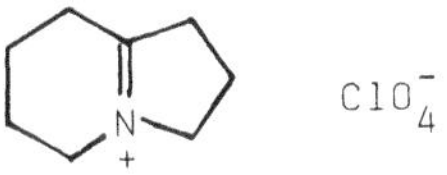

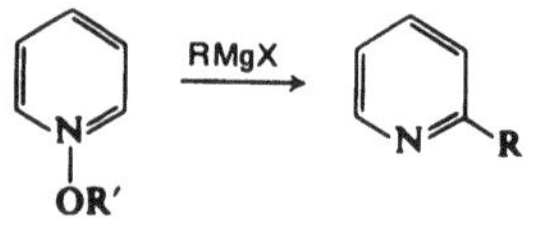

(56)   (57)

Alkyllithium reagents react with iminium salts in a manner similar to that of Grignard reagents (see Table 7) [18,131,138,145]. It should be noted that the products from addition to conjugated iminium salts occur primarily by 1,2- and not 1,4-addition [18].

In some cases, Grignard reagents cause reduction of the iminium salt [136,139,140,147].

Alkylaluminum reagents sometimes add to iminium ions also [148, 149]. Sodium trialkyl-1-propynylborates (58) react with iminium salts to form a mixture of E-(59) and Z-dialkylborylallylamines (60) [150].

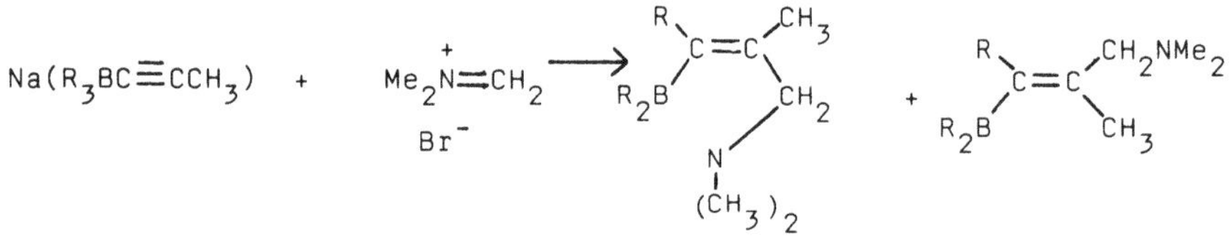

(58)   (59)   (60)

B. Addition of Hydride

The reduction of iminium salts can be achieved by a variety of methods. Some of the methods have been studied primarily on quaternary salts of aromatic bases, but the results can be extrapolated to simple iminium salts in most cases. The reagents available for reduction of iminium salts are: sodium amalgam [151], sodium hydrosulfite [152], potassium borohydride [153,154], sodium borohydride [18,22,155–163], lithium aluminum hydride [18,21,81,104,157,164,165], sodium cyanoborohydride [163,166–168], zinc-modified sodium cyanoborohydride [168], lithium tri sec-butylborohydride (L-Selectride) [163], sodium bis(2-methoxyethoxy)aluminum hydride (Red-al) [163], diborane [18,169,170], t-butylamine borane [163], carboxyhydridoferrates [171], organoaluminum and organoaluminum hydrides [149,172], zinc and hydrochloric or acetic acid [22,162,192], phosphorous acid [173], dihydropyridine [174–178], formic acid [179–190], and catalytic hydrogenation [21–23,219–221].

TABLE 7 Reaction of Iminium Salts with Lithium Reagents

Compound	Reagent	Product (yield)	Ref.
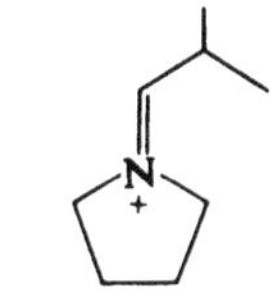	ϕ-Li	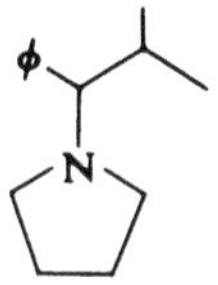 (70)	18
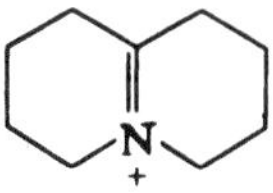	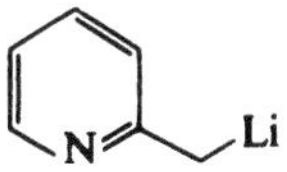	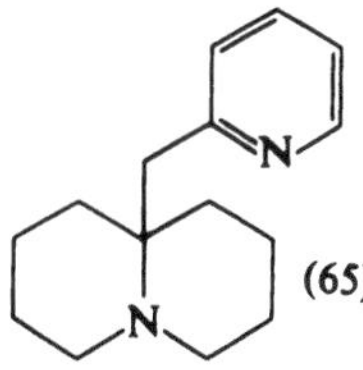 (65)	131
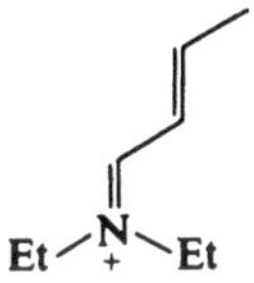	ϕ-Li	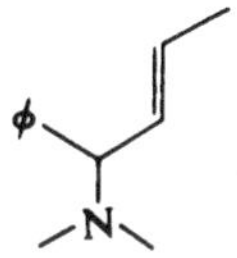 (13)	18
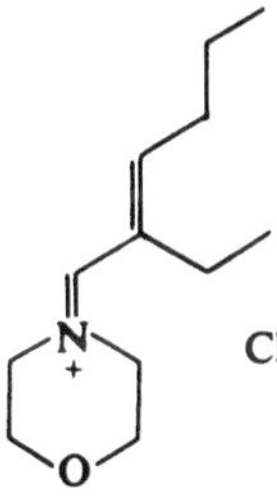	ϕ-Li	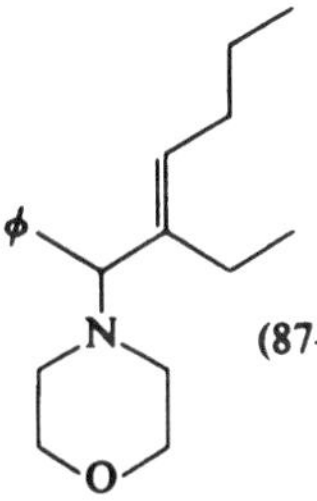 (87–97)	18

When a pyridinium salt such as (61) is treated with sodium borohydride, the final product is the tetrahydropyridine (64). The mechanism for this reaction was proposed by Katritzky [193] and experimentally verified by Anderson and Lyle [194—196]. The sequence is
visualized as reduction of the

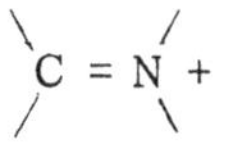

$$\text{\textbackslash}C = N \text{\textbackslash} +$$

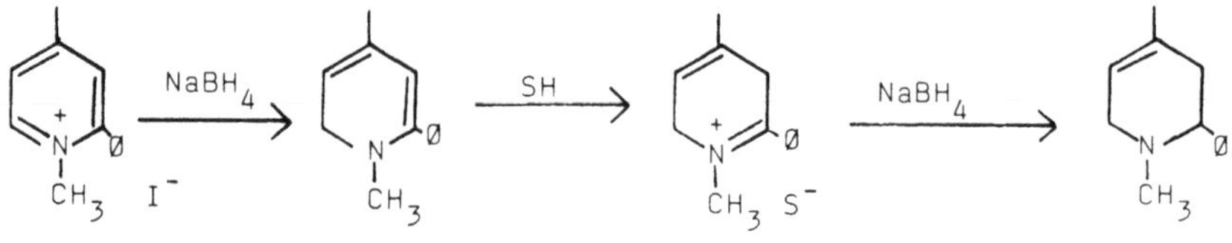

followed by protonation in the center of the dienamine system (62) to
give a second iminium salt intermediate (63) and further reduction to
give the observed tetrahydropyridine (64). The direct attack of a
proton from solvent was shown by deuterium-labeling experiments.
If only a molar equivalent of borohydride is used, a dihydropyridine
can be isolated [196,197]. Examples of isolation of the dihydroprod-
uct are numerous and can be exemplified by reduction of N-benzyl-3-
cyanopyridinium bromide (65) to the dihydro stage (66), which can be

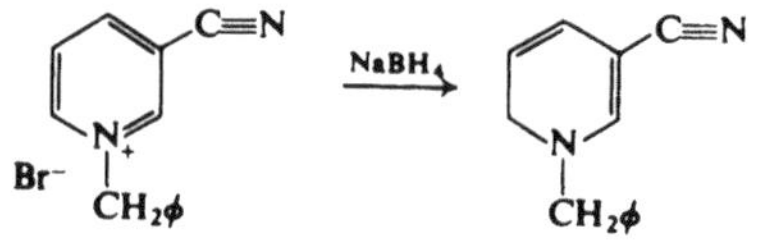

(65)   (66)

used directly in other reactions [198]. Completely reduced pyridines
or piperidines have been reported [155], as well as 1,4- rather than
1,2-additions of hydride to give 1,4-dihydropyridines [199]. Lyle
and Anderson have provided the following correlations for sodium
borohydride reduction of nitrogen heterocycles [200]: the initial attack will occur at the C = N+ carbon if there is no steric interference;
if there are severe steric interactions, attack will occur slowly or at
another electrophilic position; the resulting dihydropyridine will be
protonated by solvent provided that the position for attack is free of
substitution and that the nitrogen is not conjugated to other π-elec-
tron systems; the resulting dihydropyridine can be intercepted by
reaction with base; the resulting iminium intermediate is then further
reduced to the tetrahydropyridine.

The reduction of iminium salts with sodium borohydride is quite stereoselective [162,163,191]. For example, iminium salt 67 is re-

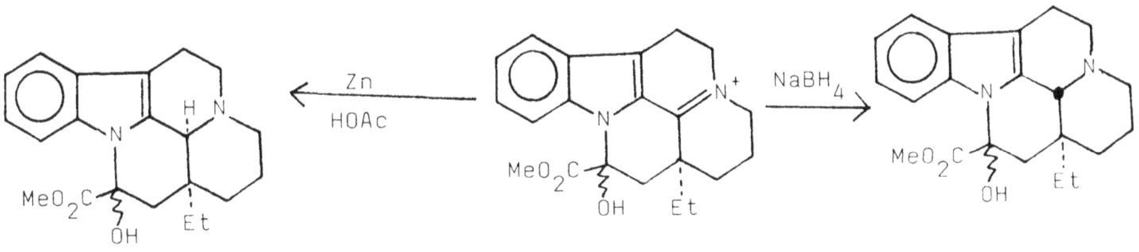

(69)   (67)   (68)

duced by sodium borohydride to give the (+)-3-epivincamines (68) [162]. On the other hand, reduction of 67 with zinc in acetic acid gave principally the stereoisomers (+)-vincamines (69). Dehydro-yohimbine perchlorate (70) is also readily reduced by zinc and hydro-chloric acid at room temperature to yield pseudoyohimbine (71) [192].

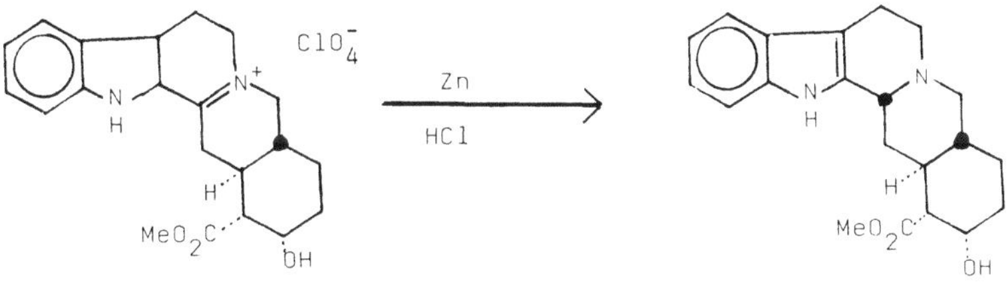

(70)   (71)

Sodium cyanoborohydride is a useful reducing agent because it is stable in acids down to pH 3 [166]. At pH 3, the hydrogen atoms in sodium cyanoborohydride can be readily exchanged for either deuterium or tritium. The iminium salt of the morpholine enamine of cyclohexanone is reduced quantitatively to the saturated amine in 15 min by NaBH$_3$CN [166]. Since the iminium ion is reduced much faster by this reagent than the carbonyl group at pH 6–8, it is possible to reductively aminate ketones or aldehydes to the corresponding sat-urated amines in a nearly neutral solution [167]. When NaBH$_4$ is used instead of NaBH$_3$CN, the major products are alcohols obtained from reduction of the carbonyl starting materials [167]. Zinc-modified cyanoborohydride (generated from NaBH$_3$CN and ZnCl$_2$ in a 2:1 molar ratio) reduces iminium salts to amines smoothly and quickly in excel-lent yields at room temperature [168].

Lithium aluminum hydride reduction of pyridinium salts is similar to sodium borohydride reduction and gives similar products, but the ratio of 1,2- and 1,4-dihydro- or tetrahydropyridines differs considerably [200]. Isoquinolinium salts are reduced by sodium borohydride or lithium aluminum hydride in a manner identical to pyridinium salts [200]. Quinolinium salts are reduced by sodium borohydride to give primarily tetrahydroquinolines [201], as shown by the conversion of 72 to 73 and 74. When lithium aluminum hydride is used, the product is usually the dihydroquinoline [202], as shown by the conversion of 75 to 76 and 77. Examples of the reduction of other aromatic salts

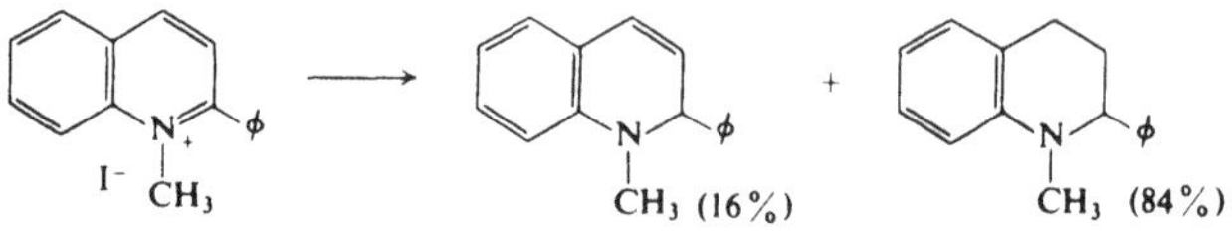

(72) → (73) CH₃ (16%) + (74) CH₃ (84%)

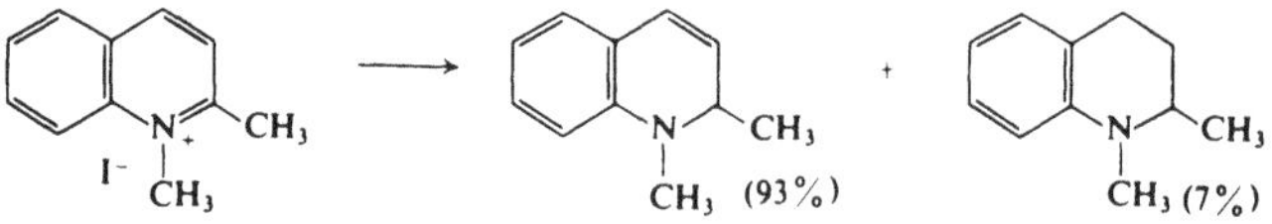

(75) → (76) CH₃ (93%) + (77) CH₃ (7%)

are given by Lyle and Anderson in their review of the subject [200] and by others [203–209].

Various complex hydrides are stereoselective in reducing 2-, 3-, or 4-alkylcyclohexyl iminium salts [163,165]. For example, using an excess of hydride, when LiAlH$_4$ is the reducing agent (i.e., MH =

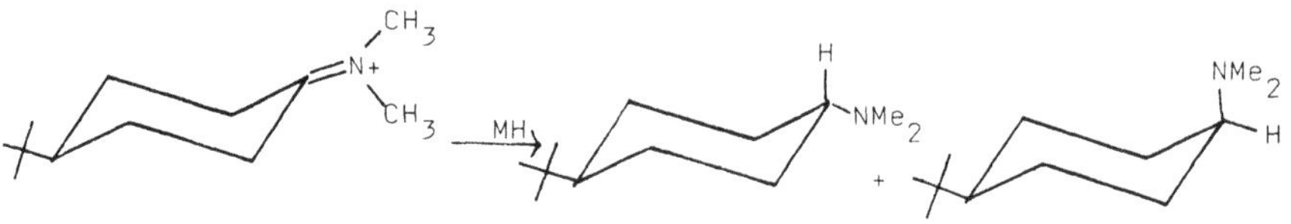

(78) (79) (80)

LiAlH$_4$), the principal product is the one resulting from axial hydride attack, that is, 79 (*trans* isomer) in a 63:37 ratio of 79:80 [165]. Other small complex hydrides such as NaBH$_4$ or NaBH$_3$CN give similar results with 3- or 4-alkylcyclohexyl iminium salts [163]. However, moderately bulky complex hydrides such as L-Selectride or Red-al attack from the equatorial direction to give primarily the *cis* product (80) [163]. This equatorial attack is true for the reduction of 2-, 3-, or 4-alkylcyclohexyl iminium salts by hindered complex hydrides.

TABLE 8 Reduction of Iminium Salts with Complex Hydrides

Compound	Product (yield, %)	Ref.
		212
		213,214
		215,216
		217
		218
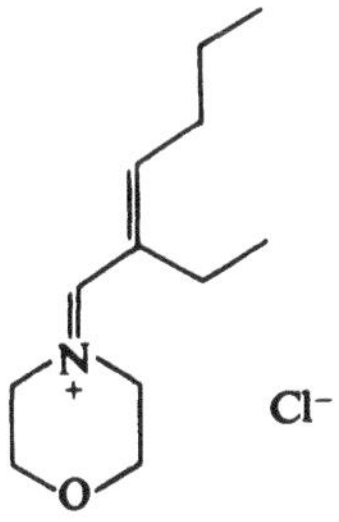	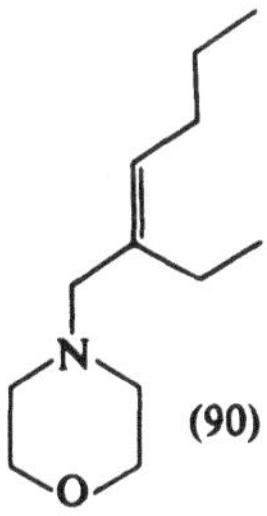 (90) 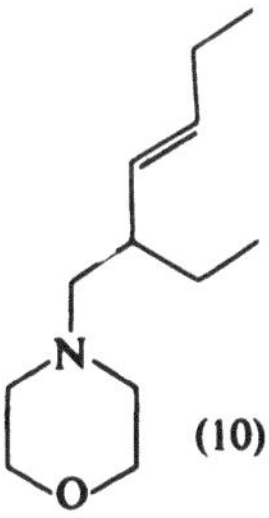 (10)	18
B₂H₆	(50) (50)	18
LAH	(97–60) (3–40)	18

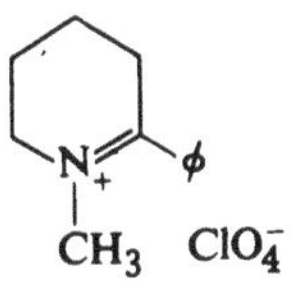

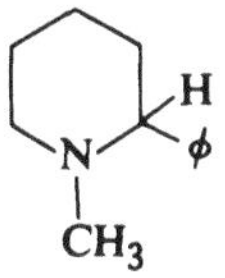

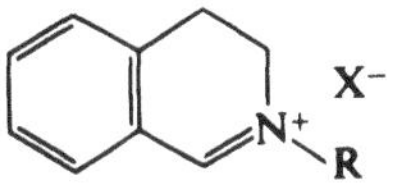

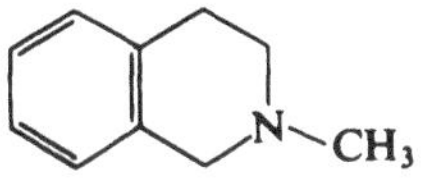

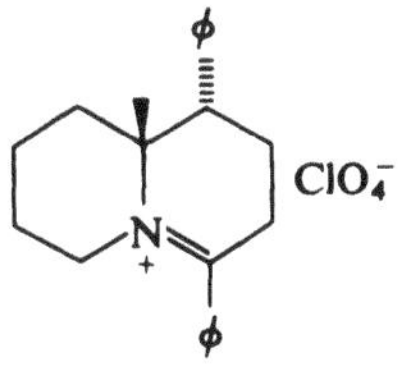

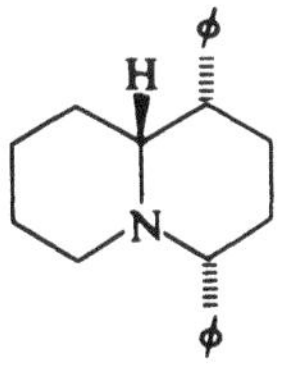

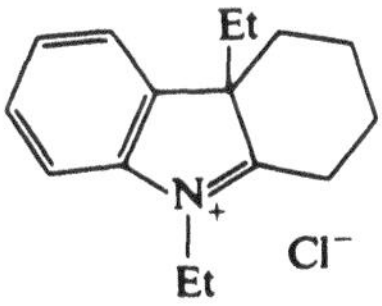

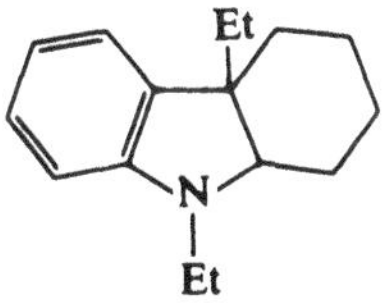

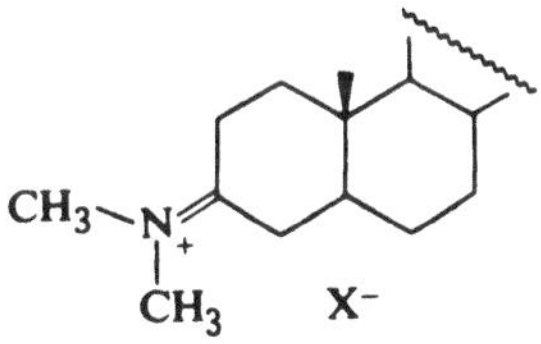

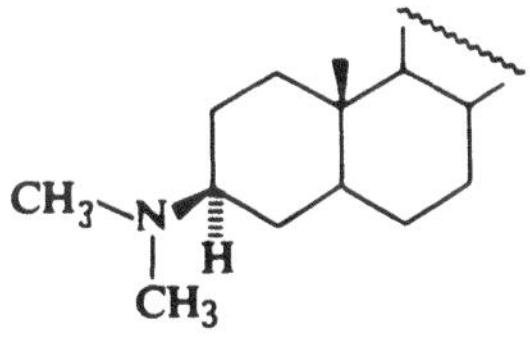

Unhindered small complex hydrides likewise give mainly the product resulting from equatorial attack of the iminium salt.

Hydride attack from the less hindered side of a molecule by complex hydrides has also been observed in bicyclo[2.2.1]heptyl iminium ions [104,210,211]. For example, reduction of N-2-bicyclo[2.2.1]-heptylidendimethylaminium perchlorate (81) with lithium aluminum hy-

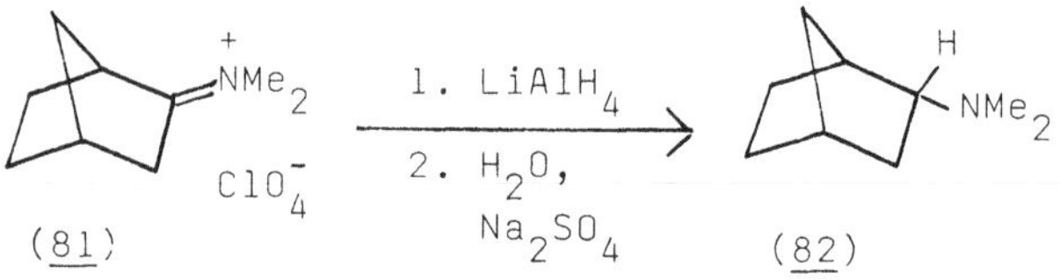

(81)   (82)

dride resulted in the formation of *endo*-saturated amine 82 because of hydride attack from the less hindered *exo*-side [211].

Reduction of iminium salts with diborane [18,169,170] and by the Meerwein-Ponndorf [18] method has been reported.

Further examples of reduction of simple iminium salts by complex hydrides are given in Table 8.

The Hantzsch ester (83) is a 1,4-dihydropyridine which is considered a good NADH model [174–178]. This ester will reduce iminium salts such as 11 to their corresponding saturated amines in yields

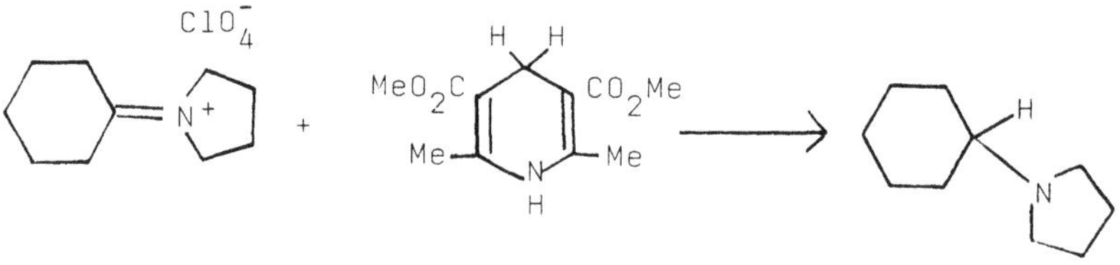

(11)   (83)

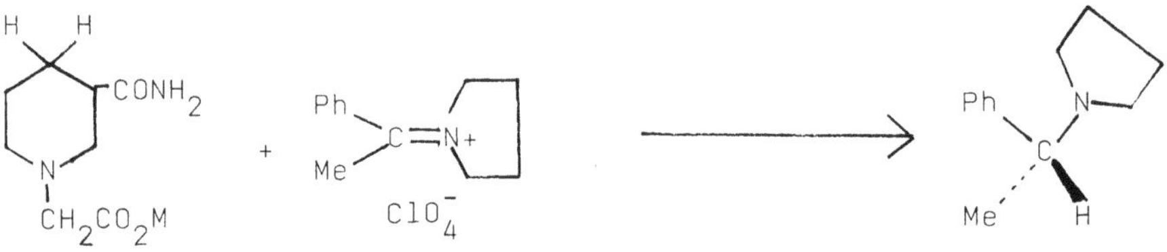

M ≡ (-)menthol
      moiety

(84)   (85)   (86)

greater than 90% [175]. The reaction can be stereospecific [174], and the use of a chiral dihydropyridine 84 with a prochiral iminium salt 85 results in an asymmetrical reaction to produce chiral amine 86 [176].

There have been only a few examples of iminium ion reduction by catalytic hydrogenation, since the reductions with complex hydrides are so easy to do in the laboratory. A report that hydrogenation using platinum oxide as catalyst of $\Delta^{5(10)}$-dehydroquinolizidinium perchlorate (10) gives quinolizidine (87) in quantitative yield indicates that such reductions should be facile [21]. Hydrogenation of an imin-

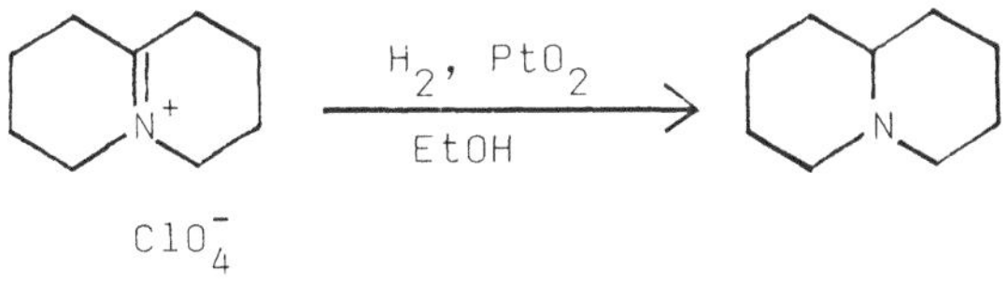

(10)   (87)

ium salt intermediate in the synthesis of vasicine used $Pd/BaSO_4$ catalyst [219]. Other examples of similar reductions are available in the literature [220,221].

The remaining major method for the reduction of the $C = N^+$ functionality is the reaction with formic acid. The first report was that of Lukes, who found [179] that thermal cleavage of 1,1-dimethyl-2-methylenepyrrolidinium formate (88) was accompanied by reduction. Lukes then explored the generality of the reaction as shown in Table

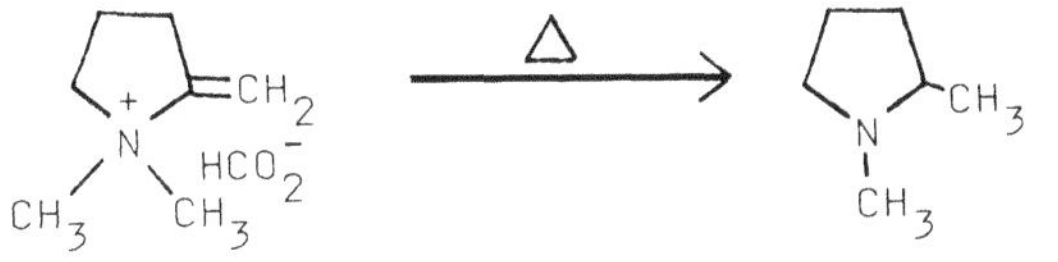

(88)

9 in a series of papers over many years [222–225]. The reduction of enamines can be carried out easily since under the condition of the reaction (i.e., formic acid), the enamine is protonated.

That the reduction with formic acid proceeds by a hydride transfer reaction was proposed by Lukes and Jizba [222] and finally prov-

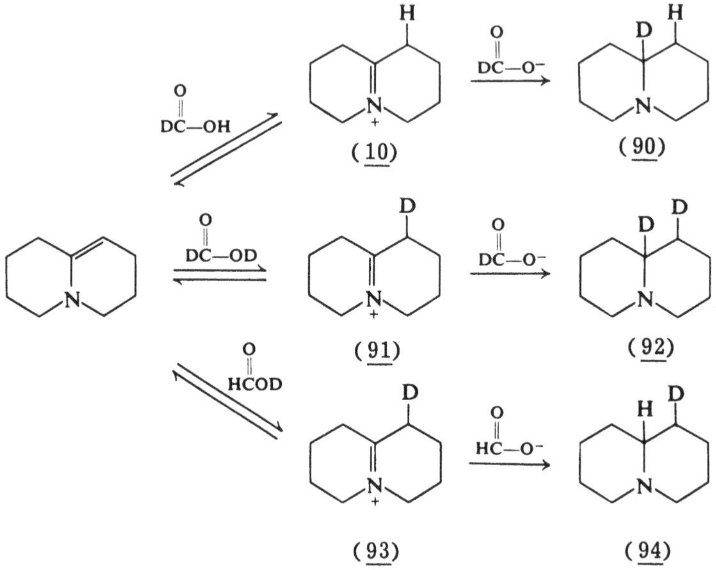

Scheme 4

en by Leonard and Sauers (see Scheme 4) [185]. The use of variously deuterated formic acid allowed Leonard and Sauers to determine
that: (a) protonation or deuteration of 89 occurs first and is a reversible process; (b) deuteration (91, 93) or protonation occurs at
the β-carbon atom; (c) reduction of hydride transfer occurs from the
formic (*d*) acid to the 10 position, exclusively. These facts are only
compatible with a mechanism as shown above in the formation of 90,
92, 94. The preformed iminium perchlorate was reduced in equal or
in better yield (73% compared to 58—65% for the enamine). The nature
of the hydride transfer agent is not known, but is presumed to be a
formate or a formate ion pair or some formate ester [185].

The reaction can be shown to have good stereospecificity in the
conversion of (-)-Δ^5-dehydrosparteine (95) and (-)-$\Delta^{5,11}$-didehydro-
sparteine (97) to (-)-sparteine (96) and (-)-α-isoparteine (98) [22,
226].

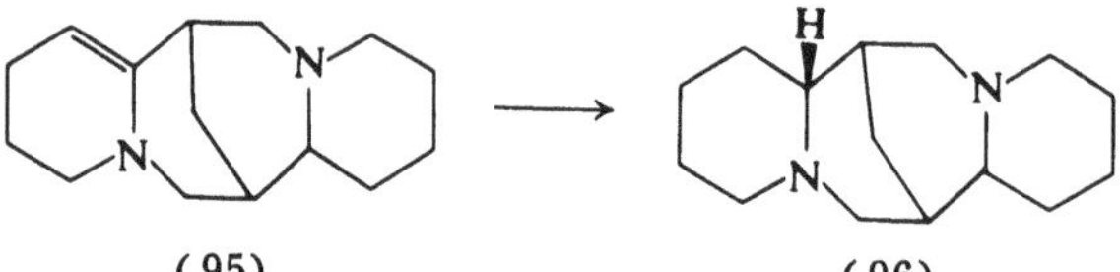

(95)   (96)

TABLE 9 Reduction of Iminium Salts with Formic Acid [222–225]

Compound	R	R′	Yield of tertiary amine, %
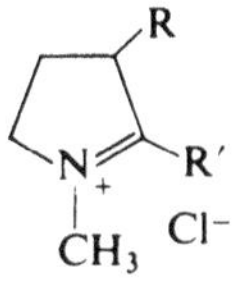 ![structure]	H	Et	91
	H	2-Pr	94
	Me	Et	78
	Me	2-Pr	77
	Me	Pr	90
	Me	Me	76
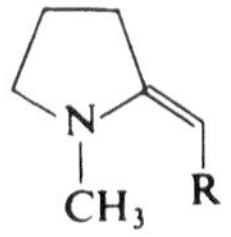 ![structure]	H	—	80
	Me	—	92
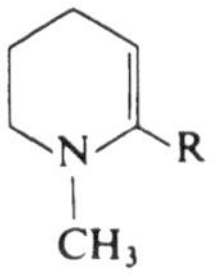 ![structure]	CH$_3$	—	91
	Et	—	95
	Pr	—	96
	Bu	—	97
	Pent	—	92
	Benz	—	96
	—(CH$_2$)$_3$CH=CH$_2$	—	77

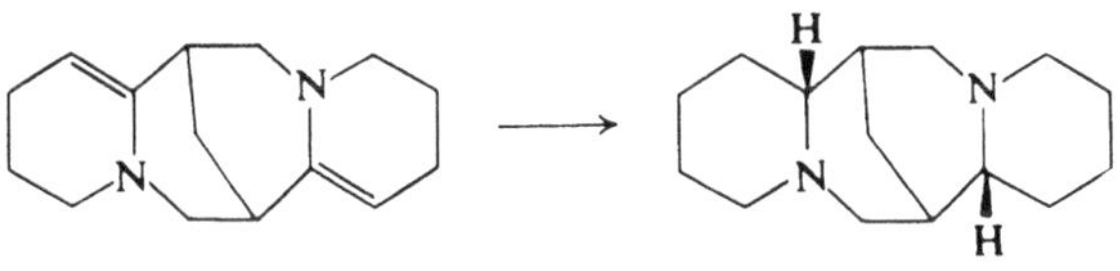

(97) → (98)

Another example [227] of selective reduction in a simpler system is shown by the conversion of 99 to 100.

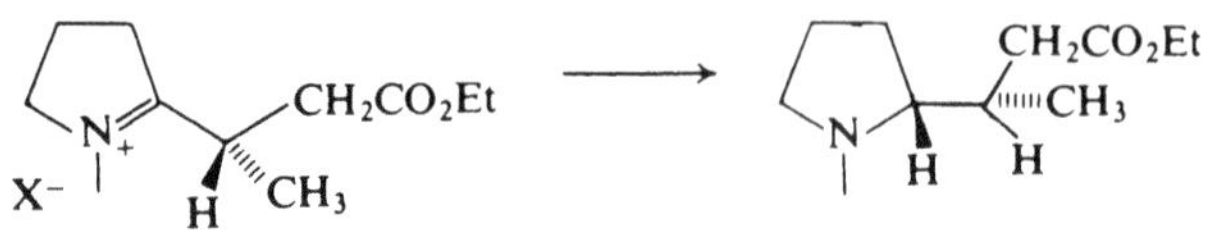

(99)   (100)

Dienamines, for example, 101, have been reported [18] to be reduced by formic acid to give a mixture of products 102 and 103 in 21%
yield.

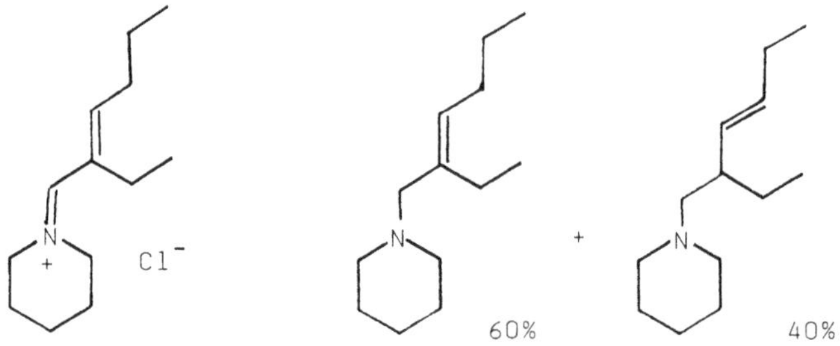

60%   40%

(101)   (102)   (103)

Lukes and Jizba [222] have reported that 1,2,6-trimethylpyridin-
ium formate (104) is reduced at the 4 position and eventually gives a
product 3-methylcyclohex-2-enone (105).

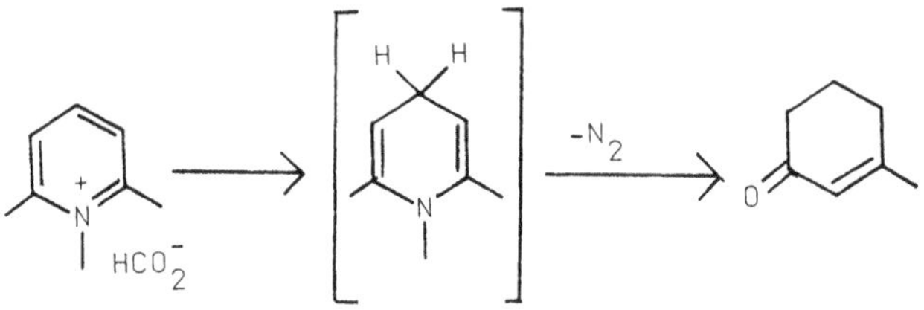

(104)   (105)

Formic acid reduction of 2-alkyl-cyclohexyl iminium salts proceeds
with high stereoselectivity [187]. For example, reduction of iminium
salt 106 with formic acid gives a 85:15 mixture of cis:trans (107:108)
isomers. This stereoselectivity is further illustrated by the reduction

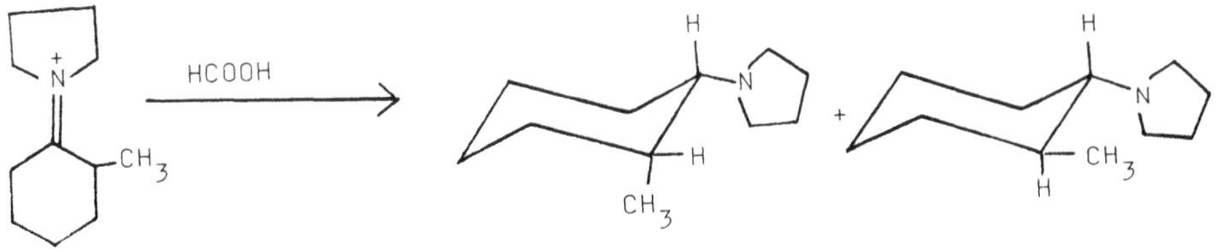

(106)   (107)   (108)

of the piperidine enamine of camphor (109) with formic acid to yield bornylpiperidine (110 and 111) in a 92:8 ratio of *endo* (110) to *exo* (111) epimers [189]. Both catalytic hydrogenation of enamine 109 and reduction of iminium salt of 109 with NaBH$_4$ are *less* stereoselective than formic acid reduction of 109 [241]. The *exo* epimer (111) is the chief product from catalytic hydrogenation, and the *endo* epimer (110) is the main product from NaBH$_4$ reduction of the iminium salt. The *endo/exo* ratio from catalytic hydrogenation is 20/80, and the *endo/exo* ratio from NaBH$_4$ reduction of the iminium salt is 79/21 [241].

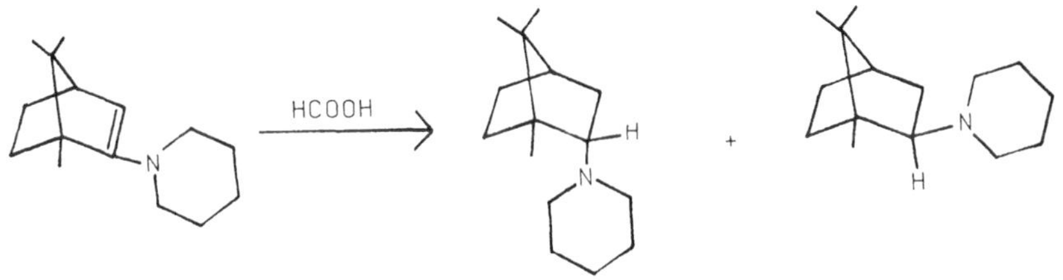

(109)   (110)   (111)

Iminium salts can be reduced with phosphorous acid [173]. The stereoselectivity of this reaction is somewhat less than that of formic acid reduction. This is surprising in view of greater bulk of phosphorous acid as compared to formic acid.

C. Addition of Diazoalkanes

Another class of nucleophilic addition reactions with iminium salts is the addition of diazoalkanes. Leonard and Jann [228—230] found that treatment of iminium perchlorates with diazomethane and other diazoalkanes yielded aziridinium salts. Treatment of an iminium salt such as N-cyclohexylidinepyrrolidinium perchlorate (11) with diazomethane yielded a new product whose structure was established by spectral and chemical means to be 5-azoniadispiro[4.0.5.1]dodecane perchlorate (112). The UV spectrum was devoid of any absorption above

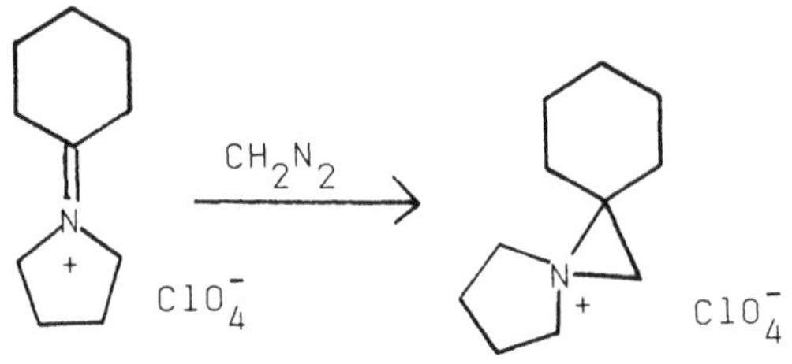

(11)   (112)

200 nm. The infrared spectrum did not show C = N+ or $\overset{+}{N}$–H. The molecular weight was consistent only with monomeric species, and the NMR spectrum showed the presence of a new singlet at δ = 3.02 whose integrated area corresponded to two hydrogen atoms. The product reacted with thiosulfate in a manner analogous to that observed for aziridinium intermediates [231,232]. The postulated mechanism (see Scheme 5) involves nucleophilic attack on the iminium moiety followed by loss of nitrogen and ring closure to give the aziridinium salt [233, 234]. It is not known whether the amine assists the elimination of

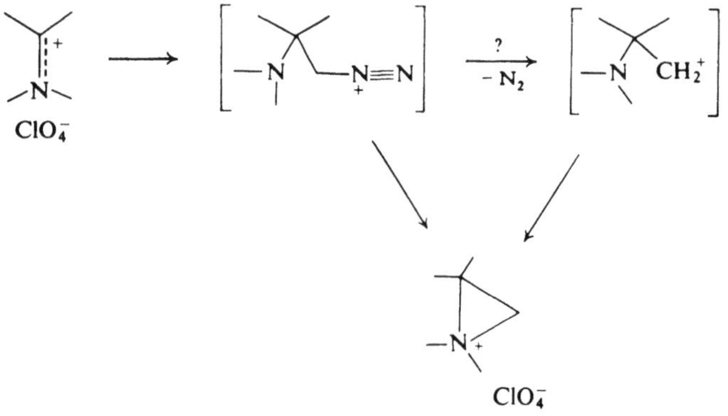

Scheme 5

the nitrogen, but that the iminium salt retains its stereochemistry has been demonstrated [233]. When a mixture of 113 and 114 of 1:5 ratio was treated with diazomethane, the ratio of 115:116 obtained in 75% yield and determined spectroscopically was still 1:5. The trans-N-isopropyl-N-methylisobutylidinium perchlorate (114) was prepared by alkylation of an aldimine salt with diazomethane and then by fur- ther reaction with diazomethane to give 116, so that the spectral as- signments for determining cis/trans ratios are on a firm ground.

(113) (114) (115) (116)

 (116)

(114)

The lifetime of the diazomethane adduct is not known, but it must be of sufficient length that carbon can compete with nitrogen in participation as a nitrogen molecule leaves [233]. The reaction of 11 with diazoethane yields 117, the normal adduct, and 118, the product from ring expansion. The structure of 118 was determined by hydrolysis to 2-methylcycloheptanone. In a similar reaction, 119 yields 112 directly without isolation of 11. A similar ring expansion has been observed for cyclooctanone derivatives [7].

(11) (117) (118)

(119)

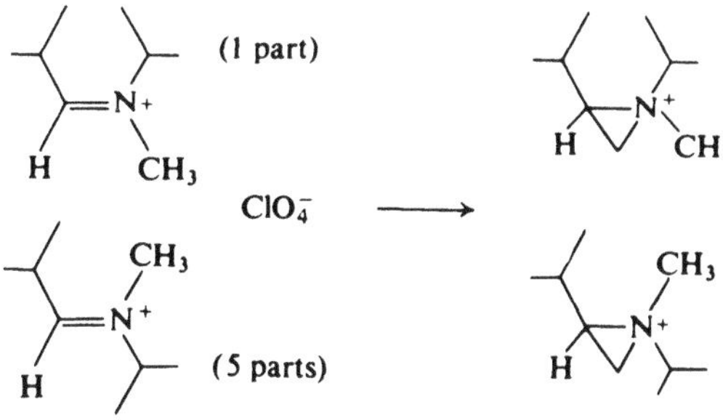

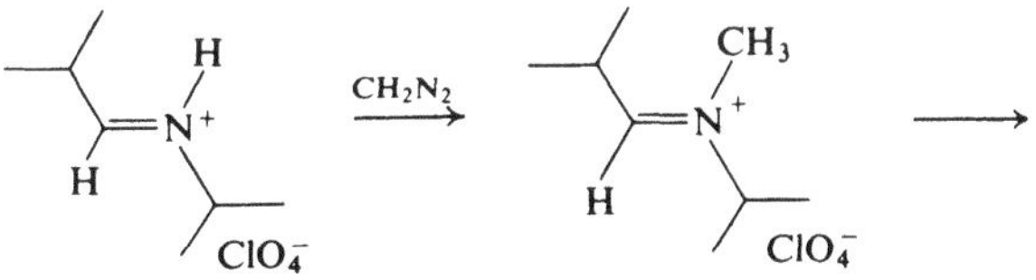

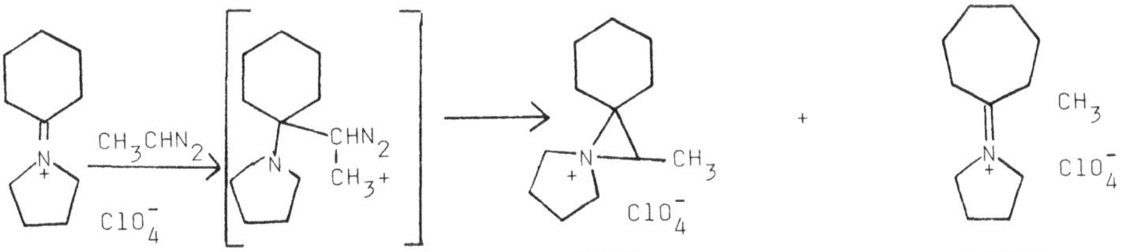

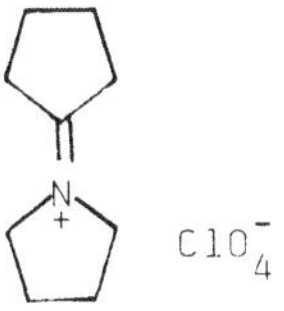

Opitz and Griesinger [7] investigated the reaction of diazomethane and iminium chlorides, but because of the nucleophilicity of the chloride ion, they obtained only β-chloroamines. The amines were assumed to result from collapse of a carbonium ion with a chloride ion and have structure 120. Further investigation [235] has shown that an aziridinium intermediate was formed first and was then opened to give β-chloroamines that were isomeric (121) with the structures (120) proposed by Opitz and Griesinger. It was also shown [235] that iminium chlorides with diazomethane give the same products as opening of corresponding aziridinium perchlorates with chloride ion. All the

(120)

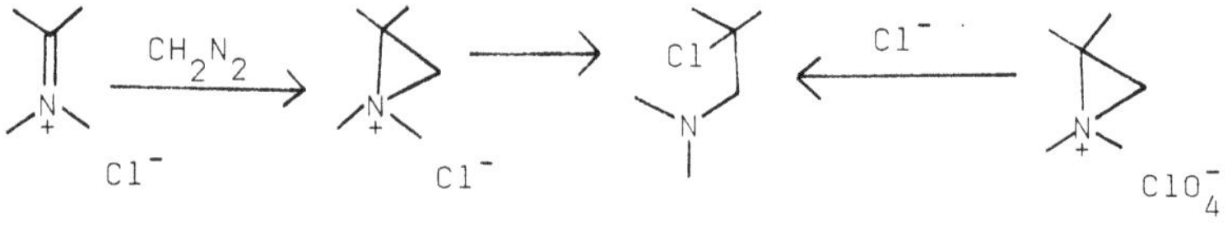

(121)

structures were established beyond a reasonable doubt by spectral methods [235]. The synthesis of β-chloroamines by reaction of α-chloroamines with diazomethane had been observed earlier, but the presence of the aziridinium intermediate had not been detected (see Scheme 6) [236].

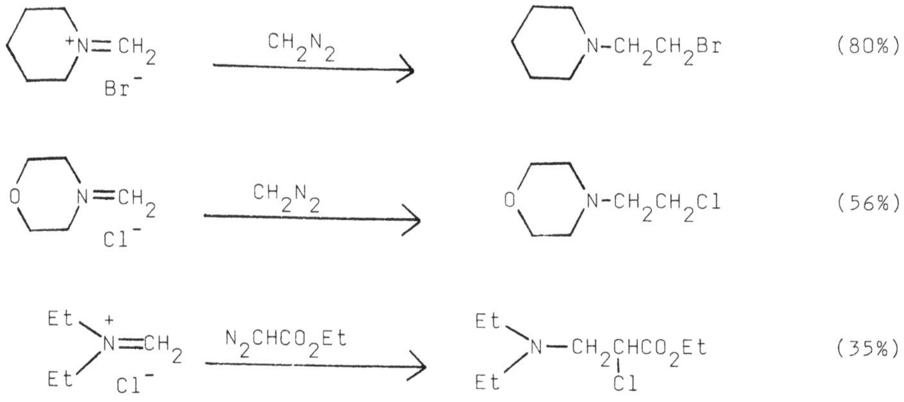

Scheme 6

A representative list of aziridinium salts prepared by reaction of iminium salts with diazomethane is given in Table 10. The reactions of aziridinium salts are many and varied, but will not be given here since their synthetic utility has been explored and reported elsewhere [233,234,237−239]. The products from the reaction of iminium chlorides and diazomethane are reported in Table 11. Many more examples are available in the literature [7].

In some cases, low product stability can lead to further reaction of the intermediate aziridinium salt to form a dimer, as shown in Scheme 7 [240].

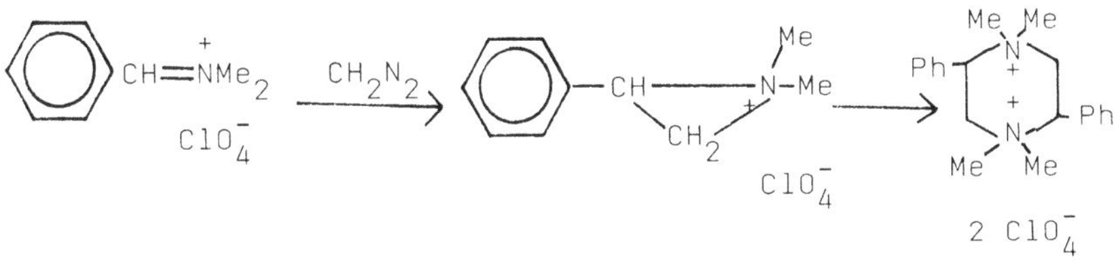

Scheme 7

D. Addition of Other Nucleophiles

The reaction of a large number of other nucleophiles with iminium salts will be discussed in this section. One can say with little reservation that almost all nucleophiles will react with iminium salts and that only the product stability will determine whether the adducts can be isolated.

The addition of cyanide was first examined by Kaufmann [242−246], who found that aromatic iminium compounds such as quinolinium methiodide (122) added potassium cyanide to give 1,4-addition product 123. The reaction has been examined more recently by several

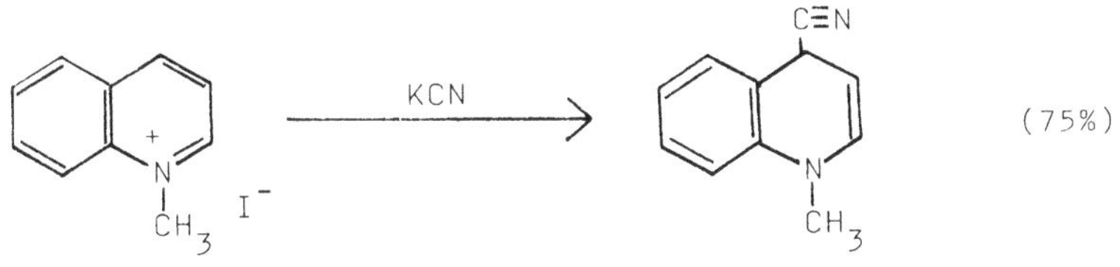

(122) (123)

groups with simple iminium salts, and the results are analogous [22−24,69,87,131,136,142,247−251,321]. Treatment of $\Delta^{1(16)}$-dehydrosparteinium perchlorate (124) with cyanide ion gave (-)-6-cyanosparteine (125) [22]. Other examples of this reaction are listed in Table 12.

TABLE 10 Reaction of Iminium Salts with Diazomethane

Iminium salt	Product (yield, %)	Ref.
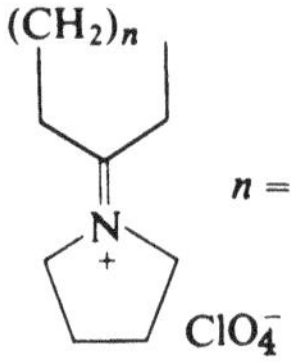 $n = 3, 4, 5$	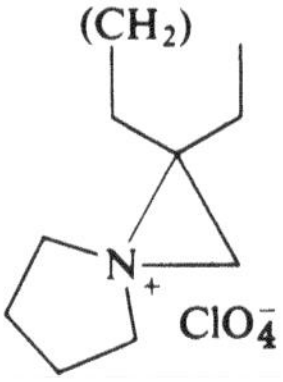 (93) (99) (79)	228–230 100
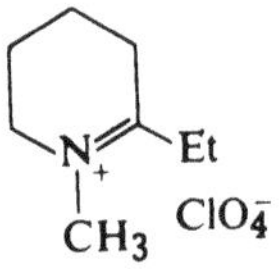	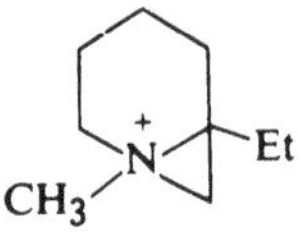 (87)	238
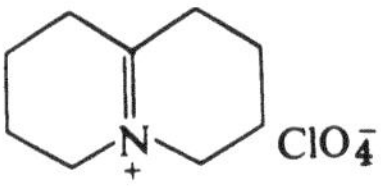	 (90)	238
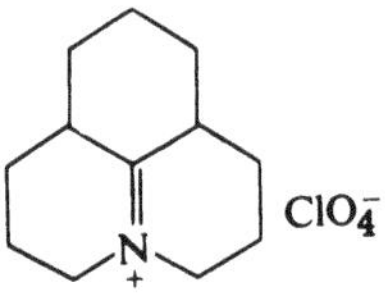	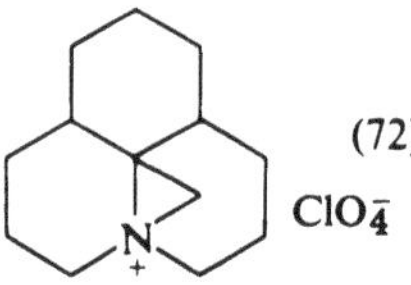 (72)	238
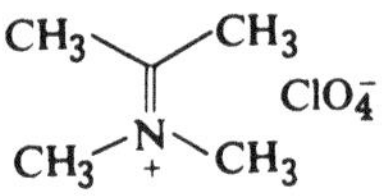	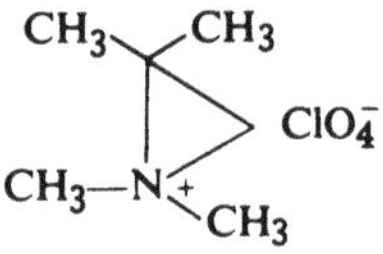 (93)	100
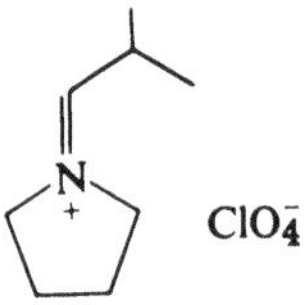	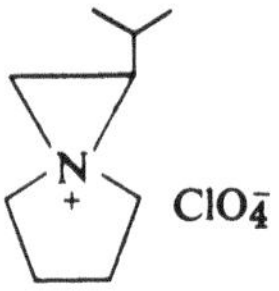 (85)	100
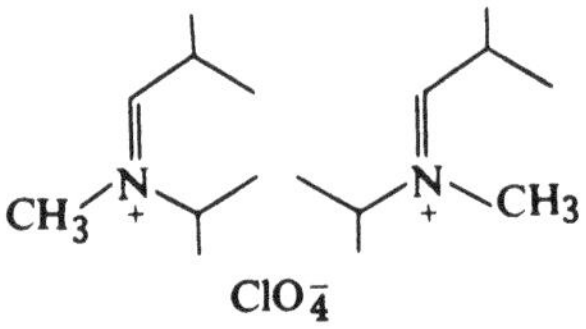	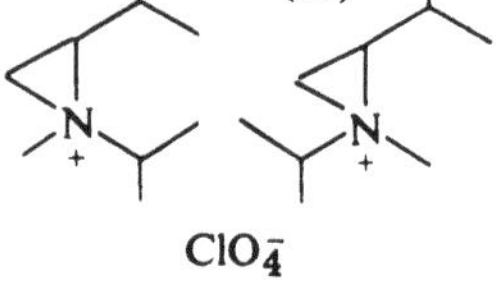 (75)	233

TABLE 11   Reaction of Iminium Chloride with Diazomethane

Compound	Product	Ref.
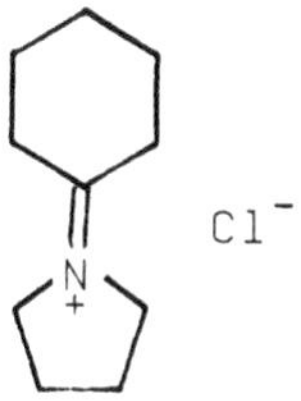	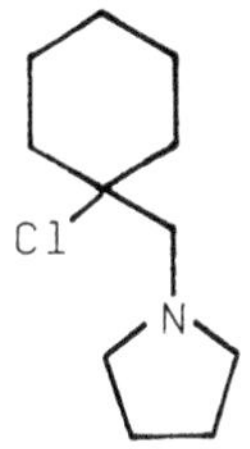	7,235
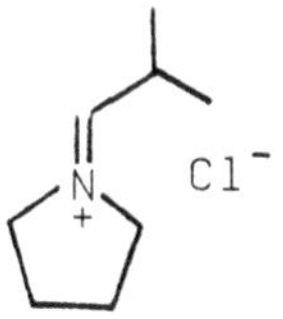	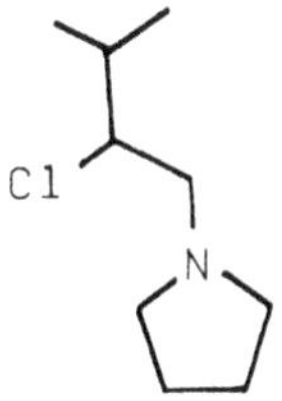	7,235

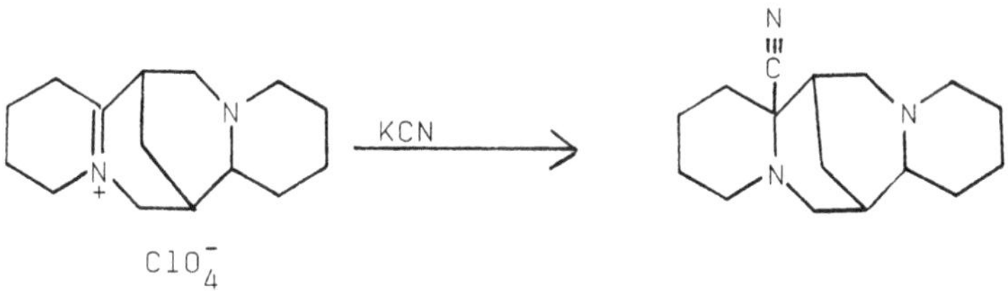

KCN →

(124)   (125)

The addition of a cyanide ion to an iminium can occur in a stereospecific manner. It has been reported that 1-alkyl-4-(N-phenylpro-
panamido)-3,4,5,6-tetrahydropyridinium salts (126) react with cyanide
ion to give the *trans* diastereoisomers (127) [252].

TABLE 12 Addition of Cyanide to Iminium Salts

Compound	Product	(yield, %)	Ref.
		(97)	23
		(82)	23
		(21)	142
		(70)	24
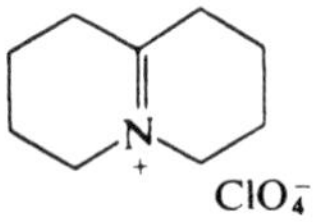	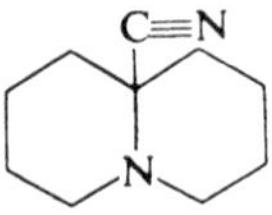	(88)	131

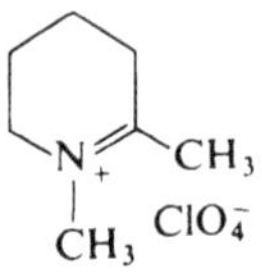

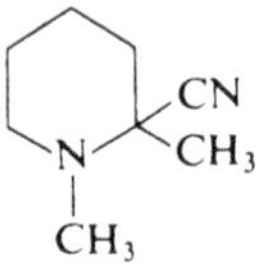

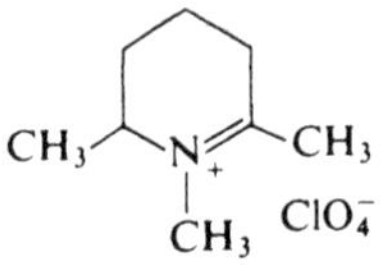

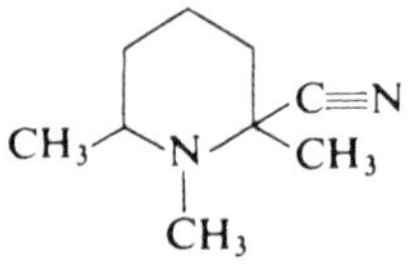

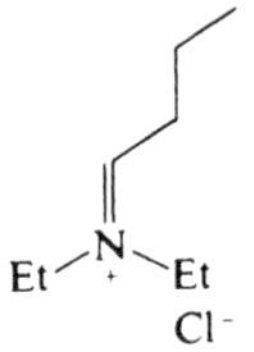

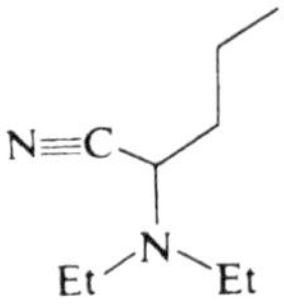

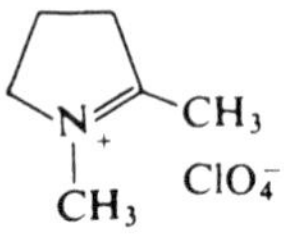

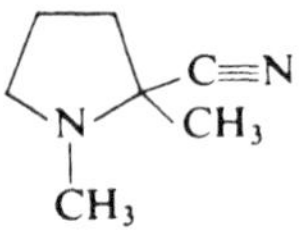

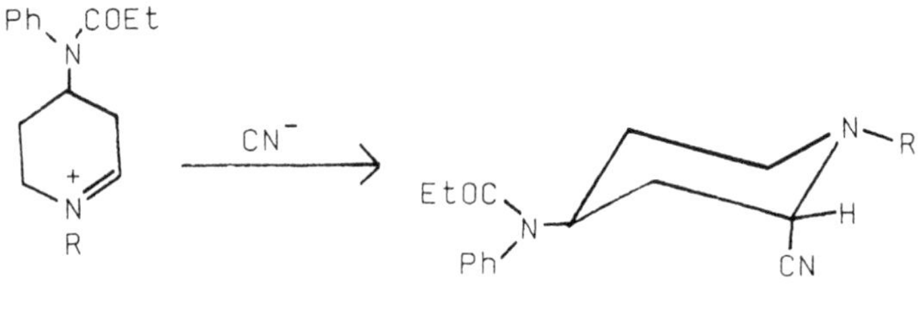

(126)   (127)

α-Cyano tertiary amines can also be synthesized by treating ena-mines with diethyl phosphorocyanidate ($(EtO)_2P(O)CN$) in tetrahy-drofuran followed by hydrolysis with aqueous 20% hydrochloric acid [253].

The cyano group in the α-cyano tertiary amine can enter into a number of reactions, including elimination to give the iminium salts (see Scheme 8) [47,131,185].

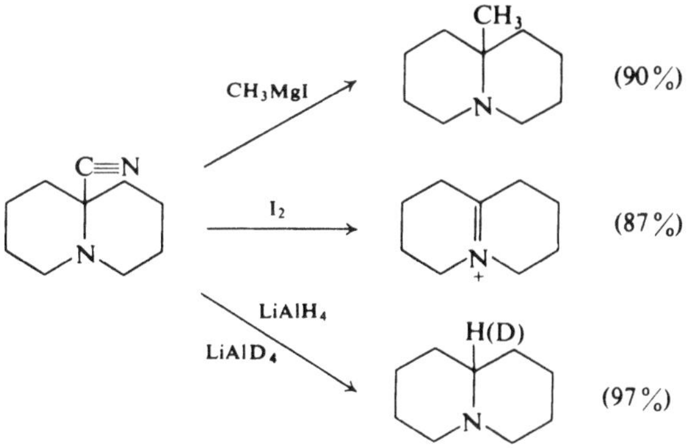

Scheme 8

Aryl- and alkylthiols [69,131,254,257], thiolacids [255,256], and α-mercapoacids [256] have been shown to react with iminium salts. The potassium salt of p-thiocresol adds to dehydroquinolizidinium salt 10 to give adduct 128 [131]. This adduct decomposes on standing in air or on treatment with dilute acids to give back the p-thiocresol.

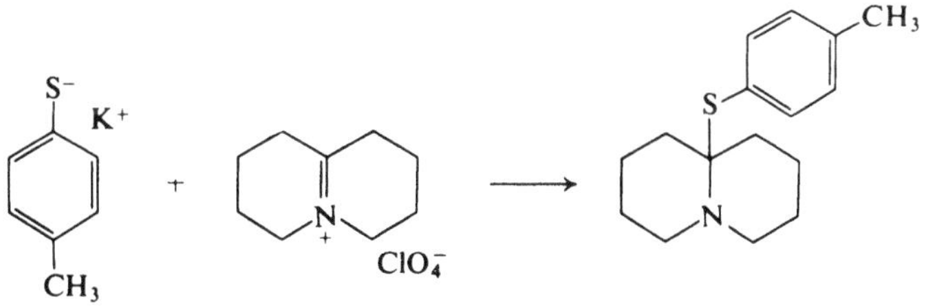

(10)   (128)

At -40°C thiolacetic acid adds to 1-(N-morpholino)cyclopentene to give addition product 129 [255,256]. This product, in turn, when allowed to come to room temperature, decomposes into N-acetylmorpholine (130) and a tarry product.

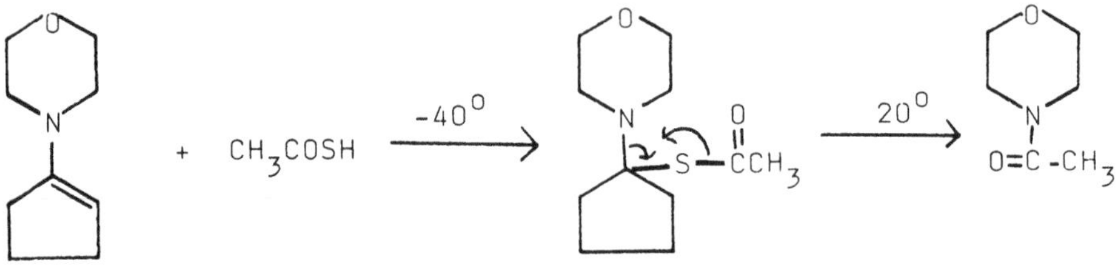

(129)   (130)

Treatment of an acetate iminium salt with trifluoroacetic acid gives a trifluoroacetate α-addition product [328].

Alcohols and alkoxides add to iminium salts to give α-alkoxyamines [131,247,257]. This is illustrated by the reaction of $\Delta^{2,3}$-dehydro-3-ethyl-3-azoniumbicyclo[3.3.1]nonane perchlorate (131) with potassium ethoxide to give N,O-acetal 132. When 132 is treated with water, dimeric etherbase 133 is formed, probably by attack of the pseudobase (α-hydroxyamine) of 131 on an iminium ion (131) [247].

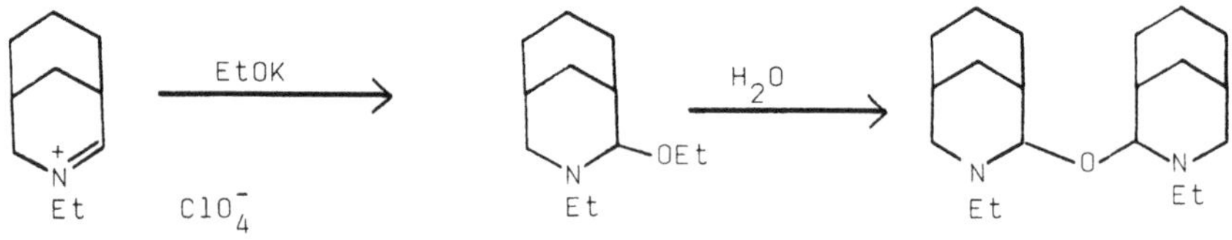

(131)   (132)   (133)

Amines have been reported to add to iminium ions either directly [69,108,257] or through conjugate addition of an enamine to an iminium ion [23,24,108,258–260,279,338]. Iminium ion 134 (an intermediate produced by protonation of dimethylaminoallene) reacts with dimethylamine to yield aminal 135 [257]. This reaction even takes place under mild conditions when the sterically crowded 2,2,6,6-tetramethylpiper-

$$CH_2=CHCH\overset{+}{=}NMe_2 \quad + \quad Me_2NH \quad \longrightarrow \quad CH_2=CHCH(NMe_2)_2$$

(134)   (135)

idine is used as the amine [257]. 1,4-Addition to conjugated iminium salts instead of the 1,2-addition shown in the example cited above has also been observed [69]. Refluxing 1-(N-piperidino)cyclopentene (138) with iminium ion 136 for 72 hr produced 137 in a 75% yield [108], obviously the result of the nucleophilic attack of an enamine on an imini-

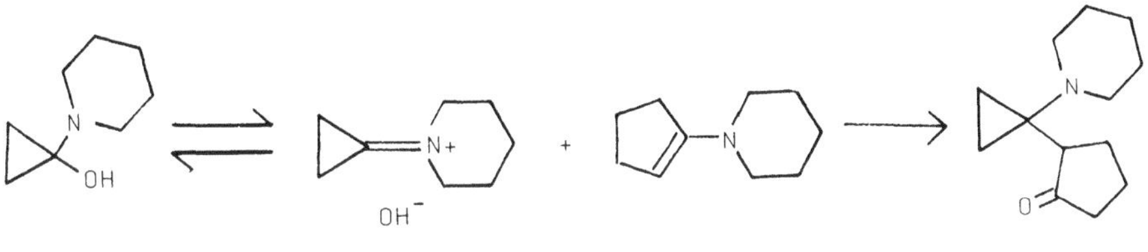

(136)   (138)   (137)

um ion. Production of 1-methyl-Δ²-tetrahydropyridine (139) by mer-curic acetate oxidation of 1-methylpiperidine is followed by attack of

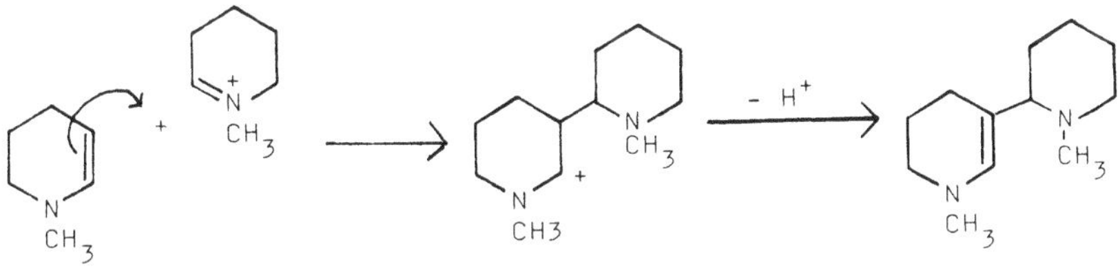

(139)   (140)

Scheme 9

an iminium ion by an enamine to give dimer 140 in a 67% yield (see Scheme 9) [23]. Pyrrole and pyrazole add to the α-carbon of an ena-mine [331,332]. Ynamines also react readily (at room temperature) in a conjugate manner with iminium salts [261]. A combination of an imine nitrogen and its β-carbon both attacking the two iminium ion carbon sites of a diiminium salt to form a pyrrole has been reported [270].

Dialkyl and trialkyl phosphites as well as phosphines have been shown to react with iminium salts (see Scheme 10) [142,146,285].

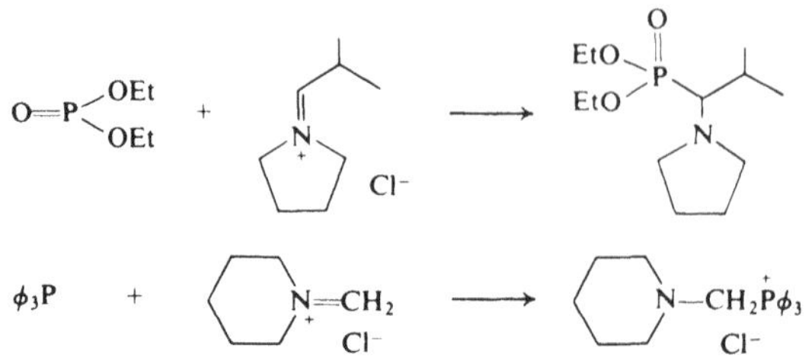

Scheme 10

Diphenylphosphine oxide adds to an enamine by, apparently, first promoting the enamine to give the iminium salt followed by nucleophilic attack on the α-carbon of the iminium ion by the diphenylphosphine oxide anion (see Scheme 10) [335–337].

Hydrolysis of iminium salts (nucleophilic attack of iminium ions by water) is covered in Chapter 3.

Azides have been shown to react with iminium salts to give addition products. The same product is obtained if the iminium salt is treated with azide ion or if the enamine is treated with hydrazoic acid [18]. The yields of the products were all very high (85–95%). The interest in this reaction centers on the fact that the azides react with isonitriles to give substituted tetrazoles (141) [142].

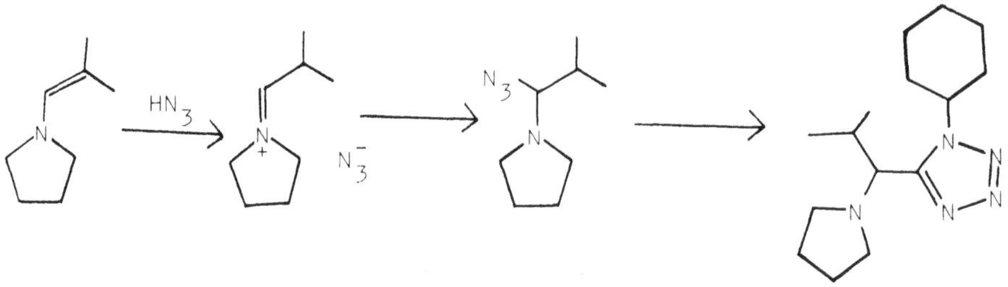

(141)

The reaction of iminium salts such as 11 with salts of trichloroacetic acid has been shown to yield amides such as 143 on hydrolysis [262]. It was suggested that the reaction proceeds by addition of dichlorocarbene to give an aziridinium intermediate (142), which was opened by trichloroacetate followed by hydrolysis to give the observed products [262]. The observed products from the reaction can

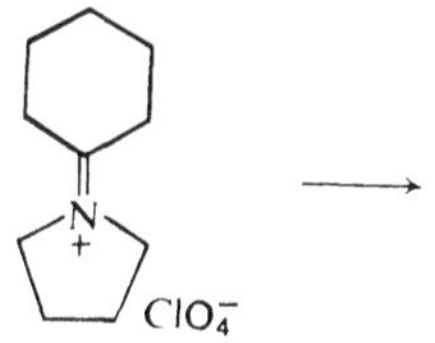

(11)

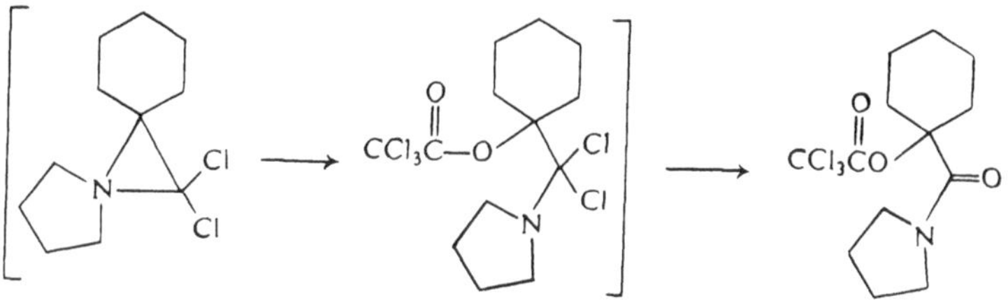

(142)   (143)

be accounted for by formation of CCl$_3$, which could add to C = N+ to give 144 [263–268] followed by displacement of chloride to give the aziridinium intermediate in a process analogous to that reported for other β-chloroamines [235]. The conversion of 142 to 143 would follow by known steps. A reaction product similar to intermediate 144

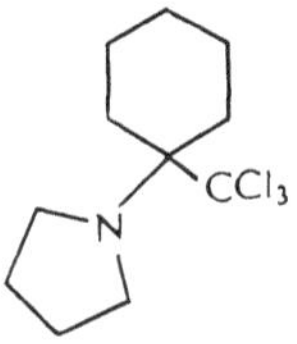

(144) 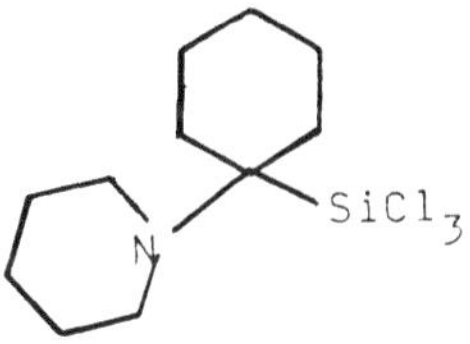   (145)

is obtained when trichlorosilane is allowed to react with 1-(N-piperidino)cyclohexene to give adduct 145 [269].

A commonly found nucleophile in organic chemistry is the carbanion stabilized by one or two α-carbonyl groups, α-nitrile groups, or α-nitro groups. This type of nucleophile has been extensively studied and has been observed to attack iminium salts. Among the types of compounds that fit into this category of nucleophile and are reported to attack iminium salts are carbanions of the following: ketones and aldehydes (including the Mannich reaction, which will be further discussed and documented below) [247,271], oxazol-5(4H)-ones

[329,330], nitroalkanes [272], β-diesters [273], β-dicarboxylic acids [274], β-ketoacids [274], cyanoacetic acid [274,275], and nitroacetic acid [276]. A reaction scheme that resulted in the

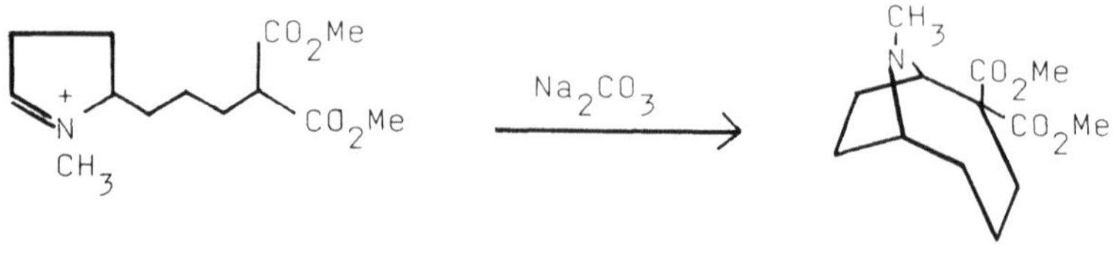

(146)   (147)

synthesis of *anatoxin a* involved the intramolecular cyclization of iminium salt 146 in basic medium to bicyclic aminodiester 147 [273]. Treatment of 1-(N-pyrrolidino)cyclohexene with cyanoacetic acid initially gives iminium salt 11 and, under conditions of refluxing dioxane

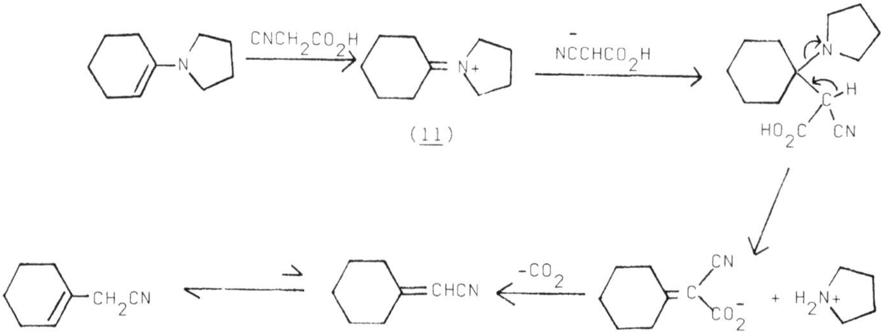

(11)

(148)

Scheme 11

solvent, produces 1-cyclohexen-1-acetonitrile (148) (see Scheme 11) in a 92% yield [274].

The Mannich reaction is a classic reaction and one of the most important methods for the α-aminoalkylation of a CH acidic compound [277,278]. The reaction involves an aldehyde (usually formaldehyde), the salt of an amine (usually a secondary amine such as dimethylamine), and a CH acid compound (usually a carbonyl compound such as a ketone). This reaction mixture is then refluxed for several days in a solvent such as ethanol, after which the product is isolated in widely varying yields. The iminium salt formed from the reaction of the amine salt and formaldehyde is postulated as a probable intermediate in this reaction. The disadvantages of this reaction are the long reaction times and the occurrence of side reactions.

However, more recently a new, improved approach to this reaction has been used which involves the prior formation of the iminium salt such as that formed between dimethylammonium ion and formaldehyde (iminium salt 26). There are three basic advantages to this

$$CH_3 \\ \diagdown \overset{+}{N} = CH_2$$
$$CH_3 \diagup \quad X^-$$

(26)

a: $X = CF_3CO_2^-$
b: $X = Cl^-$
c: $X = I^-$
d: $X = ClO_4^-$

approach: (a) faster reactions since the concentration of iminium ion is higher than that generated via equilibria; (b) lower temperatures are possible; (c) aprotic conditions may be used [67]. Because either protic or aprotic solvents can be used, regioselectivity was observed in the Mannich reaction of unsymmetrical ketones such as 3-methyl-2-butanone (149) [38,39]. When the protic trifluoroacetic acid solvent is used, the thermodynamically favor product 151 is formed almost exclusively (85/15 ratio of 151/150). If aprotic solvent acetonitrile is the solvent used (with the perchlorate iminium salt), then

$$(CH_3)_2CH\overset{O}{\overset{\|}{C}}CH_3 \;+\; (CH_3)_2\overset{+}{N}{=}CH_2 \;\longrightarrow\; (CH_3)_2\overset{O}{\overset{\|}{C}}CH_2CH_2NMe_2 \;+\; Me_2NCH_2\overset{CH_3}{\underset{CH_3}{\overset{|}{C}}}\overset{O}{\overset{\|}{C}}CH_3$$

(149)   (150)   (151)

the kinetically favored product 150 is the exclusive product formed [38,39]. Holy and co-workers [67] state that dichloromethane is the solvent of choice in these Mannich reactions because of the convenience of product isolation.

Another variable that must be considered in this modified Mannich reaction is the iminium counterion. The most commonly used anions are trifluoroacetate [38,39,62,66], chloride [43,71,280], iodide [46, 75–77], and perchlorate [38,56,57]. Recently the triflate anion (trifluoromethanesulfonate) has been used also [280].

Finally, there have been variations in the nature of the CH acidic compound and in the prior use or nonuse of various bases to remove the acidic hydrogen. A good summary review of these variations has been written by Holy and co-workers [67]. There have been reports of simple addition of the iminium salt to the CH acidic compound [38, 39,46,56,62], prior removal of a proton from the CH acidic compound using lithium diisopropylamide (LDA) followed by addition of iminium salt [281], prior formation of an enol borinate [75], prior formation of a silyl derivative either with [66,77,280] or without [76] prior formation of the carbanion with LDA or methyllithium, and prior proton removal from the CH acidic compound using potassium hydride [281]. An example of a reaction taking place under these last conditions is the aminomethylation of camphor to produce 152 in a 70% yield [67].

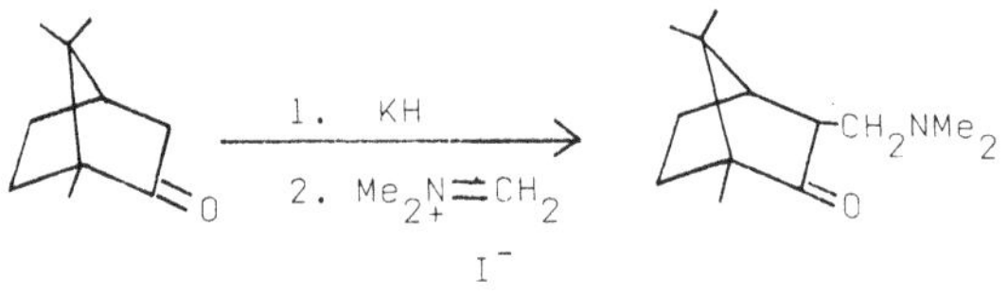

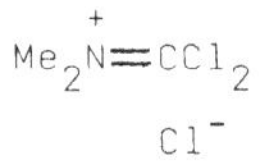

(152)   (153)

Dichloromethylenammonium salts such as dimethyl(dichloromethylene)ammonium chloride (153) have been used as Mannich reagents also [282].

IV. DEPROTONATION OF IMINIUM SALTS

A ternary iminium ion can lose a proton from an α-carbon, a β-carbon, or an α'-carbon (see Figure 1). The proton that is lost most readily is the β-proton, which is just the reversal of C-protonation of an

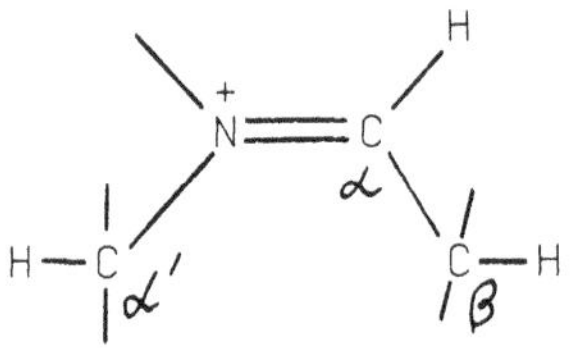

Figure 1 Protons of a ternary iminium ion.

enamine (see Section II.B and Chapter 1, Section IV.E). For example, $\Delta^{1(10)}$-dehydroquinolizidine (154) is obtained by treating its iminium perchlorate salt (10) with aqueous sodium hydroxide and extracting [21]. Similar observations have been made in other ternary iminium ion systems as well [22,34,138,283,284].

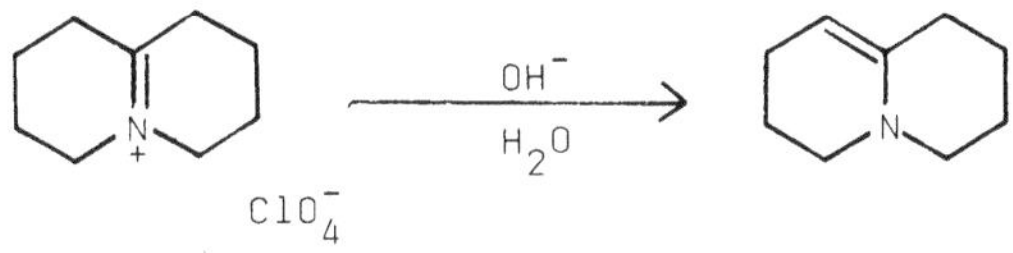

(10)   (154)

Removal of a β-proton from a ternary iminium ion can be carried out regiospecifically. This can be done by treating the asymmetrical iminium salt with a hindered primary amine such as *t*-butylamine or a moderately bulky secondary amine such as morpholine, piperidine, or diisopropylamine in pentane [19,334]. The less substituted enamine will be formed as the kinetically controlled product, quite often in quantitative yields. For example, treatment of a mixture of the

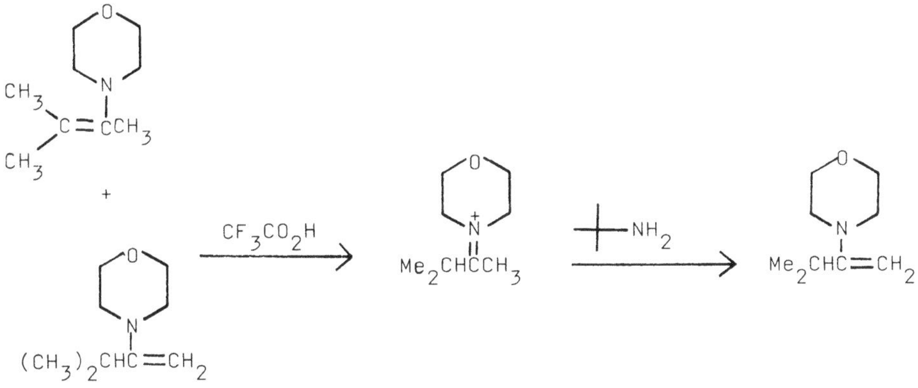

Scheme 12

two possible isomers for the morpholine enamine of 3-methyl-2-butanone with the C-protonating agent trifluoroacetic acid gives a common ternary iminium salt. Deprotonating this with *t*-butylamine then gives the least substituted enamine product since the primary β-methyl proton is kinetically more labile than the tertiary β-methine proton (see Scheme 12).

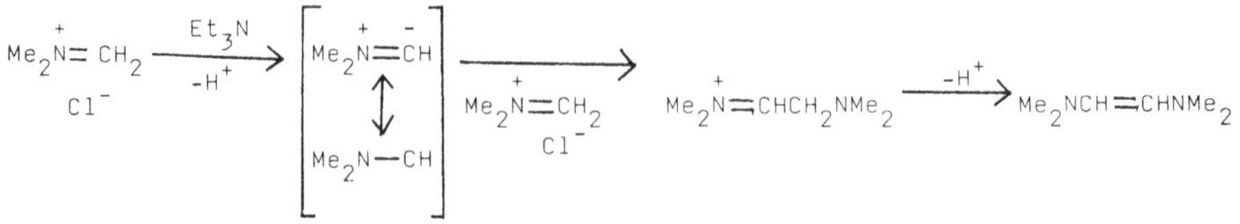

(26b) (155) (156)

The reaction of dimethyl(methylene)ammonium chloride (26b) with triethylamine gives intermediate 155, which in turn reacts with another molecule of 26b to give enediamine 156 [285]. This reaction involves the removal of an α-proton to give intermediate 155.

If there are no α- or β-protons in the ternary iminium salt, then the possibility of α'-proton removal to form an azomethine ylid [308–312] becomes important. Dimethyldiphenylmethyleniminum iodide (157)

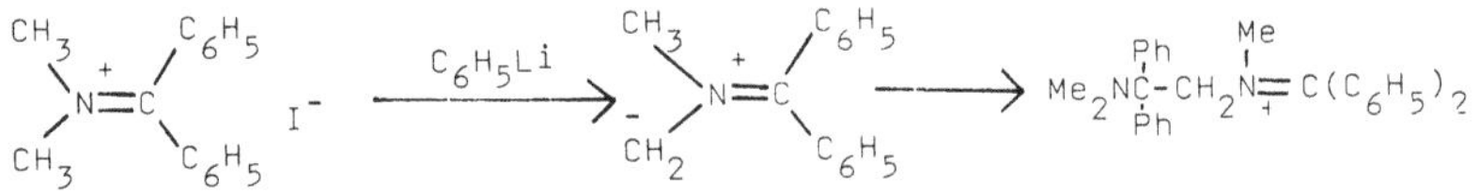

(157) (158) (159)

is an example of such an iminium salt. When it is allowed to react with phenyllithium, α'-proton removal takes place and azomethine ylid 158 is formed. This ylid then reacts with a second molecule of 157 to form dimer 159 [140].

In an effort to find a new pathway for the synthesis of aziridines, N-(benzhydrylidene)methyl-tert-butylaminium fluorosulfonate (160) was treated with the sterically hindered, nonnucleophilic strong base sodium bis(trimethylsilyl)amide (161). Initial α'-proton removal produced azomethine ylid intermediate 162 (a 1,3-dipole), which then cyclized to aziridine 163 in a near quantitative yield [286]. The main competing reaction when other bases were used was dealkylation.

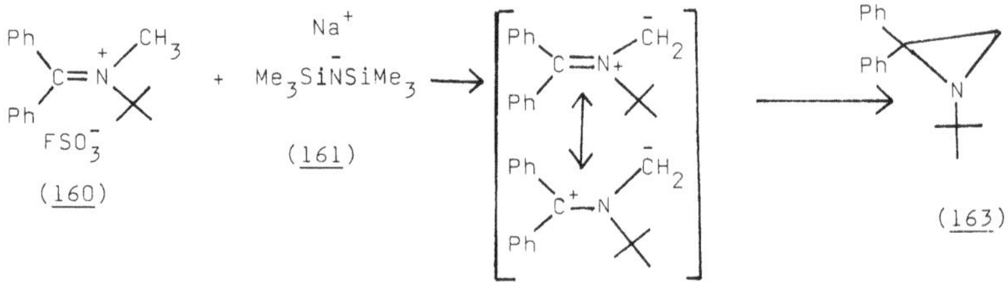

(160) (161) (162) (163)

Keteniminium salt 164a can be α'-deprotonated to azomethine ylid 165, which then can be trapped by the 1,3-dipolarophile, norbornene, to give enamine 166 [28]. This enamine then formed iminium salt 14

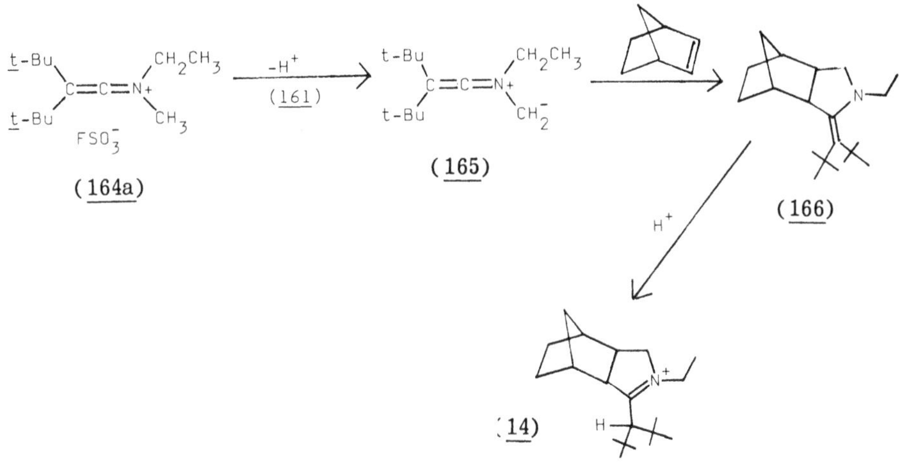

(164a)   (161)   (165)   (166)

(14)

on protonation, and this iminium salt is extremely stable toward addition, deprotonation, hydrolysis delkylation, or reduction by a wide variety of reagents [28].

Simple azomethine ylids of this type may also be produced by methods other than deprotonation. They may be produced by desilylation of iminium silated salts [306,313,314] or by treatment of a tertiary amine N-oxide with LDA (lithuim diisopropylamide) [306,315–317].

V. CYCLOADDITIONS OF IMINIUM SALTS

Keteniminium salts undergo ($_\pi 2_s$ + $_\pi 2_a$) cycloadditions [287] with alkenes to form cyclobutyl iminium salts [288–291]. For example, cyclohexene reacts with keteneiminium in 164b to form bicycloiminium salt 167 in an 88% yield [289]. Keteniminium salts undergo ($_\pi 2_s$ + $_\pi 2_a$)

$Me_2C= C =NMe_2$ +   ⟶

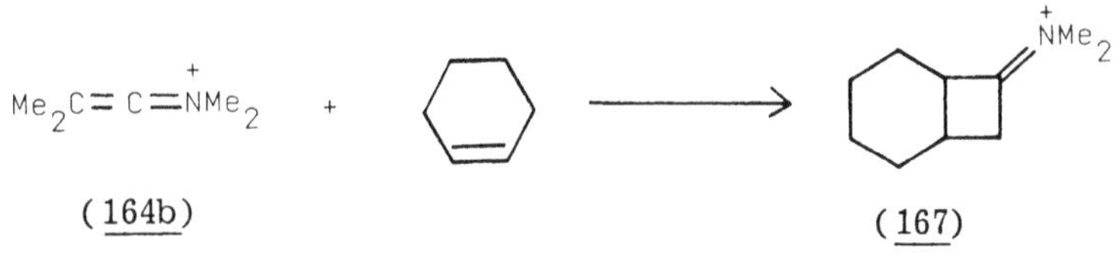

(164b)   (167)

cycloadditions with imines [292] and alkynes [293] also. Keteniminium ions do not dimerize or polymerize like ketenes or ketenimines because keteniminium ions are more electrophilic than the latter substances.

Diiminium salts such as 168 will carry out a double electrophilic

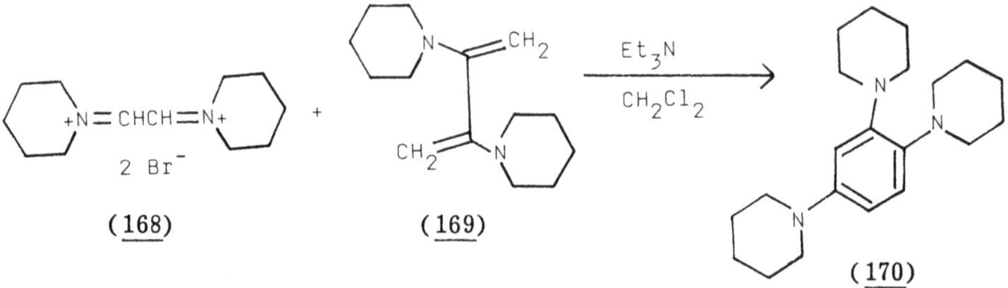

(168)   (169)   (170)

attack on a dienamine like 169 to form, after elimination of piperidine, the trisubstituted benzene 170 [294−296].

Cycloadditions of simple azomethine ylids (some formed from ternary iminium salts) with 1,3-dipolarophiles such as alkenes [28,314, 315,317], enamines [307], or alkynes [306] have been reported.

VI. PHOTOCHEMICAL REACTIONS OF IMINIUM SALTS

Iminium salts undergo photochemical reactions. This is illustrated by the photolysis of the iminium perchlorate salt of 1-(N-pyrrolidino)

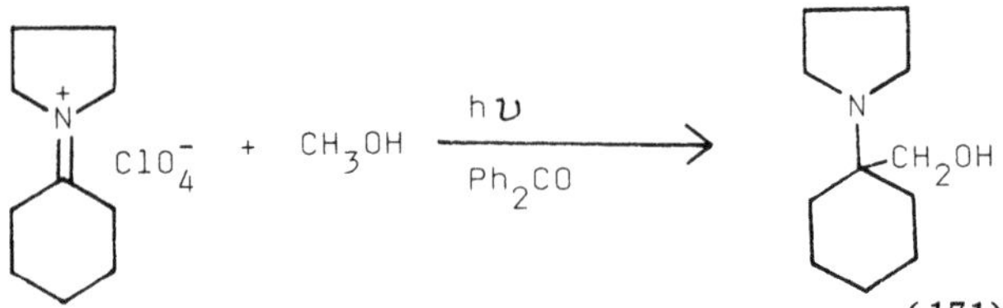

(171)

cyclohexene with methanol to produce aminoalcohol 171 in a 75% yield [102]. This is an example of sequential electron-proton transfer [304,305].

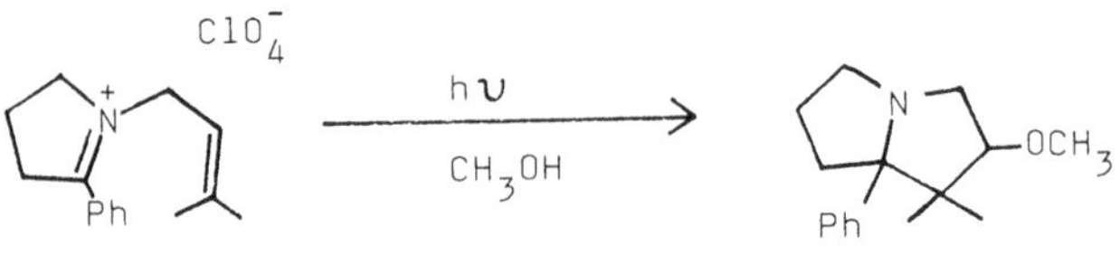

(172)   (173)

The delocalized positive charge in the iminium ion allows for its ready participation in excited-state single-electron transfer reactions with electron donors including alkenes, arenes, alcohols, and ethers

[297–303]. For example, irradiation of iminium salt 172 in methanol gave 173 in a 50% yield [299].

VII. REARRANGEMENTS OF IMINIUM SALTS

Some iminium salts undergo [3,3]-sigmatropic rearrangements (also called 2-aza-Cope rearrangements) under very mild conditions (see Scheme 13) [322–327]. For example, benzaldehyde will react with

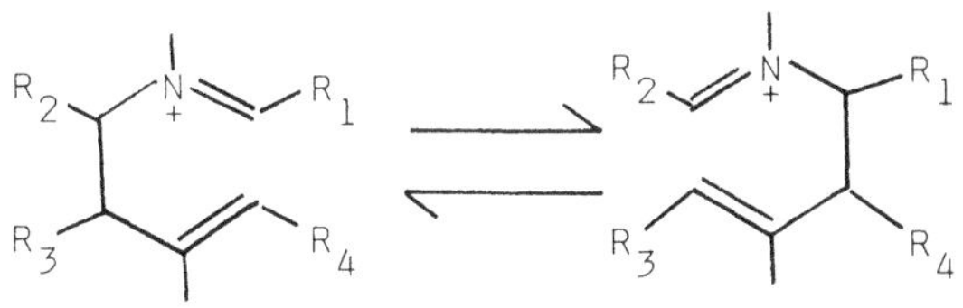

Scheme 13

ammonium salt 174 to give imminium salt 175, which rearranges to imminium salt 176. This intermediate then undergoes a Mannich condensation to pyrrolidine derivative 177, which can be hydrolyzed to 3-acetylpyrrolidine 178 in an 85% yield [324].

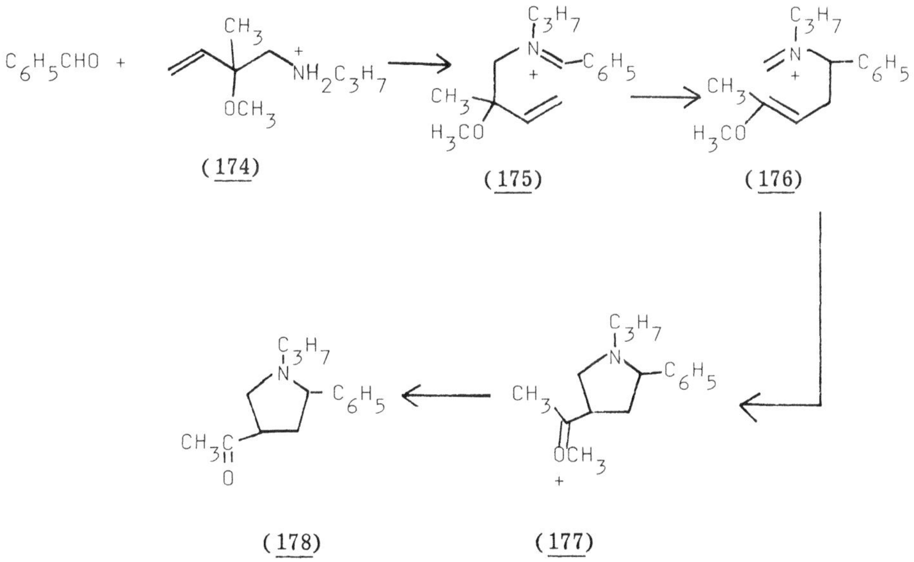

This type of rearrangement reaction of iminium salts has been extensively reviewed [323].

REFERENCES

1. H. Bohme and H. G. Viehe, eds., *Iminium Salts in Organic Chemistry*, in *Advances in Organic Chemistry* (E. C. Taylor, ed.), Vol. 9, Parts 1 and 2, Wiley-Interscience, New York, 1976 and 1979.

2. P. A. Kollman in *Iminium Salts in Organic Chemistry* (H. Bohme and H. G. Viehe, eds.), in *Advances in Organic Chemistry* (E. C. Taylor, ed.), Vol. 9, Part 1, Wiley-Interscience, New York, 1976, pp. 1−21.

3. W. J. Hehre, R. F. Stewart, and J. A. Pople, *J. Chem. Phys.*, *51*, 2651 (1969).

4. A. G. Cook, unpublished results.

5. M. J. S. Dewar and W. Thiel, *J. Am. Chem. Soc.*, *99*, 4899 (1977).

6. L. M. Trefonas, R. L. Flurry, Jr., R. Majeste, E. A. Meyers, and R. F. Copeland, *J. Am. Chem. Soc.*, *88*, 2145 (1966).

7. G. Opitz and A. Griesinger, *Liebigs Ann. Chem.*, *665*, 101 (1963).

8. E. J. Stamhuis and W. Maas, *J. Org. Chem.*, *30*, 2156 (1965).

9. J. Elguero, R. Jacquier, and G. Tarrago, *Tetrahedron Lett.*, 4719 (1965).

10. H. Matsushita, Y. Tsujino, M. Noguchi, and S. Yoshikawa, *Chem. Lett.*, 1087 (1976).

11. H. Matsushita, Y. Tsujino, M. Noguchi, and S. Yoshikawa, *Bull. Chem. Soc. Jpn.*, *50*, 1513 (1977).

12. H. Matsushita, Y. Tsujino, M. Noguchi, M. Saburi, and S. Yoshikawa, *Bull. Chem. Soc. Jpn.*, *51*, 201 (1978).

13. H. Matsushita, Y. Tsujino, M. Noguchi, M. Saburi, and S. Yoshikawa, *Bull. Chem. Soc. Jpn.*, *51*, 862 (1978).

14. M. R. Ellenberger, D. A. Dixon, and W. E. Farneth, *J. Am. Chem. Soc.*, *103*, 5377 (1981).

15. H. Bohme, *Angew. Chem.*, *68*, 224 (1956).

16. H. Bohme, H. Ellenberg, O. -E. Herboth, and W. Lehners, *Chem. Ber.*, *92*, 1608 (1959).

17. G. Opitz, H. Hellmann, and H. W. Schubert, *Liebigs Ann. Chem.*, *623*, 117 (1959).

18. G. Opitz and W. Merz, *Liebigs Ann. Chem.*, *652*, 139 (1962).

19. L. Nilsson, R. Carlson, and C. Rappe, *Acta Chem. Scand.*, *Ser. B, 30*, 271 (1976).

20. N. J. Leonard, A. S. Hay, R. W. Fulmer, and V. W. Gash, *J. Am. Chem. Soc.*, *77*, 439 (1955).

21. N. J. Leonard and V. W. Gash, *J. Am. Chem. Soc.*, *76*, 2781 (1954).

22. N. J. Leonard, P. D. Thomas, and V. W. Gash, *J. Am. Chem. Soc.*, *77*, 1552 (1955).

23. N. J. Leonard and F. P. Hauck, Jr., *J. Am. Chem. Soc.*, *79*, 5279 (1957).

24. N. J. Leonard and A. G. Cook, *J. Am. Chem. Soc.*, *81*, 5627 (1959).
25. A. G. Cook, Ph.D. thesis, University of Illinois, 1959.
26. A. T. Blomquist and E. J. Moriconi, *J. Org. Chem.*, *26*, 3761 (1961).
27. U. Edlund, *Acta Chem. Scand.*, *27*, 4027 (1973).
28. J. A. Deyrup and G. S. Kuta, *J. Org. Chem.*, *43*, 501 (1978).
29. D. A. Evans, C. H. Mitch, R. C. Thomas, D. M. Zimmerman, and R. L. Robey, *J. Am. Chem. Soc.*, *102*, 5955 (1980).
30. A. G. Schultz, R. D. Lucci, J. J. Napier, H. Kinoshita, R. Ravichandran, P. Shannon, and Y. K. Yee, *J. Org. Chem.*, *50*, 217 (1985).
31. L. Alais, R. Michelot, and B. Tchoubar, *C. R. Acad. Sci. Paris*, *273*, 261 (1971).
32. R. Carlson, L. Nilsson, C. Rappe, A. Babadjamian, and J. Metzger, *Acta Chem. Scand.*, *Ser. B.*, *32*, 85 (1978).
33. C. R. Hauser and D. Lednicer, *J. Org. Chem.*, *24*, 48 (1959).
34. G. N. Walker and D. Alkalay, *J. Org. Chem.*, *32*, 2213 (1967).
35. T. D. Stewart and W. E. Bradley, *J. Am. Chem. Soc.*, *54*, 4172 (1932).
36. H. Bohme and D. Morf, *Chem. Ber.*, *91*, 660 (1958).
37. H. Bohme and M. Haake in *Iminium Salts in Organic Chemistry* (H. Bohme and H. G. Viehe, eds.), in *Advances in Organic Chemistry* (E. C. Taylor, ed.), Vol. 9, Part 1, Wiley-Interscience, New York, 1976, pp. 117–146.
38. Y. Jasor, M. -J. Luche, M. Gaudry, and A. Marquet, *J. Chem. Soc., Chem. Commun.*, 253 (1974).
39. R. M. Coates, ed., *Organic Syntheses*, Vol. 59, Wiley, New York, 1980, p. 153.
40a. E. Jongejan, W. J. M. van Tilborg, C. H. V. Dusseau, H. Steinberg, and T. J. de Boer, *Tetrahedron Lett.*, 2359 (1972).
40b. E. Jongejan, H. Steinberg, and T. J. de Boer, *Synth. Commun.*, *4*, 11 (1974).
40c. E. Jongejan, H. Steinberg, and T. J. de Boer, *Tetrahedron Lett.*, 397 (1976).
41. H. Bohme and K. Hartke, *Chem. Ber.*, *93*, 1305 (1960).
42. G. Zinner and W. Kliegel, *Chem. Ber.*, *98*, 4036 (1965).
43. G. Kinast and L-F. Tietze, *Angew. Chem., Int. Ed. Engl.*, *15*, 239 (1976).
44. M. V. Rangaishenvi, S. V. Hiremath, and S. N. Kulkarni, *Indian J. Chem.*, *Sect. B*, *21*, 56 (1982).
45. T. A. Bryson, G. H. Bonitz, C. J. Reichel, and R. E. Dardis, *J. Org. Chem.*, *45*, 524 (1980).
46. J. Schreiber, H. Maag, N. Hashimoto, and A. Eschenmoser, *Angew. Chem., Int. Ed. Engl.*, *10*, 330 (1971).
47. H. G. Reiber and T. D. Stewart, *J. Am. Chem. Soc.*, *62*, 3026 (1940).

48. R. K. Boeckman, Jr., P. F. Jackson, and J. P. Sabatucci, *J. Am. Chem. Soc.*, *107*, 2191 (1985).

49. R. T. Dean, H. C. Padgett, and H. Rapoport, *J. Am. Chem. Soc.*, *98*, 7448 (1976).

50. R. T. Dean and H. Rapoport, *J. Org. Chem.*, *43*, 2115 (1978).

51. I. G. Csendes, Y. Y. Lee, H. C. Padgett, and H. Rapoport, *J. Org. Chem.*, *44*, 4173 (1979).

52. H. Meerwein, H. Allendorfer, P. Beckmann, F. Kunert, H. Morschel, F. Pawellek, and K. Wunderlich, *Angew. Chem.*, *70*, 211 (1958).

53. H. Meerwein, V. Hederich, H. Morschel, and K. Wunderlich, *Liebigs Ann. Chem.*, *635*, 1 (1960).

54. N. J. Bauld and Y. S. Rim, *J. Am. Chem. Soc.*, *89*, 6763 (1967).

55. H. J. Dauben, Jr. and D. F. Rhodes, *J. Am. Chem. Soc.*, *89*, 6764 (1967).

56. H. Volz and H. H. Kiltz, *Tetrahedron Lett.*, 1917 (1970).

57. H. Volz and H. H. Kiltz, *Liebigs Ann. Chem.*, *752*, 86 (1971).

58. R. Damico and C. D. Broaddus, *J. Org. Chem.*, *31*, 1607 (1966).

59. C. Jutz, *Chem. Ber.*, *97*, 2050 (1964).

60. M. Polonovski and M. Polonovski, *Bull. Soc. Chim. Fr.*, *41*, 1190 (1927).

61. A. Cave, C. Kan-Fan, P. Potier, and J. Le Men, *Tetrahedron*, *23*, 4681 (1967).

62. A. Ahond, A. Cave, C. Kan-Fan, H-P. Husson, J. de Rostolan, and P. Potier, *J. Am. Chem. Soc.*, *90*, 5622 (1968).

63. A. Ahond, A. Cave, C. Kan-Fan, and P. Potier, *Bull. Soc. Chim. Fr.*, 2707 (1970).

64. L. Alais, P. Angibeaud, and R. Michelot, *C.R. Acad. Sci. Paris*, *269C*, 150 (1969).

65. P. A. Bather, J. R. L. Smith, and R. O. C. Norman, *J. Chem. Soc. (C)*, 3060 (1971).

66. N. L. Holy and Y. F. Wang, *J. Am. Chem. Soc.*, *99*, 944 (1977).

67. N. L. Holy, R. Fowler, E. Burnett, and R. Lorenz, *Tetrahedron*, *35*, 613 (1979).

68. L. Chevolot, A. Husson, C. Kan-Fan, H-P. Husson, and P. Potier, *Bull. Soc. Chim. Fr.*, 1222 (1976).

69. D. S. Grierson, M. Harris, and H-P. Husson, *J. Am. Chem. Soc.*, *102*, 1064 (1980).

70. H. Bohme, M. Dahne, W. Lehners, and E. Ritter, *Liebigs Ann. Chem.*, *723*, 34 (1969).

71. H. Bohme, M. Hilp, L. Koch, and E. Ritter, *Chem. Ber.*, *104*, 2018 (1971).

72. E. Schmidt and F. M. Litterscheid, *Liebigs Ann. Chem.*, *337*, 37 (1904).

73. H. Bohme and E. Boll, *Chem. Ber.*, *90*, 2013 (1957).

74. H. Gross and J. Rusche, *Angew. Chem.*, *76*, 534 (1964).

75. J. Hooz and J. N. Bridson, *J. Am. Chem. Soc.*, *95*, 602 (1973).

76. S. Danishefsky, T. Kitahara, R. McKee, and P. F. Schuda, *J. Am. Chem. Soc.*, *98*, 6715 (1976).
77. S. Danishefsky, P. F. Schuda, T. Kitahara, and S. J. Etheredge, *J. Am. Chem. Soc.*, *99*, 6066 (1977).
78. M. Lamchen, W. Pugh, and A. M. Stephen, *J. Chem. Soc.*, 4418 (1954).
79. T. Zincke and W. Wurker, *Liebigs Ann. Chem.*, *338*, 133 (1905).
80. N. J. Leonard and J. V. Paukstelis, *J. Org. Chem.*, *28*, 3021 (1963).
81. A. G. Cook, W. C. Meyer, K. E. Ungrodt, and R. H. Mueller, *J. Org. Chem.*, *31*, 14 (1966).
82. T. R. Keenan and N. J. Leonard, *J. Am. Chem. Soc.*, *93*, 6567 (1971).
83. P. E. Sonnet, *J. Org. Chem.*, *37*, 925 (1972).
84. N. M. Libman, *Zh. Org. Khim.*, *3*, 1235 (1967).
85. R. Kuhn and H. Schretzmann, *Chem. Ber.*, *90*, 557 (1957).
86. H. E. Nikolajewski, S. Daehne, and B. Hirsch, *Z. Chem.*, *8*, 63 (1968).
87. J. W. Stanley, J. G. Beasley, and I. W. Mathison, *J. Org. Chem.*, *37*, 3746 (1972).
88. H. Newman and T. L. Fields, *Tetrahedron*, *28*, 4051 (1972).
89. R. Breslow, T. Eicher, A. Krebs, R. A. Peterson, and J. Posner, *J. Am. Chem. Soc.*, *87*, 1320 (1965).
90. A. Krebs and J. Breckwoldt, *Tetrahedron Lett.*, 3797 (1969).
91. A. Krebs, *Tetrahedron Lett.*, 1901 (1971).
92. A. Vilsmeier and A. Haack, *Chem. Ber.*, *60*, 119 (1927).
93. C. Jutz in *Iminium Salts in Organic Chemistry* (H. Bohme and H. G. Viehe, eds.), in *Advances in Organic Chemistry* (E. C. Taylor, ed.), Vol. 9, Part 1, Wiley-Interscience, New York, 1976, pp. 225–342.
94. G. A. Reynolds and J. A. Van Allan, *J. Org. Chem.*, *34*, 2736 (1969).
95. A. Buzas, J-P. Jacquet, and G. Lavielle, *J. Org. Chem.*, *45*, 32 (1980).
96. K. Hafner, K. H. Vopel, G. Ploss, and C. Konig, *Liebigs Ann. Chem.*, *661*, 52 (1963).
97. W. D. Emmons, ed., *Organic Syntheses*, Vol. 47, Wiley, New York, 1967, p. 52.
98. K. Hafner, K. H. Hafner, C. Konig, M. Keuder, G. Ploss, G. Schulz, E. Sturm, and K. H. Vopel, *Angew. Chem.*, *Int. Ed. Engl.*, *2*, 123 (1963).
99. E. D. Bergmann, *Chem. Rev.*, *68*, 41 (1968).
100. J. Paukstelis, Ph.D. thesis, University of Illinois, Urbana, 1964.
101. K. L. Erickson, J. Markstein, and K. Kim, *J. Org. Chem.*, *36*, 1024 (1971).

102. W. Dorscheln, H. Tiefenthaler, H. Goth, P. Cerutti, and H. Schmid, *Helv. Chim. Acta, 50*, 1759 (1967).
103. G. Opitz, H. Hellman, and H. W. Schubert, *Liebigs Ann. Chem., 623*, 112 (1959).
104. A. G. Cook, W. M. Kosman, T. A. Hecht, and W. Koehn, *J. Org. Chem., 37*, 1565 (1972).
105. J. F. Stephen and E. Marcus, *J. Org. Chem., 34*, 2535 (1969).
106. K. G. R. Sundelin, R. A. Wiley, R. S. Givens, and D. R. Rademacher, *J. Med. Chem., 16*, 235 (1973).
107. J. Marchand-Brynaert and L. Ghosez, *J. Am. Chem. Soc., 94*, 2870 (1972).
108. H. H. Wasserman and M. S. Baird, *Tetrahedron Lett.*, 1729 (1970).
109. S. F. Nelsen, C. R. Kessel, and D. J. Brien, *J. Am. Chem. Soc., 102*, 702 (1980).
110. R. Merenyi in *Iminium Salts in Organic Chemistry* (H. Bohme and H. G. Viehe, eds.), in *Advances in Organic Chemistry* (E. C. Taylor, ed.), Vol. 9, Part 1, Wiley-Interscience, New York, 1976, pp. 23–105.
111. R. T. La Londe, *Acc. Chem. Res., 13*, 39 (1980).
112. V. Prelog and O. Hafliger, *Helv. Chim. Acta, 32*, 185 (1949).
113. K-A. Kovar and U. Schwiecker, *Arch. Pharm., 307*, 384 (1974).
114. K-A. Kovar, F. Schielein, T. G. Dekker, K. Albert, and E. Breitmaier, *Tetrahedron, 35*, 2113 (1979).
115. K-A. Kovar and M. Bojadiew, *Arch. Pharm., 315*, 883 (1982).
116. M. Reinecke and L. R. Kray, *J. Org. Chem., 31*, 4215 (1966).
117. C. Rabiller, J. P. Renou, and G. J. Martin, *J. Chem. Soc., Perkin Trans., 2*, 536 (1977).
118. J-P. Gouesnard and J. Dorie, *Bull. Soc. Chim. Fr., 132* (1985).
119. M. S. Kharasch and O. Reinmuth, *Grignard Reactions of Non-Metallic Substances*, Prentice-Hall, Englewood Cliffs NJ, 1954, pp. 1251ff.
120. M. Freund, *Chem. Ber., 44*, 2346 (1911).
121. M. Freund, *Chem. Ber., 36*, 4257 (1903).
122. M. Freund, *Chem. Ber., 37*, 4666 (1904).
123. M. Freund and H. Beck, *Chem. Ber., 37*, 4679 (1904).
124. M. Freund and H. H. Reitz, *Chem. Ber., 39*, 2219 (1906).
125. M. Freund and L. Richard, *Chem. Ber., 42*, 1101 (1909).
126. D. Craig, *J. Am. Chem. Soc., 60*, 1458 (1938).
127. H. Gilman, E. A. Zoeller, and J. B. Dickey, *J. Am. Chem. Soc., 51*, 1579 (1929).
128. E. L. May and E. M. Fry, *J. Org. Chem., 22*, 1366 (1957).
129. W. Bradley and S. Jeffrey, *J. Chem. Soc.*, 2770 (1954).
130. M. Sommolet, *Compt. Rend., 183*, 302 (1926).

131. N. J. Leonard and A. S. Hay, *J. Am. Chem. Soc.*, *78*, 1984 (1956).
132. H. Bohme, E. Mundlos, and O-E. Herboth, *Chem. Ber.*, *90*, 2003 (1957).
133. K. Wiesner, Z. Valenta, A. J. Manson, and F. W. Stonner, *J. Am. Chem. Soc.*, *77*, 675 (1955).
134. R. Lukes, V. Dienstbierova, and O. Cervinka, *Chem. Listy*, *52*, 1137 (1958).
135. H. Bohme and M. Haake, *Chem. Ber.*, *100*, 3609 (1967).
136. D. Carbaret, G. Chauviere, and Z. Welvart, *Bull. Soc. Chim. Fr.*, 4457 (1969).
137. H. Bohme and W. Hover, *Liebigs Ann. Chem.*, *748*, 59 (1971).
138. H. Bohme and P. Plappert, *Chem. Ber.*, *108*, 3574 (1975).
139. H. Bohme, M. Haake, and G. Auterhoff, *Arch. Pharm. (Weinheim)*, *305*, 10 (1972).
140. H. Bohme and P. Plappert, *Chem. Ber.*, *108*, 2827 (1975).
141. C. R. Hauser, R. M. Manyk, W. R. Brassen, and P. L. Bayless, *J. Org. Chem.*, *20*, 1119 (1956).
142. G. Opitz, A. Griesinger, and H. W. Schubert, *Liebigs Ann. Chem.*, *665*, 91 (1963).
143. O. Cervinka, *Chem. Ind. (London)*, 1482 (1960).
144. O. Cervinka, *Collect. Czech. Chem. Commun.*, *28*, 536 (1963).
145. H. Bohme and H. Ellenberger, *Chem. Ber.*, *92*, 2976 (1959).
146. H. Bohme, M. Haake, and G. Auterhoff, *Arch. Pharm. (Weinheim)*, *305*, 88 (1972).
147. D. Cabaret, G. Chauviere, and Z. Welvart, *Tetrahedron Lett.*, 549 (1968).
148. H. Reinheckel, H. Gross, K. Haage, and G. Sonnek, *Chem. Ber.*, *101*, 1736 (1968).
149. A. Alberola and F. J. L. Lopez, *An. Quim*, *73*, 893 (1977).
150. P. Binger and R. Koster, *Chem. Ber.*, *108*, 395 (1975).
151. P. Karrer, G. Schwarzenbad, and G. E. Utzinger, *Helv. Chim. Acta*, *20*, 72 (1937).
152. K. Schenker and J. Bruey, *Helv. Chim. Acta*, *42*, 1960 (1959).
153. J. J. Panouse, *Compt. Rend.*, *233*, 260 (1951).
154. J. J. Panouse, *Compt. Rend.*, *233*, 1200 (1951).
155. M. Ferles, *Collect. Czech. Chem. Commun.*, *23*, 479 (1958).
156. K. Schenker, *Angew. Chem.*, *72*, 638 (1960).
157. J. W. Daly and B. Witkop, *J. Org. Chem.*, *27*, 4104 (1962).
158. J. A. Marshall and W. S. Johnson, *J. Org. Chem.*, *28*, 421 (1963).
159. J. Schmitt, J. J. Panouse, A. Hallot, P-J. Cornu, P. Comoy, and H. Pluchet, *Bull. Soc. Chim. Fr.*, 798 (1963).
160. C. Jutz, A. F. Kirschner, and R-M. Wagner, *Chem. Ber.*, *110*, 1259 (1977).
161. P. Heinstein, J. Stoeckigt, and M. H. Zenk, *Tetrahedron Lett.*, *21*, 141 (1980).

162. G. Rossey, A. Wick, and E. Wenkert, *J. Org. Chem.*, *47*, 4745 (1982).
163. R. O. Hutchins, W-Y. Su, R. Sivakumar, F. Cistone, and Y. P. Stercho, *J. Org. Chem.*, *48*, 3412 (1983).
164. M. Ferles, *Collect. Czech. Chem. Commun.*, *24*, 2221 (1959).
165. D. Cabaret, G. Chauviere, and Z. Welvart, *Tetrahedron Lett.*, 4109 (1966).
166. R. F. Borch, M. D. Bernstein, and H. D. Durst, *J. Am. Chem. Soc.*, *93*, 2897 (1971).
167. R. F. Borch in *Organic Syntheses* (H. O. House, ed.), Vol. 52, Wiley, New York, 1972, p. 124.
168. S. Kim, C. H. Oh, J. S. Ko, K. H. Ahn, and Y. J. Kim, *J. Org. Chem.*, *50*, 1927 (1985).
169. O. Cervinka and L. Hub, *Tetrahedron Lett.*, 463 (1964).
170. T. Kudo and A. Nose, *Yakugaku Zasshi*, *94*, 1475 (1974).
171. T. Mitsudo, Y. Watanabe, M. Tanaka, S. Alsuta, K. Yamamoto, and Y. Takegami, *Bull. Chem. Soc. Jpn.*, *48*, 1506 (1975).
172. E. Winterfeldt, *Synthesis*, 617 (1975).
173. D. Redmore, *J. Org. Chem.*, *43*, 992 (1978).
174. U. K. Pandit, E. F. M. Cabre, R. A. Gase, and M. J. de Nie-Sarink, *J. Chem. Soc., Chem. Commun.*, 627 (1974).
175. U. K. Pandit, R. A. Gase, F. R. M. Cabre, and M. J. de Nie-Sarink, *J. Chem. Soc., Chem. Commun.*, 211 (1975).
176. N. Baba, K. Nishiyama, J. Oda, and Y. Inouye, *Agr. Biol. Chem.*, *40*, 1441 (1976).
177. J. Stockigt, H. P. Husson, C. Kan-Fan, and M. H. Zenk, *J. Chem. Soc., Chem. Commun.*, 164 (1977).
178. P. Heinstein, J. Stockigt, and M. H. Zenk, *Tetrahedron Lett.*, *21*, 141 (1980).
179. R. Lukes, *Collect. Czech. Chem. Commun.*, *10*, 66 (1938).
180. R. Lukes and J. Priml, *Collect. Czech. Chem. Commun.*, *15*, 463 (1950).
181. R. Lukes and J. Priml, *Collect. Czech. Chem. Commun.*, *15*, 512 (1950).
182. R. Lukes and M. Ferles, *Collect. Czech. Chem. Commun.*, *22*, 121 (1957).
183. P. L. de Benneville and J. H. Macartney, *J. Am. Chem. Soc.*, *72*, 3073 (1950).
184. P. L. de Benneville, U.S. Pat. 2,578,787 (1951).
185. N. J. Leonard and R. R. Sauers, *J. Am. Chem. Soc.*, *79*, 6210 (1957).
186. O. Cervinka and O. Kriz, *Collect. Czech. Chem. Commun.*, *30*, 1700 (1965).
187. J. O. Madsen and P. E. Iversen, *Tetrahedron*, *30*, 3493 (1974).
188. W. Himmele, W. Bremser, and H. Siegel, *Angew. Chem., Int. Ed. Engl.*, *18*, 320 (1979).
189. R. Carlson and A. Nilsson, *Acta Chem. Scand., Ser. B*, *39*, 181 (1985).

190. A. Nilsson and R. Carlson, *Acta Chem. Scand., Ser. B, 39,* 187 (1985).
191. E. Wenkert and B. Wickberg, *J. Am. Chem. Soc., 87,* 1580 (1965).
192. F. L. Weisenborn and P. A. Diassi, *J. Am. Chem. Soc., 78,* 2022 (1956).
193. A. R. Katritzky, *J. Chem. Soc.,* 2586 (1955).
194. R. E. Lyle, P. S. Anderson, C. K. Spicer, and S. S. Pelosi, Abstracts of the 142nd American Chemical Society Meeting, 1962, pp. 25Q.
195. R. E. Lyle, P. S. Anderson, C. K. Spicer, S. S. Pelosi, and D. A. Nelson, *Angew. Chem., 75,* 386 (1963).
196. P. S. Anderson and R. E. Lyle, *Tetrahedron Lett.,* 153 (1964).
197. N. Kinoshita, M. Hamana, and T. Kawasaki, *Chem. Pharm. Bull. (Jpn.), 10,* 753 (1962).
198. G. Buchi, D. L. Coffen, K. Kocsis, P. E. Sonnet, and F. E. Ziegler, *J. Am. Chem. Soc., 88,* 3099 (1966).
199. M. Saunders and E. H. Gould, *J. Org. Chem., 27,* 1439 (1962).
200. R. E. Lyle and P. S. Anderson, *Adv. Heterocyclic Chem., 6,* 45 (1966).
201. R. C. Elderfield and B. H. Wark, *J. Org. Chem., 27,* 543 (1962).
202. H. Schmid and P. Karrer, *Helv. Chim. Acta, 32,* 960 (1949).
203. R. Torossian and C. Sannie, *Comp. Rend., 236,* 824 (1953).
204. R. Mirza, *J. Chem. Soc.,* 4400 (1957).
205. R. B. Woodward, F. E. Bader, A. J. Frey, and R. W. Kierstead, *Tetrahedron, 2,* 1 (1958).
206. N. A. Nelson, J. E. Ladbury, and R. S. Hsi, *J. Am. Chem. Soc., 80,* 6633 (1958).
207. W. M. Whaley and C. N. Robinson, *J. Am. Chem. Soc., 75,* 2008 (1963).
208. J. W. Huffman, *J. Am. Chem. Soc., 80,* 5193 (1958).
209. A. P. Gray, E. E. Spinner, and C. J. Cavallito, *J. Am. Chem. Soc., 76,* 2792 (1953).
210. A. G. Cook and W. M. Kosman, *Tetrahedron Lett.,* 5847 (1966).
211. A. G. Cook and C. R. Schulz, *J. Org. Chem., 32,* 473 (1967).
212. O. Cervinka, *Collect. Czech. Chem. Commun., 26,* 673 (1961).
213. M. Sasamoto, *Chem. Pharm. Bull. (Tokyo), 8,* 324 (1960).
214. M. F. Grundon, *J. Chem. Soc.,* 3010 (1959).
215. F. Bohlmann, E. Winterfeldt, P. Studt, H. Laurent, G. Boroschewski, and K. M. Keine, *Chem. Ber., 94,* 3141 (1961).
216. F. Bohlmann, E. Winterfeldt, G. Boroschewski, R. Mayer-Mader, and B. Gatscheff, *Chem. Ber., 96,* 1792 (1963).
217. P. Cerutli and H. Schmid, *Helv. Chim. Acta, 45,* 1992 (1962).
218. J. A. Marshall and W. S. Johnson, *J. Am. Chem. Soc., 84,* 1485 (1962).

219. N. J. Leonard and M. J. Martell, Jr., *Tetrahedron Lett.*, 44 (1960).
220. N. J. Leonard, R. W. Fulmer, and A. S. Hay, *J. Am. Chem. Soc.*, *78*, 3457 (1956).
221. N. J. Leonard, L. A. Miller, and P. D. Thomas, *J. Am. Chem. Soc.*, *78*, 3463 (1956).
222. R. Lukes and J. Jizba, *Chem. Listy*, *47*, 1336 (1953).
223. R. Lukes and O. Cervinka, *Chem. Listy*, *51*, 2086 (1957).
224. R. Lukes and V. Dedek, *Chem. Listy*, *51*, 2082 (1957).
225. R. Lukes and O. Cervinka, *Chem. Listy*, *51*, 2142 (1957).
226. R. Bonnett, V. M. Clark, A. Giddey, and A. Todd, *J. Chem. Soc.*, 2087 (1959).
227. O. Cervinka, *Chem. Listy*, *52*, 307 (1958).
228. N. J. Leonard and K. Jann, *J. Am. Chem. Soc.*, *83*, 6418 (1960).
229. N. J. Leonard, Abstracts of the Organic Chemistry Symposium at Bloomington, IN, 1961, pp. 1-10.
230. N. J. Leonard and K. Jann, *J. Am. Chem. Soc.*, *84*, 4806 (1962).
231. P. D. Bartlett, S. D. Ross, and C. G. Swain, *J. Am. Chem. Soc.*, *71*, 1415 (1949).
232. G. Golumbic, J. S. Fruton, and M. Bergmann, *J. Org. Chem.*, *11*, 518 (1946).
233. N. J. Leonard, *Record Chem. Progr. Kresge-Hooker Sci. Lib.*, *26*, 211 (1965).
234. D. R. Crist and N. J. Leonard, *Angew. Chem., Int. Ed. Engl.*, *8*, 962 (1969).
235. N. J. Leonard and J. V. Paukstelis, *J. Org. Chem.*, *30*, 821 (1965).
236. H. Bohme, E. Mundlos, W. Lehners, and O-R. Herboth, *Chem. Ber.*, *90*, 2008 (1957).
237. N. J. Leonard, E. F. Kiefer, and L. E. Brady, *J. Org. Chem.*, *28*, 2850 (1963).
238. N. J. Leonard, K. Jann, J. V. Paukstelis, and C. K. Steinhardt, *J. Org. Chem.*, *28*, 1499 (1963).
239. N. J. Leonard, J. V. Paukstelis, and L. E. Brady, *J. Org. Chem.*, *29*, 3383 (1964).
240. N. J. Leonard and J. A. Klainer, *J. Heterocycl. Chem.*, *8*, 215 (1971).
241. F. Bondavalli, P. Schenone, and A. Ranise, *J. Chem. Res. (S)*, *4*, 257 (1980); *J. Chem. Res. (M)*, 3256 (1980).
242. A. Kaufmann, *J. Chem. Soc.*, *114*, 187 (1918).
243. A. Kaufmann, *Chem. Ber.*, *51*, 116 (1918).
244. A. Kaufmann and A. Albertini, *Chem. Ber.*, *42*, 1999 (1909).
245. A. Kaufmann and A. Albertini, *Chem. Ber.*, *44*, 2052 (1911).
246. A. Kaufmann and A. Albertini, *Chem. Ber.*, *42*, 3776 (1909).
247. W. Schneider and H. Gotz, *Liebigs Ann. Chem.*, *653*, 85 (1962).

248. I. Murakoshi, A. Kubo, J. Saito, and J. Haginiwa, *Yakugaku Zasshi, 88,* 900 (1968).

249. I. Murakoshi, K. Takada, and J. Haginiwa, *Yakugaku Zasshi, 89,* 1661 (1969).

250. W. C. Groutas, M. Essawi, and P. S. Portoghese, *Synth. Commun., 10,* 495 (1980).

251. N. De Kimpe, R. Verhe, L. De Buyck, and N. Schamp, *Chem. Ber., 116,* 3846 (1983).

252. M. Y. H. Essawi and P. S. Portoghese, *J. Org. Chem., 48,* 2138 (1983).

253. S. Harusawa, Y. Hamada, and T. Shioiri, *Synthesis,* 716 (1979).

254. S. O. Lawesson, E. H. Larsen, and H. J. Jakobsen, *Recl. Trav. Chim., 83,* 461 (1964).

255. P. D. Klemmensen and S-O. Lawesson, *Chem. Commun.,* 205 (1968).

256. P. D. Klemmensen, J. Z. Mortensen, and S-O. Lawesson, *Tetrahedron, 26,* 4641 (1970).

257. W. Klop, P. A. A. Klusener, and L. Brandsma, *Recl. Trav. Chim. Pays-Bas, 103,* 27 (1984).

258. D. Beke, C. Szantay, and M. Barczai-Beke, *Liebigs Ann. Chem., 636,* 150 (1960).

259. R. F. Parcell, *J. Am. Chem. Soc., 81,* 2596 (1959).

260. N. J. Leonard and W. J. Musliner, *J. Org. Chem., 31,* 639 (1966).

261. R. Fuks, G. S. D. King, and H. G. Viehe, *Angew. Chem., Int. Ed. Engl., 8,* 675 (1969).

262. A. G. Cook and E. K. Fields, *J. Org. Chem., 27,* 3686 (1962).

263. A. Lukasievicz, *Tetrahedron, 20,* 1113 (1964).

264. A. Lukasievicz, *Tetrahedron, 20,* 1 (1964).

265. A. Lukasievicz, *Tetrahedron, 21,* 193 (1965).

266. A. Lukasievicz and J. Lesinska, *Tetrahedron, 21,* 3247 (1965).

267. G. H. Alt and A. J. Speziale, *J. Org. Chem., 31,* 2073 (1966).

268. G. H. Alt and A. J. Speziale, *J. Org. Chem., 31,* 1340 (1966).

269. D. C. Snyder, *J. Organomet. Chem., 301,* 137 (1986).

270. S. Baroni, R. Stradi, and M. L. Saccarello, *J. Heterocycl. Chem., 17,* 1221 (1980).

271. H. H. Wasserman and M. S. Baird, *Tetrahedron Lett.,* 3721 (1971).

272. H. H. Wasserman, M. J. Hearn, B. Haveaux, and M. Thyes, *J. Org. Chem., 41,* 153 (1976).

273. H. A. Bates and H. Rapoport, *J. Am. Chem. Soc., 101,* 1259 (1979).

274. N. Kumagaya, K. Suzuki, and M. Sekiya, *Chem. Pharm. Bull., 21,* 1601 (1973).

275. G. H. Alt and G. A. Gallegos, *J. Org. Chem., 36,* 1000 (1971).

276. W. L. F. Armarego, *J. Chem. Soc. (C)*, 986 (1969).

277. C. Mannich and W. Krosche, *Arch. Pharm.*, *250*, 647 (1912).

278. M. Tramontini, *Synthesis*, 703 (1973).

279. H. Bohme, K. Osmers, and P. Wagner, *Tetrahedron Lett.*, 2785 (1972).

280. H-U. Reissig and H. Lorey, *J. Chem. Soc., Chem. Commun.*, 269 (1986).

281. J. L. Roberst, P. S. Borromeo, and C. D. Poulter, *Tetrahedron Lett.*, 1621 (1977).

282. H. G. Viehe and Z. Janousek, *Angew. Chem., Int. Ed. Engl.*, *12*, 806 (1973).

283. O. Cervinka, *Collect. Czech. Chem. Commun.*, *25*, 1183 (1960).

284. L. Duhamel, P. Duhamel, and P. Siret, *Tetrahedron Lett.*, 3607 (1972).

285. F. Knoll and U. Krumm, *Chem. Ber.*, *104*, 31 (1971).

286. J. A. Deyrup and W. A. Szabo, *J. Org. Chem.*, *40*, 2048 (1975).

287. R. B. Woodward and R. Hoffmann, *The Conservation of Orbital Symmetry*, Academic Press, New York, 1969.

288. J. Marchand-Brynaert and L. Ghosez, *J. Am. Chem. Soc.*, *94*, 2870 (1972).

289. A. Sidani, J. Marchand-Brynaert, and L. Ghosez, *Angew. Chem., Int. Ed. Engl.*, *13*, 267 (1974).

290. I. Marko, B. Ronsmans, A-M. Hesbain-Frisque, S. Dumas, and L. Ghosez, *J. Am. Chem. Soc.*, *107*, 2192 (1985).

291. L. Ghosez and J. Marchand-Brynaert in *Iminium Salts in Organic Chemistry* (H. Bohme and H. G. Viehe, eds.), in *Advances in Organic Chemistry* (E. C. Taylor, ed.), Vol. 9, Part 1, Wiley-Interscience, New York, 1976, pp. 508–524.

292. M. De Poortere, J. Marchand-Brynaert, and L. Ghosez, *Angew. Chem., Int. Ed. Engl.*, *13*, 267 (1974).

293. C. Hoornaert, A. M. Hesbain-Frisque, and L. Ghosez, *Angew. Chem., Int. Ed. Engl.*, *14*, 569 (1975).

294. S. Baroni, E. Rivera, R. Stradi, and M. L. Saccarello, *Tetrahedron Lett.*, *21*, 889 (1980).

295. G. Crispi, P. Giacconi, E. Rossi, and R. Stradi, *Synthesis*, 787 (1982).

296. P. Giacconi, E. Rossi, R. Stradi, and R. Eccel, *Synthesis*, 789 (1982).

297. S. Mariano, *Acc. Chem. Res.*, *16*, 130 (1983).

298. P. S. Mariano, J. L. Stavinoha, and R. Swanson, *J. Am. Chem. Soc.*, *99*, 6781 (1977).

299. P. S. Mariano, J. L. Stravinoha, G. Pepe, and E. F. Meyer, Jr., *J. Am. Chem. Soc.*, *100*, 7114 (1978).

300. J. L. Stavinoha and P. S. Mariano, *J. Am. Chem. Soc.*, *103*, 3136 (1981).

301. P. S. Mariano and A. Leone, *Tetrahedron Lett.*, 4581 (1980).

302. P. S. Mariano, J. L. Stavinoha, and R. J. Swanson, *J. Am. Chem. Soc.*, *99*, 6781 (1977).

303. J. L. Stavinoha, P. S. Mariano, A. Leone-Bay, R. Swanson, and C. J. Bracken, *J. Am. Chem. Soc.*, *103*, 3148 (1981).

304. J. L. Stavinoha, E. Bay, A. Leone, and P. S. Mariano, *Tetrahedron Lett.*, 3455 (1980).

305. P. S. Mariano, J. L. Stavinoha, and E. Bay, *Tetrahedron*, *37*, 3385 (1981).

306. S. Eguchi, Y. Wakata, and T. Sasaki, *J. Chem. Res.* (*S*), 146 (1985); *J. Chem. Res.* (*M*), 1729 (1985).

307. N. S. Basketter and A. O. Plunkett, *J. Chem. Soc.*, *Chem. Commun.*, 188 (1973).

308. H. Hermann, R. Huisgen, and H. Mader, *J. Am. Chem. Soc.*, *93*, 1779 (1971).

309. J. W. Lown, *Rec. Chem. Prog.*, *32*, 51 (1971).

310. R. Huisgen, *J. Org. Chem.*, *41*, 403 (1976).

311. R. M. Kellogg, *Tetrahedron*, *32*, 2165 (1976).

312. R. Huisgen, *Angew. Chem.*, *Int. Ed. Engl.*, *19*, 947 (1980).

313. E. Vedejs and G. R. Martinez, *J. Am. Chem. Soc.*, *101*, 6452 (1979).

314. E. Vedejs and G. R. Martinez, *J. Am. Chem. Soc.*, *102*, 7993 (1980).

315. R. Beugelmans, G. Negron, and G. Roussi, *J. Chem. Soc.*, *Chem. Commun.*, 31 (1983).

316. L. Benadjilla-Iguertaira, J. Chastanet, G. Negron, and G. Roussi, *Can. J. Chem.*, *63*, 725 (1985).

317. J. Chastanet and G. Roussi, *J. Org. Chem.*, *50*, 2910 (1985).

318. R. T. LaLonde, E. Auer, C. F. Wong and V. P. Muralidharan, *J. Am. Chem. Soc.*, *93*, 2501 (1971).

319. A. Langlois, F. Gueritte, Y. Langlois, and P. Potier, *J. Am. Chem. Soc.*, *98*, 7017 (1976).

320. L. Chevolot, H-P. Husson, and P. Potier, *Tetrahedron*, *31*, 2491 (1975).

321. W. C. Groutas, M. Essawi, and P. S. Portoghese, *Synth. Commun.*, *10*, 495 (1980).

322. R. M. Horowitz and T. A. Geissman, *J. Am. Chem. Soc.*, *72*, 1518 (1950).

323. H. Heimgartner, H-J. Hansen, and H. Schmid in *Iminium Salts in Organic Chemistry* (H. Bohm and H. G. Viehe, eds.), in *Advances in Organic Chemistry* (E. C. Taylor, ed.), Vol. 9, Part 2, Wiley-Interscience, New York, 1979, pp. 655–731.

324. L. E. Overman and M. Kakimoto, *J. Am. Chem. Soc.*, *101*, 1310 (1979).

325. L. E. Overman, M. Kakimoto, and M. Okawara, *Tetrahedron Lett.*, 4041 (1979).

326. L. E. Overman, M. Kakimoto, M. Okazaki, and G. P. Meier, *J. Am. Chem. Soc.*, *105*, 6622 (1983).

327. L. E. Overman, L. T. Mendelson, and E. J. Jacobsen, *J. Am. Chem. Soc.*, *105*, 6629 (1983).

328. P. P. Lynch and P. H. Doyle, *Gazz. Chim. Ital.*, *98*, 645 (1968).

329. A. M. Knowles, A. Lawson, G. V. Boyd, and R. A. Newberry, *J. Chem. Soc.* (*C*), 598 (1971).

330. A. M. Knowles and A. Lawson, *J. Chem. Soc.*, *Perkin*, *1*, 1240 (1972).

331. O. Tsuge, M. Tashiro, and Y. Kiryu, *Chem. Lett.*, 795 (1974).

332. M. Tashiro, Y. Kiryu, and O. Tsuge, *Bull. Chem. Soc. Jpn.*, *48*, 616 (1975).

333. R. Carlson and L. Nilsson, *Acta Chem. Scand.*, *Ser. B*, *31*, 732 (1977).

334. R. Carlson, L. Nilsson, C. Rappe, A. Babadjamian, and J. Metzger, *Acta Chem. Scand.*, *Ser. B*, 85 (1978).

335. M. I. Kabachnik, T. Y. Ledved, and Y. M. Polikarpov, *Izv. Akad Nauk SSSR Ser. Khim*, 367 (1966).

336. N. L. J. M. Broekhof, P. van Elburg, and A. van der Gen, *Recl. Trav. Chim. Pays-Bas*, *103*, 312 (1984).

337. N. L. J. M. Broekhof, P. van Elburg, D. J. Hoff, and A. van der Gen, *Recl. Trav. Chim. Pays-Bas*, *103*, 317 (1984).

338. V. N. Charushin, I. Y. Postovskii, and O. N. Chupakhin, *Dokl. Akad. Nauk SSSR*, 246, 351 (1979).

7

Cycloaddition Reactions of Enamines

A. GILBERT COOK   *Valparaiso University, Valparaiso, Indiana*

I.   Introduction   347

II. Theory   348

III. Carbocycloadditions   353

     A. Electrophilic Alkenes   353
     B. Electrophilic Cyclopropanes and Cyclopropenes   381
     C. Electrophilic Alkynes   384
     D. Divalent Carbon   388
     E. Miscellaneous   389

IV. Heterocycloadditions   392

     A. Oxygen Heteroatom   392
     B. Sulfur Heteroatom   394
     C. Nitrogen Heteroatom   403

     References   415

I. INTRODUCTION

Most of the reactions described in this chapter would fall under
Huisgen's definition of a cycloaddition reaction [1]. However, some
of the reactions described would not be considered cycloaddition reactions according to this restrictive definition. Therefore, the more
liberal definition given by Baldwin will be used as a guideline, namely,

"Cycloadditions are chemical transformations giving at least one product having at least two new bonds as constitutents of a new ring" [2].
The carbon atoms of the enamine functional group alone

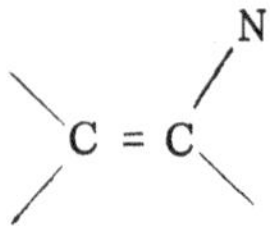

are the sites of attack and hence provide the two-carbon cyclization
bridge in most of the reactions described, although a few reactions
involve an adjacent functional group with the enamine. However,
cycloadditions to ternary iminium ions which involve the α-carbon and
the nitrogen atom or the carbons of a cumulated double bond as the
sites of attack are discussed in Chapter 6 (Sections III.C and V) and
will not be discussed here.

The types of cycloadditions discovered for enamines range through
a regular sequence, starting with divalent addition to form a cyclopropane ring, followed by 1,2 addition [3] of an alkene or an alkyne
to form a cyclobutane or a cyclobutene, then 1,3-dipolar addition with
the enamine the dipolarophile [4], and finally a Diels-Alder type of
reaction [5] with the enamine the dienophile (see Scheme 1).

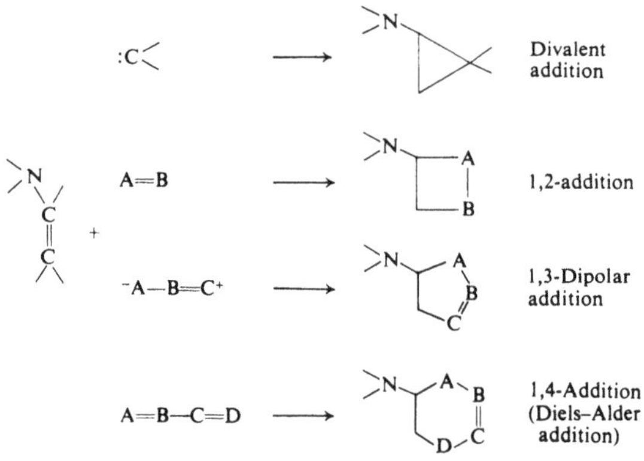

Scheme 1

II. THEORY

The mechanistic possibilities for cycloaddition reactions fall into two
broad categories, *concerted* reactions and *two-step* reactions. The
broad category of concerted reactions can be further subdivided into

synchronous reactions and *two-stage* reactions. The definitions of these four terms are as follows: a *concerted* reaction is one that takes place in a single kinetic step; a *synchronous* reaction is a concerted reaction where all the bond-making and bond-breaking processes take place in unison, having all proceeded to comparable extents in the transition state; a *two-stage* reaction is concerted but not synchronous, some of the changes in bonding taking place in the first part of the reaction, followed by the rest; a *two-step* reaction takes place in two kinetically distinct steps, via a stable intermediate [6,7]. The category of *two-step* reactions can be further subdivided according to which of the following two intermediates are formed: a *diradical* or a *zwitterion*. These are illustrated for enamines in Scheme 2 with *a* showing the concerted mechanism, *b* showing the two-step biradical mechanism, and *c* showing the two-step zwitterion mechanism.

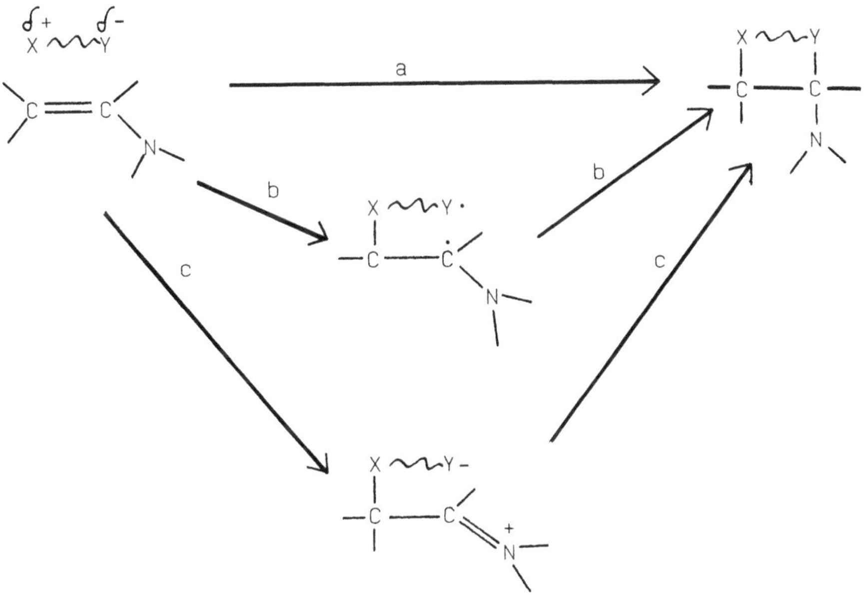

Scheme 2

A useful basis for a description of these cycloaddition reactions is the frontier molecular orbital theory [8–17]. In this theory, the most important orbitals in describing reactivities are the highest occupied molecular orbital (HOMO) of one reactant and the lowest unoccupied molecular orbital (LUMO) of the other reactant.

$$CH_2=CH-X$$

Electrophilic Alkenes (Acceptors)			Electron-rich Alkenes (Donors)	
X	Energy		Energy	X
			+2.5 ___	NMe$_2$
H ___	+1.5			

LUMO

CN,CO$_2$Me ___	0.0			
NO$_2$ ___	-0.7			

			-7.6 ___	NMe$_2$

HOMO

			-9.1 ___	OCH$_3$
H ___	-10.5			
CO$_2$Me ___	-10.5			
CN ___	-10.9			
NO$_2$ ___	-11.4			

Figure 1 HOMO and LUMO of enamine and some electrophilic olefins.

In every cycloaddition reaction (except dimerizations), one of the partners in the cycloaddition can be defined as the donor and the other as the acceptor. In considering the two reactants, I and II, if $E_{HOMO}(I) - E_{LUMO}(II) < E_{HOMO}(II) - E_{LUMO}(I)$, then reactant I is the donor and reactant II is the acceptor [18]. So the dominant interaction is between the HOMO of the donor (enamine in our case) and the LUMO of the acceptor (see Figure 1). The HOMO of a molecule can be determined from its ionization potential, and its LUMO can be determined from its electron affinity.

One can now define a spectrum of interactions between reactants in cycloadditions running from dimerizations, where the electron-donating and electron-accepting abilities of the two reactants are identical, to the case where the two reactants have extremely different electron-donating and electron-accepting abilities. In following along the spectrum from an dimerization reaction to a reaction with cycloaddends of very different electron-donating and -accepting abilities, the quantity E_{HOMO}(donor) $- E_{LUMO}$(acceptor) will decrease [18]. The former (dimerization) is a nonpolar reaction with a nonpolar transition state, and the latter is a polar reaction with a polar transition state. One can describe the transition state of these cycloadditions by resonance contributors, where D = donor, A = acceptor, DA

$$DA \leftrightarrow D^+A^-$$

is the no-bond configuration, and D^+A^- is a charge transfer configuration [18,19]. For a nonpolar reaction, the no-bond contributor (DA) is most important, whereas for a polar reaction the charge transfer contributor (D^+A^-) is most important.

The chief theoretical constraints for a concerted cycloaddition mechanism are the Woodward-Hoffmann rules [20]. Applying these rules to a [2+2] cycloaddition and a [4+2] cycloaddition, the *thermally* allowed concerted reactions are $[_\pi2_s+_\pi2_a]$ and $[_\pi4_s+_\pi2_s]$, respectively. The *photochemically* allowed concerted reactions are $[_\pi2_s+_\pi2_s]$ and $[_\pi4_s+_\pi2_a]$, respectively.

However, concerted cycloaddition reactions that violate these rules ("forbidden" reactions) have been shown to be theoretically possible using both charge transfer [19,21—23] and configuration interaction (CI) [24—28] theories. *Nonpolar concerted cycloadditions follow Woodward-Hoffmann rules, but polar concerted cycloadditions can violate these rules.* This generally will not happen with polar, concerted [4+2] cycloaddition reactions, but it usually occurs with polar, concerted [2+2] cycloaddition reactions to give (thermally) $[_\pi2_s+_\pi2_s]$ reactions [27]. In these latter reactions, both the $[_\pi2_s+_\pi2_a]$ and the $[_\pi2_s+_\pi2_s]$ concerted pathways are stabilized electronically, but the unfavorable steric features of the $[_\pi 2_s + _\pi 2_a]$ transition state detract from that transition state leaving the $[_\pi 2_s + _\pi 2_s]$ pathway [19].

Thermal [2+2] cyclodimerizations (nonpolar cycloadditions) show a $[_\pi 2_s + _\pi 2_s]$ concerted pathway that is not stabilized, and a $[_\pi 2_s + _\pi 2_a]$ concerted pathway that is only minimally stabilized. So the cyclodimerization shows a regioselectivity consistent with a two-step diradical mechanism rather than a concerted mechanism [19,30]. The two-step reaction mechanisms (involving either diradical or zwitterion intermediates) do not have the Woodward-Hoffman rules constraint which is required of concerted reaction mechanisms.

In $[_\pi 4_s + _\pi 2_s]$ concerted cycloadditions, the simultaneous closure of two new σ-bonds between the two reactants predicts a stereospecific reaction. A two-step mechanism with a biradical or zwitterion intermediate would give a nonstereospecific reaction if rotations about single bonds in the intermediate are faster than ring closure to the cycloadduct. However, if rotations are slower than a rapid ring closure, then a two-step reaction can take place stereospecifically. So a two-step reaction can also be a sterospecific reaction [29].

Conclusive proof of a two-step reaction mechanism with a zwitterion intermediate is present in a reaction with the following characteristics: It is nonstereospecific, it is strongly influenced by solvent polarity, and it has a high ρ-value [29,34]. An example of this is the very polar [2+2] cycloaddition of an enol ether with tetracyanoethylene [31].

In cycloaddition reactions, one can rationalize any set of chemical data by assuming a two-step mechanism with a diradical or zwitterion intermediate. However, the predictive power of such a reaction mechanism is poor. By using a concerted reaction mechanism, there is the possibility of great predictive power.

Enamines have very low ionization potentials and asymmetric molecular orbital coefficients (see Chapter 1, Section II), so they are good donor cycloaddends [11,32]. They undergo both [2+2] and [4+2] cycloaddition reactions. Enamines undergo polar cycloaddition reactions with various acceptor molecules, so this provides favorable conditions for the formation of a zwitterion. Since it often cannot be stated with certainty whether a given cycloaddition reaction has a concerted or a two-step mechanism, the mechanisms that follow in this chapter will be written as two-step mechanisms with zwitterion intermediates unless there is evidence to the contrary. The controversial question as to whether or not any of the concerted reactions are synchronous [6,7,33] will not be discussed.

Koikov and Bundel [286] have postulated what they call a "synchronous [3+2]-interaction" as a possible mode of formation of the zwitterion intermediate when an enamine reacts with a dienophile. This postulate is based solely on their interpretation of some MINDO/3 [287] calculations and not on any experimental evidence.

III. CARBOCYCLOADDITIONS

A. Electrophilic Alkenes

Introduction

The reactions of electrophilic alkenes (alkenes attached to electron-withdrawing groups) with enamines produce one or more of the following products: simple alkylation (2), 1,2 cycloaddition (3), and 1,4 cycloaddition (4) (see Scheme 3). Competition with C-alkylation by N-alkylation is inconsequential and therefore will be largely ignored [35,36]. A two-step ionic mechanism leading to these products necessarily involves the formation of a zwitterion intermediate (1) as the first step, which is then followed either by one of the two possible

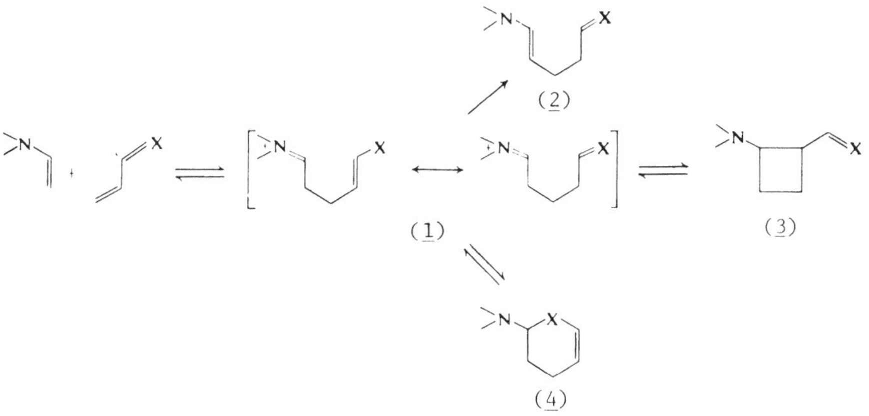

Scheme 3

cycloadditions to give a cyclic molecule or by proton elimination-addition to give a simple alkylated molecule (see Scheme 3). This proton elimination-addition has been observed to be intramolecular in one example studied [37]. It has been shown in some cases that the cyclobutyl form (3) must lie somewhere along the pathway between starting materials and simple alkylated product [38]. It probably exists as a branch that is in equilibrium with a common intermediate, namely, zwitterion 1. The 1,4 cycloaddition product (4) also seems to be in equilibrium with this zwitterion intermediate (1) (see Scheme 3) [39–42].

Conjugated with Carbonyl Group

The presence of a methyl group next to the carbonyl in the conjugated system allows still another type of cyclization to take place in addition to the two described above, namely, a 1,4 carbocycloaddition. This reaction involves proton transfer in the originally formed zwitterion intermediate from the α-methyl group to the methylene carbanion. This is followed by cyclization (see Scheme 4). Methyl vinyl ketone is a prime example of a reagent that will undergo this type of cycloaddition.

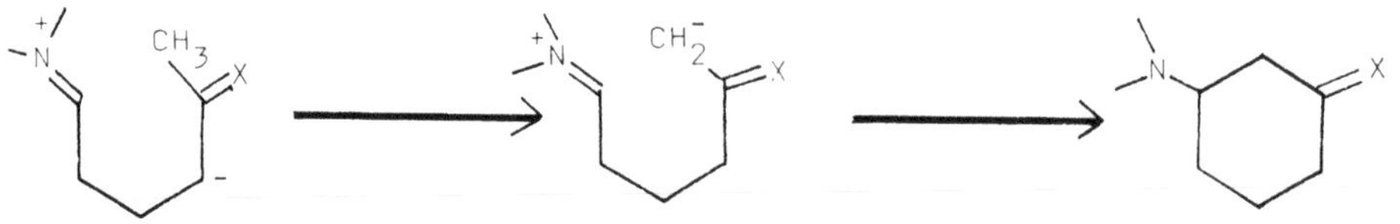

Scheme 4

The first reported cyclization involving an enamine was the 1,4 carbocycloaddition of methyl vinyl ketone with the enamine of cyclohexanone to give, after hydrolysis, $\Delta^{1,9}$-octal-2-one (5) [35,36,43, 656,657]. This reaction has been used a great deal in general synthetic sequences [44–46,662], as well as in alkaloid [47–54,74,76,77], steroid [55,75], and terpene [56–61,94] syntheses. It was reported that isomeric octalones were also formed during this reaction along with some disproportionation products [62]. Subsequently, it was determined that two isomeric enamines (6, 7) and possibly a third (8) were produced before hydrolysis along with diketone 9, but no disproportionation products were observed [63]. These "disproportionation" products may have arisen from the reduction of the enamine by some excess secondary amine [64]. This is a definite option since the oxidation product from the proposed disproportionation reaction apparently was not isolated [65]. The 1,4 cycloalkylation can be described by the following mechanism:

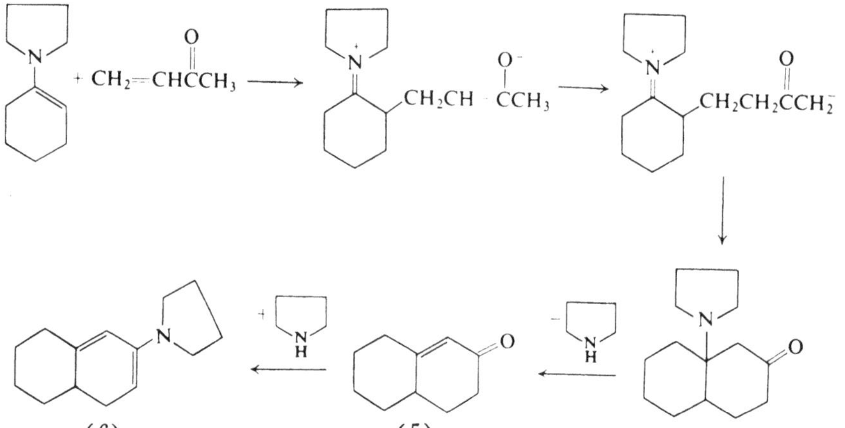

(6)   (5)

The amount of diketone 9 formed during the reaction could be enhanced by using no solvent or ethanol solvent in place of benzene
solvent. The use of ethanol solvent favors dialkylation of enamines
[36,65,66]. It was shown that enamines 6, 7, and 8 are not precur-

sors for 9. Therefore, the following (see Scheme 5) is a likely mechanism (starting with the disubstituted product):

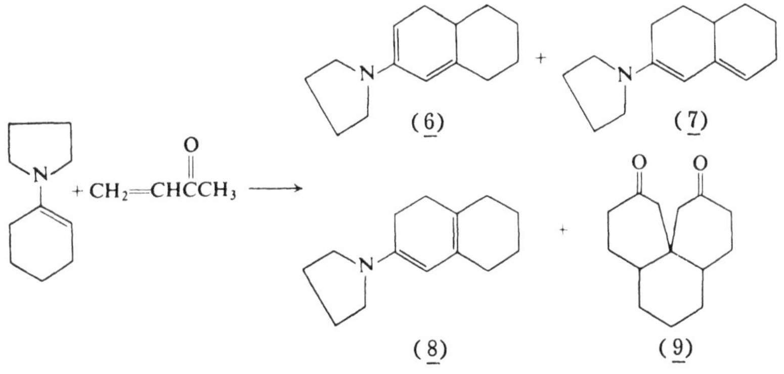

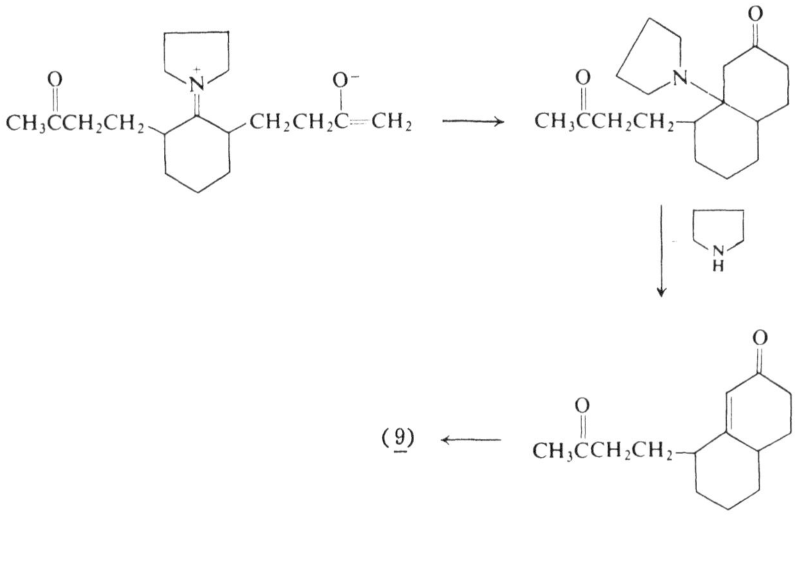

Scheme 5

The initial product formed when methyl vinyl ketone is mixed with an enamine [such as N,N-dimethylisobutenylamine (10) is the dihydropyran (11) from a 1,4 heterocycloaddition [40,67,68]. The chemical reactions that the dihydropyran undergoes indicate that it is readily equilibrated with the cyclobutane isomer 12a and zwitterion 12 [40]. Treatment of adduct 11 with phenyllithium gives cyclobutane 13, possibly via intermediate 12a [40].

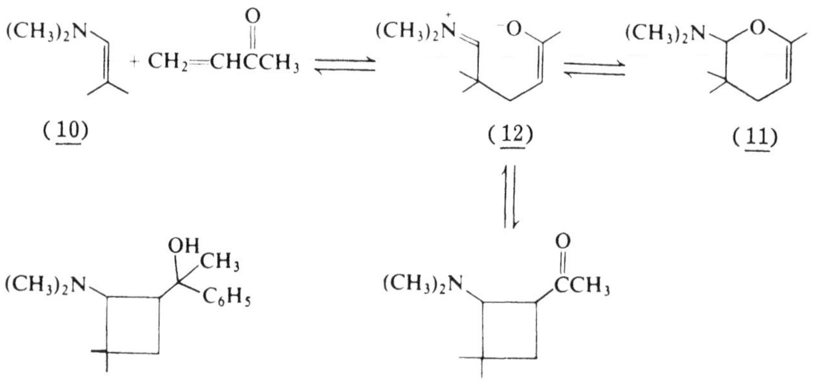

(10)   (12)   (11)

(13)   (12a)

In these reactions, the cyclobutyl ketones, such as 12a, have not been isolated in the aliphatic series because of the greater thermodynamic stability of the corresponding dihydropyrans (such as 11).

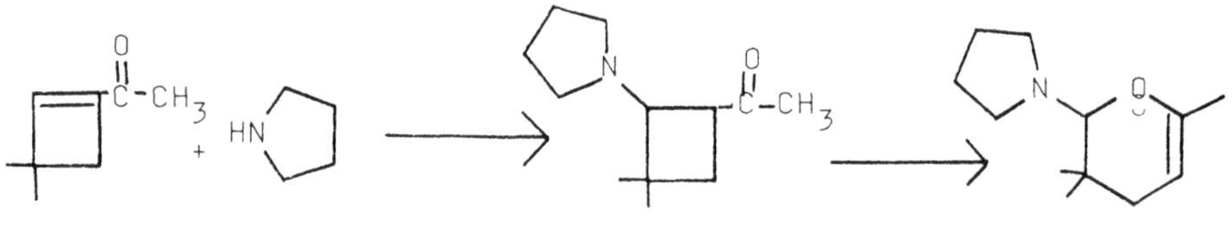

(14)   (15)

This was demonstrated by the synthesis of aminocyclobutane 14, which spontaneously decomposed into dihydropyran 15 [69]. When dihydropyran 15 is allowed to react with aqueous acid, it rearranges to carbocycle 16 [70,71]. When 16 is treated with pyrrolidine, dimerized enamine 17 is formed in a reaction similar to that ob-

(15)   (16)

(17)

served by Leonard and Musliner [72] from the pyrrolidine enamine of cyclohex-2-en-1-one. Other similar reactions have been reported [73, 265].

1,4 Cycloaddition of methyl vinyl ketone to the morpholine enamine of (+)-(R)-pulegone resulted in a pyran product which could then be treated with hydroxylamine hydrochloride in polar media to give optically active 5,6,7,8-tetrahydroquinolines [506]. Optically active alkylvinylpyridines have been synthesized using a similar method with aldehyde enamines and methyl vinyl ketone [655].

When methyl vinyl ketone or similar conjugated ketones are treated with an enamine, a 1,4 heterocycloaddition to form a pyran generally forms first, followed by rearrangement (usually brought about by aqueous acid catalysis) via 1,4 carbocycloaddition to a cyclohexyl compound. Some additional types of enamines undergoing these reactions are nitrogen heterocyclic [77–81,428], phosphorus heterocyclic [82], cycloheptanone [83], 2,7-decalinedione [84], and 2-decalone [85] enamines. The piperidine or morpholine enamine of 2,2,6,6-tetramethylpiperidin-3-one gave only the Michael addition product with methyl vinyl ketone [85]. Some of the conjugated enones that have been used other than methyl vinyl ketone are benzylidene ketones [42,87–93,189,429], phenyl vinyl ketone [85], methoxy-substituted [94] and t-butyl-carboxylate-substituted [95] methyl vinyl ketones, pyrylium salts [82,96], tropone [97], quinone methides (from phenolic Mannich bases) [86,98–100], methyl isopropenyl ketone [35], ethyl acetylacrylate [35], 2-cyclohexenone [72], and 1-acetyl-1-cyclohexene [101]. Chlorovinyl ketones yield pyrans when allowed to react with the enamines of either alicyclic ketones or aldehydes [102].

Various fluoro-substituted 3-buten-2-ones (methyl vinyl ketones) have been allowed to react with enamines to give fluorinated heterocyclic, alicyclic, or aromatic products. These substituted 3-buten-2-ones have fluoro substituents in the 4-position [103] or the 3-position [104,105], and trifluoromethyl groups at the 4-position [103,106] or the 1-position [107]. Some typical examples and yields of these reactions are shown in Scheme 6.

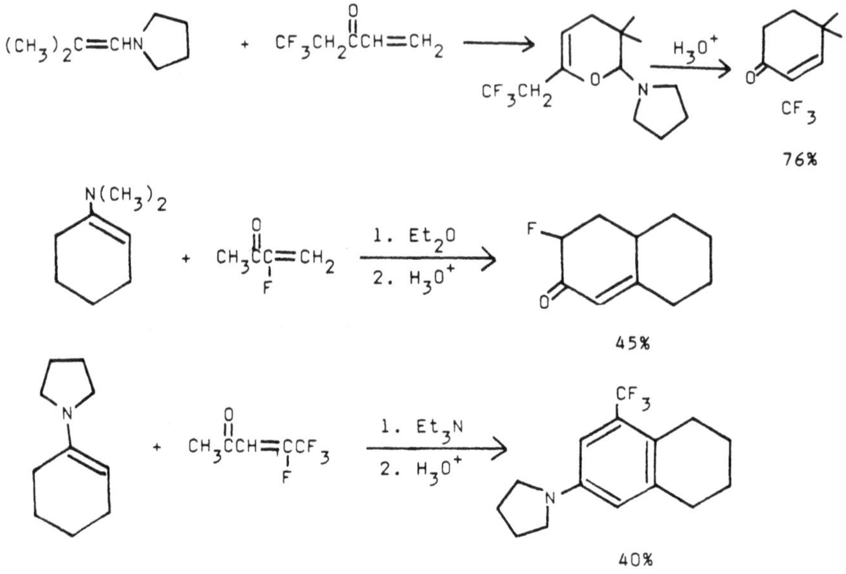

Scheme 6

The use of high pressure in these reactions helps to accelerate those that take place only very slowly [108].

Asymmetric induction has been observed in some of these cyclo-addition reactions [109–112]. For example, the pyrrolidine amide of L-proline (18) formed an enamine with 2-phenylpropanal, which in turn was allowed to react with methyl vinyl ketone, followed by hy-

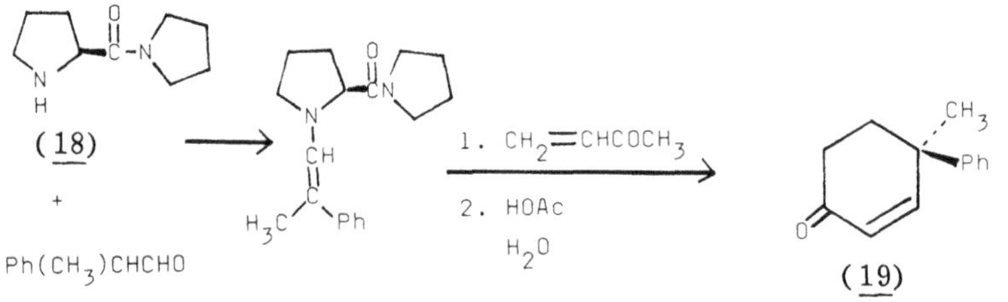

Scheme 7

drolysis, to form dextro-rotatory 4-methyl-4-phenyl-2-cyclohexenone (19) in a 48% yield and a 36.5% optical yield (see Scheme 7) [110]. This reaction was shown to be a thermodynamically controlled asymmetric synthesis.

Generally, the reaction of enamines with electrophilic olefins (such as methyl vinyl ketone and other conjugated enones) is a highly stereoselective reaction in which the enamines of 2-alkylcyclohexa-nones give rise to 2,6-disubstituted products [36]. However, 2,2-disubstituted products have also been found to form under some conditions. These reactions have been studied using methyl vinyl ketone as the electrophilic olefin [61,113]. The mechanism and factors involved in the regio- and stereoselectivity of the reactions have been extensively discussed (see Chapter 1, Section III.B) [61,113–115,664].

The sterochemistry of the 1,4 heterocycloaddition of a conjugated enone to a cyclohexyl enamine to form a 1,4-dihydropyran was studied [93]. The crystalline product of the reaction between enamine 20 and enone 21 was formed in a 72% yield. This product (22) was shown by x-ray diffraction studies to have *cis*-fusion between the cyclohexane ring and the dihydropyran ring [93]. In an earlier re-

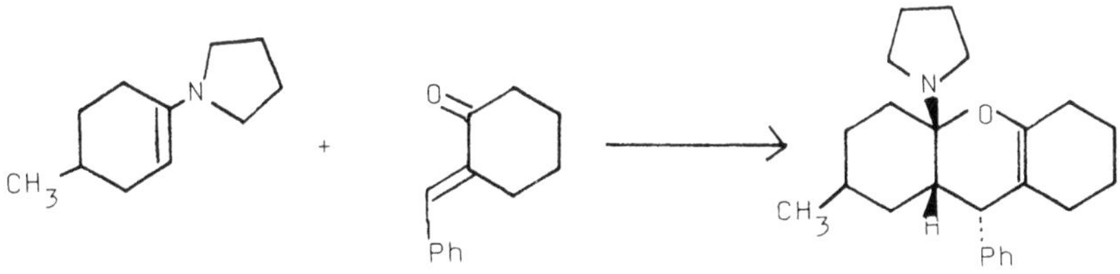

(20)   (21)   (22)

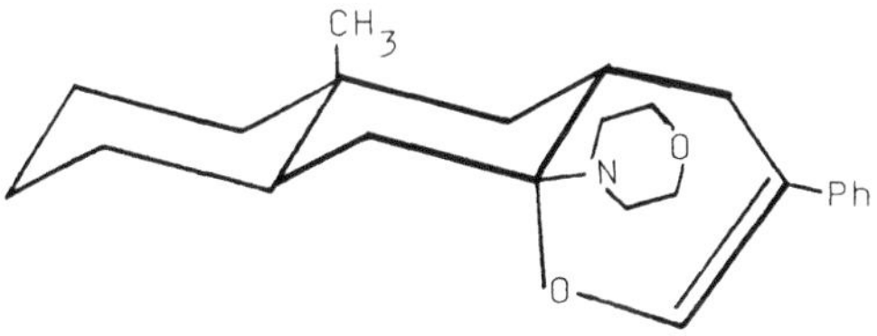

(23)

port [85], the structure of the dihydropyran adduct of the morpholine enamine of *trans*-decalin-2-one was shown to be 23 (*cis*-fusion of dihydropyran ring) by x-ray analysis.

In the case of 1,4 carbocycloaddition reactions of six-membered-ring enamines and enone, *cis*-fused products are generally found [52,84]. This is illustrated by the 1,4 carbocycloaddition of methyl vinyl ketone with enamine 24 to form *cis*-fused product 25 [52]. How-

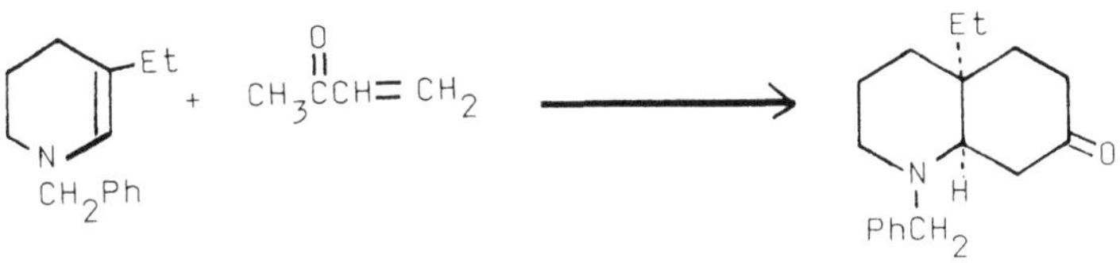

(24)   (25)

ever, under some conditions (such as a bulky substituent in the 2-position of the Δ^2-tetrahydropyridine enamine), stereoelectronic control will not allow cyclization to take place, and only a simple Michael addition takes place [116].

Heterocyclic enamines often undergo two-step "1,3 cycloaddition" with methyl vinyl ketone. This involves electrophilic attacks by an olefinic carbon and by a carbonyl carbon [117,118]. For example, 1,2-dimethyl-Δ^2-pyrroline (26), when treated with methyl vinyl ketone, produces 1,6-dimethyl-2,3,4,5-tetrahydroindole (27) [117]. The requirement that must be met so that this type of cyclization reaction can take place is that the α-position of the heterocyclic enamine be carbon substituted. This provides the possibility for an isomeric

(26)   (27)

enamine. This isomeric enamine, in the second step of the reaction, undergoes electrophilic attack by the side-chain carbonyl group.

The vinylogous 3,5-hexadien-2-one adds in a 1,4 cycloaddition with Δ^2-dehydroquinolizidine (28) to form compound (29) [119].

(28)   (29)

Acrolein (30), when allowed to react with an enamine such as the pyrrolidine enamine of cyclohexanone at room temperature followed by distillation, gives an interesting bicycloaminoketone (31) in a 75% yield [120]. This reaction has proved to be a very useful one for ring expansions and for making bicyclic systems [121]. The mechanism of this two-step 1,3 cycloaddition reaction was first studied by Untch [122]. He showed that, following the first electrophilic attack, the reaction occurred intermolecularly with transfer of the amine from the ketonic enamine to the aldehyde followed by cyclization. For cyclohexanone enamines, the initial product formed is dihydropyran (32) [123,506]. Distillation of this product produces bicyclo-

aminoketone 31, the stereochemistry of which has been studied [124].
A mixture of stereoisomeric bicycloaminoketones consisting primarily
of the *endo* isomer is obtained from this distillation when N-phenyl-

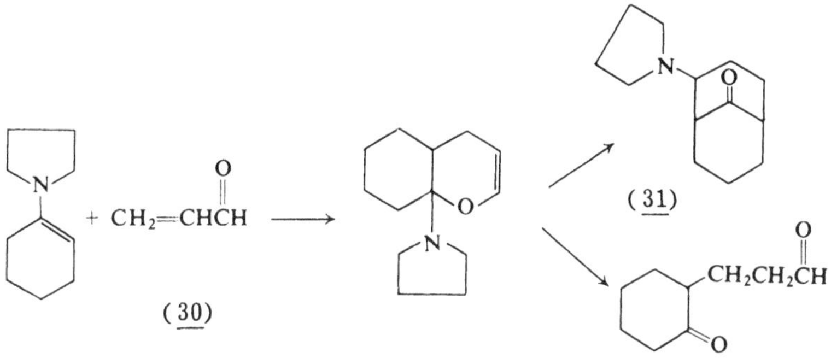

(30)

(31)

(32)   (33)

piperazine is the amine [123]. Hydrolysis of dihydropyran 32 yields
ketoaldehyde 33 [125,126]. Cyclopentanone and cycloheptanone
enamines give bicycloaminoketones directly with no dihydropyran
intermediates when treated with acrolein [123]. Enamines from acyclic
ketones also undergo a similar carbocycloaddition reaction. For example, the morpholine enamine of 3-pentanone (34) when treated with
acrolein (30) at 0°C followed by acidification produces 2,6-dimethyl-

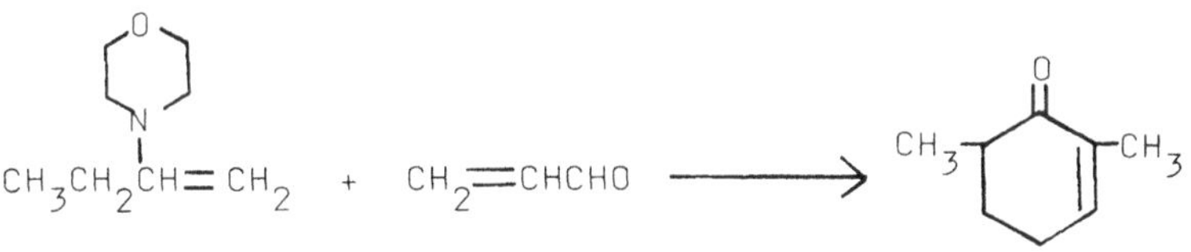

(34)   (30)   (35)

2-cyclohexenone (35) [127]. Dihydropyrans alone are found when
aldehyde enamines (either with or without β-hydrogen atoms) and
acrolein are allowed to react [39,128,404]. The electron-poor acrolein becomes the dienophile when it is allowed to react with electron-
rich dienamines. This is illustrated by the reaction between acrolein
and 1-N-pyrrolidino-2-ethyl-1,3-hexadiene (36) to give 2-N-pyrrol-
idino-3,5-diethyl-Δ^3-tetrahydrobenzaldehyde (37) [39]. A substi-

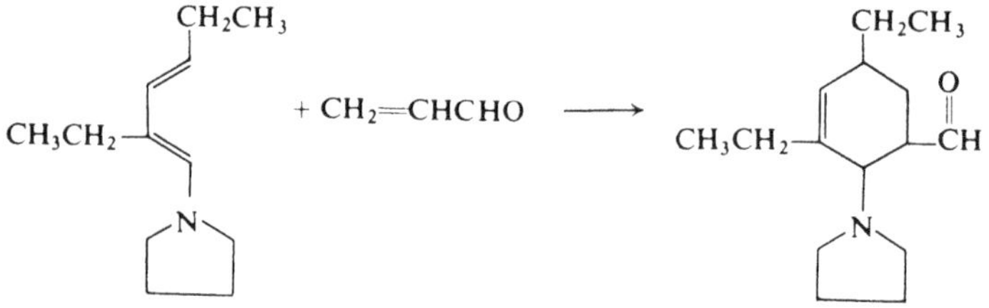

(36)   (37)

tuted α,β-unsaturated aldehyde, cinnamaldehyde, has been observed to undergo the same type of two-step 1,3 cycloaddition reaction with a cyclohexanone enamine as acrolein does, forming in this case a stereoisomeric mixture of substituted bicycloaminoketones in excellent yield [124,129–131].

Acryloyl chloride can be used to cause ring enlargement with the production of a bicyclodiketone when it is treated with a cyclohexanone enamine. This is shown by the reaction of acryloyl chloride (38) with 1-N-morpholino-1-cyclohexene (39), affording diketone 40 upon hydrolysis [132,133]. This reaction has been used to synthe-

(39)   (38)   (40)

size heterocyclic dihydropyrans [134] and pyrones [135,136], as well as carbocyclic cyclohexyl [137], indanone [138], and bicyclic [139–145] systems, some of which are used to go to adamantane derivatives [146–150].

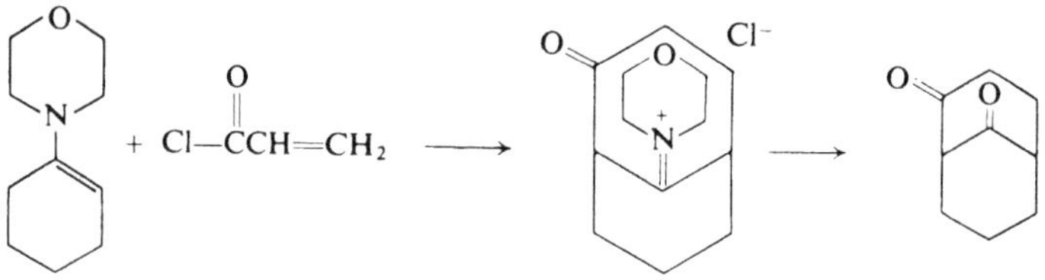

The reaction of methyl or ethyl acrylate with the enamine of an alicyclic ketone results in simple alkylation when the temperature is allowed to rise uncontrolled in the reaction mixture [36,86,113,114, 151–153]. If the reaction mixture is kept below 30°C, however, a mixture of the simple alkylated and cyclobutane (from 1,2 cycloaddition) products is obtained [151]. Upon distillation of this mixture, only starting material and simple alkylated product is obtained because of the instability of the cyclobutane adduct.

Enamines of aldehydes or acyclic ketones undergo exclusive 1,2 cycloaddition when treated with acrylate esters below 30°C [38,154, 155]. For example, treatment of N,N-dimethylisobutenylamine (10) with methyl acrylate (41) in refluxing acetonitrile gives cyclobutane

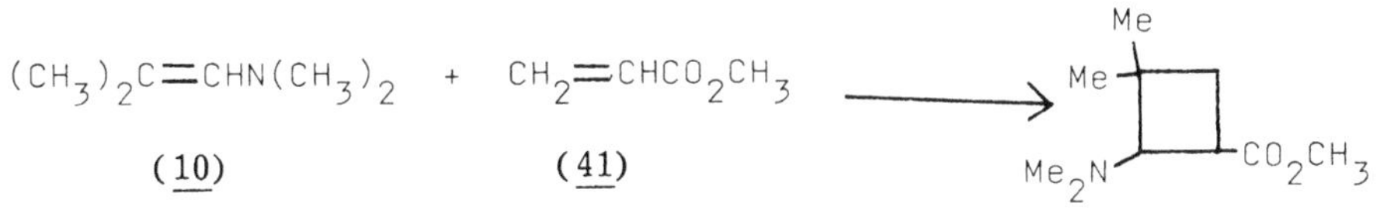

$(CH_3)_2C{=}CHN(CH_3)_2$
+
$CH_2{=}CHCO_2CH_3$
Me
Me
Me_2N
CO_2CH_3

(10) (41) (42)

42 in 91% yield [155]. Simple alkylation of enamines by electrophilic olefins depends on the presence of a β-hydrogen in the enamine. Therefore, it would be expected that the cyclobutane adduct of an enamine with no β-hydrogens and an acrylate ester should be stable with respect to the simple alkylated product. This is borne out in fact since these adducts can be distilled with no apparent decomposition [154–159]. Those adducts from enamines that have β-hydrogens decompose into starting materials and simple alkylation products when heated above about 125°C [155]. The cyclobutane adduct has been shown to lie along the pathway between starting materials and simple alkylation product [38], probably as a branch that is in equilibrium with a common zwitterion intermediate. For example, when the piperidine enamine of butyraldehyde is allowed to react with methyl acrylate (41), the 1,2 cycloadduct (44) forms initially and reversibly. Raising the temperature of the reaction mixture produces

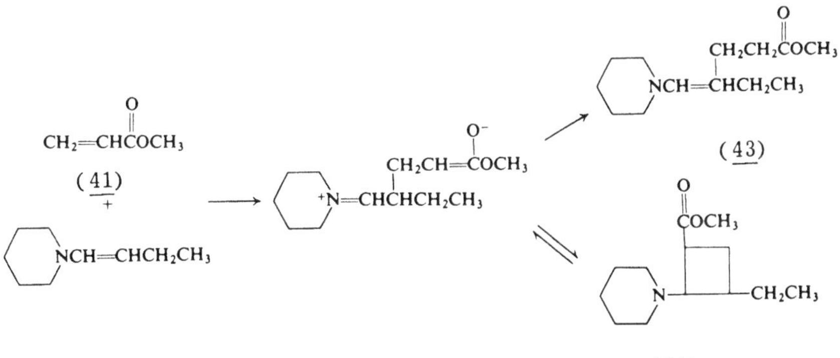

O
$CH_2{=}CHCOCH_3$
(41)
+
$NCH{=}CHCH_2CH_3$
O^-
$CH_2CH{=}COCH_3$
$^+N{=}CHCHCH_2CH_3$
O
$CH_2CH_2COCH_3$
$NCH{=}CHCH_2CH_3$
(43)
O
$COCH_3$
N
CH_2CH_3

(44)

the simple alkylated product (43) [155]. Steric requirements for 1,2 cycloaddition seem to be stringent since methyl methacrylate, methyl

crotonate, or methyl cinnamate do not form an isolatable cyclobutane
adduct with an enamine [38,155].

N-Methyl-1,2-dihydropyridine behaves as a simple enamine when
allowed to react with methyl acrylate (41) at -10°C to give cyclobutyl

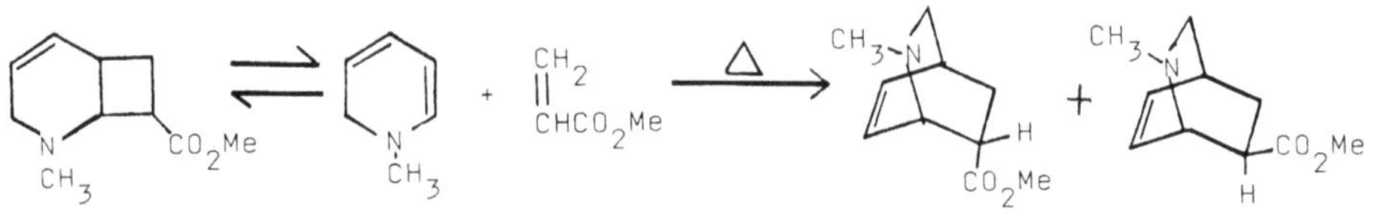

(45)   (41)   (46a)   (46b)

cycloadduct 45 in equilibrium with starting material [79]. In refluxing
benzene, the thermodynamically more stable Diels-Alder products
(46a and 46b) are formed in a 3.2-to-1 ratio [79]. So in this reaction,
N-methyl-1,2-dihydropyridine acts like a diene.

In a similar manner, diethyl maleate forms unstable 1,2 cycloadducts with enamines with β-hydrogens at temperatures below 30°C
[155]. The initial mixing of the enamine and maleate ester results in
the formation of a yellow to orange charge-transfer complex which
slowly fades [160–162], and in an exothermic reaction [151]. The maleate ester is rapidly and completely isomerized to the fumarate ester
upon contact with the reaction mixture [160], either by trace amounts
of secondary amine which might be present [161] or by the rapid and
reversible addition of the maleate ester to the enamine to form a zwitterion intermediate [41]. In the latter case, the second step of either
proton transfer or cyclization is so much slower than the first step

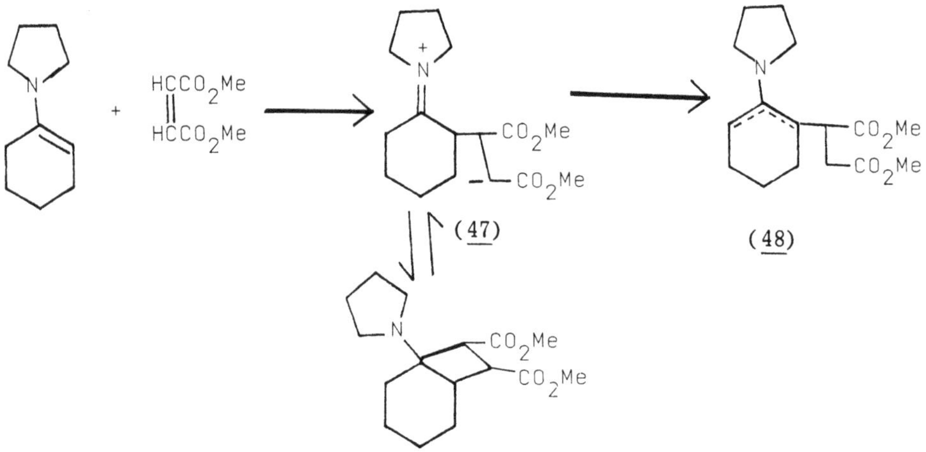

(47)   (48)

that the thermodynamically more stable fumarate ester is quantitatively formed before the second step has appreciably progressed. At
higher temperatures the equilibrium is shifted to the left (reactants)
[41,151,162], and then simple alkylated products are slowly formed
[151,160,162]. For example, dimethyl maleate, after being refluxed
for 45 min with 1-N-pyrrolidylcyclohexene, will give alkylation product 48 via zwitterion 47 in an 84% yield [160]. In these reactions,
cyclization of the zwitterion gives the *kinetically controlled* cyclobutane product, whereas proton transfer in the zwitterion yields the
thermodynamically controlled alkylation product. Enamines with no
β-hydrogens form very stable 1,2 cycloadducts with diethyl maleate
[154,155,160]. So α-N-pyrrolidylmethylenecyclohexane (49) reacts
with dimethyl maleate to form substituted cyclobutane 50 [160]. The
two adjacent carboalkoxy groups of the cyclobutane adduct have been

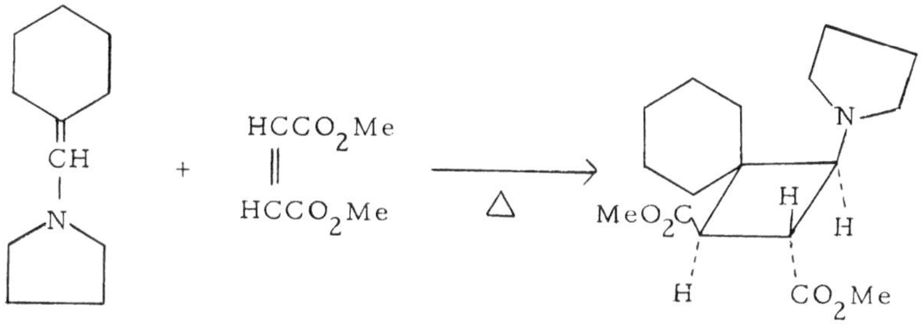

shown to be *trans* to one another [154,155,161,163]; and, using the
adduct of 1-(N-morpholino)-4-*t*-butylcyclohexene and diethylmaleate
(51), the cyclohexane and cyclobutane rings have been shown to be

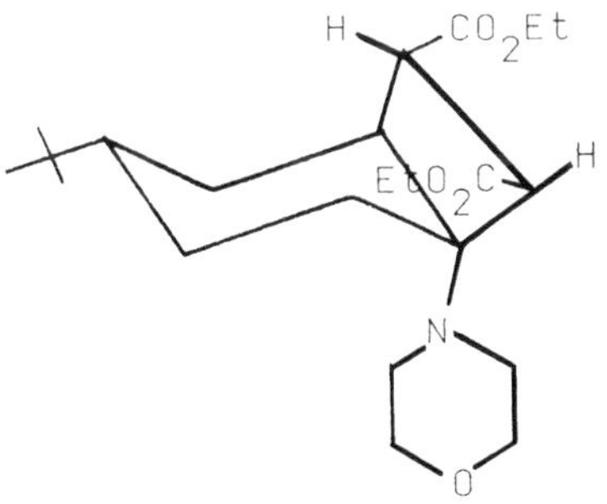

cis-fused by x-ray analysis [163]. These reactions occur much faster in polar solvents than in nonpolar solvents [155,161].

The photochemical carbocycloaddition of dimethyl fumarate with N-isobutenylpyrrolidine (52) to cyclobutane derivative 53 has been

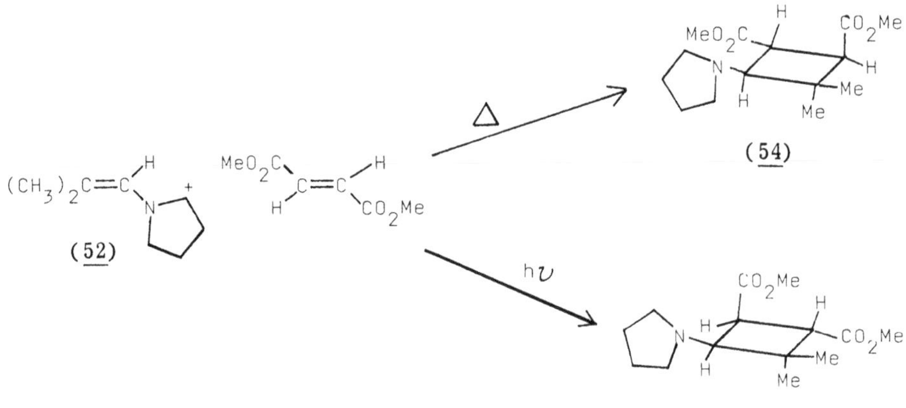

(54)

(52)

(53)

observed [161]. The photochemical reaction produced a different iso-mer, 53, from that obtained from the thermal reaction, i.e., 54. Un-like its thermal counterpart, the efficiency of the photochemical cyclo-addition decreases dramatically with increasing solvent polarity [161]. So the thermal and the photochemical reactions obviously do not occur via a common intermediate. The photochemical process probably oc-curs in nonpolar solvents via charge-transfer complex excitation to a singlet exciplex [188] and then cycloaddition, possibly a $[_{\pi}2_s + _{\pi}2_s]$ concerted reaction. This type of photochemical [2+2] cycloaddition was not observed with enamine 52 and methyl acrylate, methyl methac-rylate, or dimethyl isopropylidenemalonate; nor was it observed with dimethyl fumarate and N-isobutenylpiperidine or N,N-dimethylisobu-tenylamine [161].

Trimethyl ethylenetricarboxylate (54), when allowed to react with

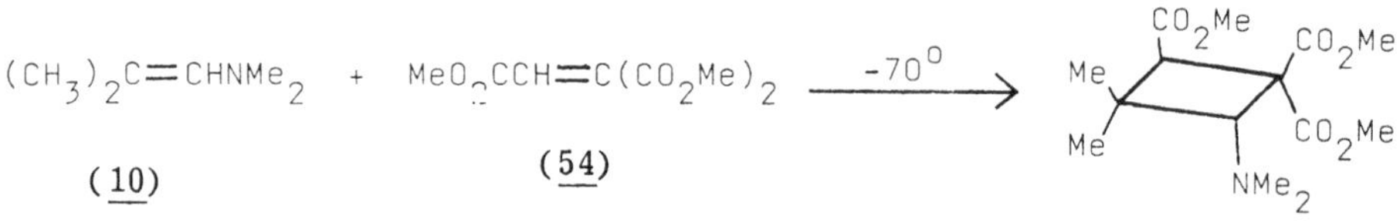

(10)   (54)   (55)

$H_2C = CHNMe_2$

(56)

N,N-dimethylisobutenylamine (10) in ether at -55°C, gave cyclobutyl product 55 in quantitative yield [162,164]. This product decomposed at room temperature [162], and when N,N-dimethylvinylamine (56) was the enamine used, only an acyclic alkylation product could be isolated, even at -55°C [162,165].

Two-step 1,4 carbocycloaddition of enamines, such as was observed with methyl vinyl ketone, is not possible with acrylate or maleate esters. This is due to the fact that, following the initial simple substitution, no side-chain carbanion is available for nucleophilic attack on the α-carbon of the iminium ion. Likewise, two-step 1,3 cycloaddition, such as that found when alicyclic enamines were treated with acrolein, is impossible with acrylate or maleate esters because transfer of the amine moiety from the original enamine to the side chain to form a new enamine just prior to the final cyclization step is not possible. That is, the reaction between a secondary amine and an ester does not produce an enamine.

If the α-position of an enamine is carbon substituted, providing the possibility of an isomeric enamine, and if the amine group and other substituent groups are sufficiently removed from the sites of electrophilic attack as to not cause any steric interference, then simple alkylation of an enamine by an acrylate ester can be followed with a second electrophilic attack by the carbonyl group on the isomeric enamine to form a two-step 1,3 cycloaddition product. α-Substituted heterocyclic enamines completely fulfill these requirements and hence undergo this type of cycloaddition with acrylate esters [118,166–169]. For example, the reaction between 1,2-dimethyl-Δ^2-tetrahydropyridine (57) and ethyl acrylate resulted in the formation of cyclization product 58 [168].

(57)   (58)

Enamines have been observed to act as both dienophiles [170–174] and dienes [172,175–179] (dienamines in this case) in one-step, Diels-Alder-type 1,4 cycloadditions with acrylate esters and their vinylogs. This is illustrated by the reaction between 1-(N-pyrrolidino)cyclohexene (46) and methyl *trans*-2,4-pentadienoate, where the enamine acts as the dienophile to give the adduct 59 [172]. In a competitive type of reaction, however, the electron-rich dienamine pref-

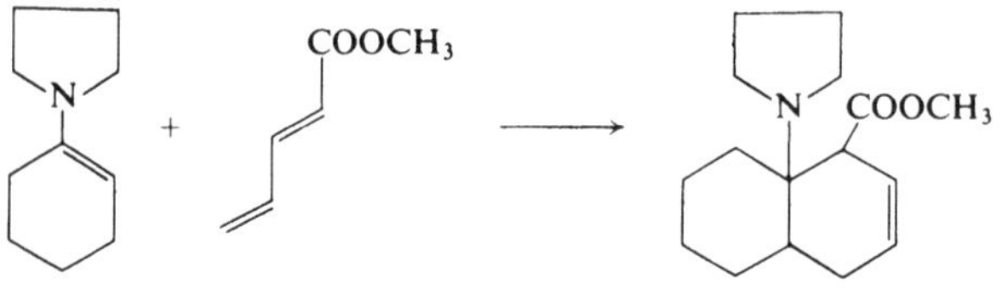
COOCH$_3$
N
+
N
COOCH$_3$

(46) (59)

erentially acts as the diene, with the electron-poor pentadienoate ester acting as the dienophile, as is shown by the reaction between

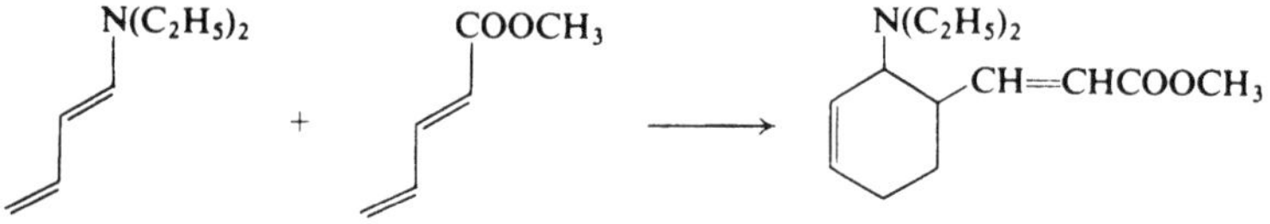
N(C$_2$H$_5$)$_2$
+
COOCH$_3$
N(C$_2$H$_5$)$_2$
CH=CHCOOCH$_3$

(60)

methyl *trans*-2,4-pentadienoate and 1-diethylamino-1,3-butadiene to give product 60 [172]. A similar reaction has been observed between α-chloroacrylonitrile and the dienamine 1-N-(morpholino)-1,3-butadiene. Upon attempted vacuum distillation of the reaction product, hydrogen cyanide and hydrogen chloride were eliminated to finally produce N-phenylmorpholine hydrochloride [178].

Sometimes a 2:1 adduct is formed between enamines and unsaturated esters instead of the usual 1:1 adduct. This is the case when N,N-dimethylisobutenylamine (10) is allowed to react with diethyl methylenemalonate (61) producing compound 63 [155]. This product

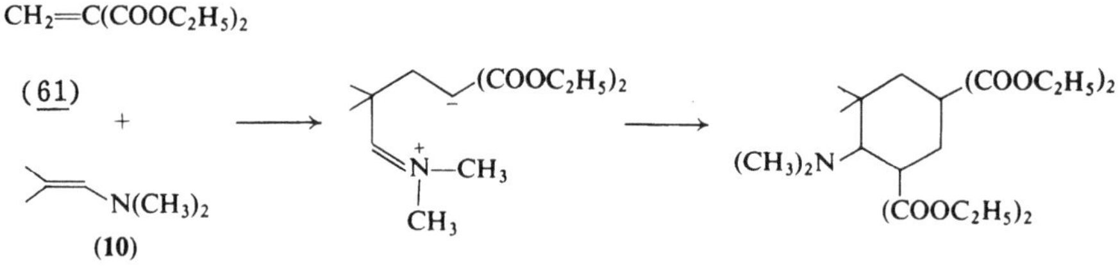
CH$_2$=C(COOC$_2$H$_5$)$_2$
(61)
+
N(CH$_3$)$_2$
(10)
(COOC$_2$H$_5$)$_2$
N—CH$_3$
CH$_3$
(COOC$_2$H$_5$)$_2$
(CH$_3$)$_2$N
(COOC$_2$H$_5$)$_2$

(10) (62) (63)

is formed even when an excess of enamine is present. The ease of formation of this 2:1 adduct is probably due to the stabilizing effect of the adjacent ester groupings on the anionic center in intermediate 62 and to the minimal steric requirements of the incoming electrophilic olefin [155].

A two-step cyclization of an enamine with an electrophilic olefin has been reported in which the first step is alkylation by an allyl

halide and the second step is alkylation by the electrophilic olefin
[179]. The reaction involves dimethyl bromomesconate (64) and 1-(N-
pyrrolidino)cyclohexene (46), which, after hydrolysis, yields bicyclic
ketodiester 65. Other cycloadditions of this type involving α-bromo-
methylacrylate esters adding to cyclohexyl enamines have been reported [180−184]. This reaction has given a good entrance into synthesis of adamantanes.

CH₃O₂C and related structures (chemical scheme)

(64)   (46)   (65)

Substituted azulenes have been produced by allowing enamines to
react with 2H-cyclohepta[b]furan-2-ones [185−187].

The addition of *p*-quinone to enamines normally produces furan
derivatives, especially when the enamine possesses a β-hydrogen (see
Section IV.A). 1,2 Cycloaddition is claimed to take place to give a
cyclobutane derivative when *p*-quinone and an enamine with no β-hydrogens are allowed to react at low temperatures [190]. However,
little evidence is reported to verify this structural assignment, and
the actual structure probably is a benzofuranol [191]. Reaction of a
dienamine (formed in situ) with *p*-quinone in the presence of acetic
acid results in two 1,4 cycloadditions to form a 2:1 dienaminequinone
carbocycle adduct [192]. Another conjugated ketone, dibenzalacetone
(66), adds to the pyrrolidine enamine of cyclohexanone (46) to yield
a bicyclo[5.3.1]undecane adduct (67) [193,194]. N-Ethyl maleimide
and N,N-dimethylaminoisobutyraldehyde add by 1,2 cycloaddition to
give a cyclobutane derivative [155,195]. Enamines can react with
2-oxoindolin-3-ylidene derivatives to form 1,2- or 1,4-cycloadducts
or Michael addition products [196−198]. Cyclopropenones also react
with enamines to give carbocyclic products [199,200] (see Section
III.B for further discussion of this).

(46)   (66)   (67)

Conjugated with Nitrile, Nitro, or Sulfone Groups

Olefins conjugated with electron-withdrawing groups other than a
carbonyl group undergo reactions with enamines in a manner similar

to the carbonyl-conjugated electrophilic alkenes described above.
Namely, they condense with an enamine to form a zwitterion intermediate from which either 1,2 cycloaddition to form a cyclobutane ring
or simple alkylation can take place.

Such an electron-withdrawing group is the nitrile. Acrylonitrile
(68) adds to enamines in a manner very similar to that of acrylate esters. 1,2 Cycloaddition of the unsaturated nitrile with the enamine
is the initial step in almost all cases [38,153]. The thermal stability
of the cyclobutane product depends on the absence or presence of a
hydrogen in the original enamine. Cyclobutane adducts derived from
enamines without β-hydrogens are thermally stable above room temperature and can be distilled. Those cyclobutane adducts obtained
from enamines with β-hydrogens are thermally unstable and will decompose on distillation to give back starting materials and/or simple
alkylated products. For example, Stork and Landesman observed
that acrylonitrile (68), when refluxed with 1-(N-pyrrolidino)cyclo-
hexene (46) in dioxane solvent, produced simple alkylated product 69
[35].

It was later noted that at low temperatures, cyclobutane adduct
70 is formed from this reaction mixture, but as the temperature is
increased, the adduct reverts back to starting material and also forms
some of simple alkylation product 69 [38,201]. The reaction between

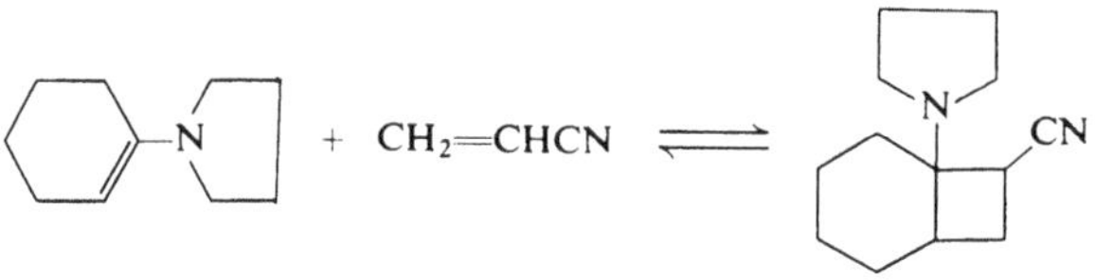

(46)   (68)   (70)

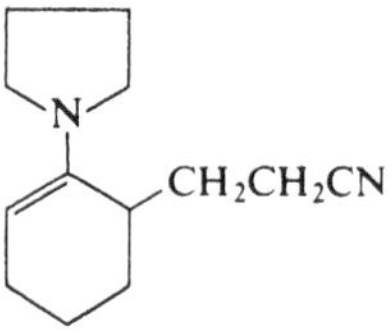

(69) 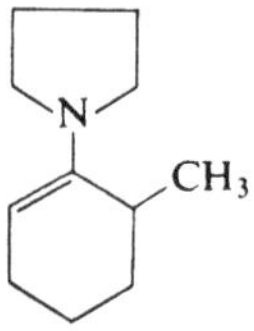   (71)

acrylonitrile (68) and 6-methyl-1-(N-pyrrolidino)cyclohexene (71) at
room temperature produced an equilibrium mixture of cyclobutane
adduct and starting materials, thus illustrating the steric sensitivity
of the 1,2 cycloaddition reaction [38,162]. On the other hand, when
N,N-dimethylaminoisobutene (10), an enamine with no β-hydrogens,

is allowed to react with fumaronitrile (72), a thermally stable cyclobutane adduct (73) is formed in a 73% yield [155]. Similar observations have been made by others [44,68,157,164,165].

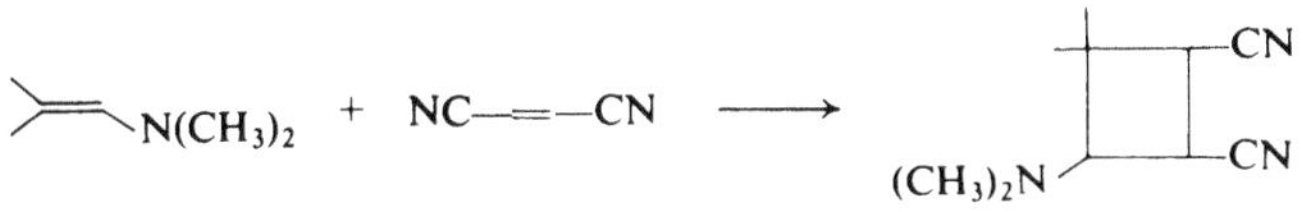

α-Chloroacrylonitrile undergoes 1,2 cycloaddition with aldehydic
enamines to give the corresponding cyclobutane adducts [178,202,
203]. However, when it is allowed to react with the enamines of cyclic ketones, quaternary chloride salts are formed by the 1,3 cycloaddition of the enamine to the electrophilic olefin.

Nitroolefins also offer the possibilities of 1,2 cycloaddition [155,
204] or simple alkylation [204–207] products when they are allowed
to react with enamines. The reaction of nitroethylene with the morpholine enamine of cyclohexanone led primarily to a cyclobutane adduct in nonpolar solvents and to a simple alkylated product in polar
solvents [204]. These products are evidently formed from kinetically
controlled reactions since they cannot be converted to the other product under the conditions in which the other product was formed, and
hence there is apparently no equilibrium set up between either of the
products and the zwitterion intermediate.

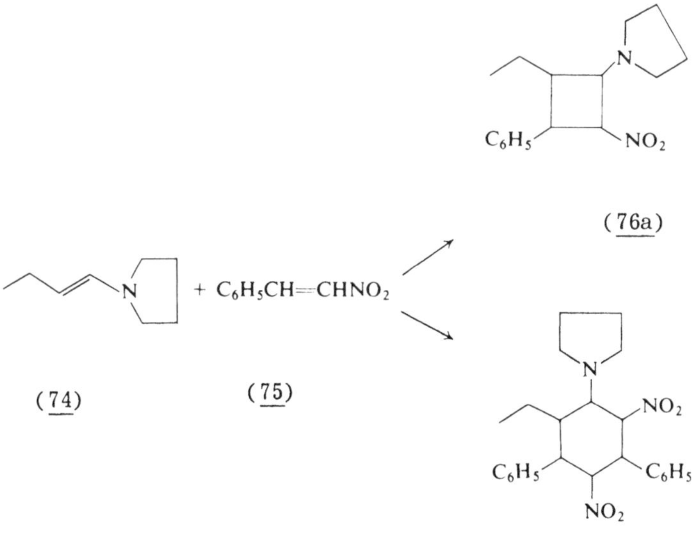

The reaction between the pyrrolidine enamine of butyraldehyde (74) and β-nitrostyrene (75) provides cyclobutane adduct 76a quantitatively in either petroleum ether or acetonitrile solvent, but in the more polar ethanol solvent, a 2:1 condensation product occurs. The structure of the product was shown to be 76b [204].

Nitroalkylation of enamines from cyclohexanones with β-nitrostyrene is very stereoselective. It attacks the enamines by an antiparallel mechanism unless there is steric hindrance in the parent enamines (see Chapter 1, Section III.B) [208,209].

Aromatic systems have been synthesized by allowing enamines to react with 2-chloronitroethene [210] or with 3-cyano-1-(p-toluenesulfonyl)-1,3-butadienes [211]. Using a method paralleling that of Nelson and Lawton (see Section III.A) [179], the reaction between (E)-

(77)

2-nitro-2-hepten-1-yl pivalate and 4-t-butyl-1-(N-(S)-2-methoxymethylpyrrolidyl)cyclohexene gives a stereoselective [3+3] carbocycloaddition product, 77, as an enantiomerically pure product in 37% yield [212].

An unusual 1,4 heterocycloaddition has been reported to take place when an enamine is treated with 2-nitropropene in ether at 0°C to form an oxazine derivative [213].

Methyl vinyl sulfone forms 1,2 cycloaddition adducts with aldehydic enamines, both with and without β-hydrogens [155]. Simple alkylation was reported to take place when phenyl vinyl sulfone was allowed to react with cyclohexanone enamines [207,214], but it has been shown that phenyl vinyl sulfone also forms cyclobutane adducts [215.217,218]. When the morpholine enamine of trans-decalin-2-one (78) is allowed to react with p-bromophenyl vinyl sulfone (79), the reaction proceeds under stereoelectronic control to give dipolar intermediate 80 (see Chapter 1, Section III.B), which then cyclizes to 81 [216]. The structure of 81 was determined by x-ray analysis.

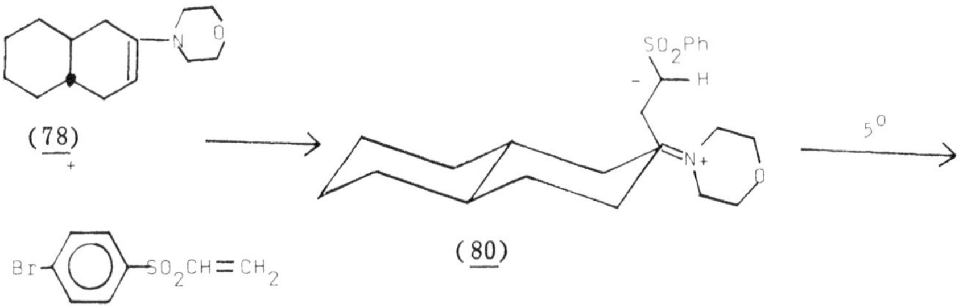

(78)

+

(79)

(80)

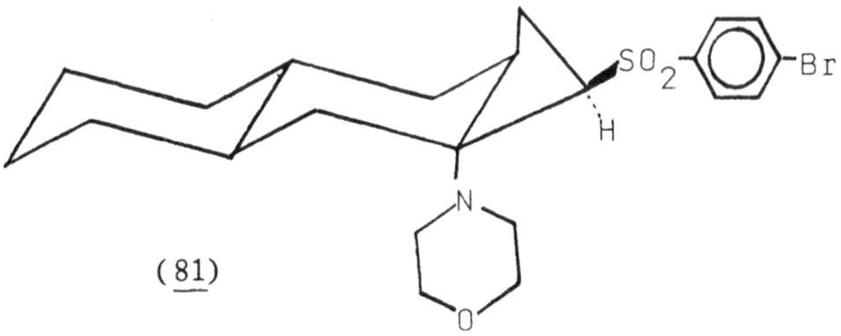

(81)

Dienamine 82a has been reported to undergo a 1,4 cycloaddition with acrylonitrile to form bicycloaminonitrile 83 in a 74% yield [219]. A report has indicated that both possible 1,4 cycloaddition adducts are obtained from the reaction of acrylonitrile with a 1:1 equilibrium mixture of the linear- and cross-conjugated isomers of dienamine 82b [176,220]. Similar adducts are formed when the dienamine is allowed to react with methyl vinyl ketone and with methyl acrylate. β-Nitrostyrene also undergoes [4+2] cycloaddition with dienamines [221,222].

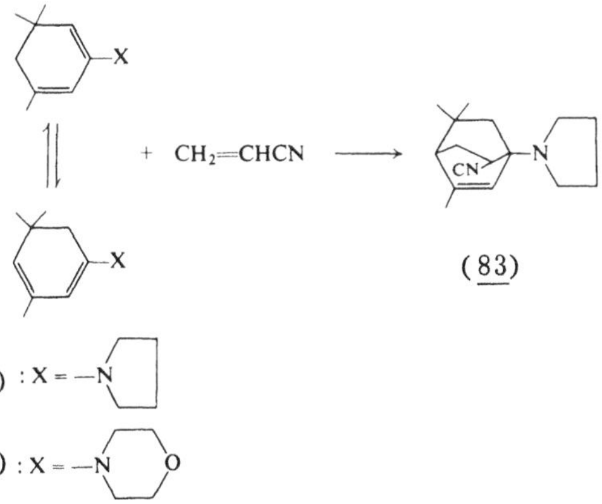

+ CH₂=CHCN ⟶

(83)

(82a) : X = —N⟨⟩

(82b) : X = —N⟨O⟩

Cumulated with Carbonyl Group

The treatment of enamines with acid halides that possess no α-hydrogens results in the simple acylation of the enamine [36,43,223–228]. If the acid halide possesses an α-hydrogen, however, ketenes are produced in situ through base-catalyzed elimination of hydrogen chloride from the acid halide. The base catalyst for this reaction may be the enamine itself or some other base introduced into the reaction mixture, such as triethylamine. However, if the ketene is produced in situ instead of externally, there still remains the possibility of a side reaction between the acid halide and the enamine other than the production of ketene [228,229].

The initial reaction between a ketene and an enamine is apparently a 1,2 cycloaddition to form an aminocyclobutanone adduct ($\underline{84}$) (see Scheme 8) [230–241]. The thermal stability of this adduct depends on the nature of substituents R_1, R_2, R_3, and R_4. The enolic forms of $\underline{84}$ can exist only if R_2 and/or R_4 is a hydrogen. If the enamine involved in the reaction is an aldehydic enamine with no β-hydrogens and the ketene involved is disubstituted (i.e., R_2, R_2, R_3, and R_4

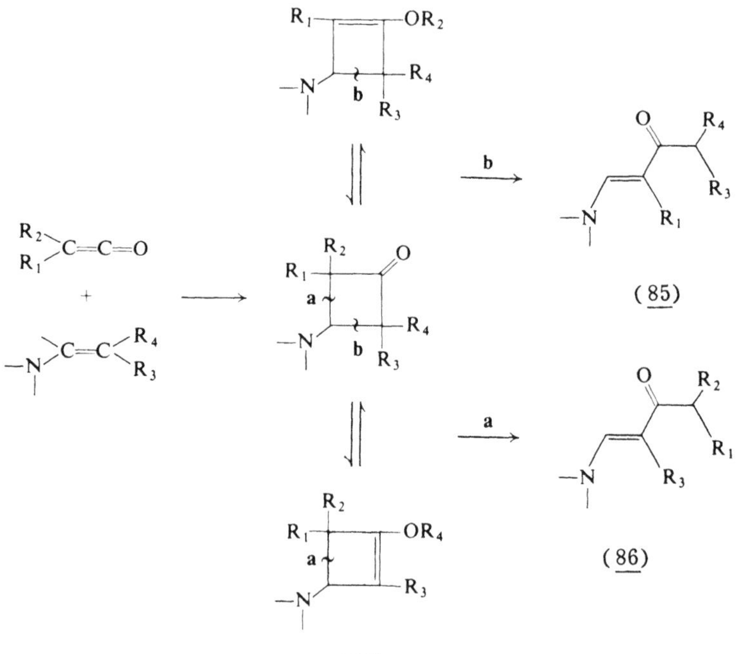

Scheme 8

are not hydrogens), then the cyclobutanone adduct is thermally stable. For example, the reaction of dimethylketene with N,N-dimethyl-
aminoisobutene (10) in isopropyl acetate solvent produces 3-(N,N-di-
methylamino)-2,2,4,4-tetramethylcyclobutanone (87) in a 64% yield,
a product that can be distilled with no decomposition [237]. Some
2:1 and 3:1 dimethylketene-enamine adducts are also formed during
this reaction. If acetonitrile is used as solvent, then larger quantities of the 2:1 and 3:1 adducts are formed at the expense of the 1:1

$$
\begin{array}{ccc}
\text{CH}_3 \\
\quad\;\; \text{C}\!\!=\!\!\text{C}\!\!=\!\!\text{O} \;+\; \text{)}\!\!=\!\!\text{CHN(CH}_3)_2 \;\longrightarrow \\
\text{CH}_3
\end{array}
$$

(10)   (87)   (88)

adduct [237]. The 2:1 adduct was identified as δ-lactone (88) [242],
which undoubtedly is formed by the further addition of a dimethylketene molecule to the initially formed zwitterion intermediate. The
structure of this product is analogous to the cyclic trimer of dimethylketene [243].

The reaction mechanism for these reactions may involve a zwitterion intermediate (89) [237] in a two-step mechanism leading to either
a cyclobutanone (87) or to a δ-lactone (88). However, two reaction

$$(CH_3)_2 C\!=\!C \overset{\displaystyle O^-}{\underset{\displaystyle Me_2CCH \,=\!=\, \overset{+}{N}(CH_3)_2}{\diagdown}}$$

(89)

mechanisms may be involved [259]. One would be a two-step mechanism leading to the δ-lactone, as described above. The other would
be a [$_\pi 2s + _\pi 2a$] concerted cycloaddition which is an allowed concerted
thermal cycloaddition according to the Woodward-Hoffmann rules [20]
(see Figure 2). Ketenes are ideal antarafacial components in reactions
of this type since they offer a minimum of steric hindrance to antarafacial addition. This is because one of the carbon atoms involved is
sp-hybridized, resulting in much less crowding in the transition state.
An indication that two different mechanisms may be involved in these
cycloadditions is given by the observation that the 1,2 cycloaddition
rate is relatively insensitive to a change of solvent polarity in the

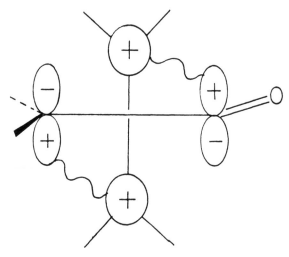

Figure 2 Concerted cycloaddition of ketene and enamine orbitals represented are HOMO (vertical) of enamine and LUMO (horizontal) of ketene.

reaction mixture, whereas the 1,4 cycloaddition rate is greatly accelerated by an increase in solvent polarity [259]. The concerted reaction between ketenes and enamines is much faster than that between ketenes and other olefins because of the electron-rich ketenophilic double bond present in enamines [260].

The reaction between an aldehydic enamine with no β-hydrogens and ketene yields a cyclobutanone adduct that is not thermally stable (R_3 and $R_4 \neq H$, $R_1 = R_2 = H$) [228,232,234]. Thermal decomposition gives just one product by way of pathway b (see Scheme 8). This is illustrated by treating 1-(N-morpholino)isobutylene with ketene to yield adduct 90, which decomposes upon heating to vinyl amide 91 [232,235]. When this enamine is allowed to react with acetyl chloride alone, only simple acylation of the enamine takes place to form hydrochloride salt 91a [228]. Cyclobutane adduct 90 was shown not to cleave and form 91a under these reaction conditions and hence is not a precursor in this reaction [228]. Apparently, simple acylation by the acid halide occurs more rapidly than ketene formation with only the weakly basic morpholine enamine acting as base catalyst [229]. Aldehydic enamines with β-hydrogens, such as 1-(N-morpholino)butene, undergo 1,2 cycloaddition with ketene to form a very unstable cyclobutane adduct ($R_2 = R_4 = H$), in this case adduct 92. This adduct rapidly changes into a mixture of vinyl amides via pathways b and a, in this case vinyl amides 93 and 94 in a 4:1 ratio [235].

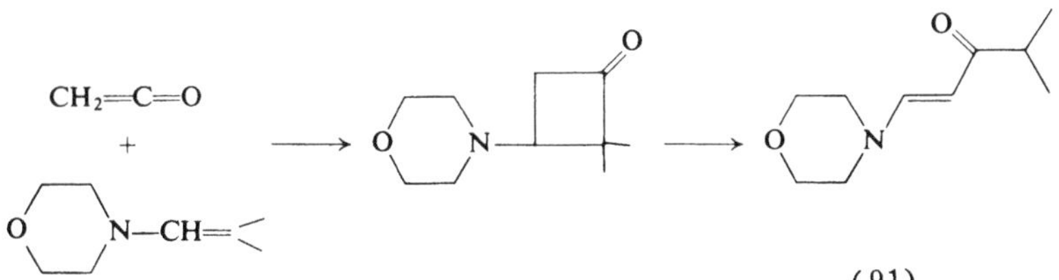

CH$_2$=C=O

+

O⟩N—CH=⟨

⟶ (structure with morpholine-N, cyclobutanone)

(90)

⟶ (structure) (91)

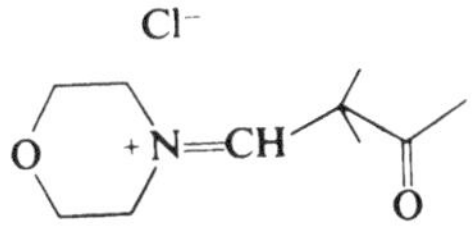

Cl⁻

(structure with morpholine +N=CH and ketone)

(91a)

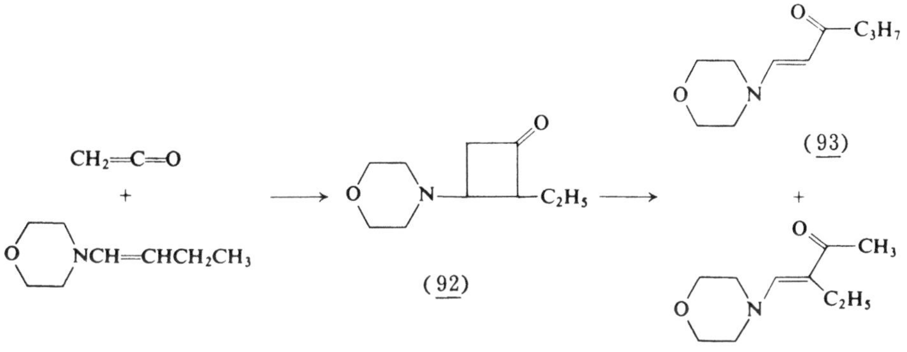

CH$_2$=C=O

+

O⟩NCH=CHCH$_2$CH$_3$

⟶ (structure) —C$_2$H$_5$

(92)

⟶

(structure with C$_3$H$_7$) (93)

+

(structure with CH$_3$, C$_2$H$_5$)

(94)

Enamines derived from cycloalkanones add to ketene to produce first a cyclobutanone adduct 95 [231,232] followed by cleavage of the cyclobutane ring at *a* or *b* (see Scheme 9). The point of cleavage of the cyclobutane ring depends on the ring size of the original cyclo-alkanone. For adducts from five- and six-membered cycloalkanone enamines (*n* = 3 and 4), decomposition takes place by pathway *a* [229, 231,232,234,235,244—257]. The adduct from the nine-membered-ring enamine (*n* = 7) is produced in very poor yields, and decomposition follows both path *a* and path *b* [229,255—257]. As the enamine ring size increases from ten-membered through fifteen-membered [229,248, 249,253,254—258], the cyclobutane ring cleavage follows path *b* and

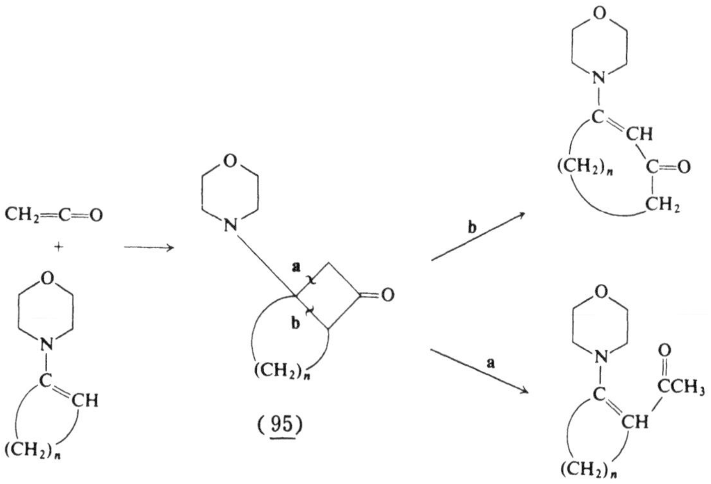

Scheme 9

the amount of product increases. Treatment of enamines obtained from ketones with excess ketene produces 2H-pyran-2-ones (see Section IV.A) [235].

Treatment of enamines with acylketenes results in a product that leads to 4-pyranones [261]. The reaction between vinylketene and enamines produces cyclohexenones and vinylcyclobutanones [262].

Others

Cyanoallene, when treated with the morpholine enamine of cyclohexanone, undergoes a 1,2 cycloaddition reaction to form 96 [263,264]. The reaction between cyanoallene and diendiamine 97 produces di-1,2 cycloaddition adduct 98 [264]. Cycloadditions of morpholine enamines

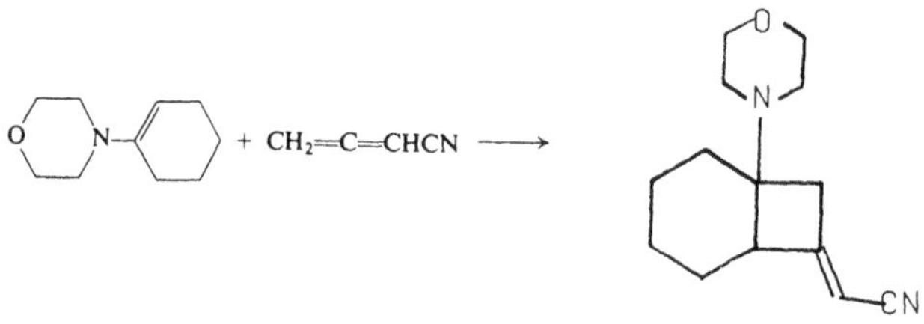

(96)

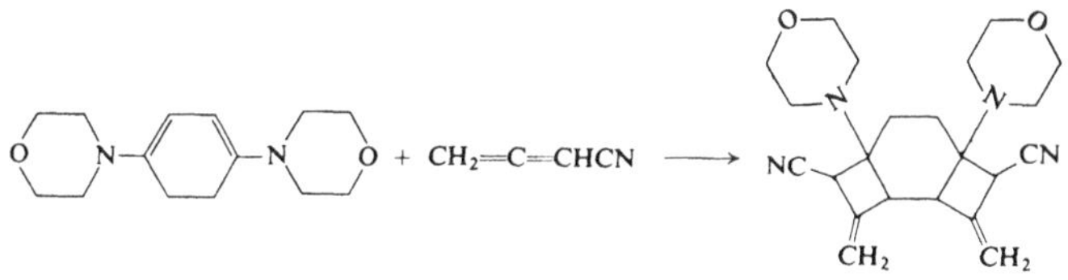

(97)   (98)

with other electron-poor allenes such as 1-benzenesulfonylallene or
1,1-bis(alkoxycarbonyl)allenes form cyclobutane products also [386].
    Cycloadditions of β-disubstituted enamines to 2- and 4-vinylpyr-
idines produce pyridylcyclobutanes in an acid-catalyzed reaction [266].
Perfluoroalkenes also add to enamines to form cyclobutane derivatives
[267]. Ethenesulfonyl fluoride reacts with enamine 10a to give cyclo-

$(CH_3)_2C=CHNR_2$ + $CH_2=CHX$

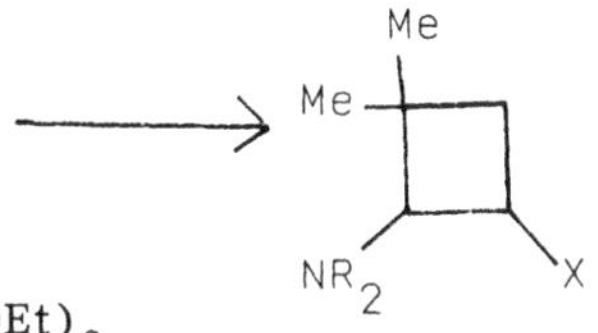

a: NR_2=morpholine $X = SO_2F$

b: NR_2=pyrrolidine   or $P(O) (OEt)_2$

(10)   (99): $X = SO_2F$

                                                  (100): $X = P(O) (OEt)_2$

butane 99a in 100% yield [268], and diethyl vinylphosphonate reacts
with enamine 10b to produce cyclobutane 100a in a 62% yield [269].
This latter dienophile also undergoes Diels-Alder reactions with dienamines [269]. Both dienophiles undergo Michael additions with
enamines.
    Dienes activated by cyano groups [34] or by a diethyl phospho-
nate group [270] undergo 1,4 cycloaddition with enamines. For example, dienedinitrile 101 reacts with enamines to produce 102. This

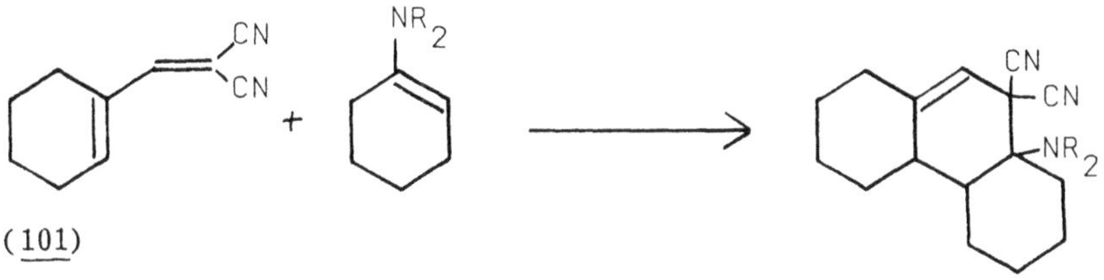

(101)   (102)

reaction goes 120 times faster in polar acetonitrile solvent than in nonpolar cyclohexane solvent [34]. Unactivated dienes have been observed to undergo 1,4 cycloadditions with enamines also [271--273]. The 4a-azonioanthracene ion (103) readily undergoes a 1,4 cycloaddition reaction with nucleophilic dienophiles such as enamines [274]. The cycloaddition is stereoselective, so that the α- and β-carbon atoms of the enamine add to the 10- and 9-positions, respectively, of the 4a-azonionanthracene ion. An unusual [3+3] cycloaddition takes place between cyclohexanone enamines and N-ethoxyquinolinium salts [275]. Fulvenes undergo both intermolecular [276,277] and intramolecular [278] cycloadditions. This latter case is a [6+2] cycloaddition in which the product was obtained in a 74% yield (see Scheme 10) [278].

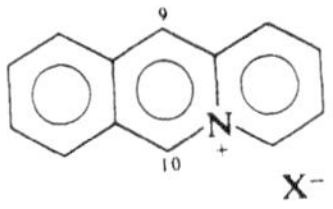

(103)

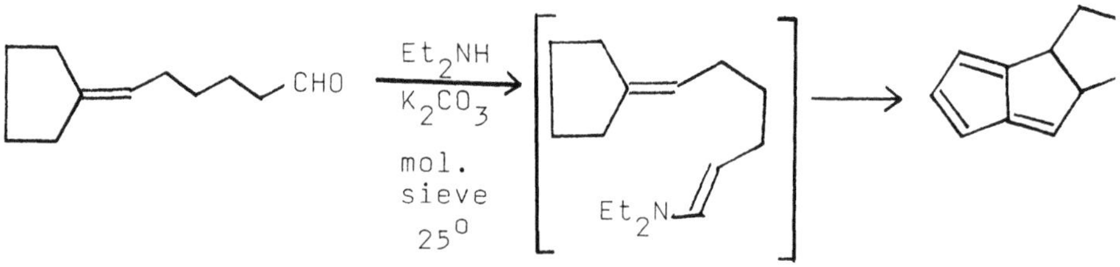

Scheme 10

Enamines of cyclohexanone react with tetracyanoethylene to give

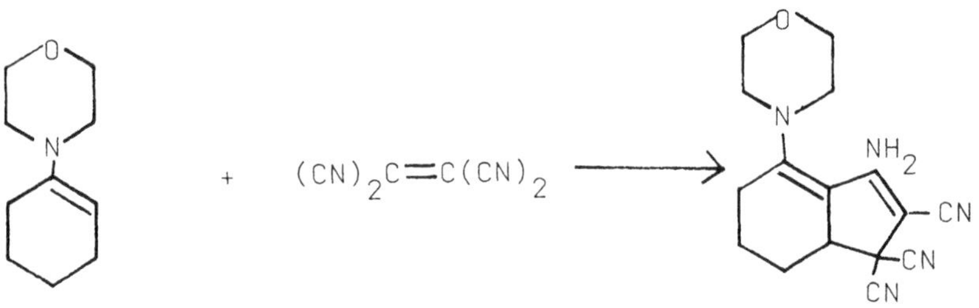

(104)

tetrahydroindene 104 in a 75% yield [279]. This is one of the first reports showing γ-carbon reactivity of an enamine.

B. Electrophilic Cyclopropanes and Cyclopropenes

The similarity between the reactions of alkenes and cyclopropanes is further demonstrated by the reactions of electrophilic cyclopropanes and cyclopropenes with enamines. Cyclopropylcyanoester, when treated with the pyrrolidine enamine of cyclohexanone, undergoes what would be a 1,2 cycloaddition in the analogous alkene case, but is actually a 1,3 cycloaddition here, to form adduct 105 [280]. A similar reaction between the heterocyclic three-membered ring, N-

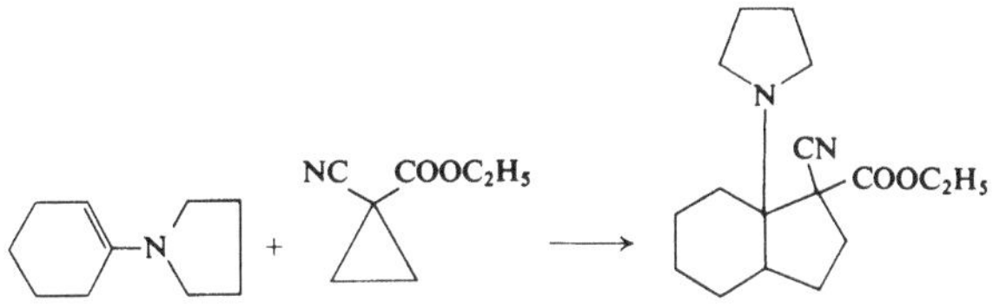

(105)

carbethoxyaziridine, and this enamine to form a heterocyclic product has been observed (see Section IV.C) [281].

This cycloaddition of an enamine to an activated cyclopropane has been observed with two nitrile groups doing the activating of the cyclopropane ring to produce adduct 106 [282]. Product 106 is

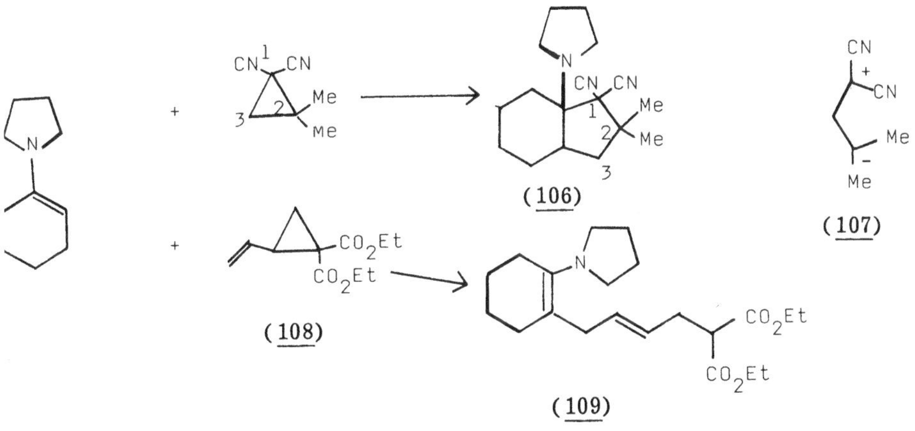

(106)

(107)

(108)

(109)

formed by an S_N2 mechanism with the nucleophilic β-carbon of the enamine attacking the 3-position of the cyclopropane ring and the α-enamine carbon adding to the cyclopropane's 1-position. If a zwitterion intermediate (107) were involved, 1,2-bond breaking would have occurred in the cyclopropane with the formation of a different isomeric product [282]. Homoconjugate addition takes place with activated cyclopropane 108 to form 109 [283–285]. In xylene solvent this reaction takes place by an S_N2' mechanism [282].

The reaction of a methylenecyclopropene molecule (triafulvene)

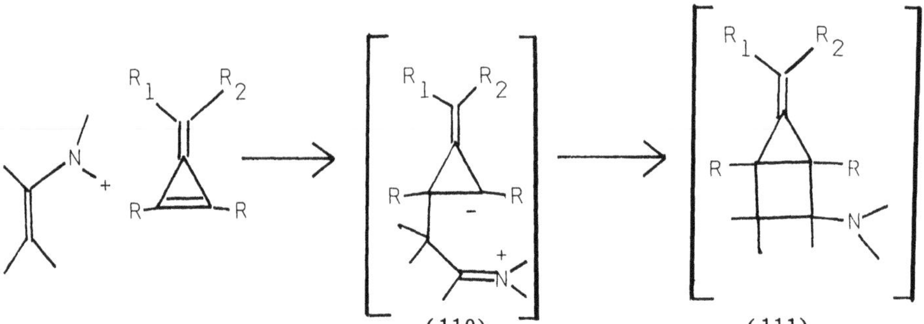

(110)   (111)

with an enamine has a relatively simple reaction sequence passing through zwitterion intermediate 110 and bicyclo intermediate 111 [288]. What product is formed from the reactive intermediate 111 depends on the substituents on the terminal carbon of the exocyclic double bond, R_1 and R_2. Dicyanomethylenecyclopropenes ($R_1 = R_2 = CN$) when treated with an enamine will have sigma-bond cleavage of both the

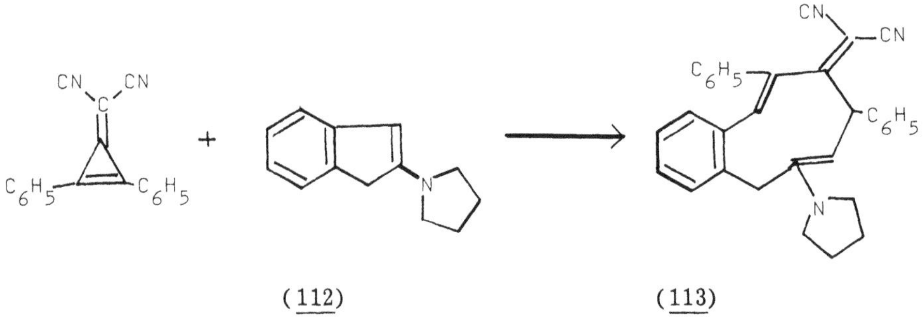

(112)   (113)

three- and four-membered ring in intermediate 111 in a conrotatory electrocyclic reaction [289,290]. For example, dicyanomethylenecyclopropene will react with enamine 112 to form ring-enlarged product 113 in an 80% yield [290]. In the case of two acyl groups being situated on the exocyclic methylene carbon ($R_1 = R_2 = -\overset{\overset{\text{O}}{\|}}{C}R'$ in intermediate 111), only the three-membered ring opens and the acyl group oxygen

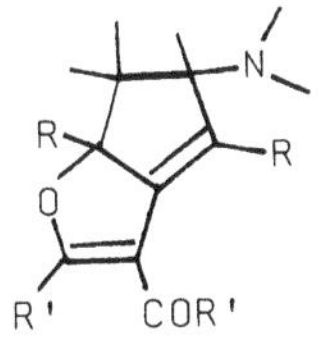

(114) 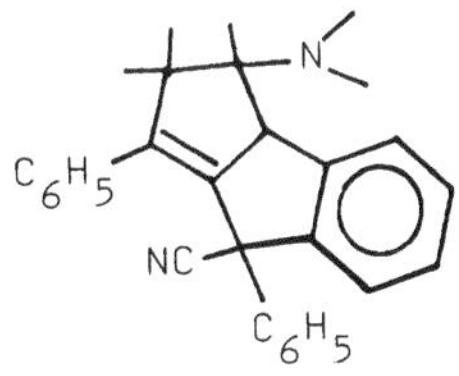   (115)

cyclizes to produce dihydrofuran 114 [291]. When R = R_1 = phenyl
and R_2 = -CN, then the three-membered ring opens and one of the
phenyl groups attacks the terminal methylene carbon yielding product
115 [292].

Cyclopropenes add to dienamines in a [2+4] cycloaddition manner

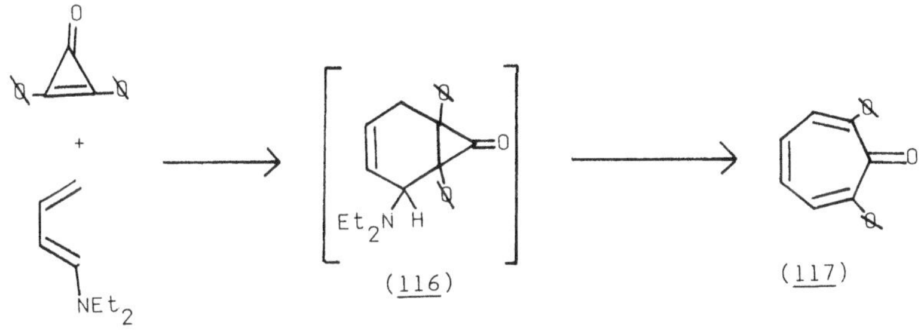

(116)   (117)

(116)   (117)

to produce a bicyclic intermediate which ring-opens. For example,
diphenylcyclopropenone reacts with 1-(N,N-diethyl)-1,3-butadiene
giving bicyclo intermediate 116, which ring-opens and eliminates diethyl amine to give 2,7-diphenyltropone (117) [200].

The first reports of the reaction between cyclopropenones and
simple enamines indicated a fairly straightforward cycloaddition reaction [199,200]. The same was reported for cyclopropenethiones and
simple enamines [293]. However, this type of reaction can give four
different products with both cyclopropenones and cyclopropenethiones
[288,294–303]. The initial intermediate is ylide 118a and betaine 118b
(see Scheme 11). The four reactions and their products are as follows: C,N-insertion (insertion of cyclopropenone into enamine C—N
bond) to give an amide (119); C,C-insertion (insertion of cyclopropenone into enamine C=C double bond) to give a β-aminodienone (120);
addition (Stevens rearrangement of ylide 115a) to give α-aminoketone

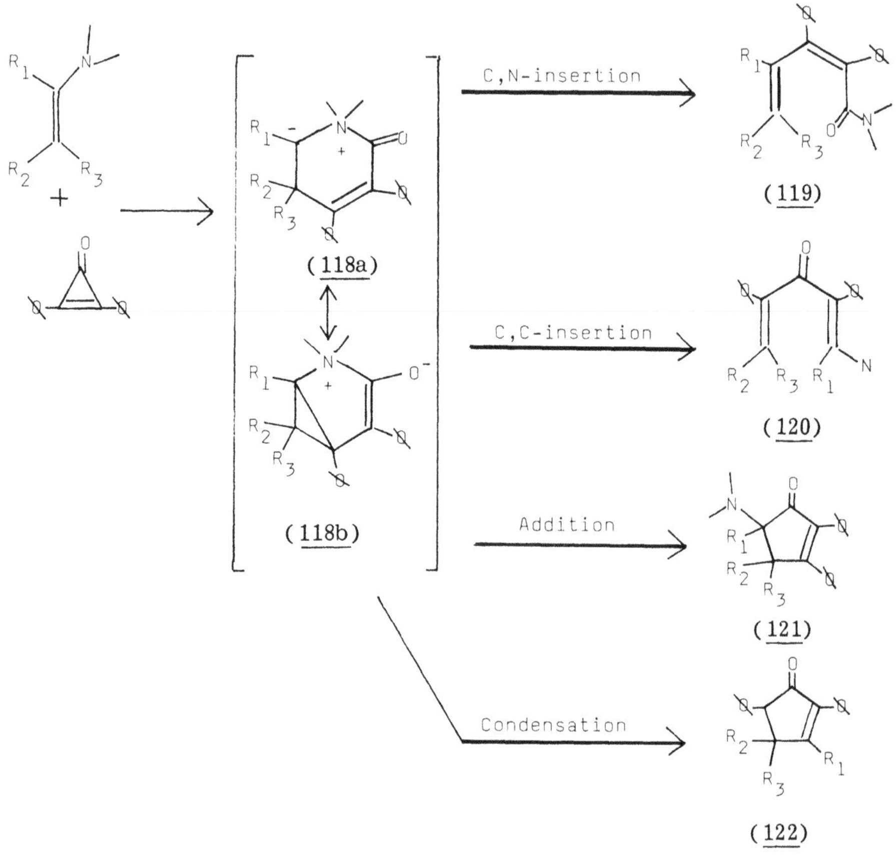

Scheme 11

(121); condensation to give a cyclopentenone (122) (see Scheme 11). With an excess of cyclopentenone, 2:1-adducts (cyclopentenone:enamine) are often formed [300].

C. Electrophilic Alkynes

Terminal alkynes with no electron-withdrawing group next to the acetylenic linkage when treated with enamines merely add across the double bonds of the enamines [304]. But electrophilic alkynes (those with an electron-withdrawing group next to the acetylenic linkage) undergo cycloaddition reactions with enamines. The most commonly used electron-withdrawing groups on the alkyne are carboxylate esters.

Solvent polarity is an important factor in determining the reaction pathway of an electrophilic alkyne and an enamine [305–313]. In an apolar solvent, dimethyl acetylenedicarboxylate undergoes initial 1,2 cycloaddition with acyclic [314,404], alicyclic [153,305,306,309,311, 314–325], or heterocyclic [326,362] enamines. The mechanism for this [2+2] cycloaddition reaction in an apolar solvent has been demonstrated to be *concerted*, giving the *cis cycloadduct* [326]. This cycloadduct then undergoes a electrocyclic ring opening. In the case of alicyclic enamines, the net result is a ring enlargement product two members larger than the original [327]. This outcome of this sequence of reactions seems to be dependent on the nature of the amine moiety in the enamine [323].

For example, treatment of 1-(N-pyrrolidyl)cyclohexene with dimethyl acetylenedicarboxylate gives cycloadduct 123 [326]. Cycloadduct 123 rapidly undergoes a symmetry-allowed conrotatory ring opening to *cis,trans*-2,8-cyclooctadiene (124). This compound in turn thermally changes to 125 at room temperature; or it photochemically changes to 126, possibly via a concerted, suprafacial [1,5] sigmatropic hydrogen shift (see Scheme 12). Compound 126 changes at room tem-

Scheme 12

perature to 125 [326]. In some cases the dienamine ring expansion product undergoes a 1,4 cycloaddition with a second molecule of dimethyl acetylenedicarboxylate [328].

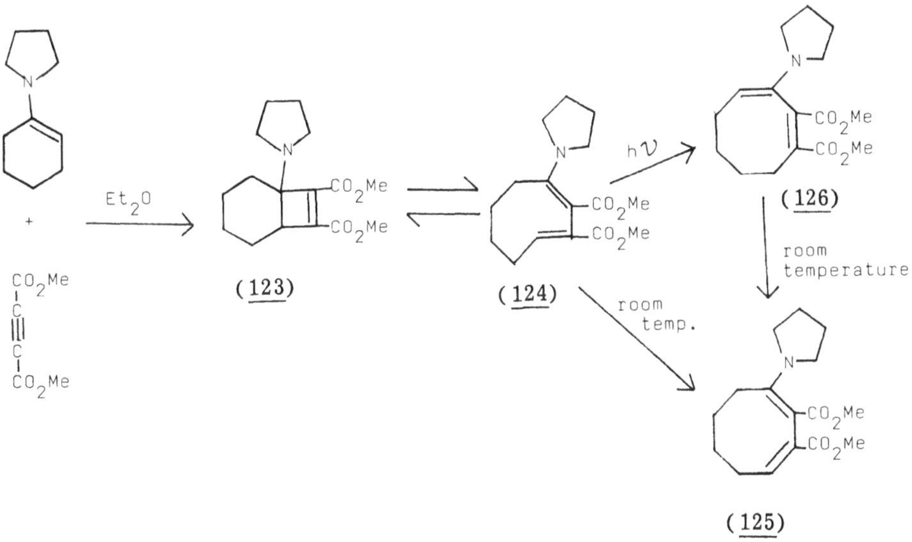

In a polar protic solvent such as methanol, the reaction of dimethyl acetylenedicarboxylate with enamines gives pyrrolizines [308–313,326]. For example, 1-(N-pyrrolidyl)cyclopentene reacts with dimethyl acetylenedicarboxylate to form pyrrolizine 127 [308,310,326].

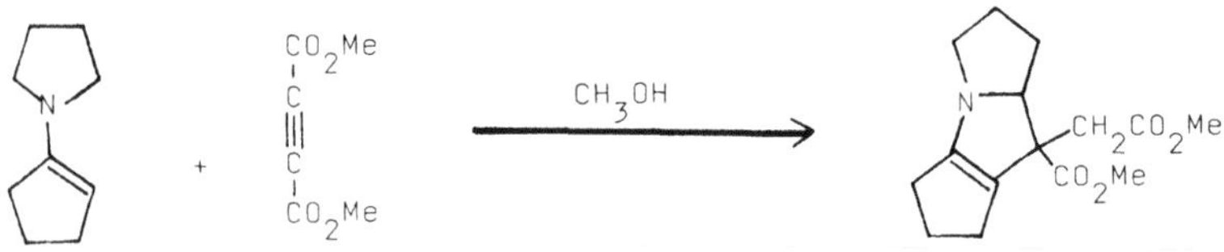

(127)

This reaction involves zwitterion intermediate 128. The mechanism

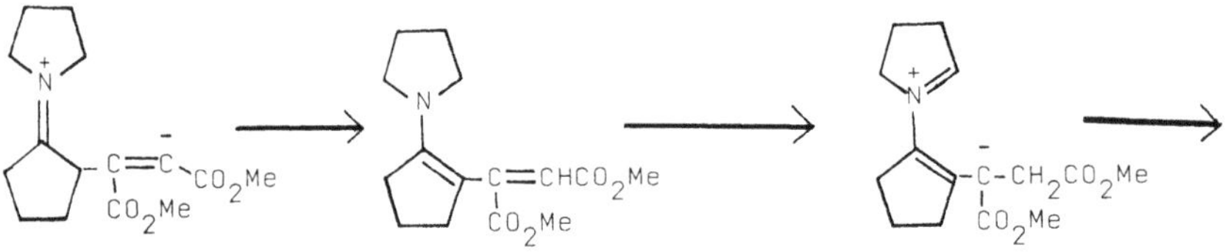

(128)   (129)   (130)

Scheme 13

is shown in Scheme 13. It involves two consecutive hydrogen transfers, the first from the solvent and the second intramolecularly [309, 313]. Protonation by the solvent of 128 produces 129, which undergoes an antarafacial [1,6] hydrogen shift to give dipolar 130 [313]. Then 130 undergoes a concerted, disrotatory electrocyclic, stereospecific, and symmetry-allowed reaction to yield pyrrolizine 127 [313, 326].

Dimethyl acetylenedicarboxylate will also undergo [2+2] cycloaddition reactions followed by ring expansions with indoles [329], pyrrolidine enamine of thiochroman-4-one [330,331], and N-methyl-1,2-dihydropyridine [79]. Treatment of some dimethylamino fulvenes with dimethyl acetylenedicarboxylate results in elimination of dimethylamine and cyclization to form azulenes [332] or pentalenes [333,334].

The photochemical reaction of diphenylacetylene with enamines to give cycloaddition products and ring expansion products has been studied [335].

The reaction of methyl propiolate with acyclic enamines produces acyclic dienamines [336], as was the case with dimethyl acetylenedi-

carboxylate. The treatment of the pyrrolidine enamines of cycloheptanone, cyclooctanone, cycloundecanone, and cyclododecanone with
methyl propiolate results in ring enlargement products [336—338,661].
When the enamines of cyclohexanone are allowed to react with methyl
propiolate, rather anomalous products are formed [336,661]. The pyrrolidine enamine of cyclopentanone forms stable 1,2 cycloaddition adduct 131 with methyl propiolate. Adduct 131 rearranges to the simple
alkylation product 132 upon standing at room temperature, and heating
131 to about 90°C causes ring expansion to 133 [315,336,665].

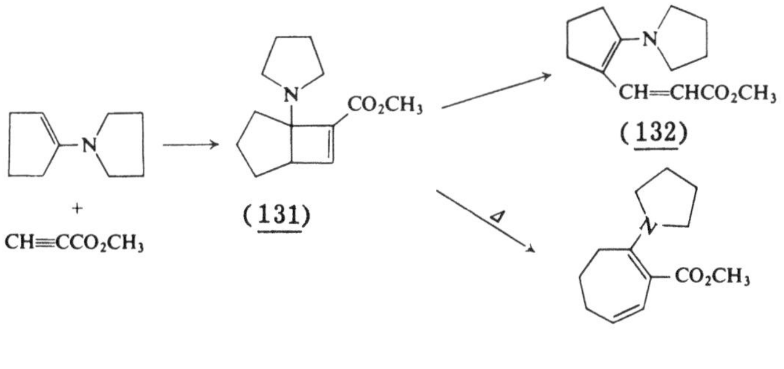

The reaction of ethyl propynoate with 1-(N,N-dimethylamino)bi-
cyclo[2.2.1]heptene in refluxing diethyl ether produces cycloadduct

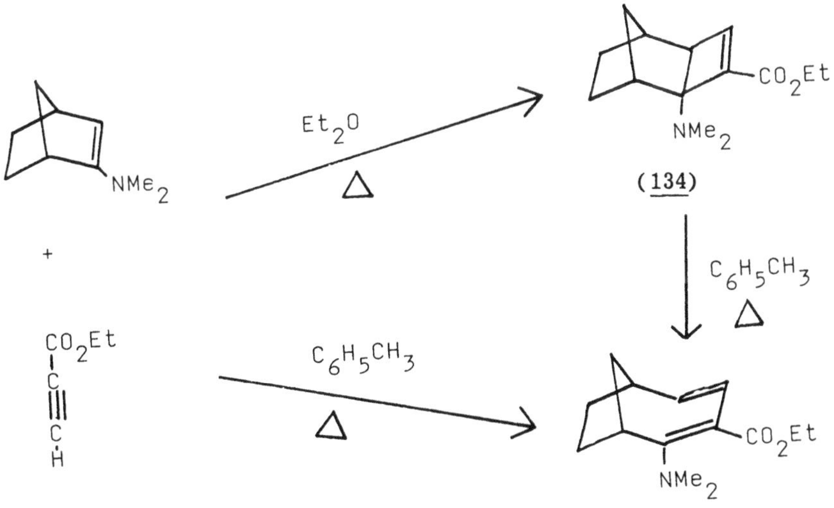

134, a remarkably stable compound that remained intact after 20 hr
in refluxing ether [339]. When 134 was refluxed in toluene solvent
or if the original reaction was carried out in refluxing toluene solvent,
then ethyl 2-(N,N-dimethylamino)bicyclo[4.2.1]nona-2,4-diene-3-
carboxylate (135) was produced.

Methyl phenylpropiolate [328] and nitroacetylenes [340] undergo
similar reactions with enamines.

The reaction of an alicyclic enamine with benzyne intermediate
yields simple arylation products and/or 1,2 cycloaddition products,
depending on the reaction conditions [341]. This is illustrated by
the reaction of 1-(N-pyrrolidino)cyclohexene with benzyne (obtained
from fluorobenzene and butyl lithium or o-bromofluorobenzene and
lithium amalgam), which produces benzocyclobutene 136 [341].

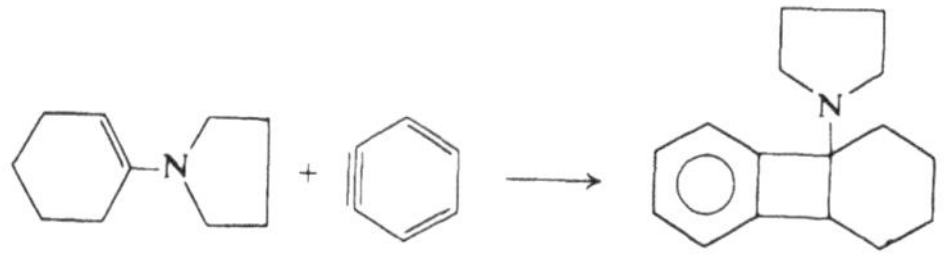

(136)

D. Divalent Carbon

Addition of dichlorocarbene to the enamine of cyclohexanone gives a
relatively stable adduct 137 (n = 4) [342−346]. Hydrolysis of this

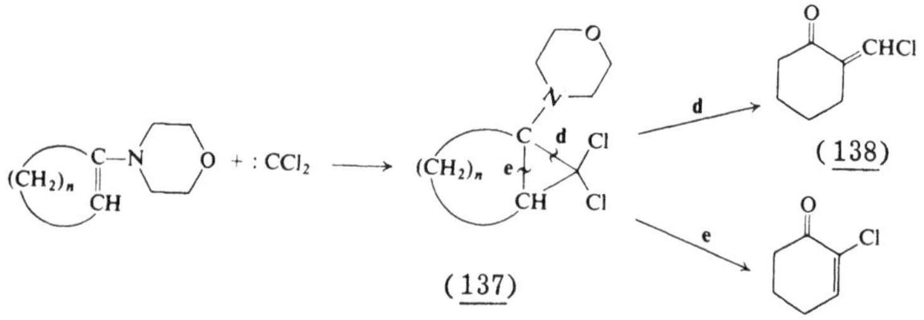

(138)

(137)

(139)

adduct causes cleavage at d and formation of ketone 138 [344]. When
the enamine of cyclopentanone is treated with dichlorocarbene, the
adduct (137, n = 3) is unstable, and cleavage at e occurs to form
ring-expanded product 139 [343,344]. Dichlorocarbene also adds to
other alicyclic enamines to form similar cycloaddition products [347−
350]. Acyclic enamines also form unstable cycloadducts with dichlo-

rocarbene, which are readily hydrolyzed through an *e*-type cleavage to produce α-choro-α,β-unsaturated aldehydes [345,346,351].

Dibromocarbene [349], phenylchlorocarbene [347—349], and fluorochlorocarbene [349,350] have all exhibited reactions with enamines similar to those observed with dichlorocarbene. A variety of conditions have been used to produce these halocarbenes. These methods include thermal decomposition of sodium trichloroacetate in 1,2-dimethoxyethane (glyme,DME) [347—350], treatment of a haloform with a strong base such as potassium *t*-butoxide or butyllithium in an aprotic solvent [343,344,347—349], treatment of a haloform in a phase transfer reaction with NaOH in water and triethylbenzylammonium halide as the transfer agent [345,346,349], and either treatment with sodium iodide in DME [349,350] or thermal decomposition [349] of phenyltrihalomethyl mercury.

Addition of a methylene group to an enamine to form aminocyclopropanes can be brought about by the use of carbenoids from diiodomethane and zinc-copper couple [352] or diethylzinc [353,354], or from *bis*(iodomethyl)zinc [355]. This can also be accomplished by allowing an enamine to react with diazomethane in the presence of cuprous halide [349,354,356].

Cycloadditions of diphenylcarbene (from catalyzed decomposition of diphenydiazomethane) [357,358], of thiocarbene [359], of methoxyphenylcarbene [from pentacarbonyl(methoxyphenylcarbene)chromium] [360,361], and of CHCO2Et (from ethyl diazoacetate and copper) [348, 349] to enamines to form cyclopropylamines have been reported.

E. Miscellaneous

Stereochemical positioning of a functional group, relative to a separate enamine moiety in the same molecule, can be done in such a manner that a simple intramolecular alkylation or acylation will cause cyclization. Such intramolecular cycloalkylations with alkyl halides have been reported [363,364,663]. Intramolecular cycloacylations of enamines with esters [365—367] and with nitriles [367—369] have also been observed.

Bifunctional molecules undergo intermolecular cyclizations with enamines through simple alkylations [370—379], metal-catalyzed alkylations [380—384], and acylations [385]. For example, the reaction between 1-(N-pyrrolidino)cyclopentene and 1,4-diiodobutane produces, after hydrolysis, ketospirans 140 and 141 [371]. A novel

(140)　　(141)

[3+2] cycloaddition reaction between α,α'-dibromoketones and morpholino enamines is catalyzed by iron carbonyls [381–384]. This is illustrated by the reaction of 1-(N-morpholino)cyclohexene with 2,4-di-

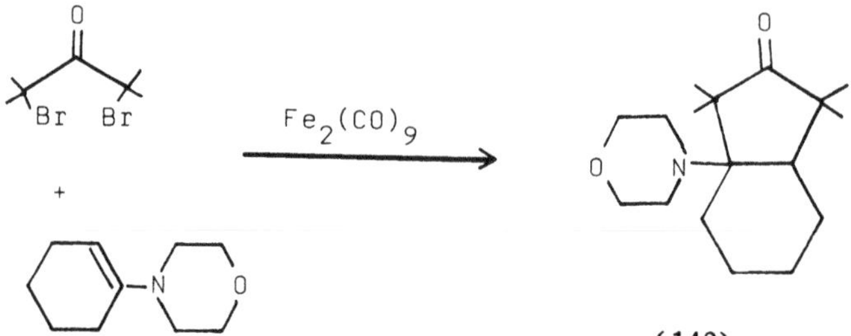

(142)

bromo-2,4-dimethylpentan-3-one to produce cycloadduct 142 in an 87% yield [383]. Treatment of enamine 143 with 1,3-dichloro-1,3-dimethoxypropane (144) resulted in a "one-pot" synthesis of 2,6-disubstituted aniline 145 [377].

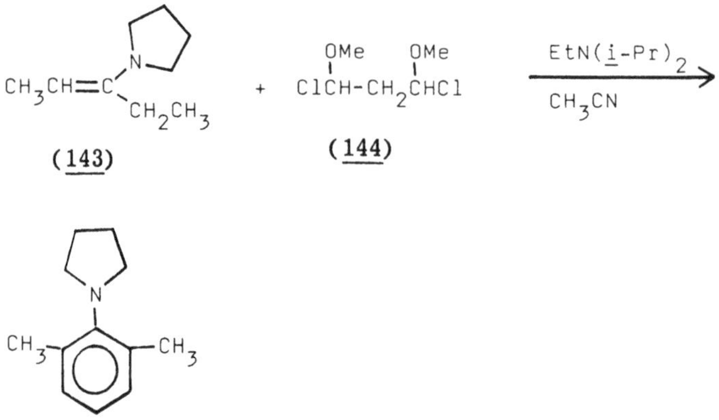

(143)   (144)

(145)

Methyl 4-trimethylsilyl-3-dialkylaminocrotonate (146) reacts with enamines from cyclic ketones (cyclopentanone through cyclooctanone) in the presence of trifluoroacetic acid (TFA) in a [4+2] cycloaddition when the dialkylamino group in 146 is pyrrolidine to give an aromatic compound (for example, 147 in Scheme 14); or the enamine adds at the γ-position of 146b when the dialkylamino group is morpholine (for

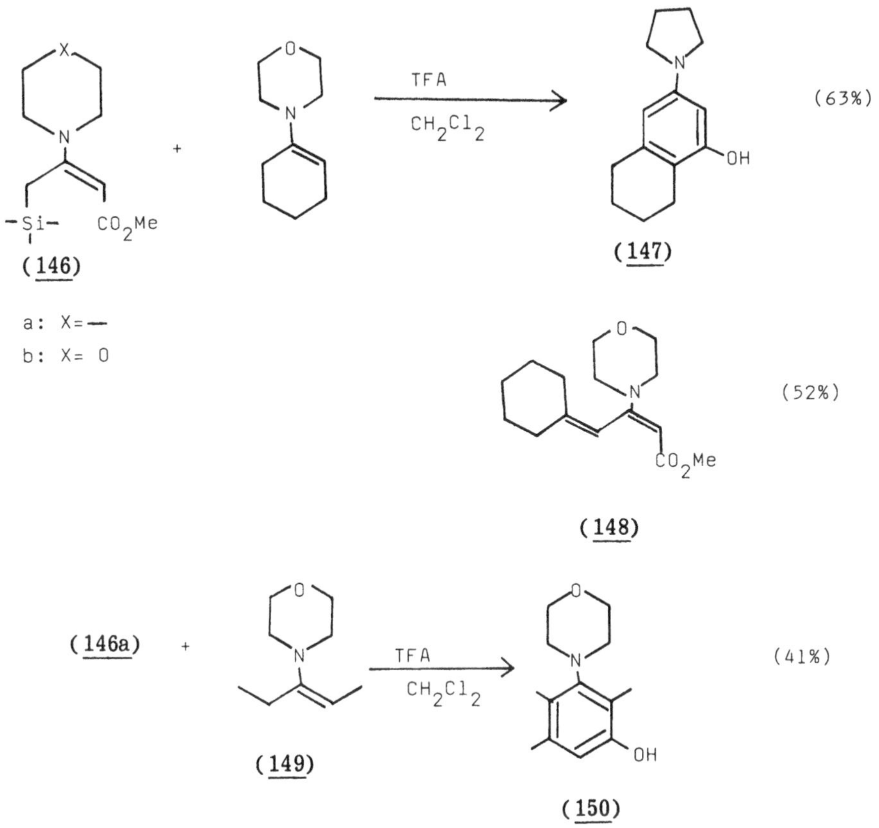

(147)

(63%)

(148)

(52%)

(150)

(41%)

Scheme 14

example, 148 in Scheme 14) [387]. Treatment of 146a with an enamine from an acyclic ketone such as 149 in the presence of TFA produces a [3+3] cycloaddition to give aromatic aminophenol 150 [387].

The reaction between *sym*-trinitrobenzene, acetone, and diethyl-amine results in the formation of a red Meisenheimer complex [388] of structure 151 by way of enamine cycloaddition to *sym*-trinitrobenzene

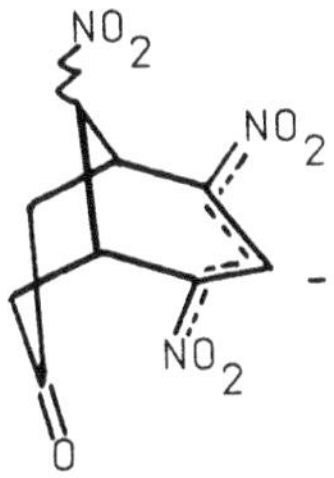

(151)

[389–391]. Treatment of 3,5-dinitroacetophenone with acetone and diethylamine gives 1-methyl-3-diethylamino-5,7-dinitronaphthalene instead of a Meisenheimer addition product [392].

Enamines containing one β-hydrogen atom react with the lactone dimer of dimethylketene to form aminocyclohexanediones [393]. Polycondensation of acetone diethyl ketal takes place by treating it with morpholine and a catalytic amount of p-toluenesulfonic acid while distilling off the ethanol formed [394–396]. The resulting spiran, bicyclo, and cyclooctadienone products differ from the known polycondensation products of acetone, and hence their formation probably involves enamine intermediates [396].

Irradiation of dienamine 152 for 3 days results in the formation of 153 through a 1,4 cycloaddition [397].

(152)   (153)

It has been reported that perfluoroisobutylene and perfluorocyclobutene undergo 1,2 cycloaddition with 1-(N-morpholino)isobutylene in ether at room temperature to give the corresponding perfluorocyclobutane derivatives [398]. Enamines of cyclic ketones produce only simple alkylation products when treated with these perfluoroolefins.

IV. HETEROCYCLOADDITIONS

A. Oxygen Heteroatom

Dihydropyrans have been produced by the 1,3 cycloaddition of methyl vinyl ketone or acrolein with enamines (see Section III.A). δ-Lactones have been formed as a side product in the reaction of dimethyl ketene with enamines [242], and as the primary products in the reaction of excess ketene with enamines derived from ketones (see Section III.A) [235].

p-Quinone (154) undergoes 1,3 cycloaddition, di-1,3 cycloaddition, or both when it is treated with enamines to form benzofuranols [191,399–404] and/or benzodifurans [405,406]. For example, when 1-(N-morpholine)-1-phenylethene (155) is allowed to react with p-quinone (154), benzodifuran 156 is formed [406]. Adduct 158 is formed

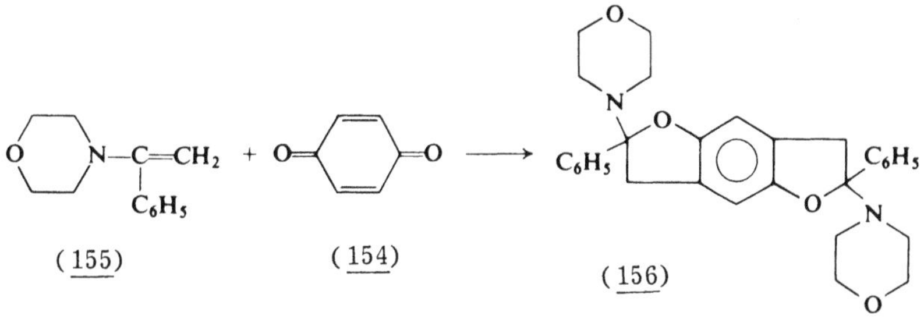

(155) (154) (156)

from the 1,4 cycloaddition of o-quinone (157) with the morpholine enamine of cyclohexanone [407]. 1,3-Dioxanes are produced when

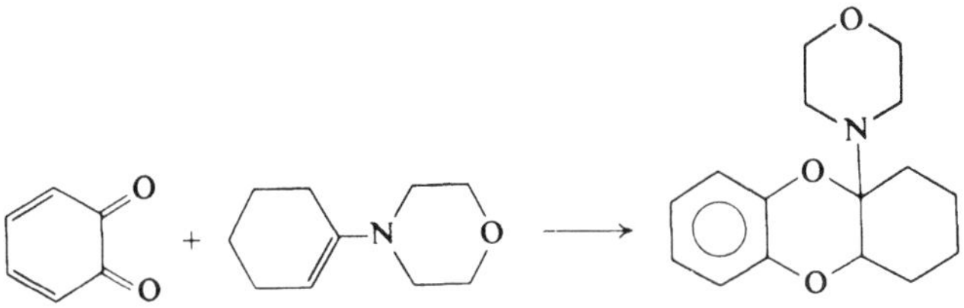

(157) (158)

the morpholine enamines of cyclohexanone and cyclopentanone are allowed to react with 2 moles of chloral [408].

Treatment of styrene oxide with cyclic enamines at elevated temperatures (about 230°C) produces α-aminotetrahydrofurans in excellent yields [409]. The reaction of cyclic enamines with α-bromoketones gave furan derivatives [410]. The photoreaction of enamines with *tris*(2,4-pentanedionato)cobalt(III) (159) gives α-aminodihydro-

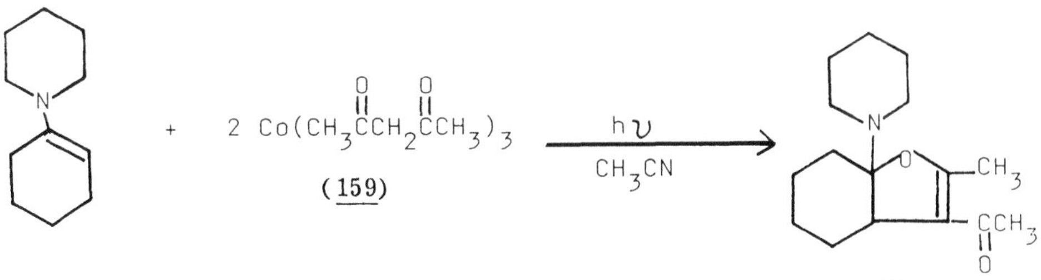

(159) (160)

furan derivative 160 [411]. α-Aminodihydrofurans are also obtained from the reactions of 2-chlorotropone with enamines [412].

The synthesis of a large number of γ-pyrones, γ-pyranols, and
α-aminopyrans from enamines has been brought about through the use
of a wide variety of bifunctional molecules [413—436]. These molecules include phenolic aldehydes and ketones [413—419], phenolic
Mannich bases [420—423], o-hydroxy-w-nitrostyrenes [424], keto and
ketal esters [425,426], acyl halides with dienamines [135], and diketene [142,258,427,434—436]. Most of these molecules have an electrophilic carbonyl group and a nucleophilic oxygen center in relative 1,4-
positions. This is illustrated by the reaction between salicylaldehyde
(161) and the morpholine enamine of cyclohexanone to give pyranol
162 in a quantitative yield [414].

(161)

(162)

Dihydro-1,3-oxazine-4-ones and -4-thiones have been produced
by treating enamines with acyl isocyanates [430,431] and acyl isothiocyanates [432], respectively. The reaction between the piperidine
enamine of dimedone and β-nitrostyrene yields a 1,2-oxazine derivative [433].

A cyclic peroxide is obtained when an enaminoester is treated with
hydrogen peroxide [437]. The dye-sensitized photooxygenation of
enamines, resulting in the ultimate formation of a ketone or aldehyde
and an amide by cleavage of the double bond, probably involves a
cyclic peroxide intermediate (see Chapter 5, Section IV.A).

B. Sulfur Heteroatom

Elemental sulfur undergoes nucleophilic attack by amines at low temperatures. Therefore, the conjugate β-position of an enamine is
sufficiently nucleophilic to attack elemental sulfur and yield thiolated
intermediate 163. When 163 is treated with phenyl isothicyanate, the
cyclic adduct 164 is formed [438]. In a similar manner, 163 produces
an aminothiazole when treated with cyanamide [439]. A thiazolidone
is formed when 163 is treated with an isocyanate (see Section IV.C)
[440]. Refluxing the morpholine enamine of cyclohexanone and elemental sulfur in benzene solvent results in the formation of hydrogenated thianthrene 165 [441].

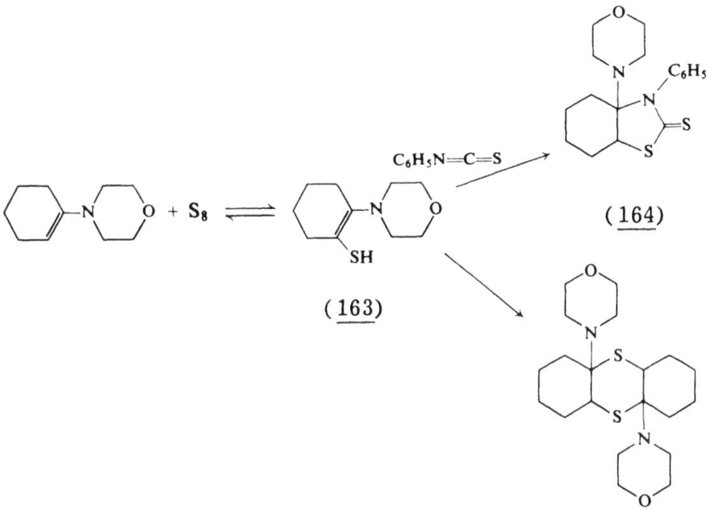

(163)

(164)

(165)

Many acyclic enamines form thioamides when they are allowed to react with elemental sulfur at room temperature in dimethylformamide solvent [442]. They also produce cyclic 1,3-dithiole by-products, which become main products at higher temperatures [439]. For example, the reaction of 1-(N-morpholino)-1-phenylethene (166) and sulfur in dimethylformamide solvent yields 1,3-dithiole 167.

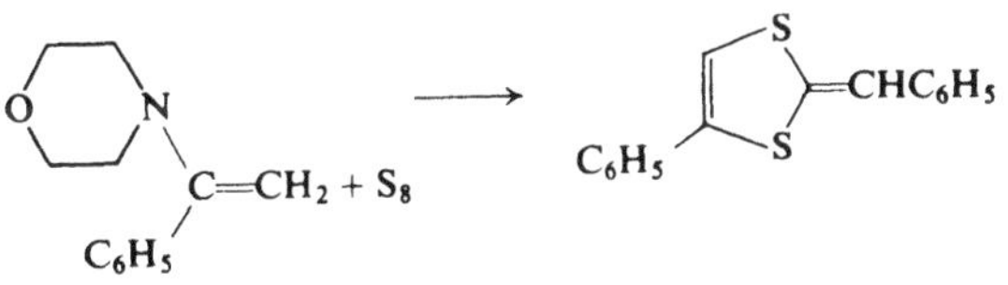

(166)

(167)

Carbon disulfide can act as an electrophilic agent with enamines at room temperature. Therefore, treatment of an enamine with both elemental sulfur and carbon disulfide in a polar solvent can result in the formation of a 3H-1,2-dithiole-3-thione (such as 168) and/or a 2H-1,3-dithiole-2-thione (such as 169) [439,443,444]. These products are the result of competing reaction paths, each of which is initiated by a different electrophilic agent, namely, sulfur and carbon disulfide. All aliphatic and many aromatic enamines give 3H-1,2-dithiole-3-thiones as the exclusive products when treated with sulfur and car-

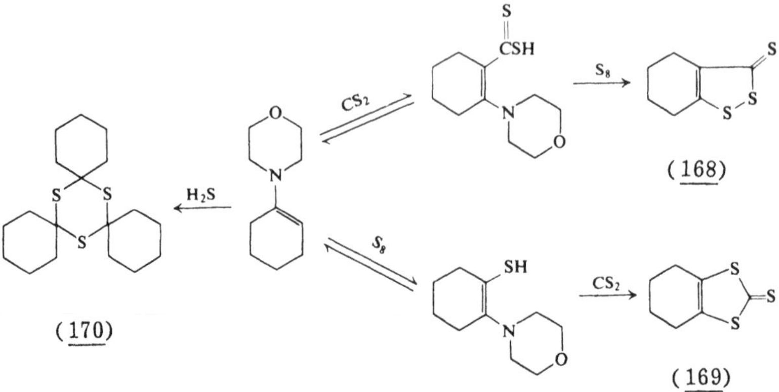

(168)

(170)

(169)

bon disulfide. Some aromatic enamines, however, yield 2H-1,3-di-
thiole-2-thione products [439]. A 1,3-dithiolethione product is also
obtained by allowing an enamine to react with thiuram disulfides and
hydrogen sulfide [439,445]. Treatment of the morpholine enamine of
cyclohexanone with hydrogen sulfide alone gives cyclic product 170
[446].

A by-product of the reaction between an enamine, elemental sulfur, and carbon disulfide is an α-dithiopyrone. This by-product is
the result of the condensation of two enamine molecules with one carbon disulfide molecule. In the case of aldehydic enamines, the reaction probably proceeds through a carbon-disulfide-catalyzed enamine
condensation to form a dienamine. Carbon disulfide then adds to the
dienamine, yielding an α-thiopyrone. This last step is illustrated by
the reaction between dienamine 171 and carbon disulfide to produce α-
dithiopyrone 172 [439]. Enamines of aliphatic methyl ketones and aro-

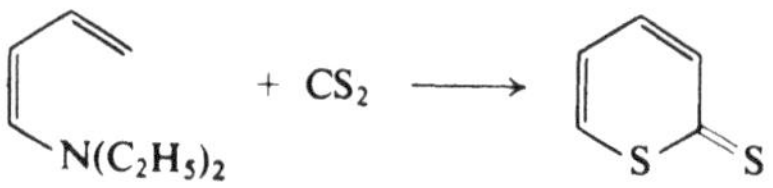

(171)   (172)

matic methyl ketones apparently add carbon disulfide first and then
condense with a second enamine molecule to give α-dithipyrones without passing through the dienamine stage [447]. Enamines of cyclic ketones do not yield α-dithiopyrone products when treated with carbon
disulfide.

The formation of α-dithiopyrone by-products during the reaction
of an enamine with elemental sulfur and carbon disulfide is enhanced

by one or a combination of the following: The carbon disulfide is allowed to stand for a long period of time with the enamine in the absence of sulfur, a high reaction temperature, and the use of a relatively nonpolar solvent [439].

α-Thiopyranthiones can be synthesized from enamines by treating the enamine with 1,2-dithiole-3-thiones [448,449]. For example, the reaction of the pyrrolidine enamine of cyclopentanone with 1,2-dithiole-3-thione 173 gives α-thiopyranthione 175 in a 45% yield, probably via intermediate zwitterions 174a and 174b [449]. Thioacetal derivatives of γ-thiopyranthione have been produced by allowing o-thioquinone methide (176) to react with an enamine to give a [4+2] cycloadduct [450,451].

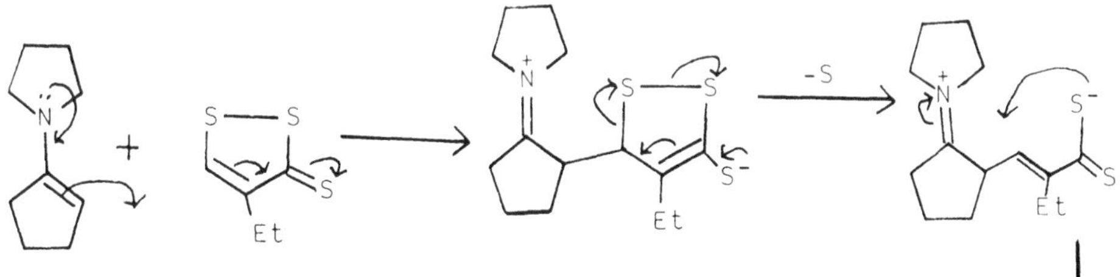

(173)   (174a)   (174b)

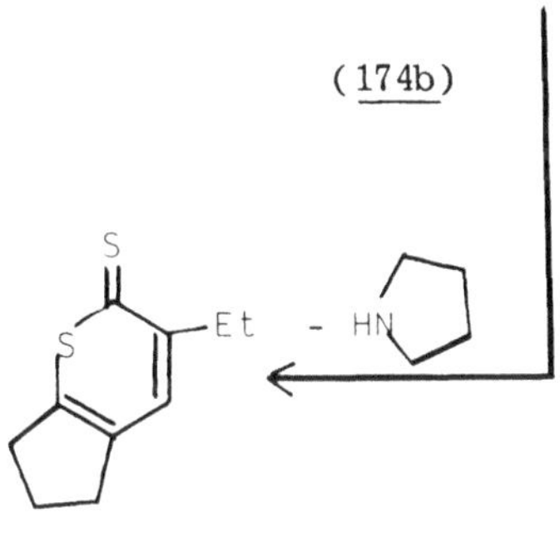

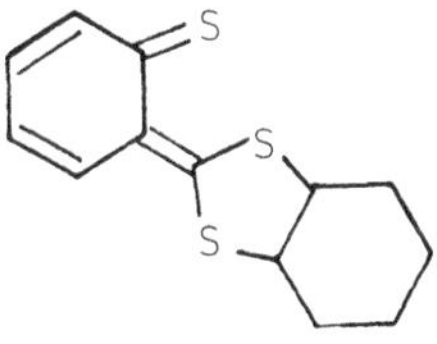

(176)

(175)

The reaction of isothiocyanates (177) with β,β-disubstituted enamines gives an equilibrium mixture of [2+2] cycloadduct 179 and zwitterion 178 (see Scheme 15) [452,453]. Which way the cyclization takes place (that is, whether thietane 179a or azetidine 179b is formed) depends on whether a sulfonyl group is present or not. In the case of cycloadduct 179b, a second mole of isothiocyanate 177b gives product 180 [453].

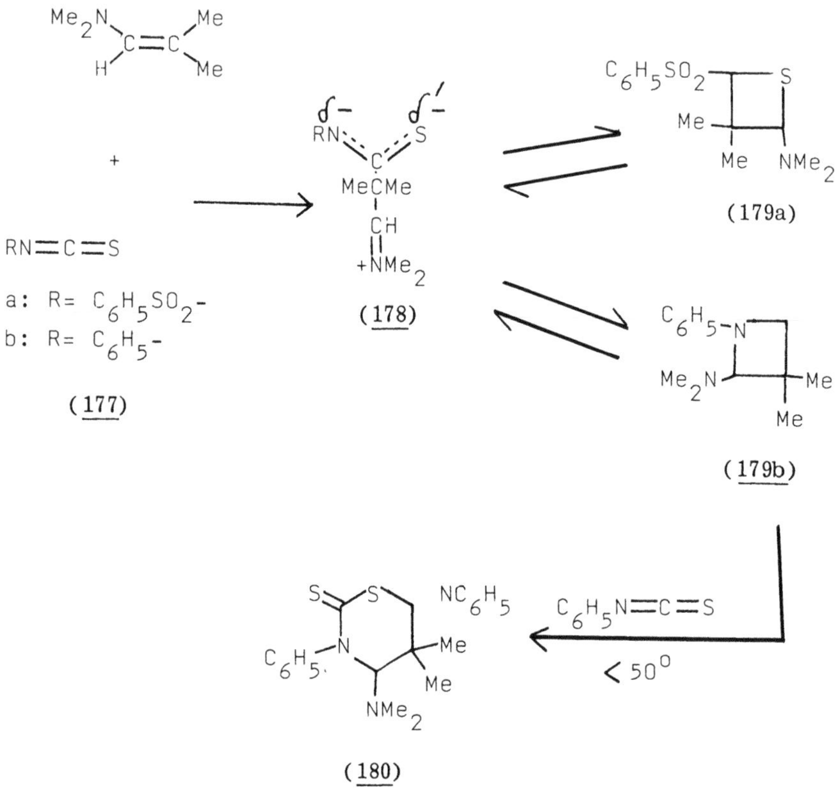

Me_2N Me
$C=C$
H Me

$+$

$RN=C=S$

a: $R= C_6H_5SO_2-$
b: $R= C_6H_5-$

(177)

$\xrightarrow{\hspace{1cm}}$

$\delta-$ $\delta'-$
$RN\cdots C\cdots S$
C
$MeCMe$
CH
$+NMe_2$

(178)

$C_6H_5SO_2$ — S
Me
Me NMe_2

(179a)

C_6H_5-N
Me_2N — Me
Me

(179b)

$S=$ S NC_6H_5
C_6H_5—N —Me
Me
NMe_2

(180)

$C_6H_5N=C=S$

$< 50°$

Scheme 15

Synthesis of five-membered sulfur heterocyclic 1,3-dithiol-2-thio-nenes from enamines, tetramethylthiourethanedisulfide, and hydrogen sulfide has been reported [454]. Azuleno thiophenes have been synthesized starting with the morpholine enamines of 3-ketotetrahydro-thiophene [455]. Tetrahydrothiophenes are the products of the reaction of thiiranimines with enamines [456].

When enamines are allowed to react with 2,3-diphenylthiirene 1,1-dioxide (181), simple cycloaddition can take place (e.g., cyclo-adduct 182), or ring expansion to a medium- or large-size sulfur-con-taining heterocycle can occur [457]. When a solution of 2,2-dimethyl cyclopropanium ion in fluorosulfonic acid is warmed to about 35°C, cycloaddition of sulfur trioxide takes place to give a six-membered sultone [458].

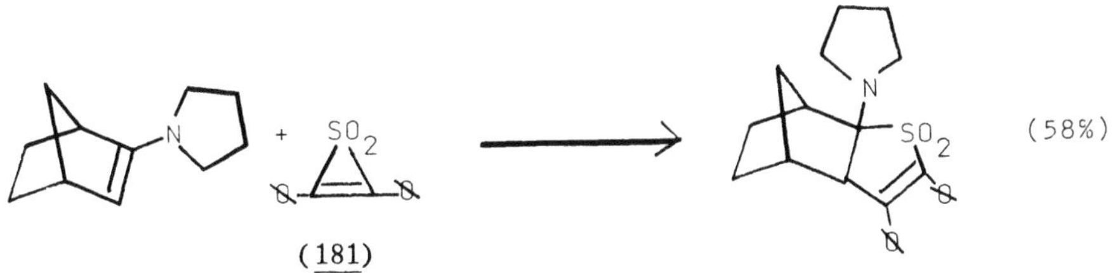

(181)

(182)

(58%)

Sulfonyl imides, RN = SO_2, are isoelectronic with sulfur trioxide and sulfenes. They can be produced in solution with enamines by treating N-alkylsufamoyl chloride with triethylamine [459]. With enamines having no β-hydrogens, a four-membered cycloadduct is formed by way of a zwitterion intermediate (see Scheme 16) [459].

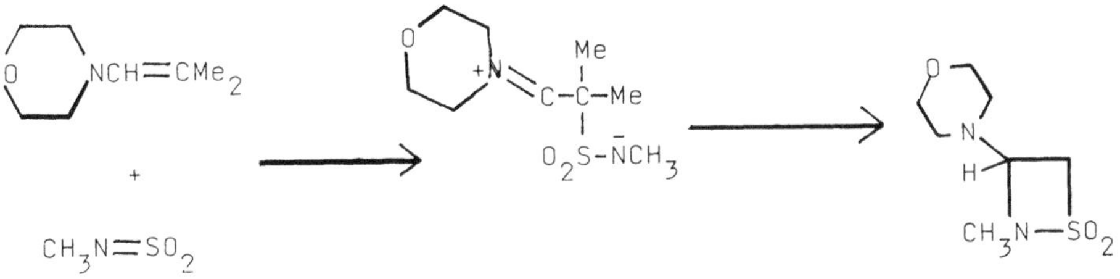

Scheme 16

Thiete 1,1-dioxide (183) reacts with the dimethylamine enamine of isobutryaldehyde to give 2-thiabicyclo[2.2.0]hexane 184 in a 60% yield [460]. With dienamines [4+2], cycloaddition takes place [460].

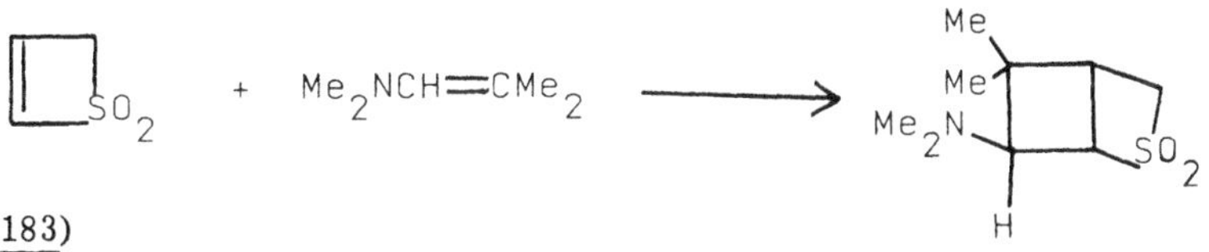

(183)

(184)

Sulfenes (R_2C = SO_2) are the sulfonyl analogs of ketene, and they can readily be generated by various means, including treatment of sulfonyl chlorides with bases, such as triethylamine treatment of α-chlorosulfenic acid with triethylamine [463], treatment of alkylsul-

fonyl fluorides with phosphoranes [494], and treatment of diazometh-
anes with sulfur dioxide [461,462]. It has been observed that sulfene
(usually generated in situ from methanesulfonyl chloride and triethylamine) undergoes 1,2 cycloaddition with enamines to form a β-amino-
thietane dioxide regardless of whether a β hydrogen is present in the
original enamine or not [222,353,404,463–486]. This is illustrated by

(185)

(186)

the reaction of enamine 185 with sulfene to produce adduct 186 in an
80% yield [466]. The product 186 was also observed in an 18% yield
from the reaction of diazomethane, sulfur dioxide, and enamine 185
[487]. It was demonstrated that this cyclization reaction must involve
sulfene adding to the enamine directly and not acylation of the enamine
by methanesulfonyl chloride followed by cyclization [465,468,487].
However, in some reactions acylation products were found along with
cyclization products [465,468,473]; in others the acylation product
was almost the exclusive product [467,488]. It is not entirely clear
whether the simple acylation product forms by direct acylation of the
enamine by a sulfonyl chloride, as is the case with phenyl sulfonyl
chloride, where no sulfene intermediate is possible [489], or whether
a cyclic sulfone or at least a zwitterion intermediate arising from sulfene addition is involved in the simple acylation. The latter appears
to be a strong possibility [488].

The cycloaddition of sulfene to bicyclo[2.2.1]heptyl enamines is
stereospecific, addition coming from the exo side [476,490]. However,
the steric preference of cis and trans isomers relative to the four-
membered ring generated does not seem as strong, at least in the
case of the addition of chlorosulfene ($ClCH = SO_2$) to bicyclic enamines, where a mixture of stereoisomers is obtained [491].

Nevertheless, the cis isomer is formed predominantly over the
trans isomer in the sulfene [2+2] cycloadducts with enamines even
though the trans isomer is thermodynamically more stable [462,474,
483]. The enamines of 4-t-butylcyclohexanone (187) show low stereoselectivity with methyl sulfenes [486]. Asymmetric induction has been
observed in the cycloaddition reaction of methyl sulfene to individual
enantiomers of chiral enamines [480].

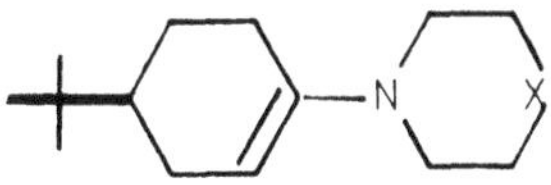

(187)   X= O-, CH₂-, —

The mechanism of the cycloaddition of sulfenes to enamines does not involve a concerted process in many if not all cases, but rather a two-step process in which a zwitterion is the initially formed intermediate [492,493].

Dienamines undergo 1,4 cycloaddition with sulfenes as well as 1,2 cycloaddition [222,353,477,482,495]. For example, 1-(N,N-diethylamino)butadiene (171), when treated with sulfene (generated from methanesulfonyl chloride and triethylamine), produces 1,4 cycloadduct 189 in an 18% yield and di-1,2-cycloadduct 190 in a 60% yield [495]. Cycloadduct 189 was shown not to be the precursor for 190 by treating 189 with excess sulfene and recovering the starting material unchanged [495]. This reaction probably takes place by way of zwitterion 188, which can close in either a 1,4 or 3,4 manner to form cycloadducts 189 and 191, respectively. The 3,4 cycloaddition would then be followed by a 1,2 cycloaddition of a second mole of sulfene to form 190. Cycloadduct 190 must form in the 3,4 cycloaddition followed by a 1,2 cycloaddition sequence rather than the reverse sequence since sulfenes undergo cycloaddition only in the presence of an electron-rich olefinic center [493]. Such a center is present as an enamine in 191, but it is not present in 192.

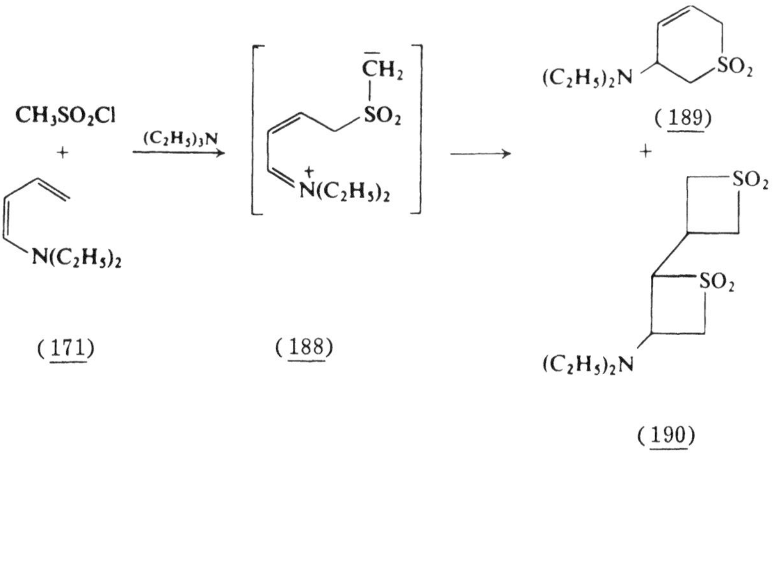

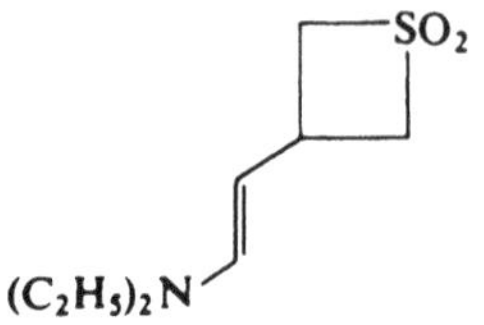

(191) 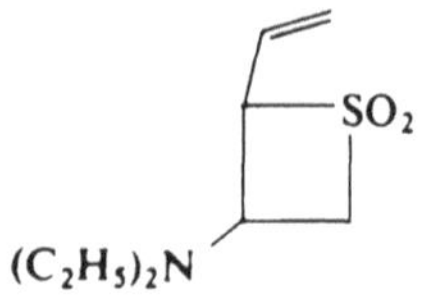   (192)

1,3-bis(Dimethylamino)-1-alkenes undergo similar reactions with sulfenes because of the possibility of elimination of the elements of dimethylamine to form a dienamine [493,496]. These 1,3-diaminoalkenes, when treated with sulfenes, also yield other products which are formed primarily because of the presence of a nonenamine tertiary nitrogen at C−3, which can compete with the neighboring enamine system for the electrophilic sulfene [492,493,497].

Enaminoketones undergo 1,4 cycloadditions with sulfene [258,498]. This is illustrated by the reaction of enamine 193 with sulfene to form sulfone 194 in an 80% yield [497,499].

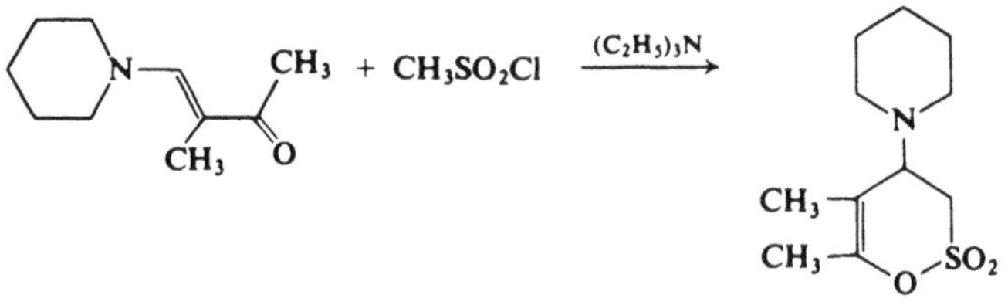

(193)   (194)

α,β-Unsaturated sulfenes, obtained by treatment of allyl sulfonyl chlorides with triethylamine, have been observed to undergo only 1,2 cycloaddition with enamines [500,501]. Simple sulfonylation products also resulted from these reactions.

Disulfenes react with 2 moles of enamine to produce a double 1,2 cycloaddition adduct [502].

The reaction of benzoylsulfene with enamines produces 2:1 cycloadducts [503].

Stable sulfenes have been isolated by treating methanesulfonyl chloride with triethylamine or trimethylamine in acetonitrile solvent at -40°C [474,504,505]. These stable sulfenes undergo 1,2 cycloaddition with enamines to form the expected thietanes (trimethylenesulfones).

C. Nitrogen Heteroatom

1,2 and 1,4 Cycloadditions

Treatment of enamines possessing β-hydrogens with isocyanates or thioisocyanates results in the formation of the corresponding carboxamides or thiocarboxamides [507–511]. The reaction of an enamine possessing no β-hydrogen with phenylisocyanate at room temperature produces β-lactam 196 by 1,2 cycloaddition [512,513], probably via zwitterion intermediate 195. At higher temperatures a second mole of phenylisocyanate adds to 195, yielding aminohydrouracil 197 [514, 515]. A similar reaction is observed with phenylisothiocyanate [516] and with sulfonylisocyanates [517]. Treatment of enamines with o-phenylene diisothiocyanate produces thiazine ring systems [518]. When another reagent is present, such as sulfur [440] or cyclohexylisonitrile (199) [519], cyclization to form five-membered rings takes

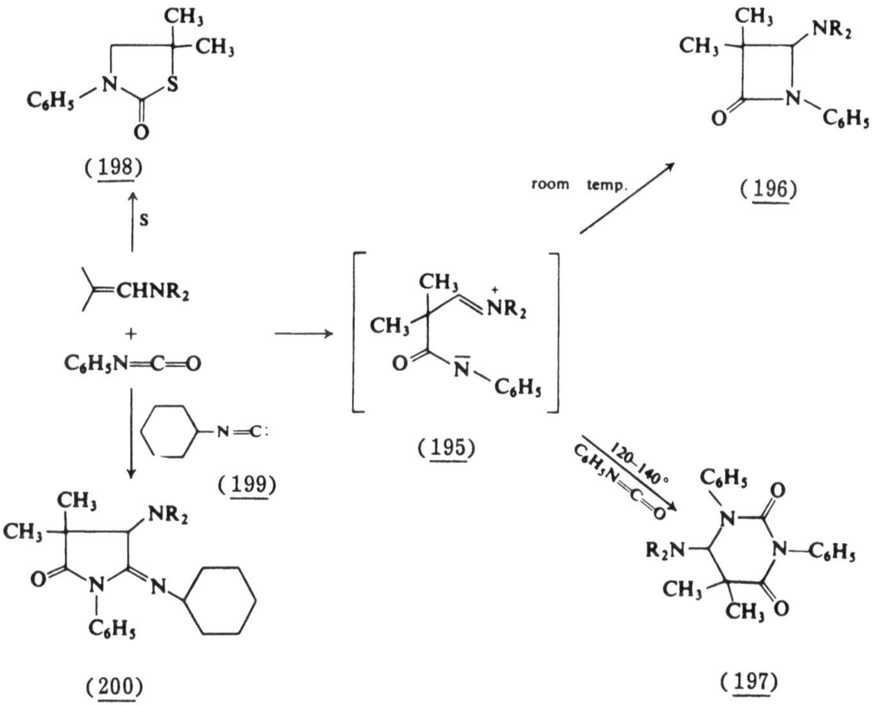

place, in these cases 1,3-thiazolidin-2-one (198) and pyrrolidone (200) respectively.

The 1,4 cycloaddition reaction between enamines possessing β-hydrogens and isocyanates or isothiocyanates which are conjugated to a group such as a carbonyl, aromatic ring, or an alkene is outlined

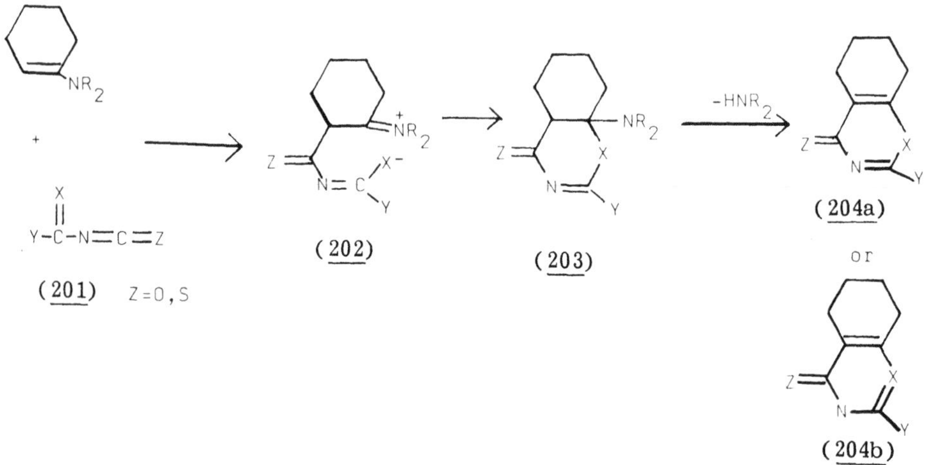

(201) (202) (203) (204a) (204b)

Scheme 17

in Scheme 17. When the conjugating group is a carbonyl (X = 0), then the final product has been observed to be 203 in some cases (NR_2 = morpholino; Z = 0; Y = C_6H_5, CCl_3, or Cl) [520,521], and 204a in other cases (NR_2 = morpholino; Z = S; Y = C_6H_5) [522]. When the conjugating group is an amidinoyl group (Y = N; X = NAr), then 204a (Z = S; NR_2 = morpholino or piperidino) is formed [523]. An aromatic conjugating group (XCY = C_6H_5 or RC_6H_4) results in 204b (Z = 0; NR_2 = morpholino) being formed after treating intermediate 202 with strong acid [510,511,524–530]. Using an alkene conjugating group (X = C), the interaction of an enamine with this conjugated isocyanate gives 204b (Z = O; NR_2 = pyrrolidino or morpholino) [531]. For example, refluxing 1-isocyanato-1-cyclohexene (205) with the

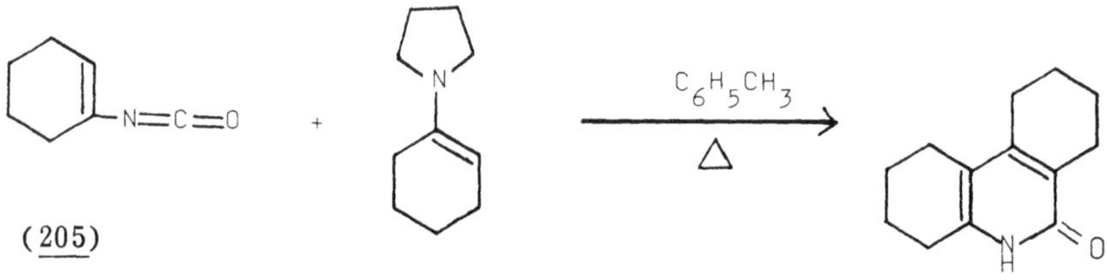

(205) (206)

pyrrolidine enamine of cyclohexanone in toluene solvent produces octahydrophenanthridinone 206 in an 81% yield [531].

The reaction between an enamine and (ethoxycarbonyl)nitrene (207) usually gives simple addition products [532,533], but in the case of the N-methylaniline enamine of cyclohexanone, the product is a relatively unstable aziridine 208 [533].

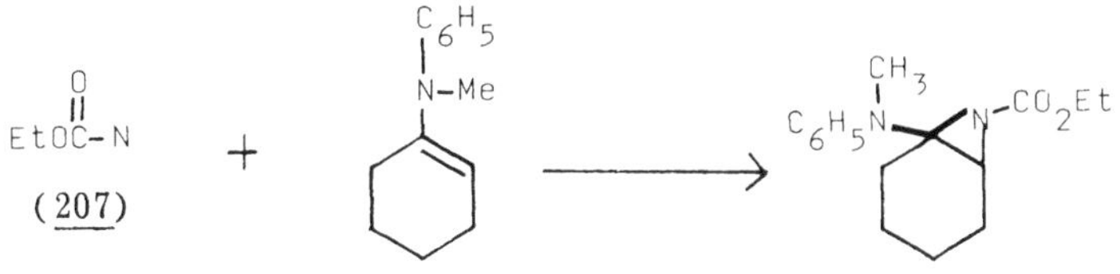

(207)

(208)

A pseudo 1,2 cycloaddition (actually a 1,3 cycloaddition, but may be considered a 1,2 type if a three-membered ring is considered analogous to an alkene) is observed when the pyrrolidine enamine of cyclohexanone is allowed to react with N-carbethoxyaziridine to produce octahydroindole (210) [281]. Octahydroindoles and pyrrolidines can also be produced through the intramolecular alkylation of the enamines of certain haloketourethanes [534]. Treatment of 2H-azirines (209) with enamines causes initial [2+2] cycloaddition to give an azabicyclo[2.1.0] system followed by bond breaking to a monocyclic dihydropyrrole [535,536].

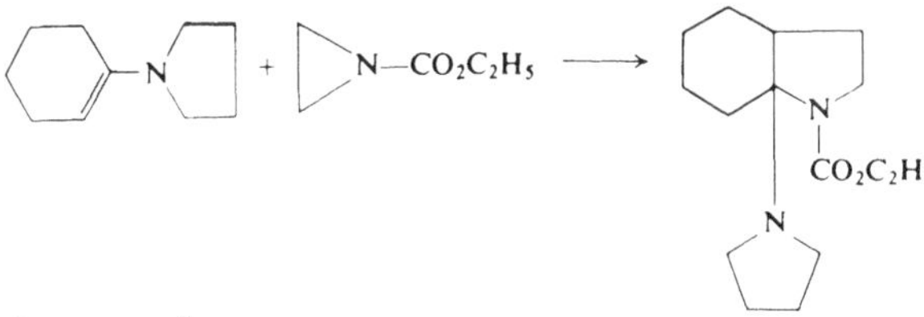

(210)

(209)

Imines initially undergo 1,2 cycloadditions with enamines in acid media, whereas in neutral media only normal additions are observed [537–539]. In acid media (acetic acid or p-toluenesulfonic acid in methanol) with enamines having an α-hydrogen, 1,4 cycloaddition will take place with benzylideneaniline to form 1,2,3,4-tetrahydroquinoline derivative 211 [540]. The acid catalyst is necessary because

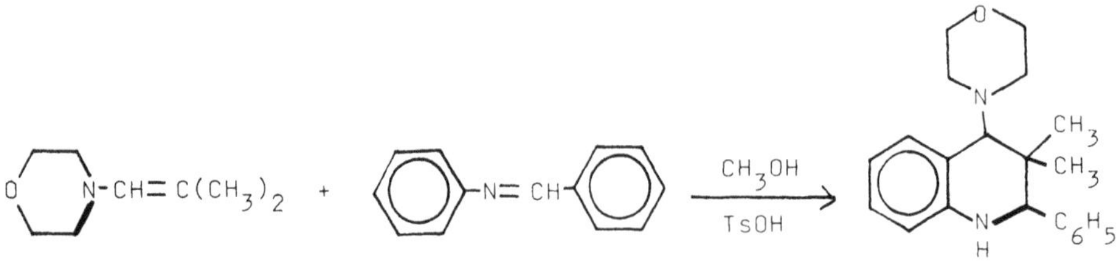

(211)

the iminium salt formed from the acid is the electrophile which attacks the aromatic ring to complete the cyclization.

A unique synthetic route to 3,5-disubstituted pyridine derivatives involves the initial 1,2 cycloaddition of N-methylene-*tert*-butylamine

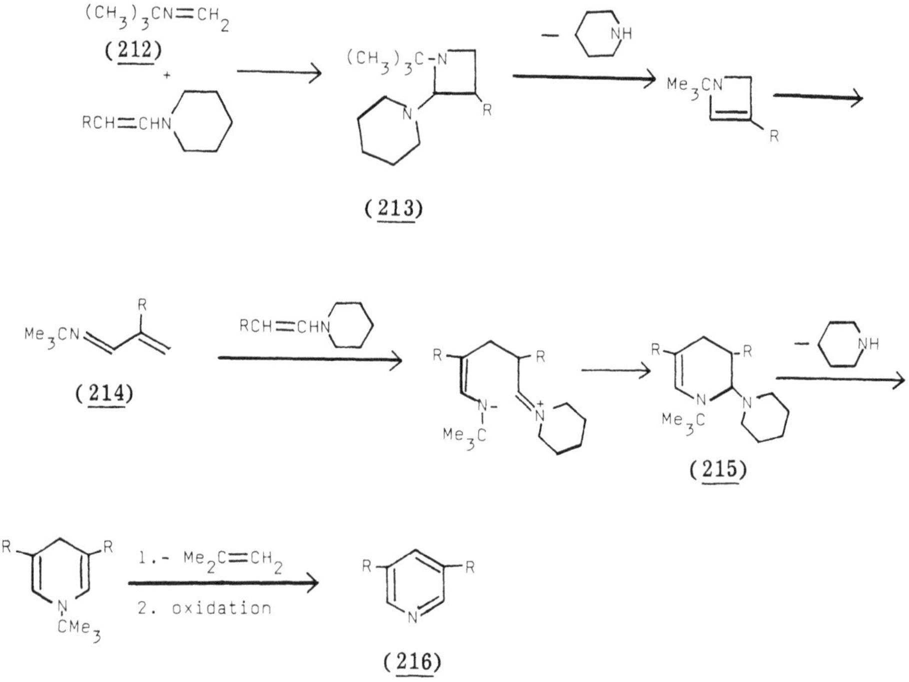

Scheme 18

(212) with an enamine to give intermediate 213 (see Scheme 18) [541–543]. This intermediate in turn loses piperidine and ring-opens to intermediate 214. Condensation of this imine with a second mole of

enamine produces tetrahydropyridine 215. Loss of piperidine and
isobutylene along with oxidation then gives 3,4-disubstituted pyridine in yields ranging from 67 to 87% [542,543]. Using a variation of
this method, unsymmetrically 3,5-disubstituted pyridines were also
synthesized [543].

Enamines have been reported to undergo 1,4 cycloadditions with
heterocyclic compounds. The reaction of isobenzofuroxan (217) with
the morpholine enamine of cyclohexanone results in a 1,4 cycloaddition
to form quinoxaline-di-N-oxide 218 [544—547]. Quinone dibenzenesulfonimide has been found to undergo 1,4 cycloaddition with enamines

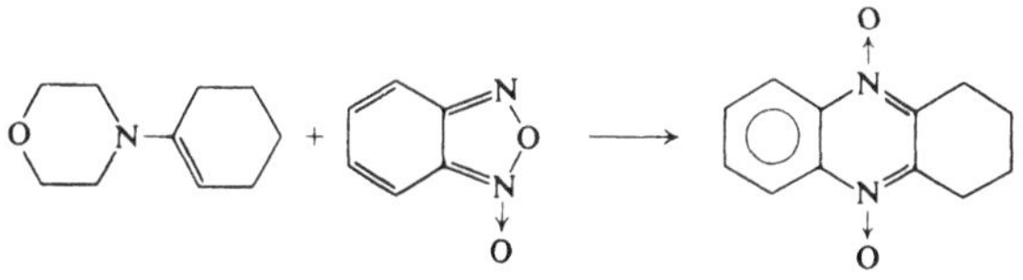

(217)   (218)

in a manner similar to that of *p*-quinone (see Section III.A) [341].
Substituted pyridines have been produced by the ring cleavage of
1,2-oxazolium perchlorates [548] or 1,3-oxazin-4-ones [549,550] with
enamines. A 1,5 cycloaddition to an enamine takes place as the β-carbon of an enamine attacks the 2-carbon of a 1,3-oxazolidine breaking
the carbon-oxygen bond heterolytically in the presence of trifluoroacetic acid, and then the oxygen adds to the enamine's α-carbon to
give a 1,4-oxazepine [551].

5-Nitropyrimidine (219) undergoes a 1,4 cycloaddition with an

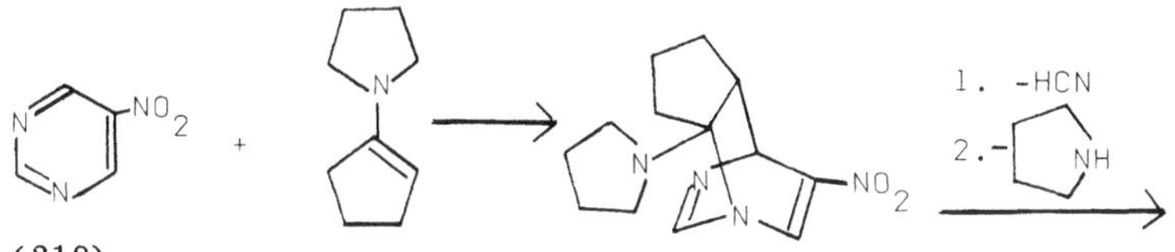

(219)   (220)

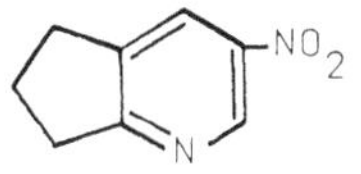

(221)

enamine to give cycloadduct 220. This molecule then undergoes a
reverse Diels-Alder reaction with subsequent loss of pyrrolidine to

give pyridine <u>221</u> [552]. In some similar reactions, both 1,2,4-tri-
azine and 1,3,5-triazine have been used as electron-deficient heterocycles to react with electron-rich enamines in an "inverse electron
demand" Diels-Alder reaction [635] in a regiospecific manner to give
pyridines [636—638] and pyrimidines [639]. This is illustrated in
Scheme 19.

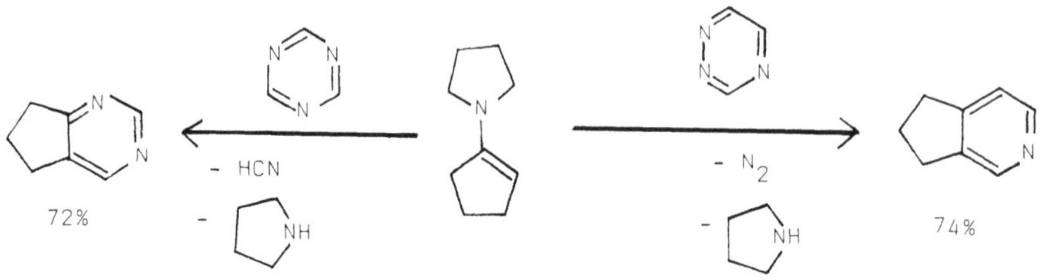

Scheme 19

Enamines have also been reported to undergo 1,4 cycloadditions
with bifunctional acyclic compounds. Various enamines react with
3-bromopropylamine hydrobromide to form tetrahydropyridines [553,
554]. Acrylamide, when heated with an enamine in the presence of
p-toluenesulfonyl chloride, gives good yields of 3,4-dihydro-2-pyri-
done [555]. α-Cyanocinnamide produces a 1,4 cycloadduct when
treated with an enamine [556]. α-Bromoximes react with enamines to
give 5,6-dihydro-4H-1,2-oxazines [557—561]. An oxazine derivative
is also formed by the 1,4 cycloaddition of 2-nitropropene to an enamine [213]. The reactions between the enamines of methylcyclohex-
anones and ω-bromoacetophenone semicarbazone produce tetrahydro-
pyridazine derivatives [562]. The reaction of N-(phenylsulfonyl)-
benzohydrazonoyl chloride with enamines gives substituted pyrazoles
[563]. Treating enamines with N-haloamidines yields 1,4-dihydro-5-
triazine derivatives [563] or 4,5-dihydroimidazoles [564,565].

The reactions of 1-nitroso-2-naphthol with the morpholine enamines of cyclohexanone and propanal gave dihydro-1,4-oxazine derivatives [566]. Acylation of N-arylenamines with phosgene yields 4-
chloroquinolinium salts by final electrophilic attack on the aromatic
ring by the intermediate acylenamine [567]. Enamines will react with
2 moles of trifluoroacetonitrile to produce trifluoromethyl-substituted
pyrimidines [568,569], as will the intermediate 1,3-diazabuta-1,3-
dienes [570].

Sulfonylcarbodiimides react with enamines to form triazolidines
[640]. Hexafluoroacetone azine undergoes a [2+2] cycloaddition with
enamines without β-hydrogen atoms to give azetidine derivatives
[641]. Treatment of cyclohexene enamines with phenylcarbamoyldi-

imide produces either 1,2,4-triazin-3-one derivatives or spirotria-
zolidine derivatives, depending on whether the pyrrolidine enamine
was used, or whether the morpholine or piperidine enamine was used
[642]. 1,3,4-Oxadiazine derivatives are obtained in a regiospecific
way when cyclohexanone enamines are allowed to react with acyl aryl-
diazenes [643–645]. The reaction between an enamine from cyclohexanone and diacyldiimides is a 1,4 cycloaddition to form a 1,3,4-benzo-
oxadiazine derivative [646–648]. Azodicarbonyl compounds when
treated with enamines undergo either 1,2 cycloadditions to form 1,2-
diazetidines or 1,4 cycloadditions to form 1,3,4-oxadiazino compounds
[659,660].

The nonoxidative photocyclization of N-aryl enamines derived
from cyclic or acyclic ketones proceeds under mild conditions to produce 2,3-dihydroindole derivatives [322,571–573] or 1,2-dihydroquin-
oline derivatives [574,575] in good yields. For example, irradiation
of N-methylanilinocyclohexene (222) gave N-methyl-*trans*-hexahydro-
carbazole (224) in a 55% yield (see Scheme 20) [573]. This enamine
could not be cyclized thermally [573]. Irradiation of the N-methyl-

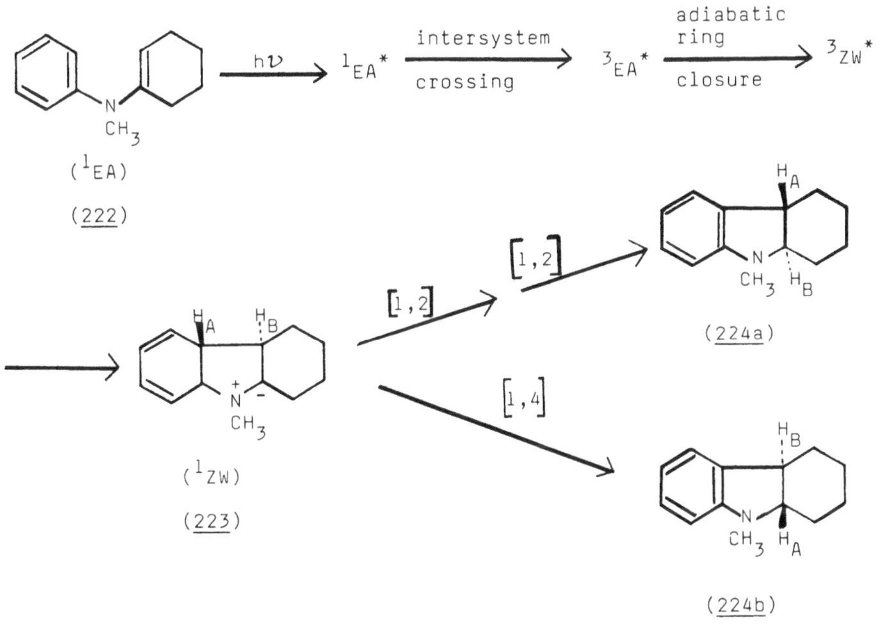

Scheme 20

aniline enamines of cycloheptanone and cyclopentanone produced similar products in yields of 75 and 54%, respectively [573]. The ring

closure of this reaction ($^3EA^*$ to $^3ZW^*$) must involve a conrotatory electrocyclical reaction of a divinylamine which is allowed for this photochemical reaction [20]. This results in a *trans* geometry for the product. It was shown by flash photolysis experiments that a colored zwitterion or ylide ground-state intermediate (^1ZW, 223; λ_{max}520 nm, lifetime τ2.3 msec) is formed during the reaction [576,577]. So the mechanism (see Scheme 20 [577]) involves the initial excitation of enamine 222 to excited state $^1E^*$ followed by intersystem crossing to $^3EA^*$. An adiabatic ring closure to triplet-state zwitterion $^3ZW^*$ takes place followed by deactivation to singlet zwitterion 223. At this point, either a thermally allowed sigmatropic [1,4] hydrogen shift takes place to give product 224b, or a series of two sigmatropic [1,2] hydrogen shifts takes place to give product 224a, both of which are *trans* (see Scheme 10). It has been shown that there is a temperature-dependent kinetic isotopic effect (largely governed by a proton tunnel process) when H_A is replaced by deuterium [578,579]. This effect is absent when H_{BA} is replaced by deuterium. This isotope effect (kH/kD) ranges from 10 to 10^4, depending on the temperature. So the final step of the mechanism must be a sigmatropic [1,4] hydrogen shift [579].

Asymmetrical photocyclization of this type of enamine by circularly polarized light yields optically active dihydroindole derivatives [580]. Molecules in which a carbonyl group is attached to a vinyl group cross-conjugated to the nitrogen, such as in 225, give good

(225)   (226)

yields of products (226) [581]. Owing to the presence of the carbonyl group, relatively low-energy radiation may be used, and this carbonyl group also provides stabilization for the zwitterion intermediate.

1,3-Dipolar Cycloadditions

1,3 Cycloaddition across the double bond of an enamine with a 1,3-dipole (a species described by zwitterionic octet structures) results in the loss of formal charges to give an uncharged five-membered ring

[582−585]. The terminal centers of a 1,3-dipole can be both nucleophilic and electrophilic. This ambivalence of the 1,3-dipole is best
seen in the sextet resonance contributing forms of these 1,3-dipoles

Contributing

Octet

Forms

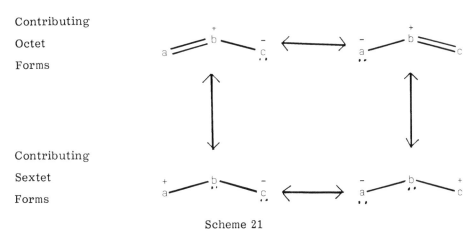

Contributing

Sextet

Forms

Scheme 21

(see Scheme 21). Several reactions of these 1,3-dipolar cycloaddition
reactions have been observed in which an enamine acts as the dipolarophile. The types of dipolar reactants that have been reported to
undergo this type of 1,3 cycloaddition with dipolarophilic enamines
are listed below.

azomethine ylide	$C{+}-\bar{N}-Cl^-$	$\leftrightarrow$	$^-C-\overset{+}{N}{=}C$
nitrones	$\overset{+}{C}-\bar{N}-\bar{O}l^-$	$\leftrightarrow$	$C{=}\overset{+}{N}-\bar{O}l^-$
nitrilimines	$-\overset{+}{C}{=}N-\bar{N}^-$	$\leftrightarrow$	$-C{\equiv}\overset{+}{N}-\bar{N}^-$
nitrile oxides	$-\overset{+}{C}{=}N-\bar{O}l^-$	$\leftrightarrow$	$-C{\equiv}\overset{+}{N}-\bar{O}l^-$
azides	$l\overset{+}{N}{=}N-\bar{N}^-$	$\leftrightarrow$	$lN{\equiv}\overset{+}{N}-\bar{N}^-$
diazoalkanes	$l\overset{+}{N}{=}\bar{N}-C^-$	$\leftrightarrow$	$lN{\equiv}\overset{+}{N}-C^-$

An azomethine ylide 1,3 dipole, isoquinolinium diethoxycarbonyl-
methylide, has been reported to undergo 1,3 cycloaddition with enamines [633].

The cycloaddition of nitrones to enamines results in the formation
of an isoxazolidine [586−590]. The reaction of 1-(N-pyrrolidino)-1-

phenyl-ethylene (227) with nitrone 228 producing isoxazolidine 229
illustrates this type of cycloaddition [587]. In some cases the orientation of cycloaddition is reversed [591].

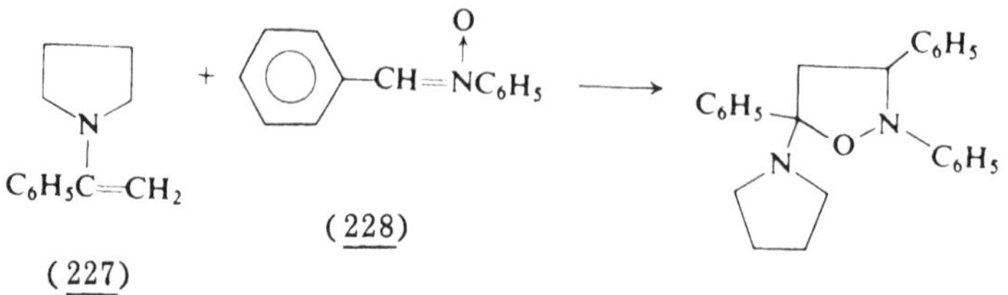

(227) (228) (229)

Nitrile imines can be produced by treating halogenated hydrazones
with a base such as triethylamine. These nitrile imines undergo 1,3
cycloaddition with enamines to form pyrazoles [592–595]. This is
shown by the reaction of the pyrrolidine enamine of cyclohexanone
with diphenylnitrilimine to form diphenyltetrahydroisoindazole (230)
[593]. Treatment of halohydroxamic acids with a base yields nitrile
oxides, which in turn add to the dipolarophilic enamines to produce
isoxazoles [593,596,597,599–602]. Therefore, benzonitrile oxide adds
to the pyrrolidine enamine of cyclohexanone to yield aminodihydroisox-
azole 231 [593]. A dioxazole adduct is formed when 1,2-bis(N,N-di-
ethylamino)ethylene is allowed to react with terephthalonitrile oxide
[598].

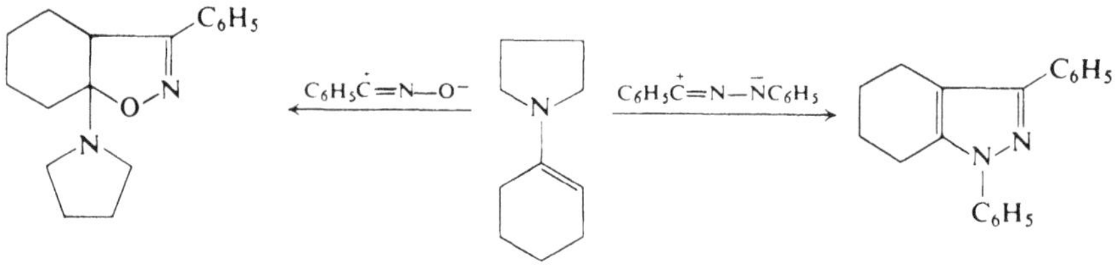

(231) (230)

Azides can use enamines as dipolarophiles for 1,3 cycloadditions
to form triazolines. These azides can be formate ester azides [603],
phenyl azides [599,604–620], arylsulfonyl azides [599,608–610,621–
623,634], benzoylazides [624,625], vinyl azides [626], hydrazonoyl
azides [627], or phosphoryl azides [628–630,654]. The last type of
azide has been used to very successfully carry out ring contractions
of ketones via their enamine [654]. For example, the reaction between phenyl azide and the piperidine enamine of propionaldehyde
(232) gives 1-phenyl-4-methyl-5-(1-piperidino)-4,5-dihydro-1,2,3-
triazole (233), exclusively, in a 53% yield [607]. None of the isomeric

Table 1 Rates of Cycloaddition of Phenylazide to Some Alkenes [611]
(Benzene Solvent, 25°C)

Alkene dipolarophile	$k_2 \times 10^7$ (1/mol-sec)
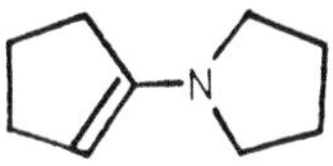	115,000
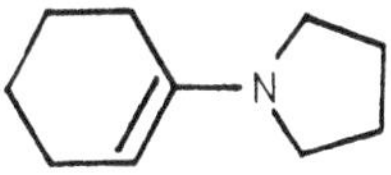	9,930
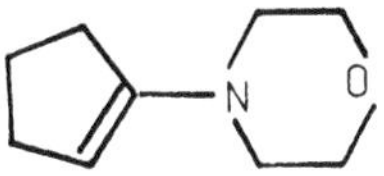	2,580
	254
	2.4
 OEt	0.49

1-phenyl-5-methyl product was formed. This indicates that the products formed from this reaction are the results of electronic control

$$C_6H_5N_3 + \langle\;\rangle NCH{=}CHCH_3 \longrightarrow$$

(232)   (233)

[607]. The negative end of the azide dipole (the phenyl-substituted nitrogen) is directed to the β-position of the enamine. In general, this reaction is regiospecific in that the terminal nitrogen atom of the 1,3 dipole bearing the substituent is directed to the α-carbon atom of enamine dipolarophile [616].

Enamines add azides faster than any other category of alkenes (see Table 1). Using frontier molecular orbital theory for describing 1,3 dipole cycloadditions to electron-rich olefins such as enamines, the most important interaction is the interaction between LUMO of the 1,3 dipole and HOMO of the dipolarophile [17,631]. So this means the lower the energy of the LUMO of the azide, and the higher the energy of the HOMO of the enamine, the faster the reaction. It has been observed that the relative rates of enamines with aryl azides are pyrrolidino > piperidino > morpholino [611,616,626]. These observations correspond to the relative energies of the HOMOs of the enamines (see IP$_1$s in Chapter 1, Table 1). An electron-withdrawing substituent on the enamine dipolarophile slows down the reaction, whereas an electron-withdrawing substituent on the 1,3 dipole accelerates the reaction.

The mechanism of the cycloaddition of phenyl azide to norbornene. has been shown to involve a concerted mechanism with a charge imbalance in the transition state [632]. In a similar manner, the cycloaddition of phenyl azide to enamines apparently proceeds by a concerted mechanism [611,612]. This is shown by a rather large negative entropy of activation (-36 entropy units for 1-(N-morpholino)cyclopentene in benzene solvent at 25°C), indicative of a highly ordered transition state. Varying solvents from those of small dielectric constants to those of large dielectric constants has only a small effect on the relative rates of the cycloaddition reaction. This also supports a concerted reaction mechanism. However, a very large positive Hammett ρ value of +2.5 for 1-(N-pyrrolidino)cyclohexene in benzene solvent at 25°C seems to militate against a concerted mechanism since it indicates a charge separation in the transition state [611,616].

These results can be correlated by describing the mechanism as concerted (but not synchronous) with formation of transition state 234 and the formation of bond *a* being further developed than the formation of bond *b*; that is, δ- or δ+ is greater than δ'- or δ'+.

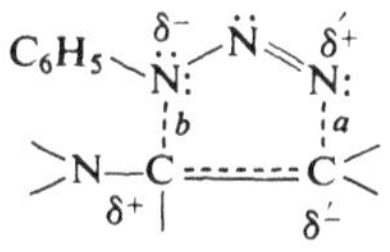

(234)

This discrepancy in the extent of bond formation between a and b is greater with enamines as dipolarophiles than with unsubstituted olefins as dipolarophiles [607,611,616].

The reaction of cyanogen azide with enamines of cyclic ketones to yield a cyanoamidine with one less member in the carbocyclic ring represents a potentially valuable method of ring contraction under mild conditions [633]. The reaction probably proceeds first by 1,3 cycloaddition of the azide to the enamine followed by rearrangement and elimination of a molecule of nitrogen.

Enamines do not react with diazomethane, but they undergo 1,3-dipolar cycloaddition with α-diazo carbonyl compounds [649–651,658]. For example, ethyl diazoacetate reacts with the pyrrolidine enamine

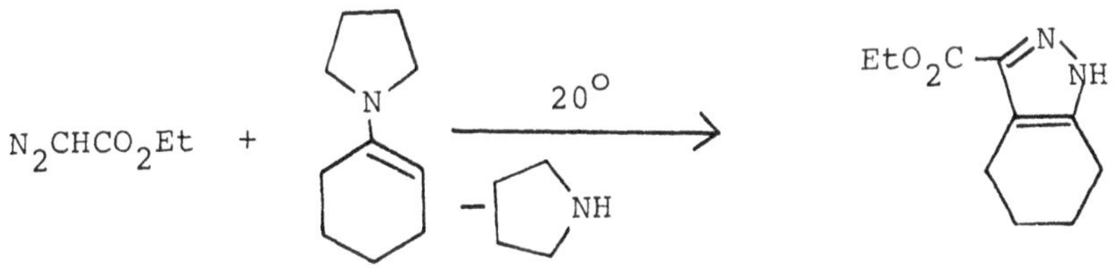

(235)

of cyclohexanone to produce adduct 235 [650]. Azoalkenes react with enamines to afford criss-cross cycloadducts [652,653].

REFERENCES

1. R. Huisgen, *Agnew. Chem., Int. Ed. Engl.*, 7, 321 (1968).
2. J. E. Baldwin, *J. Org. Chem.*, 32, 2438 (1967).
3. L. L. Muller and J. Hamer, *1,2-Cycloaddition Reactions*, Wiley-Interscience, New York, 1967.
4. R. Huisgen, *Proc. Chem. Soc.*, 351 (1961).
5. J. Hamer, ed., *1,4-Cycloaddition Reactions*, Academic Press, New York, 1967.
6. M. J. S. Dewar and A. B. Pierini, *J. Am. Chem. Soc.*, 106, 203 (1984).

7. M. J. S. Dewar, *J. Am. Chem. Soc.*, *106*, 209 (1984).
8. K. Fukui, *Chem. Forsch.*, *15*, 1 (1970).
9. K. Fukui, *Acc. Chem. Res.*, *4*, 57 (1971).
10. W. C. Herndon, *Chem. Rev.*, *72*, 157 (1972).
11. K. N. Houk, *J. Am. Chem. Soc.*, *95*, 4092 (1973).
12. K. N. Houk, *Acc. Chem. Res.*, *8*, 361 (1975).
13. R. Sustmann, *Pure Appl. Chem.*, *40*, 569 (1974).
14. K. Fukui, *Theory of Orientation and Steroselection in Reactivity and Structure Concepts in Organic Chemistry*, Vol. 2, Springer, Berlin, 1975.
15. N. D. Epiotis, *Theory of Organic Reactions in Reactivity and Structure Concepts in Organic Chemistry*, Vol. 5, Springer, Berlin, 1978.
16. N. T. Anh, E. Canadell, and O. Einsenstein, *Tetrahedron, 34*, 2283 (1978).
17. K. N. Houk in *Pericyclic Reactions* (A. P. Marchand and R. E. Lehr, eds.), Vol. 2, Academic Press, New York, 1977, pp. 181–271.
18. N. D. Epiotis, *J. Am. Chem. Soc.*, *94*, 1924 (1972).
19. N. D. Epiotis, *Angew. Chem., Int. Ed. Engl.*, *13*, 751 (1974).
20. R. B. Woodward and R. Hoffmann, *The Conservation of Orbital Symmetry*, Academic Press, 1970.
21. N. D. Epiotis, *J. Am. Chem. Soc.*, *94*, 1935 (1972).
22. N. D. Epiotis, *J. Am. Chem. Soc.*, *94*, 1941 (1972).
23. N. D. Epiotis, *J. Am. Chem. Soc.*, *94*, 1946 (1972).
24. N. D. Epiotis, *J. Am. Chem. Soc.*, *95*, 1191 (1973).
25. N. D. Epiotis, *J. Am. Chem. Soc.*, *95*, 1200 (1973).
26. N. D. Epiotis, *J. Am. Chem. Soc.*, *95*, 1206 (1973).
27. N. D. Epiotis, *J. Am. Chem. Soc.*, *95*, 1214 (1973).
28. N. D. Epiotis, R. Yates, D. Carlberg, and F. Bernardi, *J. Am. Chem. Soc.*, *98*, 453 (1976).
29. J. Sauer and R. Sustmann, *Angew. Chem., Int. Ed. Engl.*, *19*, 779 (1980).
30. N. D. Epiotis, *J. Am. Chem. Res.*, *95*, 5624 (1973).
31. R. Huisgen, *Acc. Chem. Res.*, *10*, 117 (1977).
32. G. Desimoni and G. Tacconi, *Chem. Rev.*, *75*, 651 (1975).
33. K. N. Houk, Y-T. Lin, and F. K. Brown, *J. Am. Chem. Soc.*, *108*, 554 (1986).
34. R. Gompper, *Angew. Chem., Int. Ed. Engl.*, *8*, 312 (1969).
35. G. Stork and H. Landesman, *J. Am. Chem. Soc.*, *78*, 5128 (1956).
36. G. Stork, A. Brizzolara, H. Landesman, J. Szmuskovicz, and R. Terrell, *J. Am. Chem. Soc.*, *85*, 207 (1963).
37. U. K. Pandit and H. O. Huisman, *Tetrahedron Lett.*, 3901 (1967).
38. I. Fleming and J. Harley-Mason, *J. Chem. Soc.*, 2165 (1964).
39. G. Opitz and H. Holtmann, *Liebigs Ann. Chem.*, *684*, 79 (1965).

40. I. Fleming and M. H. Karger, *J. Chem. Soc. (C)*, 226 (1967).
41. A. Risaliti, E. Valentin, and M. Forchiassin, *J. Chem. Soc., Chem. Commun.*, 233 (1969).
42. F. P. Colonna, S. Fatutta, A. Risaliti, and C. Russo, *J. Chem. Soc., (C)*, 2377 (1970).
43. G. Stork, R. Terrell, and J. Szmuszkovicz, *J. Am. Chem. Soc., 76*, 2029 (1954).
44. H. Christol, D. Lafont, and F. Plenat, *Bull. Soc. Chim. Fr.*, 1176 (1966).
45. R. L. N. Harris, F. Komitsky, Jr., and C. Djerassi, *J. Am. Chem. Soc., 89*, 4765 (1967).
46. J. Davey, B. R. T. Keene, and G. Mannering, *J. Chem. Soc. (C), 120* (1967).
47. T. J. Curphey and H. L. Kim, *Tetrahedron Lett.*, 1441 (1968).
48. D. A. Evans, *Tetrahedron Lett.*, 1573 (1969).
49. R. V. Stevens and M. P. Wentland, *J. Am. Chem. Soc., 90*, 5580 (1968).
50. R. V. Stevens and M. P. Wentland, *J. Chem. Soc., Chem. Commun.*, 1104 (1968).
51. R. V. Stevens and L. E. DuPree, Jr., *J. Chem. Soc., Chem. Commun.*, 1585 (1970).
52. R. V. Stevens, *Acc. Chem. Res., 10*, 193 (1977).
53. S. L. Keely, Jr., A. J. Martinez, and F. C. Tahk, *Tetrahedron, 26*, 4729 (1970).
54. C. P. Forbes, G. L. Wenteler, and A. Wiechers, *Tetrahedron, 34*, 487 (1978).
55. P. Houdewind, J. C. L. Armande, and U. K. Pandit, *Tetrahedron Lett.*, 591 (1974).
56. G. D. Joshi and S. N. Kulkarni, *Indian J. Chem., 3*, 91 (1965).
57. S. H. Mashraqui and G. K. Trivedi, *Indian J. Chem., Section B*, 305 (1977).
58. R. M. Coates and J. E. Shaw, *J. Am. Chem. Soc., 92*, 5657 (1970).
59. M. Shibasaki, S. Terashima, and S. Yamada, *Chem. Pharm. Bull., 23*, 272 (1975).
60. W. Pickenhagen and C. Starkemann, *Nouv. J. Chim., 2*, 365 (1978).
61. W. M. B. Konst, J. G. Witteveen, and H. Boelens, *Tetrahedron, 32*, 1415 (1976).
62. R. L. Augustine and H. V. Cortez, *Chem. Ind. (London)*, 490 (1963).
63. H. O. House, B. M. Trost, R. W. Magin, R. G. Carlson, R. W. Franck, and G. H. Rasmusson, *J. Org. Chem., 30*, 2513 (1965).
64. A. G. Cook and C. R. Schulz, *J. Org. Chem., 32*, 473 (1967).
65. T. L. Westman, R. Paredes, and W. S. Brey, Jr., *J. Org. Chem., 28*, 3512 (1963).

66. T. L. Westman and A. E. Kober, *J. Org. Chem.*, *29*, 2448 (1964).

67. M. Brown, *J. Org. Chem.*, *33*, 162 (1968).

68. T. Oishi, M. Ochiai, M. Nagai, and Y. Ban, *Tetrahedron Lett.*, 497 (1968).

69. A. Venot and G. Adrian, *Tetrahedron Lett.*, 4663 (1972).

70. J. W. Lewis and P. L. Myers, *J. Chem. Soc.* (*C*), 753 (1971).

71. Y. Chan and W. W. Epstein, *Organic Syntheses* (A. Brossi, ed.), Vol. 53, Wiley, New York, 1973, p. 48.

72. N. J. Leonard and W. J. Musliner, *J. Org. Chem.*, *31*, 639 (1966).

73. Y. K. Singh and R. B. Rao, *Chem. Letters*, 653 (1979).

74. G. Stork and R. N. Guthikonda, *J. Am. Chem. Soc.*, *94*, 5109 (1972).

75. J. Tsuji, I. Shimizu, H. Suzuki, and Y. Naito, *J. Am. Chem. Soc.*, *101*, 5070 (1979).

76. H. Nishimura, T. Takabatake, K. Kaku, A. Seo, and J. Mizutani, *Agric. Biol. Chem.*, *45*, 1861 (1981).

77. F. Bohlman, H-J. Muller, and D. Schumann, *Chem. Ber.*, *106*, 3026 (1973).

78. J. W. Lewis and P. A. Mayor, *J. Chem. Soc.* (*C*), 1074 (1970).

79. B. Weinstein, L-C. C. Lin, and F. W. Fowler, *J. Org. Chem.*, *45*, 1657 (1980).

80. G. Oszbach, D. Szabo, and M. E. Vitai, *Acta Chi... Acad. Sci. Hung.*, *95*, 273 (1977).

81. G. Oszbach, D. Szabo, and M. E. Vitai, *Acta Chim. Acad. Sci. Hung.*, *101*, 119 (1979).

82. G. Markl and H. Baier, *Tetrahedron Lett.*, 4439 (1972).

83. J. V. Silverton, M. Ziffer, and H. Ziffer, *J. Org. Chem.*, *44*, 3959 (1979).

84. F. De Pessemier, P. Vanhee, and D. Tavernier, *Bull. Soc. Chim. Belg.*, *86*, 551 (1977).

85. M. Forchiassin, A. Risaliti, C. Russo, M. Calligaris, and G. Pitacco, *J. Chem. Soc. Perkin I*, 660 (1974).

86. E. G. Rozantsev, M. Dagonneau, E. S. Kagan, V. I. Mikhailov, and V. D. Sholle, *J. Chem. Res.* (*S*), 260 (1979); *J. Chem. Res.* (*M*), 2901 (1979).

87. G. Tacconi and G. Desimoni, *Gazz. Chim. Ital.*, *98*, 1314 (1968).

88. R. B. Rao and G. V. Bhide, *Chem. and Ind.*, 653 (1970).

89. J. W. Lewis and P. L. Myers, *Chem. and Ind.*, 1625 (1970).

90. K. K. Prasad and V. M. Girijavallabhan, *Chem. and Ind.*, 426 (1971).

91. J. W. Lewis, P. L. Myers, J. A. Ormerod, and I. A. Selby, *J. Chem. Soc., Perkin Trans.*, *I*, 1549 (1972).

92. O. Tsuge and I. Shinkai, *Nippon Kagaku Zasshi*, *92*, 263 (1971).

93. G. Oszbach, D. Szabo, G. Argay, A. Kalman, and A. Neszmelyi, *J. Chem. Soc., Perkin Trans.*, *II*, 447 (1983).

94. G. L. Lange, D. J. Wallace, and S. So, *J. Org. Chem.*, *44*, 3066 (1979).
95. S. Ohta, A. Shimabayashi, S. Hatano, and M. Okamoto, *Synthesis*, 715 (1983).
96. G. Markl and H. Baier, *Tetrahedron Lett.*, 4379 (1968).
97. M. Oda, M. Funamizu, and Y. Kitahara, *J. Chem. Soc.*, *Chem. Commun.*, 737 (1969).
98. M. Von Strandmann, M. P. Cohen, and J. Shavel, Jr., *J. Org. Chem.*, *30*, 3240 (1965).
99. M. Von Strandmann, M. P. Cohen, and J. Shavel, Jr., *Tetrahedron Lett.*, 3103 (1965).
100. M. Von Strandmann, M. P. Cohen, and J. Shavel, Jr., *J. Heterocycl. Chem.*, *7*, 1311 (1970).
101. L. H. Hellberg and M. F. Stough, III, *Acta. Chem. Scand.*, *21*, 1368 (1967).
102. W. Schroth and G. Fischer, *Angew. Chem.*, *Int. Ed. Engl.*, *2*, 394 (1963).
103. H. Molines, M. Tordeux, and C. Wakselman, *Bull. Soc. Chim. Fr.*, II-367 (1982).
104. H. Molines and C. Wakselman, *J. Chem. Soc.*, *Chem. Commun.*, 232 (1975).
105. H. Molines and C. Wakselman, *Tetrahedron*, *32*, 2099 (1976).
106. H. Molines and C. Wakselman, *J. Fluorine Chem.*, *16*, 97 (1980).
107. H. Molines and C. Wakselman, *J. Chem. Soc. Perkin I*, 1114 (1980).
108. W. G. Dauben and A. P. Kozikowski, *J. Am. Chem. Soc.*, *96*, 3664 (1974).
109. T. Igarashi, J. Oda, and Y. Inouye, *Bull. Inst. Chem. Res.*, *Kyoto Univ.*, *50*, 222 (1972).
110. G. Otani and S. Yamada, *Chem. Pharm. Bull.*, *21*, 2112 (1973).
111. G. Otani and S. Yamada, *Chem. Pharm. Bull.*, *21*, 2125 (1973).
112. T. Sone, K. Hiroi, and S. Yamada, *Chem. Pharm. Bull.*, *21*, 2331 (1973).
113. J. W. Huffman, C. D. Rowe, and F. J. Matthews, *J. Org. Chem.*, *47*, 1438 (1982).
114. P. W. Hickmott and N. F. Firrell, *J. Chem. Soc.*, *Perkin Trans.*, *I*, 340 (1978).
115. P. W. Hickmott, *Tetrahedron*, *38*, 1975 (1982).
116. R. V. Stevens, *Acc. Chem. Res.*, *17*, 289 (1984).
117. R. E. Ireland, *Chem. Ind.* (*London*), 979 (1958).
118. F. Bohlmann and O. Schmidt, *Chem. Ber.*, *97*, 1354 (1964).
119. F. Bohlmann, D. Schumann, and E. Bauerschmidt, *Chem. Ber.*, *100*, 542 (1967).
120. G. Stork and H. K. Landesman, *J. Am. Chem. Soc.*, *78*, 5129 (1956).
121. A. Z. Britten and J. O'Sullivan, *Tetrahedron*, *29*, 1331 (1973).

122. K. G. Untch, Ph.D. thesis, Columbia Univ., New York, 1959.
123. R. N. Schut and T. M. H. Liu, *J. Org. Chem.*, 30, 2845 (1965).
124. R. A. Appleton, K. H. Baggaley, C. Egan, J. M. Davies, S. H. Graham, and D. O. Lewis, *J. Chem. Soc. (C)*, 2032 (1968).
125. A. C. Cope, D. L. Nealy, P. Scheiner, and G. Wood, *J. Am. Chem. Soc.*, 87, 3130 (1965).
126. R. D. Allan, B. G. Cordiner, and R. J. Wells, *Tetrahedron Lett.*, 6055 (1968).
127. H. Fischer, U.S. Pat. 4,188,341 (1980); *CA*, 92, 198085x (1980).
128. G. Opitz and I. Loschmann, *Angew. Chem.*, 72, 523 (1960).
129. V. Dressler and K. Bodendorf, *Tetrahedron Lett.*, 4243 (1967).
130. V. Dressler and K. Bodendorf, *Liebig's Ann. Chem.*, 720, 71 (1968).
131. R. L. Autrey and F. C. Tahk, *Tetrahedron*, 24, 3337 (1968).
132. P. W. Hickmott, *Proc. Chem. Soc.*, 287 (1964).
133. P. W. Hickmott, *Tetrahedron*, 23, 3151 (1967).
134. P. W. Hickmott and B. J. Hopkins, *J. Chem. Soc. (C)*, 2918 (1968).
135. P. W. Hickmott and C. T. Yoxall, *J. Chem. Soc. (C)*, 1829 (1971).
136. P. W. Hickmott, B. J. Hopkins, and C. T. Yoxall, *Tetrahedron Lett.*, 2519 (1970).
137. J. R. Hargreaves, P. W. Hickmott, and B. J. Hopkins, *J. Chem. Soc. (C)*, 2599 (1968).
138. P. W. Hickmott, B. J. Hopkins, G. Sheppard, and D. J. Barraclough, *J. Chem. Soc., Perkin Trans.*, I, 1639 (1972).
139. N. F. Firrell and P. W. Hickmott, *J. Chem. Soc. (C)*, 2320 (1968).
140. J. R. Hargreaves, P. W. Hickmott, and B. J. Hopkins, *J. Chem. Soc. (C)*, 592 (1969).
141. N. F. Firrell, P. W. Hickmott, and B. J. Hopkins, *J. Chem. Soc. (C)*, 1477 (1970).
142. P. W. Hickmott, G. J. Miles, G. Sheppard, R. Urbani, and C. T. Yoxall, *J. Chem. Soc., Perkin Trans.*, I, 1514 (1973).
143. P. W. Hickmott, P. J. Cox, and G. A. Sim, *J. Chem. Soc., Perkin Trans.*, I, 2544 (1974).
144. P. W. Hickmott, K. N. Woodward, and R. Urbani, *J. Chem. Soc. Perkin I*, 1885 (1975).
145. E. N. Aredova, V. V. Sevostyanova, M. M. Krayushkin, S. S. Novikov, and N. F. Karpenko, *Izv. Akad. Nauk SSSR, Ser. Khim.*, 1408 (1976).
146. P. W. Hickmott, H. Suschitzky, and R. Urbani, *J. Chem. Soc. Perkin I*, 2063 (1973).
147. D. Gravel and S. Rahal, *Can. J. Chem.*, 53, 2671 (1975).
148. S. A. Ahmed and P. W. Hickmott, *J. Chem. Soc., Perkin Trans.*, I, 2180 (1979).

149. S. A. Ahmed, P. W. Hickmott, and S. Wood, *S. Afr. J. Chem.*, *38*, 55 (1984).
150. P. W. Hickmott, M. G. Ahmed, S. A. Ahmed, S. Wood, and M. Kapon, *J. Chem. Soc., Perkin Trans.*, *I*, 2559 (1985).
151. K. C. Brannock, R. D. Burpitt, V. W. Goodlett, and J. G. Thweatt, *J. Org. Chem.*, *29*, 813 (1964).
152. L. Birkofer and C. D. Barnikel, *Chem. Ber.*, *91*, 1998 (1958).
153. H. Mazarguil and A. Lattes, *Bull. Soc. Chim. Fr.*, 3874 (1972).
154. K. C. Brannock, A. Bell, R. D. Burpitt, and C. A. Kelly, *J. Org. Chem.*, *26*, 625 (1961).
155. K. C. Brannock, A. Bell, R. D. Burpitt, and C. A. Kelly, *J. Org. Chem.*, *29*, 801 (1964).
156. H. Christol and F. Plenat, *Bull. Soc. Chim. Fr.*, 1132 (1963).
157. H. Christol, D. Lafont, and F. Plenat, *Bull. Soc. Chim. Fr.*, 3947 (1966).
158. A. G. Wilson and L. Weintraub, U.S. Pat. 3,133,924; *CA, 61*, 2986f (1964).
159. W. C. Agosta and D. K. Herron, *J. Org. Chem.*, *34*, 2782 (1969).
160. A. G. Cook, Ph.D. thesis, University of Illinois, Urbana, 1959.
161. F. D. Lewis, T-I. Ho, and R. J. DeVoe, *J. Org. Chem.*, *45*, 5283 (1980).
162. H. K. Hall, Jr. and P. Ykman, *J. Am. Chem. Soc.*, *97*, 800 (1975).
163. S. Bruckner, G. Pitacco, and E. Valentin, *J. Chem. Soc., Perkin Trans.*, *II*, 1804 (1975).
164. H. K. Hall, Jr. and P. Ykman, *J. Chem. Soc., Chem. Commun.*, 587 (1974).
165. H. K. Hall, Jr., M. Abdelkader, and M. E. Glogowski, *J. Org. Chem.*, *47*, 3691 (1982).
166. O. Cervinka, *Chem. Ind. (London)*, 1129 (1959).
167. O. Cervinka, *Collect. Czech. Chem. Commun.*, *25*, 1183 (1960).
168. O. Cervinka, *Collect. Czech. Chem. Commun.*, *25*, 1174 (1960).
169. O. Cervinka, *Collect. Czech. Chem. Commun.*, *25*, 2675 (1960).
170. S. Danishefsky and R. Cunningham, *J. Org. Chem.*, *30*, 3676 (1965).
171. S. Danishefsky and R. Cavanaugh, *J. Org. Chem.*, *33*, 2959 (1968).
172. G. A. Berchtold, J. Ciabattoni, and A. A. Tunick, *J. Org. Chem.*, *30*, 3679 (1965).
173. F. Bohlmann and O. Schmidt, *Chem. Ber.*, *99*, 1652 (1966).
174. H. L. Gingrich, D. M. Roush, and W. A. Van Saun, *J. Org. Chem.*, *48*, 4869 (1983).
175. S. Hunig and H. Kahanek, *Chem. Ber.*, *90*, 238 (1957).
176. A. J. Birch and E. G. Hutchinson, *J. Chem. Soc., Perkin Trans.*, *I*, 1757 (1973).

177. L. E. Overman, C. B. Petty, and R. J. Doedens, *J. Org. Chem.*, *44*, 4183 (1979).

178. J. O. Madsen and S-O. Lawesson, *Tetrahedron, 24*, 3369 (1968).

179. P. R. Nelson and R. G. Lawton, *J. Am. Chem. Soc.*, *88*, 3884 (1966).

180. H. Stetter and H. G. Thomas, *Chem. Ber.*, *101*, 1115 (1968).

181. H. Stetter, H. G. Thomas, and K. Meyer, *Chem. Ber.*, *103*, 863 (1970).

182. H. Stetter and K. Komorowski, *Chem. Ber.*, *104*, 75 (1971).

183. W. N. Speckamp, J. Dijkink, and H. O. Huisman, *J. Chem. Soc., Chem. Commun.*, 196 (1970).

184. A. W. J. D. Dekkers, W. N. Speckamp, and H. O. Huisman, *Tetrahedron Lett.*, 489 (1971).

185. P. W. Yang, M. Yasunami, and K. Takase, *Tetrahedron Lett.*, 4275 (1971).

186. M. Yasunami, P. W. Yang, Y. Kondo, Y. Noro, and K. Takase, *Chem. Letters*, 167 (1980).

187. M. Yasunami, A. Chen, Y. Noro, and K. Takase, *Chem. Letters*, 555 (1981).

188. F. D. Lewis, *Acc. Chem. Res.*, *12*, 152 (1979).

189. J. W. Lewis, P. L. Myers, and M. J. Readhead, *J. Chem. Soc. (C)*, 771 (1970).

190. V. I. Shvedov and A. N. Grinev, *Zh. Org. Khim, 1*, 1125 (1965).

191. K. C. Brannock, R. D. Burpitt, H. E. Davis, H. S. Pridgen, and J. G. Thweatt, *J. Org. Chem.*, *29*, 2579 (1964).

192. G. Domschke, *Chem. Ber.*, *99*, 934 (1966).

193. H. A. P. de Jongh, F. J. Gerhartl, and H. Wynberg, *J. Org. Chem.*, *30*, 1409 (1965).

194. T. V. Rao, L. Anandan, H. H. Mathur, and G. K. Trivedi, *Indian J. Chem.*, *Sect. B, 22B*, 864 (1983).

195. R. H. Rynbrandt, *J. Heterocycl. Chem.*, *11*, 787 (1974).

196. G. Tacconi, A. Gamba, F. Marinone, and G. Desimoni, *Tetrahedron, 27*, 561 (1971).

197. G. Tacconi, F. Marinone, A. Gamba, and G. Desimoni, *Tetrahedron, 28*, 1517 (1972).

198. G. Tacconi, A. G. Invernizzi, and G. Desimoni, *J. Chem. Soc., Perkin Trans., I*, 1872 (1976).

199. J. Ciabattoni and G. A. Berchtold, *J. Am. Chem. Soc.*, *87*, 1404 (1965).

200. J. Ciabattoni and G. A. Berchtold, *J. Org. Chem.*, *31*, 1336 (1966).

201. S. Penades, H. Kisch, K. Tortschanoff, P. Margaretha, and O. E. Polansky, *Monatsh. Chem.*, *104*, 447 (1973).

202. J. O. Madsen and S-O. Lawesson, *Bull. Soc. Chim. Belg.*, *85*, 805 (1976).

203. H. Kolind-Andersen and S-O. Lawesson, *Bull. Soc. Chim. Belg.*, *86*, 543 (1977).
204. M. E. Kuehne and L. Foley, *J. Org. Chem.*, *30*, 4280 (1965).
205. M. C. Moorjani and G. K. Trivedi, *Indian J. Chem.*, *16B*, 405 (1978).
206. A. Risaliti, M. Forchiassin, and E. Valentin, *Tetrahedron Lett.*, 6331 (1966).
207. A. Risaliti, S. Fatutta, M. Forchiassin, and E. Valentin, *Tetrahedron Lett.*, 1821 (1966).
208. E. Valentin, G. Pitacco, and F. P. Colonna, *Tetrahedron Lett.*, 2837 (1972).
209. F. P. Colonna, E. Valentin, G. Pitacco, and A. Risaliti, *Tetrahedron*, *29*, 3011 (1973).
210. H. G. Viehe and R. Verbruggen, *Chimia*, 352 (1975).
211. Y. Masuyama, H. Yamazaki, Y. Toyoda, and Y. Kurusu, *Synthesis*, 964 (1985).
212. D. Seebach, G. Calderari, W. L. Meyer, A. Merritt, and L. Odermann, *Chimia*, *39*, 183 (1985).
213. A. Risaliti, M. Forchiassin, and E. Valentin, *Tetrahedron*, *24*, 1889 (1968).
214. A. Risaliti, S. Fatutta, and M. Forchiassin, *Tetrahedron*, *23*, 1451 (1967).
215. J. Elguero, R. Jacquier, and G. Tarrago, *Bull. Soc. Chim. Fr.*, 1149 (1968).
216. M. Calligaris, M. Forchiassin, A. Risaliti, and C. Russo, *Gazz. Chim. Ital.*, *105*, 689 (1975).
217. A. Risaliti, S. Fatutta, M. Forchiassin, and C. Russo, *Ric. Sci.*, *38*, 827 (1968).
218. F. Johnson, L. G. Duquette, A. Whitehead, and L. C. Dorman, *Tetrahedron*, *30*, 3241 (1974).
219. G. Opitz and W. Merz, *Liebigs Ann. Chem.*, *652*, 139 (1962).
220. H. Nozaki, T. Yamaguti, S. Ueda, and K. Kondo, *Tetrahedron*, *24*, 1445 (1968).
221. G. Pitacco, A. Risaliti, M. L. Trevisan, and E. Valentin, *Tetrahedron*, *33*, 3145 (1977).
222. F. Benedetti, G. Pitacco, and E. Valentin, *Tetrahedron*, *35*, 2293 (1979).
223. G. H. Alt and A. J. Speziale, *J. Org. Chem.*, *29*, 798 (1964).
224. G. H. Alt, *J. Org. Chem.*, *31*, 2384 (1966).
225. R. D. Campbell and W. L. Harmer, *J. Org. Chem.*, *28*, 379 (1963).
226. R. D. Campbell and J. A. Jung, *J. Org. Chem.*, *30*, 3711 (1965).
227. W. L. Harmer, M.S. thesis, University of Iowa, Ames, 1962.
228. T. I. Nukai and R. Yoshizawa, *J. Org. Chem.*, *32*, 404 (1967).
229. S. Hunig and H. Hoch, *Tetrahedron Lett.*, 5215 (1966).

230. G. Opitz, H. Adolph, M. Kleemann, and F. Zimmermann, *Angew Chem.*, *73*, 654 (1961).
231. G. Opitz, H. Adolph, M. Kleemann, and F. Zimmermann, *Angew. Chem.*, *Int. Ed. Engl.*, *1*, 51 (1962).
232. G. Opitz and F. Zimmermann, *Liebigs Ann. Chem.*, *662*, 178 (1963).
233. G. Opitz and M. Kleemann, *Liebigs Ann. Chem.*, *665*, 114 (1963).
234. G. A. Berchtold, G. R. Harvey, and G. E. Wilson, Jr., *J. Org. Chem.*, *26*, 4776 (1961).
235. G. A. Berchtold, G. R. Harvey, and G. E. Wilson, Jr., *J. Org. Chem.*, *30*, 2642 (1965).
236. R. H. Hasels and J. C. Martin, *J. Org., Chem.*, *26*, 4775 (1961).
237. R. H. Hasek, and J. C. Martin, *J. Org. Chem.*, *28*, 1468 (1963).
238. R. H. Hasek, P. G. Gott, and J. C. Martin, *J. Org. Chem.*, *31*, 1931 (1966).
239. R. Huisgen, L. A. Feiler, and P. Otto, *Tetrahedron Lett.*, 4485 (1968).
240. P. Otto, L. A. Feiler, and R. Huisgen, *Angew. Chem.*, *Int. Ed. Engl.*, *7*, 737 (1968).
241. L. A. Feiler and R. Huisgen, *Chem. Ber.*, *102*, 3428 (1969).
242. J. C. Martin, P. G. Gott, and H. U. Hostettler, *J. Org. Chem.*, *32*, 1654 (1967).
243. R. D. Clark, *J. Org. Chem.*, *32*, 399 (1967).
244. S. Hunig, E. Benzing, and E. Lucke, *Chem. Ber.*, *90*, 2833 (1957).
245. S. Hunig, E. Lucke, and E. Benzing, *Chem. Ber.*, *91*, 129 (1950).
246. S. Hunig and W. Lendle, *Chem. Ber.*, *93*, 909, 913 (1960).
247. S. Hunig and M. Salzwedel, *Chem. Ber.*, *99*, 823 (1966).
248. H. J. Buysch and S. Hunig, *Angew. Chem.*, *Int. Ed. Engl.*, *5*, 128 (1966).
249. S. Hunig, H. J. Buysch, H. Hoch, and W. Lendle, *Chem. Ber.*, *100*, 3996 (1967).
250. B. Eistert and R. Wessendorg, *Chem. Ber.*, *94*, 2590 (1961).
251. F. Johnson, U.S. Pat. 3,153,650; *CA*, *62*, 1669g (1965).
252. H. Stetter, H. Held, and A. Schulte-Oestrich, *Chem. Ber.*, *95*, 1687 (1962).
253. A. Kirrmann and C. Wakselman, *C. R. Seances Acad. Sci.*, *261*, 759 (1965).
254. C. Wakselmann, *Bull. Soc. Chim. Fr.*, 3766 (1967).
255. A. Kirrmann and C. Wakselman, *Bull. Soc. Chim. Fr.*, 3766 (1967).
256. S. Hunig and H. Hoch, *Chem. Ber.*, *105*, 2216 (1972).

257. H. Hoch and S. Hunig, *Chem. Ber.*, *105*, 2660 (1972).
258. S. Hunig and H. Hoch, *Chem. Ber.*, *105*, 2197 (1972).
259. R. Huisgen and P. Otto, *J. Am. Chem. Soc.*, *91*, 5922 (1969).
260. R. Huisgen, L. A. Feiler, and P. Otto, *Chem. Ber.*, *102*, 3444 (1969).
261. L. Capuano, T. Tammer, and R. Zander, *Chem. Ber.*, *109*, 3497 (1976).
262. J. M. Berge, M. Rey, and A. S. Dreiding, *Helv. Chim. Acta*, *65*, 2230 (1982).
263. J. E. Baldwin, R. H. Fleming, and D. M. Simmons, *J. Org. Chem.*, *37*, 3963 (1972).
264. W. Ried and W. Kaeppeler, *Liebigs Ann. Chem.*, *687*, 183 (1965).
265. M. Miocque, M. D. d'Engenieres, and O. Lafont, *Tetrahedron Lett.*, 2133 (1976).
266. D. E. Heitmeier, J. T. Hortenstine, Jr., and A. P. Gray, *J. Org. Chem.*, *36*, 1449 (1971).
267. N. I. Delyagina, E. Y. Pervova, and I. L. Knunyants, *Dokl. Akad. Nauk SSSR*, *176*, 93 (1967).
268. J. J. Krutak, R. D. Burpitt, W. H. Moore, and J. A. Hyatt, *J. Org. Chem.*, *44*, 3847 (1979).
269. S. D. Darling and N. Subramanian, *Tetrahedron Lett.*, 3279 (1975).
270. S. D. Darling and N. Subramanian, *J. Org. Chem.*, *40*, 2851 (1975).
271. S. F. Martin, T. Chou, and C. Tu, *Tetrahedron Lett.*, 3823 (1979).
272. S. F. Martin, S. R. Desai, G. W. Phillips, and A. C. Miller, *J. Am. Chem. Soc.*, *102*, 3294 (1980).
273. S. F. Martin, C. Tu, and T. Chou, *J. Am. Chem. Soc.*, *102*, 5274 (1980).
274. D. L. Fields, T. H. Regan, and J. C. Dignan, *J. Org. Chem.*, *33*, 390 (1968).
275. H. Noda, K. Narimatsu, M. Hamana, and I. Ueda, *Hukusokan Kagaku Toronkai Koen Yoshishu*, *8th*, 144 (1975).
276. M. Oda, H. Tani, and Y. Kitahara, *J. Chem. Soc. Chem. Comun.*, 739 (1969).
277. M. Oda and Y. Kitahara, *Synthesis*, 367 (1971).
278. T. Wu and K. N. Houk, *J. Am. Chem. Soc.*, *107*, 5308 (1985).
279. C. Hubschwerien, J-P. Fleury, and H. Fritz, *Helv. Chim. Acta*, *60*, 2576 (1977).
280. J. E. Dolfini, K. Menich, and P. Corliss, *Tetrahedron Lett.*, 4421 (1966).
281. J. E. Dolfini and J. D. Simpson, *J. Am. Chem. Soc.*, *87*, 4381 (1965).
282. W. F. Berkowitz and S. C. Grenetz, *J. Org. Chem.*, *41*, 10 (1976).

283. D. Danishefsky, G. Rovnyak, and R. Cavenaugh, *J. Chem. Soc., Chem. Commun.*, 636 (1969).

284. S. Danishefsky and G. Rovnyak, *J. Chem. Soc., Chem. Commun.*, 820 (1972).

285. S. Danishefsky and G. Rovnyak, *J. Chem. Soc., Chem. Commun.*, 821 (1972).

286. L. N. Koikov and Y. G. Bundel, *Vestn. Mosk. Univ., Ser. 2: Khim.*, *26*, 86 (1985).

287. R. C. Bingham, M. J. S. Dewar, and D. H. Lo, *J. Am. Chem. Soc.*, *97*, 1285 (1975).

288. T. Eicher and J. L. Weber, *Topics Current Chem.*, *57*, 1 (1975).

289. J. Ciabattoni and E. C. Nathan, III, *J. Am. Chem. Soc.*, *89*, 3081 (1967).

290. O. Tsuge, S. Okita, M. Noguchi, and S. Kanemasa, *Chem. Letters*, 847 (1982).

291. T. Eicher and T. Born, *Liebigs Ann. Chem.*, *762*, 127 (1972).

292. O. Tsuge, S. Okita, M. Noguchi, and S. Kanemasa, *Chem. Letters*, 993 (1982).

293. J. W. Lown and T. W. Maloney, *Chem. Ind. (London)*, 870 (1970).

294. M. A. Steinfels and A. S. Dreiding, *Helv. Chim. Acta, 55*, 702 (1972).

295. V. Bilinski, M. A. Steinfels, and A. S. Dreiding, *Helv. Chim. Acta, 55*, 702 (1972).

296. V. Bilinski and A. S. Dreiding, *Helv. Chim. Acta, 55*, 1271 (1972).

297. T. Eicher and S. Bohm, *Tetrahedron Lett.*, 2603 (1972).

298. T. Eicher and S. Bohm, *Tetrahedron Lett.*, 3965 (1972).

299. T. Eicher and S. Bohm, *Chem. Ber.*, *107*, 2186 (1974).

300. T. Eicher and S. Bohm, *Chem. Ber.*, *107*, 2215 (1974).

301. T. Eicher and S. Bohm, *Chem. Ber.*, *107*, 2238 (1974).

302. M. H. Rosen, I. Fengler, and G. Bonet, *Tetrahedron Lett.*, 949 (1973).

303. O. Tsuge, S. Okita, M. Noguchi, and H. Watanabe, *Chem. Letters*, 1439 (1981).

304. K. C. Brannock, R. D. Burpitt, and J. G. Thweatt, *J. Org. Chem.*, *28*, 1462 (1963).

305. D. N. Reinhoudt and C. G. Kouwenhoven, *Tetrahedron, 30*, 2093 (1974).

306. D. N. Reinhoudt and C. G. Kouwenhoven, *Tetrahedron, 30*, 2431 (1974).

307. D. N. Reinhoudt, W. P. Trompenaars, and J. Geevers, *Tetrahedron Lett.*, 4777 (1976).

308. D. N. Reinhoudt, J. Geevers, and W. P. Trompenaars, *Tetrahedron Lett.*, 1351 (1978).

309. D. N. Reinhoudt, J. Geevers, W. P. Trompenaars, S. Harkema, and G. J. van Hummel, *J. Org. Chem.*, *46*, 424 (1981).
310. W. Verboom, G. W. Visser, W. P. Trompenaars, D. N. Reinhoudt, S. Harkema, and G. J. van Hummel, *Tetrahedron*, *37*, 3525 (1981).
311. G. W. Visser, W. Verboom, W. P. Trompenaars, and D. N. Reinhoudt, *Tetrahedron Lett.*, 1217 (1982).
312. G. W. Visser, W. Verboom, P. H. Benders, and D. N. Reinhoudt, *J. Chem. Soc., Chem. Commun.*, 669 (1982).
313. D. N. Reinhoudt, G. W. Visser, W. Verboom, P. H. Benders, and M. L. M. Pennings, *J. Am. Chem. Soc.*, *105*, 4775 (1983).
314. K. C. Brannock, R. D. Burpitt, V. W. Goodlett, and J. G. Thweatt, *J. Org. Chem.*, *28*, 1464 (1963).
315. C. F. Huebner, L. Dorfman, M. M. Robinson, E. Donoghue, W. G. Pierson, and P. Strachan, *J. Org. Chem.*, *28*, 3134 (1963).
316. G. A. Berchtold and G. F. Uhlig, *J. Org. Chem.*, *28*, 1459 (1963).
317. P. Schenone and G. Minardi, *Gazz. Chim. Ital.*, *100*, 945 (1970).
318. J. A. Hirsch and F. J. Cross, *J. Org. Chem.*, *36*, 955 (1971).
319. A. J. Birch and E. G. Hutchinson, *J. Chem. Soc. (C)*, 3671 (1971).
320. D. N. Reinhoudt and C. G. Kouwenhoven, *Tetrahedron Lett.*, 5203 (1972).
321. D. N. Reinhoudt and C. G. Leliveld, *Tetrahedron Lett.*, 3119 (1972).
322. M. Riviere, N. Paillous, and M. A. Lattes, *Bull. Soc. Chim. Fr.*, 1911 (1974).
323. T. Fex, J. Froborg, G. Magnusson, and S. Thoren, *J. Org. Chem.*, *41*, 3518 (1976).
324. D. Becker, L. R. Hughes, and R. A. Raphael, *J. Chem. Soc., Perkin Trans.*, *I*, 1674 (1977).
325. A. Mondon and G. Aumann, *Chem. Ber.*, *105*, 1459 (1972).
326. D. N. Reinhoudt, W. Verboom, G. W. Visser, W. P. Tompenaars, S. Harkema, and G. J. van Hummel, *J. Am. Chem. Soc.*, *106*, 1341 (1984).
327. D. N. Reinhoudt, *Adv. Heterocycl. Chem.*, *21*, 253 (1977).
328. A. K. Bose, G. Mina, M. S. Manhas, and E. Rzucidlo, *Tetrahedron Lett.*, 1467 (1963).
329. R. M. Acheson, R. M. Letcher, and G. Procter, *J. Chem. Soc., Perkin Trans.*, *I*, 535 (1980).
330. B. Lamm and C-J. Aurell, *Acta Chem. Scand. B.*, *36*, 435 (1982).
331. B. Lamm, L. Andersen, and C-J. Aurell, *Tetrahedron Lett.*, *24*, 2413 (1983).
332. R. W. Alder and G. Whittaker, *J. Chem. Soc., Chem. Commun.*, 776 (1971).

333. M. Suda and K. Hafner, *Tetrahedron Lett.*, 2449 (1977).

334. M. Suda and K. Hafner, *Tetrahedron Lett.*, 2453 (1977).

335. N. Miyamoto and H. Nozaki, *Bull. Chem. Soc. Jpn.*, 46, 1257 (1973).

336. K. C. Brannock, R. D. Burpitt, V. W. Goodlett, and J. Thweatt, *J. Org. Chem.*, 29, 818 (1964).

337. W. E. Parham and R. J. Sperley, *J. Org. Chem.*, 32, 926 (1967).

338. R. D. Burpitt and J. G. Thweatt, *Organic Syntheses* (P. Yates, ed.), Vol. 48, Wiley, New York, 1968, p. 56.

339. D. W. Boerth and F. A. Van-Catledge, *J. Org. Chem.*, 40, 3319 (1975).

340. V. Jager and H. G. Viehe, *Angew. Chem.*, 82, 836 (1970).

341. M. E. Kuehne, *J. Am. Chem. Soc.*, 84, 837 (1962).

342. G. Stork, Enamine Symposium, 140th National Meeting of the American Chemical Society, Chicago, Sept. 1961.

343. M. Ohno, *Tetrahedron Lett.*, 1753 (1963).

344. J. Wolinsky, D. Chan, and R. Novak, *Chem. Ind.* (*London*), 720 (1965).

345. M. Makosza and A. Kacprowicz, *Bull. Acad. Pol. Sci., Ser. Sci. Chim.*, 22, 467 (1974).

346. J. Graefe, M. Adler, and M. Muehlstaedt, *Z. Chem.*, 15, 14 (1975).

347. U. K. Pandit, S. A. G. de Graaf, C. T. Braams, and J. S. T. Raaphorst, *Recl. Trav. Chim. Pays-Bas*, 91, 799 (1972).

348. S. A. G. de Graaf and U. K. Pandit, *Tetrahedron*, 29, 2141 (1973).

349. S. A. G. de Graaf and U. K. Pandit, *Tetrahedron*, 29, 4263 (1973).

350. S. A. G. de Graaf and U. K. Pandit, *Tetrahedron*, 30, 1115 (1974).

351. E. Elkik and P. Vandescal, *C. R. Seances Acad. Sci.*, 261, 1015 (1965).

352. E. P. Blanchard, H. E. Simmons, and J. S. Taylor, *J. Org. Chem.*, 30, 4321 (1965).

353. M. E. Kuehne and G. DiVincenzo, *J. Org. Chem.*, 37, 1023 (1972).

354. M. E. Kuehne and J. C. King, *J. Org. Chem.*, 38, 304 (1973).

355. S. Wittig and F. Wingler, *Chem. Ber.*, 97, 2146 (1964).

356. D. L. Muck and E. R. Wilson, *J. Org. Chem.*, 33, 419 (1968).

357. H. Nozaki, S. Moriuti, M. Yamabe, and R. Noyori, *Tetrahedron Lett.*, 59 (1966).

358. H. Nozaki, H. Takaya, S. Moriuti, and R. Noyori, *Tetrahedron*, 24, 3655 (1968).

359. R. H. Rynbrandt and F. E. Dutton, *J. Org. Chem.*, 40, 2282 (1975).

360. B. Dorrer, E. O. Fischer, and W. Kalbfus, *J. Organomet. Chem.*, *81*, C20 (1974).
361. K. H. Dotz and I. Pruskil, *Chem. Ber.*, *114*, 1980 (1981).
362. G. Dannhardt and R. Obergrusberger, *Arch. Pharm.* (*Weinheim*), *314*, 787 (1981).
363. A. I. Meyers and J. C. Sircar, *Tetrahedron, 23*, 785 (1927).
364. J. C. Sircar and A. I. Meyers, *J. Org. Chem.*, *32*, 1248 (1967).
365. N. A. Nelson, J. E. Ladbury, and R. S. P. Hsi, *J. Am. Chem. Soc.*, *80*, 6633 (1958).
366. W. Sobotka, W. N. Beverung, G. G. Munoz, J. C. Sircar, and A. I. Meyers, *J. Org. Chem.*, *30*, 3667 (1965).
367. A. I. Meyers, A. H. Reine, J. C. Sircar, K. B. Rao, S. Singh, H. Weidmann, and M. Fitzpatrick, *J. Heterocycl. Chem.*, 5, 151 (1968).
368. A. I. Meyers and J. C. Sircar, *J. Org. Chem.*, *32*, 1250 (1967).
369. A. I. Meyers, J. C. Sircar, and S. Singh, *J. Heterocycl. Chem.*, *4*, 461 (1967).
370. G. Opitz and H. Mildenberger, *Liebigs Ann. Chem.*, *650*, 115 (1961).
371. H. Krieger, H. Ruolsalainen, and J. Montin, *Chem. Ber.*, *99*, 3715 (1967).
372. M. Winn and F. G. Bordwell, *J. Org. Chem.*, *32*, 1610 (1967).
373. J. M. McEuen, R. P. Nelson, and R. G. Lawton, *J. Org. Chem.*, *35*, 690 (1970).
374. H. Stetter, K-D. Ramsch, and K. Elfert, *Liebigs Ann. Chem.*, 1322 (1974).
375. W. C. Still, *Synthesis*, 453 (1976).
376. J. A. Peters, J. M. van der Toorn, and H. van Bekkum, *Tetrahedron, 30*, 633 (1974).
377. P. Camps, C. Jaime, and J. Molas, *Tetrahedron Lett.*, *22*, 2487 (1981).
378. P. Camps and C. Jaime, *Tetrahedron, 36*, 393 (1980).
379. P. Camps and C. Jaime, *Org. Magn. Reson.*, *14*, 177 (1980).
380. H. Onoue, I. Moritani, and S-I. Murahashi, *Tetrahedron Lett.*, 121 (1973).
381. R. Noyori, K. Yokoyama, S. Makino, and Y. Hayakawa, *J. Am. Chem. Soc.*, *94*, 1772 (1972).
382. R. Noyori, F. Shimizu, K. Fukuta, H. Takaya, and Y. Hayakawa, *J. Am. Chem. Soc.*, *99*, 5196 (1977).
383. Y. Hayakawa, K. Yokoyama, and R. Noyori, *J. Am. Chem. Soc.*, *100*, 1799 (1978).
384. R. Noyori, K. Yokoyama, and Y. Hayakawa, *Organic Syntheses*, (W. A. Sheppard, ed.), Vol. 58, Wiley, 1978, p. 56.
385. M. Muhlstadt and J. Reimer, *Z. Chem.*, *4*, 70 (1964).
386. R. Gompper and D. Lach, *Angew. Chem., Int. Ed. Engl.*, *12*, 567 (1973).
387. G. K. Kang and T. H. Chan, *Can. J. Chem.*, *63*, 3102 (1985).

388. E. Buncel, A. R. Norris, and K. E. Russell, *Chem. Soc. Q. Rev.*, *22*, 123 (1968).
389. M. J. Strauss and H. Schran, *J. Am. Chem. Soc.*, *91*, 3974 (1969).
390. M. J. Strauss, T. C. Jensen, H. Schran, and K. O'Conner, *J. Org. Chem.*, *35*, 383 (1970).
391. H. Schran and M. J. Strauss, *J. Org. Chem.*, *36*, 856 (1971).
392. S. R. Alpha, *J. Org. Chem.*, *38*, 3136 (1973).
393. J. C. Martin, R. D. Burpitt, and H. U. Hostettler, *J. Org. Chem.*, *32*, 210 (1967).
394. G. Bianchetti, D. Pocar, P. D. Croce, G. G. Gallo, and A. Vigevani, *Tetrahedron Lett.*, 1637 (1966).
395. G. Bianchetti, P. D. Croce, D. Pocar, R. Stradi, and G. G. Gallo, *Gazz. Chim. Ital.*, *97*, 564 (1967).
396. G. Bianchetti, D. Pocar, R. Stradi, P. D. Croce, and A. Vigevani, *Gazz. Chim. Ital.*, *97*, 872 (1967).
397. L. A. Paquette, *Tetrahedron Lett.*, 2027 (1963).
398. N. I. Delyagina, E. Y. Perova, and I. L. Knunyants, *Dokl. Akad. Nauk. SSSR*, *176*, 93 (1967).
399. G. Domschke, *Z. Chem.*, *4*, 29 (1964).
400. G. Domschke, *J. Prakt. Chem.*, *32*, 144 (1966).
401. G. R. Allen, Jr., *J. Org. Chem.*, *33*, 3346 (1968).
402. A. G. Makhsumov, I. T. Turdimukhamedova, and A. Safev, *Dokl. Akad. Nauk Uzb. SSR*, *28*, 44 (1971).
403. R. Cassis, R. Tapia, and J. A. Valderrama, *J. Heterocycl. Chem.*, *21*, 869 (1984).
404. L. Birkofer and W. Quittman, *Chem. Ber.*, *119*, 257 (1986).
405. G. Domschke, *J. Prakt. Chem.*, *32*, 140 (1966).
406. G. Domschke, *Chem. Ber.*, *99*, 930 (1966).
407. W. Ried and E. Torok, *Liebigs Ann. Chem.*, *687*, 187 (1965).
408. C. Nolde and S-O. Lawesson, *Bull. Soc. Chem. Belg.*, *86*, 313 (1977).
409. P. Jakobsen and S-O. Lawesson, *Tetrahedron*, *24*, 3671 (1968).
410. M. S. Manhas, J. W. Brown, and U. K. Pandit, *Heterocycles*, *3*, 117 (1975).
411. T. Sato and K. Watanabe, *Chem. Letters*, 1499 (1983).
412. M. Oda and Y. Kitahara, *Synthesis*, 368 (1971).
413. L. A. Paquette, *Tetrahedron Lett.*, 1291 (1965).
414. L. A. Paquette and H. Stucki, *J. Org. Chem.*, *31*, 1232 (1966).
415. M. S. Manhas and J. R. McCoy, *J. Chem. Soc. (C)*, 1419 (1969).
416. S. Tobinaga, N. Takeuchi, and H. Nakagawa, *J. Chem. Soc., Chem. Commun.*, 890 (1972).
417. H. Sliwa and G. Cordonnier, *J. Heterocycl. Chem.*, *14*, 169 (1977).
418. F. M. Dean and R. S. Varma, *Tetrahedron Lett.*, 2113 (1981).

419. F. M. Dean and R. S. Varma, *J. Chem. Soc., Perkin Trans.,* *I*, 1193 (1982).

420. M. von Strandtmann, M. P. Cohen, and J. Shavel, Jr., *Tetrahedron Lett.*, 3103 (1965).

421. D. Blondeau and H. Sliwa, *C. R. Seances Acad. Sci.*, 947 (1975).

422. E. S. Kagan, V. I. Mikhailov, V. D. Sholle, V. A. Smirnov, and E. G. Rozantsev, *Izv. Akad. Nauk SSSR, Ser. Khim.*, 1668 (1978).

423. D. Blondeau and H. Sliwa, *J. Chem. Res. (S)*, 2 (1979); *J. Chem. Res. (M)*, 117 (1979).

424. S. Klutchko, A. C. Sonntag, M. von Strandtmann, and J. Shavel, Jr., *J. Org. Chem.*, *38*, 3049 (1973).

425. G. V. Boyd and D. Hewson, *J. Chem. Soc., Chem. Commun.*, 536 (1965).

426. R. S. Monson, *J. Heterocycl. Chem.*, *13*, 893 (1976).

427. F. Eiden and K. T. Wanner, *Arch. Pharm.*, *317*, 958 (1984).

428. G. Tacconi, F. Marinone, and G. Desimoni, *Gazz. Chim. Ital.*, *100*, 284 (1970).

429. F. Eiden and G. Felbermeir, *Arch. Pharm.*, *317*, 861 (1984).

430. B. A. Arbuzov and N. N. Zobova, *Synthesis*, 461 (1974).

431. R. Lattrell, *Liebigs Ann. Chem.*, *722*, 142 (1969).

432. M. Uher, J. Foltin, and L. Floch, *Coll. Czech. Chem. Commun.*, *46*, 2696 (1981).

433. T. V. Rao, V. S. Ekkundi, and G. Kumar, *J. Chem. Res. (S)*, 116 (1986).

434. B. B. Millward, *J. Chem. Soc.*, 26 (1960).

435. S. Hunig, *Angew. Chem.*, *71*, 312 (1959).

436. S. Hunig, E. Benzing, and K. Hubner, *Chem. Ber.*, *94*, 486 (1961).

437. A. Riecke, E. Schmitz, and E. Grundermann, *Angew. Chem.*, *72*, 635 (1960).

438. H. Hartmann and R. Mayer, *Z. Chem.*, *5*, 152 (1965).

439. R. Mayer and K. Gewald, *Angew. Chem., Int. Ed. Engl.*, *6*, 294 (1967).

440. K. Ley and R. Nast, *Angew. Chem., Int. Ed. Engl.*, *4*, 519 (1965).

441. Y. Nomura and Y. Takeuchi, *Bull. Chem. Soc. Jpn.*, *36*, 1044 (1963).

442. R. Mayer and H. Wehl, *Angew. Chem., Int. Ed. Engl.*, *3*, 705 (1964).

443. J. Fabian, K. Gewald, and R. Mayer, *Angew. Chem., Int. Ed. Engl.*, *2*, 45 (1963).

444. R. Mayer, P. Wittig, J. Fabian, and R. Heitmuller, *Chem. Ber.*, *97*, 654 (1964).

445. E. Fanghaenel, *Z. Chem.*, *4*, 41 (1964).

446. C. Djerassi and B. Tursch, *J. Org. Chem.*, *27*, 1041 (1962).
447. J. P. Sauve and N. Lozac'h, *Bull. Soc. Chim. Fr.*, 2016 (1970).
448. F. Ishii, M. Stavaux, and N. Lozac'h, *Tetrahedron Lett.*, 1473 (1975).
449. F. Ishii, M. Stavaux, and N. Lozac'h, *Bull. Soc. Chim. Fr.*, 1142 (1977).
450. F. Ishii, R. Okazaki, and N. Inamoto, *Tetrahedron Lett.*, 4283 (1976).
451. R. Okazaki, F. Ishii, and N. Inamoto, *Bull. Chem. Soc. Jpn.*, *51*, 309 (1978).
452. E. Schaumann, S. Sieveking, and W. Walter, *Tetrahedron*, *30*, 4147 (1974).
453. E. Schaumann, H-G. Bauch, S. Sieveking, and G. Adiwidijaja, *Chem. Ber.*, *116*, 55 (1983).
454. E. Fanghanel, *J. Prakt. Chem.*, *317*, 123 (1975).
455. K. Fujimori, T. Fujita, K. Yamane, M. Yasunami, and K. Takase, *Chem. Letters*, 1721 (1983).
456. J-P. Declercq, G. Germain, and M. Van Meerssche, *Tetrahedron Lett.*, 1819 (1979).
457. M. H. Rosen and G. Bonet, *J. Org. Chem.*, *39*, 3805 (1974).
458. E. Jongejan, H. Steinberg, and T. J. de Boer, *Tetrahedron Lett.*, 397 (1976).
459. T. Nagai, T. Shingaki, M. Inagaki, and T. Ohshima, *Bull. Chem. Soc. Jpn.*, *52* 1102 (1979).
460. L. A. Paquette, R. W. Houser, and M. Rosen, *J. Org. Chem.*, *35*, 905 (1970).
461. T. J. Wallace, *Q. Rev. (London)*, *20*, 67 (1966).
462. T. Tanabe and T. Nagai, *Bull. Chem. Soc. Jpn.*, *50*, 1179 (1977).
463. J. F. King and R. P. Beatson, *J. Chem. Soc., Chem. Commun.*, 663 (1970).
464. G. Stork and I. J. Borowitz, *J. Am. Chem. Soc.*, *84*, 313 (1962). (1962).
465. I. J. Borowitz, *J. Am. Chem. Soc.*, *86*, 1146 (1964).
466. G. Opitz and H. Adolph, *Angew. Chem., Int. Ed. Engl.*, *1*, 113 (1962).
467. G. Opitz and H. Schempp, and H. Adolph, *Liebigs Ann. Chem.*, *684*, 92 (1965).
468. F. Fusco, R. Rossi, and S. Maiorana, *Chim. Ind. (Milan)*, *44*, 873 (1962).
469. W. E. Truce, J. R. Norell, J. E. Richman, and J. P. Walsh, *Tetrahedron Lett.*, 1677 (1963).
470. D. C. Dittmer and F. A. Davis, *J. Org. Chem.*, *29*, 3131 (1964).
471. L. A. Paquette, *J. Org. Chem.*, *30*, 629 (1965).
472. L. A. Paquette, M. Rosen, and H. Stucki, *J. Org. Chem.*, *33*, 3020 (1968).

473. J. N. Wells and F. S. Abbott, *J. Med. Chem.*, *9*, 489 (1966).
474. G. Opitz, *Angew. Chem., Int. Ed. Engl.*, *7*, 646 (1968).
475. P. L-F. Chang and D. C. Dittmer, *J. Org. Chem.*, *34*, 2791 (1969).
476. S. F. Stephen and E. Marcus, *J. Org. Chem.*, *34*, 2535 (1969).
477. L. A. Paquette and R. W. Begland, *J. Org. Chem.*, *34*, 2896 (1969).
478. L. A. Paquette, J. P. Freeman, and R. W. Houser, *J. Org. Chem.*, *34*, 2901 (1969).
479. H. Mazarguil and A. Lattes, *Bull. Soc. Chim. Fr.*, 3713 (1969).
480. L. A. Paquette, J. P. Freeman, and S. Maiorana, *Tetrahedron*, *27*, 2599 (1971).
481. C. T. Goralski and T. E. Evans, *J. Org. Chem.*, *37*, 2080 (1972).
482. H. Mazarguil and A. Lattes, *Bull. Soc. Chim. Fr.*, 500 (1974).
483. W. E. Truce and J. F. Rach, *J. Org. Chem.*, *39*, 1109 (1974).
484. A. Etienne and B. Desmazieres, *J. Chem. Res. (S)*, 484 (1978); *J. Chem. Res. (M)*, 5501 (1978).
485. R. A. Ferri, G. Pitacco, and E. Valentin, *Tetrahedron*, *34*, 2537 (1978).
486. P. Bradamante, M. Forchiassin, G. Pitacco, C. Russo, and E. Valentin, *J. Heterocycl. Chem.*, *19*, 985 (1982).
487. G. Opitz and K. Fisher, *Z. Naturforsch. B*, *18*, 775 (1963).
488. J. J. Looker, *J. Org. Chem.*, *31*, 2973 (1966).
489. M. E. Kuehne, *J. Org. Chem.*, *28*, 2124 (1963).
490. L. A. Paquette, *J. Org. Chem.*, *29*, 2851 (1964).
491. L. A. Paquette, *J. Org. Chem.*, *29*, 2854 (1964).
492. L. A. Paquette and M. Rosen, *Tetrahedron Lett.*, 703 (1967).
493. L. A. Paquette and M. Rosen, *J. Am. Chem. Soc.*, *89*, 4102 (1967).
494. B. A. Reith, J. Strating, and A. M. van Leusen, *J. Org. Chem.*, *39*, 2728 (1974).
495. G. Opitz and F. Schweinsberg, *Angew. Chem., Int. Ed. Engl.*, *4*, 786 (1965).
496. L. A. Paquette and M. Rosen, *Tetrahedron Lett.*, 311 (1966).
497. G. Opitz and E. Tempel, *Angew. Chem., Int. Ed. Engl.*, *3*, 754 (1964).
498. A. Gandini, P. Schenone, and G. Bignardi, *Monastsh.*, *98*, 1518 (1967).
499. G. Opitz and E. Tempel, *Liebigs Ann. Chem.*, *699*, 68 (1966).
500. S. Maiorana and G. Pagani, *Chim. Ind. (Milan)*, *48*, 1193 (1966).
501. S. Bradamante, S. Maiorana, and G. Pagani, *J. Chem. Soc., Perkin Trans.*, *I*, 282 (1972).
502. T. Nagai, H. Namikoshi, and N. Tokura, *Tetrahedron Lett.*, 4329 (1968).

503. O. Tsuge, S. Iwanami, and S. Hagio, *Bull. Chem. Soc. Jpn.*, *45*, 237 (1972).

504. G. Opitz, M. Kleeman, D. Beucher, G. Walz, and K. Rieth, *Angew. Chem., Int. Ed. Engl.*, *5*, 594 (1966).

505. G. Opitz and D. Buecher, *Tetrahedron Lett.*, 5263 (1966).

506. F. Soccolini, G. Chelucci, and C. Botteghi, *J. Heterocycl. Chem.*, *21*, 1001 (1984).

507. G. A. Berchtold, *J. Org. Chem.*, *26*, 3043 (1961).

508. R. Fusco, G. Bianchetti, and S. Rossi, *Gazz. Chim. Ital.*, *91*, 825 (1961).

509. S. Hunig, K. Hubner, and E. Benzing, *Chem. Ber.*, *95*, 926 (1962).

510. W. Ried and W. Kaeppeler, *Liebigs Ann. Chem.*, *673*, 132 (1964).

511. W. Ried and W. Kaeppeler, *Liebigs Ann. Chem.*, *688*, 177 (1965).

512. M. Perelman and S. A. Mizsak, *J. Am. Chem. Soc.*, *84*, 4988 (1962).

513. G. Opitz and J. Koch, *Angew. Chem., Int. Ed. Engl.*, *2*, 152 (1963).

514. A. K. Bose and G. Mina, *J. Org. Chem.*, *30*, 812 (1965).

515. L. Capuano, M. Kussler, and H. R. Kirn, *Ann. Univ. Sarav. Math.-Naturwiss. Fak.*, 7 (1981).

516. R. Oda, A. Miyasu, and M. Okano, *Nippon Kagaku Zasshi*, *88*, 96 (1967).

517. E. Schaumann, S. Sieveking, and W. Walter, *Tetrahedron Lett.*, 209 (1974).

518. A. W. Faull, D. Griffiths, R. Hull, and T. P. Seden, *J. Chem. Soc., Perkin Trans.*, I, 2587 (1980).

519. K. Ley, U. Eholzer, and R. Nast, *Angew. Chem., Int. Ed. Engl.*, *4*, 519 (1965).

520. B. A. Arbuzov, N. N. Zobova, and F. B. Balabanova, *Izv. Akad. Nauk SSSR, Ser. Khim.*, 2086 (1972).

521. L. A. Lazukina, V. I. Gorbatenko, L. F. Lur'e, and V. P. Kukhar, *Zh. Org. Khim.*, *13*, 290 (1977).

522. S. Hunig and K. Hubner, *Chem. Ber.*, *95*, 937 (1962).

523. S. Stankovsky, *Coll. Czech. Chem. Commun.*, *48*, 3575 (1983).

524. A. K. Bose, M. S. Manhas, V. V. Rao, C. T. Chen, I. R. Trehan, S. D. Sharma, and S. G. Amin, *J. Heterocycl. Chem.*, *8*, 1091 (1971).

525. A. K. Bose, M. S. Manhas, S. D. Sharma, S. G. Amin, and H. P. S. Chawla, *Syn. Commun.*, *1*, 33 (1971).

526. A. K. Bose, M. S. Manhas, V. V. Rao, I. R. Trehan, S. D. Sharma, and H. P. S. Chawla, *Tetrahedron*, *28*, 2931 (1972).

527. M. S. Manhas, C. T. Chen, V. V. Rao, I. R. Trehan, and S. D. Sharma, *J. Chem. Soc., Perkin Trans.*, I, 2119 (1972).

528. S. D. Sharma and V. Rani, *Indian J. Chem.*, *Sect. B*, *14*, 132 (1976).
529. S. D. Sharma, P. K. Gupta, and A. L. Gauba, *Indian J. Chem.*, *Sect. B.*, *15*, 960 (1977).
530. S. D. Sharma, S. Manuja, and A. L. Gauba, *Indian J. Chem.*, *Sect. B*, *19*, 246 (1980).
531. J. H. Rigby and N. Balasubramonian, *J. Org. Chem.*, *49*, 4569 (1984).
532. L. Pellacani, P. Pulcini, and P. A. Tardella, *J. Org. Chem.*, *47*, 5023 (1982).
533. S. Fioravanti, M. A. Loreto, L. Pellaconi, and P. A. Tardella, *J. Org. Chem.*, *50*, 5365 (1985).
534. J. E. Dolfini and D. M. Dolfini, *Tetrahedron Lett.*, 2053 (1965).
535. G. L'abbe, P. V. Stappen, and J-P. Dekerk, *J. Chem. Soc.*, *Chem. Commun.*, 784 (1982).
536. K. W. Law, T-F. Lai, M. P. Sammes, A. R. Katritzky, and T. C. W. Mak, *J. Chem. Soc.*, *Perkin Trans.*, *I*, 111 (1984).
537. S. Tomoda, Y. Takeuchi, and Y. Nomura, *Tetrahedron Lett.*, 3549 (1969).
538. Y. Nomura, S. Tomoda, and Y. Takeuchi, *Chem. Letters*, 79 (1972).
539. Y. Nomura, M. Kimura, T. Shibata, Y. Takeuchi, and S. Tomoda, *Bull. Chem. Soc. Jpn.*, *55*, 3343 (1982).
540. Y. Nomura, M. Kimura, Y. Takeuchi, and S. Tomoda, *Chem. Letters*, 267 (1978).
541. M. Komatsu, H. Ohgishi, Y. Ohshiro, and T. Agawa, *Tetrahedron Lett.*, 4589 (1976).
542. M. Komatsu, H. Ohgishi, S. Takamatsu, Y. Ohshiro, and T. Agawa, *Angew. Chem.*, *Int. Ed. Engl.*, *21*, 213 (1982).
543. M. Komatsu, S. Takamatsu, M. Uesaka, S. Yamamoto, Y. Ohshiro, and T. Agawa, *J. Org. Chem.*, *49*, 2691 (1984).
544. M. J. Haddadin and C. H. Issidorides, *Tetrahedron Lett.*, 3253 (1965).
545. J. W. McFarland, *J. Org. Chem.*, *36*, 1842 (1971).
546. N. A. Mufarrij, M. J. Haddadin, C. H. Issidorides, J. W. McFarland, and J. D. Johnson, *J. Chem. Soc.*, *Perkin Trans.*, *I*, 965 (1972).
547. P. Devi, J. S. Sandhu, and G. Thyagarajan, *J. Chem. Soc.*, *Chem. Commun.*, 710 (1979).
548. I. Adachi, *Chem. Pharm. Bull.*, *17*, 2209 (1969).
549. W. Steglich and O. Hollitzer, *Angew. Chem.*, *Int. Ed. Engl.*, *12*, 495 (1973).
550. T. Kato, Y. Yamamoto, and M. Kondo, *Heterocycles*, 293 (1975).
551. H. Griengl, G. Prischl, and A. Bleikolm, *Liebigs Ann. Chem.*, 1573 (1980).

552. V. N. Charushin and H. C. van der Plas, *Tetrahedron Lett.*, 23, 3965 (1982).
553. R. F. Parcell and F. P. Hauck, Jr., *J. Org. Chem.*, 28, 3468 (1963).
554. L. H. Hellberg, R. J. Milligan, and R. N. Wilke, *J. Chem. Soc. (C)*, 35 (1970).
555. I. Ninomiya, T. Naito, S. Higuchi, and T. Mori, *J. Chem. Soc., Chem. Commun.*, 457 (1971).
556. H. Person and A. Foucaud, *C. R. Seances Acad. Sci.*, 265, 1007 (1967).
557. A. Maccioni, E. Marongiu, and G. Bianchetti, *Gazz. Chim. Ital.*, 100, 288 (1970).
558. P. Bravo, G. Gaudiano, P. P. Ponti, and A. Umani-Ronchi, *Tetrahedron*, 26, 1315 (1970).
559. S. Nakanishi, Y. Shirai, K. Takahashi, and Y. Otsuji, *Chem. Letters*, 869 (1981).
560. G. Cerioni, A. Maccioni, P. P. Piras, and A. Plumitallo, *Gazz. Chim. Ital.*, 111, 391 (1981).
561. M. Cannas, M. T. Cocco, A. Maccioni, G. Marongiu, and A. Plumitallo, *Tetrahedron*, 40, 2041 (1984).
562. M. T. Cocco, A. Maccioni, and A. Plumitallo, *Gazz. Chim. Ital.*, 114, 521 (1984).
563. L. Citerio, D. Pocar, R. Stradi, and B. Gioia, *Tetrahedron*, 35, 69 (1979).
564. L. Citerio, D. Pocar, R. Stradi, and B. Gioia, *J. Chem. Soc., Perkin Trans., I*, 309 (1978).
565. L. Citerio, D. Pocar, M. L. Saccarello, and R. Stradi, *Tetrahedron*, 35, 2453 (1979).
566. J. W. Lewis, P. L. Myers, and J. A. Ormerod, *J. Chem. Soc., Perkin Trans., I*, 1129 (1973).
567. H. Ahlbrecht and C. Vonderheid, *Chem. Ber.*, 108, 2300 (1975).
568. K. Burger, F. Hein, U. Wasssmuth, and H. Krist, *Synthesis*, 904 (1981).
569. K. Burger, U. Wasmuth, F. Hein, and S. Rottegger, *Liebigs Ann. Chem.*, 991 (1984).
570. K. Burger, U. Wassmuth, B. Forster, and S. Penninger, *Z. Naturforsch. B*, 39, 1442 (1984).
571. O. L. Chapman and G. L. Eian, *J. Am. Chem. Soc.*, 90, 5329 (1968).
572. A. Bloom and J. C. Clardy, *J. Chem. Soc., Chem. Commun.*, 531 (1970).
573. O. L. Chapman, G. L. Eian, A. Bloom, and J. Clardy, *J. Am. Chem. Soc.*, 93, 2918 (1971).
574. A. Rahman and M. Ghazala, *Heterocycles*, 16, 261 (1981).
575. A. Rahman and M. Ghazala, *J. Chem. Soc., Perkin Trans., I*, 59 (1982).

576. K. H. Grellmann, W. Kuhnle, and T. Wolff, *Z. Phys. Chem. (Frankfurt am Main)*, *101*, 295 (1976).

577. T. Wolff and R. Waffenschmidt, *J. Am. Chem. Soc.*, *102*, 6098 (1980).

578. K. H. Grellmann, U. Schmitt, and H. Weller, *Chem. Phys. Lett.*, *88*, 40 (1982).

579. U. Baron, G. Bartelt, A. Eychmuller, K. H. Grellmann, U. Schmitt, E. Tauer, and H. Weller, *J. Photochem.*, *28*, 187 (1985).

580. J. F. Nicoud and H. B. Kagan, *Isr. J. Chem.*, *15*, 78 (1977).

581. A. G. Schultz and C-K. Sha, *Tetrahedron*, *36*, 1757 (1980).

582. R. A. Firestone, *J. Org. Chem.*, *33*, 2285 (1968).

583. R. Huisgen, *J. Org. Chem.*, *33*, 2291 (1968).

584. R. Huisgen, *J. Org. Chem.*, *41*, 403 (1976).

585. A. Padwa, ed., *1,3-Dipolar Cycloaddition Chemistry*, Vols. 1 and 2, Wiley-Interscience, New York, 1984.

586. Y. Nomura, F. Furasaki, and Y. Takeuchi, *Bull. Chem. Soc. Jpn.*, *40*, 1740 (1967).

587. O. Tsuge, M. Tashiro, and Y. Nishihara, *Tetrahedron Lett.*, 3769 (1967).

588. O. Tsuge, M. Tashiro, and Y. Nishihara, *Nippon Kagaku Zasshi*, *92*, 72 (1971).

589. J. P. Freeman, M. J. Haddadin, and J. F. Hansen, *J. Org. Chem.*, *44*, 4978 (1979).

590. H. Suschitzky, B. J. Wakefield, and J. P. Whitten, *J. Chem. Soc., Perkin Trans., I*, 2709 (1980).

591. Y. Nomura, F. Furusaki, and Y. Takeuchi, *Bull. Chem. Soc. Jpn.*, *43*, 1913 (1970).

592. R. Fusco, G. Bianchetti, and D. Pocar, *Gazz. Chim. Ital.*, *91*, 1233 (1961).

593. M. E. Kuehne, S. J. Weaver, and P. Franz, *J. Org. Chem.*, *29*, 1582 (1964).

594. R. Huisgen, R. Sustmann, and G. Wallbillich, *Chem. Ber.*, *100*, 1786 (1967).

595. S. Kitane, T. Kabula, J. Vebrel, and B. Laude, *Tetrahedron Lett.*, *22*, 1217 (1981).

596. G. Bianchetti, D. Pocar, and P. D. Croce, *Gazz. Chim. Ital.*, *93*, 1714 (1963).

597. G. Bianchetti, D. Pocar, and P. D. Croce, *Gazz. Chim. Ital.*, *93*, 1726 (1963).

598. A. Halleux and H. G. Viehe, *J. Chem. Soc. (C)*, 1726 (1968).

599. G. Bianchetti, D. Pocar, and R. Stradi, *Gazz. Chim. Ital.*, *100*, 726 (1970).

600. P. Caramella and E. Cereda, *Synthesis*, 433 (1971).

601. H. Noda, T. Yamamori, M. Yoshida, and M. Hamana, *Heterocycles*, *4*, 453 (1976).

602. A. Krutosikova, J. Kovac, M. Dandarova, and M. Valentiny, *Coll. Czech. Chem. Commun.*, *43*, 288 (1978).

603. G. Nathansohn, E. Testa, and N. DiMola, *Experientia*, *18*, 57 (1962).

604. R. Fusco, G. Bianchetti, and D. Pocar, *Gazz. Chim. Ital.*, *91*, 849 (1961).

605. R. Fusco, G. Bianchetti, and D. Pocar, and R. Ugo, *Gazz. Chim. Ital.*, *92*, 1040 (1962).

606. D. Pocar, G. Bianchetti, and P. D. Croce, *Chem. Ber.*, 97 1225 (1964).

607. M. E. Munk and Y. K. Kim, *J. Am. Chem. Soc.*, *86*, 2213 (1964).

608. G. Bianchetti, P. D. Croce, and D. Pocar, *Tetrahedron Lett.*, 2039 (1965).

609. D. Pocar, G. Bianchetti, and P. D. Croce, *Gazz. Chim. Ital.*, *95*, 1220 (1965).

610. S. Maiorana, D. Pocar, and P. D. Croce, *Tetrahedron Lett.*, 6043 (1966).

611. R. Huisgen, G. Szeimies, and L. Mobius, *Chem. Ber.*, *100*, 2494 (1967).

612. G. L'Abbe, *Ind. Chem. Belge*, *32*, 541 (1967); *CA*, *67*, 116272c (1967).

613. A. S. Bailey and J. E. White, *J. Chem. Soc. (B)*, 819 (1966).

614. G. Bianchetti, R. Stradi, and D. Pocar, *J. Chem. Soc., Perkin Trans.*, *I*, 997 (1972).

615. R. Sustmann, *Tetrahedron Lett.*, 963 (1974).

616. M. K. Meilahn, B. Cox, and M. E. Munk, *J. Org. Chem.*, *40*, 819 (1975).

617. L. Citerio, M. L. Saccarello, and R. Stradi, *Synthesis*, 305 (1979).

618. N. Almirante, M. L. Gelmi, P. Marelli, and D. Pocar, *Tetrahedron*, *42*, 57 (1986).

619. N. Almirante, M. Ballabio. G. Bianchetti, A. Cambiaghi, and D. Pocar, *J. Chem. Res. (S)*, 132 (1986); *J. Chem. Res. (M)*, 1233 (1986).

620. D. Pocar, R. Stradi, and G. Bianchetti, *Gazz. Chim. Ital.*, *100*, 1135 (1970).

621. R. Fusco, G. Bianchetti, D. Pocar, and R. Ugo, *Chem. Ber.*, *96*, 802 (1963).

622. D. Pocar and P. Trimarco, *J. Chem. Soc., Perkin Trans.*, *I*, 622 (1976).

623. P. D. Croce and R. Stradi, *Tetrahedron*, *33*, 865 (1977).

624. E. Fanghaenel, *Z. Chem.*, *3*, 309 (1963).

625. R. D. Burpitt and V. W. Goodett, *J. Org. Chem.*, *30*, 4308 (1965).

626. Y. Nomura, Y. Takeuchi, S. Tomoda, and M. M. Ito, *Bull. Chem. Soc. Jpn.*, *54*, 261 (1981).

627. L. Bruche, L. Garanti, and G. Zecchi, *J. Chem. Soc., Perkin Trans., I,* 1427 (1984).
628. S. Yamada, Y. Hamada, K. Ninomiya, and T. Shioiri, *Tetrahedron Lett.,* 4749 (1976).
629. T. Shioiri and N. Kawai, *J. Org. Chem., 43,* 2936 (1978).
630. S. Kitane, M. Berrada, J. Vebrel, and B. Laude, *Bull. Soc. Chim. Belg., 94,* 163 (1985).
631. R. Sustmann and H. Trill, *Angew. Chem., Int. Ed. Engl., 11,* 838 (1972).
632. P. Scheiner, J. H. Schomaker, S. Deming, W. J. Libbey, and G. P. Nowack, *J. Am. Chem. Soc., 87,* 306 (1965).
633. N. S. Basketter and A. O. Plunkett, *J. Chem. Soc., Chem. Commun.,* 188 (1973).
634. S. Hunig, G. Kiesslich, F. Linhart, and H. Schlaf, *Liebigs Ann. Chem., 752,* 182 (1971).
635. W. E. Bachmann and N. C. Deno, *J. Am. Chem. Soc., 71,* 3062 (1949).
636. D. L. Boger and J. S. Panek, *J. Org. Chem., 46,* 2179 (1981).
637. D. L. Boger, J. S. Panek, and M. M. Meier, *J. Org. Chem., 47,* 895 (1982).
638. E. C. Taylor, S. R. Fletcher, and S. Fitzjohn, *J. Org. Chem., 50,* 1010 (1985).
639. D. L. Boger, J. Schumacher, M. D. Mullican, M. Patel, and J. S. Panek, *J. Org. Chem., 47,* 2673 (1982).
640. G. L'abbe, E. Van Loock, R. Albert, S. Toppet, G. Verheist, and G. Smets, *J. Am. Chem. Soc., 96,* 3973 (1974).
641. K. Burger and F. Hein, *Liebigs Ann. Chem.,* 853 (1982).
642. M. Forchiassin and C. Russo, *Gazz. Chim. Ital., 114,* 243 (1984).
643. A. Bigotto, M. Forchiassin, A. Risaliti, and C. Russo, *Tetrahedron Lett.,* 4761 (1979).
644. M. Forchiassin, A. Risaliti, and C. Russo, *Tetrahedron, 37,* 2921 (1981).
645. F. Benedetti, M. Forchiassin, C. Russo, and A. Risaliti, *Gazz. Chim. Ital., 115,* 663 (1985).
646. L. Marchetti and G. Tosi, *Tetrahedron Lett.,* 3071 (1971).
647. L. Marchetti, E. F. Serantoni, R. Mongiorgi, and L. R. Di Sanseverino, *Gazz. Chim. Ital., 103,* 615 (1973).
648. L. Marchetti, *J. Chem. Soc., Perkin Trans., II,* 382 (1978).
649. F. Piozzi, A. Umani-Ronchi, and L. Merlini, *Gazz. Chim. Ital., 95,* 814 (1965).
650. R. Huisgen and H-U. Reissig, *Angew. Chem., Int. Ed. Engl., 18,* 330 (1979).
651. R. Huisgen, W. Bihlmaier, and H-U. Reissig, *Angew. Chem., Int. Ed. Engl., 18,* 331 (1979).
652. S. Sommer, *Angew. Chem., Int. Ed. Engl., 18,* 695 (1979).

653. A. G. Schultz, W. K. Hagman, and M. Shen, *Tetrahedron Lett.*, 2965 (1979).

654. Y. M. Hamada and T. Shioiri, *Organic Syntheses* (M. F. Semmelhack, ed.), Vol. 62, Wiley, New York, 1984, p. 191.

655. C. Botteghi, G. Caccia, and S. Gladiali, *Syn. Commun.*, *9*, 69 (1979).

656. R. L. Augustine and J. A. Caputo, *Organic Syntheses* (W. G. Dauben, ed.), Vol. 45, Wiley, 1965, p. 80.

657. J. M. Watson, U.S. Pat. 3,954,890 (1976); *CA*, *85*, 46260t (1976).

658. R. Huisgen, H-U. Reissig, H. Huber, and S. Voss, *Tetrahedron Lett.*, 2987 (1979).

659. J. Firl and S. Sommer, *Tetrahedron Lett.*, 1133, 1137 (1969).

660. J. Firl and S. Sommer, *Tetrahedron Lett.*, 4193 (1971).

661. G. J. M. Vos, P. H. Benders, D. N. Reinhoudt, R. J. M. Egberink, S. Harkema, and G. J. van Hummel, *J. Org. Chem.*, *51*, 2004 (1986).

662. J. C. Aumiller and J. A. Whittle, *J. Org. Chem.*, *41*, 2955 (1976).

663. M. Haslanger, S. Zawacky, and R. G. Lawton, *J. Org. Chem.*, *41*, 1807 (1976).

664. S. L. Schreiber, H. V. Meyers, and K. B. Wiberg, *J. Am. Chem. Soc.*, *108*, 8274 (1986).

665. G. Bengtson, S. Keyaniyan, and A. de Meijere, *Chem. Ber.*, *119*, 3607 (1986).

8
Heterocyclic Enamines

OTAKAR ČERVINKA *Prague Institute of Chemical Technology, Prague, Czechoslovakia*

I.	Introduction	442
II.	Preparation of Heterocyclic Enamines	443
	A. Condensation Reactions	443
	B. Reduction and Elimination Methods	451
	C. Isomerization of Tetrahydropyridines	456
	D. Preparation of Enamino Ketones	458
III.	Structure and Physicochemical Properties	460
	A. Pyrrolines and Piperideines	460
	B. Enamines of 1-Azabicycloalkanes	463
	C. Enamino Ketones	465
	D. Enamines That Cannot Exhibit Mesomerism	466
	E. Tautomerism of Enamines	467
	F. Structure of Enamine Salts	469
IV.	Reactions of Heterocyclic Enamines	473
	A. Reactions of Electrophilic Reagents with the Double Bond of Enamines	474
	B. Reactions of Enamine Salts with Nucleophilic Reagents	486
V.	Importance of Heterocyclic Enamines for the Synthesis of Alkaloids	496
	References	514

This chapter covers the period to 1984. As in the first edition of this book, we are specifically excluding pyrroles, indoles, partial dihydropyridines, exoenamine tautomers of oxazolines, and imidazolines. Enamine systems further conjugated with electron acceptor groups (enaminones, enaminonitriles, enamides) have not been included in this survey.

I. INTRODUCTION

The term "heterocyclic enamines" designates compounds containing an enamine grouping

$$-N-\overset{|}{C}=\overset{|}{C}-$$

in the molecule in which all three atoms are part of a heterocyclic system [1]. Typical behavior of enamines has been mainly observed for compounds possessing a tertiary nitrogen atom [2,3]. In this group are included variously substituted dehydroderivatives of pyrrolidine (pyrrolines) (1), piperidine (piperideines) (2), rarely the derivatives of dehydro-1-azacycloalkanes with more than six-membered rings, and finally, compounds containing some of these basic skeletons: enamines of quinolizidine (3), indolizidine (4), pyrrolizidine (5), and their benzoderivatives, which occur in a large number of alkaloids.

(1)   (2)   (3)   (4)   (5)

Analogous compounds with a secondary amino group (α,β-unsaturated secondary amines) can, in principle, exist in either the form of imines (6) or the tautomeric form of enamines (7). As they practically occur and react in the former structure, it is more convenient to use the group designation "imines."

(6)   (7)

II. PREPARATION OF HETEROCYCLIC ENAMINES

A. Condensation Reactions

Preparation of Heterocyclic Enamines from γ- and δ-Amino Ketones

2-Alkyl-Δ^1-pyrrolines ($\underline{8}$, n = 1) and 2-alkyl-Δ^1-piperideines ($\underline{8}$, n = 2) are readily formed by the methods used to prepare γ- and δ-amino ketones [4−6]. The reaction of corresponding halogeno ketones with ammonia belongs to the classical reactions of this type.

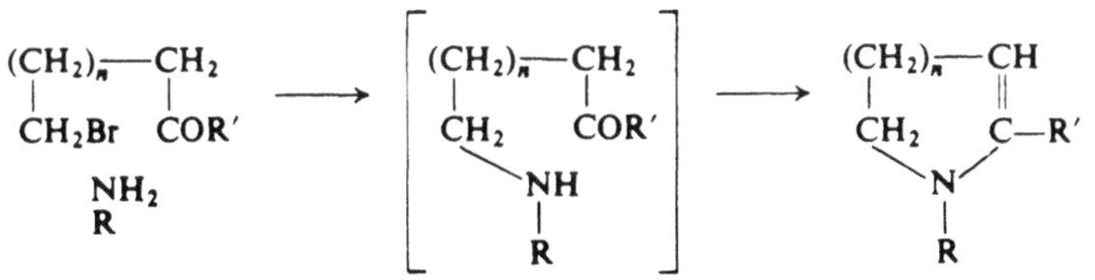

$$(\underline{8})$$

By use of potassium phthalimide we can isolate the intermediate.

α-Oxo-δ-aminovaleric ($\underline{9}$, n = 1) and α-oxo-ε-aminocaproic acids ($\underline{9}$, n = 2) readily yield Δ^1-pyrroline-2-carboxylic acid ($\underline{10}$, n = 1) and Δ^1-piperideine-2-carboxylic acids ($\underline{10}$, n = 2), respectively [7−10]. The equilibrium of the acids with their cyclic forms was observed in water solutions [11,12].

$$(\underline{9}) \qquad (\underline{10})$$

Phenols and phenol ethers can be acylated with γ- or δ-amino acids in the presence of polyphosphoric acid to form 2-aryl-Δ^1-pyrrolines ($\underline{11}$, n = 1) or 2-aryl-Δ^1-piperideines ($\underline{11}$, n = 2), respectively [13].

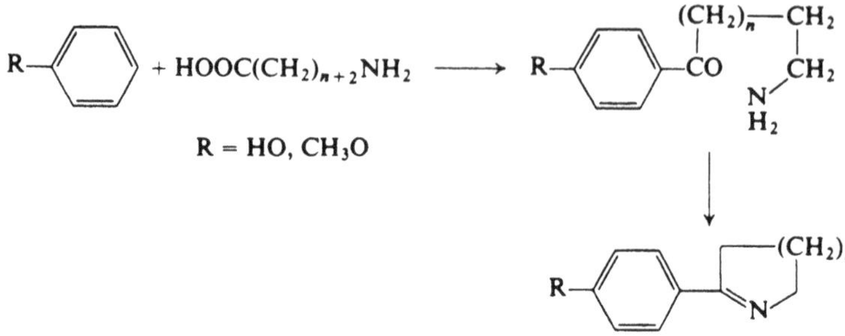

(11)

The use of primary amines instead of ammonia affords 1,2-dialkyl-Δ^2-pyrrolines or 1,2-dialkyl-Δ^2-piperideines. Amino ketones with a primary amino group are intermediates in the reduction of γ-nitropropylalkyl ketones [14,15] or δ-nitrobutylalkyl ketones [16–18] by catalytic hydrogenation over Raney nickel or with zinc and hydrochloric acid (Scheme 1).

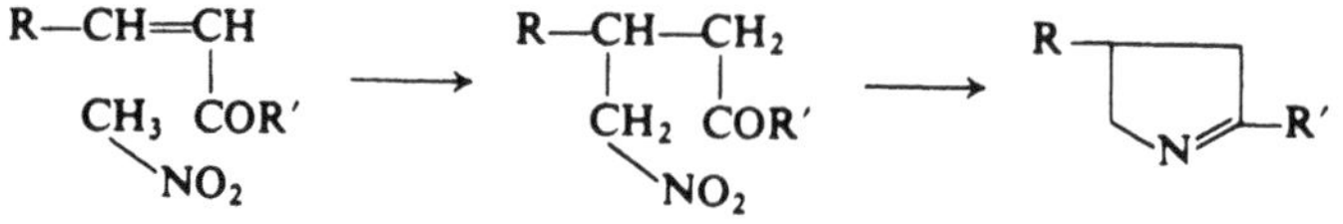

Scheme 1

Δ^1-Pyrroline-N-oxides (12) are sometimes isolated when using zinc ammonium chloride [19,20], iron-sulfuric acid [14], or hydrazine-Raney nickel [21] as a reducing agent. During the reduction, dimerization has been often observed [22].

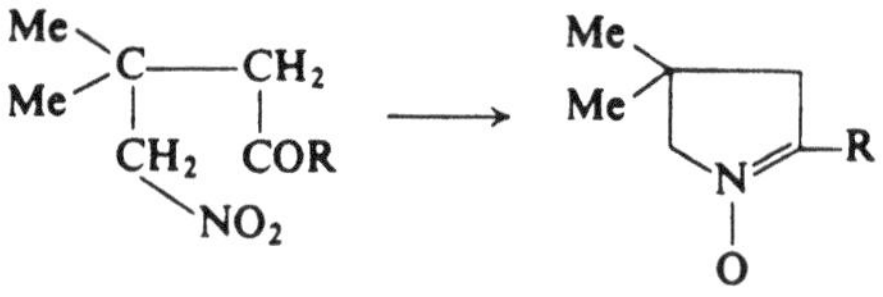

(12)

Δ^1-Pyrrolines have been isolated from the hydrogenation products of γ-ketonitriles [23–26] and in a large number of reactions during which enamino ketones are formed as intermediates. The preparation of pyrrolines from anhydro-5-hydroxyoxazolinium hydroxides (13, R, R = Ph, R = Me) is also important [27]. By the reaction of 13 with styrene, 1-methyl-2,3,5-triphenyl-Δ^2-pyrroline (14) is formed.

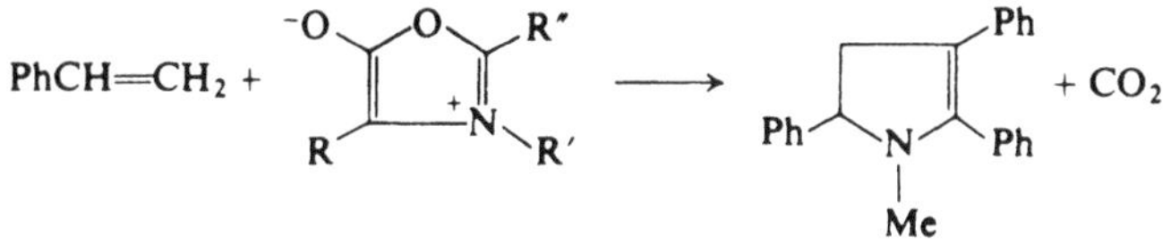

$$PhCH{=}CH_2 \;+\; \underset{(13)}{\underset{R}{\overset{-O}{\diagdown}}\underset{}{\Big\langle}\overset{O}{\diagup}\underset{N^+}{\overset{R'}{\diagdown}}\underset{R'}{\diagup}} \;\longrightarrow\; \underset{(14)}{\underset{Ph}{\diagup}\underset{N}{\diagdown}\underset{\underset{Me}{|}}{}\underset{Ph}{\diagup}\overset{Ph}{\diagdown}} \;+\; CO_2$$

(13)   (14)

When the 1-monoximes or dioximes of 4-acetyl-1-tetralones are hydrogenated in the presence of palladium, mixtures of diastereoisomeric 1-aminotetralones are formed. The *cis*-aminoketone isomers readily form dehydrobenzoisoquinuclideines (3,4-disubstituted-1,4-dihydro-1,4-ethanoisoquinolines). Quaternary immonium salts prepared from these bicyclical imines are then converted by bases to bicyclical enamines 2,4-disubstituted-3-alkylidene-1,4-ethano-1,2,3,4-tetrahydroisoquinolines [28].

Preparation of Heterocyclic Enamines by Means of Organometallic Reagents

Δ^1-Pyrrolines and Δ^1-piperideines are formed on treatment of γ- and δ-halogenonitriles [29–34], respectively, with Grignard reagents. As for the reaction mechanism, it is probable that the pyrrolines form mainly by a direct thermal displacement of the imine and only to a small extent by rearrangement of the cyclopropylketimine intermediate [35,36] (Scheme 2). It has been reported that treatment of the cyclopropylnitrile with phenyllithium yields an isolatable ketimine [37–39]. Simple thermal rearrangement of the ketimine does not occur at tem-

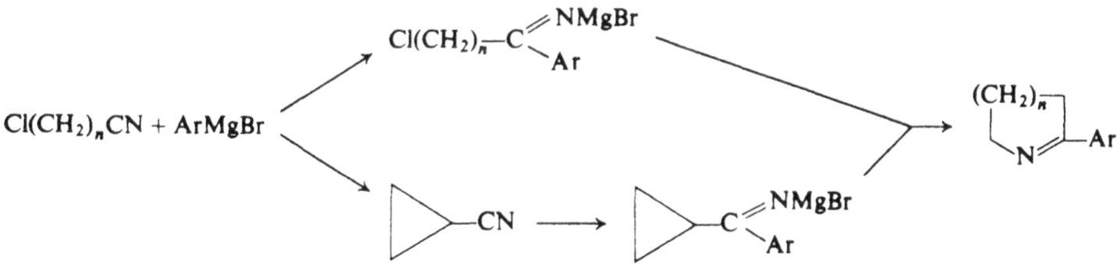

$$Cl(CH_2)_nCN + ArMgBr$$

$$Cl(CH_2)_n\!-\!C{\lessgtr}\genfrac{}{}{0pt}{}{NMgBr}{Ar}$$

$$\triangleright\!-\!CN \;\longrightarrow\; \triangleright\!-\!C{\lessgtr}\genfrac{}{}{0pt}{}{NMgBr}{Ar}$$

$$\underset{N}{\overset{(CH_2)_n}{\diagdown}}\!-\!Ar$$

Scheme 2

peratures up to 200°C, but acid-catalyzed thermal rearrangement does occur, producing 2-phenyl-Δ^1-pyrroline in good yield. The quaternary methiodide salt of a ketimine can be thermally rearranged to a Δ^2-pyrroline [37–41].

For synthesis of other alkaloids, the requisite cyclopropylimines were prepared by bisalkylation of the appropriately substituted benzyl cyanide [42]. Cyanides could be selectively reduced in high yield to the corresponding aldehydes by employing diisobutylaluminium hydride (Scheme 3).

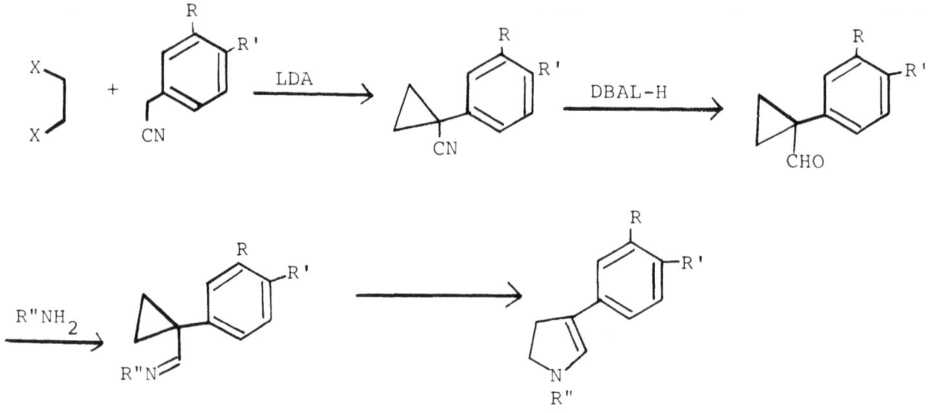
X
X
R
R'
CN
LDA
CN
DBAL-H
CHO
R"NH2
R"N
N
R"

Scheme 3

Treatment of these aldehydes with an appropriate primary amine completes the synthesis of cyclopropyl imines. This rearrangement is not a purely thermal process—an acid catalyst is required. Catalytic amounts of anhydrous halohydrogen acids induce the rearrangement upon heating [43,44]. By contrast, fluoroborate or perchlorate salts fail to catalyze the rearrangement under similar conditions, proving that the anion must be nucleophilic (Scheme 4).

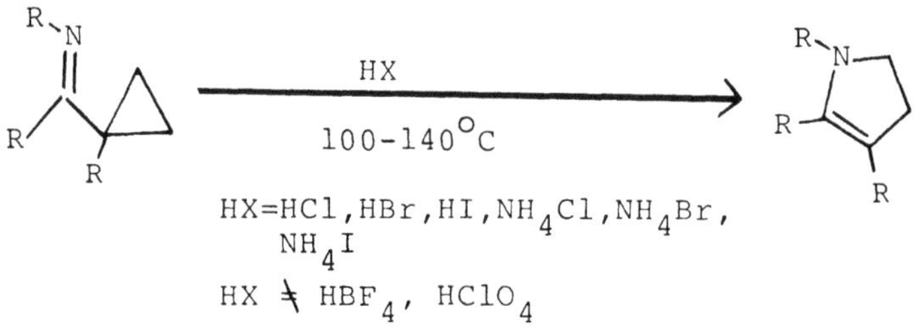
R
N
HX
100-140°C
HX=HCl,HBr,HI,NH4Cl,NH4Br,
NH4I
HX ≠ HBF4, HClO4

Scheme 4

Rearrangement afforded the 2-pyrrolines in high yield. However, in contrast to the cyclopropyl imine rearrangement, the acid-catalyzed rearrangement of various cyclobutyl imines proved to be sluggish and much less general [45].

2-Alkyl-Δ^1-piperideines were obtained on treatment of imino ethers with Grignard reagent [46,47].

Lukeš studied the reaction of N-methyl lactams with Grignard reagents. With the five- [48–51] and six-membered [52–56] rings, 2,2-dialkylated bases (16, $n = 1,2$) are formed as by-product in addition to the 1-methyl-2-alkyl pyrrolines (15, $n = 1$) or 1-methyl-2-alkyl piperideines (15, $n = 2$). Aromatic Grignard reagents afford only the unsaturated bases, probably because of steric factors [57,58]. Separation of enamines and 2,2-dialkylated amines from each other can be easily achieved since the perchlorates of the enamines and the picrates of 2,2-dialkylated bases crystallize readily. Therefore, enamines can be isolated as crystalline perchlorates and the 2,2-dialkylated bases as crystalline picrates. Some authors who repeated the reactions isolated only pyrrolines [59,60] or, by contrast, 2,2-dialkylated bases [61]. This can be explained by use of unsuitable isolation techniques by the authors.

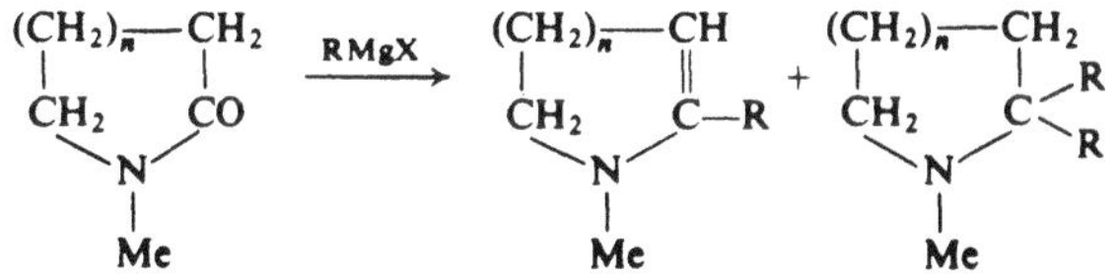

$$(\underline{15}) \qquad (\underline{16})$$

The reaction was also carried out with variously substituted [62–64] and nonmethylated [65] lactams. Treatment of 1-methyl-2-piperidone with phenylmagnesium bromide and subsequent reaction with acetic anhydride gave the corresponding acetate in small yield [66]. This indicates that the carbinolamine salt is an intermediate in this reaction which, on liberation, affords another tautomeric form. In

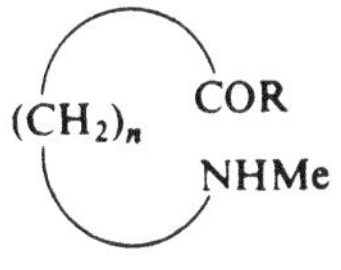

$$(\underline{17}) \qquad\qquad (\underline{18}) \qquad\qquad (\underline{19})$$

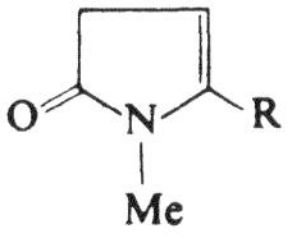

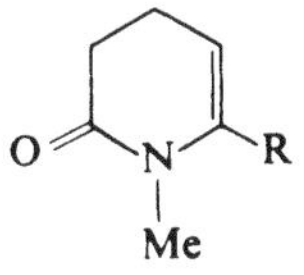

the five- and six-membered series, this tautomeric form is a cyclic enamine. Lactams with larger rings, i.e., seven- [67,68], eight- [69], nine- [70], 11-, and 13-membered [71] lactams, yield only acyclic amino ketones (17, n = 5–11) when treated with Grignard reagents. Treatment of a 1-naphthyl-substituted seven- or 13-membered ring with Grignard reagents produces both the enamine and the amino ketone [72]. Treatment of imides of dicarboxylic acids with organometallic compound forms cyclic enamines such as 18 and 19 [73–78].

Evans [79] described the general method for the synthesis of a variety of cyclic enamines from structurally diverse imine anions. The imine was added to a cooled (−30°C) solution of lithium diisopropylamide in tetrahydrofuran; after 15 min 1-chloro-2-iodoethane (1-chloro-3-iodopropane) was added in one lot [79–81] (Scheme 5).

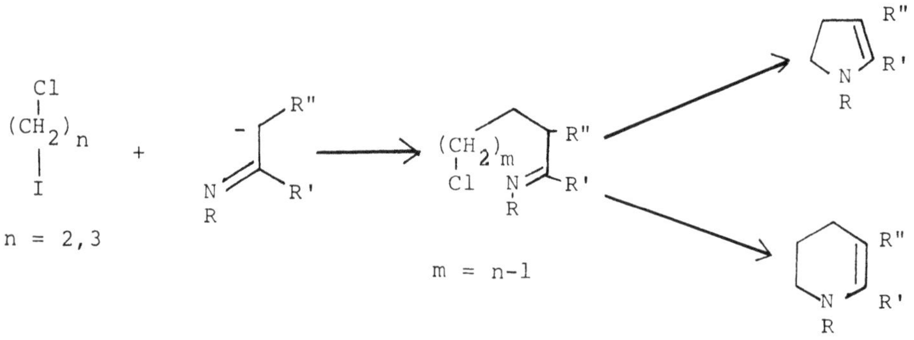

n = 2,3

m = n-1

Scheme 5

2-Methyl-3,4,5,6-tetrahydropyridine (20) and $\Delta^{1(9)}$-octahydroquinoline (21) could be transformed cleanly into the cyclical bases (22, 23).

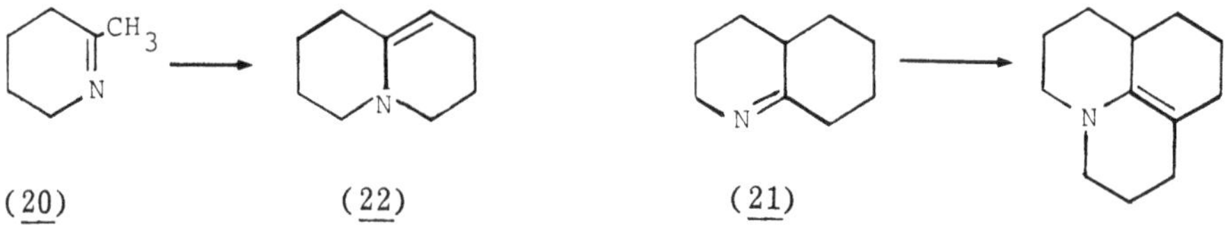

(20) (22) (21)

(23)

The methylimide of 2-methylcyclohexanone afforded an equal mixture of 24 and 25. The only attempt at the synthesis of 2-pyrroline

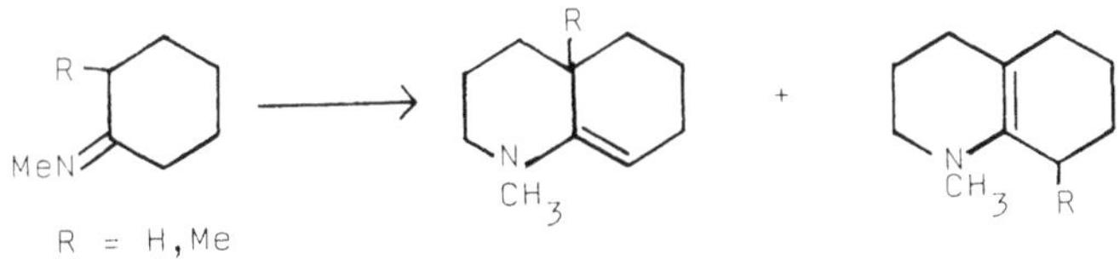

(24)   (25)

derivative is represented by the transformation of enamine (26) into the tetrahydrobenzene indole (27).

(26)   (27)

Preparation of Heterocyclic Enamines by Means of Claisen Condensation of Lactams

The reactivity of the methylene group adjacent to the lactam group affords the possibility of a Claisen condensation. Thus, treatment of 2-pyrrolidone or 2-piperidone with ethyl oxalate leads to the Δ^1-pyrrolinecarboxylic [82] and Δ^1-piperideine-2-carboxylic acids [83], respectively. N-Methyl lactams furnish N-methyl derivatives [84,85] (Scheme 6).

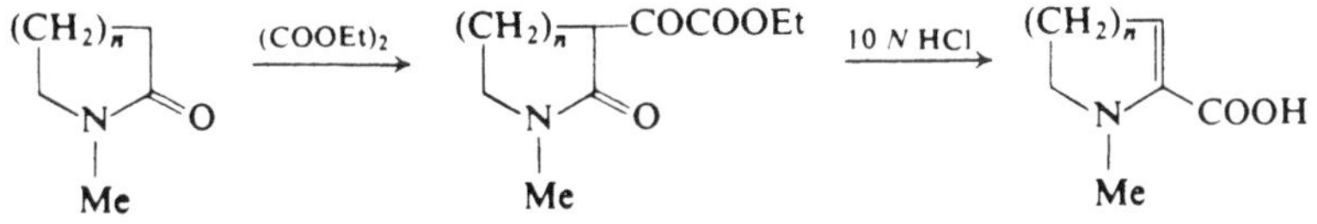

Scheme 6

In some cases, the products of Claisen condensation of two molecules of 2-pyrrolidones (28) can be hydrolyzed to 2-alkylidenepyrrolidines (29) [86].

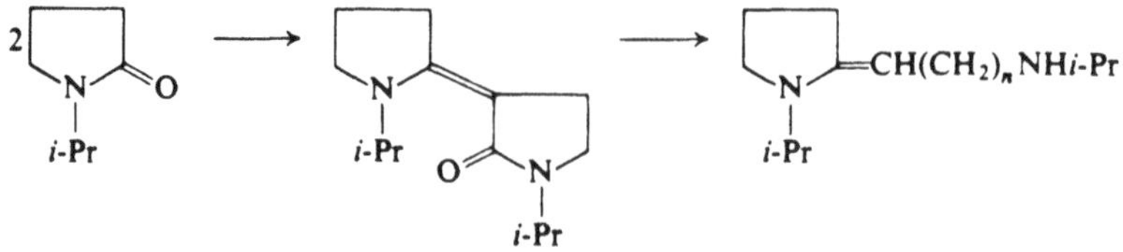

(28)   (29)

The treatment of esters of aromatic acids with 1-alkyl-2-pyrrol-
idones and 1-alkyl-2-piperidones is an extremely useful method for
the preparation of simple pyrrolines and piperideines, respectively.
The 1-alkyl-3-aryl-2-pyrrolidones (30, $n = 1$) and 1-alkyl-3-aryl-2-
piperidones (30, $n = 2$) thus formed are cleaved by the action of
concentrated hydrochloric acid to 1-alkyl-2-aryl-Δ^2-pyrrolines [87]
(31, $n = 1$) and 1-alkyl-2-aryl-Δ^2-piperideines [88] (31, $n = 2$),
respectively. After blocking the secondary nitrogen atom in a lactam
by means of acylation, one can prepare 2-aryl-Δ^1-pyrrolines and

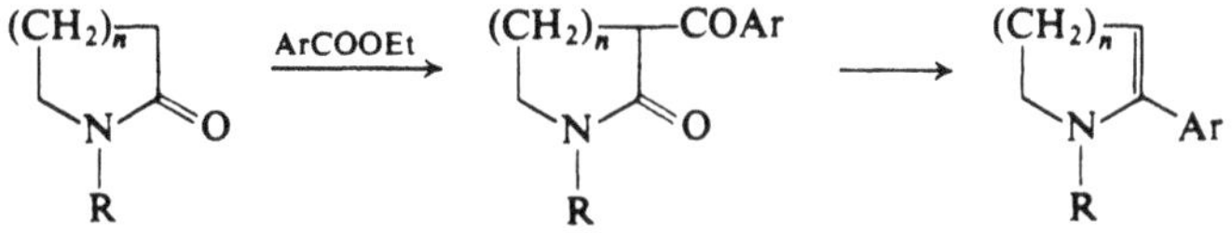

(30)   (31)

2-aryl-Δ^1-piperideines by the same reaction route [89—91]. Treatment of ethyl formate with 1-methyl-2-piperidone yields 1-methyl-2-
piperideine (32) and, subsequently, dimer (33) in alkaline medium
[92,93].

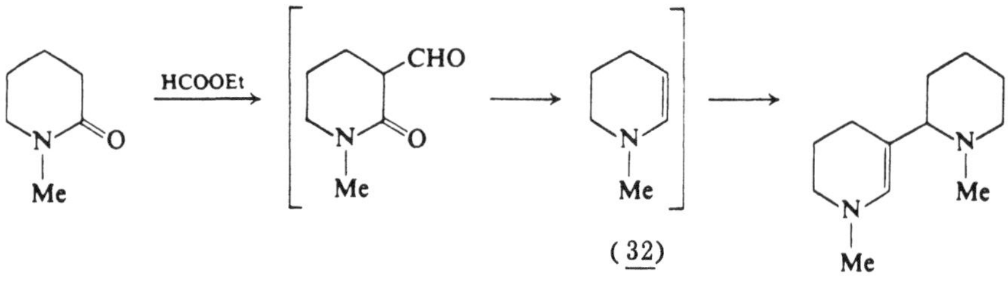

(32)

(33)

B. Reduction and Elimination Methods

Partial hydrogenation of pyrrole derivatives and partial dehydrogenation of pyrrolidines afford Δ^1-pyrrolines [94–96]. However, because
of the complex nature of the reaction, it is of little preparative value.
The same is true for isomerization of Δ^3-pyrrolines to Δ^1-pyrrolines
[97]. A photodehydrogenation of 2,6-dimethylpiperidine (34) has
been observed affording 2,6-dimethyl-3,4,5,6-tetrahydropyridine
(35) in a good yield [98].

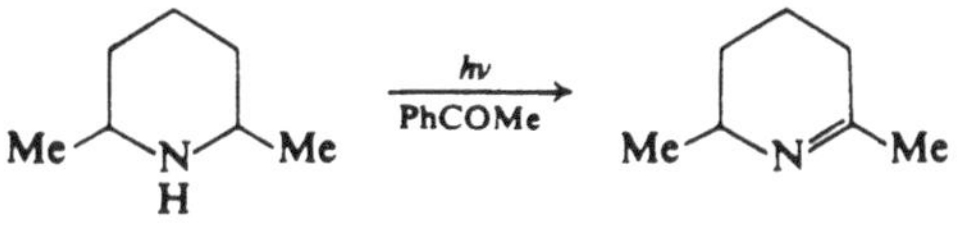

(34)   (35)

The 1,2-dihydro derivative is formed by reduction of pyridine
with LiAlH$_4$ [99]. Analogous reduction with sodium in 95% alcohol
affords the 1,4-dihydro derivative. Monomeric N-trimethylsilyl-1,2,
3,4-tetrahydro (36) and the corresponding 1,2-dihydro (37) and
1,4-dihydro derivatives (38) of pyridine have been prepared by
treatment of pyridine with trimethylsilane in the presence of palladium
[100]. Pyridines with a carbonyl function (exemplified by ketones,
an ester, and an amide) in the 3 position can be hydrogenated to corresponding 1,4,5,6-tetrahydropyridine derivatives in good yield [101].

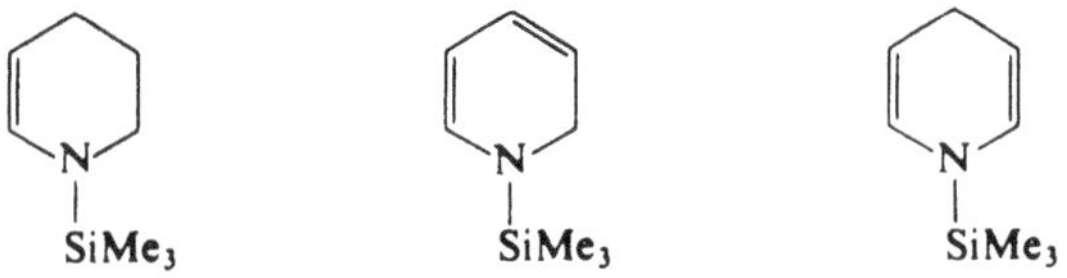

(36)   (37)   (38)

Reduction of quaternary pyridine salts with sodium amalgam or
sodium hydrosulfite affords 1,2-dihydro [102] and 1,4-dihydro derivatives, respectively. From the preparative point of view, partial hydrogenation of quaternary pyridine salts in alkaline media to give
substituted 1-methyl-Δ^2-piperideines is very important [103,104].
Formation of 1,2-dihydroisoquinolines was observed in the reduction
of quaternary isoquinoline salts with sodium hydrosulfite [105], lithium aluminium hydride [106,107], dialkylaluminohydrides [108], or
on treatment with Grignard reagent [109]. The reduction of quino-

line with either lithium or sodium in ammonia produces only 1,4-dihy-droquinoline [110]. 1,2- and 1,4-dihydro derivatives were also formed from 3,5-dicyanopyridines by reduction with sodium borohy-dride [111] or by reaction with Grignard reagent [112]. Formation of enamines was observed in the reduction of 1-methyl-2-piperidone with sodium in ethanol [113] or lithium aluminium hydride [114]. Electroreduction of N-methylglutarimide likewise produces enamines [115,116].

Δ^1-Pyrroline has been prepared in low yield by oxidation of pro-line with sodium hypochlorite [83], persulfate [117], and periodate [118]. Δ^1-Pyrroline and Δ^1-piperideine are products of enzymatic oxidation via deamination of putrescine and cadaverine or ornithine and lysine, respectively [119,120]. This process plays an important part in metabolism and in the biosynthesis of various heterocyclic compounds, especially of alkaloids.

The best procedure for the preparation of Δ^1-pyrroline (39, n = 1) and Δ^1-piperideine (39, n = 2) consists of dehydrohalogenation of N-chloropyrrolidine and N-chloropiperidine, respectively, by means of potassium hydroxide [121].

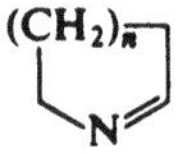

(39)

Selective reduction of lactams with diisobutylaluminiumhydride affords endocyclic enamines [122,123] (Scheme 7).

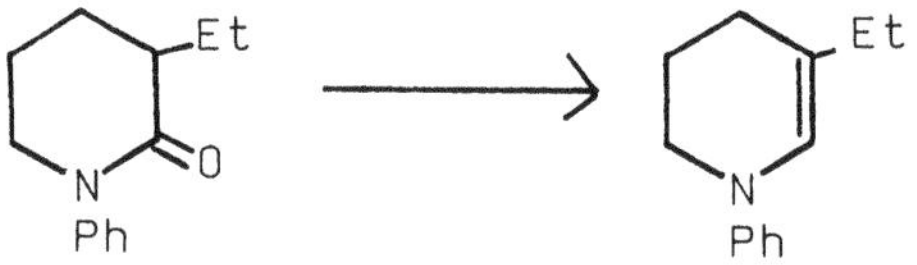

Scheme 7

1,2-Dihydropyridines are intermediates in the synthesis of pyri-dines from mixtures of aldehydes or ketones and NH_3 in the liquid phase [124].

Δ^1-Piperideine-N-oxide was obtained along with a dimeric product by oxidation of N-hydroxypiperidines with mercuric acetate or potas-sium ferricyanide [125–127]. Δ^1-Pyrroline-N-oxide is formed by oxidation of N-ethylpyrrolidine with hydrogen peroxide with simultaneous formation of ethylene [128].

Another convenient method for the preparation of tertiary enamines involves the dehydrogenation of saturated bases with mercuric acetate [129–134]. A *trans*-1,2 elimination occurs, which requires an antiperiplanar position of the nitrogen-free electron pair and the eliminated atom. A preferential elimination of the hydrogen atom from the tertiary carbon atom is supposed. Overoxidation can be avoided by adding disodium ethylenediaminotetraacetate to the reaction mixture [135,178].

Bohlmann [136–139] observed that an infrared absorption band between 2700 and 2800 cm^{-1} is characteristic of a piperidine derivative possessing at least two axial carbon-hydrogen bonds in antiperiplanar position to the free-electron pair on the nitrogen atom. The possibility of forming an enamine by dehydrogenation can be determined by this test. Compounds that do not fulfill this condition cannot usually be dehydrogenated [59,140,141]. Thus, for example, yohimbine can be dehydrogenated by mercuric acetate, whereas reserpine and pseudoyohimbine do not react [142]. The quinolizidine [143] enamines (Scheme 8), 1-azabicyclo(4,3,0)-nonane, 1-azabicyclo(5,3,0)decane, 1-azabicyclo(5,4,0)undecane, and 1-azabicyclo(5,5,0)dodecane, have been prepared in this manner [130,144].

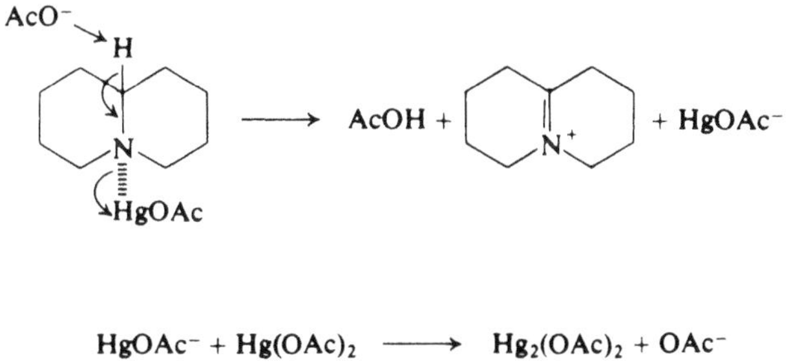

$$HgOAc^- + Hg(OAc)_2 \longrightarrow Hg_2(OAc)_2 + OAc^-$$

Scheme 8

The dehydrogenation of 4-aryl quinolizidines is very interesting, too. The double bond of the salts is formed in the $\Delta^{9,10}$ position and not in the expected $\Delta^{4,10}$ position [145]. In several cases, hydroxylation takes place in the dehydrogenation of 1-methylquinolizidine [133], especially of *cis*- and *trans*-1-methyldecahydroquinolines [146, 147] (Scheme 9).

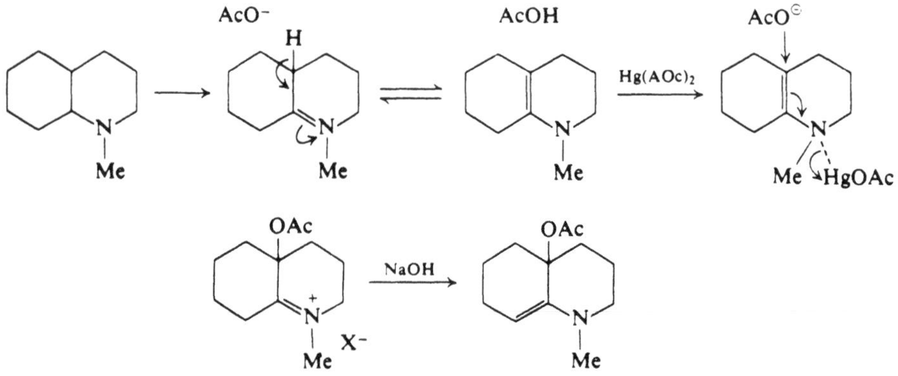

AcO⁻ → H (structure) ⇌ AcOH (structure) → Hg(AOc)₂ → AcO⊖ (structure) Me ↘HgOAc

OAc (structure) → NaOH → OAc (structure)

Scheme 9

5α,22αH,25βH-Solanidanole-3 (40) is one of several complex compounds containing an indolizidine skeleton which can be dehydrogenated by mercuric acetate as well as by N-bromosuccinimide, yielding in this case a mixture of immonium salts, namely Δ22(N)-5α,25βH-solanidenole-3β (41) and Δ16(N)-5α,22αH,25βH-solanidenole-3β (42) [148].

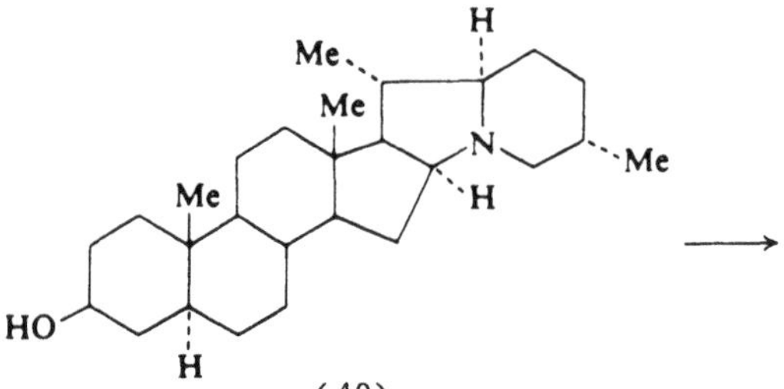

(40)

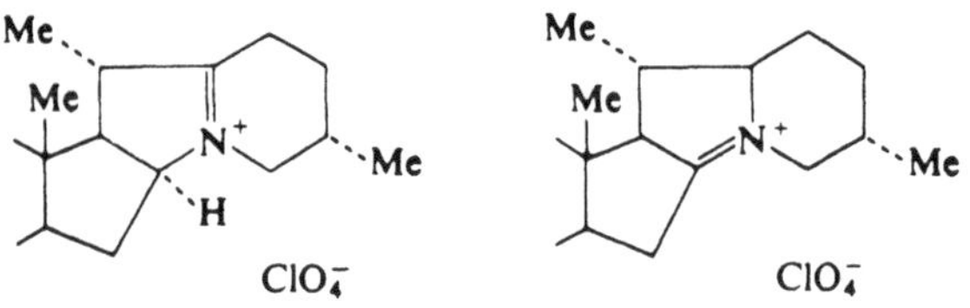

(41)

(42)

The 1,2-, 1,2,5-, 1,3,4-, and 1,2,5-substituted pyrrolidines afford the corresponding pyrrolines very readily by oxidation with mercuric acetate. In the case of 1,2,2-trimethylpyrrolidine (43), the formation of a double bond involving the unsubstituted α-carbon atom is followed by dimerization of the intermediate (44) to 1,5,5-trimethyl-3-(1',5',5'-trimethyl-2'-pyrrolidyl)-Δ2-pyrroline (45) [149]. The formation of oligomers is a frequent complication in the preparation of enamines. Dehydrogenation of 1-methylpyrrolidine (46) gives dimer (47) in addition to a trimer which is identical to a product obtained by

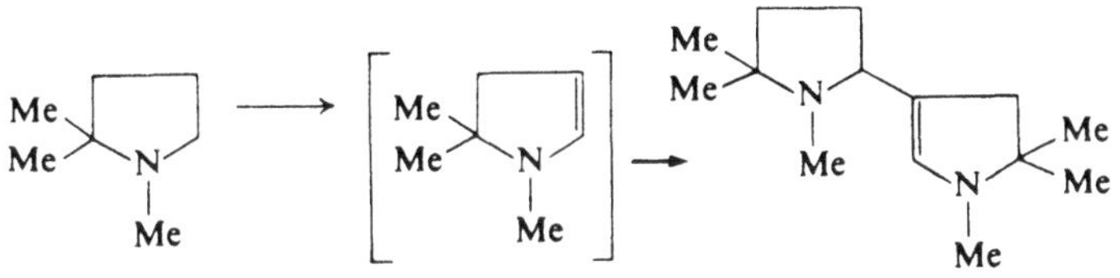

(43) (44) (45)

the reduction of N-methylpyrrole with zinc and hydrochloric acid [149,150]. A dimer is formed by analogous dehydrogenation of 1,3-

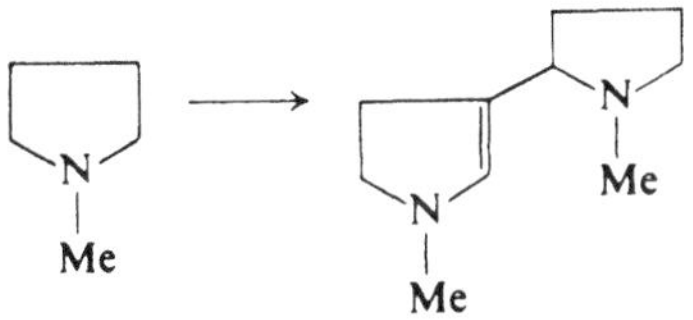

(46) (47)

dimethylpyrrolidine. In the same manner, 1,3,4-trimethylpyrrolidine is dimerized and oxidized to 2-(1',3',4'-trimethyl-2'-pyrrolidyl)pyrrole [149].

Dehydrogenation of 1-methyl-1-azacycloheptane through azacyclononane, followed by treatment with hydrogen sulfide, gave the trithiane derivatives only [151]. These results give further evidence about the instability of enamines with medium-sized rings.

Dehydrogenation of amino alcohols of type 48 affords bicyclic compounds (49), the formation of which can be explained by nucleophilic attack of the hydroxyl group on the formed enamine salt [152, 153].

(48) (49)

Several cyclic [154] and bicyclic [155] enamines were so prepared.

C. Isomerization of Tetrahydropyridines

Since N-alkylpyridinium salts are easily prepared and smoothly reduced by borohydride in protic solvents to tetrahydropyridines with
the double bond at the 3,4-position, i.e., to cyclic allylamines, then
if cyclic allylamine could be isomerized to cyclic enamine, a very useful simple route to these intermediates would be available. No such
isomerization is recorded in a 1970 review of 3-piperideine chemistry
[156].

Since that review, there have been a few reported examples of
preparative allyl-amine-enamine isomerizations. Thus, 1-methyl-1,2-
dihydro-1-benzazocine (50) was converted into enamine (51) by reaction of t-BuOK-DMSO-room temperature [157]. 1-Methyl-1,2-dihy-
droquinoline (52) was isomerized under the same conditions into
enamine (53). The indoloquinolizine (54) was isomerized to (55) by
base and t-BuOK-DMSO-100°C [158,159].

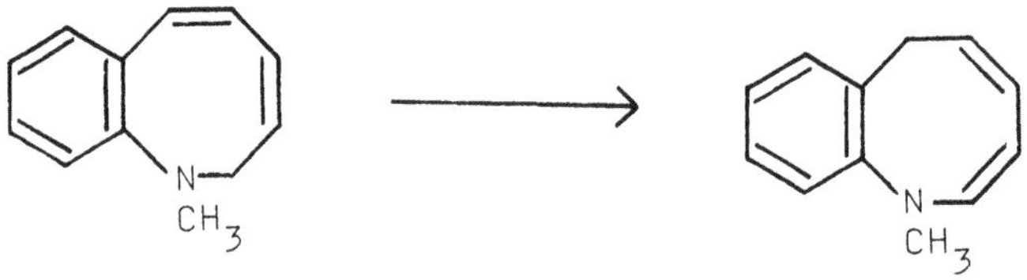

(50)

(51)

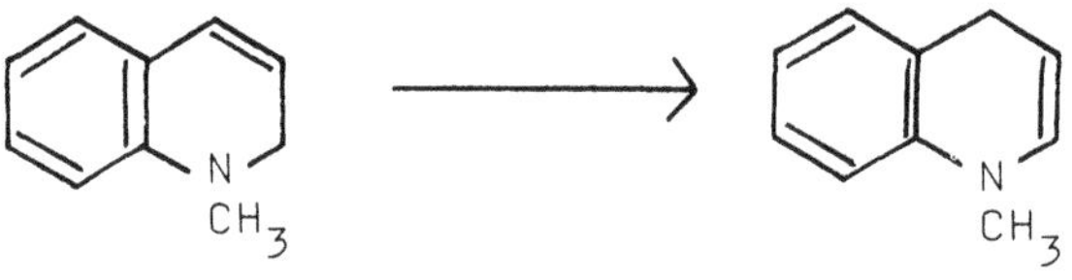

(52)

(53)

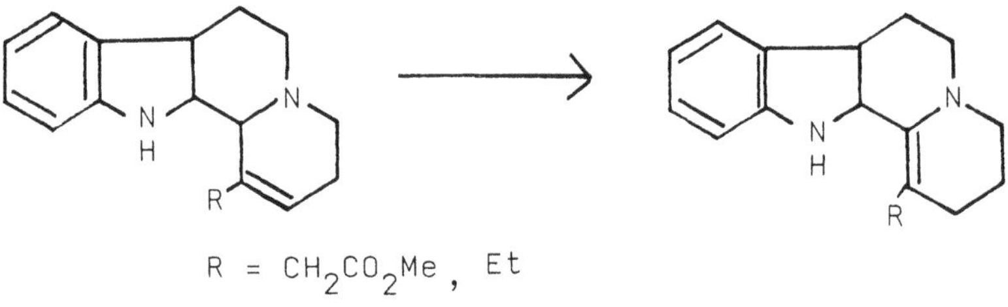

R = CH₂CO₂Me, Et

(54)

(55)

Beeken and Fowler [160] have observed that N-methyl-1,2,3,4-tetrahydropyridin (57) can be prepared by treatment of N-methyl-1,2,3,6-tetrahydropyridine (56) with 1.0 M potassium tert-butoxide in dimethyl sulfoxide. The pure enamine can be isolated by the addition of water to the reaction mixture, followed by extraction with pentane. However, the enamine is extremely unstable and is stable only if stored in the refrigerator (5°C) over potassium hydroxide pellets.

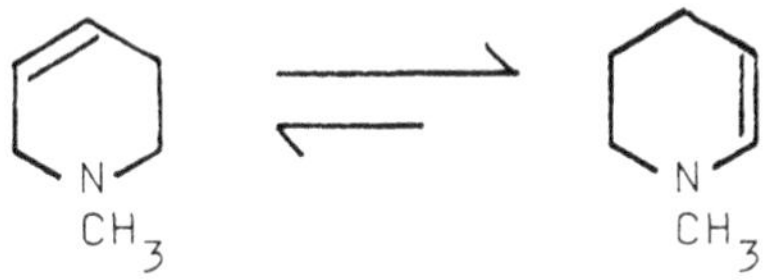

(56)   (57)

Direct observation by proton and carbon-13 magnetic resonance of the reaction mixture confirms that the equilibrium strongly favors the enamine structure. The free-energy difference between them is probably at least 4.0 kcal/mol. The most stable conformation of the enamine was assumed to be half-chair.

Martinez and Joule [161] were unable to directly observe or isolate the vinyl amine. The main product they isolated from their base-catalyzed isomerization was N,N-dimethyl-1,4,5,6-tetrahydroanabasine [162] (Scheme 10).

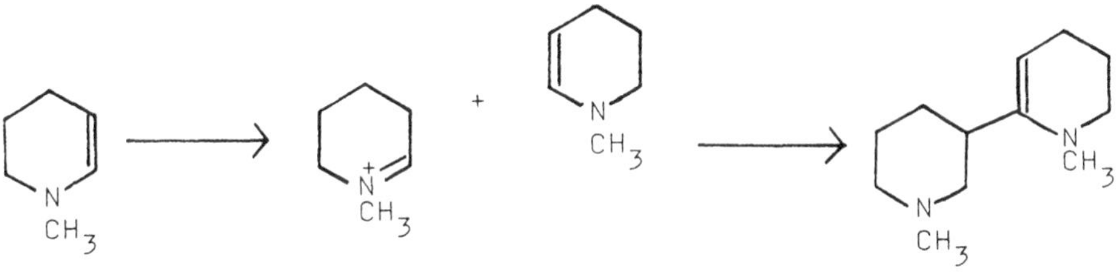

Scheme 10

Besselièvre, Beugelmans, and Husson [163] reported that the isomerization of cyclic allylamine (58) to the enamine (59) could be achieved by a photocatalyzed process believed to involve initial *trans-cis* isomerization and abstraction of C_2 hydrogen as a radical by photoexcited ester oxygen and its transfer thereby to C_4. It was also noted that 58 was not transformed into the enamine by methoxide treatment.

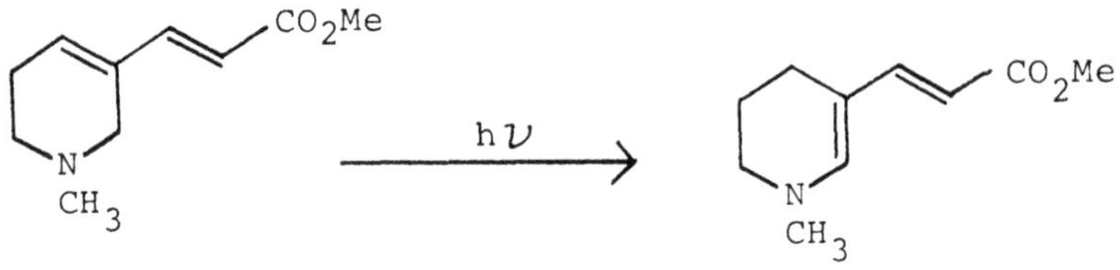

(58)

(59)

Eisner and Sadeghi [164] described the isomerization of dihydro-pyridine with $RhCl_3$-$(C_6H_5)_3P$ complex in benzene.

The experiments clearly show that 2-piperideines (enamine isomer) are thermodynamically more stable than their 3-piperideine(allylamine) isomers and can be produced from them by equilibrative transformation. The enamine conjugation is probably worth more than about 2.5 kcal/mol^{-1}.

D. Preparation of Enamino Ketones

Acylated alkyl aminoisobutyrylmalonates (60) can be easily converted to 3-oxo-Δ^2-pyrrolines (61) [165, 166].

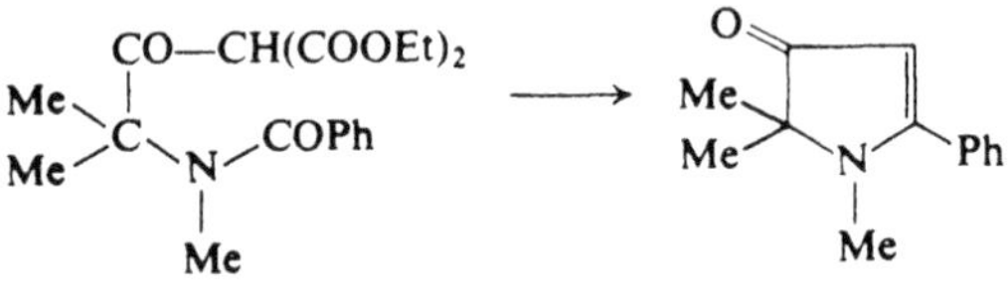

(60)

(61)

The general method for preparation of heterocyclic enamino ke-tones, amide vinylogs, consists of a cyclization (Scheme 11). The most convenient technique involves heating the starting substance in acetonitrile in the presence of silver perchlorate [167−169].

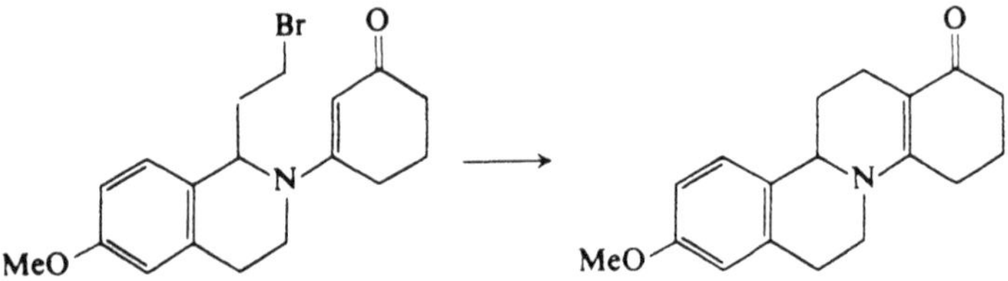

Scheme 11

Photocyclization [170–175] of N-haloaryl-substituted enaminones (Scheme 12) and related enamides (Scheme 13) has been employed in the synthesis of a variety of heterocyclic compounds. The reactions, in most cases, are efficient and thus useful in the preparation of complex structures found in natural products or their precursors.

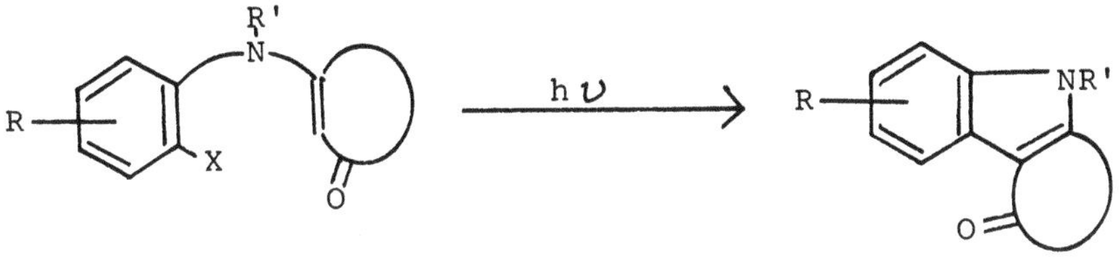

Scheme 12

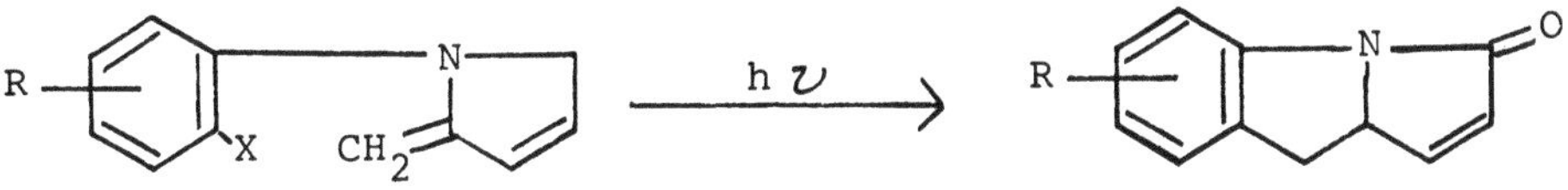

Scheme 13

Enaminoketones were obtained from lactam acetals by reaction with active methylene compounds [176]. Heterocyclic β-enaminoesters are versatile synthons in the preparation of new condensed heterocyclic systems [177]. Tautomeric equilibrium for the enaminone has been investigated by Grob and Wilkens [209]. CD and ORD data indicated that in enaminoketones (62) (R = Ph, PhCH$_2$, n = 1,2,3) hyperconjugation existed between the enaminoketone and aryl chromophores [179].

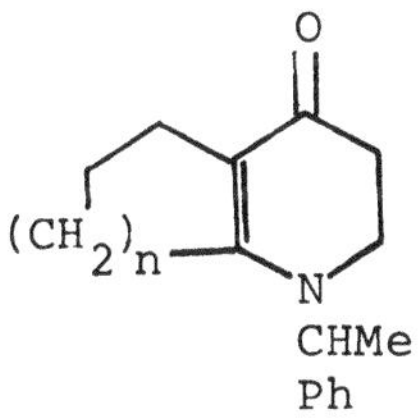

(62)

The preparation of enamine ketones by addition of α,β-unsaturated ketones to enamines is described in Chapter 4.

III. STRUCTURE AND PHYSICOCHEMICAL PROPERTIES

The presence of an enamine grouping in a molecule makes possible several interconvertible structures. The application of physicochemical methods has been of great importance for determining the actual structure. Unsaturated amines with the double bond separated from the nitrogen atom by one saturated carbon atom do not show behavior different from that of other organic bases, and the character of the double bond corresponds to that of other unsaturated compounds. The shift of a double bond to the α,β position with respect to nitrogen atom leads, by contrast, to the formation of a new reactive grouping in which the nitrogen-free electron pair is conjugated with the π electrons of the double bond. The mesomeric character of an enamine grouping is then exemplified by the fact that reaction may occur on either the nitrogen or β-carbon atom of the grouping, an increased thermodynamic basicity of the molecule, and a change in the spectral properties of the double bond.

A. Pyrrolines and Piperideines

Secondary

As pointed out in the introduction, if one of the substituents on the nitrogen atom is a hydrogen atom, tautomeric equilibrium between enamino and imino forms strongly favors the latter form [18,180,181]. According to physicochemical measurements, the occurrence of simply substituted Δ^2-pyrrolines and Δ^2-piperideines is highly improbable. The formulation of this type of compound with a double bond in the Δ^2 position (used mainly by early authors) was of formal meaning only, having no experimental evidence [182-184].

A study of infrared and NMR spectra makes it possible to distinguish between both tautomeric structures [185-187]. A single strong absorption band at 1620 cm^{-1}, attributable to the carbon-nitrogen double bond, has been found in the infrared spectra of 2-alkyl pyrrolines (63, R = alkyl) and 2-aryl pyrrolines [188] (63, R = aryl), but absorption at 3300 cm^{-1} due to the presence of the N−H group was not present. The estimation of an active hydrogen, which is negative in every reported case, leads to the same conclusions [184,

(63)

189]. The determination was even negative with 2,3-diphenyl-Δ^1-pyrroline (64), where the probability of the Δ^2-structure stabilization ought to be higher [190]. It is, nevertheless, expected that conju-

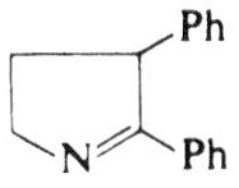

(64)

gation would support the enamine structure. According to spectral data, 2-benzylpyrroline exists as a mixture of both imino and enamino tautomers, the enamino form possessing an exocyclic double bond, whereas the negative estimation of active hydrogen points to its existence solely in Δ^1-form [184].

The piperideine derivatives have not been studied as extensively as the analogous pyrrolines [191,192]. The imino structure has been established, for example, for the alkaloid γ-coniceine (65) [186]. The great influence of conjugation on the structure is seen with 1-(α-picolyl)-6,7-methylenedioxy-3,4-dihydroisoquinoline (66) possessing an enamine structure, whereas the analogous 1-methyl derivative (67) possesses an imine structure, according to infrared spectra [192, 193].

In contrast to the five-membered-ring compound, conformational factors would be expected to influence the equilibrium between the imine and enamine forms in the case of the six-membered-ring piperideine derivatives [194].

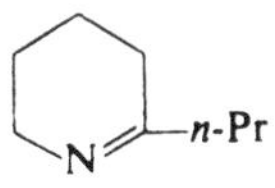

(65) 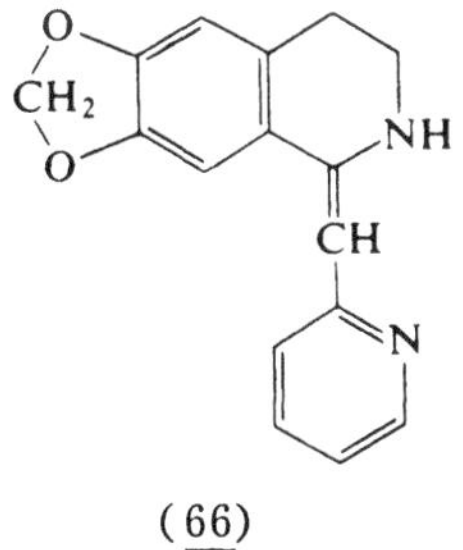   (66) 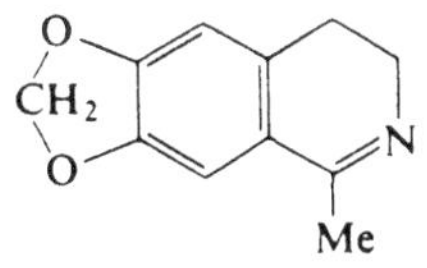   (67)

Tertiary

Tertiary pyrrolines (68, $n = 1$) and piperideines (68, $n = 2$) (if R = H and the enamine can exist in the monomeric form, or if R = aryl) evidently possess an endocyclic Δ^2-double bond [93,195,196]. The stretching frequency of the double bond can be lowered to 1620—1635 cm^{-1} by conjugation with an aromatic substituent. The double bond of an analogous compound with aliphatic substituents in position 2 may occupy either the endo or the exo position. Lukeš and co-workers [197] have shown that the majority of the five-membered-ring compounds, traditionally formulated with the double bond in a Δ^2-position, possess the structure of 2-alkylidene derivatives (69), with an exocyclic double bond and infrared absorption at 1627 cm^{-1}. Only the 1,2-dimethyl derivative (70) is actually a Δ^2-pyrroline, absorbing at 1632 cm^{-1}. For comparison, 1,3,3-trimethyl-2-methylene pyrrolidine (71) with an unambiguous exocyclic double bond has been prepared [63].

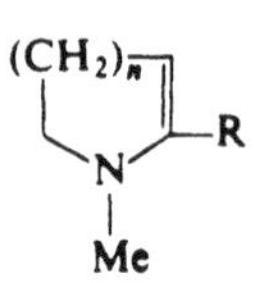

(68) 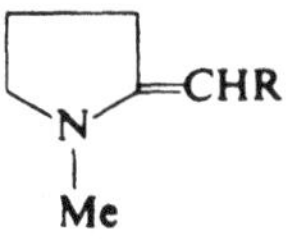   (69) 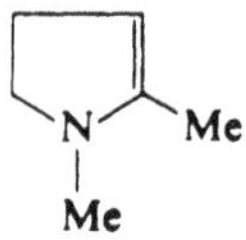   (70) 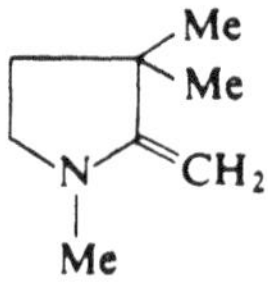   (71)

The ultraviolet spectra were also used for determination of the pyrroline structure [1,198—200]. They exhibit a bathochromic shift to 225—235 nm, caused by the auxochromic action of the nitrogen-free electron pair which is in conjugation with π-electrons of the enamine double bond [201,202].

More complex compounds containing enamine grouping, e.g., holarrhena alkaloids such as conkurchine and conessidine, possess an endocyclic rather than exocyclic double bond [199]. On the other hand, 1-methyl-2-alkylpiperideines (72) possess a fixed endocyclic double bond [203,204] ($\nu_{C=C}$ 1635—1645 cm^{-1}), probably because of higher endocyclic double-bond stability in six-membered rings [205, 206].

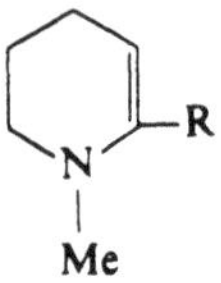

(72)

2-Piperideines are thermodynamically more stable than their 3-piperideine isomers.

A number of methods are available for the synthesis of the ring systems of isomeric 1,2- (73) and 1,4-dihydropyridines (74), and

(73)   (74)

much is known about their chemistry. However, little information is available concerning their relative stabilities. The effect of substituents on the relative stabilities of the 1,2- and 1,4-dihydropyridine is unknown. HMO claculations on the π-system of the dihydropyridine ring system indicate the 1,2-dihydropyridine system is more stable.

Fowler [207] prepared both N-methyldihydropyridines by reduction of the N-carbomethoxydihydropyridines with lithium aluminium hydride. Treatment of either isomer with 1.0 M potassium tert-butoxide in dimethyl sulfoxide at 91.6°C produces an equilibrium mixture containing 7.7% of N-methyl-1,2-dihydropyridine. If statistical factors are taken into consideration, the N-methyl-1,4-dihydropyridine is 2.29 ± 0.01 kcal/mol more stable than 1,2-isomer at this temperature. This assumption is consistent with the observation that the unsubstituted 1,4-dihydropyridine is a remarkably stable enamine.

B. Enamines of 1-Azabicycloalkanes

Enamines derived from 1-azabicycloalkanes, readily accessible by mercuric acetate oxidation of saturated bases [130], have been extensively studied [131—133]. Since an immonium salt is formed during dehydrogenation, the composition of the liberated enamine mixture shows the relative stability of the various possible isomers. The study of infrared and NMR spectra has shown that the position of the enamine double bond is determined by factors similar to those determining the relative stability of simple olefins.

1. In the case of enamines that can exist in two different isomeric forms (for example, indolizidine derivatives), the equilibrium is strongly in favor of the isomer containing the double bond in the endo position to a six-membered ring and in the exo position to a five-membered ring.

The enamine formed by dehydrogenation of indolizidine was considered to be a mixture of $\Delta^{1,9}$- (75) and Δ^8- (76) isomers because of infrared spectra [144]. According to the NMR spectrum, the Δ^8-isomer is the major constituent. This is demonstrated by comparison of this spectrum with the spectra of compounds (77) and (78) containing fixed double bonds.

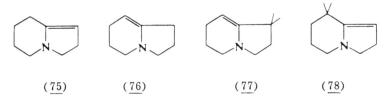

(75)   (76)   (77)   (78)

The 8-methyl derivative (79) and 1-ethyl derivative (80)
were shown to possess mainly the Δ^8-structure, whereas the
1-methyl derivative contains the Δ^8-structure in the ratio
2:1.

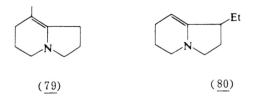

(79)   (80)

2. Substituents stabilize the double bond. The enamine of 1-
    methylquinolizidine exists as a mixture of $\Delta^{1,10}$-isomer (81)
    and Δ^9-isomer (82) in a 2:1 ratio.

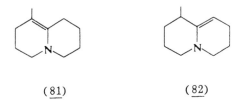

(81)   (82)

3. Formation of enamine stabilized by conjugation is preferred.
    Dehydrogenation of 1-phenylindolizidine affords the $\Delta^{1,9}$-1-
    phenyl isomer (83) only.

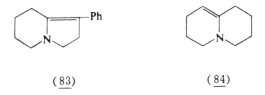

(83)   (84)

The infrared spectrum of $\Delta^{1,10}$-dehydroquinolizidine (84) exhibits
an absorption maximum at 1652 cm^{-1}, and the ultraviolet spectrum

shows a maximum at λ_{max} 228 nm (ε_{max} 5600). NMR spectra of all compounds in question possess four proton-absorption regions. Vinyl protons $H-C=C$ exhibit broad singlets and sometimes a poorly re-solved triplet centered at τ 5.65–5.98 ppm. Unusually high chemical shifts can be explained by considering mesomeric contributors. Pro-tons of methylene groups attached to a nitrogen atom $-CH_2-N-$ lead to broad triplets or multiplets centered at τ 7.1–7.3 ppm. Methyl group protons $CH_3-C=$ exhibit peaks at τ 8.4–8.5 and 8.8–9.0 ppm, typical for an allylic grouping. The other protons produce very broad multiplets at τ 7.5–8.5 ppm.

C. Enamino Ketones

Structural analogy of aliphatic amino ketones can be found in the heterocyclic series. A simple example of such compounds is $\Delta^{8,9}$-octahydro-7-quinolinone [208], which, as a vinylog of an amide, can possess enonamine-enolimine tautomerism [209].

The infrared, ultraviolet, and NMR spectra of $\Delta^{8,9}$-octahydro-7-quinolinone were compared with the corresponding spectra of the N-ethyl and O-ethyl derivatives in order to determine whether it is in the enamine (85), enolimine (86), or ketimine (87) form.

(85)   (86)   (87)

The infrared spectrum exhibits absorption maxima at 3405 cm^{-1} (NH), 1610 cm^{-1} (C=O), and 1580 cm^{-1} (C=C), without any band at 3610 cm^{-1} owing to enolic hydroxyl group. Spectra of corresponding N-ethyl derivatives contain absorption maxima at 1600 cm^{-1} (C=O) and 1550 cm^{-1} (C=C), whereas those of O-ethyl derivatives show a broad band at 1628 cm^{-1} owing to the presence of C=N and C=C double bonds. In the ultraviolet spectrum of the unsubstituted com-pound, an absorption maximum at λ_{max} 298 nm (log ε 4.49) was ob-served; the N-ethyl derivative showed a maximum at λ_{max} 304 nm (log ε 4.97) and the O-ethyl derivative showed quite different maxi-mum at λ_{max} 245 nm (log ε 4.21). The NMR spectrum consists of a singlet due to vinyl protons with chemical shift δ 5.18 ppm (the same singlet is exhibited by the N-ethyl derivative at δ 5.21 ppm and the O-ethyl derivative at δ 5.33 ppm) and a broad singlet at δ 5.9 ppm due to the NH-group proton. These data indicate unambiguous enamine structure (85). The study of spectra of Δ9,10-octahydro-5-quinolinone leads to a similar conclusion.

D. Enamines That Cannot Exhibit Mesomerism

Mesomerism involving polarized and nonpolarized contributing enamine forms influences the enamine's spectral properties and chemical reactivity. For mesomerism to be present, a planar arrangement is required for the three atoms of enamine grouping and the five atoms immediately bound to this system. If this condition is not fulfilled, full interaction of the π-electrons of the double bond with the free electron pair on the nitrogen atoms is impossible. Enamines in which mesomerism is inhibited do not show the properties characteristic of enamines, and only the mutual electrostatic interaction of the double bond and lone electron pair of the nitrogen atom can be observed. Such steric hindrance of mesomerism occurs mainly in polycyclic systems.

The simplest examples of this type of compound are enamines derived from the quinuclidine skeleton (88). The formulation of enamines of quinuclidine in a mesomeric form would violate Bredt's rule. Actually, the ultraviolet spectrum of 2,3-benzoquinuclidine shows that there exists no interaction of aromatic ring π-electrons and the nitrogen-free electron pair [200,210]. The overlap of the olefinic π-orbital and the lone-pair orbital on nitrogen is precluded.

(88)　　　　　　　(89)　　　　　　　(90)

Similar behavior can be observed even in the case of substituted quinuclideines [211]. Neostrychnine (89) serves as an example of more complex compounds which show spectra differing from those of other enamines. The ultraviolet spectrum of this compound exhibits no bathochromic shift, and its basicity is considerably decreased [199,212,213] (pK_a in methylcellosolve at 20°C is 3.8, whereas the analogous saturated compound has a pK_a under the same conditions of 7.45, and a compound with the double bond further removed, strychnine, has a pK_a of 7.37). As another example, the ultraviolet spectrum of trimethyl conkurchine (90) shows the same absorption maxima as a saturated tertiary amine (λ_{max} in ether, about 213 nm).

E. Tautomerism of Enamines

The study of structure and reactivity of tertiary heterocyclic enamines is associated with the problem of equilibrium of the cyclic enamine form (91) and the tautomeric hydration products [214,215]: quaternary hydroxide (92), pseudobase (so-called carbinolamine) (93), and an opened form of amino aldehyde or amino ketone (94).

The position of the equilibrium is determined not only by ring size and polar and steric factors but also by the environment of the molecule. The experimental evidence for the existence of three tautomeric forms has been based on the study of their reactivity and, to a lesser degree, on physicochemical measurements [216–218]. Often the existence of the corresponding carbinolamine or its acyclic tautomeric form in addition to the basic dehydrated form is quite important.

Five- (91, $n = 1$) and six-membered (91, $n = 2$) enamines substituted in position 2 generally exist in the cyclic form. Lukeš and coworkers observed that partial ring opening occurs with the pyrroline [197] or piperideine [203,204] derivatives by atmospheric moisture. This leads to the formation of amino ketones, which can be detected

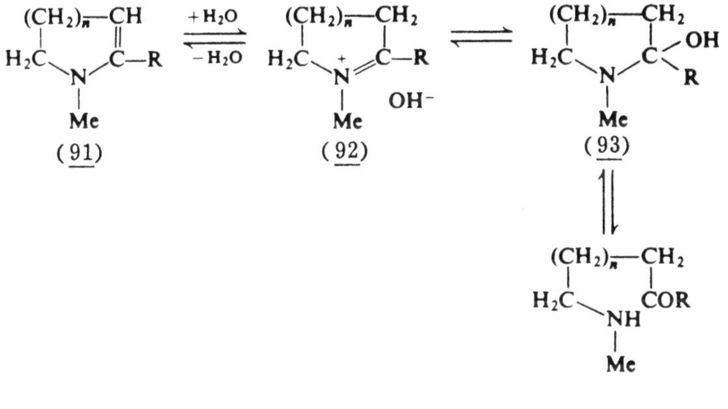

by the ketonic carbonyl absorption at 1705–1710 cm^{-1} in the infrared spectra. Higher basicity in water solution indicates the presence of quaternary hydroxide (92) (pK_a of 1,2-dimethyl-Δ^2-pyrroline is, for example, 11.94, and that of the corresponding saturated compound is 10.23). The analogous six-membered compound has pK_a 11.43, and the corresponding saturated base, 10.26 [1,51]. Introduction of a double bond in the α,β position of primary and secondary amines causes, by contrast, a decrease of basicity [31,219]. The five- and six-membered enamines unsubstituted in position 2 generally exist in cyclic form also.

On the other hand, the cyclic form of the analogous 7- to 13-membered compounds is energetically disadvantageous, and easy formation of amino ketones is encountered. In accordance with this, the compound unsubstituted in position 2 (1-methyl-1-aza-2-cyclooctene, -nonene, and -decene) can react as acyclic amino aldehydes [217]. In the case of enamines bearing an aromatic ring in position 2, especially with 7- and 13-membered rings, a higher stability of the cyclic form can be expected. Therefore, there is a good possibility for isolation of the cyclic form.

On treatment of 1-naphthylmagnesiumbromide with corresponding N-methyllactams, cyclic enamines 1-methyl-2-α-naphthyl-1-aza-cycloheptene (91, n = 3) and 1-methyl-2-α-naphthyl-1-aza-cyclotridecene (91, n = 9) have been prepared. Infrared spectra of the enamines exhibit absorption maxima in the region of C=C double-bond vibrational frequencies at 1625—1630 cm^{-1}. These maxima correspond to double bonds in conjugation with an aromatic ring. Salts of these enamines undergo ring opening in alkalidene media to produce open amino ketones 6-methylamino-1-α-naphthyl-1-hexanone (94, n = 3) and 12-methylamino-1-α-naphthyl-1-dodecanone (94, n = 9).

Compound 95 belongs to a special group that forms cyclic aldehyde ammonia (96) (cotarnine) by interaction of the secondary amine group with the aldehyde group. This aldehyde ammonia can be considered to be a pseudobase.

(95)   (96)

The importance of ring size holds also for tautomerism of Δ^2-pyrrol-5-ones and Δ^2-dihydro-6-pyridones. Whereas the former compounds behave as cyclic 1-methyl-2-alkyl-2-hydroxy-5-pyrrolidones [220] (97) or, on distillation, as the dehydrated 1-methyl-2-alkyl-Δ^2-pyrrolones (98), the latter compounds exist as acyclic N-methylamides of δ-oxo-acids (99), as shown by infrared spectroscopy [221]. The dehydration of 99 during distillation to form 1-methyl-2-alkyl-Δ^2-dihydro-6-pyridones (100) is achieved only with difficulty

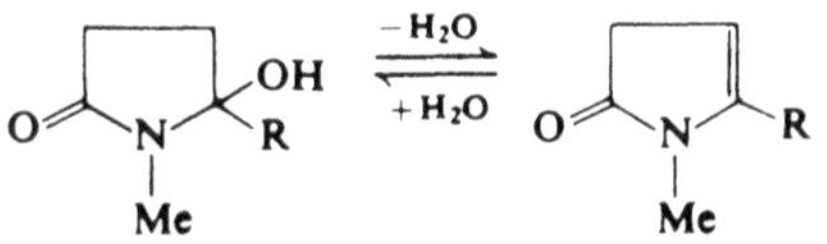

(97)   (98)

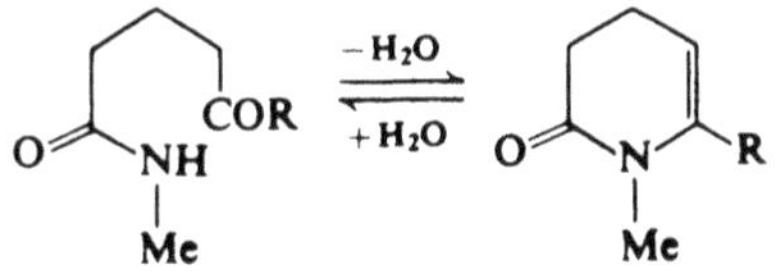

(99)   (100)

In this area of keto amino reactions, transannular cyclization reactions between the ketonic carbonyl group and the tertiary amino nitrogen atom in medium-sized rings are of great interest [216,222–224]. In addition to the alkaloid cryptopine (101), which is the most usual example, there are a large number of other simple examples. The main driving force for the cyclization of 101 upon acidification to compound 102 is the tendency to relieve the nonbonding interactions present in the medium-sized ring and to form a conformationally more favorable arrangement. This transannular reaction corresponds to the tautomeric equilibrium between carbinolamine and amino ketone.

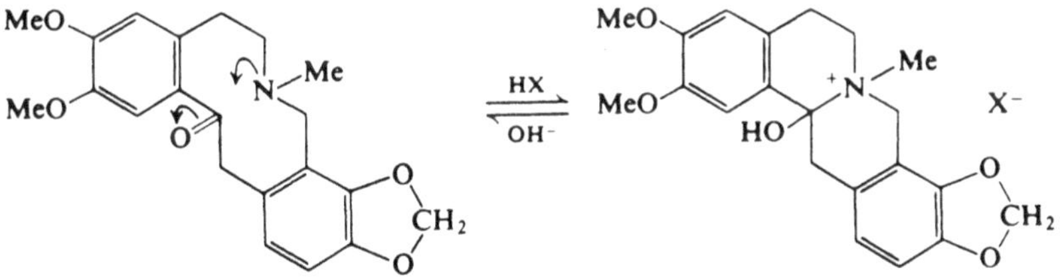

(101)   (102)

F. Structure of Enamine Salts

Physicochemical investigations of enamines and their salts have shown that the addition of a proton occurs almost exclusively at the β-carbon atom of the enamine grouping. This means that salts of pyrrolines (103), piperideines (104), and enamines of 1-azabicycloalkanes (105) possesses immonium structures.

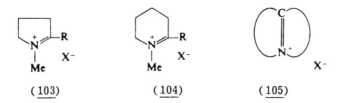

(103)   (104)   (105)

Structural differences between the free enamine and its salt are accompanied by several spectral changes. The presence of these spectral changes can serve as evidence of the presence of an enamine grouping in the molecule. Usually the presence of the new immonium chromophore is indicated by a marked shift of the absorption maximum in the double-bond stretching region to a higher frequency ($20-50$ cm^{-1}) from that present in the free enamine [193,225,226]. This shift, which is characteristic of enamines, in contrast to β, -unsaturated tertiary amines [212], was long considered to be an unambiguous criterion of the presence of enamine grouping in the molecule [133]. However, it has been shown recently in a few cases that structural change

$$-\underset{|}{C}=\underset{|}{C}-\underset{|}{\bar{N}}- \rightarrow -\underset{|}{C}H-\underset{|}{C}=\overset{+}{\underset{|}{N}}-$$

found, for example, when enamines 106 and 107 are treated with acid to form salts 108 and 109, respectively (immonium structure of which follows from interpretation of NMR spectra), is not accompanied by any increase of characteristic frequency [195].

This shift in the infrared spectrum is practically independent of the anion properties.

(106)   (107)   (108)   (109)

The study of NMR spectra [227] shows that all the Δ^1-pyrrolinium and the Δ^1-piperideinium salts exist as such. The NMR spectra of the pyrrolinium derivatives are especially clearly resolved. The chemical shift for the ring protons in the 5, 3, and 4 position lies to increasing field in this order, as is to be expected. The signals of the methylene groups in the 3 and 5 position are triplets, evidently splitting from the 4 position. The piperideine spectra are less regular; although multiplets for the 3- and 6-methylene groups can be distinguished, the 4- and 5-proton signals are merged. Also, the signals for the 3- and 6-methylene groups are in the form of broad peaks representing unresolved multiplets. This difference is evidently connected with the different geometry of the five- and six-membered rings.

The immonium salts derived from 1-azabicycloalkanes have very characteristic NMR spectra [133,134], as illustrated by the spectrum of $\Delta^{4,9}$-indolizinium perchlorate. Assignments of the peaks at $\tau =$ 5.85 and 6.30 ppm to the $-CH_2-\overset{+}{N}=$ groups and those at $\tau = 6.80$ and 7.21 ppm to the $-CH_2-C=$ groups were based on their relative areas (two protons each) and on the previous observation that the chemical shift of the former type of proton is at lower field than that of the latter.

The ultraviolet absorption at λ_{max} 222–232 nm is comparable only with immonium structure [228]. No active hydrogen (Zerewitinov) was present in the immonium salts [1,229], and no deformation vibrations of nitrogen-hydrogen linkage were detected [228].

Cases where proton is localized on the nitrogen atom and an ammonium salt is formed are exceptional. Salts of 1,4,4-trimethyl-Δ^2-piperidine (110), which consist of a mixture of immonium (111) and ammonium (112) salts, serve as an example [1].

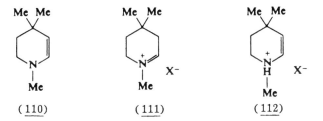

(110)   (111)   (112)

With imines, formation of salts is accompanied by characteristic spectral changes [193]: (a) a bathochromic shift in the ultraviolet region by as much as 50 nm, according to compound type and to properties of any auxochrome present, and (b) a high-frequency shift of the

$$\underset{/}{\overset{\backslash}{}}C=\overset{+}{\underset{|}{N}}-$$

stretching vibration in the infrared region. The imine salts possess an active hydrogen, whereas their quaternization products exhibit the same spectral properties as the enamine salts [230].

Enamines in which the double-bond shift is sterically prevented afford only the ammonium salts. Their spectra in the C=C stretching vibration region does not differ greatly from that of the free amine spectrum [212]. For example, neostrychnine [199] has $\nu_{C=C}$ at 1666 cm^{-1} and its perchlorate at 1665 cm^{-1}. Salts of quinuclideine (113) and the polycyclic alkaloid trimethylconkurchine have similar properties.

The salts of some enamines crystallize as hydrates. In such cases it is possible that they are derived from either the tautomeric carbinolamine or the amino ketone forms. Amino ketone salts (114, n = 5,11) can serve as examples. The proton resonance spectra of 114 show that these salts exist in the open-chain forms in trifluoroacetic acid solution, rather than in the ring-closed forms (115, n = 5,11). The spectrum of the 6-methylamino-phenylhexanone cation shows a multiplet at about 2.15 ppm for phenyl and a triplet for the N-methyl centered at 7.0 ppm and overlapped by signals for the methylene protons at about 8.2 ppm. The spectrum of 114 (n = 11) was similar. These assignments were confirmed by determination of the spectrum in deuterium oxide. Here the N-methyl group of 114 showed a sharp singlet at about 7.4 ppm since the splitting in $-\overset{+}{N}D_2Me$ was much reduced from that of the undeuterated compound.

$$PhCO(CH_2)_n\overset{+}{N}H_2Me \quad ClO_4^-$$

(113)   (114)   (115)

On the other hand, there have been isolated salts of either the acyclic amino ketone form or the cyclic enamine form, namely: 6-methylamino-1-α-naphthyl-1-hexanone (116, n = 5) and 12-methylamino-1-α-naphthyl-1-dodecanone (116, n = 11), or 1-methyl-2-α-naphthyl-1-aza-cycloheptene (117, n = 5) and 1-methyl-2-α-naphthyl-1-aza-2-cyclotridecene (117, n = 11), respectively [72].

$$\alpha C_{10}H_7CO(CH_2)_n\overset{+}{N}H_2Me$$

(116)

(117)

Schöpf et al. [231,232] observed that Δ^1-tetrahydroanabasine salts contain a molecule of water or methanol. According to infrared spectra, they exist as 2-hydroxy- or 2-methoxy-3-(2-piperidyl)piperidine salts (118). Salt (120), obtained by a transannular cyclization reaction taking place on neutralization of bicyclic amino ketone (119), also belongs to this group [222].

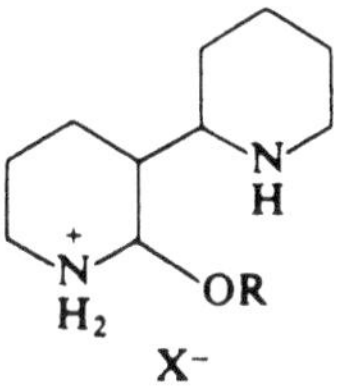

(118)

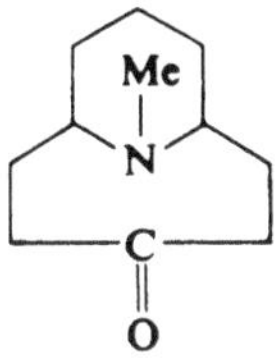

(119)

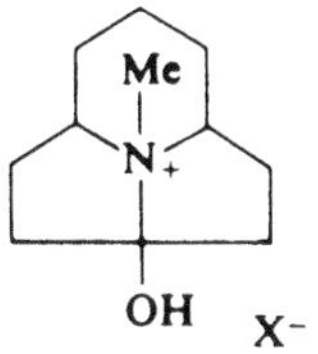

(120)

If other groups capable of conjugation are adjacent to the enamine grouping, they can also participate in the salt formation. Thus, for example, β-amino-α,β-unsaturated ketones can undergo protonation on the carbonyl oxygen atom as well as possible protonation at the carbon and nitrogen atoms. Salts of $\Delta^{8,9}$-octahydro-7-quinolone (85) have their proton situated on the oxygen atom (121). The evidence for this structural assignment comes mainly from the hypsochromic shift in the ultraviolet absorption spectrum; the free base exhibits λ_{max} 298 nm (log ε 4.53) and the salt λ_{max} 280 nm (log ε 4.30). Salts of tertiary enamino ketone (122) are formed in a similar manner.

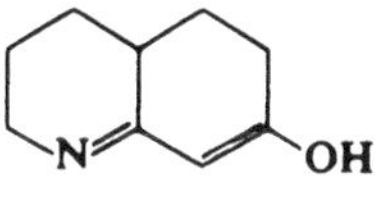

(121)

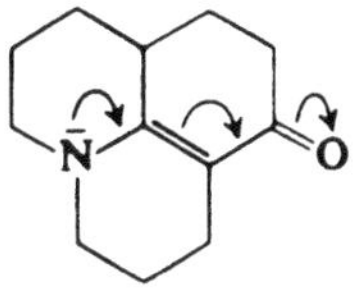

(122)

IV. REACTIONS OF HETEROCYCLIC ENAMINES

The double bonds of either enamines or their salts readily undergo many reactions. We shall divide the reactions of heterocyclic enamines on the basis of the mechanism involved.

A. Reactions of Electrophilic Reagents with the Double Bond of Enamines

Since there are two available sites for electrophilic attack in an enamine, the electrophile can add to the nitrogen atom to form an ammonium salt, or it can add to the β position to form an immonium salt.

Alkylation and Acylation

All enamines do not react in the same way. Both reactive sites are
available for electrophilic alkylation. Whether the alkylation occurs
on the nitrogen or the carbon atom depends on the reactivity of the
alkylation reagent, the structure of the enamine, and, finally, the
polarity of the solvent. Aliphatic alkylating reagents exhibit a
greater tendency to react with the nitrogen atom to form quaternary
ammonium salts. The more reactive alkyl halides, such as allyl halides, α-halogeno ketones, and α-halogeno esters, would, by contrast,
react mainly with the β-carbon atom of the enamine grouping.

The first reported alkylations at the β-carbon atom of a heterocyclic enamine were observed with the alkylations of dihydroberber-
ine (123) [233,234] and 1,3,3-trimethyl-2-2-methyleneindoline (125)
[235,236] to yield monomethylated products (124) and 1,3-trimethyl-
2-isopropylideneindoline (126), respectively.

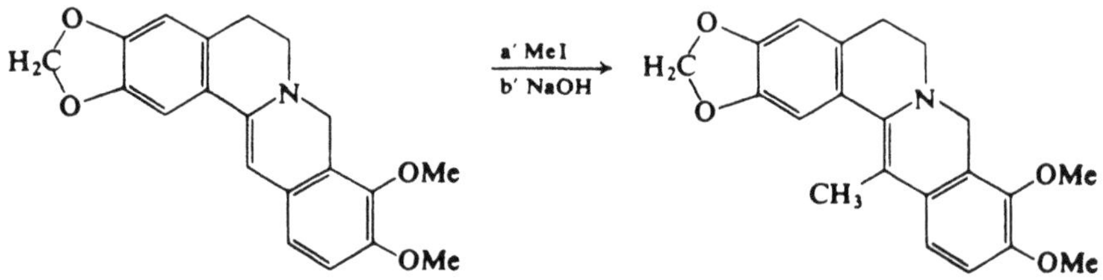

(123)   (124)

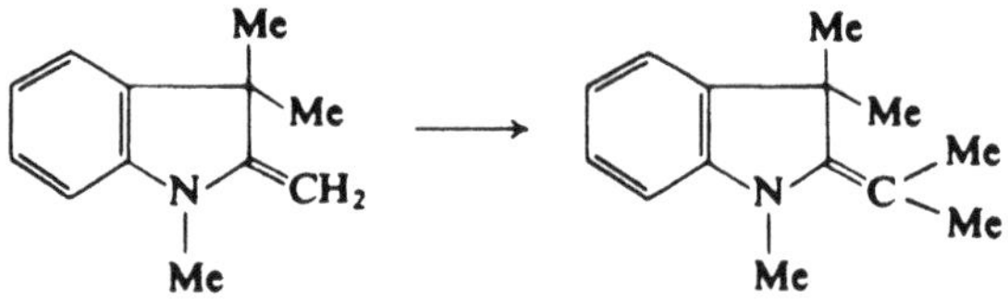

(125)   (126)

C-alkylation was used in the corydaline synthesis [237]. Lukeš
and Dědek [238] obtained on methylation of 1-methyl-2-ethylidene-
pyrrolidine a C-alkylation product, i.e., 1-methyl-2-isopropyl-Δ^2-

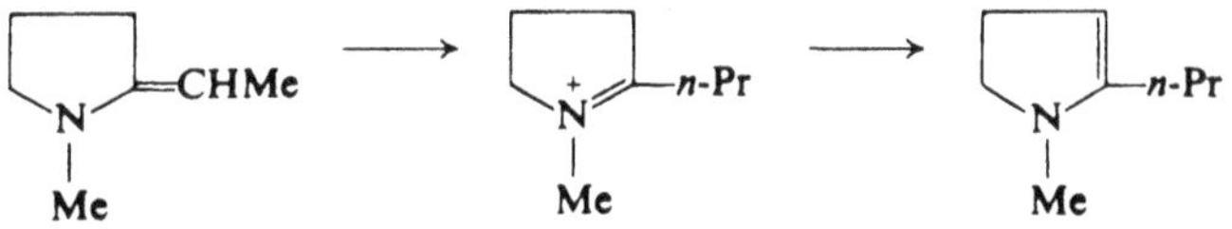

(127)

pyrroline (127). Alkylation of the same enamine with ethyl bromoacetate was the first synthetic step in the preparation of D,L-pseudo-
heliotridane [239].

Quaternary ammonium salts of pyrrolines (121) can be prepared
only indirectly [240]. Addition of bromide to 1-dimethylamino-4-pen-
tene followed by removal of hydrogen bromide afforded, depending
on the dehydrohalogenation conditions, quaternary bromides derived
from either 1,2-dimethyl-Δ^2-pyrroline (128) or 1-methyl-2-methylene-
pyrrolidine (129) (Scheme 14).

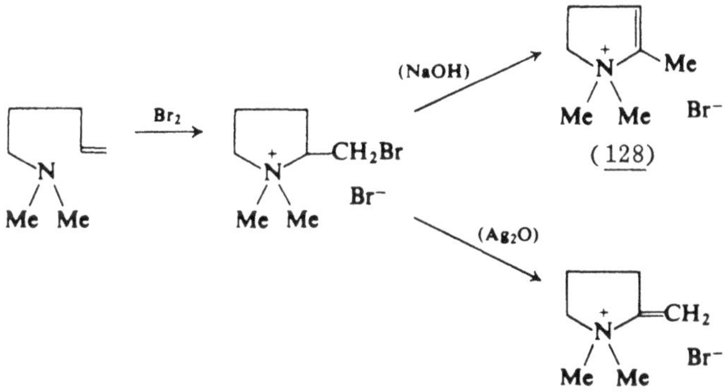

Scheme 14

It has not been hitherto unequivocally proven whether the β-
alkylation product arises directly or by isomerization of the initially
formed products of N-alkylation. Červinka, Fábryová, Chudobová
[241] used this quaternary salt in studying the course of isomerization of an N-alkylated salt to a C-alkylated product. The isomerization was carried out by heating the quaternary salt briefly above its
melting point in a closed flask. In order to identify all the reaction
products, the isomerization mixture was reduced with formic acid to
saturated bases, which were identified by gas-liquid chromatographic
comparison with authentic standards. It was found that the thermal

isomerization of the compound (129) led to a mixture of salts of unsaturated bases which on reduction with formic acid afforded 1,2-di-
methylpyrrolidine, 1-methyl-2-ethylpyrrolidine, and 1,2,3-trimethyl-
pyrrolidine. In order to decide whether the thermal isomerization is
an intramolecular or an intermolecular process, workers studied the
analogous reaction of 1,1-diethyl-2-methyl-2-pyrrolinium bromide
(130), which after similar workup afforded a mixture of 1-ethyl-2-
methylpyrrolidine (131), 1-ethyl-2-propylpyrrolidine (132), and a
base which probably was 1,3-diethyl-2-methylpyrrolidine. Finally,
thermal isomerization of a mixture of both the quaternary salts (128,
130) was carried out, which led to the pyrrolidines (133 a–g).

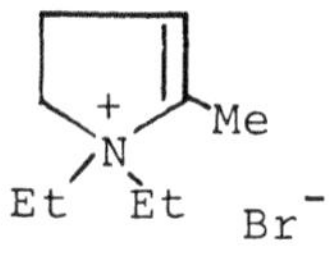

(130) 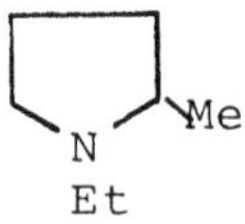   (131) 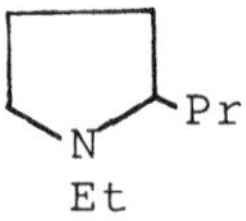   (132)

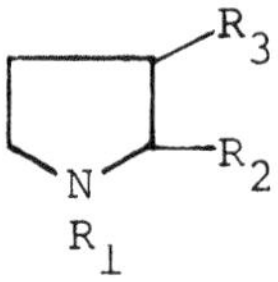

(133)

a) $R^1 = R^2 = Me$; $R^3 = H$
b) $R^1 = Me$; $R^2 = Et$; $R^3 = H$
c) $R^1 = R^2 = R^3 = Me$
d) $R^1 = Et$; $R^2 = Me$; $R^3 = H$
e) $R^1 = Et$; $R^2 = Pr$; $R^3 = H$
f) $R^1 = Me$; $R^2 = Pr$; $R^3 = H$
g) $R^1 = R^2 = Et$; $R^3 = H$

Hofmann degradation of 1,1-dimethyl-2-methylenepyrrolidinium hydroxide furnishes dimethylamine and dimethyl-3-pentynylamine [242]. 1,1,4,4-Tetramethyltetrahydropyridinium hydroxide was obtained from 1-dimethylamino-4,5-dibromopentane by means of silver oxide. Hofmann degradation of the product gives 1,4,4-trimethyl-Δ^2-tetrahydropyridine [243].

A study of methylation of 1-azabicycloalkane enamines shows the complexity of the alkylation reaction. Treatment of $\Delta^{1,10}$-dehydroquinolizidine (134) with methyliodide (129, 131) gives a mixture of three products (135, 136, and 137) containing 83% of the quaternary

(134) (135) (136) (137)

(138) (139) (140) (141)

ammonium salt (136). 1-Methyl-$\Delta^{1,10}$-dehydroquinolizidine (81) affords ammonium salt (138) only. Upon methylation of Δ^8-dehydroindolizidine (76), dialkylated compounds (140 and 141) are formed in addition to C-monomethylated product (139). Compound 140 is accessible also by methylation of 8-methyl-Δ^8-dehydroquinolizidine (131).

It is noteworthy that only in the case of dehydroquinolizidine derivatives does monomethylation produce the N-alkylated product. The formation of dialkylated products can be explained by a disproportionation reaction of the monoalkylated immonium salt caused by either the basicity of the starting enamine or some base added to the reaction mixture (most often potassium carbonate) and subsequent alkylation of the monoalkylated enamine. Reinecke and Kray [131] try to explain the different behavior of $\Delta^{1,10}$-dehydroquinolizidine and Δ^8-dehydroquinolizidine derivatives by the difference in energies of N- and C-alkylation transition states because of the presence of I-strain.

In the case of methylation of Δ^8-dehydroindolizidine on nitrogen, the orbitals of nitrogen in such a planar system would have to rehy-

bridize from trigonal to tetragonal configuration, which is not advan-
tageous for nitrogen as a part of a five-membered ring, because of
I-strain. The analogous β-carbon atom (of the enamine system) hybridization is more favorable in the transition state since this atom
is solely a part of a six-membered ring [244]. With quinolizidine
enamines, where the nitrogen atom is a part of two six-membered
rings, the sp^3 rehybridization state of N-methylation does not require
a substantial increase of activation energy for the reaction. It is
important to point out that more reactive halides, such as allylbro-
mide, also react with the β-carbon atom of the $\Delta^{1,10}$-dehydroquino-
lizidine enamine grouping [245]. Alkylation to the β-carbon atom of
enamine grouping of berberine [246] and harmaline [247] was observed.

When the immonium form of the enamine is precluded sterically,
enamines are alkylated solely on the nitrogen atom. Methylation of
neostrychnine with methyliodide proceeds in this manner, as well as
that of Δ^5-tetrahydrojulolidine (142) [248], which affords only the
N-methylated product (143).

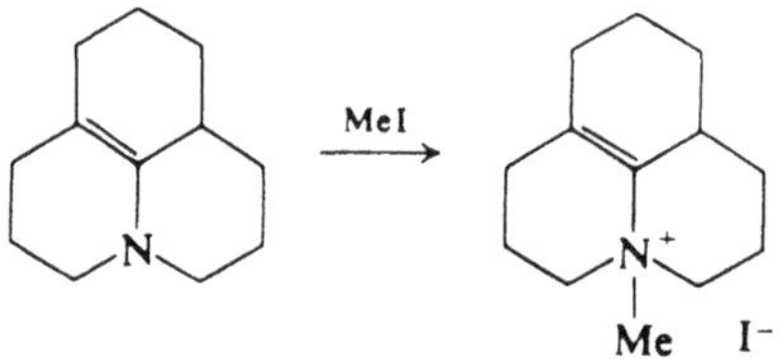

(142)

(143)

An explanation of the exclusive N-methylation of 1,2-dimethyl-
Δ^2-piperideine by means of methyliodide is more difficult. Pyrrolines
and piperideines that are not alkylated on the nitrogen atom afford
only quaternary ammonium salts on alkylation [249–251], for example,
144.

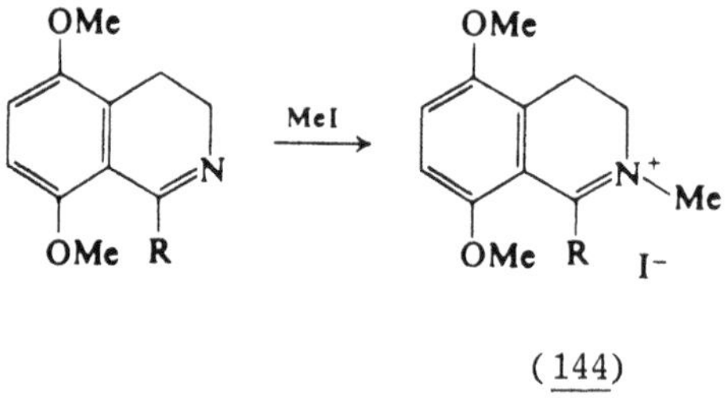

(144)

Secondary enamino ketones such as $\Delta^{8,9}$-octahydro-7-quinolone (85) furnish a mixture of N- and C-alkylated bases (145 and 146, respectively), on treatment with ethyliodide [209]. Alkylation of tertiary enamino ketones, for example, 122, proceeds exclusively on the oxygen-atom, forming product (147) [252].

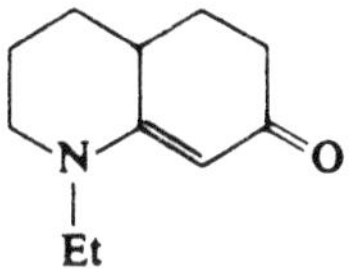

(145) 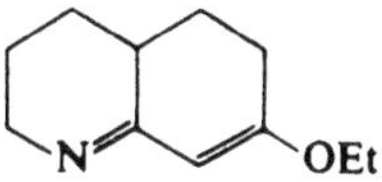   (146)

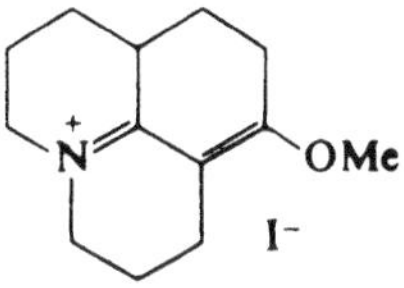

(147)

Quarternizing 1-pyrrolines gave 1-pyrrolinium salts, which were deprotonated with base to the tertiary pyrrolidine enamines [253].

Acylation of heterocyclic enamines is to a great extent similar to alkylation, and usually occurs on the β-carbon atom of the enamine grouping.

Acylation of 1,3,3-trimethyl-2-methyleneindoline (125) leads on basification to 1,3,3-trimethyl-2-acylmethyleneindoline (148) [254]. Reaction with diketene affords the corresponding β-diketo compound (149).

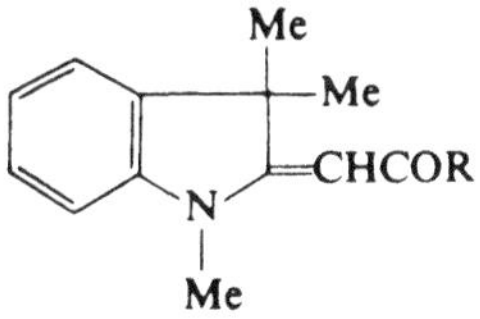

(148) 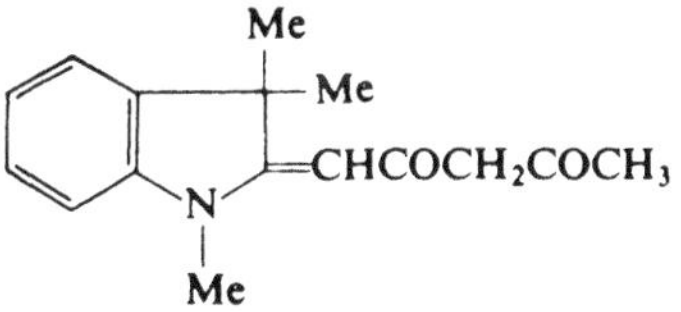   (149)

1,2-Dimethyl-3-acetylpiperidine (150) has been prepared by acetylation of 1,2-dimethyl-Δ^2-piperideine, followed by hydrogenation (Scheme 15) [203].

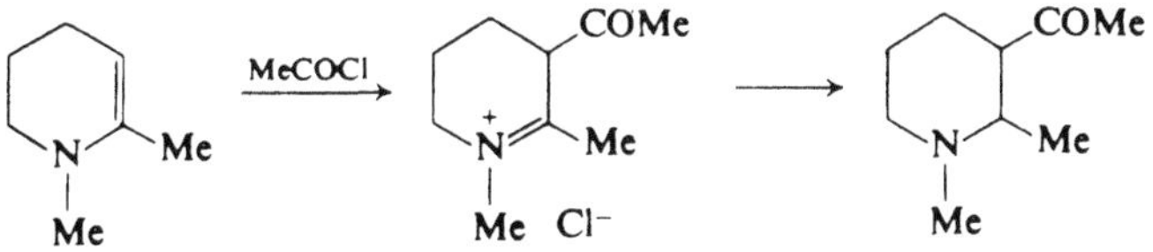

Scheme 15

Treatment of $\Delta^{1,10}$-dehydroquinolizidine with ethyl chloroformate furnishes on basification 1-carbethoxy-$\Delta^{1,10}$-dehydroquinolizidine (151).

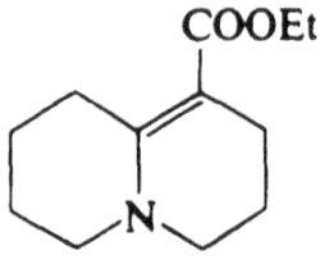

(151)

By acylation of 2-methyl-1,2-dihydroisoquinoline (152) with 3,4-dimethoxyphenacylchloride, a C-4-alkylated product (153) is formed [255].

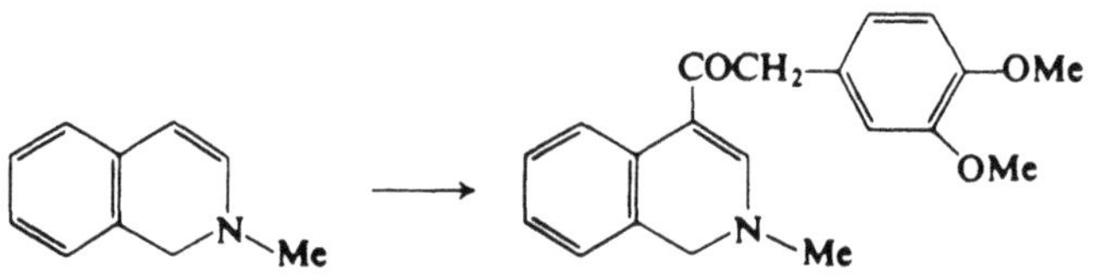

(152)   (153)

A methylene base formed from quinaldine ethiodide, 1-ethyl-2-methylene-1,2-dihydroisoquinoline (154), exhibits a number of reactions characteristic of enamines [254,256]. On treatment with benzoylchloride, a dialkylated product (155) is produced by C- and subsequent O-benzoylation [257].

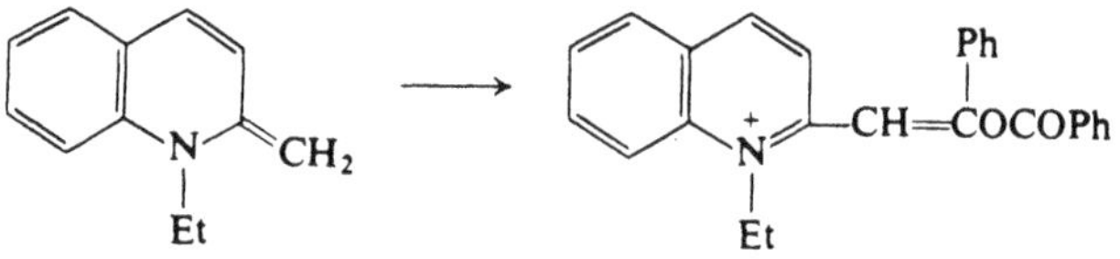

(154)

(155)

Reaction of 2-alkyl-Δ^1-pyrrolines and 2-alkyl-Δ^1-piperideines with acid chlorides leads to ring opening and formation of N-acylated amino ketones (156, n = 1,2) [258–260]. Ketene reacts with Δ^1-piperideine to form a tricyclic derivative (157) [261].

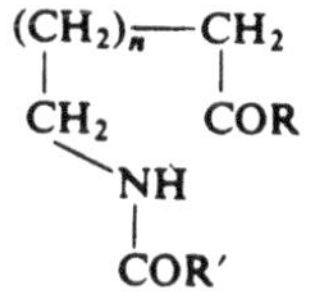

(CH₂)ₙ—CH₂
| |
CH₂   COR
   \
    NH
    |
    COR′

(156)

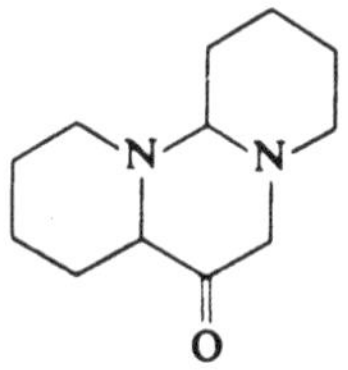

(157)

Me-1,2,3,5,6,7,7a,8-octahydro-8-(methoxycarbonyl)cyclopenta-[b]-pyrrolizine-8-acetate reacted with (CF₃CO)₂O to give trifluoro-acetylated compound [262].

Reactions of Heterocyclic Enamines with
α, β-Unsaturated Compounds

Enamines react readily with compounds containing a double bond acti-vated by electronegative groups. Addition of acrolein to 1-methyl-2-ethylidenepyrrolidine, followed by dehydrogenation, leads to 1,7-dimethylindole (158) (Scheme 16) [263].

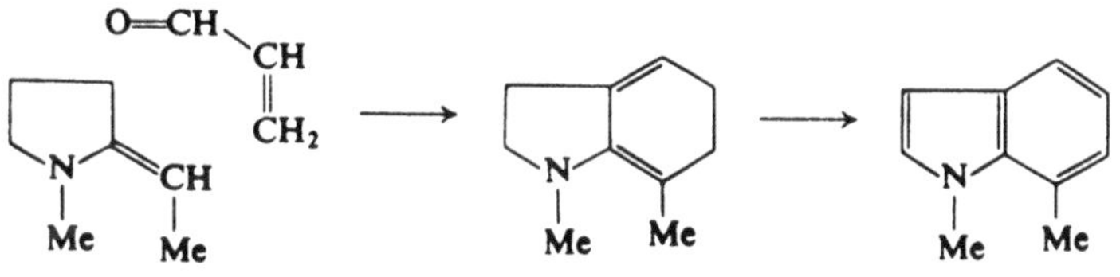

(158)

Scheme 16

In a similar addition to 1-methyl-2-alkyl-Δ²-piperideines, 1-meth-yl-8-alkyl-1,2,3,4-tetrahydroquinolines (159) were obtained (Scheme 17) [203].

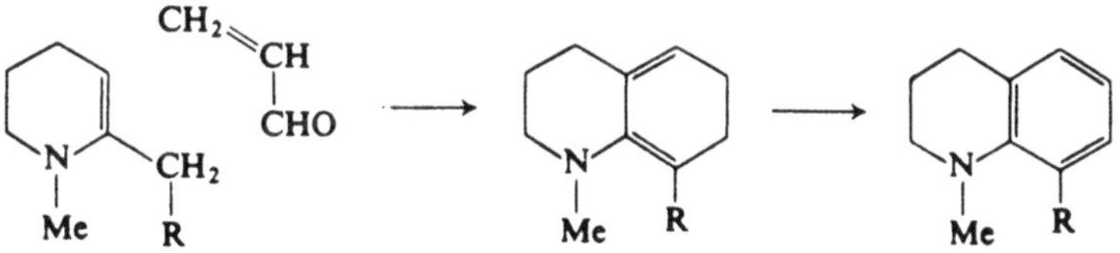

(159)

Scheme 17

Bohlmann and Schmidt [245] reported the reaction of $\Delta^{1,10}$-dehydroquinolizidine with methyl vinyl ketone and with propargyl aldehyde forming a partially saturated derivative of julolidine (160) and julolidine (161), respectively. Compound 160 can be prepared also by mercuric acetate dehydrogenation of ketone (162), which is formed by condensation of 1-bromoethylquinolizidine with ethyl acetoacetate (Scheme 18).

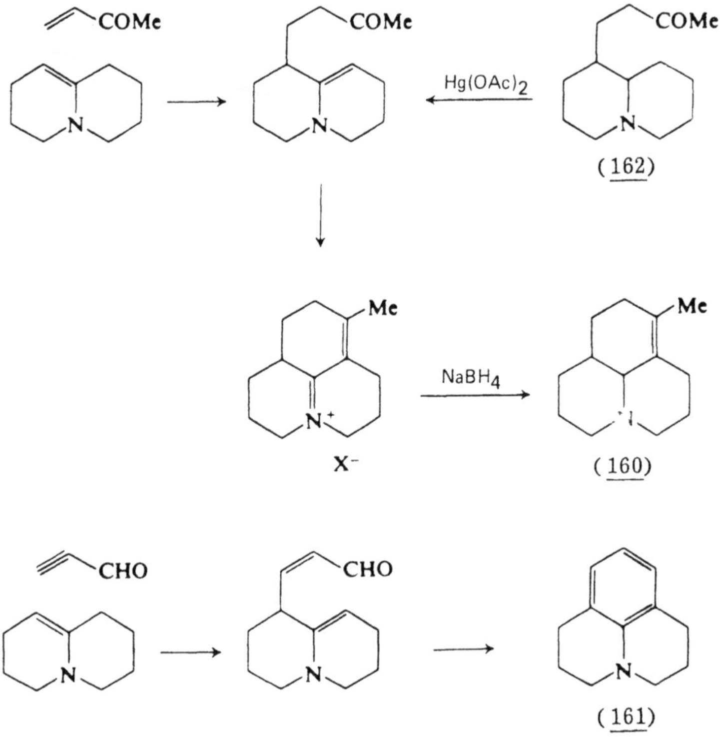

Scheme 18

The addition of ethyl acrylate to 1,2-dimethyl-Δ^2-piperideine [203], 1-methyl-2-ethyl-Δ^2-piperideine [204], and 1,2-dimethyl-Δ^2-pyrrolidine [264,265] occurs, yielding both possible enamine structures (163 and 164, n = 1,2).

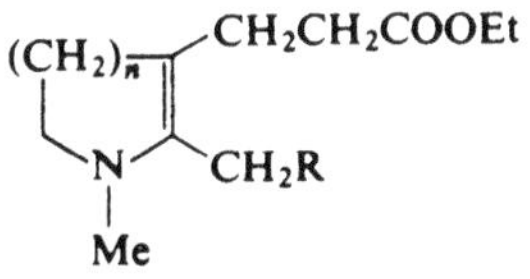

(163)

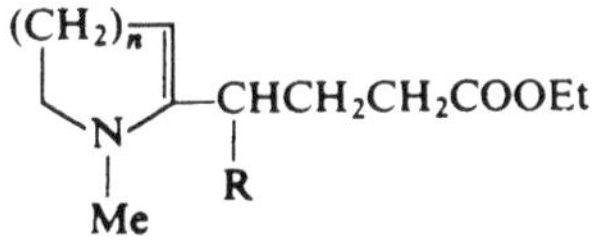

(164)

Addition to 1,2-dimethyl-Δ^2-piperideine or 1,2-dimethyl-Δ^2-pyr-roline is followed by intramolecular alkylation by the ester group as a side reaction to give 165 and 166 (n 1,2, respectively).

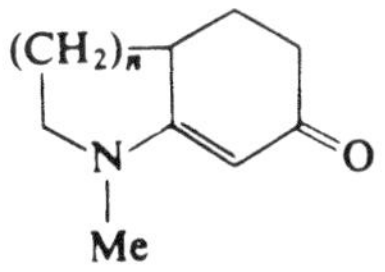

(165)

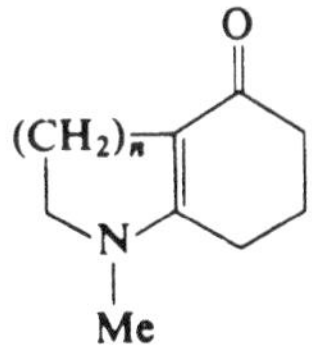

(166)

Cyclization products (167) and (168) are the main products in the reaction of ethyl acrylate or ethyl glutaconate with $\Delta^{1,10}$-dehy-droquinolizidine [245]. On the other hand, the addition of butadiene carboxylic acid leads to a mixture of products [266].

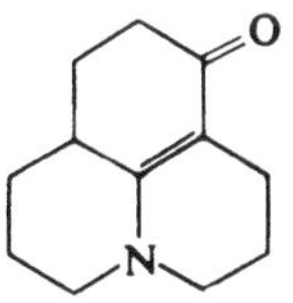

(167)

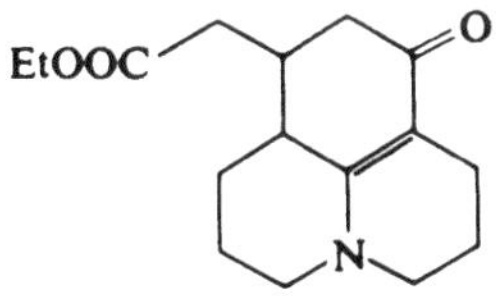

(168)

An interesting addition of ethyl acrylate has been reported in the case of 1-methyl-2-ethylidenepyrrolidine. An unsaturated amino ketone (169) is formed, which rearranges to 1,7-dimethyloctahydro-indole (170) on reduction with formic acid, as established by dehydro-genation to 1,7-dimethylindole (Scheme 19) [265].

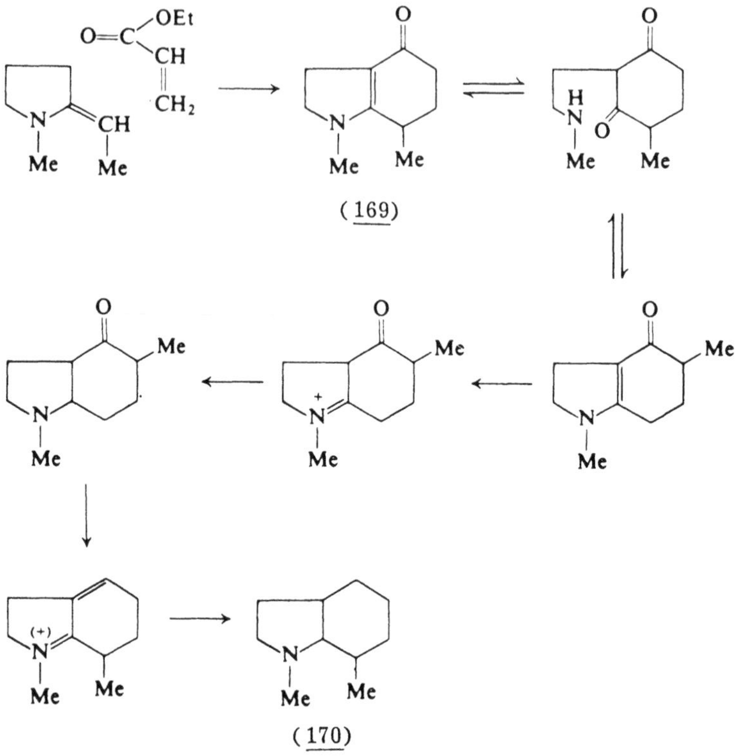

(169)

(170)

Scheme 19

Imines also react with α,β-unsaturated aldehydes or ketones [267–269]. 3,4-Dihydroisoquinoline reacts, for example, with methyl vinyl ketone to give cyclic ketone (171) [270,271].

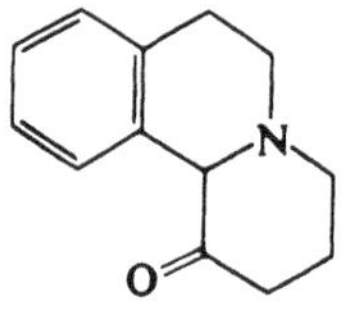

(171)

Reactions of Heterocyclic Enamines with Other Electrophilic Reagents

Enamines are generally very sensitive to oxidation [106,272]. By standing in the air, they become brown and afford an undefinable mass. Imines, by contrast, form hydroperoxides with atmospheric oxygen, which may be isolated [191,273—277]. $\Delta^{1,9}$-Octahydroquinoline (172) affords a crystalline hydroperoxide (173), which may be reduced to 10-hydroxy-$\Delta^{1,9}$-octahydroquinoline (174) or hydrolyzed to a cyclic oxolactam (175). Reactions with many analogous compounds have been reported [278—280].

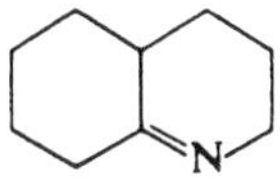

OOH   OH   O

(172) 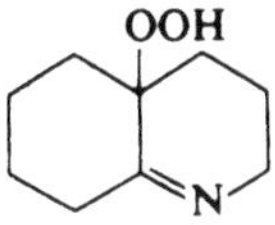   (173) 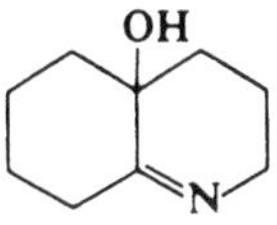   (174)

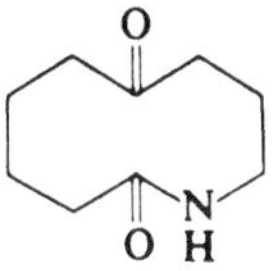

(175)

By means of perbenzoic acid oxidation, a bicyclic oxazirane (177) [281,282] is formed from 5,5-dimethyl-4-phenyl-Δ^1-pyrroline (176).

(176)   (177)

Dibenzoylperoxide oxidation of $\Delta^{1,10}$-dehydroquinolizidine affords an immonium salt, which can be reduced with sodium borohydride to 1-benzoyloxyquinolizidine. Treatment of the salt with base liberates 1-benzoyloxy-$\Delta^{9,10}$-dehydroquinolizidine [269].

A dimer is formed by the action of hydrogen peroxide on the quaternary salt of 3,4-dihydroisoquinoline [283]. The other similar reactions are of small importance.

Photooxygenation of 1-benzyl-3,4-dihydroisoquinolines gives the tautomeric enaminohydroperoxide, which readily loses a molecule of water, giving the oxidized product [284].

The fact that 2-N-substituted pyrazolines containing a

$$-\underset{|}{C}=\bar{N}-\bar{N}-R$$

grouping in the molecule react in a manner that is typical of enamines is very interesting [285].

B. Reactions of Enamine Salts with
Nucleophilic Reagents

Reactions at the carbon-nitrogen double bond of imminium salts are
analogous to nucleophilic reactions at the carbonyl group of aldehydes
and ketones. This is why free enamines do not react with nucleophilic reagents, whereas their salts can undergo such reactions.

Reduction

Tertiary heterocyclic enamines are reduced with metals in acidic media
[182] or electrolytically [286,287], and their salts are reduced with
lithium aluminium hydride or sodium borohydride [288,289] to the
corresponding saturated amines.

Reduction of 1-methyl-2-alkyl-Δ^1-pyrroline and 1-methyl-2-alkyl-
Δ^1-piperideine perchlorates with complex hydrides prepared in situ
by partial decomposition of lithium aluminium hydride with the optically active alcohols (−)-menthol and (−)-borneol affords partially
optically active 1-methyl-2-alkyl pyrrolidines (178, n = 1) and 1-
methyl-2-alkyl piperideines (178, n = 2), respectively [290,291].

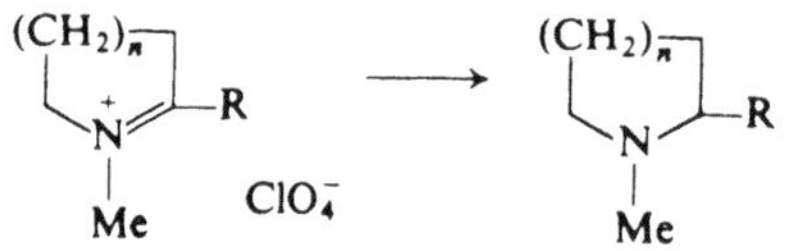

(178)

$\Delta^{1,10}$-Dehydroquinolizidine reacts with the enantiomeric (−)-
and (+)-menthyl chloroformates forming (−)- and (+)-menthoxycar-
bonyl-$\Delta^{1,10}$-dehydroquinolizidines. These can be reduced as such
or in the form of their immonium salts with sodium borohydride to
(−)- and (+)-1-menthoxycarbonylquinolizidines, which give (+)- and
(−)-lupinin, respectively, on reduction with lithium aluminum hydride [292]. The optical yield of the asymmetrical reduction is about
10%.

The intermediate formation of iminium salts is postulated in the
reduction of α-amino ketones by the Clemmensen method, occurring

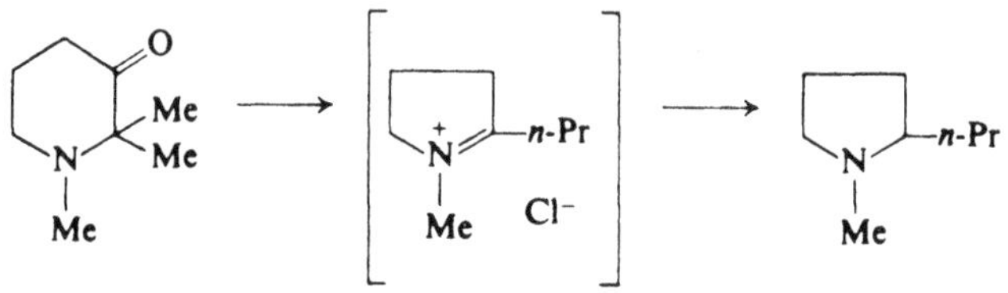

(179)

(180)

with concomitant ring enlargement or contraction [293–295]. Reduction of 1,2,2-trimethyl-3-piperidone (179) in this manner gave 1-methyl-2-isopropylpyrrolidine (180).

Enamines are also reduced with formic acid [296]. Distillation of 1,2-dimethyl-Δ^1-pyrroline formate (181) affords 1,2-dimethylpyrrolidine [297]. The reaction is usually carried out by heating of the enamine salt with formic acid. Potassium formate can be added to increase the temperature of the reaction mixture.

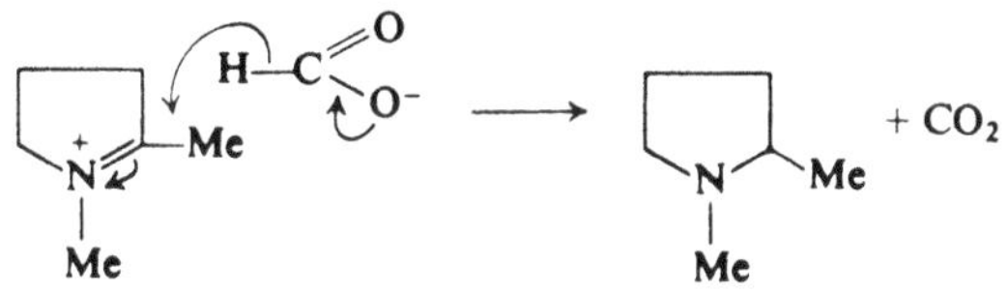

(181)

This method has been used for the reduction of 1-methyl-2-alkyl-Δ^1-pyrrolinium and 1-methyl-2-alkyl-Δ^1-piperideinium salts by Lukeš et al. [51,298–300] and for the reduction of more complex bases containing the dehydroquinolizidine skeleton by Leonard et al. [301]. The formic acid reduction may be satisfactorily explained by addition of a hydride ion, or an equivalent particle formed from the formate anion, to the β-carbon atom of the enamine [302], as shown in Scheme 20.

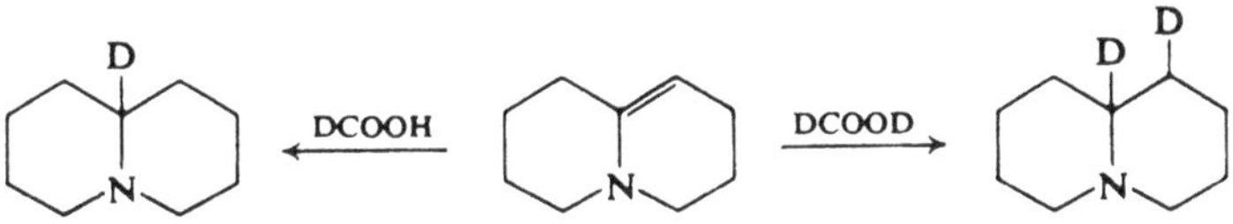

Scheme 20

1,2-Dihydro and 1,4-dihydro derivatives are formed as intermediates in the reduction of quaternary pyridine salts and their homologs with sodium borohydride or formic acid. A proton is added to the present enamine grouping, and the formed immonium salts are reduced to the 1-methyl-1,2,5,6-tetrahydropyridine derivatives (182) and to completely saturated compounds (183) [303] (Scheme 21).

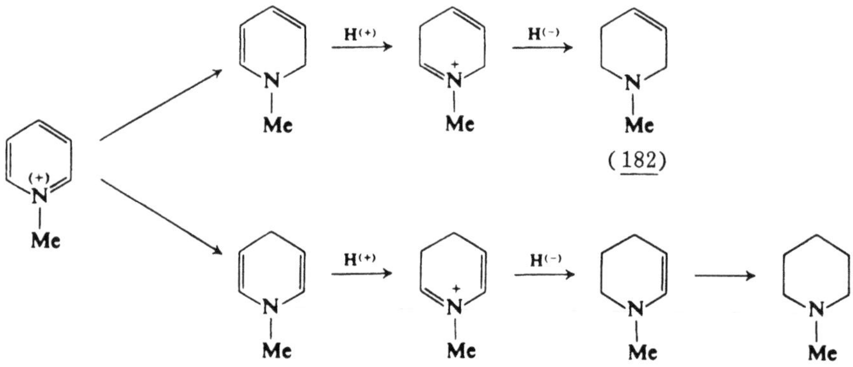

(182)

(183)

Scheme 21

The formic acid reduction has great stereospecificity. Reduction of (−)-Δ^5-dehydrosparteine (184) and (−)-Δ^{5,11}-didehydrosparteine affords (−)-sparteine (185) and (−)-α-isosparteine, respectively [301].

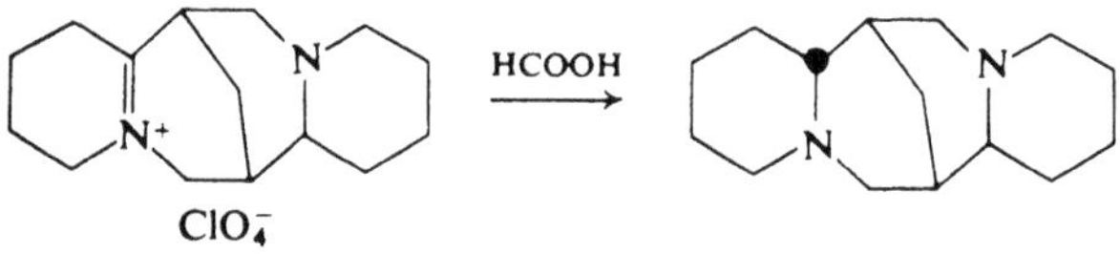

(185)

(184)

Reduction of the quaternary immonium salt (186), obtained by treatment of 1-methyl-2-ethylidenepyrrolidine with ethyl bromoacetate, by means of either sodium borohydride or formic acid, leads to (−)-erythro-2-(2-N-methylpyrrolidyl)butyric acid (187), in agreement with Cram's rule [239].

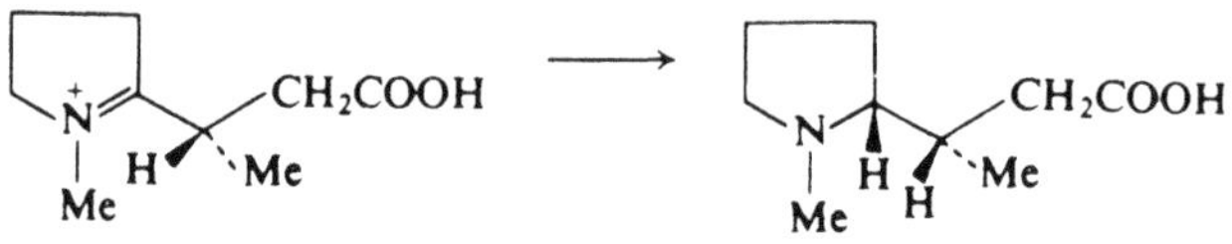

(186)

(187)

In both cases, the hydride ion approaches the double bond from the sterically more accessible side of the molecule. Reduction of imines by metals and acids, electrolytically or by formic acid, gives saturated secondary amines [47,304].

Reactions of Enamine Salts with Organometallic Compounds

Organolithium and organomagnesium compounds react with enamine salts to give amines substituted on the α-carbon atoms. The treatment of $\Delta^{5,10}$-dehydroquinolizidinium perchlorate (188) with alkylmagnesium halides gives 9-alkylated quinolizidines (189) [301,305].

(188)   (189)

Formation of 1-methyl-2,2-dialkylpyrrolidines (190) has been observed on treatment of 1-methyl-2-alkyl-Δ^1-pyrrolinium perchlorates with alkylmagnesium halides [306].

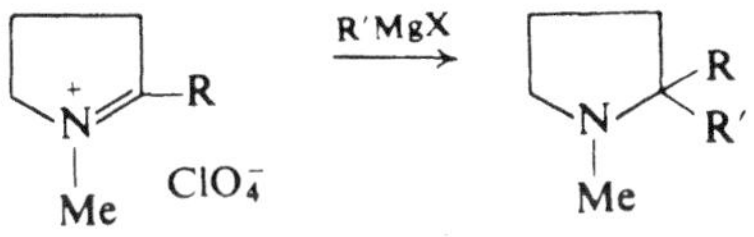

(190)

Δ^1-Pyrrolines and Δ^1-piperideines do not generally react with Grignard reagents [189,307]. The addition complex reverts to the starting amine when treated with water during the hydrolysis step. In some cases, the Grignard reagent causes proton removal. This is followed by condensation of the anion thus formed with a second molecule of the Δ^1-pyrroline [304].

The alkaloids cotarnine [308], hydrastinine [310], and berberinal [309], each possessing a grouping formed by interaction of an aldehyde with a secondary amino group in their molecule, are unusual. The Grignard reaction of free base (191) does not occur as readily as that of the corresponding salt (192). Both reactions lead to the alkylated product (193). For example, only 50% of hydrastinine reacts and 50% is regenerated, whereas hydrastinine hydrochloride reacts almost quantitatively [310]. The salt undoubtedly contains a C=N double bond. In the case of the free base, the presence of a C=N double bond was not proven, and the reaction probably occurs by direct cleavage of the C−OH bond.

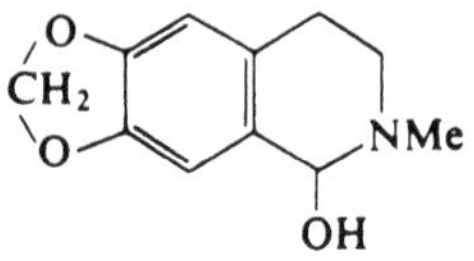

(191) 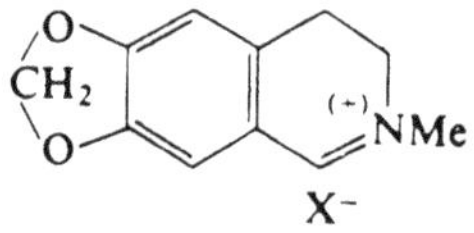 (192)

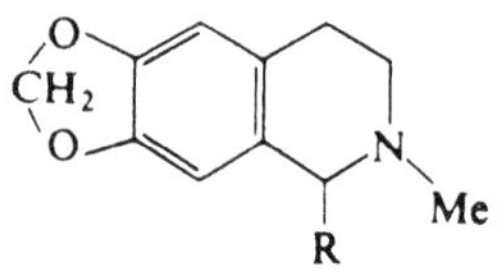

(193)

Reaction of organometallic compounds with enamine salts have been successfully used for the synthesis of some natural products [305]. Thus, reaction of the immonium salt of O-alkylated enamino ketone (147) with isobutyllithium affords compound 194.

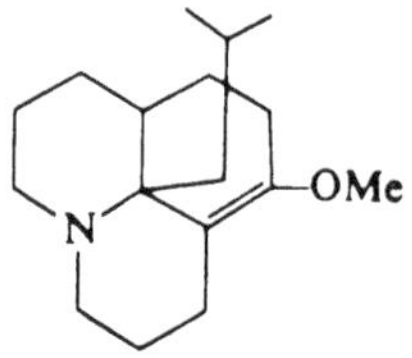

(194)

Reactions of Enamine Salts with Other Nucleophilic Reagents

Enamine salts react with many nucleophilic reagents. The reaction with the cyanide ion is noteworthy. 1-Methyl-2-ethyl-2-cyanopyrrolidine (195) is formed on treatment of alkali cyanide with 1-methyl-2-ethyl-Δ1-pyrrolinium perchlorate (191). The reduction of the tertiary nitrile (195) with lithium aluminum hydride in ether gives 1-methyl-2-ethylpyrrolidine. Hydrogen cyanide can be removed by treatment

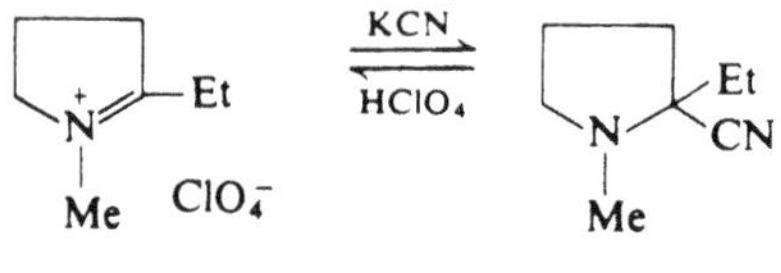

(195)

with acids with reformation of 1-methyl-2-ethyl-Δ^1-pyrrolinium salts.
On the other hand, addition of hydrogen cyanide to Δ^1-pyrrolines
yields stable nitriles, which can be easily saponified to acids or reduced to amines [304] (Scheme 22).

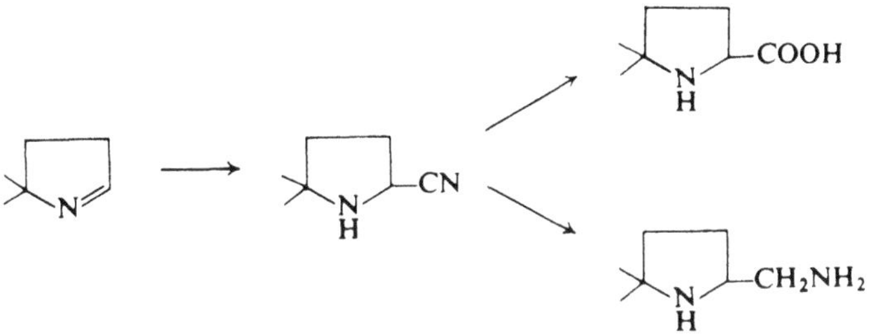

Scheme 22

Δ^1-Pyrroline-N-oxides when unsubstituted in the 2 position readily add hydrogen cyanide. The 1-hydroxy-2-cyanopyrrolidines thus
formed undergo oxidation to 2-cyano-Δ^1-pyrroline-N-oxides.

Treatment of cotarnine and similar compounds with hydrogen cyanide, alkoxides, mercaptides, hydroxylamine, hydrazine, and amines
has been reported to give 1-substituted derivatives of 1,2,3-tetrahy-
droisoquinoline (196, R = CN, OR, SP, NHOH, NHNH$_2$, NHR) [311–
314].

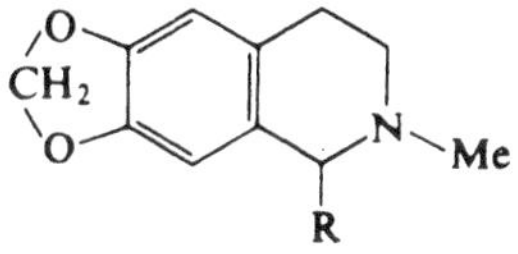

(196)

The cyclization reaction of some substituted 1,2-dihydroisoquin-
olines is of interest [315]. The reduction of papaverine with tin and
hydrochloric acid affords the 1,2-dihydro compound in the form of
immonium salt (197), which then undergoes a cyclization reaction in
the acidic medium to give compound 198, called pavine [316].

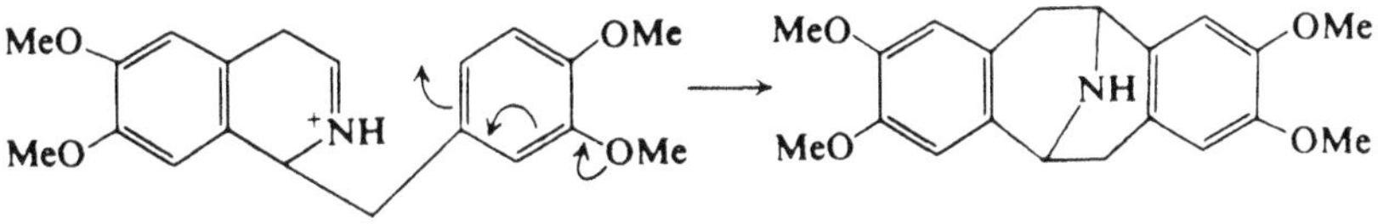

(197)   (198)

On treatment of N-methylpapaverine, formed by the lithium aluminum hydride reduction of papaverine methiodide with phosphoric
acid, N-methylpavine is formed, which is identical with the racemic
alkaloid argemonine. This reaction was used for the synthesis of
the alkaloid (+)-coreximine (199) [317] and similar compounds containing the protoberberine grouping in the molecule [318,319].

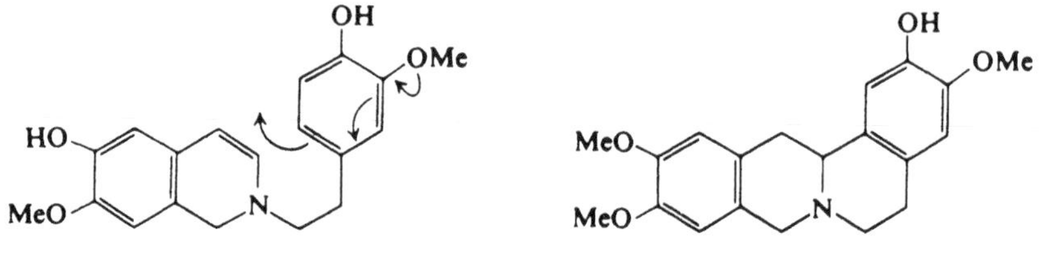

(199)

Knabe et al. [320−323] later observed that 6,7-dimethoxy-2-
methyl-1,2-dihydroisoquinolines (200), possessing either a free or a
substituted benzyl group in position 1, readily rearrange to 3,4-
dihydroisoquinoline salts (201) on treatment with dilute acids.

2-Methyl-1,2-dihydropapaverine (200, R = OMe) rearranges to
the 2-methyl-3-(3,4-dimethoxybenzyl)-6,7-dimethoxy-3,4-dihydro-
isoquinolinium salt (201, R = OMe) under very mild conditions (treatment with 2% hydrochloric acid). A similar rearrangement of 1-(3,4-
methylenedioxybenzyl)-2-methyl-6,7-dimethoxyisoquinoline (200, R,
R = −O−CH$_2$−O−) affords 3-(3,4-methylenedioxybenzyl)-6,7-dimeth-
oxy-3,4-dihydroisoquinolinium chloride (201, R,R = O−CH$_2$−O−).
The reaction was shown to be an allylic rearrangement with internal
return [324,325].

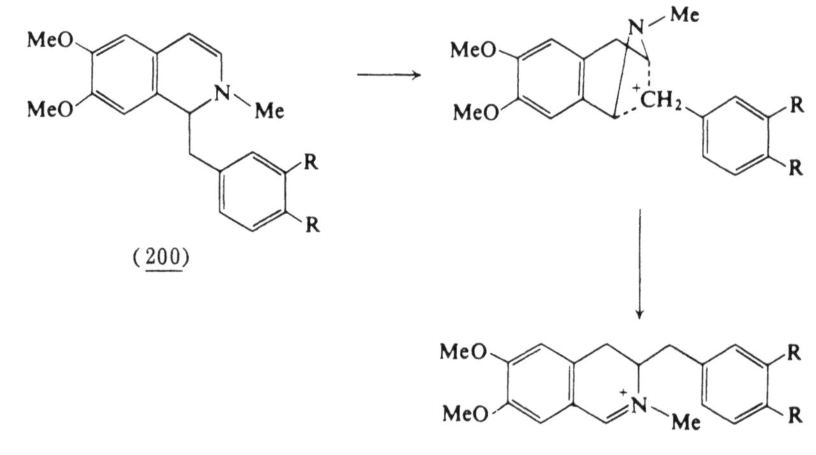

(200)

(201)

The disproportionation reaction of isoquinolinium salts to 1,2-di-hydroisoquinolines and isocarbostyril derivatives (Scheme 23) was used by Brown and Dyke for the synthesis of berberine and 8-oxo-berberine derivatives [326–328].

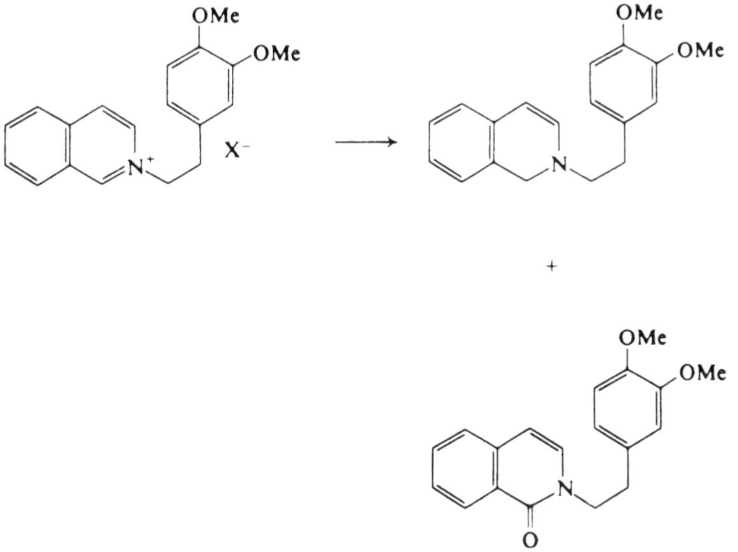

Scheme 23

The mercuric acetate dehydrogenation of carbomethoxydihydro-cleavacine (202) yields immonium salt (203), which undergoes trans-annular cyclization to give a mixture of coronaridine (205, R = Et, R_1 = H) and dihydrocantharanthine (205, R = H, R_1 = Et). The reaction is accompanied by a similar cyclization [329] to the β-position of the indole nucleus forming compound 206 via 204.

From the preparative point of view, reactions of heterocyclic aromatic compounds with nucleophilic reagents are very important, especially the reactions of their quaternary salts containing a formal grouping in the molecule.

With pyridine derivatives and compounds containing the pyridine ring, some reactions are of particular interest [131,333,334]: reaction with organometallic compounds (organomagnesium or organolith-ium); reduction with sodium in ethanol, with complex hydrides, or with formic acid; and reactions with hydroxyl or cyanide ions [330–332] and with some organic compounds containing the reactive methylene group. The N-oxides of these heterocyclic bases [335,336], especially their quaternary salts, react in a similar manner [337,338].

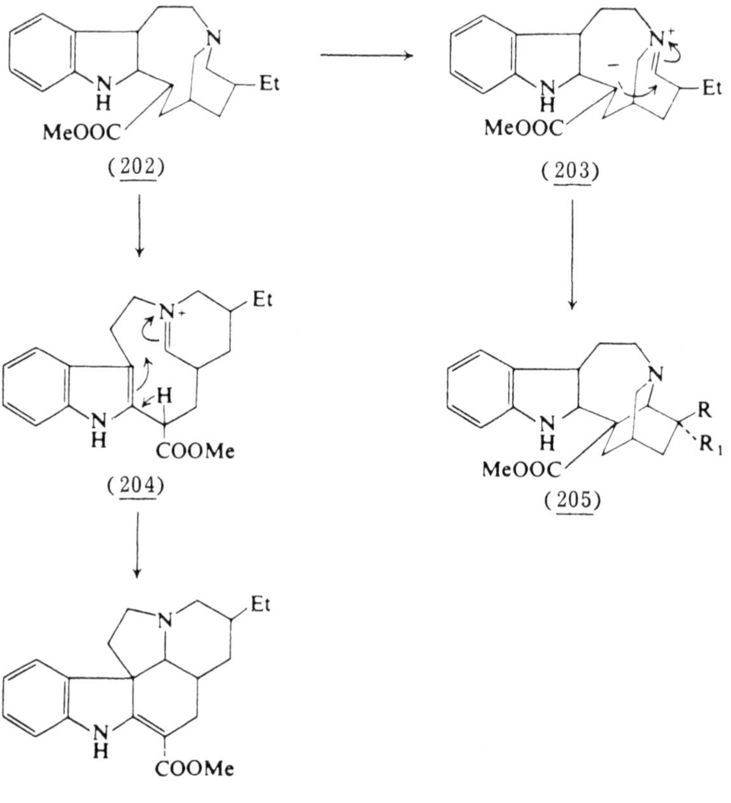

(202)

(203)

(204)

(205)

(206)

Červinka and co-workers [339] studied (3,3)-sigmatropic reaction of the heterocyclic enamines 1-methyl-2-(3-butenyl)-2-pyrroline (207) and 1-methyl-2-(3-butenyl)-2-piperideine (208), which contain a 1,5-hexadiene system. Thermal isomerization of 1-methyl-2-(3-butenyl) pyrrolidine afforded 1,2-dimethyl-3-allyl-2-pyrroline (209). Analogous reaction with 1-methyl-2-(3-butenyl)-2-piperideine gave quantitatively 1,2-dimethyl-3-allyl-2-piperideine (210).

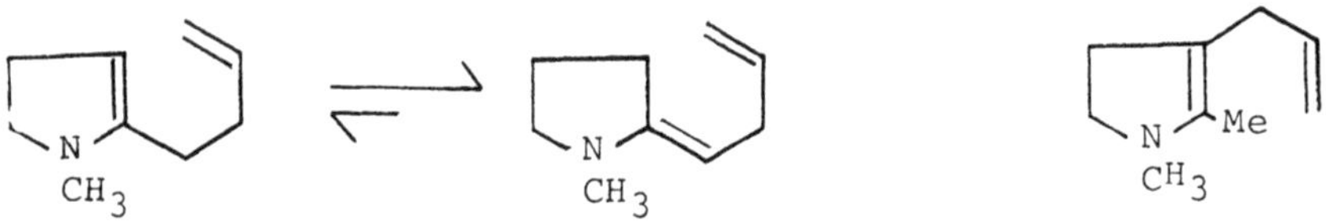

(207)

(211)

(209)

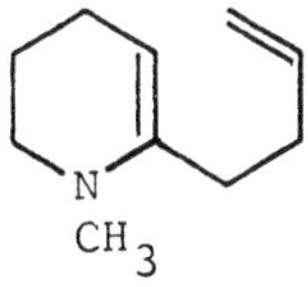

(208)

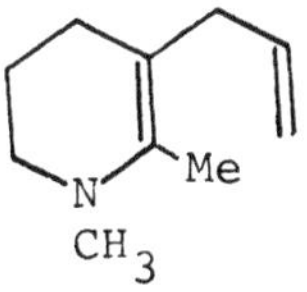

(210)

The incomplete isomerization in the five-membered series can be explained by dominant population of the tautomer with exocyclic double bond (211). On the other hand, the exclusive presence of the endocyclic double bond in the piperideine favors the rearrangement.

When the enamine (212) was brominated in methylene chloride and the reaction mixture was treated with aqueous ammonium hydroxide at room temperature, the 10-oxohomobenzomorphinan (213) was obtained in 81% yield [340].

A possible mechanism of this reaction may be represented by the sequence of reactions shown in Scheme 24.

Attack of OH⁻ to the initially formed bromo iminium bromide would give intermediate 214, which may undergo rearrangement to 213, presumably in a concerted manner.

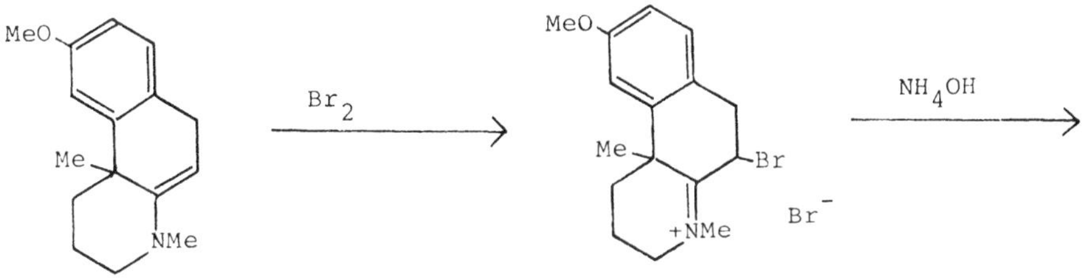

(212)

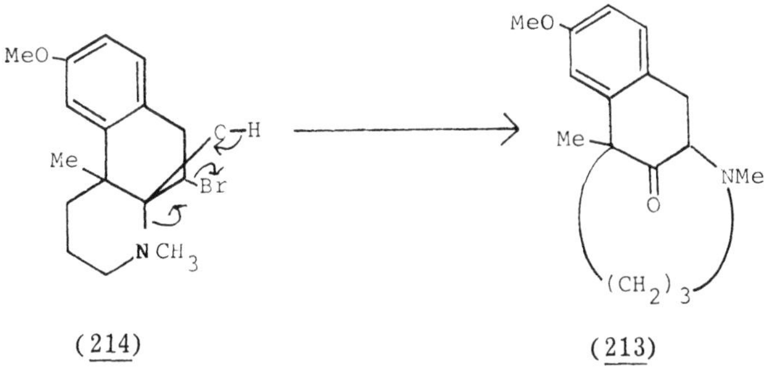

Br_2

NH_4OH

(214)

(213)

Scheme 24

Since the original discovery that the benzyl group migrates from C_1 to C_3 when a 1-benzyl-1,2-dihydroisoquinoline is treated with hot, dilute mineral acid, a substantial amount of work has been reported with a series of 1-substituted-1,2-dihydroisoquinoline derivatives [341–348] (Scheme 25).

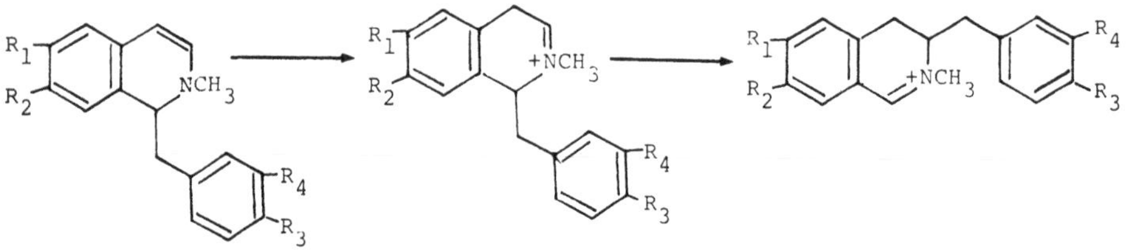

Scheme 25

Orbital symmetry analysis, together with the kinetic data, strongly supports the concept of a concerted bimolecular exchange reaction for the rearrangement of a 1-benzyl-1,2-dihydroisoquinoline into 3-benzyl-3,4-dihydroisoquinoline derivative.

V. IMPORTANCE OF HETEROCYCLIC ENAMINES FOR THE SYNTHESIS OF ALKALOIDS

The successful determination of the structure and stereochemistry of alkaloids, followed by their unequivocal synthesis, is not the only goal of organic chemists studying these natural products. The importance of solving the problem of their biogenesis in a living organism, of defining their relation to other compounds in the organism, and at the same time of getting acquainted with their importance for the plant is now becoming more apparent.

The first step in this study has involved experiments that synthesize alkaloids in vitro under quasicellular conditions, using reactions that can proceed in the living cell and compounds that actually occur in the cell or are supposed to be intermediates in the plant metabolism. Such syntheses are designated as syntheses under "physiological conditions."

Aldol reactions of enamines with reactive methylene groups constitute the basic step in the Robinson theory of alkaloid biogenesis. The theory has been modified by Schöpf, Woodward, Wenkert, and other chemists. As a result of their studies, we can obtain a satisfactory image of the biogenesis even of some very complex alkaloids. In the synthesis of alkaloids, two main groups of reactions take part: degradation reactions of amino acids or sugars to give starting materials, and synthetic processes proceeding in the cell to form these alkaloids.

Heterocyclic enamines Δ^1-pyrroline and Δ^1-piperideine are the precursors of compounds containing the pyrrolidine or piperidine rings in the molecule. Such compounds and their N-methylated analogs are believed to originate from arginine and lysine [349] by metabolic conversion. Under cellular conditions, the proper reaction with an active methylene compound proceeds via an aldehyde ammonia, which is in equilibrium with other possible tautomeric forms. It is necessary to admit the involvement of the corresponding α-ketoacid [12,350] instead of an enamine. The α-ketoacid constitutes an intermediate state in the degradation of an amino acid to an aldehyde. α-Ketoacids or suitably substituted aromatic compounds may function as components in active methylene reactions (Scheme 26).

The synthetic process proceeding under physiological conditions can be imitated in vitro with the object of establishing the validity of biogenetic hypotheses [351] as well as finding new potential routes for the synthesis of pharmaceuticals [352].

The aldol reactions of enamines may be formally considered to proceed via acyclic amino aldehyde or amino ketone forms, in spite of the fact that the cyclic enamine forms can also take part in aldol reactions.

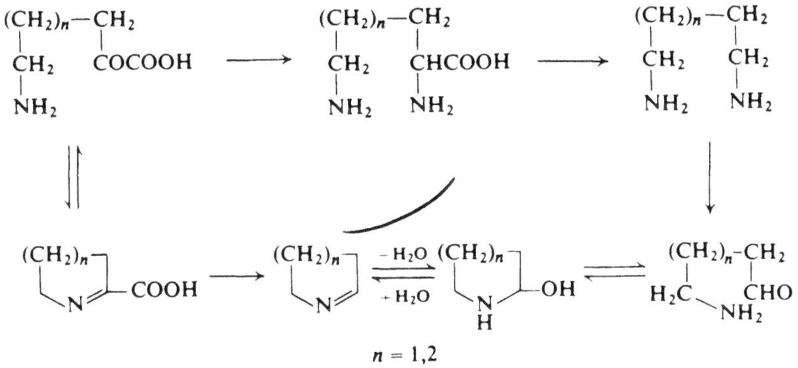

Scheme 26

The simplest compounds, Δ^1-pyrroline and Δ^1-piperideine, do not exist in the monomeric form. Schöpf et al. [353] described two geometrical isomers of Δ^1-piperideine trimer and called them α- and β-tripiperideines (215). An equilibrium exists between Δ^1-piperideine and both trimers which, therefore, react as typical aldehyde ammonia.

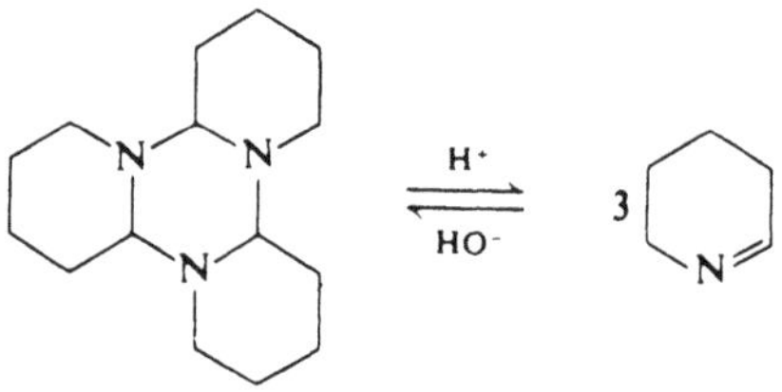

(215)

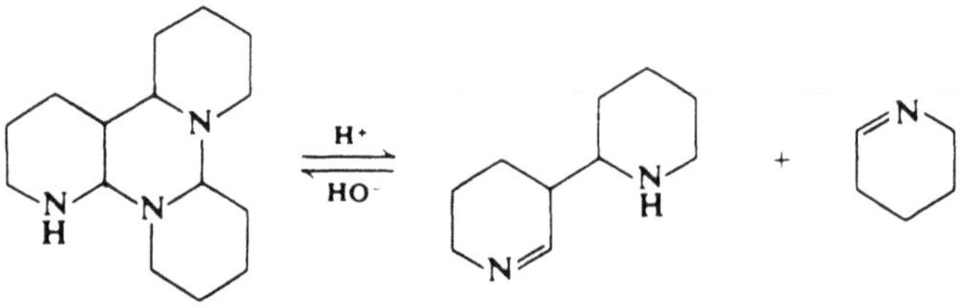

(216)   (217)

The trimer rearranges at pH 9–10 in an almost quantitative yield to isotripiperideine (216), which, in turn, is in equilibrium with tetrahydroanabasine (217) and Δ^1-piperideine.

Isotripiperideine and α-tripiperideine structures differ from each other in a new C—C bond formed in the tripiperideine by an aldol reaction [354, 355]. In aqueous media at pH 2–13, two molecules of Δ^1-piperideine yield tetrahydroanabasine [355].

The last isomer, the so-called aldotripiperideine (218), is obtained by the action of acid catalysts on α-tripiperideine at its boiling point [356, 357], or in aqueous solution at pH 9.2 and 100°C. Further aldol reaction between tetrahydroanabasine and Δ^1-piperideine obviously occurs. Hydrogenolysis of this compound gives dihydroaldotripiperideine (219), which is convertible into matridine (220), a reduction product of the alkaloid matrine.

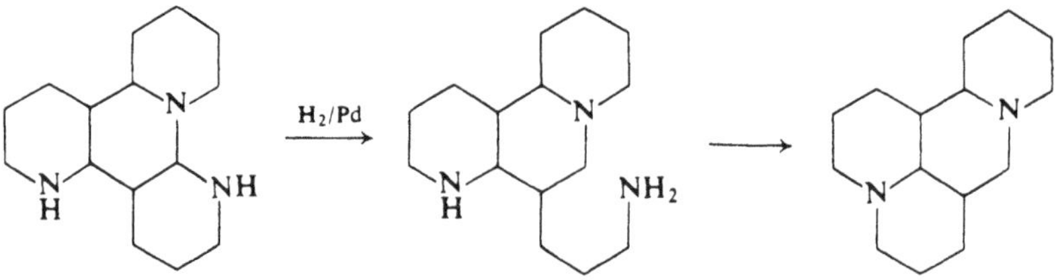

(218)   (219)   (220)

Curiously, neither Δ^1-piperideine nor tetrahydroanabasine undergoes aldolization in strongly acidic or strongly alkaline media; the reaction occurs only at pH 2–13, when both the free base and its salt are present [353]. This relation between the rate of aldolization and pH indicates that aldolization occurs by condensation of the methylene group of the immonium salt with the free base.

Δ^1-Pyrroline affords a trimer [358], tripyrroline (221). In the five-membered series, only 3-(2-pyrrolidyl)-Δ^1-pyrroline (222), a dimeric aldolization product, was always obtained under the reaction conditions causing the formation of isotripiperideine or even aldotripiperidine in the six-membered series. Product 222 can be prepared by allowing an aqueous solution of pyrroline to stand at pH 7.

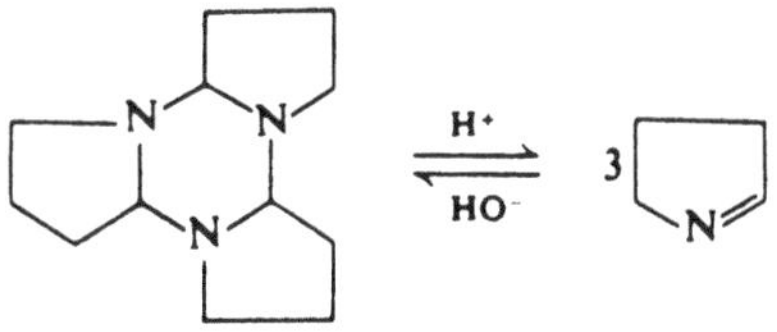

(221)

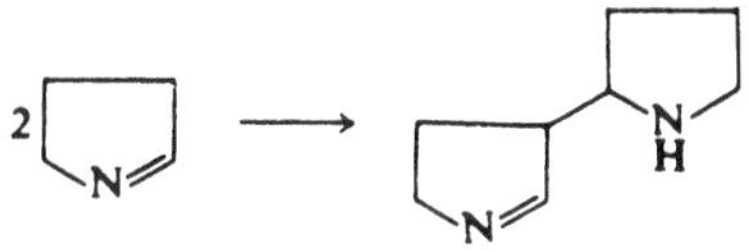

(222)

Condensation of Δ^1-pyrroline with pyrrole readily affords 2-(2-pyrrolidyl)pyrrole [96]. The dimerizations of some derivatives of Δ^1-piperideine, e.g., Δ^1-pyrroline and Δ^1-piperideine-2-carboxylic acids, take a similar course [359].

The N-substituted bases also undergo dimerization when the 2 position is free. Reduction of N-substituted five- and six-membered lactams and imides with various reducing reagents gives rise to various amounts of higher-boiling bases. It has been found that these are dimerization products of intermediate 1-alkyl-Δ^2-pyrrolines and 1-alkyl piperideines [149]. Thus, for example, 3-(1,2-dimethylpiperidyl)-1-methylpiperideine, i.e., N,N'-dimethyl-Δ^2-tetrahydroanabasine (223), is formed by reduction of 1-methyl-2-piperidone with sodium in ethanol. The same dimer was later obtained by Leonard and Hauck [1] on dehydrogenation of N-methylpiperidine with mercuric acetate, and by Schöpf on partial hydrogenation of the N-methylpyridinium salts.

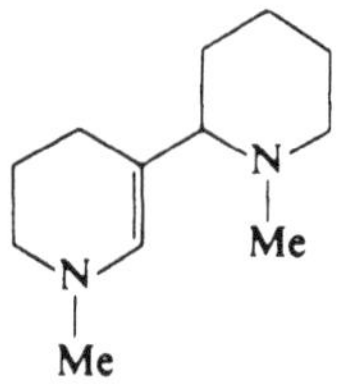

(223)

The only stable monomeric form of 1-methyl-Δ^2-piperideine is as the immonium salt.

The dimer of 1-methyl-Δ^2-pyrroline (47) was obtained by reduction of N-methylpyrrole with zinc and hydrochloric acid [150] and, together with the trimer, by mercuric acetate dehydrogenation of N-methylpyrrolidine [149]. Δ^1-Pyrroline-N-oxides form dimers in a similar manner [360]. Treatment of 1,2-dimethyl-Δ^2-piperideine with formaldehyde, producing 1-methyl-3-acetylpiperidine [361], serves as an example of a mixed aldol reaction (Scheme 27).

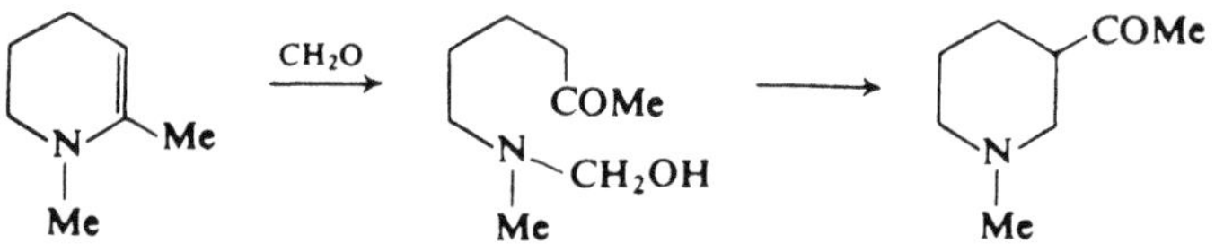

Scheme 27

The synthesis of some simple alkaloids under so-called physiological conditions has been made possible by development of methods of preparation of Δ^1-pyrroline, Δ^1-piperideine, and their N-methyl derivatives and by knowledge of their basic conversions.

Anet et al. [362] obtained the alkaloids hygrine (224) and kusk-hygrine (225) in a very good yield by treatment of γ-methylamino-butyraldehyde with acetoacetic or acetonedicarbolic acids at pH 7. The same reaction was later accomplished by Galinovsky et al. [363–365], who prepared the starting aldehyde by partial reduction of 1-methyl-2-pyrrolidone with lithium alluminum hydride. They used acetonedicarboxylic acid for the synthesis of both alkaloids and showed that a mixture of both alkaloids is formed, the composition of which depends on the ratio of components.

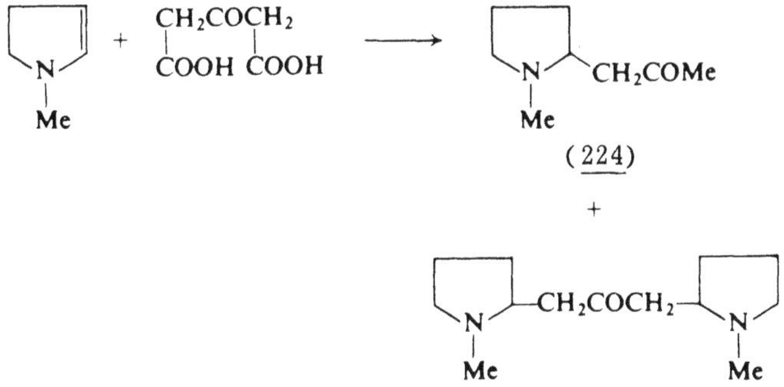

(224)

+

(225)

Norhygrine has been prepared by Schöpf from Δ^1-pyrroline. Goldschmidt [366] synthesized 1-carbethoxy-2,3-dioxopyrrolizidine (226) by condensation of Δ^1-pyrroline with diethylester of oxosuccinic acid.

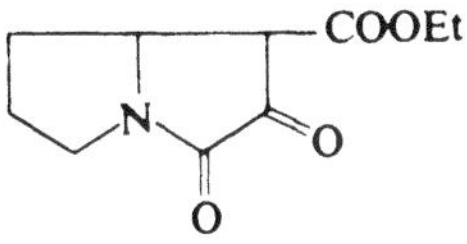

(226)

In the six-membered series, the alkaloids of *Punica granatum*, isopelletierine and methylisopelletierine, have been obtained by treatment of enamines with acetoacetic acid. Isopelletierine (227, R = H) was prepared also by Schöpf et al. from Δ^1-piperideine [367–369]. The reversibility of aldol dimerization [142,149] of enamines has been established by the synthesis of methylisopelletierine (227, R = Me] from dimethyltetrahydroanabasine, accomplished by Lukeš and Kovář [116] (Scheme 28).

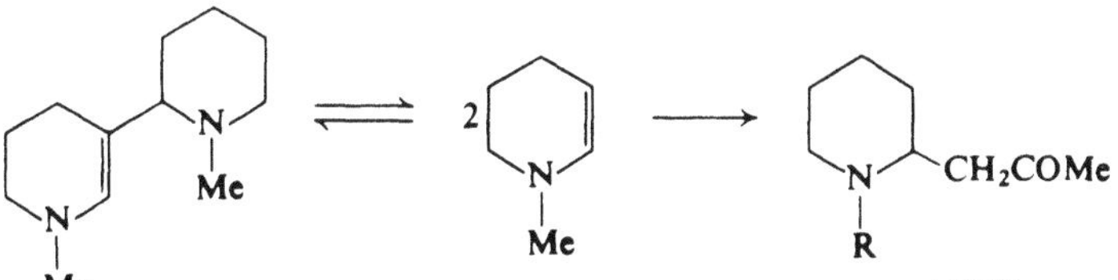

(227)

Scheme 28

The acetylation of isotripiperideine by means of a ketone in non-polar media affords a compound that decomposes in acidic media to Δ^1-piperideine and a monoacetyl derivative of the enamine form of tetrahydroanabasine (228). This monoacetyl derivative is identical with the alkaloid amodendrine [370]. A similar acylation with cinnamoylchloride affords the alkaloid orensine (229) [371], the optically active form of which is the natural alkaloid adenocarpine [372]. The hydrolysis of alkaloid santiaguine gives α-truxilic acid [372].

From a vast number of alkaloids containing the Δ^1-pyrroline or Δ^1-piperideine ring in the molecule, alkaloids myosmine (230) and lobinaline (231) [373] of Lobelia cardinalis L. may be mentioned.

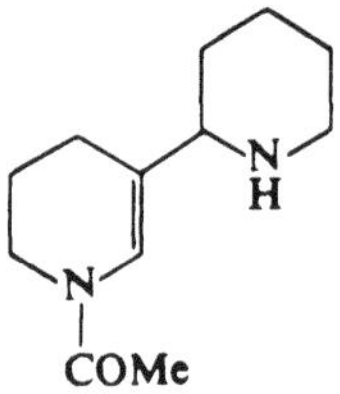

COMe

(228)

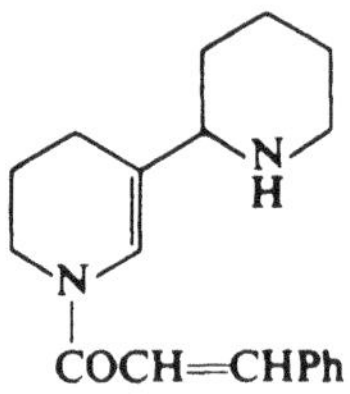

COCH=CHPh

(229)

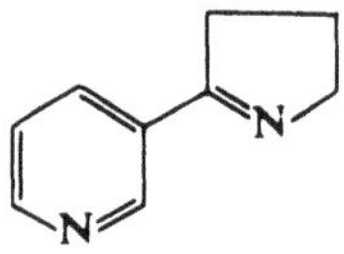

(230)

Ph

Ph

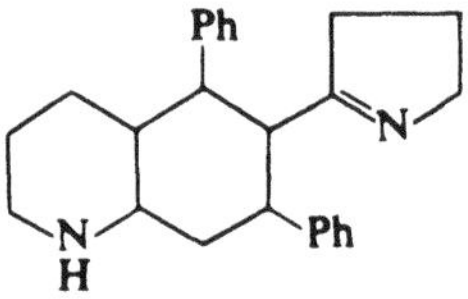

(231)

The reaction of 2-(α-pyridyl)alkylmalonic acid with Δ^1-piperideine, leading to formation of 3-(α-pyridyl)quinolizidine-1-carboxylic acid

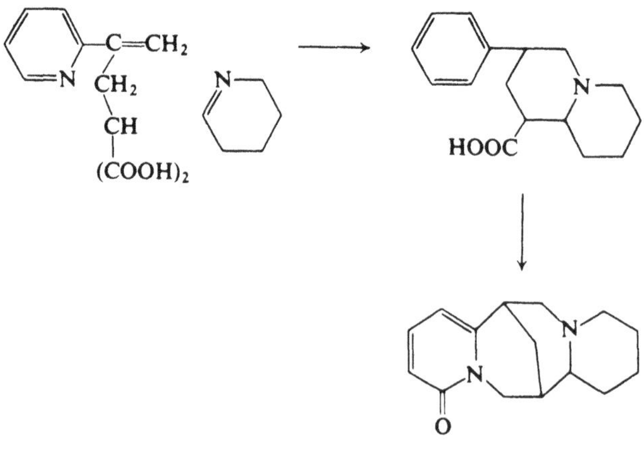

Scheme 29

on dexarboxylation, has been used by van Tamelen and Foltz [374] for synthesis of the alkaloid lupanine (Scheme 29). An elegant synthesis of matrine has been accomplished by Bohlmann et al. [375].

The alkaloid sedamine has been prepared under physiological conditions also [376]. Aldolization of the immonium salt (233), obtained by mercuric acetate dehydrogenation of (2S,8R)-(−)-sedamine (232), with benzoylacetic acid produces (2S,6R,8S)-(−)-8,10-diphenyllobe-linol [i.e., (−)-lobeline] (234). This determines its absolute configuration [377,378].

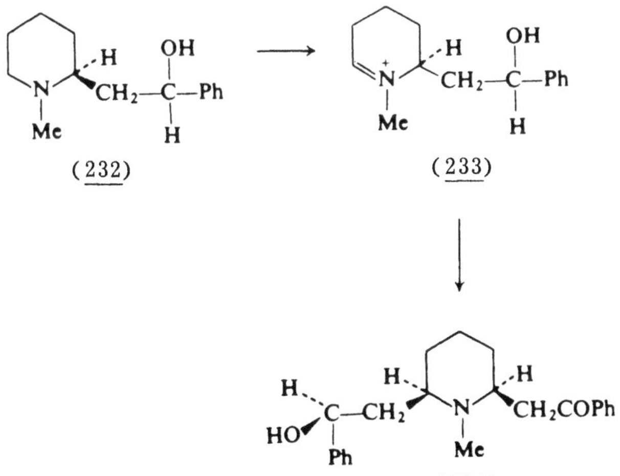

Condensation of Δ^1-pyrroline with o-aminobenzaldehyde leads to dihydroquinazolinium salt (235) on acidification, which in turn can be reduced to desoxyvasicine (236).

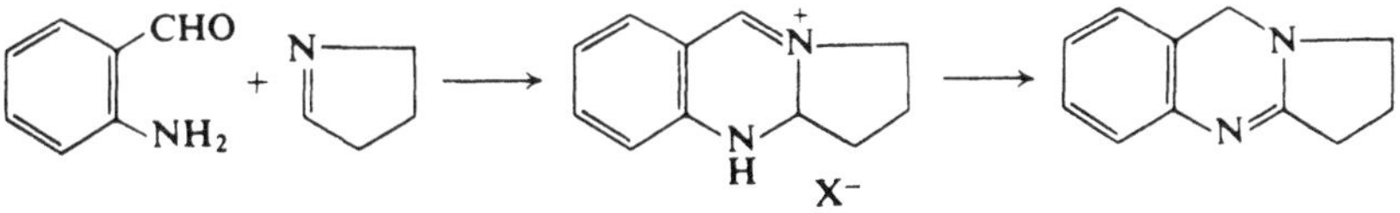

The synthesis of alkaloids from dihydronorharmane, condensation of which with o-aminobenzaldehyde gives rutaecarpine (237) [379–381], is of particular interest.

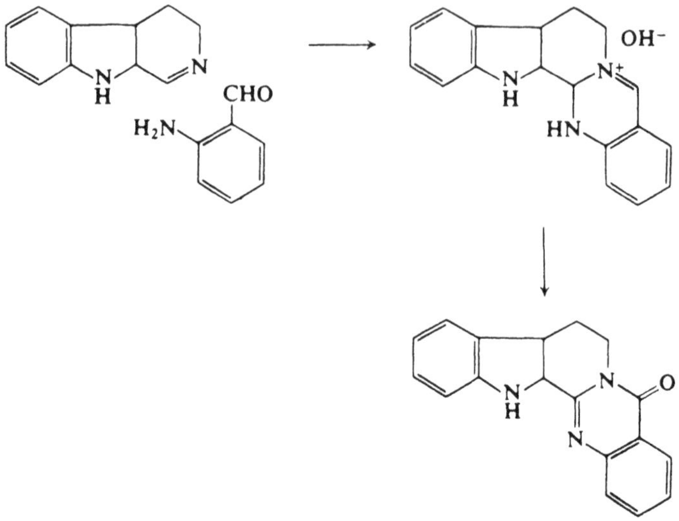

(237)

All heterocyclic enamines readily undergo condensation with o-aminobenzaldehyde. The quinoxaline derivatives thus formed have a characteristic yellow color. Therefore, this reaction can serve as evidence of the presence of an enamine in plants [353,367].

A characteristic of all the above reactions is that the yield of the aldolization product depends on the pH of the reaction mixture [382], the maximum yield usually occurring near pH 7. Such reactions have been carried out in vitro in dilute aqueous buffer under so-called physiological conditions, i.e., conditions attainable in the living cell. Although this oversimplified technique for the study of alkaloid biogenesis is now being abandoned in favor of experiments in vivo with labeled precursors, such reactions are still of interest to organic chemists.

Partial hydrogenation of the quaternary pyridinium salts in the presence of triethylamine on palladium in methanol has been used for the synthesis of a large number of alkaloids. The tetrahydropyridine derivatives thus formed undergo various cyclization reactions in acidic media [103].

Hydrogenation of t-butyl nicotinate methobromide, followed by hydrolysis of the 1-methyl-3-tert-butoxycarbonyl-1,4,5,6-tetrahydropyridine product (238) in the presence of indole, affords, on decarboxylation, the β-substituted derivative (239) [383]. The formation of compounds containing the β-carboline grouping in the molecule takes a similar course [104,384,385] (Scheme 30).

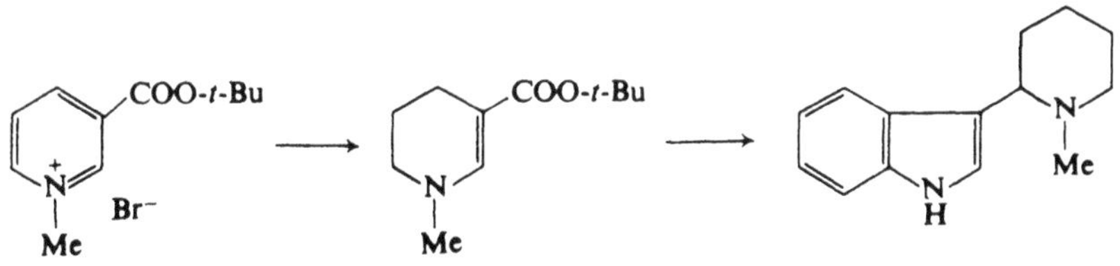

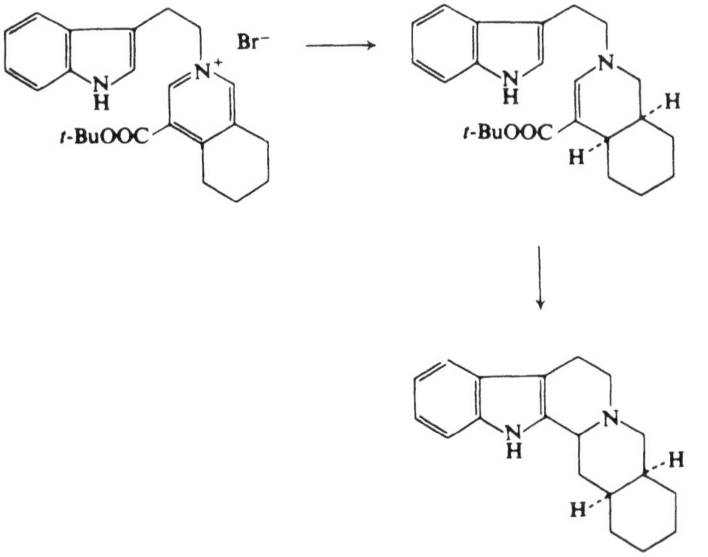

Scheme 30

Enamine chemistry has become one of the principal new research areas of synthetic organic methodology during the past 20 years. Synthesis of natural products and related compounds illustrates the usefulness of enamine chemistry [386].

The rearrangement of the cyclopropylketimines and N-methyl-imides of cyclopropylaldehydes to cyclic enamines under acidic conditions was applied to synthesis of myosmine (240) (Scheme 31), apo-ferrosamine (241) [387,388], labeled nicotine (242) [389], shihunine (243) [390].

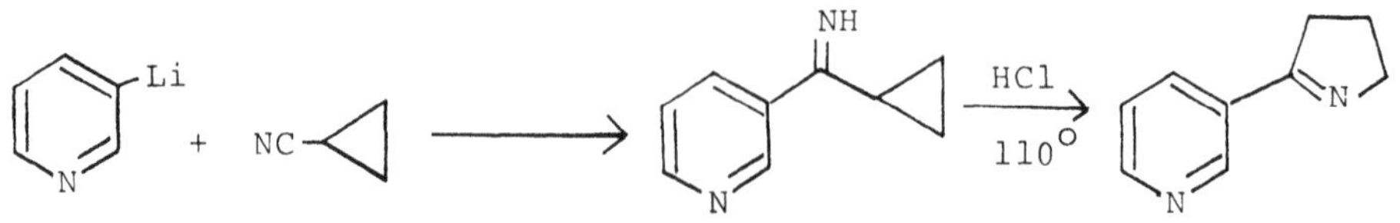

(240)

Scheme 31

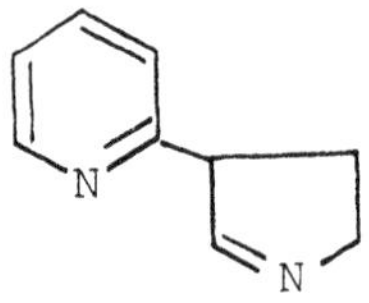

(241) 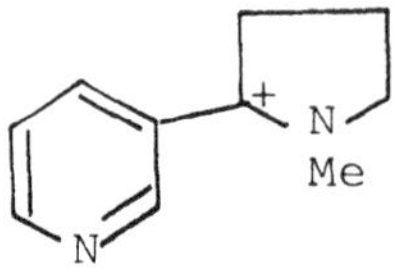   (242) 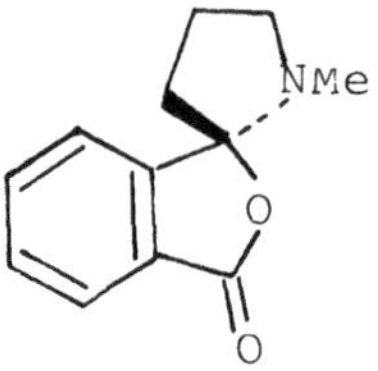   (243)

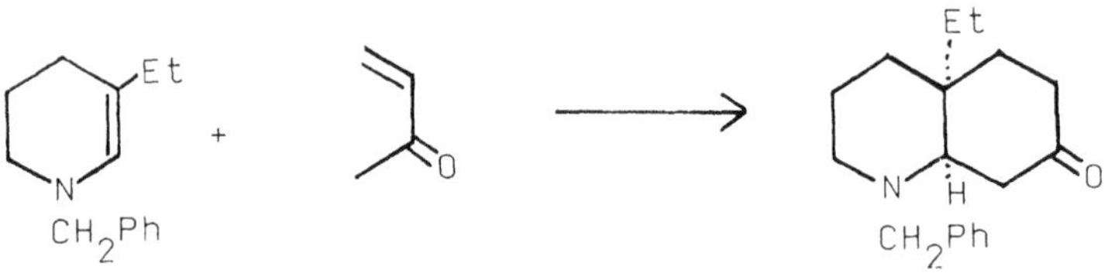

Scheme 32

An extension to the N-methylimines of β-tetralone-derived cyclo-propylketone to give 1-methyl-2,3-dihydrobenzo[e] indoles has also been reported.

2-Pyrrolines and 2-piperideines are valuable intermediates in alkaloid synthesis [44,391–394] in that they can provide the means for C—C bonding with an attracting electrophile at C_3 or via a C_3 protonated immonium salt, with a nucleophile at C_2

Annulation of 2-pyrrolines and 2-piperideines with methyl vinyl ketone, or derivatives thereof, has found wide application in the total synthesis of alkaloids. Invariably, annulation of these endocyclic enamines affords exclusively *cis*-fused hydroindolones or hydroquin-olones. The latter case is particularly intriguing. Because it is well established that hydroquinolones prefer thermodynamically to exist as the corresponding *trans* isomer, it is clear that this process is kinetically controlled [419].

By analogy with axial attack of electrophiles to cyclohexene systems, one would expect electrophilic additions to endocyclic enamines to occur on the same face of the molecule as the lone electron pair on nitrogen (Scheme 32).

Heating of endocyclic enamine (244) with methyl vinyl ketone in refluxing ethylene glycol afforded a 55% yield of (±)-mesembrine (245) [395–400] (Scheme 33). Similarly joubertiamine [401] was obtained.

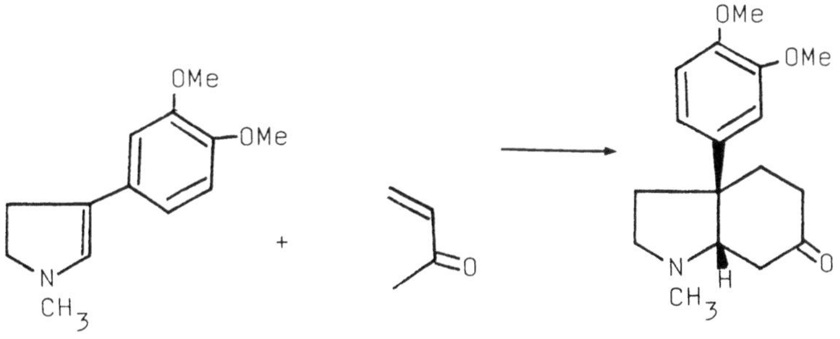

(244) (245)

Scheme 33

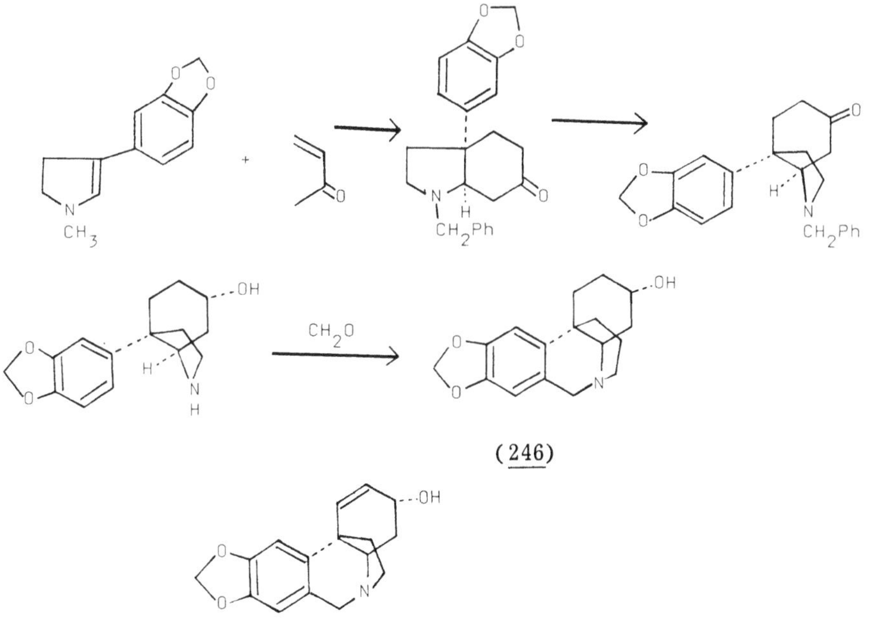

(246)

(247)

Scheme 34

Many Amaryllidaceae alkaloids, such as elwesine (246) and crinine (247), also incorporate a hydroindolone nucleus into their skeletons (Scheme 34) [402,403].

Unusual 3-phenylthio-2-pyrroline serves as a relatively stable equivalent of the unsubstituted enamine, which is notoriously unstable. Application of this particular variation to the indolizidine alkaloids δ-conicein (248), ipalbidine (249), and septicine (250) illustrated further the utility of the 3-phenylthio-2-pyrroline synthon in the synthesis of functionalized indolizidine and pyrrolizidine nuclei (Scheme 35).

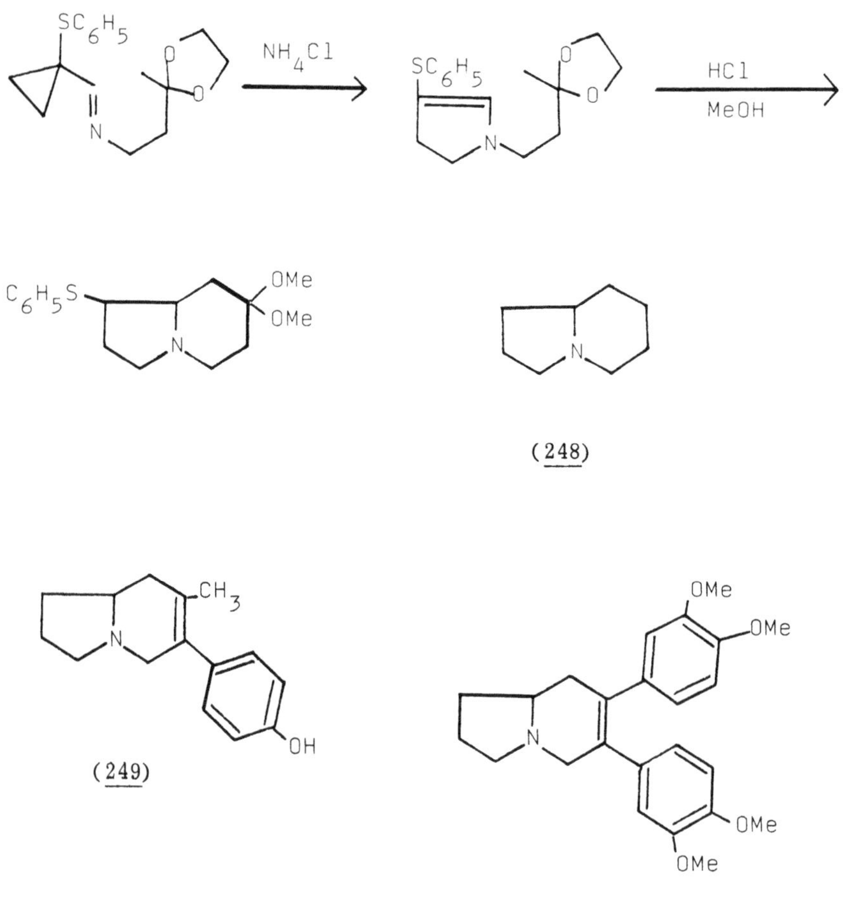

(248)

(249)

(250)

Scheme 35

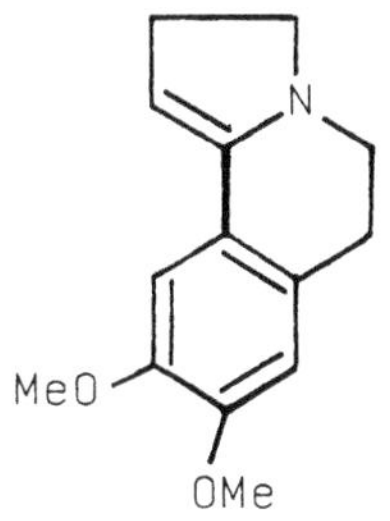

(251)

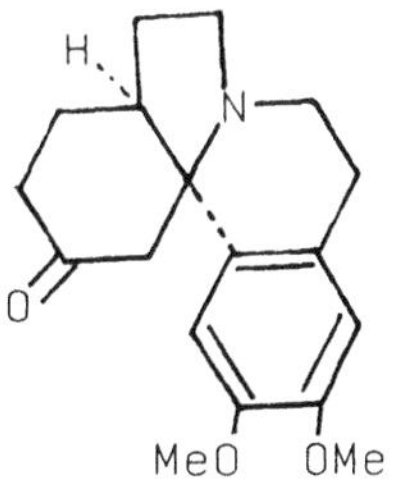

(252)

The annulation of endocyclic enamine (251) with methyl vinyl ketone affords the basic skeleton (252) of several of Erythrina alkaloids.

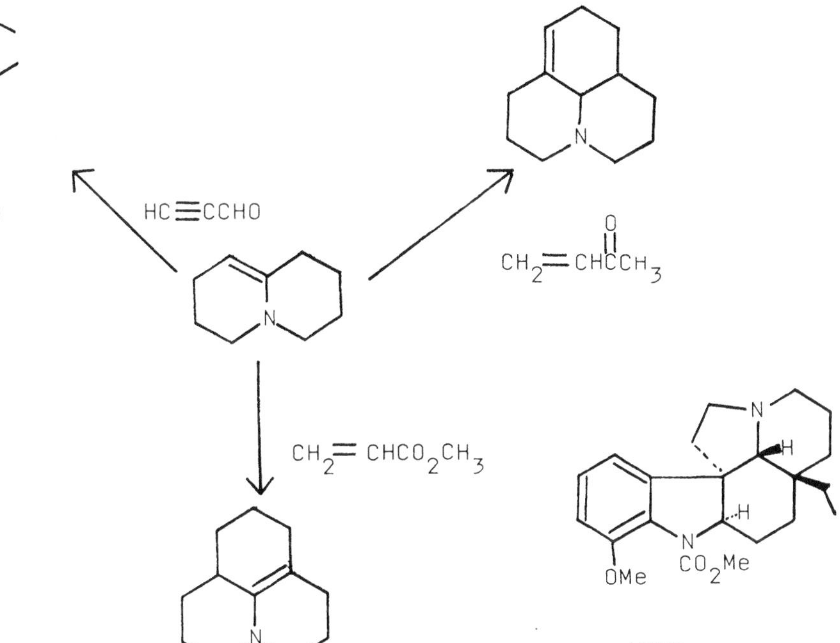

Scheme 36

The alkylation of heterocyclic enamines, which are readily accessible by mercury(II) acetate oxidations, with acrylic esters, methyl vinyl ketone, or acetylenic aldehydes provide a flexible entry into polycyclic alkaloid synthesis (Scheme 36). Thus routes to julolidine (253) and spidospermine (254) become available [404,405].

Acylation of 1,9a-dehydrochinolizidine with ethyl chloroformate gave vinylogous urethane, which could be converted to lupinine (255) [404,406] (Scheme 37).

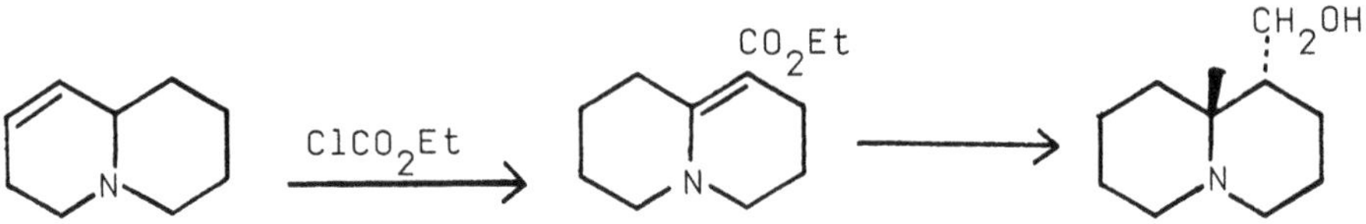

(255)

Scheme 37

The cyclic immonium salts arising from oxidation of tertiary amines with mercury(II) acetate were useful in syntheses of quebrachamine [407], dihydrocleavamine [408], and aspidospermidine [409, 410]. Oxidation of 3-(2-piperidinoethyl) indoles also provides a model for the synthesis of yohimbinoid alkaloids [411].

The formation of enamines by reduction of isoquinolinium salts with lithium aluminum hydride provided intermediates for acid-catalyzed cyclizations into aryl substituents and thus syntheses of nor-coralydine (256) [412], cureximine [412], N-methylpavine (257) [413], thalisopavine [414], and berberine [415].

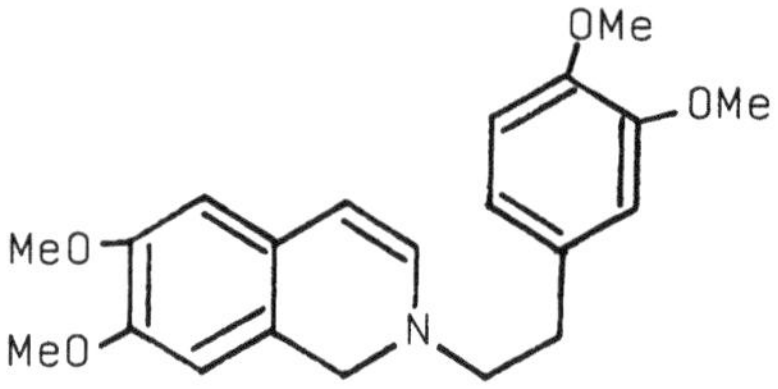

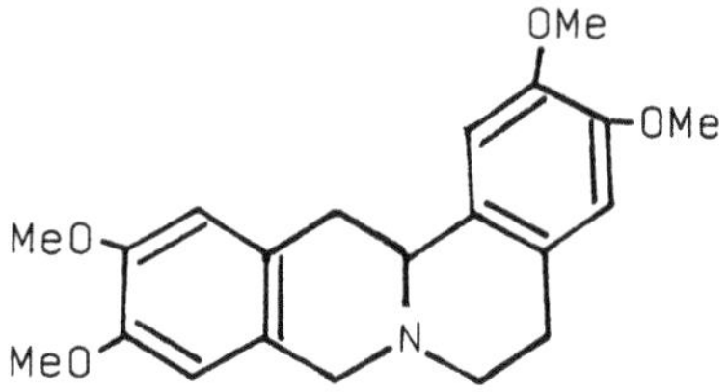

(256)

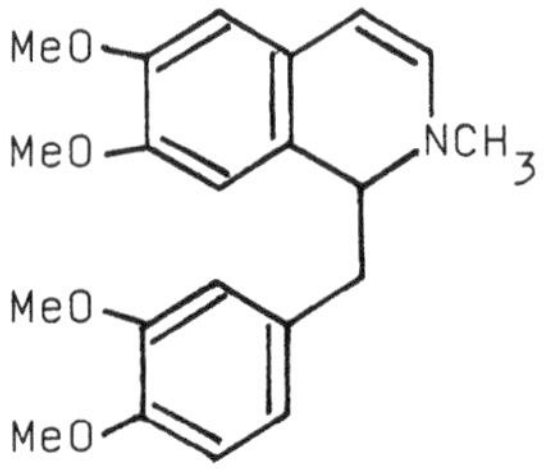

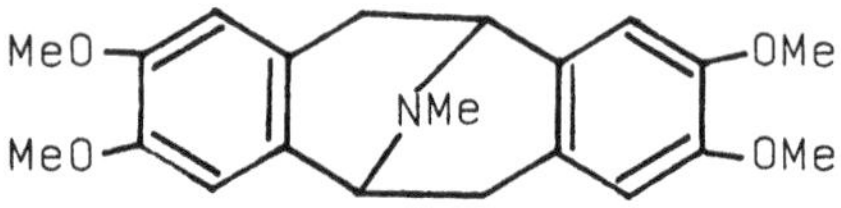

(257)

Cyclizations of 3-(2-vinylphenyl)-substituted 1,2-dihydroisoquin-
olines by irradiation [416] or treatment with acid [417,418] led to
sanquinarin (258) and epicryptopirubin chloride.

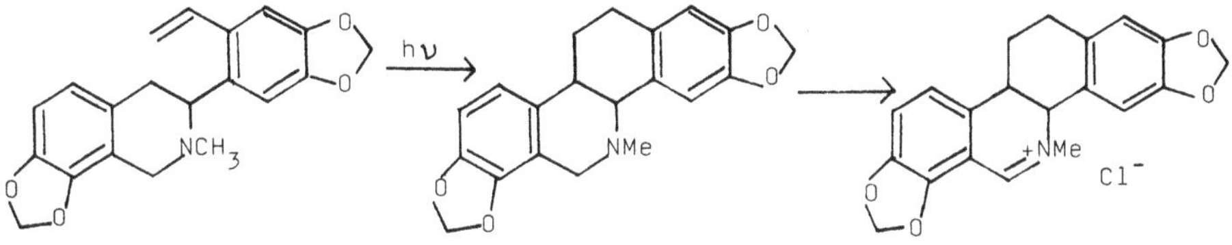

(258)

Cyclic enamines have been obtained as intermediates for alkaloid
syntheses by the Bischler-Napieralsky reaction. Their use in subsequent reactions with electrophilic alkenes is illustrated in synthesis
of 15,16-dimethoxy-3-oxoerythrinane (259) [420], analogs of vincina-
mine [421], and emetine.

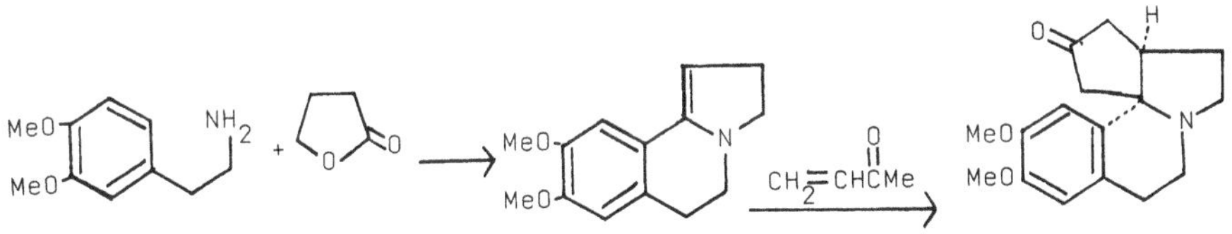

(259)

Dihydropyridines play an essential role in biochemistry and have
many applications in pharmacology [422]. They are indispensable
intermediates in the synthesis of natural products, mainly alkaloids.
These aspects, as well as the preparation and the chemical or physical properties of dihydropyridines, have been exhaustively dealt with
in several recent reviews [423–426].

Methods of synthesizing dihydropyridines can be classified into
ring-forming processes and ring transformations. The first general
principle is best represented by the versatile Hantzsch synthesis.
This method conveniently affords 1,4-dihydro- or 3,4-dihydro-, but
not 1,2-dihydropyridines. Recently, the preparation of N-acyl-1,2-
dihydropyridines by a thermal electrocyclic process was reported by
Wyle and Fowler [427]. The second principle is illustrated by partial
reduction of pyridines or pyridinium salts and nucleophilic addition
of organometallic reagents to the same species of 1,2- or 1,4-dihydro-
pyridines.

Potassium borohydride reduction of a pyridinium salt to a tetra-hydropyridine and rearrangement of the allylic amine to an enamine provided an intermediate that could be converted to dasycarpidone (260) by cyclization of the immonium salt to the 3-position of the indole [428] (Scheme 38).

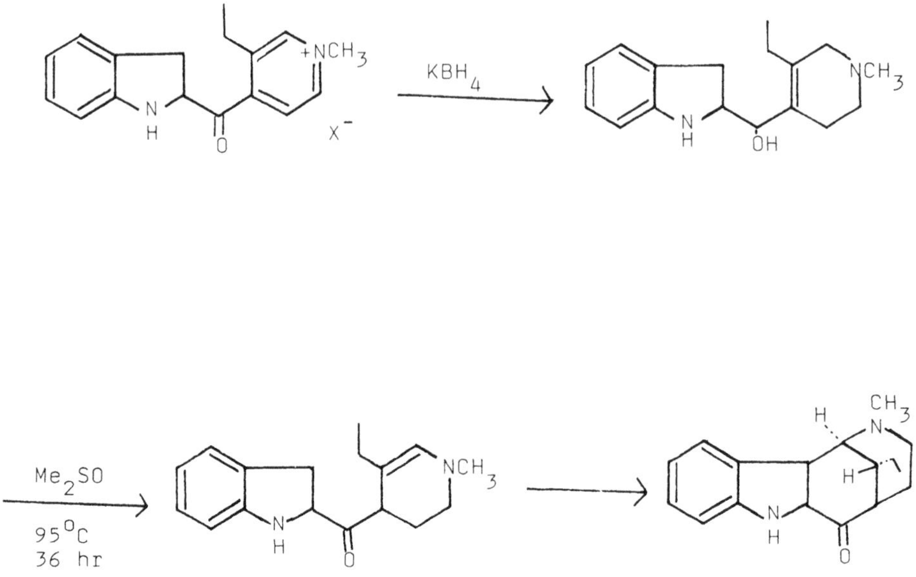

Scheme 38

On the other hand, hydride reductions of pyridinium salts under controlled conditions to dihydropyridines gave cyclic dienamines, which could be condensed with methyl vinyl ketone [429] or akrylo-nitrile [430] to isoquinuclidines, the unique structural moiety of the iboga alkaloids.

Catalytic reductions of 3-acylpyridinium salts to tetrahydropyri-dines, where the enamine function is conjugated with a carbonyl group, also gave immonium intermediates with acids, which could undergo cyclization reaction [431].

Thus closure to the indole 2-position is seen in syntheses of cor-ynantheidine (261) (Scheme 39), eburnamonine, and lupinine [431].

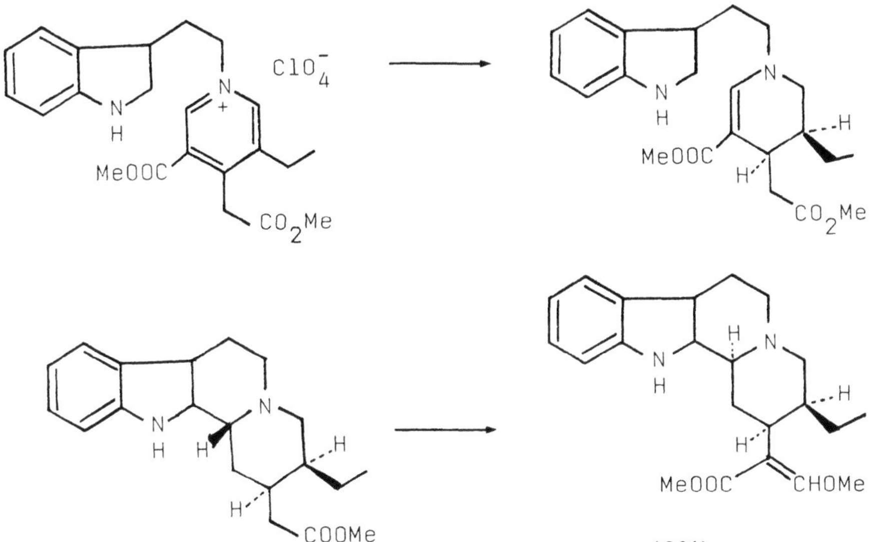

Scheme 39

Cyclic enamines are intermediates in the syntheses of azasteroids [432–435].

In order to introduce angular substituents, the alkylation of the vinylogous lactams at the nitrogen-bearing carbon with Grignard reagents or organolithium compounds was studied (Scheme 40) [436, 437].

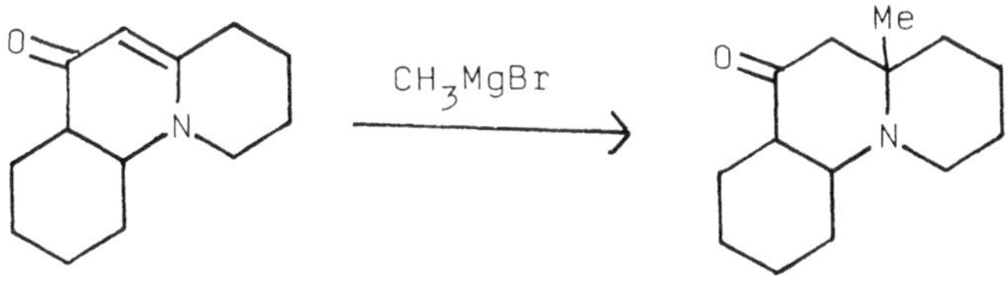

Scheme 40

Intramolecular condensations of amine and carbonyl components were also possible through reductive cyclizations and used in syntheses of quebrachamine [438].

Solvent dependence was found for the O- versus C-alkylation of vinylogous amides [439]. Intramolecular C-alkylation of a vinylogous amide was used in oxocrinane (262) syntheses [440] (Scheme 41).

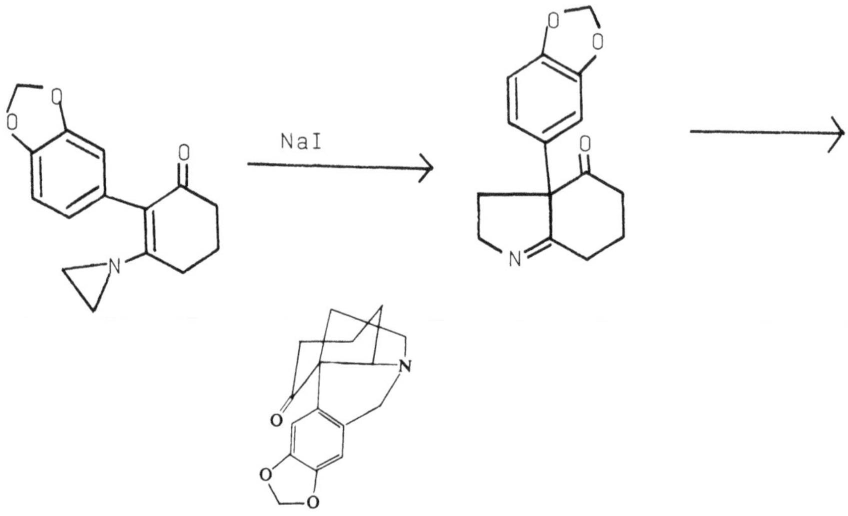

(262)

Scheme 41

REFERENCES

1. N. J. Leonard and F. P. Hauck, Jr., *J. Am. Chem. Soc.*, *79*, 5279 (1957).
2. K. Bláha and O. Červinka, *Adv. Heterocyclic Chem.*, 6, 147 (1966).
3. S. F. Dyke, *The Chemistry of Enamines*, Cambridge University Press, 1973.
4. J. B. Cloke, *J. Am. Chem. Soc.*, *51*, 1174 (1929).
5. A. Wohl, *Ber.*, *34*, 1914 (1901).
6. S. Gabriel, *Ber.*, *41*, 2010 (1908).
7. L. Macholán, *Naturwiss.*, *46*, 357 (1959).
8. L. Macholán, *Chem. Listy*, *50*, 1818 (1956); *Collection Czech. Chem. Commun.*, *22*, 479 (1957).
9. L. Macholán and L. Skurský, *Chem. Listy*, *49*, 1385 (1955).
10. L. Skurský and L. Macholán, *Chem. Listy*, *51*, 774 (1957); *Collection Czech. Chem. Commun.*, *23*, 150 (1958).
11. L. Macholán and E. Svátek, *Collection Czech. Chem. Commun.*, *25*, 2564 (1960).
12. L. Macholán, *Chem. Listy*, *51*, 2122 (1957); *Collection Czech. Chem. Commun.*, *24*, 550 (1959).
13. W. Koller and P. Schlack, *Ber.*, *96*, 93 (1963).
14. M. L. Stein and A. Burger, *J. Am. Chem. Soc.*, *79*, 154 (1957).

15. M. C. Kloetzel, F. L. Chubb, and J. L. Pinkus, *J. Am. Chem. Soc.*, *80*, 5773 (1958).
16. J. Dhout and J. P. Wibaut, *Rec. Trav. Chim.*, *63*, 81 (1944).
17. E. Profft, F. Runge, and A. Jumax, *J. Prakt. Chem.*, *1*, 57 (1955).
18. A. Sonn, *Ber.*, *68*, 148 (1935); *72*, 2150 (1939).
19. R. Bomsett, R. F. C. Brown, V. M. Clark, I. O. Sutherland, and A. Todd, *J. Chem. Soc.*, 2094 (1959).
20. R. F. C. Brown, V. M. Clark, and A. Todd, *Proc. Chem. Soc. (London)*, 97 (1957).
21. M. C. Kloetzel, F. L. Chubb, R. Gobran, and J. L. Pinkus, *J. Am. Chem. Soc.*, *83*, 1128 (1961).
22. J. B. Barat and D. St. C. Black, *Chem. Commun.*, 902 (1966).
23. J. H. Burckhalter and J. M. Shart, *J. Org. Chem.*, *23*, 1281 (1958).
24. E. B. Knott, *J. Chem. Soc.*, 186 (1948).
25. H. Ruppe and F. Gisiger, *Helv. Chim. Acta, 8*, 338 (1925).
26. H. Richtzenhain, G. Daum, and R. Modic, *Ann.*, *691*, 88 (1966).
27. H. Gotthardt, R. Huisgen, and F. C. Schaefer, *Tetrahedron Lett.*, 487 (1964).
28. G. N. Walter and D. Alkalay, *J. Org. Chem.*, *32*, 2213 (1967).
29. P. Lipp and H. Seeles, *Ber.*, *62*, 2456 (1929).
30. L. C. Craig, H. Bulbrook, and R. M. Mixon, *J. Am. Chem. Soc.*, *53*, 1831 (1931).
31. D. F. Starr, H. Bulbrook, and R. M. Mixon, *J. Am Chem. Soc.*, *54*, 3971 (1932).
32. J. B. Cloke, E. Stehr, T. R. Steadman, and L. C. Westcott, *J. Am. Chem. Soc.*, *67*, 1587 (1945).
33. J. B. Cloke and O. Ayers, *J. Am. Chem. Soc.*, *56*, 2144 (1934).
34. R. Salathiel, J. M. Burch, and R. M. Mixon, *J. Am. Chem. Soc.*, *59*, 5361 (1937).
35. J. B. Cloke, L. H. Baer, J. M. Robbins, and G. E. Smith, *J. Am. Chem. Soc.*, *67*, 2155 (1945).
36. J. V. Murray and J. B. Cloke, *J. Am. Chem. Soc.*, *68*, 126 (1946).
37. R. V. Stevens and M. C. Ellis, *Tetrahedron Lett.*, 5185 (1967).
38. R. V. Stevens, M. C. Ellis, and M. P. Wentland, *J. Am. Chem. Soc.*, *90*, 5576 (1968).
39. R. V. Stevens and M. P. Wentland, *J. Am. Chem. Soc.*, *90*, 5580 (1968).
40. S. L. Keely, Jr., and F. C. T. Tahk, *Chem. Commun.*, 441 (1968).
41. S. L. Keely, Jr., and F. C. T. Tahk, *J. Am. Chem. Soc.*, *90*, 5584 (1968).
42. R. V. Stevens and M. P. Wentland, *Tetrahedron Lett.*, 2613 (1968).

43. R. V. Stevens, *Strategies and Tactics in Organic Synthesis*, Academic Press, New York 1984.
44. R. V. Stevens, *Account Chem. Res.*, *10*, 193 (1977); *17*, 289 (1984).
45. R. V. Stevens and J. T. Shen, *Chem. Commun.*, 682 (1975).
46. O. Červinka, *Collection Czech. Chem. Commun.*, *24*, 1146 (1959); *Chem. Listy*, *52*, 1145 (1958).
47. R. Lukeš and O. Červinka, *Chem. Listy*, *52*, 83 (1958); *Collection Czech. Chem. Commun.*, *24*, 1846 (1959).
48. R. Lukeš, *Collection Czech. Chem. Commun.*, *2*, 531 (1931); *Chem. Listy*, *27*, 121 (1933).
49. L. C. Craig, *J. Am. Chem. Soc.*, *55*, 295 (1933).
50. R. Lukeš and Z. Veselý, *Collection Czech. Chem. Commun.*, *24*, 944 (1959); *Chem. Listy*, *52*, 1299 (1958).
51. R. Lukeš and O. Červinka, *Collection Czech. Chem. Commun.*, *26*, 1893 (1961).
52. R. Lukeš, *Collection Czech. Chem. Commun.*, *4*, 181 (1932).
53. R. Lukeš and M. Smetáčková, *Collection Czech. Chem. Commun.*, *6*, 231 (1934); *Chem. Listy*, *29*, 316 (1935).
54. R. Lukeš and O. Grossmann, *Collection Czech. Chem. Commun.*, *8*, 533 (1936).
55. R. Lukeš and F. Šorm, *Chem. Listy*, *36*, 282 (1942); *Collection Czech. Chem. Commun.*, *12*, 356 (1947).
56. R. Lukeš and Z. Veselý, *Collection Czech. Chem Commun.*, *24*, 2318 (1959); *Chem. Listy*, *52*, 1608 (1958).
57. K. Winterfeld and E. Hoffmann, *Arch. Pharm.*, *275*, 526 (1937).
58. K. Winterfeld and P. Pethon, *Ber.*, *82*, 156 (1949).
59. L. C. Craig, *J. Am. Chem. Soc.*, *55*, 2543 (1933).
60. R. Adams and J. E. Mahan, *J. Am. Chem. Soc.*, *64*, 2588 (1942).
61. R. Adams and E. F. Rogers, *J. Am. Chem. Soc.*, *63*, 228 (1941).
62. R. Lukeš and M. Večeřa, *Collection Czech. Chem. Commun.*, *19*, 263 (1954); *Chem. Listy*, *47*, 541 (1953).
63. R. Lukeš and V. Dědek, *Chem. Listy*, *51*, 2074 (1957); *Collection Czech. Chem. Commun.*, *24*, 391 (1959).
64. V. Dědek and M. Bárta, *Sb. Vysoké Školy Chem-Technol. Praha Odděl. Fak. Anorg. Org. Technol.*, *C8*, 89 (1966).
65. R. Lukeš, F. Šorm, and Z. Arnold, *Collection Czech. Chem. Commun.*, *12*, 641 (1947); *Chem. Listy*, *41*, 250 (1947).
66. J. Lee, A. Ziering, S. D. Heineman, and L. Berger, *J. Org. Chem.*, *12*, 885 (1947).
67. R. Lukeš and K. Smolek, *Collection Czech. Chem. Commun.*, *11*, 506 (1939).
68. R. Lukeš, V. Dudek, D. Sedlakova, and I. Koran, *Collection Czech. Chem. Commun.*, *26*, 1105 (1961).
69. R. Lukeš and J. Dobáš, *Collection Czech. Chem. Commun.*, *15*, 303 (1950).

70. R. Lukeš and J. Málek, *Collection Czech. Chem. Commun.*, *16*, 23 (1951); *Chem. Listy*, *45*, 72 (1951).
71. R. Lukeš and L. Karličková, *Collection Czech. Chem. Commun.*, *26*, 2245 (1961).
72. O. Červinka and L. Hub, *Collection Czech. Chem. Commun.*, *30*, 3111 (1965).
73. R. Lukeš and V. Prelog, *Chem. Listy*, *24*, 251 (1930); *Collection Czech. Chem. Commun.*, *1*, 282 (1929).
74. R. Lukeš and K. Smolek, *Collection Czech. Chem. Commun.*, *7*, 476 (1935); *Chem. Listy*, *30*, 185 (1936).
75. R. Lukeš and J. Gorocholinskij, *Collection Czech. Chem. Commun.*, *8*, 223 (1936).
76. R. Lukeš and M. Černý, *Collection Czech. Chem. Commun.*, *23*, 497 (1958); *Chem. Listy*, *51*, 1327 (1957).
77. R. Lukeš and M. Černý, *Chem. Listy*, *51*, 1862 (1957); *Collection Czech. Chem. Commun.*, *23*, 946 (1958).
78. F. Michael and W. Flitsch, *Ber.*, *94*, 1749 (1961).
79. D. A. Evans, *J. Am. Chem. Soc.*, *92*, 7593 (1970).
80. D. A. Evans and L. A. Domier, *Org. Syntheses*, *53*, 1849 (1973).
81. D. Bruno, G. Lesma, and G. Palmisano, *Chem. Commun.*, 860 (1980).
82. K. Hasse and A. Wieland, *Ber.*, *93*, 1686 (1960).
83. K. Langheld, *Ber.*, *42*, 392 (1909).
84. F. Korte, K. H. Büchel, H. Mader, G. Romer, and H. H. Schulze, *Ber.*, *95*, 2424 (1962).
85. K. H. Büchel and F. Korte, *Ber.*, *97*, 2453 (1964).
86. F. Korte and A. K. Bocz, *Ber.*, *99*, 1918 (1966).
87. F. Korte and H. J. Schulze-Steinen, *Ber.*, *95*, 2444 (1962).
88. E. Späth and H. Bretschneider, *Ber.*, *61*, 327 (1928).
89. E. Späth, J. P. Wibaut, and F. Kesztler, *Ber.*, *71*, 100 (1938).
90. S. Brandange and L. Lindblom, *Acta Chem. Scand. Ser. B.*, *30*, 93 (1976).
91. E. Lecte and S. A. S. Leete, *J. Org. Chem.*, *43*, 2122 (1978).
92. C. Schöpf and H. L. De Wall, *Ber.*, *89*, 915 (1956).
93. K. H. Büchel and F. Korte, *Ber.*, *95*, 2465 (1962).
94. J. P. Wibaut and W. Proost, *Rec. Trav. Chim.*, *52*, 333 (1933).
95. H. P. L. Gitels and J. P. Wibaut, *Rec. Trav. Chim.*, *59*, 1091 (1940).
96. D. W. Fuhlhage and C. A. Van der Werf, *J. Am. Chem. Soc.*, *80*, 6249 (1958).
97. G. G. Evans, *J. Am. Chem. Soc.*, *73*, 5230 (1951).
98. O. Červinka and O. Křiž, *Z. Chem.*, *5*, 190 (1967).
99. F. Bohlmann, *Ber.*, *85*, 390 (1952).
100. N. C. Cook and J. E. Lyons, *J. Am. Chem. Soc.*, *88*, 3396 (1966).
101. K. Schenker and J. Druey, *Helv. Chim. Acta*, *42*, 1960 (1959).

102. P. Karrer, G. Schwarzenbach, and G. E. Vizinger, *Helv. Chim. Acta, 20*, 72 (1937).

103. C. Schöpf, G. Herbert, R. Rausch, and G. Schröder, *Angew. Chem., 69*, 391 (1957).

104. E. Wenkert, K. G. Dave, and F. Haglid, *J. Am. Chem. Soc., 87*, 5461 (1965).

105. P. Karrer, *Helv. Chim. Acta, 21*, 223 (1938).

106. H. Schmid and P. Karrer, *Helv. Chim. Acta, 32*, 960 (1949).

107. T. Kametani and K. Fukumoto, *Yakugaku Zasshi, 80*, 1288 (1960); *CA, 55*, 3589f (1961).

108. W. P. Newmann, *Angew. Chem., 70*, 401 (1958).

109. W. Bradley and S. Jeffrey, *J. Chem. Soc.*, 2770 (1954).

110. A. J. Birch and P. G. Lehman, *Tetrahedron Lett.*, 2395 (1974).

111. J. Kuthan and E. Janečková, *Collection Czech. Chem. Commun., 28*, 1654 (1964).

112. R. Lukeš and J. Kuthan, *Collection Czech. Chem. Commun., 26*, 1845 (1961).

113. R. Lukeš and J. Kovář, *Chem. Listy, 48*, 404 (1954); *Collection Czech. Chem. Commun., 19*, 1215 (1954).

114. F. Galinovsky and R. Weiser, *Experientia, 6*, 377 (1950).

115. R. Lukeš and M. Smetáčková, *Collection Czech. Chem. Commun., 5*, 61 (1931).

116. R. Lukeš and J. Kovář, *Chem. Listy, 48*, 692 (1954); *Collection Czech. Chem. Commun., 19*, 1227 (1954).

117. K. Lang, *Z. Physiol. Chem., 241*, 68 (1936).

118. L. Shurský, *Z. Naturforsch., 146*, 473 (1959).

119. P. J. G. Mann and W. R. Smithies, *Biochem. J., 61*, 89 (1955).

120. W. B. Jacoby and J. Fredericks, *J. Biol. Chem., 234*, 2145 (1959).

121. C. Schöpf, *Angew. Chem., 59*, 174 (1947).

122. R. V. Stevens, R. K. Mehra, and R. L. Zimmerman, *Chem. Commun.*, 1968 (1969).

123. F. Bohlmann, H. J. Mueller, and D. Schumann, *Chem. Ber., 106*, 3026 (1973).

124. J. I. Grayson and R. Dinkel, *Helv. Chim. Acta, 67*, 2100 (1984).

125. J. Thesing and H. Meyer, *Ber., 89*, 2159 (1956).

126. J. Thesing and H. Meyer, *Ann., 609*, 46 (1957).

127. J. Thesing and W. Sirrenberg, *Ber., 91*, 1987 (1958).

128. J. Thesing and W. Sirrenberg, *Ber., 92*, 1748 (1959).

129. N. J. Leonard, A. S. Hay, R. W. Fulmer, and V. W. Gash, *J. Am. Chem. Soc., 77*, 439 (1955).

130. M. G. Reinecke and L. R. Kray, *J. Org. Chem., 29*, 1736 (1964).

131. M. G. Reinecke and L. R. Kray, *J. Org. Chem., 30*, 3671 (1965).

132. E. Wenkert and B. Wickberg, *J. Am. Chem. Soc., 84*, 4914 (1962).

133. M. G. Reinecke and L. R. Kray, *J. Org. Chem.*, *31*, 4215 (1966).
134. W. Schneider, R. Dillmenn, and H. J. Dechow, *Arch. Pharm.*, *299*, 397 (1966); *CA*, *65*, 3832 (1966).
135. J. Knabe, *Arch. Pharm.*, *292*, 416 (1959).
136. F. Bohlmann, *Angew. Chem.*, *69*, 641 (1957).
137. F. Bohlmann, W. Weise, H. Sander, H. G. Hanke, and E. Winterfeld, *Ber.*, *90*, 653 (1957).
138. F. Bohlmann, W. Weise, D. Rahtz, and C. Arndt, *Ber.*, *90*, 2176 (1958).
139. F. Bohlmann, D. Schumann, and H. Schulz, *Tetrahedron Lett.*, 173 (1965).
140. N. J. Leonard and D. F. Morrow, *J. Am. Chem. Soc.*, *80*, 371 (1958).
141. F. Bohlmann, *Ber.*, *92*, 1798 (1959).
142. F. L. Weisenborn and P. A. Diassi, *J. Am. Chem. Soc.*, *78*, 2023 (1956).
143. N. J. Leonard, R. W. Fulmer, and A. S. Hay, *J. Am. Chem. Soc.*, *78*, 3457 (1956).
144. N. J. Leonard, W. J. Middleton, P. D. Thomas, and D. Choudhury, *J. Org. Chem.*, *21*, 344 (1956).
145. F. Mohlmann and P. Strehlke, *Tetrahedron Lett.*, 167 (1965).
146. N. J. Leonard, L. A. Miller, and P. D. Thomas, *J. Am. Chem. Soc.*, *78*, 3463 (1956).
147. F. Bohlmann, D. Schumann, V. Friesea, and E. Poetsch, *Ber.*, *99*, 3358 (1966).
148. K. Schreiber and Ch. Horstmann, *Ber.*, *99*, 3183 (1966).
149. N. J. Leonard and A. G. Cook, *J. Am. Chem. Soc.*, *81*, 5627 (1959).
150. R. Lukeš, J. Plešek, and J. Trojánek, *Collection Czech. Chem. Commun.*, *24*, 1987 (1959).
151. N. J. Leonard and W. K. Musker, *J. Am. Chem. Soc.*, *81*, 5631 (1959).
152. N. J. Leonard, K. Conrow, and R. R. Sauers, *J. Am. Chem. Soc.*, *80*, 5185 (1958).
153. N. J. Leonard and W. K. Musker, *J. Am. Chem. Soc.*, *82*, 5148 (1960).
154. J. R. Mahajan, B. J. Nunes, H. C. Aranjo, and G. A. L. Ferreira, *J. Chem. Res. Synop.*, 284 (1979).
155. O. T. Suge, S. Otika, M. Noguchi, and S. Kanemasa, *Chem. Lett.*, 867 (1982).
156. M. Ferles and J. Pliml, *Adv. Heterocyclic. Chem.*, *12*, 43 (1970).
157. R. M. Coates and E. F. Johnson, *J. Am. Chem. Soc.*, *93*, 4016 (1971).

158. H. P. Husson, L. Chevolet, Y. Langlois, C. Thal, and P. Pitier, *Chem. Commun.*, 930 (1972).
159. L. Chevelet, A. Husson, C. Kan-Fan, H. P. Husson, and P. Pitier, *Bull. Soc. Chim. France*, 1222 (1976).
160. P. Beeken and F. W. Fowler, *J. Org. Chem.*, 45, 1336 (1980).
161. J. Martinez and J. A. Joule, *Tetrahedron*, 34, 3027 (1978).
162. L. Libit and R. Hoffmann, *J. Am. Chem. Soc.*, 96, 1370 (1974).
163. C. Besselièvre, R. Beugelmans, and H. P. Husson, *Tetrahedron Lett.*, 3447 (1976).
164. U. Eisner and M. M. Sadeghi, *Tetrahedron Lett.*, 299 (1978).
165. S. Gabriel, *Ber.*, 46, 1358 (1913).
166. E. Immendörfer, *Ber.*, 48, 612 (1915).
167. H. Dugas, R. A. Alison, Z. Valenta, K. Wiesner, and C. M. Wong, *Tetrahedron Lett.*, 1279 (1965).
168. A. I. Meyers and J. C. Sircar, *Tetrahedron*, 23, 785 (1967).
169. A. I. Meyers and J. C. Sircar, *J. Org. Chem.*, 32, 1250 (1967).
170. T. T. Harding and P. S. Mariano, *J. Org. Chem.*, 47, 482 (1982).
171. H. Iido, S. Aoyagi, and C. Kibayashi, *J. Chem. Soc., Perkin Trans.*, 1, 120 (1977).
172. H. Iido, Y. Yuasa, and C. Kibayashi, *Chem. Commun.*, 766 (1978).
173. H. Iido, Y. Yuasa, and C. Kibayashi, *J. Org. Chem.*, 44, 1236 (1979).
174. H. O. Bernard and V. Smiekus, *Tetrahedron Lett.*, 4867 (1971).
175. I. Ise and V. Smiekus, *Chem. Commun.*, 505 (1976).
176. V. Virmini, M. B. Nigam, P. C. Jain, and N. Anand, *Indian J. Chem. Sect. B*, 17B, 472 (1979).
177. H. Wamhoff, *Lect. Heterocycl. Chem.*, 5, 561 (1980).
178. J. Knabe, *Pharmazie*, 20, 741 (1965).
179. V. M. Potapov, G. V. Grishina, and G. N. Koval, *Khim. Geterotsikl. Soedin.*, 1286 (1976).
180. J. B. Cloke and T. S. Leary, *J. Am. Chem. Soc.*, 67, 1249 (1945).
181. A. Seher, *Arch. Pharm.*, 284, 371 (1951); *CA*, 47, 2123 (1953).
182. R. Hilscher, *Ber.*, 31, 277 (1898).
183. S. Gabriel, *Ber.*, 42, 1238 (1909).
184. J. H. Burchalter and J. H. Short, *J. Org. Chem.*, 23, 1278 (1958).
185. J. A. Pople, W. G. Schneider, and H. J. Bernstein, *High Resolution Nuclear Magnetic Resonance*, McGraw-Hill, New York, 1959, p. 281.
186. H. C. Beyerman, M. van Leeuwen, J. Schmidt, and A. van Veer, *Rec. Trav. Chim.*, 80, 513 (1961).
187. K. H. Büchel and F. Korte, *Ber.*, 95, 2460 (1962).
188. C. R. Eddy and A. Eisner, *Anal. Chem.*, 26, 1428 (1954).
189. M. C. Kloetzel, J. L. Pinkus, and R. M. Washburn, *J. Am. Chem. Soc.*, 79, 4222 (1957).

190. P. M. Magiumity and J. B. Cloke, *J. Am. Chem. Soc.*, *93*, 49 (1951).
191. L. A. Cohen and B. Witkop, *J. Am. Chem. Soc.*, *77*, 6595 (1955).
192. B. Witkop, *J. Am. Chem. Soc.*, *78*, 2873 (1956).
193. B. Witkop, *Experientia*, *10*, 420 (1954).
194. S. Baldwin, *J. Org. Chem.*, *26*, 3288 (1961).
195. N. J. Leonard and V. W. Gash, *J. Am. Chem. Soc.*, *76*, 2781 (1954).
196. K. H. Büchel and F. Korte, *Ber.*, *95*, 2438 (1962).
197. R. Lukeš, V. Dědek, and L. Novotný, *Chem. Listy*, *52*, 654 (1958); *Collection Czech. Chem. Commun.*, *24*, 1117 (1959).
198. N. J. Leonard and D. M. Locke, *J. Am. Chem. Soc.*, *77*, 437 (1955).
199. R. Tschesche and G. Snatzke, *Ber.*, *90*, 579 (1957).
200. B. M. Wepster, *Rec. Trav. Chim.*, *71*, 1159 (1958).
201. K. Bowden, E. A. Brande, E. R. H. Jones, and B. C. L. Weedon, *J. Chem. Soc.*, 45 (1946).
202. K. Bowden, E. A. Brande, and E. R. H. Jones, *J. Chem. Soc.*, 948 (1946).
203. O. Červinka, *Collection Czech. Chem. Commun.*, *25*, 1174 (1960).
204. O. Červinka, *Collection Czech. Chem. Commun.*, *25*, 2675 (1960).
205. H. C. Brown, *J. Org. Chem.*, *22*, 439 (1957).
206. R. B. Turner and R. H. Garner, *J. Am. Chem. Soc.*, *80*, 1424 (1958).
207. F. W. Fowler, *J. Am. Chem. Soc.*, *94*, 5926 (1972).
208. C. A. Grob and H. J. Wilkens, *Helv. Chim. Acta*, *48*, 808 (1965).
209. C. A. Grob and H. J. Wilkens, *Helv. Chim. Acta*, *50*, 725 (1967).
210. C. A. Grob, A. Kauser, and E. Renk, *Helv. Chim. Acta*, *40*, 2170 (1957).
211. C. A. Grob, A. Kaiser, and E. Renk, *Chem. Ind. (London)*, 598 (1957).
212. V. Prelog and O. Häfliger, *Helv. Chim. Acta*, *32*, 1851 (1949).
213. N. J. Leonard, D. F. Morrow, and M. T. Rogers, *J. Am. Chem. Soc.*, *79*, 5476 (1957).
214. J. Gadamer, *J. Prakt. Chem.*, *84*, 817 (1911).
215. D. Béke, *Adv. Heterocyclic Chem.*, *1*, 167 (1963).
216. N. J. Leonard, J. A. Adamcik, C. Djerassi, and O. Halpern, *J. Am. Chem. Soc.*, *80*, 4858 (1958).
217. N. J. Leonard and W. K. Musker, *J. Am. Chem. Soc.*, *81*, 5631 (1959).
218. D. Dvornik and O. E. Edwards, *Tetrahedron*, *14*, 54 (1961).

219. J. Elguero, R. Jacquier, and G. Tarrago, *Tetrahedron Lett.*, 4719 (1965).

220. R. Lukeš and Z. Linhartová, *Collection Czech. Chem. Commun.*, 25, 502 (1960).

221. R. Lukeš, A. Fábryová, S. Doležal, and L. Novotný, *Collection Czech. Chem. Commun.*, 25, 1063 (1960).

222. N. J. Leonard and M. Oki, *J. Am. Chem. Soc.*, 77, 6239 (1955).

223. N. J. Leonard and M. Oki, *J. Am. Chem. Soc.*, 77, 6241 (1955).

224. N. J. Leonard, *Record Chem. Progr. Kresge-Hooker Sci. Lib.*, 17, 243 (1956).

225. B. Witkop and J. B. Patrick, *J. Am. Chem. Soc.*, 75, 4474 (1953).

226. O. E. Edwards, F. H. Clarke, and B. Douglas, *Can. J. Chem.*, 32, 235 (1954).

227. O. Červinka, A. R. Katritzky, and F. J. Swinbourne, *Collection Czech. Chem. Commun.*, 30, 1736 (1965).

228. G. Opitz, H. Hellmann, and H. W. Schubert, *Ann.*, 623, 117 (1959).

229. N. J. Leonard, A. S. Hay, R. W. Fulmer, and V. W. Gash, *J. Am. Chem. Soc.*, 77, 439 (1955).

230. J. D. S. Goulden, *J. Chem. Soc.*, 997 (1953).

231. C. Schöpf, H. Koop, and G. Werner, *Ber.*, 93, 2457 (1960).

232. C. Schöpf, F. Braun, H. Koop, and G. Werner, *Ann.*, 658, 156 (1962).

233. J. Gadamer, *Arch. Pharm.*, 248, 680 (1911).

234. M. Freund and K. Flwischer, *Ann.*, 409, 188 (1915).

235. C. Zatti and A. Ferratini, *Ber.*, 23, 2302 (1890).

236. G. Plaucher, *Ber.*, 31, 1488 (1898).

237. F. Bruckhausen, *Arch. Pharm.*, 261, 28 (1922).

238. R. Lukeš and V. Dědek, *Chem. Listy*, 51, 2059 (1957); *Collection Czech. Chem. Commun.*, 23, 2046 (1958).

239. O. Červinka, *Chem. Listy*, 52, 307 (1958); *Collection Czech. Chem. Commun.*, 24, 1880 (1959).

240. R. Lukeš and O. Červinka, *Chem. Listy*, 47, 392 (1953).

241. O. Červinka, A. Fábryová, and H. Chudobová, *Collection Czech. Chem. Commun.*, 43, 884 (1978).

242. R. Lukeš, J. Pliml, and J. Trojánek, *Collection Czech. Chem.* 24, 3109 (1959); *Chem. Listy*, 52, 1603 (1958).

243. R. Lukeš and J. Hofman, *Ber.*, 93, 2556 (1960).

244. H. C. Brown, J. H. Brewster, and H. Schechter, *J. Am. Chem. Soc.*, 76, 467 (1954).

245. F. Bohlmann and O. Schmidt, *Ber.*, 97, 1354 (1964).

246. N. Viswanathan, V. Balakrishnan, *Indian J. Chem. Sect. B.*, 1100 (1978).

247. A. ur Rehman and M. Ghazala, *Heterocycles*, 16, 261 (1981).

248. N. J. Leonard, C. K. Steinhardt, and C. Lee, *J. Org. Chem.*, 27, 4027 (1962).

249. R. Griot and T. Wagner-Jauregg, *Helv. Chim. Acta, 42*, 605 (1959).
250. M. Lora-Tamayo, R. Madroñero, and M. Stud, *Ber.*, *95*, 2176 (1962).
251. D. Béke and L. Töke, *Ber.*, *95*, 2123 (1962).
252. Z. Valenta, P. Deslongchamps, R. A. Ellison, and K. Weisner, *J. Am. Chem. Soc.*, *86*, 2533 (1964).
253. G. Dannhardt and R. Obergrusberger, *Arch. Pharm.*, *314*, 787 (1981).
254. M. Coenen, *Angew. Chem.*, *61*, 11 (1949).
255. M. Sainsbury, S. F. Dyke, and A. R. Marshall, *Tetrahedron*, *22*, 2445 (1966).
256. F. H. Hamer, R. J. Rathbone, and B. S. Winton, *J. Chem. Soc.*, 954 (1947).
257. G. H. Alt, *J. Org. Chem.*, *31*, 2384 (1966).
258. M. C. Koetzel and J. L. Pinkus, *J. Am. Chem. Soc.*, *80*, 2332 (1958).
259. S. Gabriel and J. Colman, *Ber.*, *41*, 513 (1908).
260. A. Lipp, *Ann.*, *289*, 173 (1896).
261. J. Thesing and K. Hofmann, *Ber.*, *90*, 229 (1957).
262. W. Vorboom, D. N. Reinhoudt, S. Harmeka, and G. J. van Hummel, *J. Org. Chem.*, *47*, 3339 (1982).
263. R. E. Ireland, *Chem. Ind.* (*London*), 979 (1958).
264. O. Červinka, *Chem. Ind.* (*London*), 1129 (1959).
265. O. Červinka, *Collection Czech. Chem. Commun.*, *25*, 1183 (1960).
266. F. Bohlmann, D. Schumann, and O. Schmidt, *Ber.*, *99*, 1652 (1966).
267. H. T. Openshau and N. Whittaker, *J. Chem. Soc.*, 1449 (1963).
268. M. von Standtmann, M. P. Cohen, and J. Shavel, Jr., *J. Org. Chem.*, *31*, 797 (1966).
269. F. Bohlmann and P. Herbert, *Ber.*, *99*, 3362 (1966).
270. D. Béke and C. Szahtay, *Ber.*, *95*, 2132 (1962).
271. C. Szántay, L. Töke, K. Honty, and G. Kalaus, *J. Org. Chem.*, *32*, 423 (1967).
272. T. Kametani, K. Kukumoto, and T. Katagi, *Chem. Pharm. Bull.*, *7*, 567 (1959).
273. B. Witkop, *Ann.*, *558*, 98 (1947).
274. B. Witkop, *Bull. Soc. Chim. France*, 423 (1954).
275. B. Witkop, *J. Am. Chem. Soc.*, *76*, 5597 (1954).
276. B. Witkop, *J. Am. Chem. Soc.*, *72*, 2311 (1950).
277. C. L. Stevens and R. J. Gasser, *J. Am. Chem. Soc.*, *79*, 6057 (1957).
278. B. Witkop, *J. Am. Chem. Soc.*, *72*, 1428 (1950).
279. B. Witkop, J. B. Patrick, and M. Rosenbaum, *J. Am. Chem. Soc.*, *73*, 2641 (1951).
280. W. I. Taylor, *Proc. Chem. Soc.*, 247 (1962).

281. W. D. Emmons, *J. Am. Chem. Soc.*, *78*, 6208 (1956).
282. R. Bonnett, V. M. Clark, and A. Todd, *J. Chem. Soc.*, 2102 (1959).
283. A. Reiche, E. Höft, and H. Schultze, *Ber.*, *97*, 195 (1964).
284. N. H. Martin and CH. W. Jefford, *Helv. Chim. Acta*, *65*, 762 (1982).
285. P. Bouchet, J. E. Elguero, and R. Jacquier, *Tetrahedron Lett.*, 3317 (1964).
286. J. Tafel and M. Stern, *Ber.*, *33*, 2224 (1900).
287. R. Lukeš, *Collection Czech. Chem. Commun.*, *4*, 351 (1932); *Chem. Listy*, *27*, 1 (1933).
288. N. G. Gaylord, *Reduction with Complex Metal Hydrides*, Wiley-Interscience, New York, 1956, p. 781.
289. J. A. Marshall and W. S. Johnson, *J. Org. Chem.*, *28*, 421 (1963).
290. O. Červinka, *Collection Czech. Chem. Commun.*, *26*, 673 (1961).
291. O. Červinka, *Collection Czech. Chem. Commun.*, *30*, 2403 (1965).
292. S. I. Goldberg and I. Ragade, *J. Org. Chem.*, *32*, 1046 (1967).
293. N. J. Leonard and E. Barthel, *J. Am. Chem. Soc.*, *72*, 3632 (1950).
294. N. J. Leonard, J. W. Curry, and J. J. Sagura, *J. Am. Chem. Soc.*, *75*, 6249 (1953).
295. N. J. Leonard and R. C. Sentz, *J. Am. Chem. Soc.*, *74*, 1704 (1952).
296. P. L. de Benneville and J. H. Macartney, *J. Am. Chem. Soc.*, *72*, 3073 (1950).
297. R. Lukeš, *Collection Czech. Chem. Commun.*, *10*, 66 (1938).
298. R. Lukeš and O. Červinka, *Collection Czech. Chem. Commun.*, *23*, 1336 (1958); *Chem. Listy*, *51*, 2086 (1957).
299. R. Lukeš and V. Dědek, *Collection Czech. Chem. Commun.*, *23*, 2053 (1958); *Chem. Listy*, *51*, 2082 (1957).
300. R. Lukeš and O. Červinka, *Collection Czech. Chem. Commun.*, *24*, 309 (1959); *Chem. Listy*, *51*, 2142 (1957).
301. N. J. Leonard, P. D. Thomas, and V. W. Gash, *J. Am. Chem. Soc.*, *77*, 1552 (1955).
302. N. J. Leonard and R. R. Sauers, *J. Am. Chem. Soc.*, *79*, 6210 (1957).
303. O. Červinka and O. Křiž, *Collection Czech. Chem. Commun.*, *30*, 1700 (1965).
304. R. Bonnet, V. M. Clark, A. Giddey, and A. Todd, *J. Chem. Soc.*, 2087 (1959).
305. K. Wiesner, Z. Valenta, A. J. Manson, and F. W. Stonner, *J. Am. Chem. Soc.*, *77*, 675 (1955).
306. R. Lukeš, V. Dienstbierová, and O. Červinka, *Collection Czech. Chem. Commun.*, *24*, 428 (1959); *Chem. Listy*, *52*, 1137 (1958).

307. R. Lukeš and M. Černý, *Collection Czech. Chem. Commun.*, *26*, 2886 (1961).
308. M. Freund, *Ber.*, *37*, 3334 (1904).
309. M. Freund and H. Beck, *Ber.*, *37*, 4679 (1904).
310. M. Freund and K. Lederer, *Ber.*, *44*, 2356 (1911).
311. A. Kaufmann, *Ber.*, *51*, 116 (1918).
312. A. Kaufmann, *J. Chem. Soc.*, *114*, 187 (1918).
313. N. J. Leonard and A. S. Hay, *J. Am. Chem. Soc.*, *78*, 1984 (1956).
314. D. Béke, C. Szantay, and M. Bárczai-Beke, *Ann.*, *636*, 150 (1960).
315. S. F. Dyke and M. Sainsbury, *Tetrahedron*, *21*, 1907 (1965); *Tetrahedron Lett.*, 1545 (1964).
316. A. R. Battersby and R. Binks, *J. Chem. Soc.*, 2888 (1955).
317. A. R. Battersby, D. J. Le Count, S. Garratt, and R. I. Thrift, *Tetrahedron*, *14*, 46 (1961).
318. J. W. Huffman and E. C. Miller, *J. Org. Chem.*, *25*, 90 (1960).
319. D. W. Brown and S. F. Dyke, *Tetrahedron Lett.*, 3587 (1964).
320. J. Knabe and J. Kubitz, *Angew. Chem. Intern. Ed. Engl.*, *2*, 689 (1963); *Angew. Chem.*, *75*, 981 (1963).
321. J. Knabe and J. Kubitz, *Arch. Pharm.*, *297*, 129 (1964).
322. J. Knabe and N. Ruppenthal, *Arch. Pharm.*, *297*, 141 (1964).
323. J. Knabe and N. Ruppenthal, *Arch. Pharm.*, *299*, 159 (1966).
324. J. Knabe and N. Ruppenthal, *Naturwiss.*, *51*, 482 (1964).
325. J. Knabe and K. Detering, *Ber.*, *9*, 2873 (1966).
326. D. W. Brown and S. F. Dyke, *Tetrahedron Lett.*, 3587 (1964).
327. D. W. Brown and S. F. Dyke, *Tetrahedron*, *22*, 2429 (1966).
328. D. W. Brown and S. F. Dyke, *Tetrahedron*, *22*, 2437 (1966).
329. J. P. Kutney, R. T. Brown, and E. Piers, *J. Am. Chem. Soc.*, *86*, 2287 (1964).
330. A. Kaufmann and A. Albertini, *Ber.*, *42*, 3776 (1909).
331. A. Kaufmann and A. Widmer, *Ber.*, *44*, 2058 (1911).
332. W. E. Feely and E. M. Beavers, *J. Am. Chem. Soc.*, *81*, 4004 (1959).
333. N. J. Leonard, H. A. de Walt, and G. W. Leubner, *J. Am. Chem. Soc.*, *73*, 3325 (1951).
334. N. J. Leonard and R. L. Foster, *J. Am. Chem. Soc.*, *74*, 2110 (1952).
335. O. Červinka, *Chem. Ind. (London)*, 1482 (1960); *Collection Czech. Chem. Commun.*, *27*, 567 (1962).
336. O. Červinka, A. Fábryová, and L. Matouchová, *Collection Czech. Chem. Commun.*, *28*, 536 (1963).
337. S. Takahashi and M. Kans, *Tetrahedron Lett.*, 3789 (1965).
338. R. Eisenthal and A. R. Katritzky, *Tetrahedron*, *21*, 2205 (1965).
339. O. Červinka, A. Fábryová, J. Josef, V. Sermek, and S. Smrčková, *Collection Czech. Chem. Commun.*, *48*, 3407 (1983).

340. M. Takeda, H. Inoue, M. Konda, S. Saito, and H. Kugita, *J. Org. Chem.*, *37*, 677 (1972).
341. R. G. Kinsman, A. W. C. White, and S. F. Dyke, *Tetrahedron*, *31*, 449 (1975).
342. R. G. Kinsman and S. F. Dyke, *Tetrahedron*, *35*, 857 (1979).
343. J. Knabe, W. Krause, and K. Sierock, *Arch. Pharm.*, *303*, 255 (1970).
344. J. Knabe and H. Powilleit, *Arch. Pharm.*, *303*, 37 (1970); *304*, 52 (1971).
345. J. Knabe, R. Dorr, S. F. Dyke, and R. G. Kinsman, *Tetrahedron Lett.*, 5373 (1972).
346. J. Knabe and R. Dorr, *Arch. Pharm.*, *306*, 784 (1973).
347. R. G. Kinsman, S. F. Dyke, and J. Mead, *Tetrahedron*, *29*, 4303 (1974).
348. S. F. Dyke, *Advances in Heterocyclic Chemistry*, vol. 14, Academic Press, New York, 1974, p. 279.
349. K. Mothes, *Pharmazie*, *14*, 121 (1959).
350. G. Hahn and F. Rumph, *Ber.*, *71*, 2141 (1938).
351. C. Schöpf and G. Lehmann, *Ann. Chem.*, *518*, 1 (1935).
352. E. Jucker, *Chimia (Aarau)*, *9*, 199 (1955).
353. C. Schöpf, A. Komzak, F. Braun, and E. Jacobi, *Ann.*, *559*, 1 (1948).
354. C. Schöpf, H. Arm, and H. Krimm, *Ber.*, *84*, 690 (1951).
355. C. Schöpf, H. Arm, and F. Braun, *Ber.*, *85*, 937 (1952).
356. C. Schöpf, *Angew. Chem.*, *62*, 452 (1950).
357. C. Schöpf, H. Arm, G. Benz, and H. Krimm, *Naturwiss*, *38*, 186 (1951).
358. A. Lüttringhaus, J. Jander, and R. Schneider, *Ber.*, *92*, 1756 (1959).
359. P. Hermann, *Ber.*, *94*, 442 (1961).
360. R. F. C. Brown, V. M. Clark, I. O. Sutherland, and A. Todd, *J. Chem. Soc.*, 2109 (1959).
361. R. Robinson, *J. Chem. Soc.*, *111*, 876 (1917); 1079 (1936).
362. E. Anet, G. K. Huges, and E. Ritchie, *Nature*, *163*, 298 (1949).
363. F. Galinovsky, A. Wagner, and R. Weiser, *Monatsh.*, *82*, 551 (1951).
364. M. Pailer, *Chemiker. Ztg.*, *51*, 24 (1950).
365. F. Galinovsky and H. Zuber, *Monatsh.*, *84*, 798 (1953).
366. B. M. Goldschmidt, *J. Org. Chem.*, *27*, 4057 (1962).
367. C. Schöpf, *Angew. Chem.*, *61*, 31 (1949).
368. C. Schöpf, F. Braun, K. Burkhardt, G. Dummer, and H. Müller, *Ann.*, *626*, 123 (1959).
369. E. Anet, G. K. Huges, and E. Ritchie, *Nature*, *164*, 501 (1949); *Australian J. Sci. Res.*, *A3*, 336 (1950).
370. C. Schöpf and F. Braun, *Naturwiss.*, *36*, 377 (1949).
371. C. Schöpf and K. Kreibich, *Naturwiss*, *41*, 335 (1954).

372. M. J. Ribas and A. N. Blanco, *Anales Real Soc. Espan. Fi. Quin. (Madrid)*, *B 57*, 781 (1962); *CA*, *58*, 5738c (1963).

373. M. M. Robinson, W. G. Pierson, L. Dorfman, B. F. Lambert, and R. A. Lucas, *J. Org. Chem.*, *31*, 3206, 3213, 3220 (1966).

374. E. E. van Tamelen and R. L. Foltz, *J. Am. Chem. Soc.*, *82*, 2400 (1960).

375. F. Bohlmann, E. Winterfeld, and V. Friese, *Ber.*, *96*, 2251 (1963).

376. R. Lukeš, J. Kovář, K. Bláha, and J. Koubek, *Collection Czech. Chem. Commun.*, *21*, 1324 (1956).

377. C. Schöpf and E. Müller, *Ann.*, *687*, 242 (1965).

378. C. Schöpf and E. Schenkenberger, *Ann.*, *682*, 206 (1965).

379. C. Schöpf and F. Oechler, *Ann.*, *523*, 1 (1936).

380. N. J. Leonard and M. J. Martell, *Tetrahedron Lett.*, 44 (1960).

381. C. Schöpf and H. Steuer, *Ann.*, *558*, 124 (1947).

382. C. Schöpf, *Angew. Chem.*, *50*, 779 (1937).

383. E. E. van Tamelen and G. G. Knapp, *J. Am. Chem. Soc.*, *77*, 1860 (1955).

384. E. Wenkert and D. K. Roychandhuri, *J. Am. Chem. Soc.*, *80*, 1613 (1958).

385. W. I. Taylor, *Indole Alkaloids. An Introduction to the Enamine Chemistry of Natural Products*, Pergamon Press, New York, 1966.

386. M. E. Keuhne, *Synthesis*, 510 (1970).

387. R. V. Stevens, M. C. Ellis, *Tetrahedron Lett.*, 5185 (1967).

388. R. V. Stevens, M. C. Ellis, and M. P. Wentland, *J. Am. Chem. Soc.*, *90*, 5576 (1968).

389. R. A. Comes, M. T. Core, M. D. Edmons, W. B. Edwards, and R. W. Jenkins, *J. Labelled Compd.*, *9*, 253 (1973).

390. E. Breuer, S. Zbaida, *Tetrahedron*, *31*, 499 (1975).

391. E. Wenkert, *Accounts Chem. Res.*, *1*, 78 (1968).

392. S. J. Martinez and J. A. Joule, *Chem. Commun.*, 818 (1976).

393. M. S. Allen, A. J. Gaskell, and J. A. Joule, *Chem. Commun.*, 736 (1971).

394. T. J. Curphey, H. L. Kim, *Tetrahedron Lett.*, 1441 (1968).

395. S. L. Keely, A. J. Martinez, and F. C. Tahk, *Tetrahedron Lett.*, 2763 (1969).

396. R. V. Stevens and M. P. Wentland, *Tetrahedron Lett.*, 2613 (1968).

397. R. V. Stevens and M. P. Wentland, *J. Am. Chem. Soc.*, *90*, 5580 (1968).

398. R. V. Stevens, P. M. Lesko, and R. Lapalme, *J. Org. Chem.*, *40*, 3495 (1975).

399. S. L. Keely, Jr., and F. C. Tahk, *J. Am. Chem. Soc.*, *90*, 5580 (1968).

400. T. J. Curphey and H. L. Kim, *Tetrahedron Lett.*, 1441 (1968).

401. R. V. Stevens and J. T. Lai, *J. Org. Chem.*, *37*, 2138 (1972).
402. R. V. Stevens, L. E. Du Pree, Jr., and P. L. Loewenstein, *J. Org. Chem.*, *37*, 977 (1972).
403. R. V. Stevens and L. E. Du Pree, Jr., *Chem. Commun.*, 1585 (1970).
404. J. Bohlmann and O. Schmidt, *Chem. Ber.*, *97*, 1354 (1964).
405. M. E. Kuhne and C. Bayha, *Tetrahedron Lett.*, 1311 (1966).
406. S. I. Goldberg and I. Ragade, *J. Org. Chem.*, *32*, 1046 (1967).
407. J. P. Kutney, N. Abdurahman, P. L. Lequesne, E. Piers, and I. Vlattas, *J. Am. Chem. Soc.*, *88*, 3656 (1966).
408. P. Kutney, W. J. Cretney, P. Lequesne, B. McKague, and E. Piers, *J. Am. Chem. Soc.*, *88*, 4756 (1966).
409. A. Camerman, N. Camerman, J. P. Kutney, E. Piers, and J. Trotter, *Tetrahedron Lett.*, 637 (1965).
410. J. P. Kutney and E. Piers, *J. Am. Chem. Soc.*, *86*, 953 (1964).
411. T. Kimura and Y. Ban, *Chem. Pharm. Bull.* (*Tokyo*), *17*, 297 (1969).
412. A. R. Battersby, D. J. Lecount, S. Garratt, and R. I. Thrift, *Tetrahedron*, *14*, 46 (1961).
413. A. R. Battersby and R. Binks, *J. Chem. Soc.*, 2888 (1955).
414. S. M. Kupchan and A. Yoshitake, *J. Org. Chem.*, *34*, 1062 (1960).
415. M. Sainsbury, D. W. Brown, S. F. Dyke, and G. Hardy, *Tetrahedron*, *25*, 1881 (1969).
416. M. Onda, K. Yonezawa, and K. Albe, *Chem. Pharm. Bull.* (*Tokyo*), *17*, 404 (1969).
417. S. F. Dyke and D. W. Brown, *Tetrahedron*, *24*, 1455 (1968).
418. M. Onda, K. Albe, and K. Yonezawa, *Chem. Pharm. Bull.* (*Tokyo*), *16*, 2005 (1968).
419. R. K. Mehra and R. L. Zimmerman, *Chem. Commun.*, 877 (1969).
420. R. V. Stevens and M. P. Wentland, *Chem. Commun.*, 1104 (1968).
421. R. N. Schut, F. E. Ward, and T. J. Leipzig, *J. Org. Chem.*, *34*, 330 (1969).
422. F. Bossert, H. Meyer, E. Wehinger, *Angew. Chem. Int. Ed.*, 762 (1981).
423. V. Eismer and J. Kuthan, *Chem. Rev.*, *72*, 1 (1972).
424. J. Kuthan and A. Kurfürst, *Ind. Eng. Chem. Prod. Res. Dev.*, *21*, 191 (198ě).
425. D. M. Stout and A. I. Meyers, *Chem. Rev.*, *82*, 223 (1982).
426. S. Blechart, *Nach. Chem. Tech. Lab.*, *28*, 651 (1980).
427. H. J. Wyle and F. W. Fowler, *J. Org. Chem.*, *49*, 4025 (1984).
428. A. Jackson, A. J. Gaskell, N. D. V. Wilson, and J. A. Joule, *Chem. Commun.*, 364 (1968).
429. G. Buchi, D. L. Coffew, K. Kocsis, P. E. Sonnet, and F. E. Ziegler, *J. Am. Chem. Soc.*, *87*, 2073 (1965).
430. M. Ikezaki, T. Wykamatsu, and Y. Ban, *Chem. Commun.*, 88 (1969).

431. E. Wenkert, *Accounts Chem. Res.*, *1*, 78 (1968).
432. A. I. Meyers, *Heterocyclic Chem.*, 5, 151 (1968).
433. A. I. Meyers and W. N. Beverung, *Chem. Commun.*, 877 (1968).
434. A. I. Meyers, J. C. Sircar, *Tetrahedron*, *23*, 785 (1967).
435. J. C. Sircar, A. I. Mayers, *J. Org. Chem.*, *32*, 1248 (1967).
436. A. I. Meyers and S. Singh, *Tetrahedron*, *25*, 4161 (1969).
437. Z. Horii, K. Morihawa, and I. Ninomiya, *Chem. Pharm. Bull. (Tokyo)*, *17*, 846 (1969).
438. F. E. Ziegler and P. A. Zoretic, *Tetrahedron Lett.*, 2639 (1968).
439. A. I. Meyers, A. H. Reine, and R. Goult, *J. Org. Chem.*, *34*, 698 (1969).
440. G. L. Smith and H. W. Whitlock, *Tetrahedron Lett.*, 2711 (1966).

9

Applications of Enamines to Synthesis of Natural Products

GOWRIKUMAR GADAMASETTI AND MARTIN E. KUEHNE
The University of Vermont, Burlington, Vermont

	Abbreviations	533
I.	β-Acylation Reactions: 1-11	534
II.	β-Alkylation Reactions: 12-74	543
III.	β-Arylation Reactions, Including Photocyclizations: 75-95	590
IV.	Alkylations at the β'-Position: 96-97	604
V.	δ-Alkylation Reactions of Dienamines: 98-106	605
VI.	N-Alkylations: 107-109	612
VII.	Addition to the α-Position (or γ-Position for Dienamines): 110-125	614
VIII.	α-Arylation Reactions and γ-Arylations for Dienamines: 126-144	625
IX.	Oxidation Reactions, Including Reactions with Electrophilic Sulfur and Halogen: 145-169	640
X.	Reduction Reactions: 170-196	657
XI.	Miscellaneous Examples of Enamine Generation: 197-211	676
	Addendum: 212-213	688
	References Corresponding to Each Reaction Sample	690

Gilbert Stork's seminal studies on the use of enamine derivatives for α-alkylation of aldehydes and ketones awakened an intense interest in their extensions and in analogous electrophilic substitutions of enamines.* Within fifteen years, over 700 reports of enamine alkylactions, acylations, arylations, oxidations, and other reactions appeared in the literature.† A large number of applications to syntheses of natural products could also be noted at that time.‡

Since use of a synthetic tool for the generation of final goal compounds surely is a good measure of the utility of that synthetic method, one can obtain a fair assessment of its overall value and dispersion from a survey of its employment in the spectrum of goal directed syntheses. In contrast to reaction focused reports, synthetic applications show not only what a reaction can do, but that it is governed primarily by what it must do to meet a synthetic demand. A collection of synthetic applications of enamine chemistry may also lead one to recognize the solution to an analogous synthetic problem at hand.

Enamine reactions have been discussed in detail in the foregoing chapters and in the earlier volume of this series. It is hoped that perusal of the following collection of synthetic applications of enamines, presented in the visual impact for recognition mode, will be of help in reaching other synthetic goals. Also included in this collection are some examples of reactions of enamides and of vinylogous amides, where the transformations parallel the formal objectives of enamine transformations. Lexicographic adherance to an enamine definition was deemed less important to the mission of this chapter than provision of more extensive synthetic assistance.

The synthetic examples are presented in terms of the key enamine reaction step, i.e., β-alkylation, acylation, arylation or oxidation, or reaction at the α (iminium) carbon after an initial protonation, or after alternative reactions at the enamine β-position. The method of enamine generation is generally not shown in structural form but it is given in terms of the reaction principle, i.e., f (formation): *condensation* (meaning reaction of a carbonyl component with a secondary amine), or *oxidation* (typically indicating reaction of a tertiary amine with mercuric acetate). The reactants for the enamine formation should thus become obvious from the shown enamine structure in each example.

---

*G. Stock, R. Terrell, J. Szmuskocicz, *J. Am. Chem. Soc.*, 76, 2029 (1954).
†M. E. Kuehne in "Enamines in Organic Synthesis," Chap. 8, Enamines (G. Cook, ed.) Dekker, New York (1968).
‡M. E. Kuehne, "Applications of Enamines to Syntheses of Natural Products," *Synthesis*, 510 (1970).

ABBREVIATIONS

abs	absolute
Ac	acetyl
anh	anhydrous
aq	aqueous
Ar	aryl
atoms	atmosphere
Bu	butyl
°C	degree Celsius
cat	catalytic
conc	concentrated
crystn	crystallization
DBU	1,5-azabicyclo [5,4,0]-undecene-5
DCC	dicyclohexyl carbodimide
DIBAL	diisobutylaluminum hydride
dil	dilute
DME	1,2-dimethoxyethane
DMF	dimethyl formamide
DMS	dimethyl sulfide
DMSO	dimethyl sulfoxide
DNP	2,4-dinitrophenyl
eq	equivalent
Et	ethyl
extrn	extraction
f	formation
h	hour
hv	photolysis
HMPA	hexamethyl phosphoramide
i	iso
LAH	lithium aluminum hydride
LBH	lithium borohydride
LDA	lithium diisopropylamide
liq,lq	liquid
m	meta
M	molar
mCPBA	meta-chloro perbenzoic acid
Me	methyl
min	minute
Ms	mesyl
MVK	methyl vinyl ketone
N	normality
NBH	sodium borohydride
NCS	N-chloro succinimide
nm	nanometers
o	ortho

oxidn   oxidation
p   para
Ph   phenyl
Pr   propyl
Press   pressure
psi   pounds per square inch
PTSA   para toluene sulfonic acid
Py   pyridine
Redn   reduction
t   tertiary
TFA   trifluoro acetic acid
THF   tetrahydrofuran
Tos, Ts   tosyl

I. β-ACYLATION REACTIONS: 1—11

1

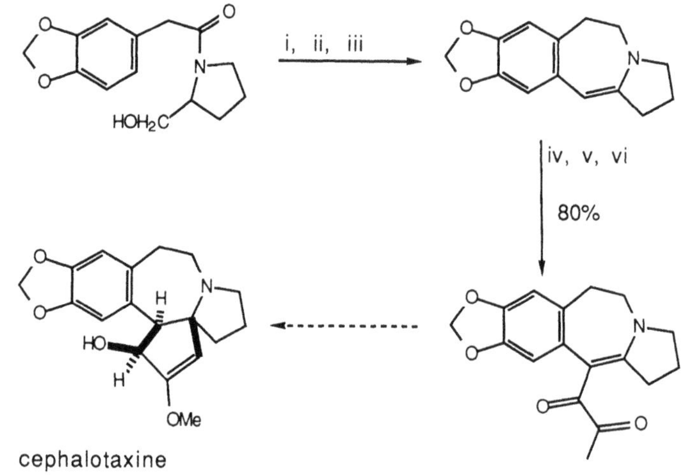

cephalotaxine

f. ß-hydroxyamide

i. DCC, Cl_2CHCO_2H, DMSO ii. $BF_3.Et_2O$, $CHCl_3$ iii. LAH, THF
iv. $MeCH(OAc)COCl$, $NaHCO_3$, MeCN v. K_2CO_3, MeOH
vi. PbO_2, PhMe or $MeCOCO_2Et$

From J. Auerbach and S. M. Weinreb, *J. Am. Chem. Soc.*, **94**, 7172 (1972).

2

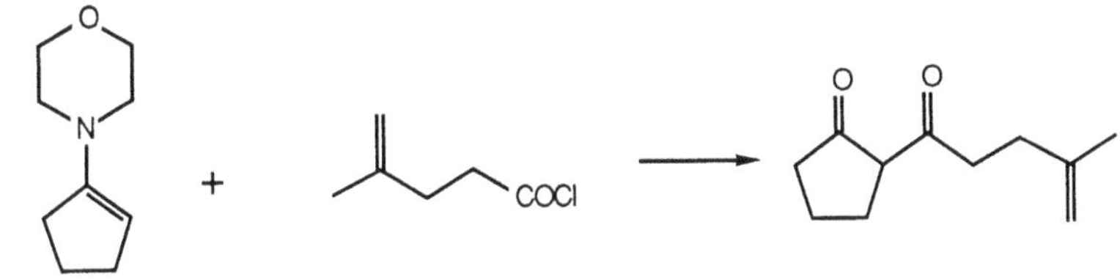

From G. Pattenden and G. M. Robertson, *Tet. Letters*, 399 (1986).

3

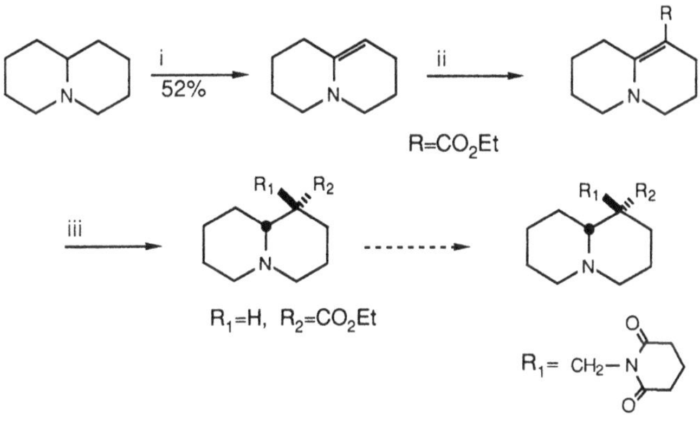

R=CO₂Et

R=CO$_2$Et

R$_1$=H, R$_2$=CO$_2$Et

R$_1$= CH$_2$—N

R$_2$=H

i. Hg(OAc)$_2$ ii. ClCO$_2$Et iii. NaBH$_4$

f. oxidation

From S. I. Goldberg and I. Ragade, *J. Org. Chem.*, **32**, 1046 (1967)
and S. I. Goldberg and A. H. Lipkins, *J. Org. Chem.*, **35**, 242 (1970).

4

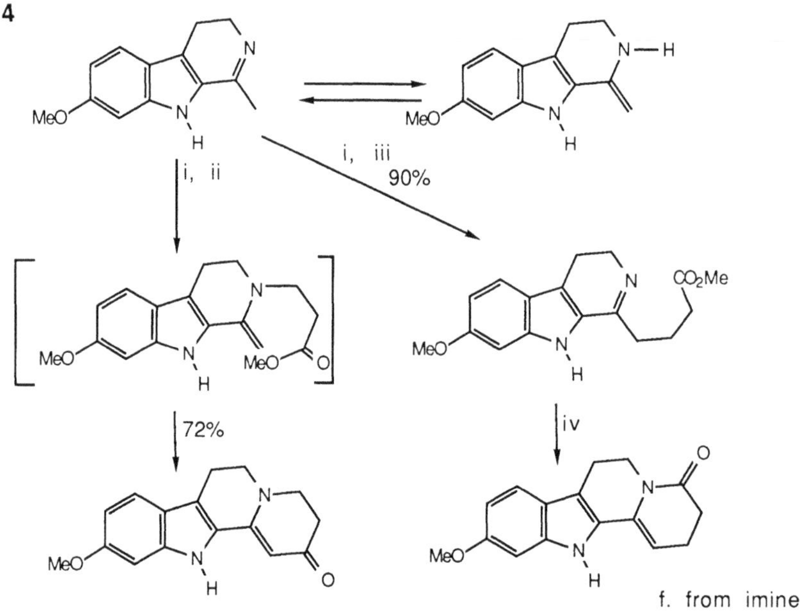

f. from imine

i. Methyl acrylate ii. PhH-Anh. MeOH, 24h, reflux   iii. MeOH-PhH, 25°C, 72h
iv. MeOH, reflux

From Atta-Ur-Rahman, *J. Chem. Soc.*, *Perkin Trans. I*, 731, 736 (1972).

5

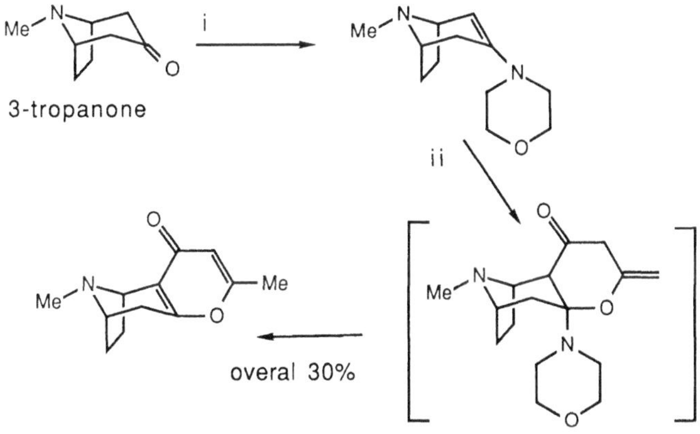
Me
N
O
3-tropanone
i
ii
overal 30%

(±)-isobellendine

f. condensation

i. morpholine, TsOH,PhH, reflux, 20h ii. diketene, CH_2Cl_2, 20°C

From M. Lounasmaa, T. Langenskiöld, and C. Holmberg, *Tet. Letters*, *22*, 5179 (1981).

6

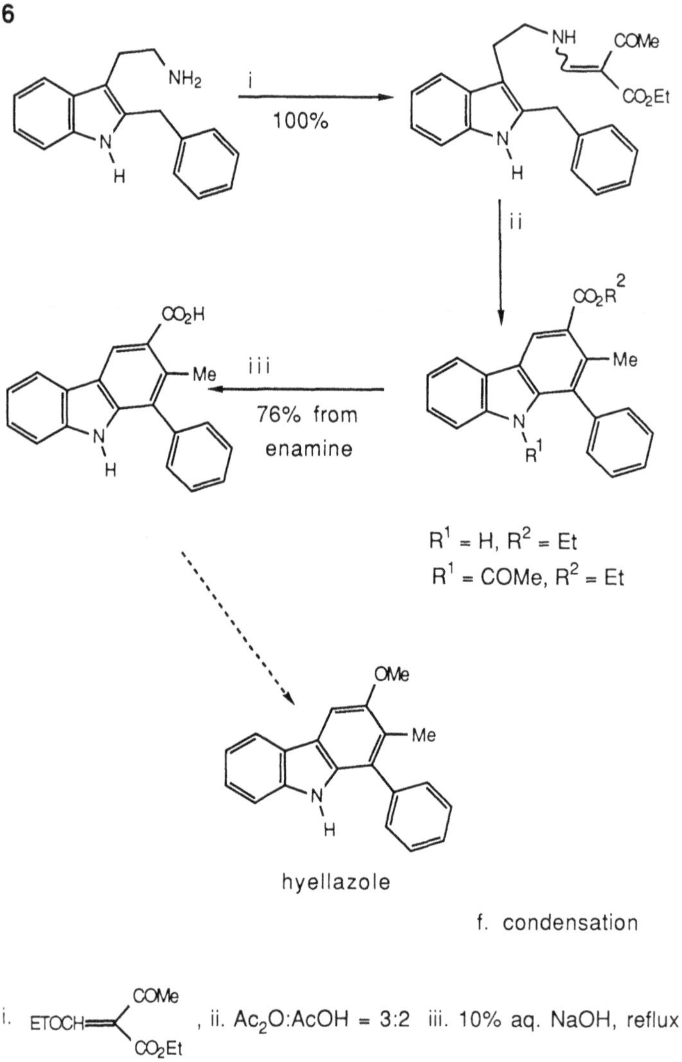
NH2
i
100%
NH
COMe
CO2Et
N
H
ii
CO2H
Me
iii
76% from
enamine
CO2R2
Me
N
R1
R1 = H, R2 = Et
R1 = COMe, R2 = Et
OMe
Me
hyellazole
f. condensation
i. ETOCH= COMe CO2Et, ii. Ac2O:AcOH = 3:2 iii. 10% aq. NaOH, reflux

From S. Takano, Y. Suzuki, and K. Ogasawara, *Heterocycles, 16*, 1479 (1981).

7

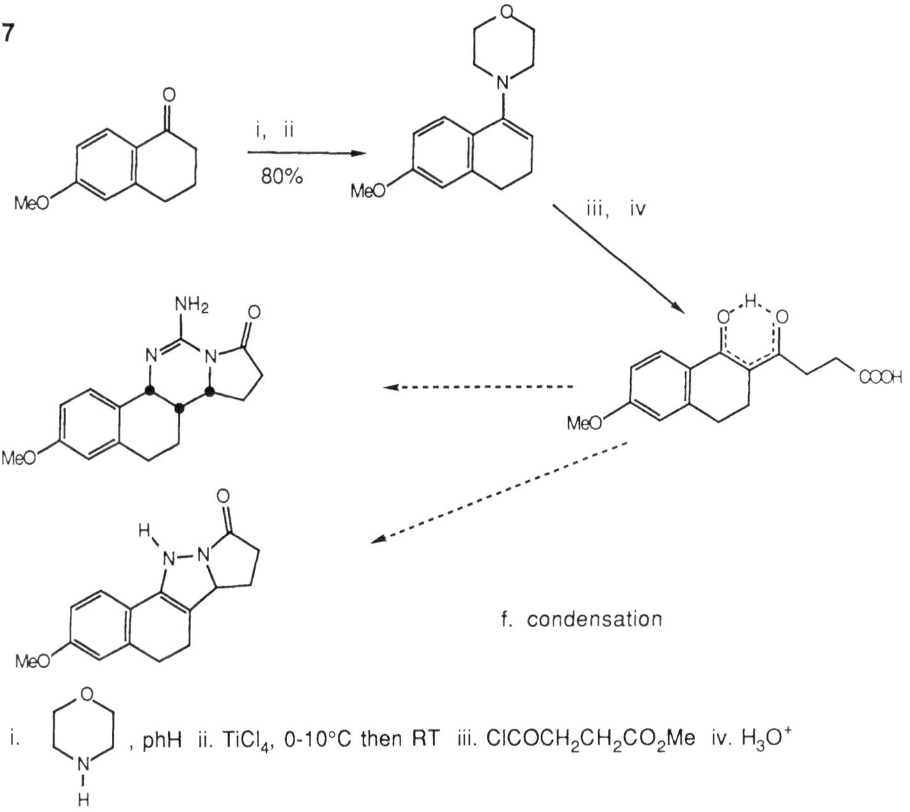

i, ii
80%

iii, iv

NH₂

f. condensation

i. ![morpholine] , phH   ii. TiCl₄, 0-10°C then RT   iii. ClCOCH₂CH₂CO₂Me   iv. H₃O⁺

From U. K. Pandit, F. A. Van der Vlugt, and A. C. Van Dalen, *Tet. Letters*, 3693 (1969).

8

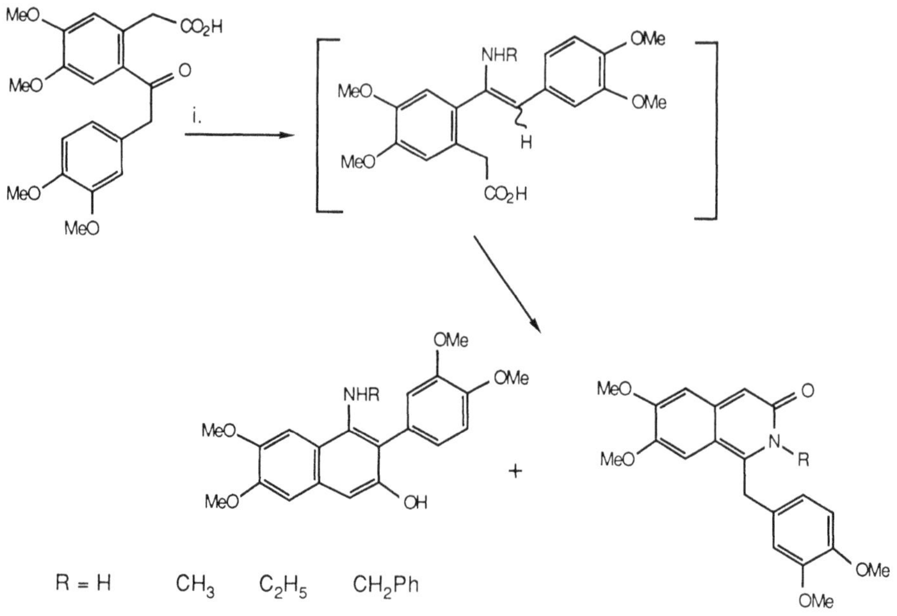

R =	H	CH₃	C₂H₅	CH₂Ph
Yield (%)	25	6	30	28

i. RNH_2, H_3O^+

From W. E. Kreighbaum, W. F. Kavanaugh, and W. T. Comes, *J. Heterocyclic Chem.*, *10*, 317 (1973).

9

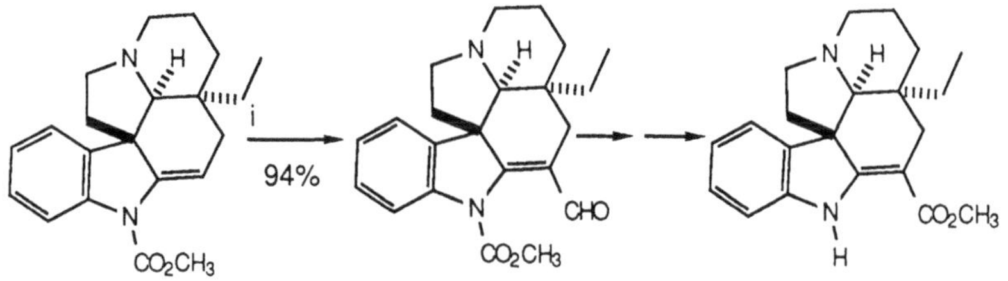

94%

vincadifformine

i. Vilsmeier-Haack reaction

From K. Yoshida, S. Nomura and Y. Ban, *Tetrahedron*, *41*, 5495 (1985).

10

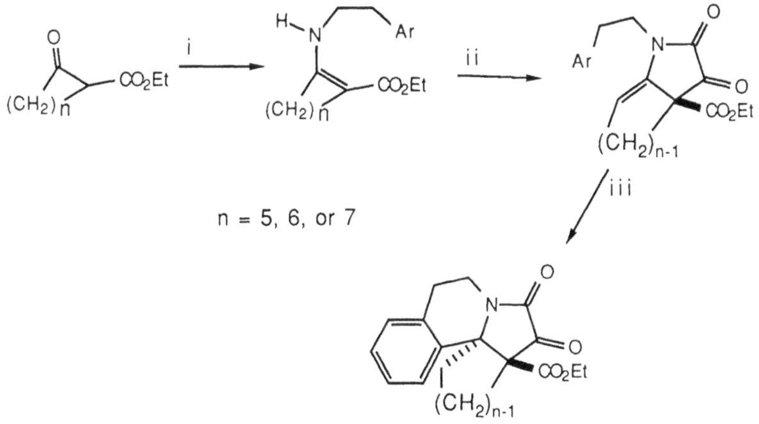

n = 5, 6, or 7

f. condensation

i. $Ar(CH_2)_2NH_2$ ii. $(COCl)_2$ iii. PPA

From Y. Tsuda, Y. Sakai, N. Kashiwaaba, T. Sano, J. Toda, and K. Isobe, *Heterocycles*, *16*, 189 (1981).

11

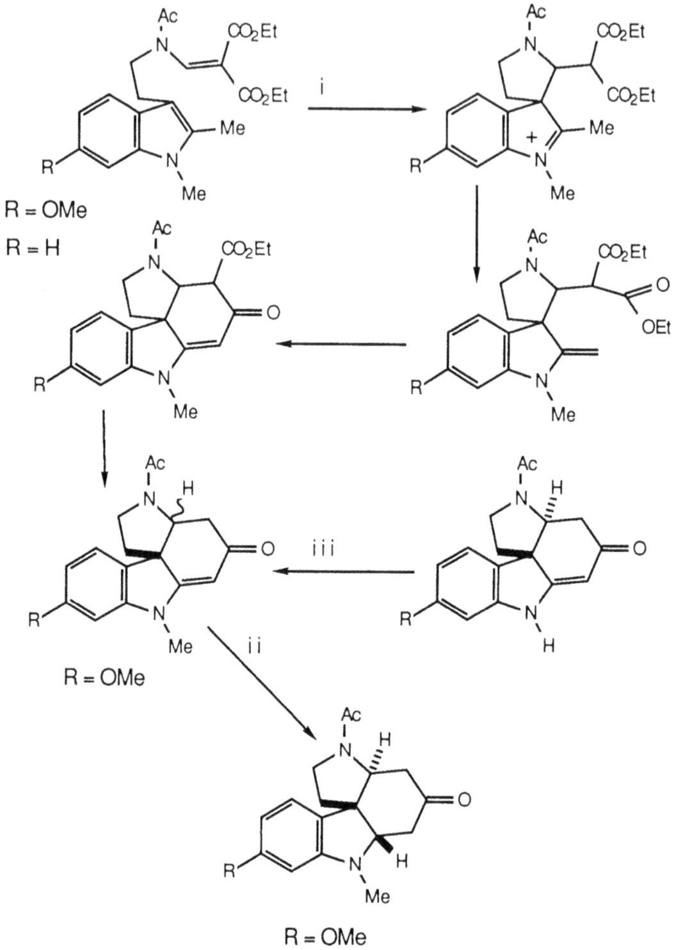

R = OMe
R = H

R = OMe

R = OMe

f. indole alkylation

i. AcOH, Ac$_2$O, reflux, 72h   ii. Li, NH$_3$ iii. NaH, MeI, DMF

From S. Takano, K. Shishido, M. Sato, and K. Ogasawara, *Hetero-cycles*, *6*, 1699 (1977).

II. β-ALKYLATION REACTIONS: 12—74

Reactions with alkyl halides, carbenoids, aldehydes, ketones, imonium compounds, enones, acrylates and Diels-Alder type reactions.

In addition to the examples given in this section see also examples: 4, 80, 103 (Sec V.), 145.

12

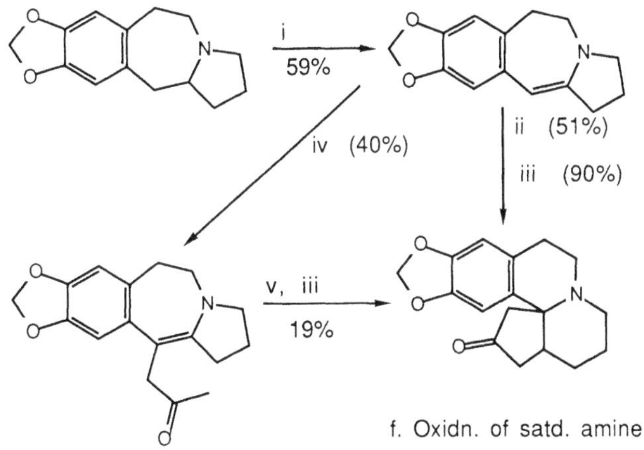

f. Oxidn. of satd. amine

i. Hg(OAc)$_2$ ii. BrCH$_2$COCH$_2$CO$_2$Et , MeCN iii. 5% H$_2$SO$_4$-H$_2$

iv. BrCH$_2$COMe, MeCN, heat v. Pyrrolidine-TsOH, Mol. Sieves

From L. J. Dolby, S. J. Nelson, and D. Senkovich, *J. Org. Chem.*, 37, 3691 (1972).

13

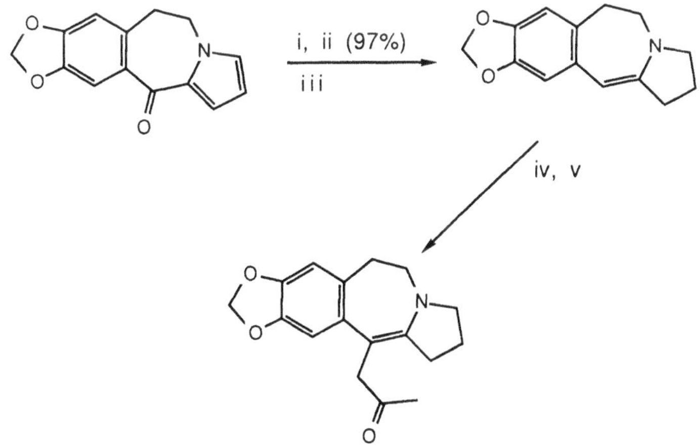

i. NBH ii. H$_2$-Rh/C-EtOH-HClO$_4$ iii. Hg(OAc)$_2$ iv. CH≡CCH$_2$Br v. Hg^{2+}

From B. Weinstein and A. R. Craig, *J. Org. Chem.*, 41, 875 (1976).

14

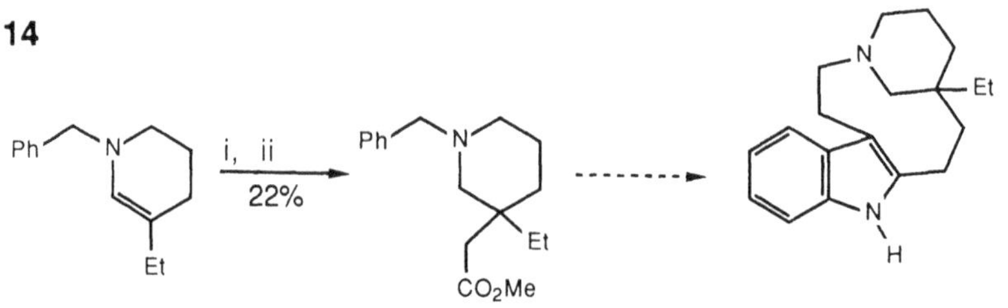

i. BrCH₂CO₂Me ii. NBH

From F. E. Ziegler, J. A. Kloek, and P. A. Zoretic, *J. Am. Chem. Soc.*, **91**, 2342 (1969).

15

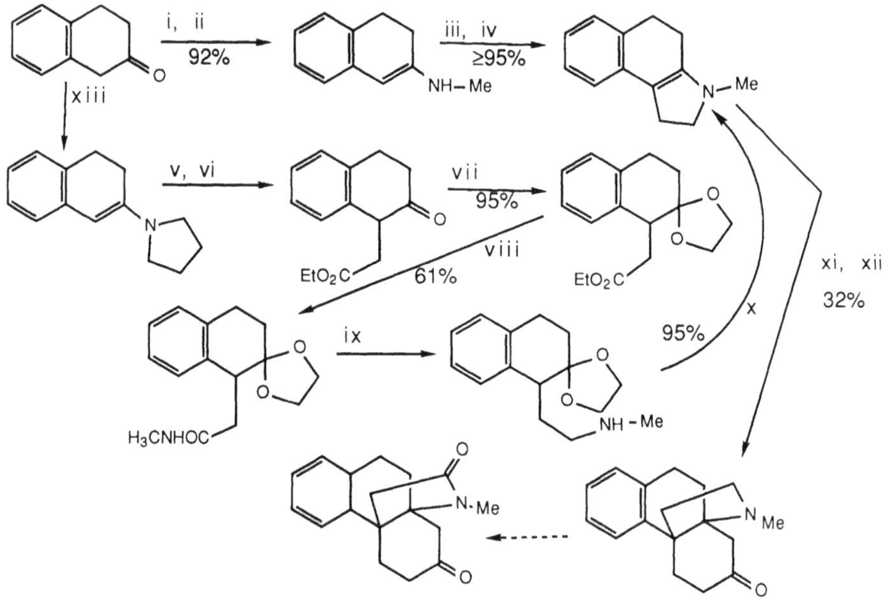

i. MeNH₂ ii. TiCl₄ iii. i-PrMgCl, THF iv. BrCH₂CH₂Cl v. BrCH₂CO₂Et vi. H₂O
vii. PTSA, HOCH₂CH₂OH, PhH viii. MeNH₂ ix. LAH, THF x. 3M H₂SO₄, EtOH

xi. [CH₂=CHCO] , MeCN, RT xii. AcOH, 80°C xiii. [pyrrolidine] , PhH

From D. A. Evans, C. A. Bryan, and G. M. Wahl, *J. Org. Chem.*, **35**, 4122 (1970).

16

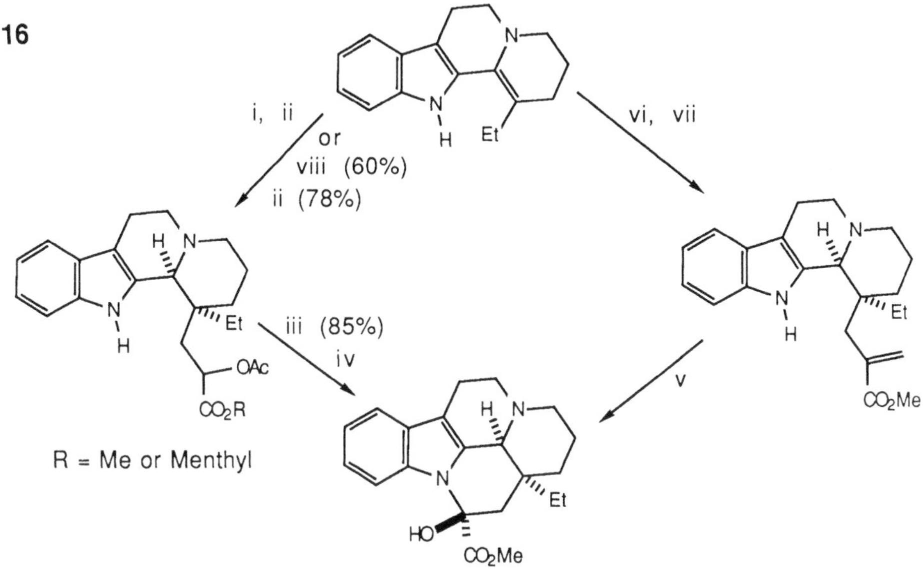

i, ii
or
viii (60%)
ii (78%)

vi, vii

iii (85%)
iv

R = Me or Menthyl

v

HO

CO₂Me

(±)-vincamine or 40% optical purity

i. $CH_2=C(OAc)CO_2Me$ ii. H_2, Pd-MeOH iii. MeONa-MeOH iv. Ag_2CO_3-

Celite, PhH, heat v. OsO_4, HIO_4 vi. $BrCH_2C(CO_2Me)=CH_2$ vii. NBH

viii. $CH_2=C(OAc)CO_2$Menthyl

From Cs. Szantay, L. Szabo, and Gy. Kalaus, *Tet. Letters*, 191 (1973), Cs. Szantay, L. Szabo, and Gy. Kalaus, *Tetrahedron*, *33*, 1803 (1977), and C. Thal, T. Sevenet, H. P. Husson and P. Potier, *Compt. rend.*, *275 C*, 1295 (1972).

17

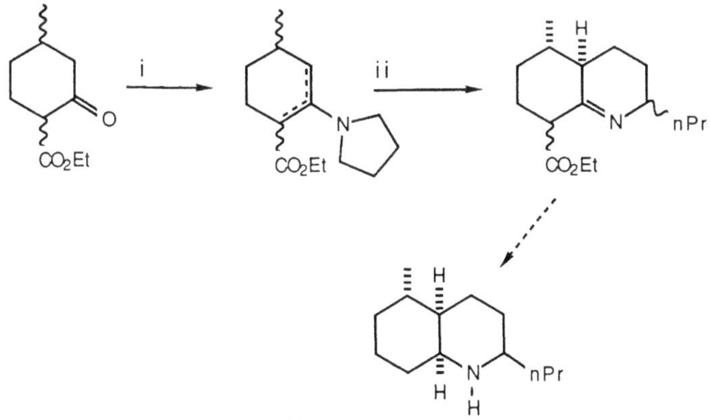

i

ii

nPr

i. pyrrolidine ii. 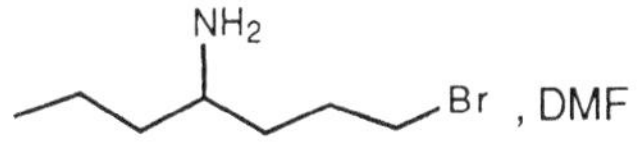 NH₂ ... Br , DMF

From G. Habermehl, H. Andres, and B. Witkop, *Naturwiss.*, *62*, 345 (1975).

18

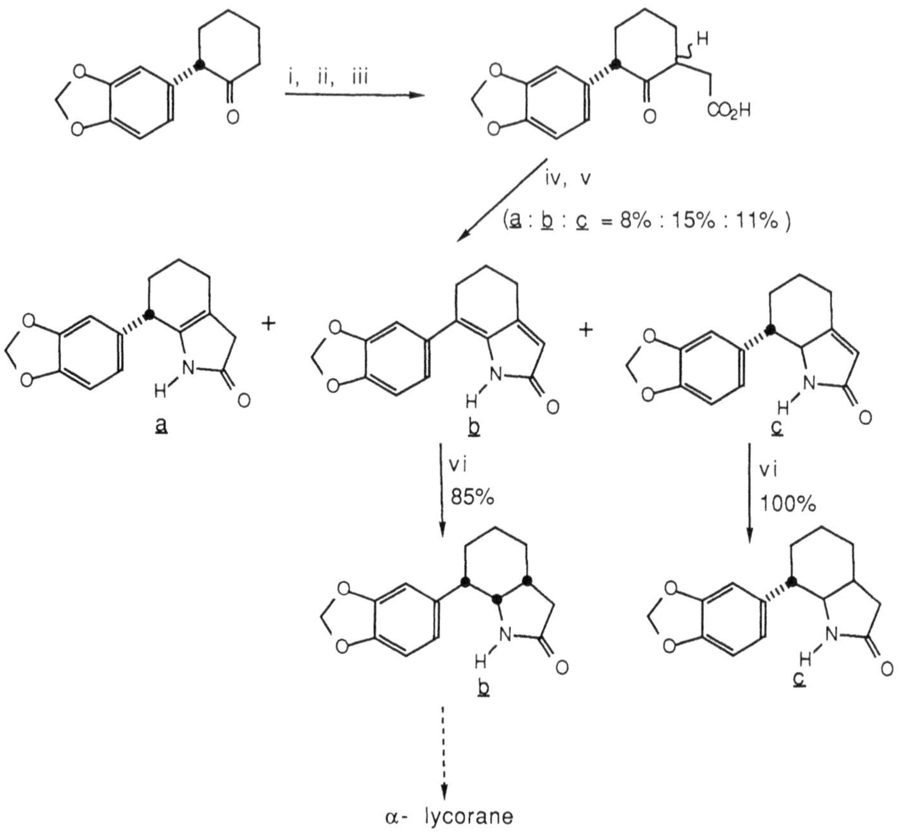

i, ii, iii

iv, v

($\underline{a}$: $\underline{b}$: $\underline{c}$ = 8% : 15% : 11%)

$\underline{a}$ + $\underline{b}$ + $\underline{c}$

vi
85%

vi
100%

$\underline{b}$ $\underline{c}$

α- lycorane

f. condensation

i. pyrrolidine, PhH, reflux   ii. BrCH$_2$CO$_2$Me   iii. aq KOH, reflux   iv. NH$_2$OH
v. Zn, AcOH, reflux   vi. Adams cat. H$_2$, CH$_3$OH

From B. Umezawa, O. Hoshino, S. Sawaki, S. Sato, and N. Numao, *J. Org. Che.*, **42**, 4272 (1977).

19

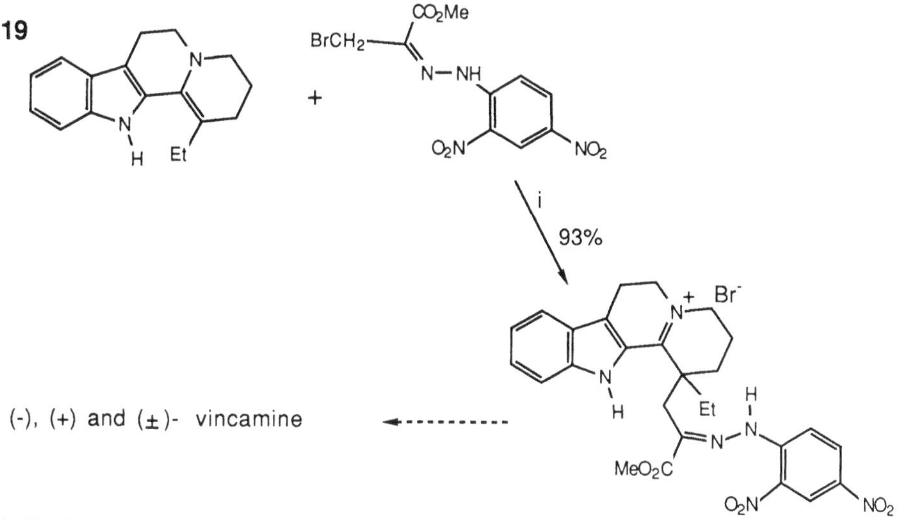

i. Et₃N, EtOAc

From G. Rossey, A. Wick, and E. Wenkert, *J. Org. Chem.*, 47, 4745 (1982).

20

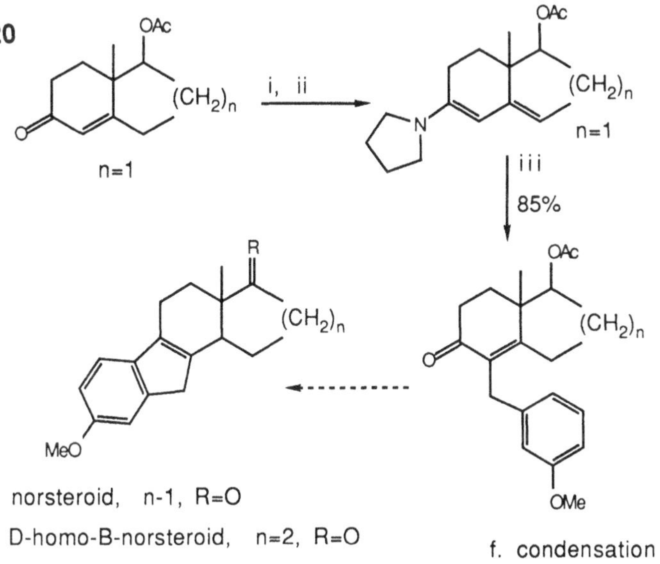

norsteroid, n-1, R=O

D-homo-B-norsteroid, n=2, R=O

f. condensation

i. pyrrolidine, phH   ii. TiCl₄, 0-10°C then RT   iii.

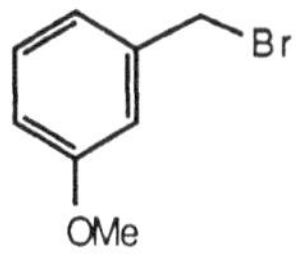

, DMF, reflux

From F. Le Goffic, A. Gouyette, and A. Ahond, *Compt. rend.*, 274 C, 2008 (1972).

21

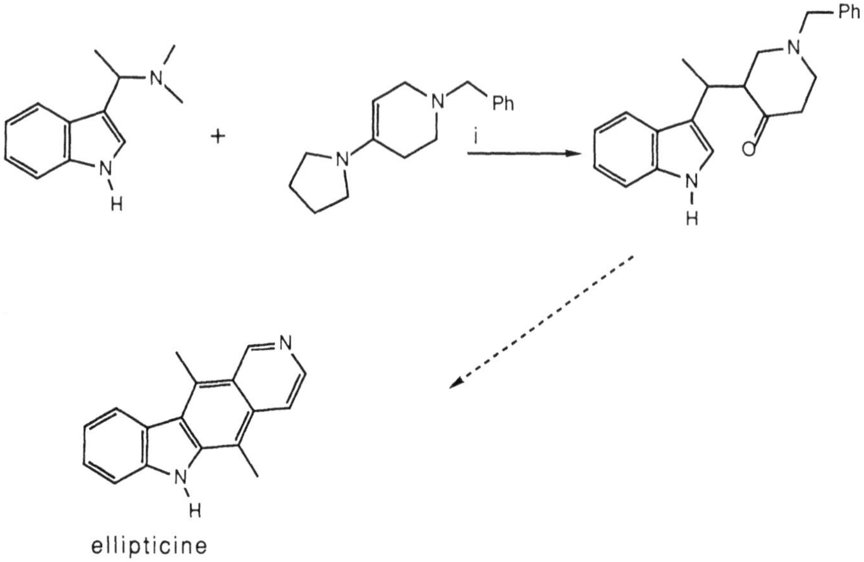

ellipticine

i. dioxane, heat

From U. K. Pandit, K. de Jonge, K. Erhart, and H. O. Huisman, *Tet. Letters*, 1207 (1969).

22

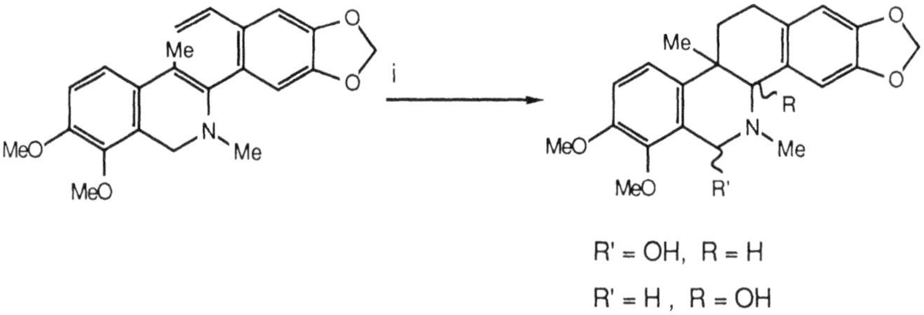

R' = OH, R = H

R' = H, R = OH

i. hv, benzene

From M. Onda, K. Yuasa, J. Okada, K. Katoaka, and K. Abe, *Chem. Pharm. Bull.* (Japan), *21*, 1333 (1973).

23

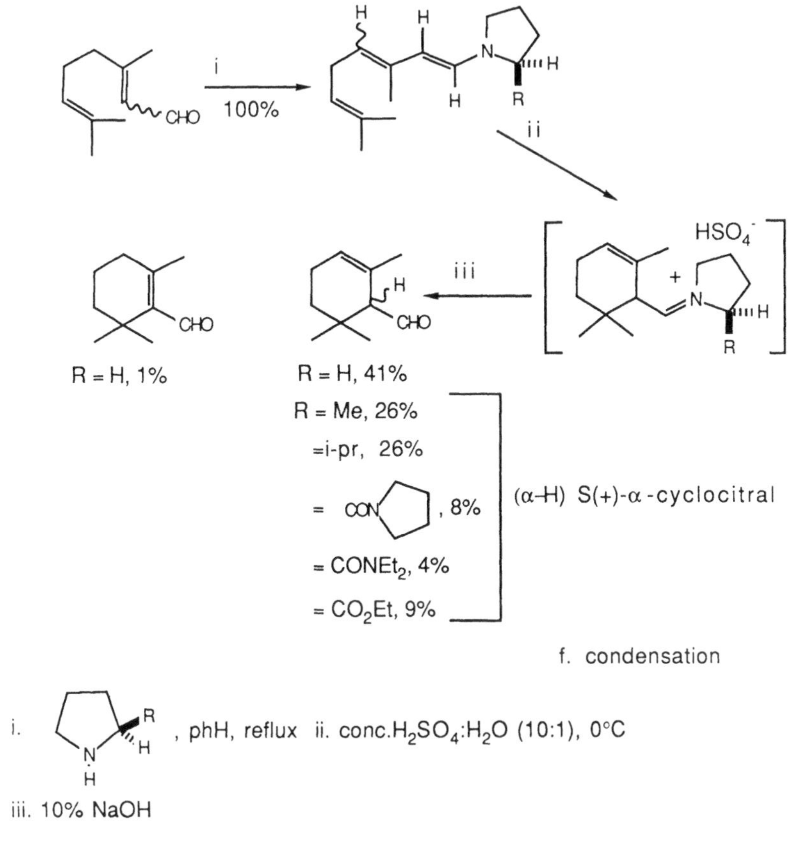

R = H, 1%

R = H, 41%
R = Me, 26%
=i-pr, 26%

= CON⟨⟩ , 8%   (α–H) S(+)-α-cyclocitral

= CONEt₂, 4%

= CO₂Et, 9%

f. condensation

i. ⟨pyrrolidine structure⟩ , phH, reflux   ii. conc.H₂SO₄:H₂O (10:1), 0°C

iii. 10% NaOH

From S. Yamada, M. Shibasaki and S. Terashima, *Tet. Letters*, 381, 377 (1973).

24

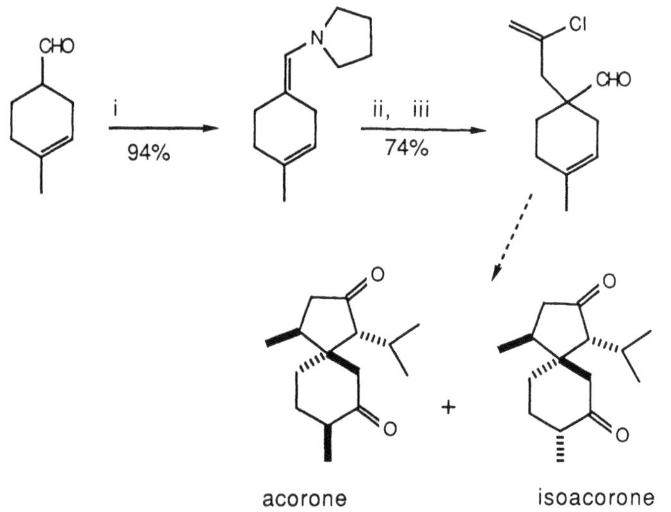
CHO
i
94%
N
ii, iii
74%
Cl
CHO
O
O
+
O
O
acorone
isoacorone

f. condensation

i. pyrrolidine, PTSA, PhH, reflux ii. $ICH_2CH(Cl)=CH_2$, CH_3CN, heat
iii. H_2O

From S. F. Martin and T. S. Chou, *J. Org. Chem.*, *43*, 1027 (1978).

25

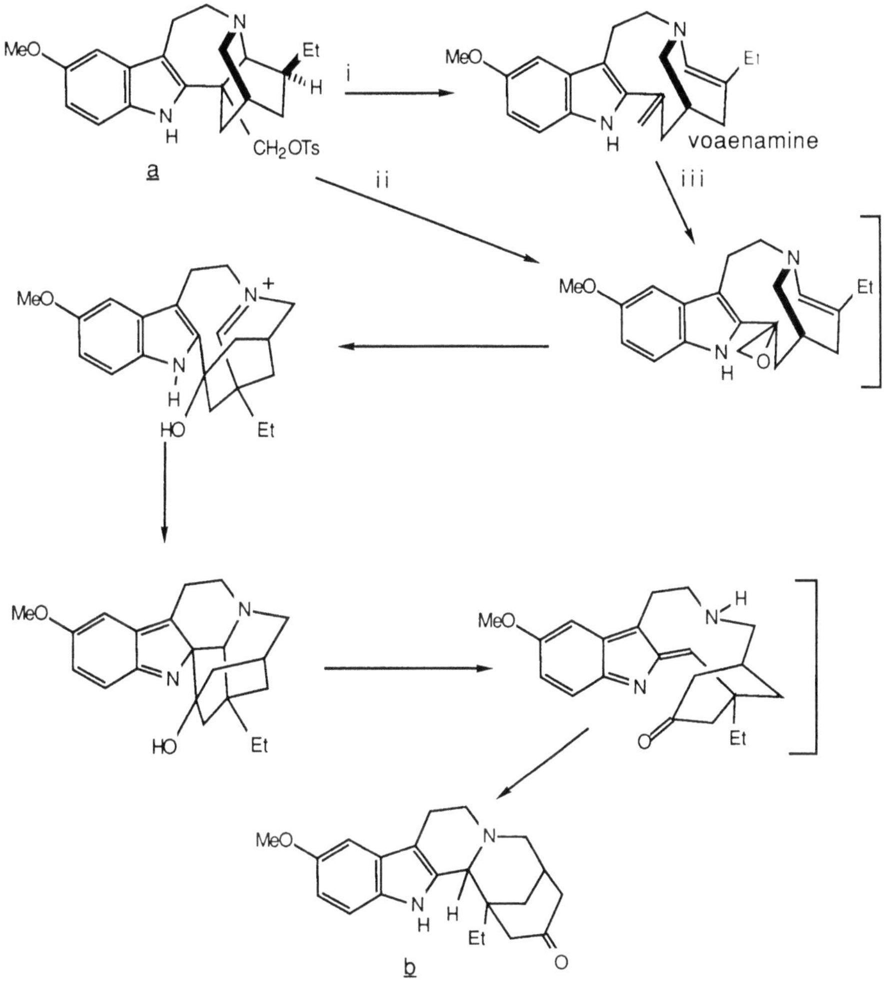

voaenamine

f. fragmentation reaction

i. py   ii. MeOH-py-O$_2$ iii. mCPBA, CH$_2$Cl$_2$

$$\underline{a} \xrightarrow[39\%]{ii} \underline{b}$$

$$\underline{a} \xrightarrow[76\%]{iii} \underline{b}$$

From Y. Morita, M. Hesse, Y. Renner, and H. Schmid, *Helv. Chim. Acta.* *59*, 532 (1976).

26

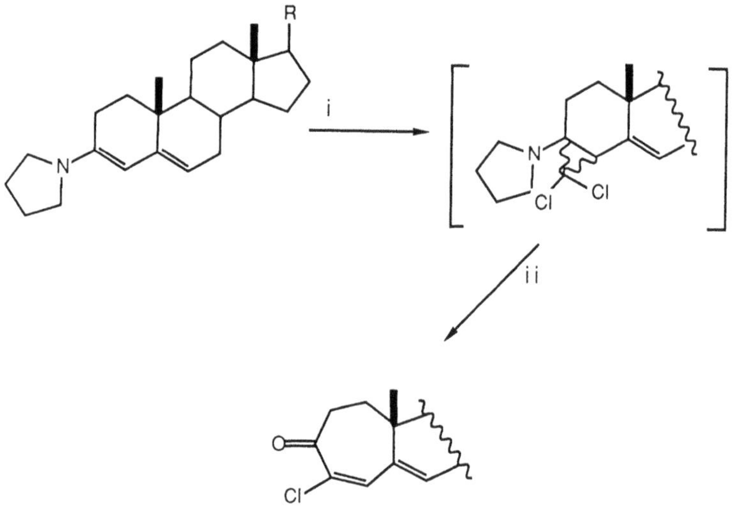

R = C$_8$H$_{17}$, 31%

R = OAc, 34%

i. :CCl$_2$ ii. H$_2$O

From U. K. Pandit and S. A. G. de Graaf, *J. Chem. Soc., Chem. Commun.*, 381 (1970).

27

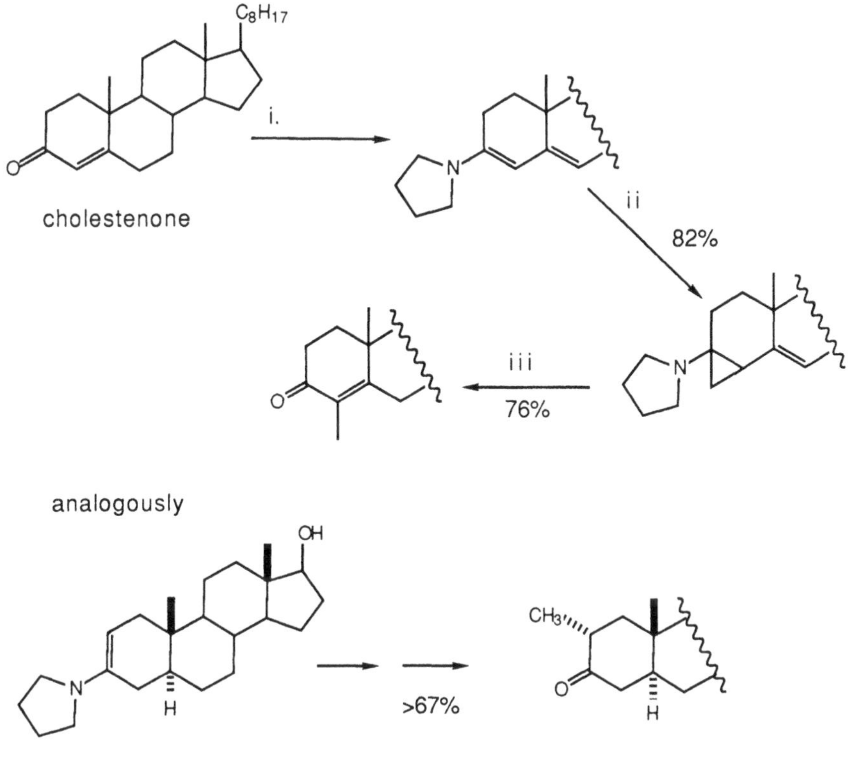
C8H17
i.
cholestenone
N
ii
82%
iii
76%
O
analogously
OH
N
H
>67%
CH3
O
H

f. condensation

i. pyrrolidine ii. CH_2I_2, Et_2Zn or CH_2N_2, CuCl
iii. 90% aq. MeOH, 160-165°C, 2.5h

From M. E. Kuehne and J. C. King, *J. Org. Chem.*, *38*, 304 (1973).

28

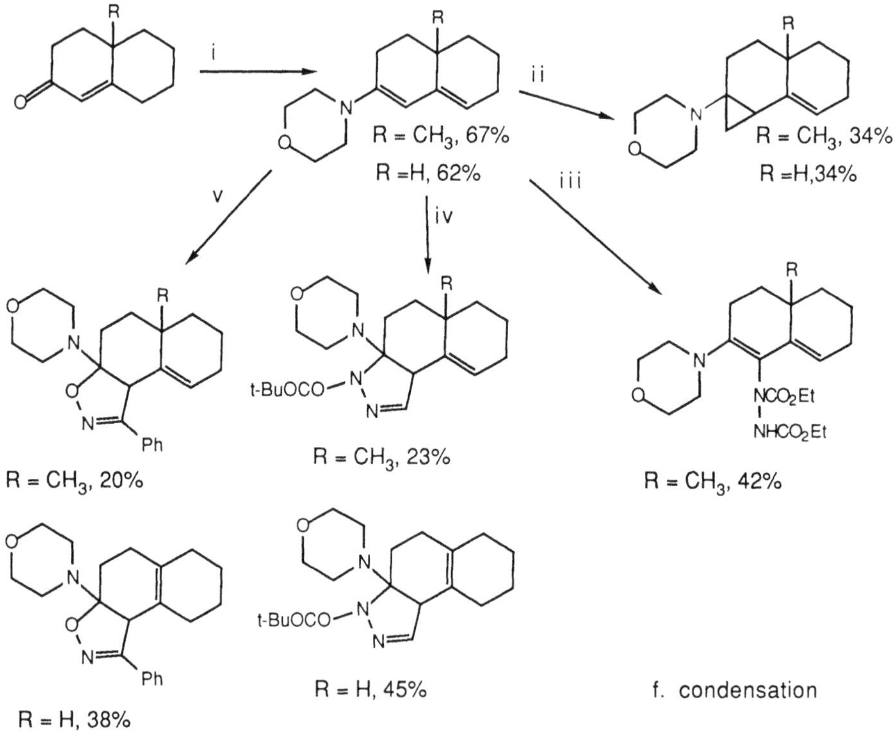

i.

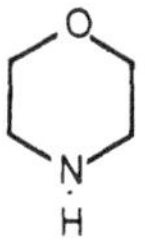

, ii. CH$_2$I$_2$, ZnEt$_2$, PhH, 0°C, 30 min, NH$_4$OH iii. EtO$_2$CN=NCO$_2$Et,

THF, 0°C (2h), 24h (RT) iv. t-BuOCON$_3$, RT, dark, N$_2$ atmos, 56h

iv. PhC(Cl)=NOH ⟶ (PhC≡N$^+$-O$^-$), dry PhH, 2h, 0°C, N$_2$ atmos

From M. E. Kuehne, G. Di Vincenzo, *J. Org. Chem.*, *37* 1023 (1972).

29

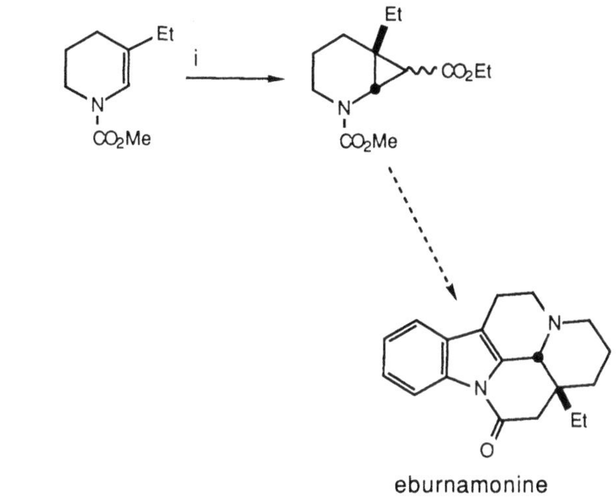

eburnamonine

i. N$_2$CHCO$_2$Et, Cu

From E. Wenkert, T. Hudlicky, and H. D. H. Showalter, *J. Am. Chem. Soc.*, *100*, 4893 (1978) and E. Wenkert, B. L. Buckwalter, and S. S. Sathe, *Synth. Commun.*, *3*, 61 (1973).

30

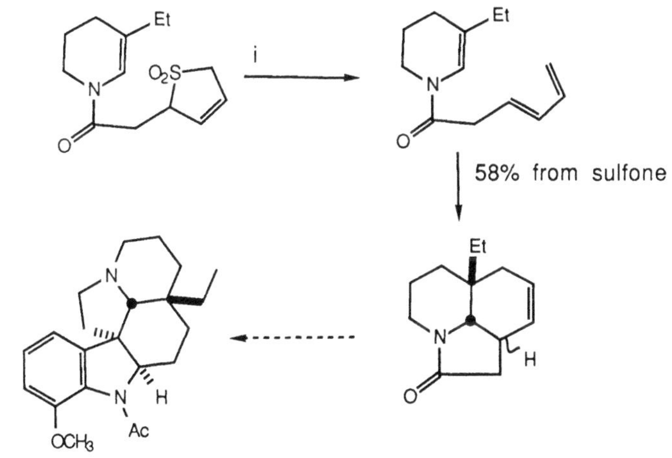

58% from sulfone

(±) aspidospermine

i. 600°C

From S. F. Martin, S. R. Desai, G. W. Phillips, and A. C. Miller, *J. Am. Chem. Soc.*, *102*, 3294 (1980).

31

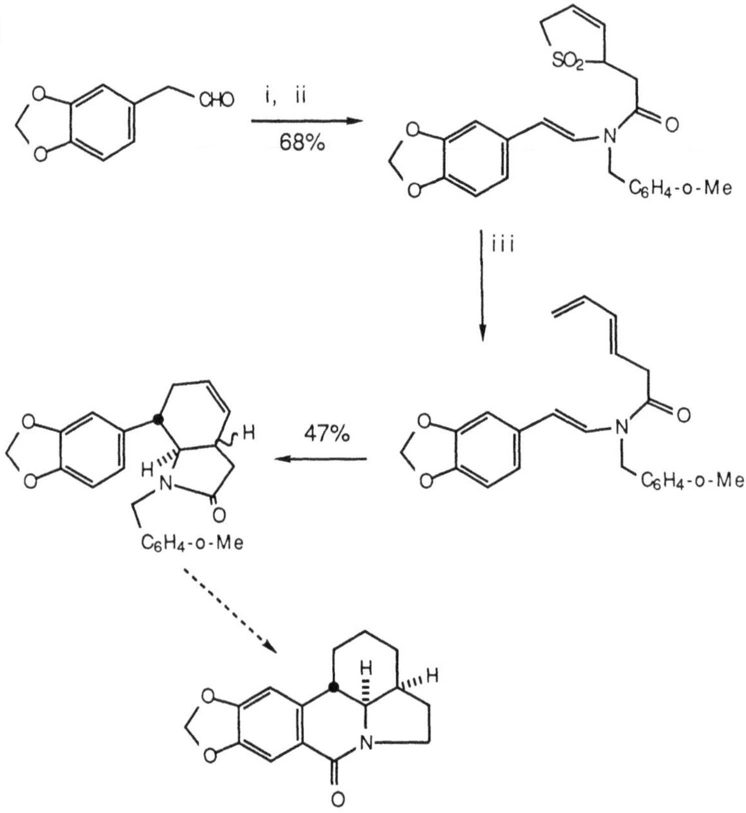

f. imine acylation

i. p-MeOCH₂NH₂, PhMe, MgSO₄ at 0°C   ii. 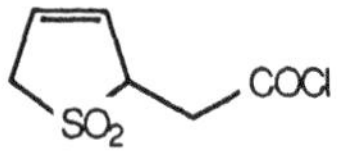   , PhNEt₂, -78°C to 25°C
iii. xylene, heat

From S. F. Martin and C. Y. Tu, *J. Org. Chem.*, **46**, 3763 (1981).

32

f. condensation

i. butadiene, heat   ii. Pd/C, H_2

From Y. Tsuda, Y. Sakai, N. Kashiwaaba, T. Sano, J. Toda, and
K. Isobe, *Heterocycles*, *16*, 189 (1981).

33

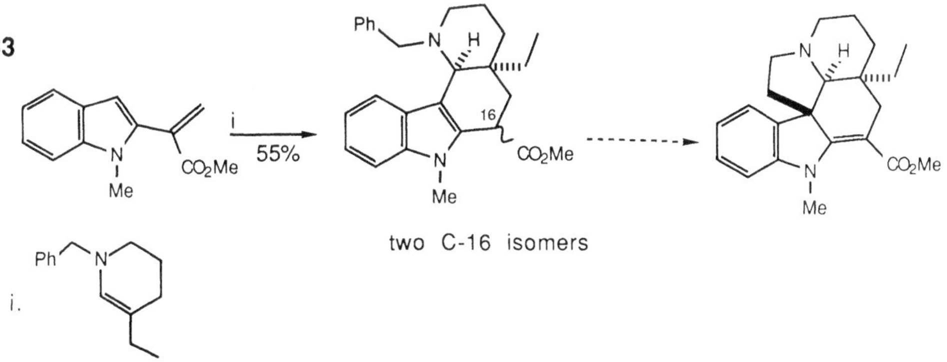

two C-16 isomers

i.

From F. E. Ziegler, E. B. Spitzner, *J. Am. Chem. Soc.*, *92*, 3492
(1970).

34

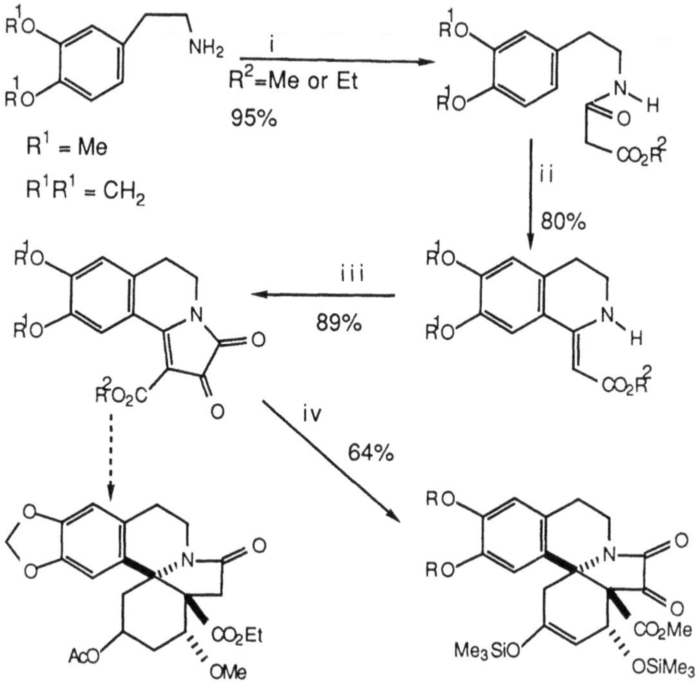

f. imine acylation

From T. Sano, J. Toda, N. Kashiwaba, Y. Tsuda, and Y. Iitaka, *Heterocycles*, *16*, 1151 (1981).

35

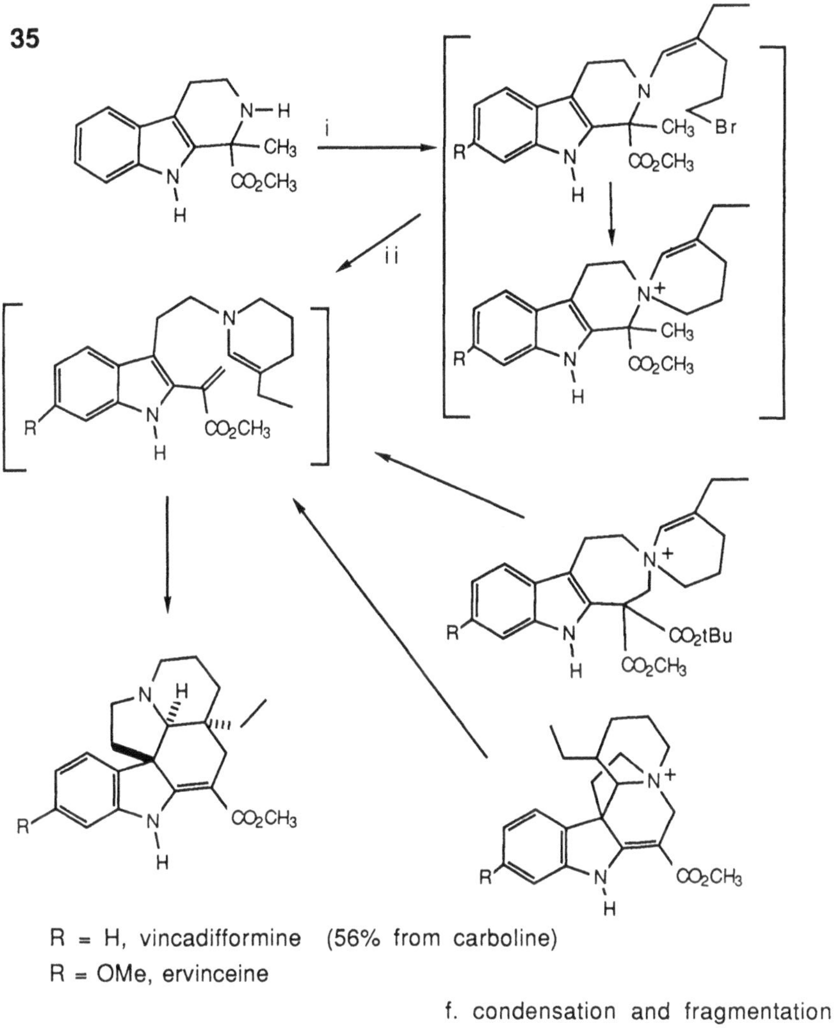

N—H
CH3
CO2CH3
H
i
ii
R
N
CH3
Br
CO2CH3
N+
CO2tBu
R = H, vincadifformine (56% from carboline)
R = OMe, ervinceine
f. condensation and fragmentation

i. $BrCH(Et)(CH_2)_3CHO$, TsOH, PhMe, reflux ii. DBU, heat

From M. E. Kuehne, D. M. Roland, and R. Hafter, *J. Org. Chem.*, *43*, 3705 (1978), M. E. Kuehne, T. H. Matsko, J. C. Bohnert, and C. L. Kirkemo, *J. Org. Chem.*, *44*, 1063 (1979), and M. E. Kuehne, J. A. Huebner, and T. H. Matsko, *J. Org. Chem.*, *44*, 2477 (1979).

36

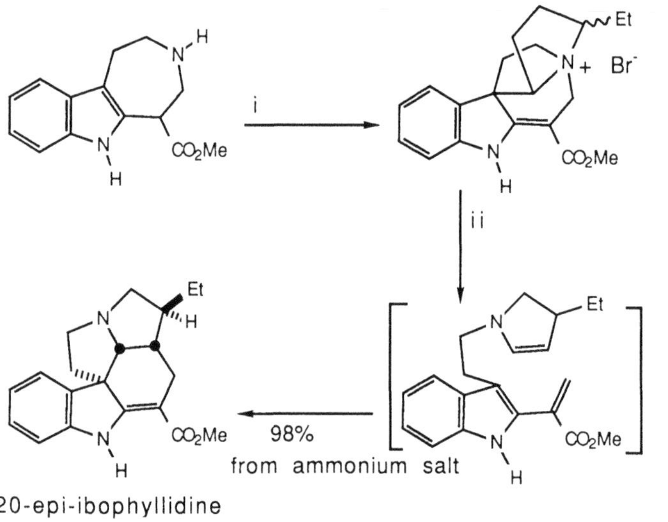

N-H
CO2Me
N
H
i
Et
N+ Br-
ii
Et
98%
from ammonium salt
20-epi-ibophyllidine

f. fragmentation

i. 4-bromohexanal, THF, 60°C ii. NEt_3, MeOH

From M. E. Kuehne and J. C. Bohnert, *J. Org. Chem.*, *46*, 3443 (1981) and M. E. Kuehne, F. J. Okuniewicz, C. L. Kirkemo, and J. C. Bohnert, *J. Org. Chem.*, *47*, 1335 (1982).

37

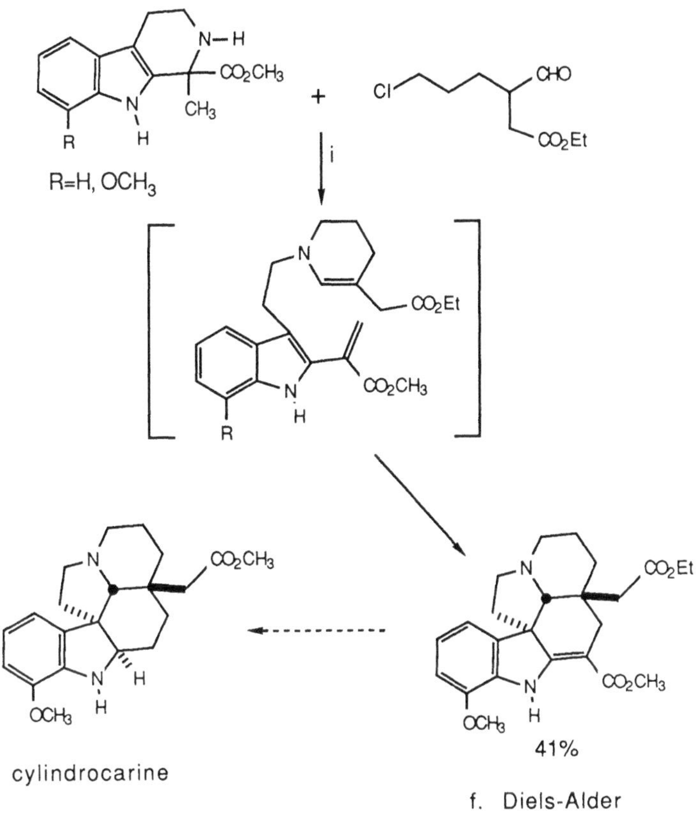

cylindrocarine

41%

f. Diels-Alder

i. 110°C, toluene, 110h, TosOH

From J. P. Brennan and J. E. Saxton, *Tetrahedron*, *42*, 6719 (1986).

38

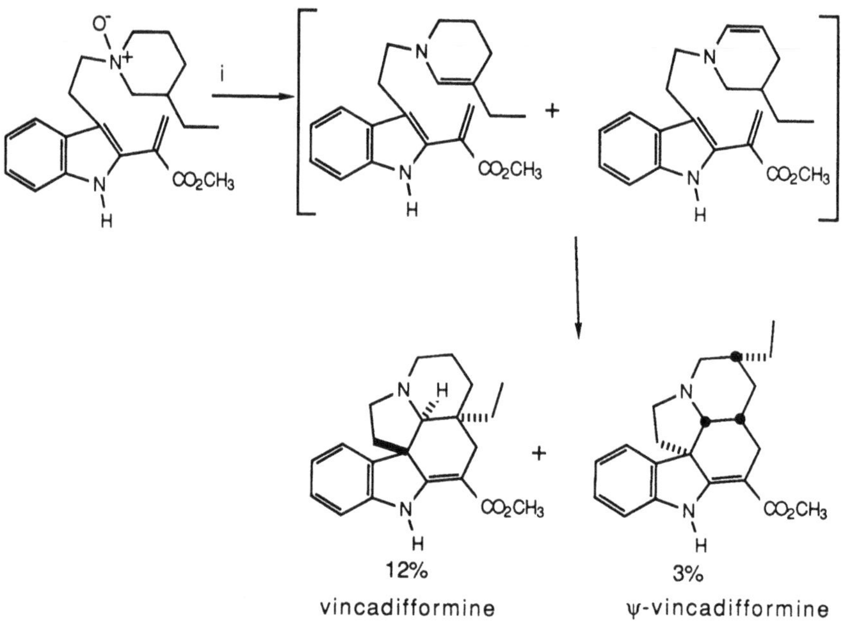

12%

vincadifformine

3%

ψ-vincadifformine

f. N-oxide elimination

i. Ac₂O, Py, RT, 1h

From Gy. Kalaus, M. Kiss, M. Kajtar-Peredy, J. Brlick, L. Szabo, and Cs. Szantay, *Heterocycles*, *23*, 2783 (1985).

39

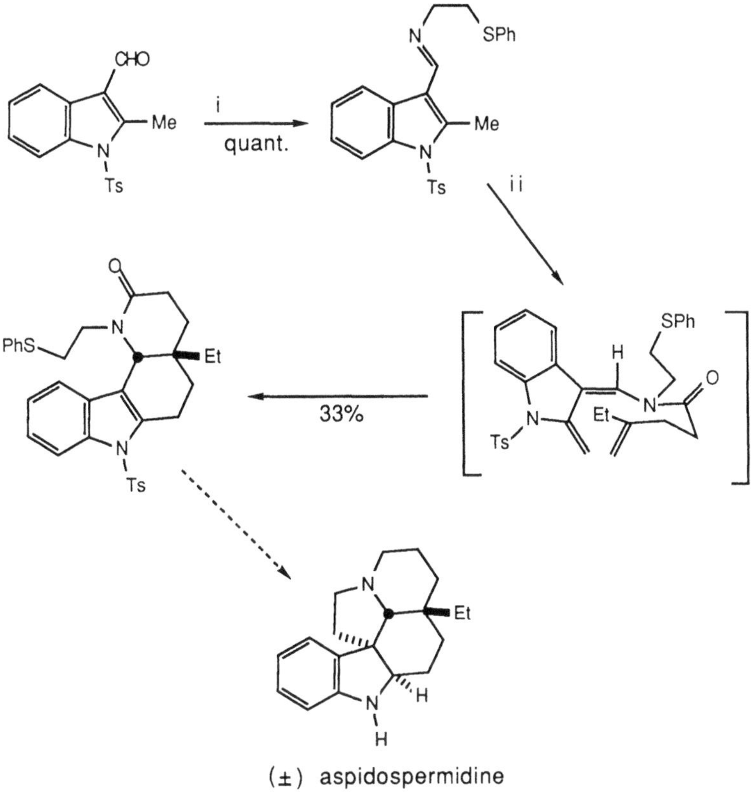

(±) aspidospermidine

f. imine acylation

i. PhSCH₂CH₂NH₂ ii. [structure: Et...O...O...OEt with CH₂] , PhCl, 140°C

i. $PhSCH_2CH_2NH_2$ ii.

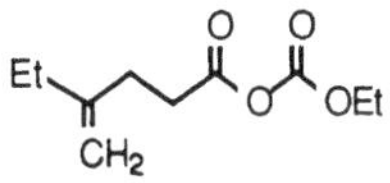

From T. Gallagher, P. Magnus, and J. C. Huffman, *J. Am. Chem. Soc.*, *104*, 1140 (1982).

40

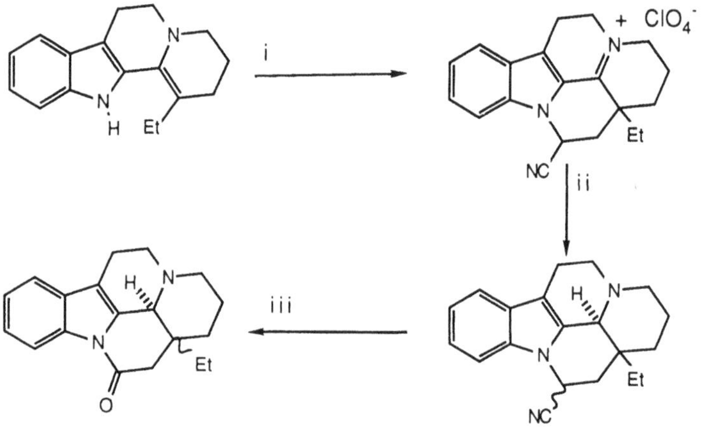

eburnamonine, 20   -Et
epieburnamonine, 20ß-Et

i. CH₂=C(Cl)CN, CH₂Cl₂ ii. Zn-EtOH,HCl iii. C₆H₅NPr'Li-THF, HMPA-O₂, -78°C

From A. Buzas, C. Herisson, and C. Lavielle, *Compt. rend.*, 283, C, 763 (1976).

41

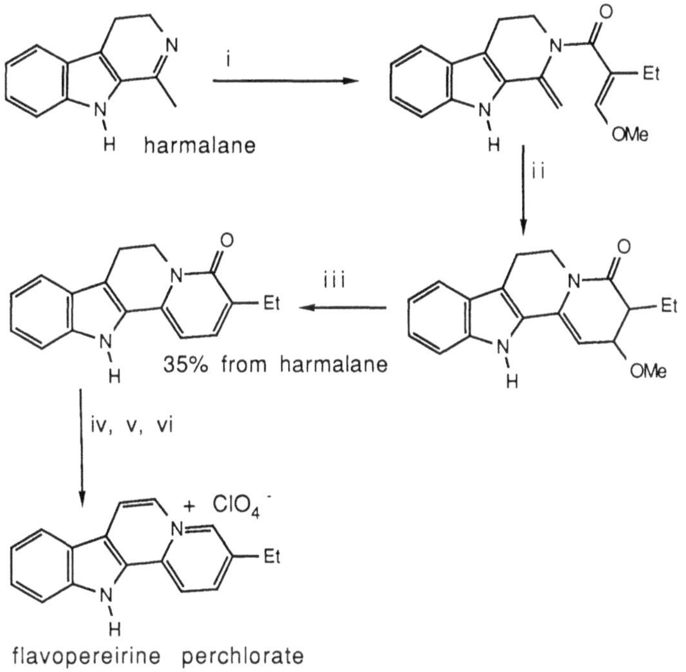

harmalane

35% from harmalane

iv, v, vi

$+ ClO_4^-$

flavopereirine perchlorate

i. MeOCH=C(Et)COCl ii. hv, PhH, RT iii. 10% HCl-H₂O iv. LAH v. Pd/C, 280-300°C vi. HClO₄

2.

harmalane $\xrightarrow{\text{i}}$ $\xrightarrow{\text{ii}}$ yohimbane
epiyohimbane
alloyohimbane

27-35% combined yields
from harmalane

i.

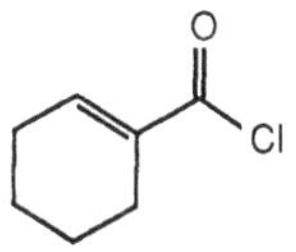

ii. similar steps as above

From I. Ninomiya, Y. Tada, T. Kiguchi, O. Yamamoto, and T. Naito, *Heterocycles*, **9**, 1527 (1978).

42

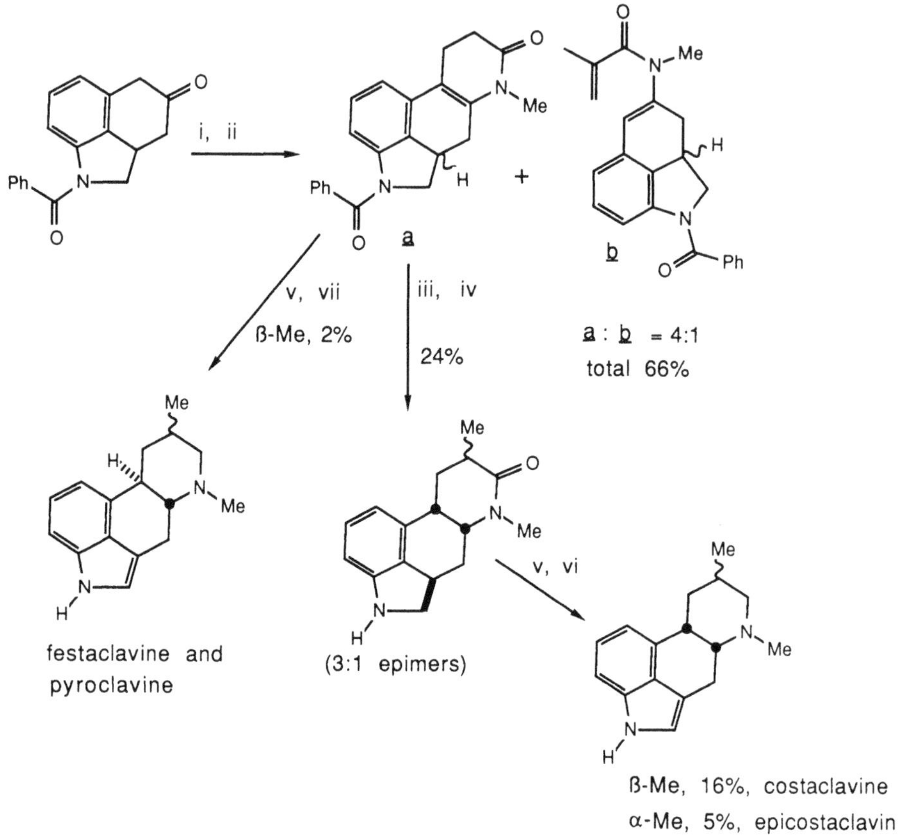

i, ii
Ph
N
O
Me
H
+
a
b
v, vii
ß-Me, 2%
iii, iv
24%
a : b = 4:1
total 66%
festaclavine and
pyroclavine
(3:1 epimers)
v, vi
ß-Me, 16%, costaclavine
α-Me, 5%, epicostaclavin

i. $MeNH_2$ ii. CH_2=C(Me)COCl iii. H_2-PtO_2 iv. HCl v. LAH
vi. MnO_2 vii. Na-NH_3

From I. Ninomiya, J. Kiguchi, *J. Chem. Soc., Chem. Comm.*, 624 (1976).

43

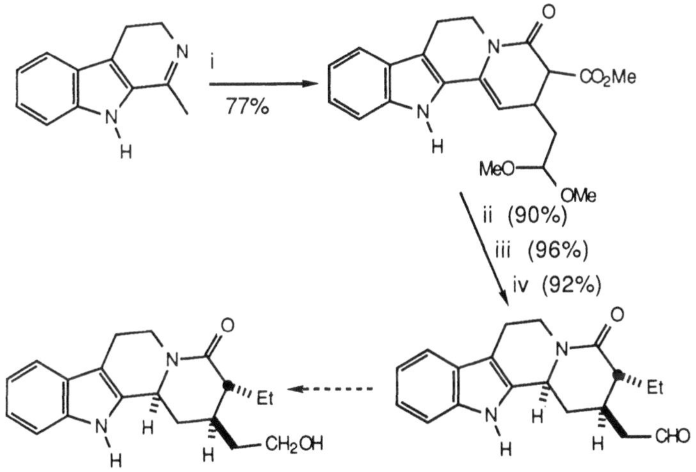

(±)-dihydrocorynantheol

f. imine acylation

i.

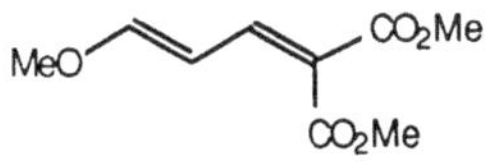

MeO—⟍⟋⟍⟋⟍—CO₂Me with CO₂Me , MeOH, 3 days , RT, then heat 24h ii. NaH, EtI, DMF

iii. H₂, PtO₂, MeOH iv. NaOMe, H₂O, MeOH

From T. Kametani, N. Kanaya, H. Hino, S. P. Huang, and M. Ihara, *Heterocycles*, *14*, 1771 (1980).

44

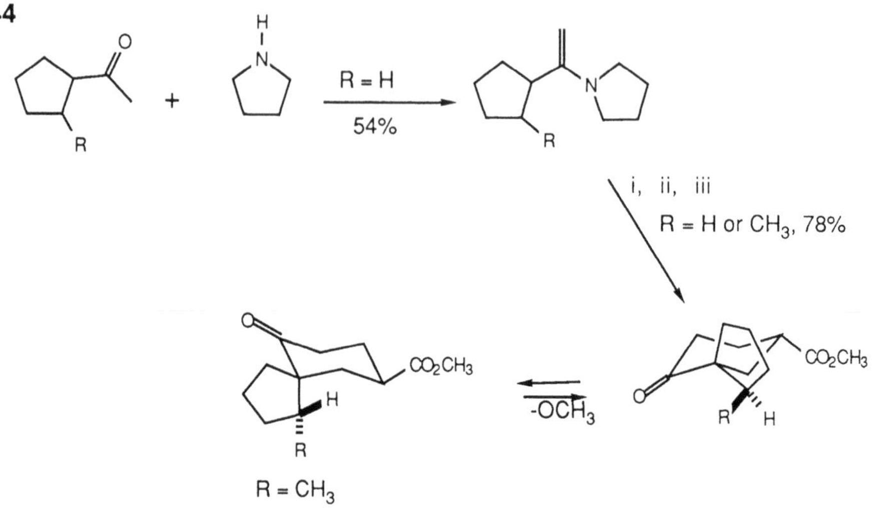

i, ii, iii

R = H or CH₃, 78%

R = CH₃

f. condensation

i. 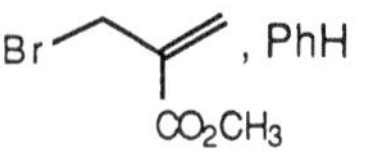 Br—⟍, PhH ii. Et₃N, CH₃CN, reflux iii. aq. AcOH
   CO₂CH₃

45

ii

30%

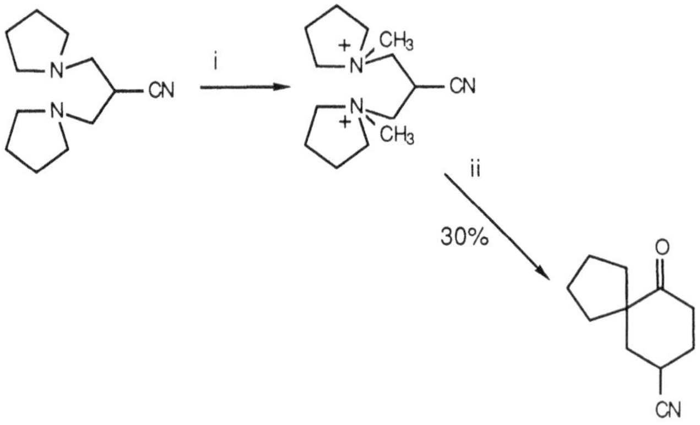

f. condensation

i. CH₃SO₃C₆H₅CH₃ ii. 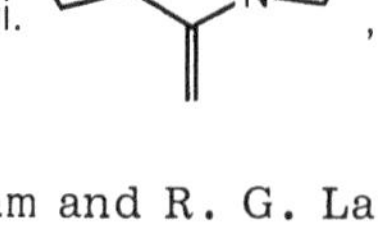 [structure], CH₃CN

From D. J. Dunham and R. G. Lawton, J. Am. Chem. Soc., 93, 2074; 2075 (1971).

46

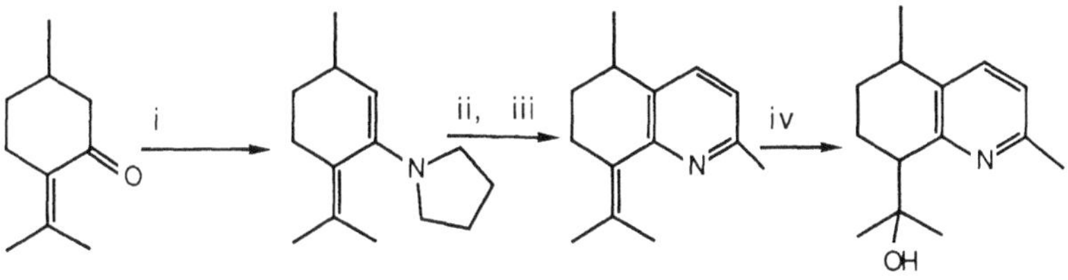

i. pyrrolidine ii. MeCOCH=CH$_2$ iii. NH$_2$OH iv. H$^+$, H$_2$O

From P. Teisseire, B. Shimizu, M. Plattier, B. Corbier, and P. Rouillier, *Recherches*, 19, 241 (CA 1975, 83, 28395v) (1974).

47

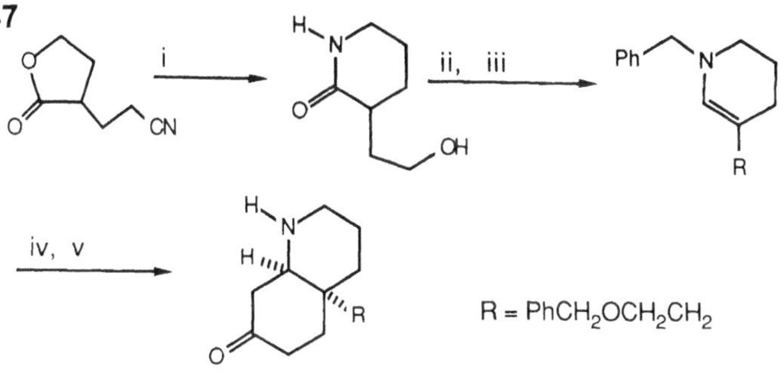

R = PhCH$_2$OCH$_2$CH$_2$

f. reduction

i. H$_2$-Ni ii. PhCH$_2$Br, MeSOCH$_2$K-DMSO iii. DIBAL iv. MVK , glycol, heat
v. H$_2$-Pd/C

From R. V. Stevens, R. K. Mehra, and R. L. Zimmerman, *J. Chem. Soc., Chem. Commun.*, 877 (1969).

48

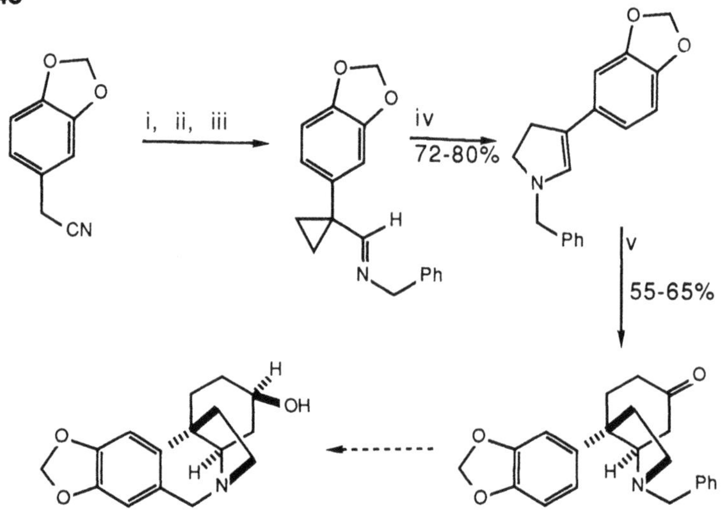

i. (CH₂Br)₂/LiNH₂ ii. i-Bu₂AlH iii. PhCH₂NH₂, CaCl₂, PhH
iv. NH₄Cl, heat v. MVK

From R. V. Stevens and L. E. Du Pree, jun., *J. Chem. Soc., Chem. Comm.* 1585 (1970).

49

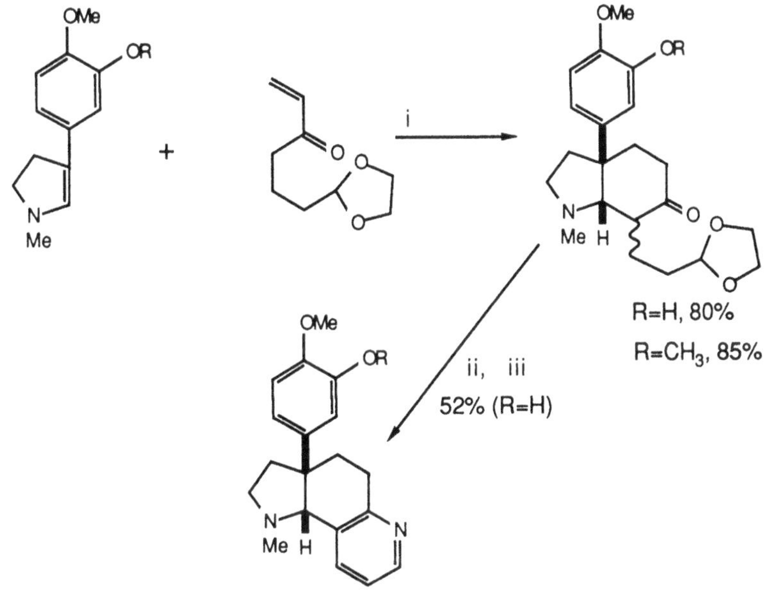

i. MeCN, reflux   ii. PTSA-dioxane, reflux   iii. NH₂OH.HCl, abs EtOH, reflux

From C. P. Forbes, G. L. Wenteler, and A. Wiechers, *Tetrahedron* 34, 487 (1978).

50

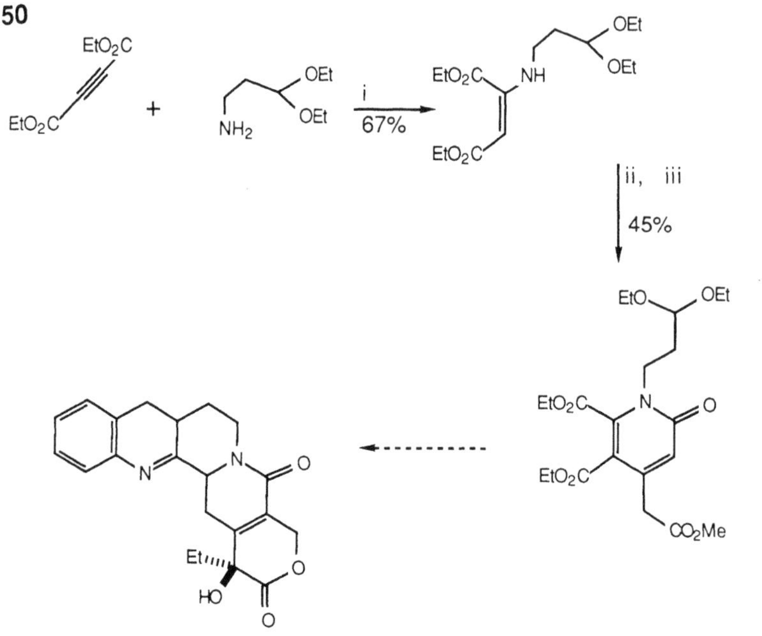

(±)-camptothecin

i. ether ii. MeO₂CCH=C=CHCO₂Me iii. MeOH, Et₃N, RT

From R. Volkmann, S. Danishefsky, J. Eggler, and D. M. Solomon, *J. Am. Chem. Soc.*, **93**, 5576 (1971).

51

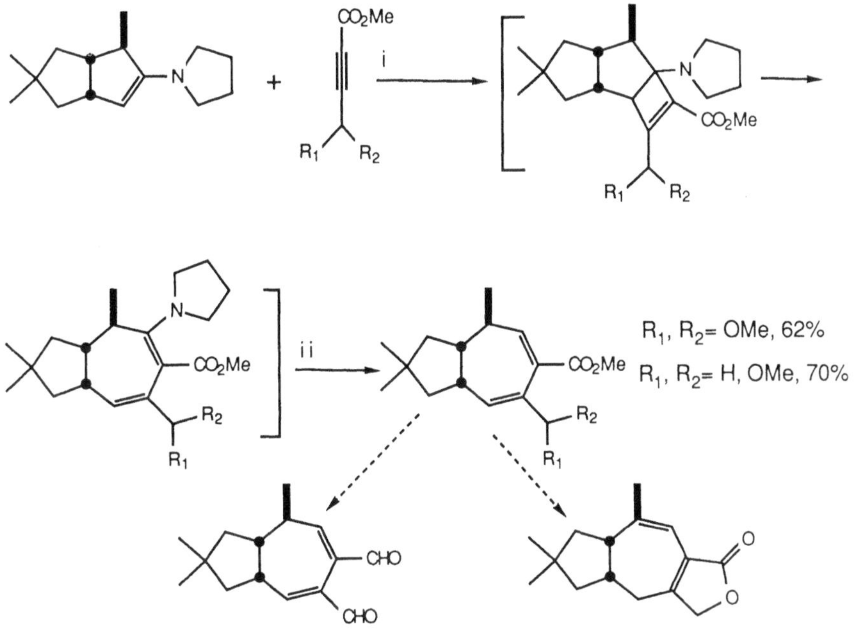

R₁, R₂= OMe, 62%
R₁, R₂= H, OMe, 70%

±-velleral

±-pyrovellorolactone

f. condensation

i. PhCH₃, heat ii. BH₃.THF

From J. Froborg and G. Magnusson, *J. Am. Chem. Soc.*, *100* 6728 (1978).

52

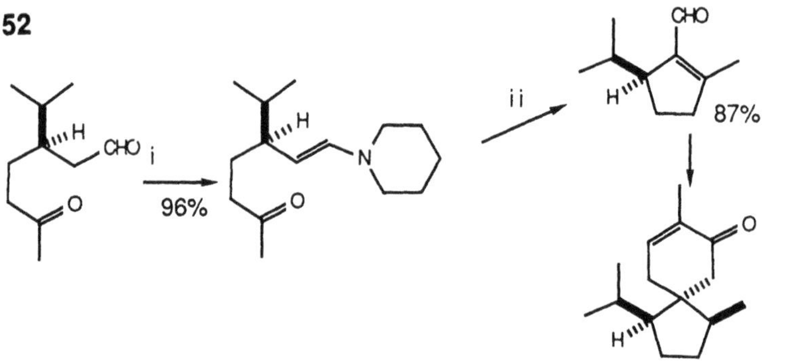

acorenone B

f. condensation

i. piperidone, ether, 0°C   ii. AcOH, ether, reflux, 0.5h

From J. D. White, *J. Am. Chem. Soc.*, *103*, 1813 (1981).

53

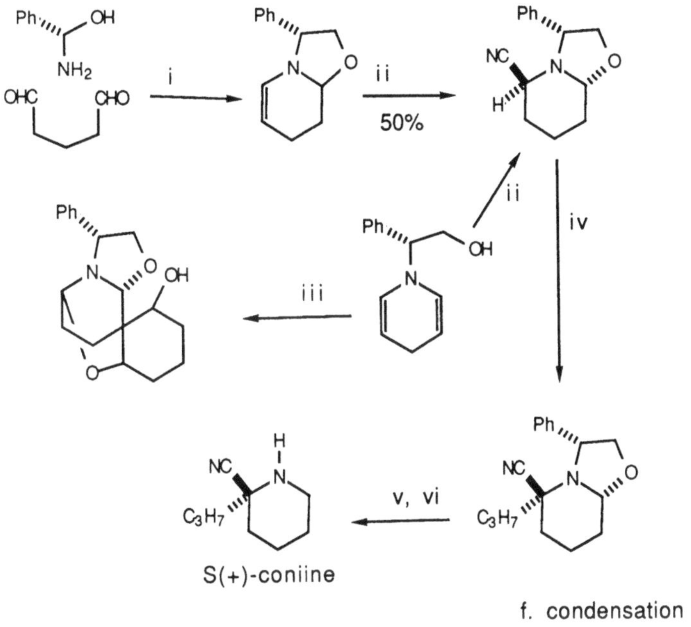

f. condensation

i. pH=3, citric acid ii. KCN iii. OHC(CH₂)₃CHO iv. LDA, C₃H₇Br
v. NBH, MeOH vi. H₂/Pd

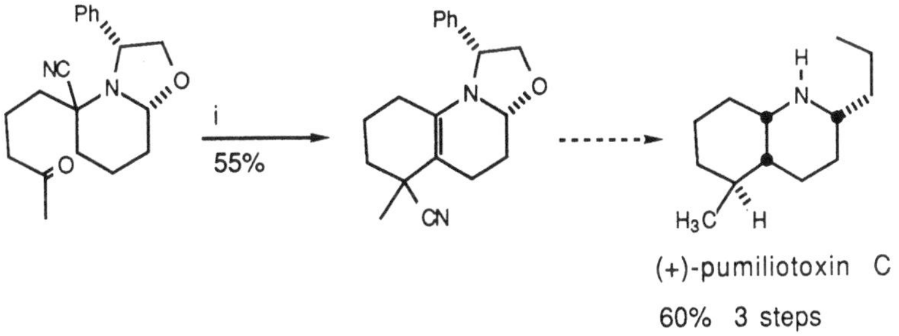

(+)-pumiliotoxin C
60% 3 steps

i. Al₂O₃, CH₂Cl₂

From H. P. Husson, *J. Nat. Prod.*, **48**, 894 (1985).

54

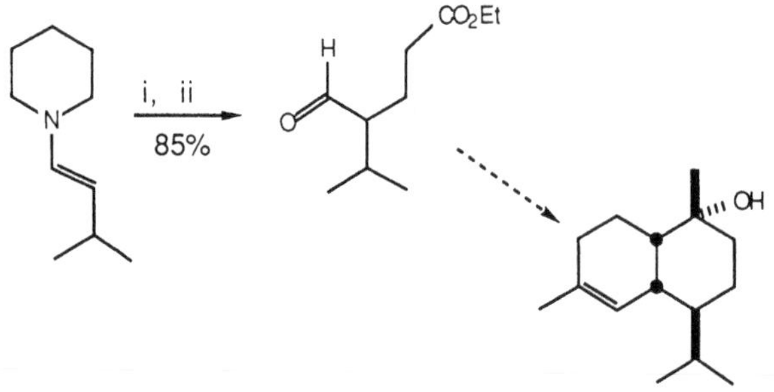

(±)-torreyol

i. ethylacrylate   ii. H_3O^+

From D. F. Taber and B. P. Gunn, *J. Am. Chem. Soc.*, *101*, 3992 (1979).

55

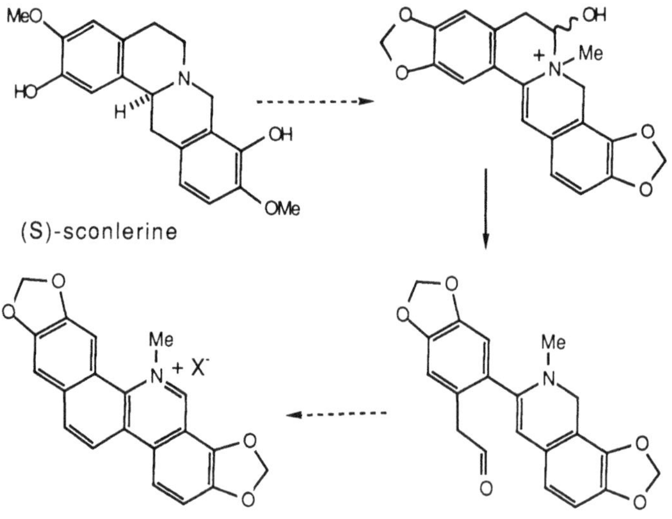

(S)-sconlerine

sanguinarine

From A. R. Battersby, R. J. Francis, M. Hirst, E. A. Ruveda, and J. Stannton, *J. Chem. Soc.*, *Perkin I*, 1140 (1975) and from A. R. Battersby, J. Stannton, H. R. Wiltshire, R. J. Francis, and R. Sonthgate, *J. Chem. Soc.*, *Perkin I*, 1147 (1975).

56

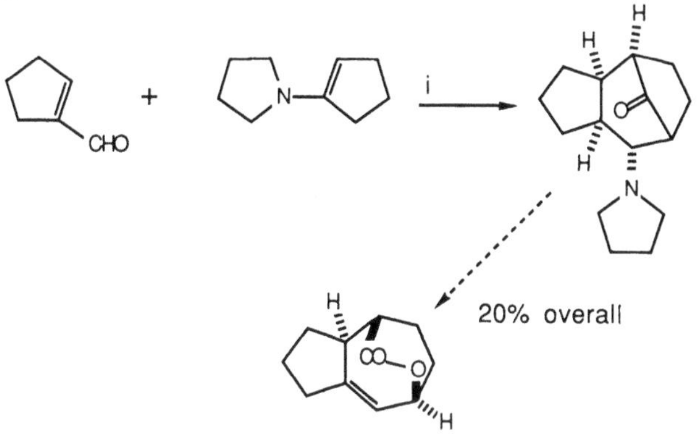

20% overall

i. dioxane, RT, 20h

From J. B. Hendrickson and R. K. Boeckman Jr., *J. Am. Chem. Soc.*, *93*, 1307 (1971).

57

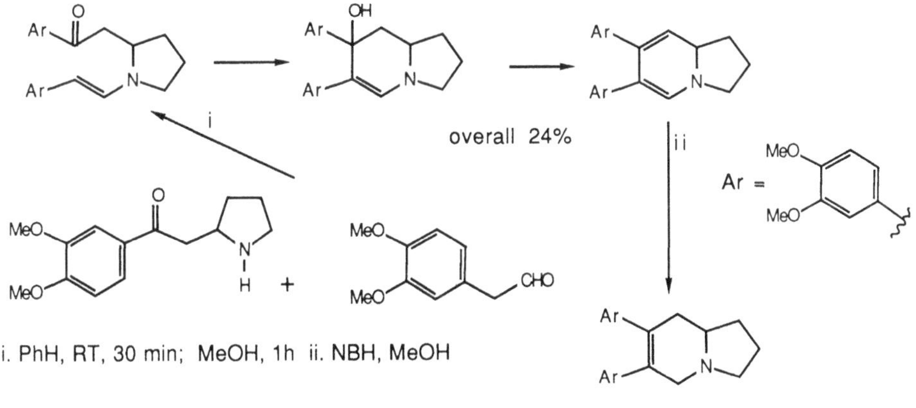

overall 24%

Ar =

septicine

i. PhH, RT, 30 min; MeOH, 1h ii. NBH, MeOH

From R. B. Herbert, F. B. Jackson, and I. T. Nicolson, *J. Chem. Soc.*, *Chem. Commun.*, 450 (1976).

58

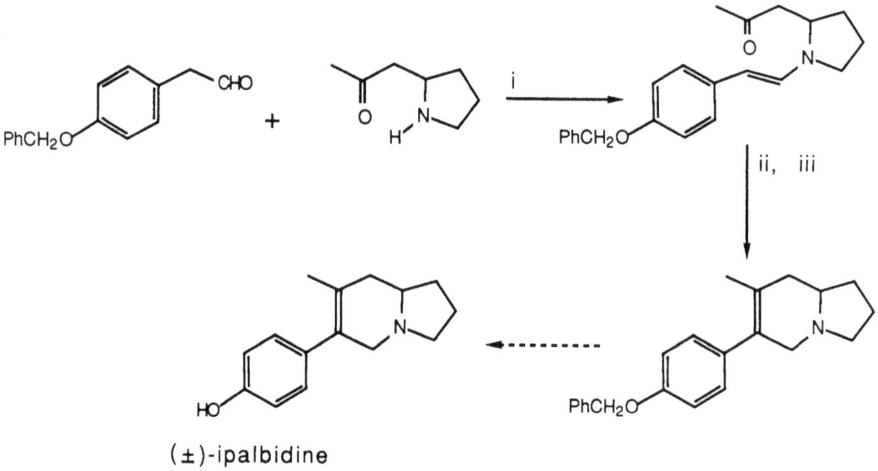

(±)-ipalbidine

f. condensation

i. PhH, RT ii. MeOH iii. NBH, Pr^{i}OH

From S. H. Hedges and R. B. Herbert, *J. Chem. Res.* (M), 413 (1979).

59

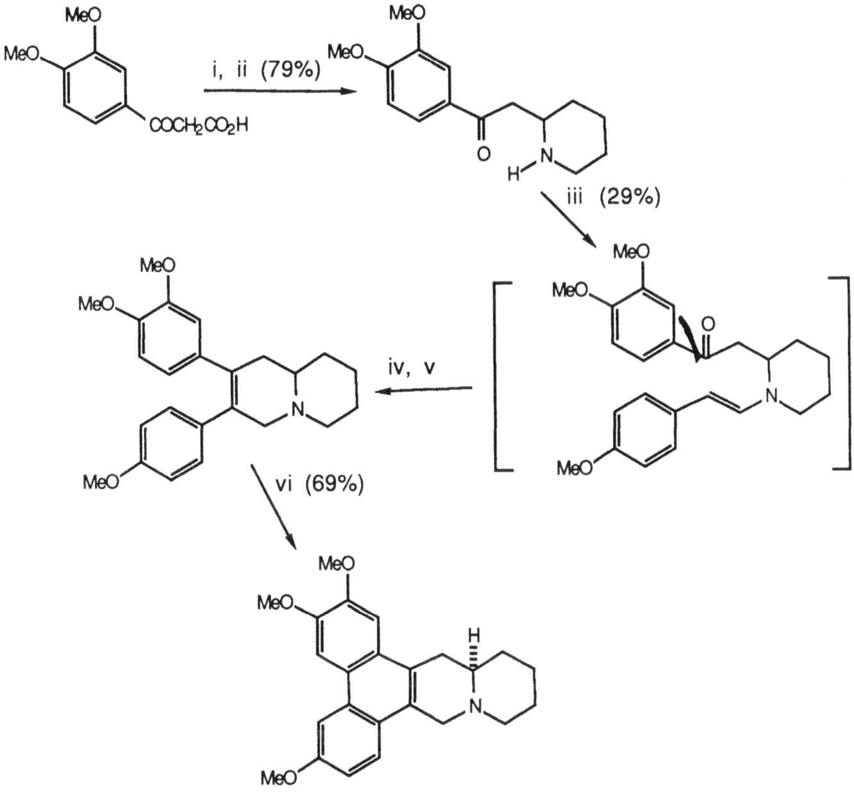

(±)-cryptopleurine

f. condensation

i. cadaverine   ii. diamine oxidase, pH=7   iii. p-MeOC$_6$H$_4$CH$_2$CHO
iv. SnCl$_4$, TiCl$_4$, PhH, MgI$_2$, Et$_2$O, v. NBH, Pr^{i}OH
vi. (F$_3$CCO$_2$)$_3$Tl, F$_3$CCO$_2$H

From R. B. Herbert, *J. Chem. Soc., Chem. Commun.*, 794 (1978).

60

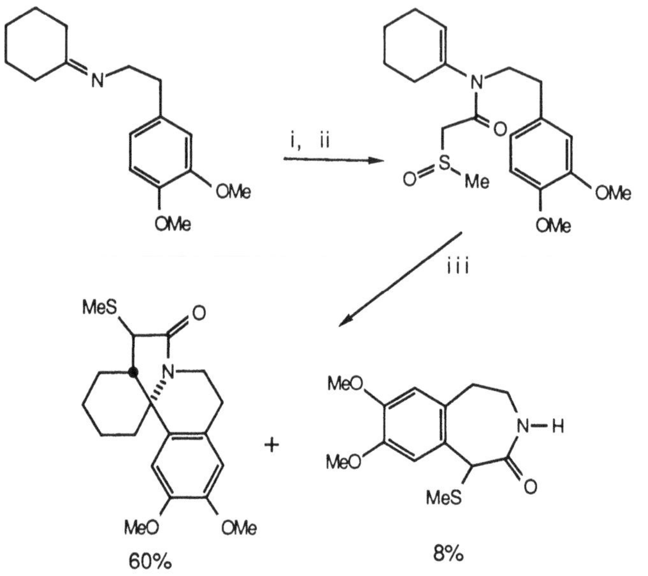

60%

8%

f. imine acylation

i. (MeSCH$_2$CO)$_2$O, Py   ii. NaIO$_4$ iii. p-TsOH (2 eq), CH$_2$Cl$_2$ boiling

From Y. Tamura, H. Maeda, S. Akai, and H. Ishibashi, *Tet. Lett.*, 23, 2209 (1982).

61

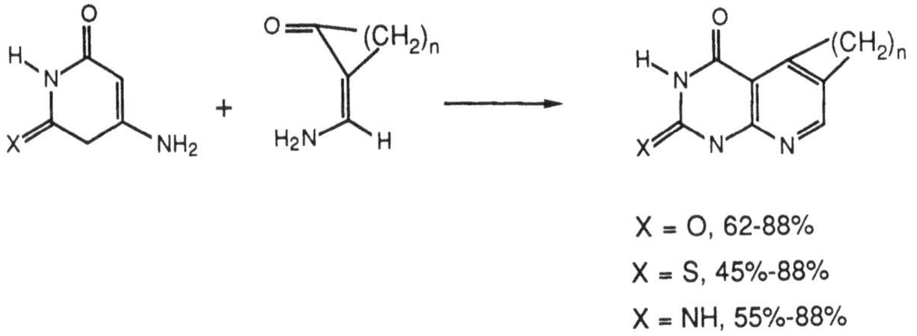

X = O, 62-88%

X = S, 45%-88%

X = NH, 55%-88%

f. condensation

From E. Stark and E. Breitmaier, *Tetrahedron 29*, 2209 (1973).

62

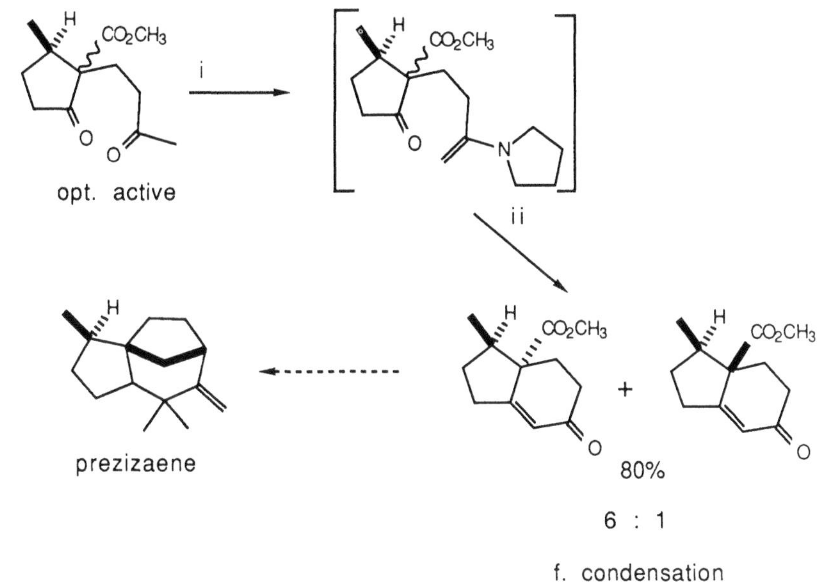
CO2CH3
H
i
O
opt. active
ii
prezizaene
80%
6 : 1
f. condensation

i. pyrrolidine, toluene,heat ii. AcOH/ NaOAc, H_2O

From R. M. Coates, J. Org. Chem., *45*, 5430 (1980).

63

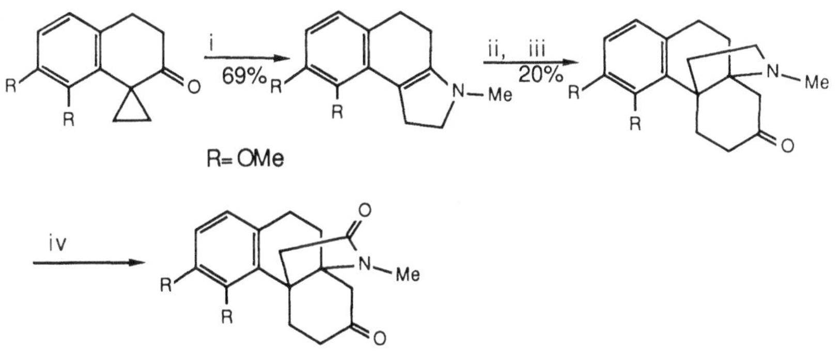
i
69%
ii, iii
20%
N—Me
R
R=OMe
iv
O

i. $MeNH_2$, PhH, 110°C ii. CO , RT to 40°C, iii. AcOH, 40°C iv. $KMnO_4$

From S. L.Keeley, Jun., A. J. Martinez, and F. C. Tahk, *Tetrahedron*, *26*, 4729 (1970).

64

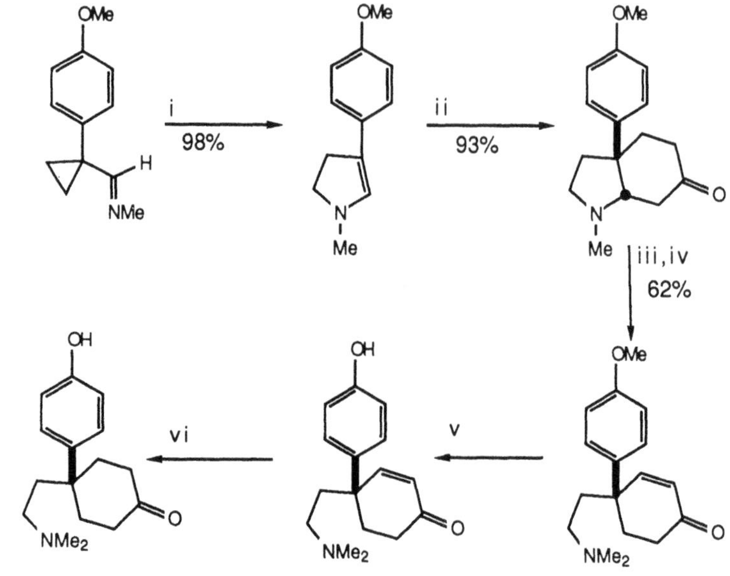

(±)-dihydrojoubertiamine

f. cyclopropylamine

i. NH₄Cl, 140°C, N₂ ii. MVK, MeCN, reflux iii. MeI, reflux

iv. 0.5 KOH v. 48% HBr vi. H₂/Pd-C, MeOH

From R. V. Stevens and J. T. Lai, *J. Org. Che.*, *37*, 2138 (1972).

65

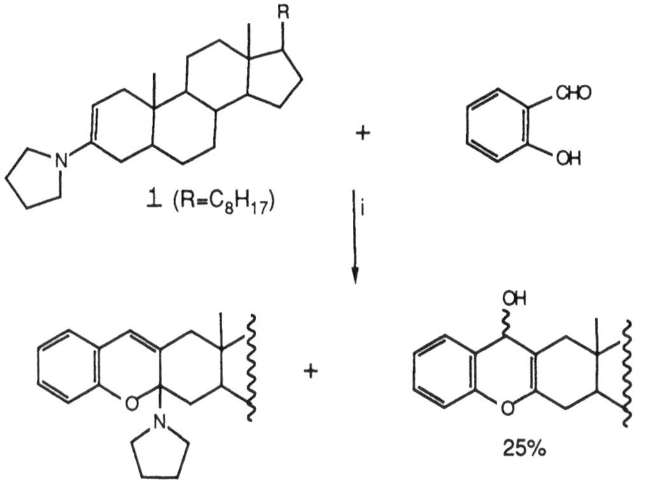

1 (R=C₈H₁₇)

+

i

70%

cholestane

+

25%

4'-hydroxy-4'H-1'-benzopyrano
(3',2':2,3)-cholestane

i. PhH, reflux, 3 days

From M. S. Manhas and J. R. McRoy, *J. Chem. Soc.*, 1419 (1969).

66

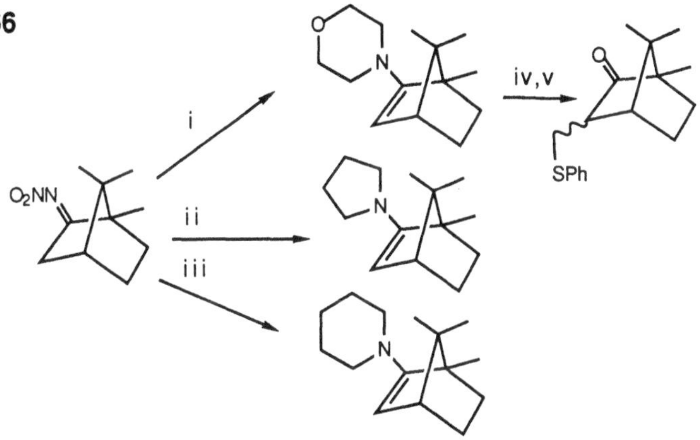

iv,v

i

ii

iii

f. condensation

i. morpholine, PhH ii. Pyrrolidine, PhH iii. piperidine, PhH
iv. PhSCH₂NMe₂ v. HCl, EtOH

From F. Bondavalli, P. Schenone, and A. Ranise, *Synthesis*, 830 (1981).

67

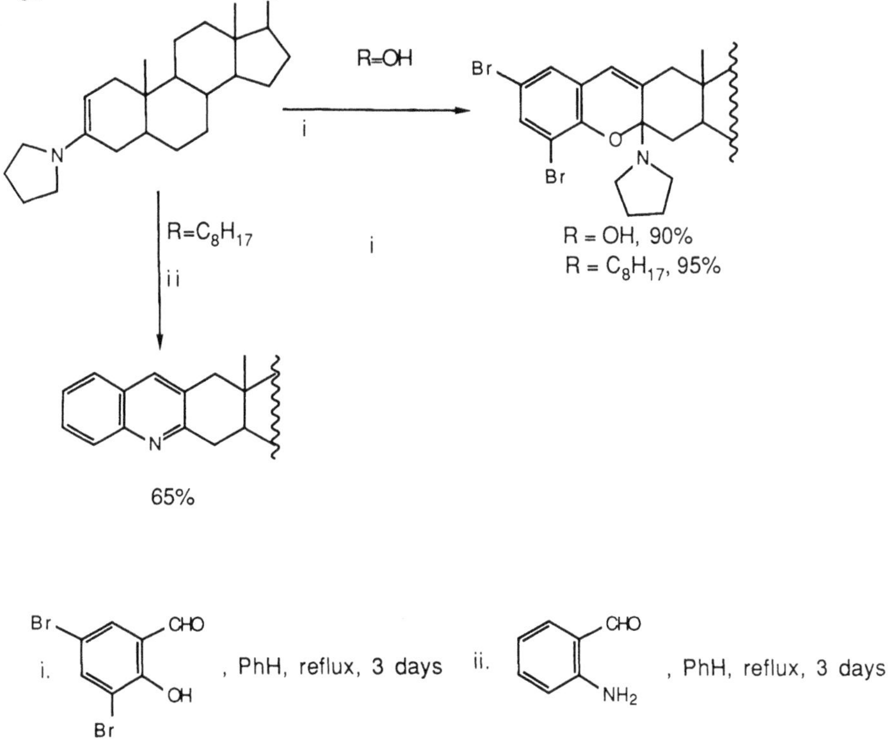

R=OH

R = OH, 90%
R = C_8H_{17}, 95%

R=C_8H_{17}

i

ii

65%

i. , PhH, reflux, 3 days   ii. , PhH, reflux, 3 days

From M. S. Manhas and J. R. McRoy, *J. Chem. Soc.*, 1419 (1969).

68

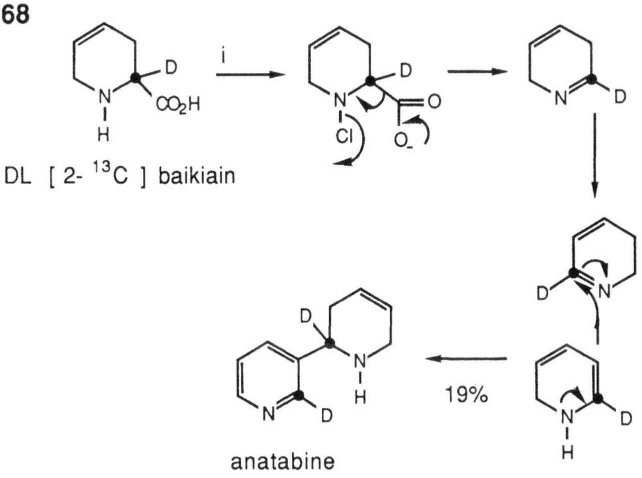

DL [2- ^{13}C] baikiain

anatabine

19%

→ indicates labelled carbon

i. NaOCl, pH=10

From E. Leete and M. E. Mueller, *J. Am. Chem. Soc.*, *104*, 6440 (1982).

69

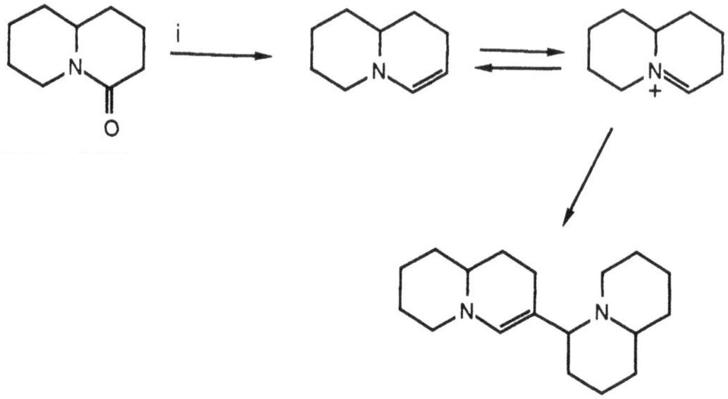

i. LAH

f. reduction of lactam

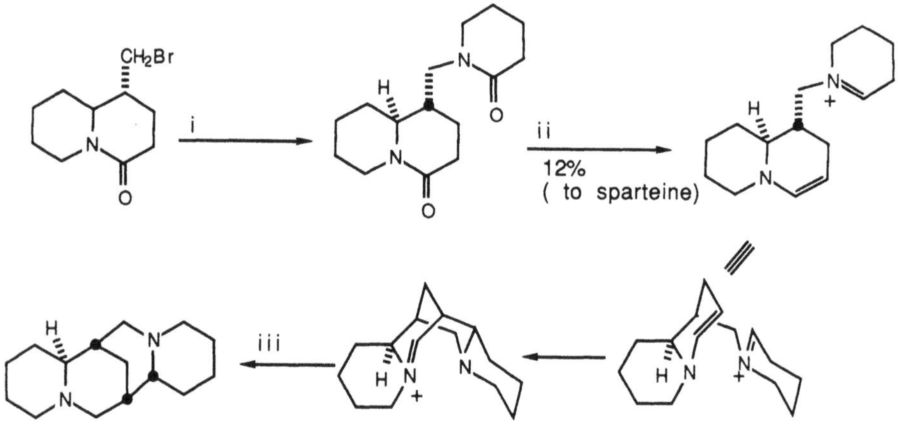

12%
(to sparteine)

sparteine

f reduction of lactam

i. 2-pipyridone, base cat. ii. DIBAL iii. NBH

From F. Bohlmann, H. J. Müller, and D. Schumann, *Chem. Ber.*, *106*, 3026 (1973).

70

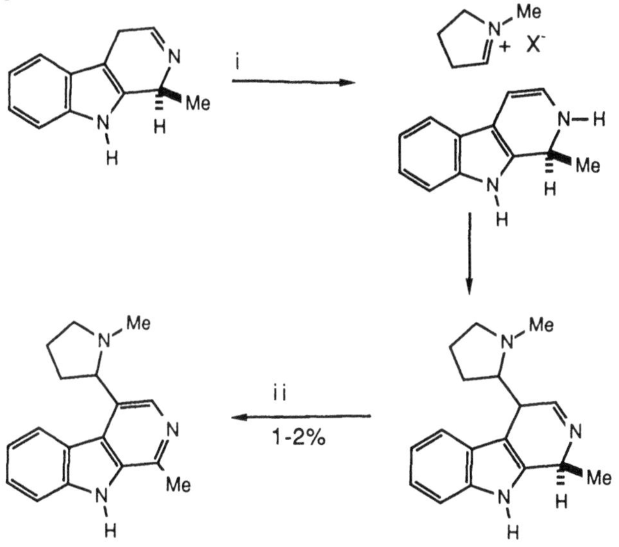

brevicolline

i. tautomeric shift   ii. autoxidn.

From E. Leete, *J. Chem. Soc., Chem. Commun.* 821 (1979).

71

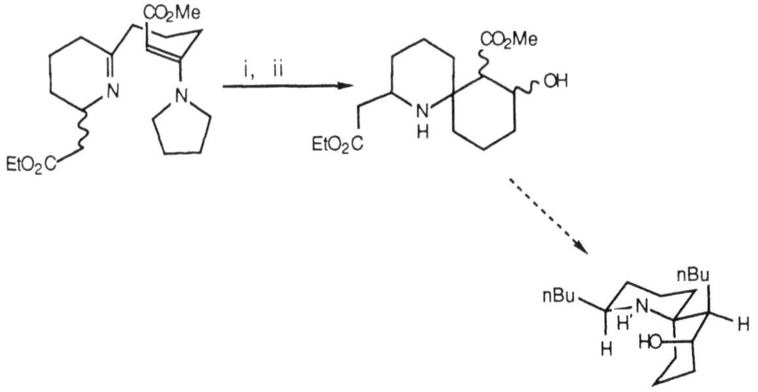

2,7-epiperhydrohistrionicotoxin

i. PTSA, H_2O, 0°C, 1h   ii. NBH

From E. J. Corey, Y. Uyeda, and R. A. Ruden, *Tet. Letters.* 4347 (1975).

72

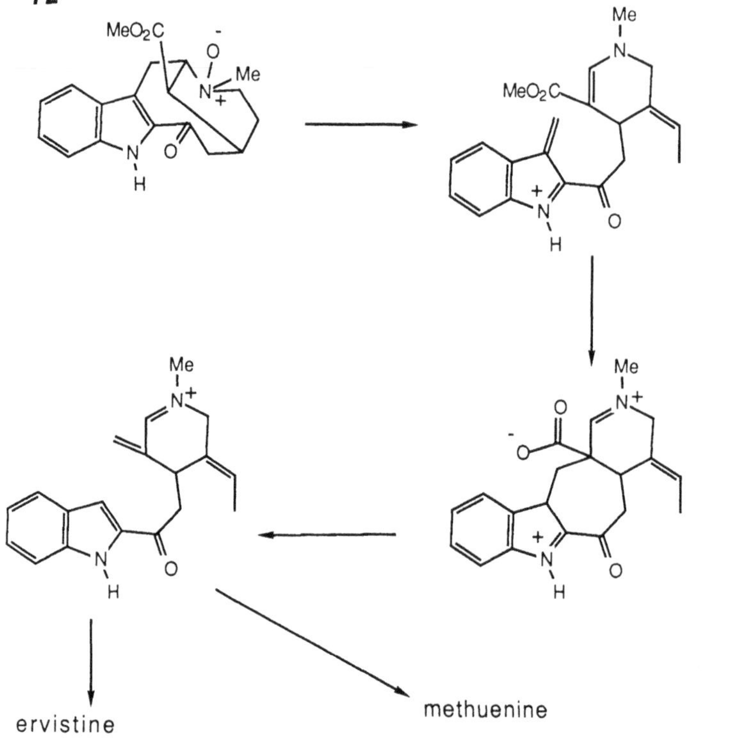

ervistine

methuenine

f. fragmentation

From M. Andrierntsiferana, R. Besselievre, C. Riche, and H. P.
Husson, *Tet. Letters*, 2587 (1977).

73

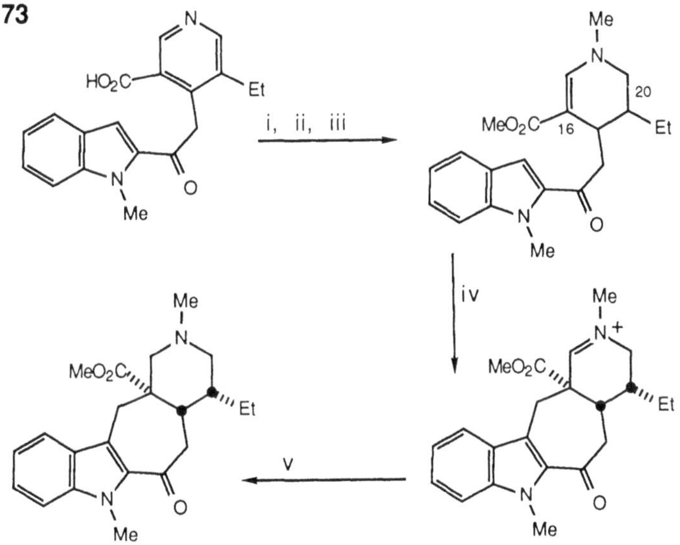

N$_a$-methyl-16,20-epi-ervatamine

yields 10% from enamine

f. reduction

i. CH$_2$N$_2$ ii. MeI iii. H$_2$, Pd/C, pH=6.8 iv. Me$_2$N$^+$=CH$_2$ CF$_3$CO$_2^-$
v. NaBH$_3$CN

From H. P. Husson, K. Bannal, R. Freire, B. Mompon, and F. A. M. Reis, *Tetrahedron, 34,* 1363 (1978).

74

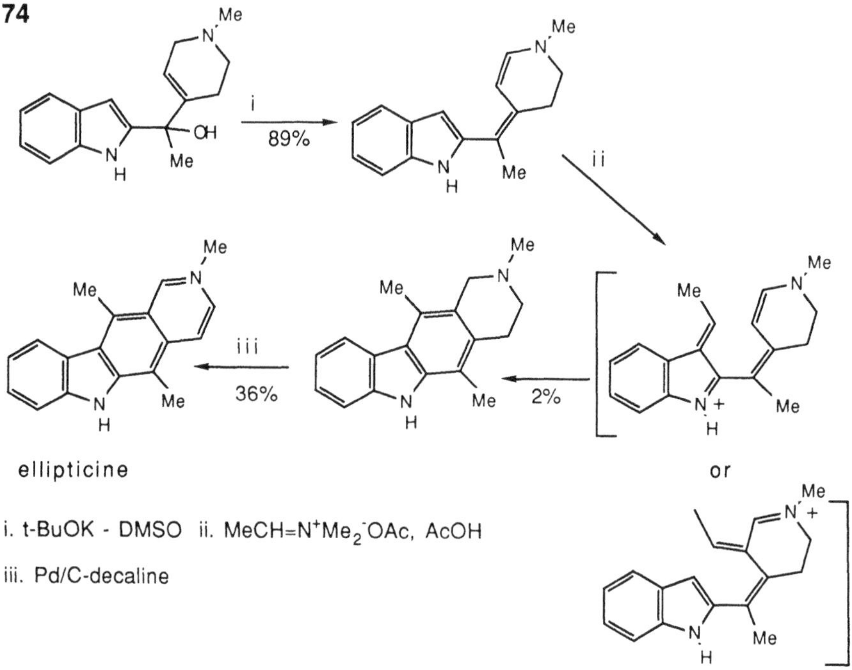

ellipticine

i. t-BuOK - DMSO ii. MeCH=N⁺Me₂⁻OAc, AcOH

iii. Pd/C-decaline

From R. Besselievre, C. Thal, H. P. Husson, and P. Potier, *J. Chem. Soc.*, *Chem. Commun.*, 90 (1975).

III. β-ARYLATION REACTIONS, INCLUDING PHOTOCYCLIZATIONS: 75—95

75

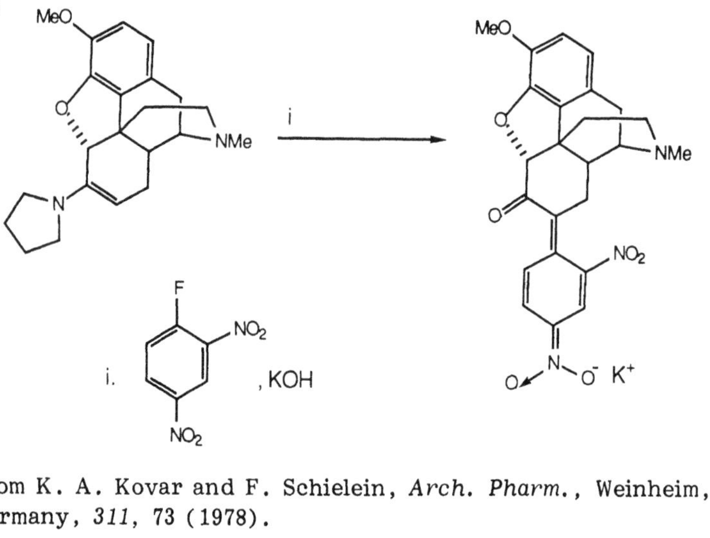

From K. A. Kovar and F. Schielein, *Arch. Pharm.*, Weinheim, Germany, *311*, 73 (1978).

76

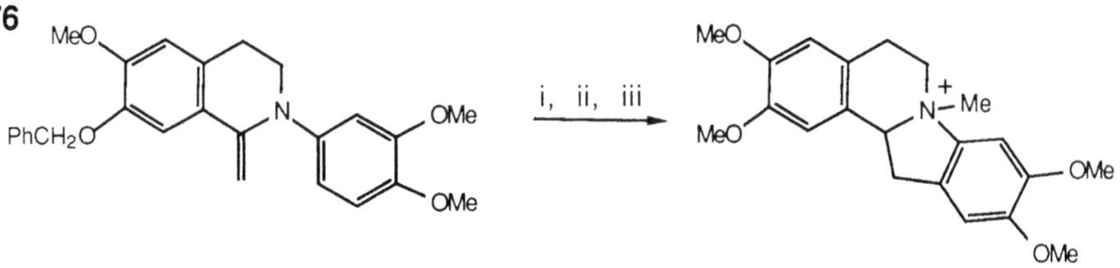

i, ii, iii

i. hv   ii. redn.   iii. (N,O)-methylation

From I. Ninomiya, J. Yasui, and T. Kiguchi, *Heterocycles*, *6*, 1855 (1977).

77

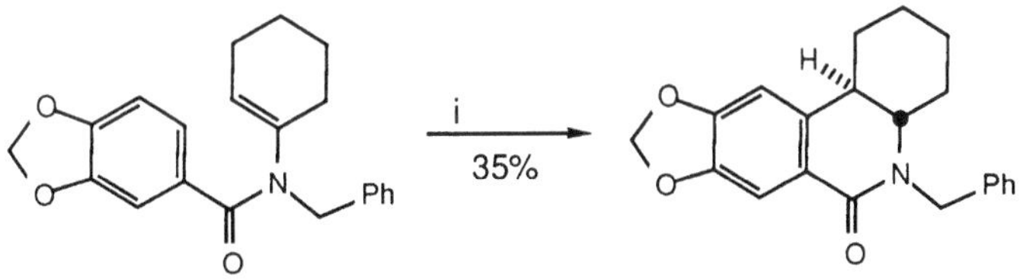

i

35%

i. hv

From I. Ninomiya, T. Naito, and T. Kiguchi, *Tet. Letters*, *51*, 4451 (1970).

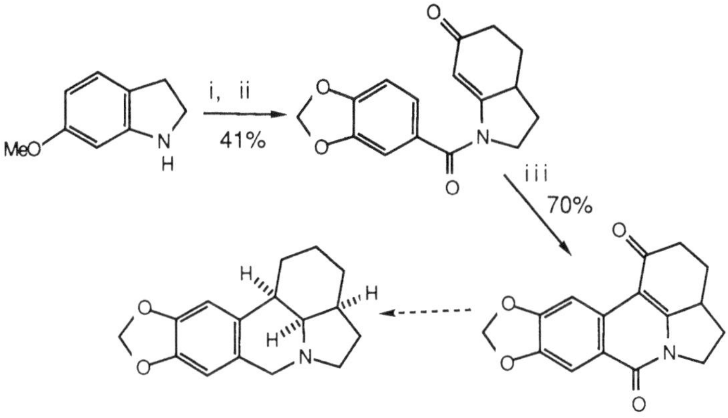

i, ii

41%

iii

70%

i. Li/lqd NH₃, THF, MeOH   ii. 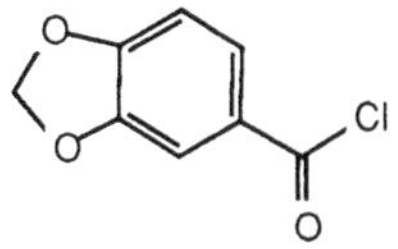 ,   aq alkaline soln.   iii. hv, THF

From H. Iida, S. Aoyagi and C. Kibayashi, *J. Chem. Soc.*, *Chem. Commun.*, 499 (1974) and H. Iida, S. Aoyagi and C. Kibayashi, *J. Chem. Soc.*, *Perkin Trans. I*, 2502 (1975).

78

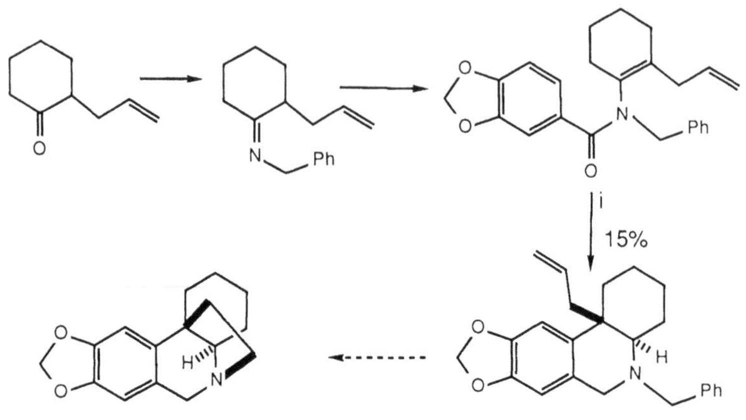
O
N
Ph
O
O
N
Ph
O
i
15%
O
O
H
N
O
O
H
N
Ph

i. hv. MeOH

From I. Ninomiya, T. Naito, and T. Kiguchi, *J. Chem. Soc., Chem. Commun.*, 1669 (1970).

79

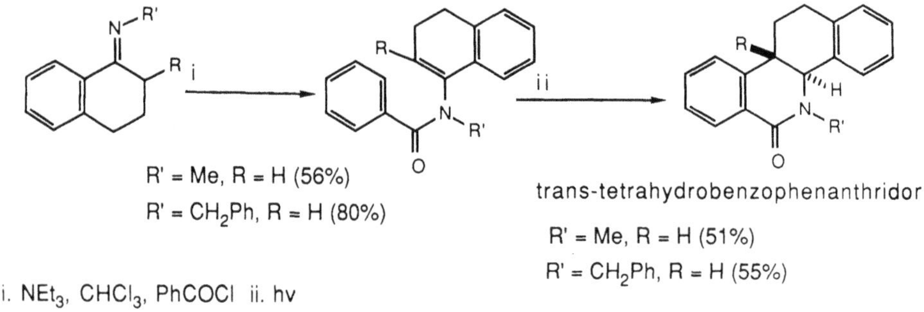
N
R'
R
i
R
N
R'
O
ii
R
H
N
R'
O
R' = Me, R = H (56%)
R' = CH2Ph, R = H (80%)
trans-tetrahydrobenzophenanthridor
R' = Me, R = H (51%)
R' = CH2Ph, R = H (55%)

i. NEt_3, $CHCl_3$, PhCOCl ii. hv

From I. Ninomiya, T. Naito; T. Kiguchi, and T. Mori, *J. Chem. Soc., Perkin Trans. I*, 1696 (1973).

80

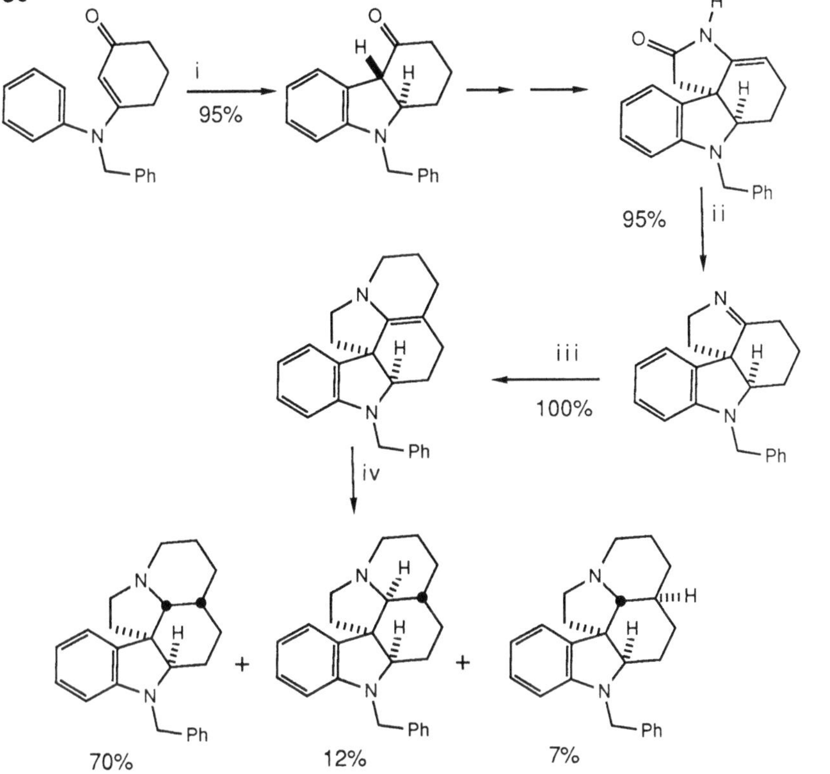

i. hv , PhH ii. LAH, THF iii. t-Bu.Li, -20°C, Br(CH$_2$)$_3$CI, -70°C to 0°C to RT,
iv. H$_2$/Pt-Al$_2$O$_3$, HCl/EtOH, 40 psi

From J.-C. Gramain, H. P. Husson, and Y. Troin, *J. Org. Chem.*, *50*, 5517 (1985).

81

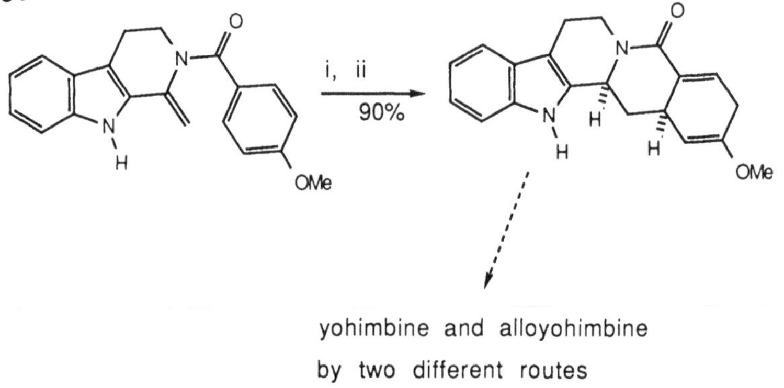
i, ii
90%
OMe
yohimbine and alloyohimbine
by two different routes

i. hv, MeCN ii. NBH, MeOH

From T. Naito, Y. Tada, Y. Nishiguchi, and I. Ninomiya, *Heterocycles 18*, 213 (1982) and O. Miyata, Y. Hirata, T. Naito, and I. Ninomiya, *J. Chem. Soc., Chem. Commun.*, 1231 (1983).

82

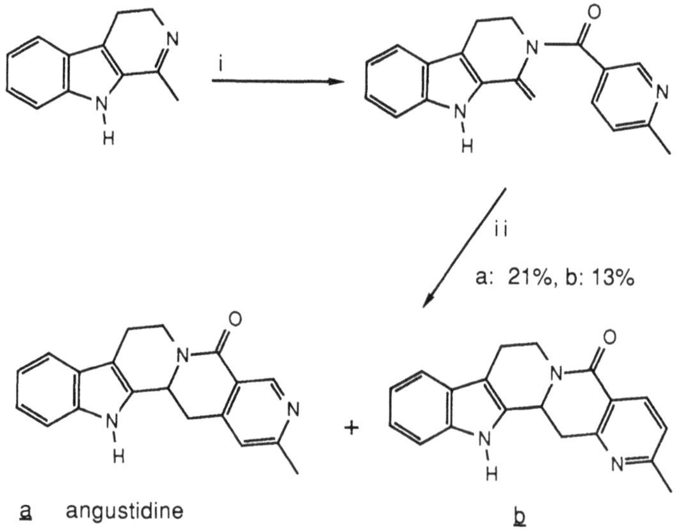
i
ii
a: 21%, b: 13%
a angustidine
b

i. 6-methyl-pyrid-3-yl-COCl ii. hv, MeOH, 8h

From I. Ninomiya, H. Takasugi, and T. Naito, *J. Chem. Soc., Chem. Commun.*, 732 (1973).

83

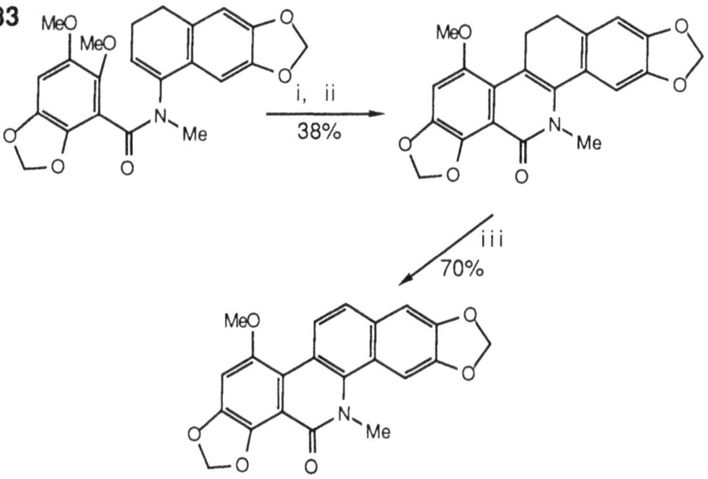

MeO
MeO
i, ii
38%
iii
70%

i. hν ii. conc. HCl iii. 30% Pd/C, p-cymene

From H. Ishii, K. Harada, T. Ishida, E. Ueda, and K. Nakajama, *Tet. Letters*, 319 (1975).

84

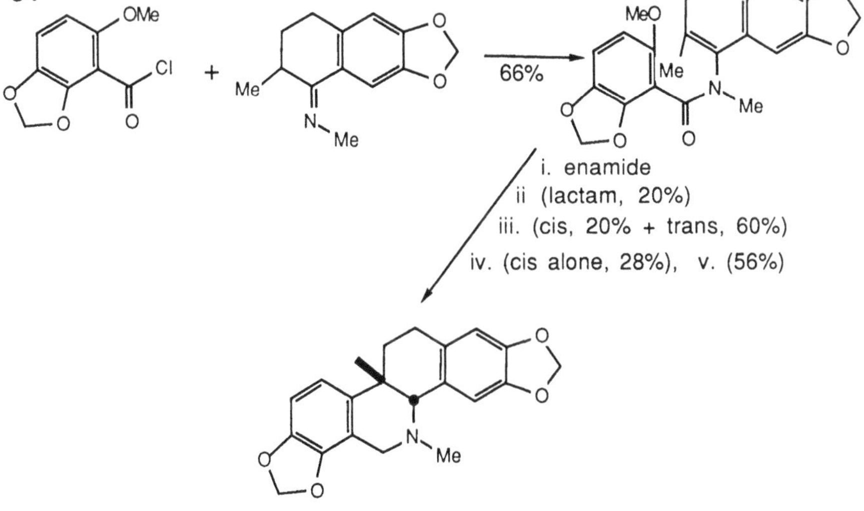

OMe
Cl
66%
MeO
i. enamide
ii (lactam, 20%)
iii. (cis, 20% + trans, 60%)
iv. (cis alone, 28%), v. (56%)

deoxycorynoline

ii. hν iii. H_2, Pd/C iv. DDQ v. LAH

From I. Ninomiya, O. Yamamoto, and T. Naito, *Heterocycles*, 4, 743 (1976).

From I. Ninomiya, O. Yamamoto, and T. Naito, *J. Chem. Soc., Chem. Commun.*, 437 (1976).

85

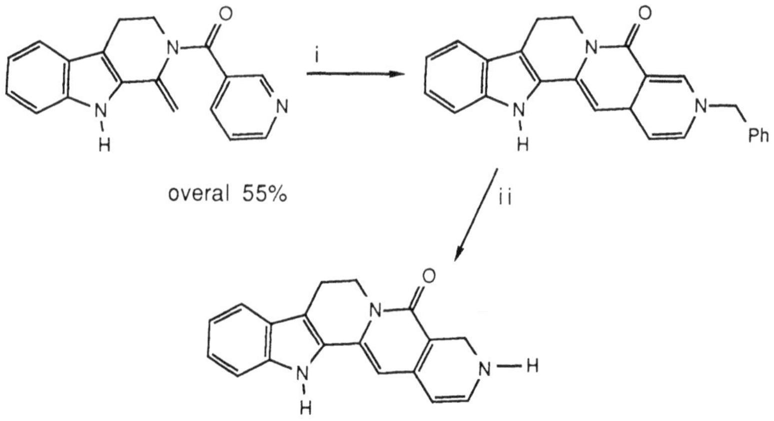

overal 55%

i. PhCH₂Br   ii. PhMe-AcOH-AcONa, heat

From M. Sainsbury and N. Uttley, *J. Chem. Soc., Chem. Commun.*, 319 (1977).

86

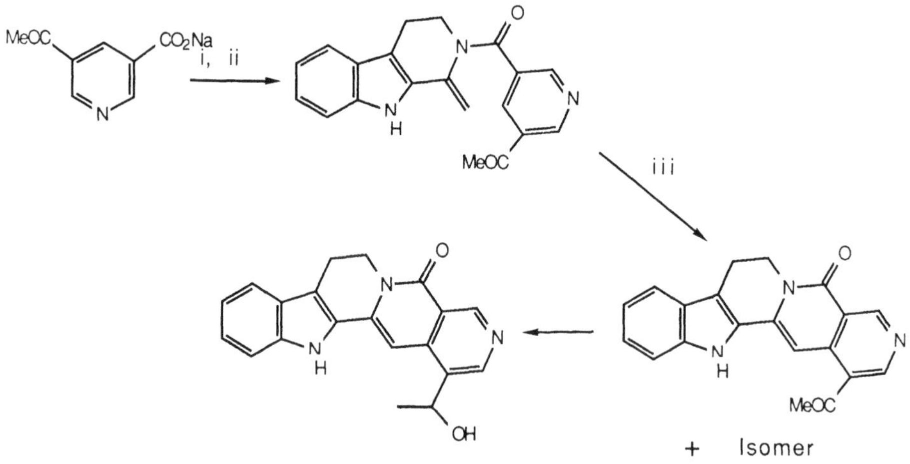

+ Isomer

i. (COCl)₂   ii. harmalan   iii. hv, MeOH, RT

From I. Ninomiya and T. Naito, *Heterocycles*, 2, 607 (1974).

87

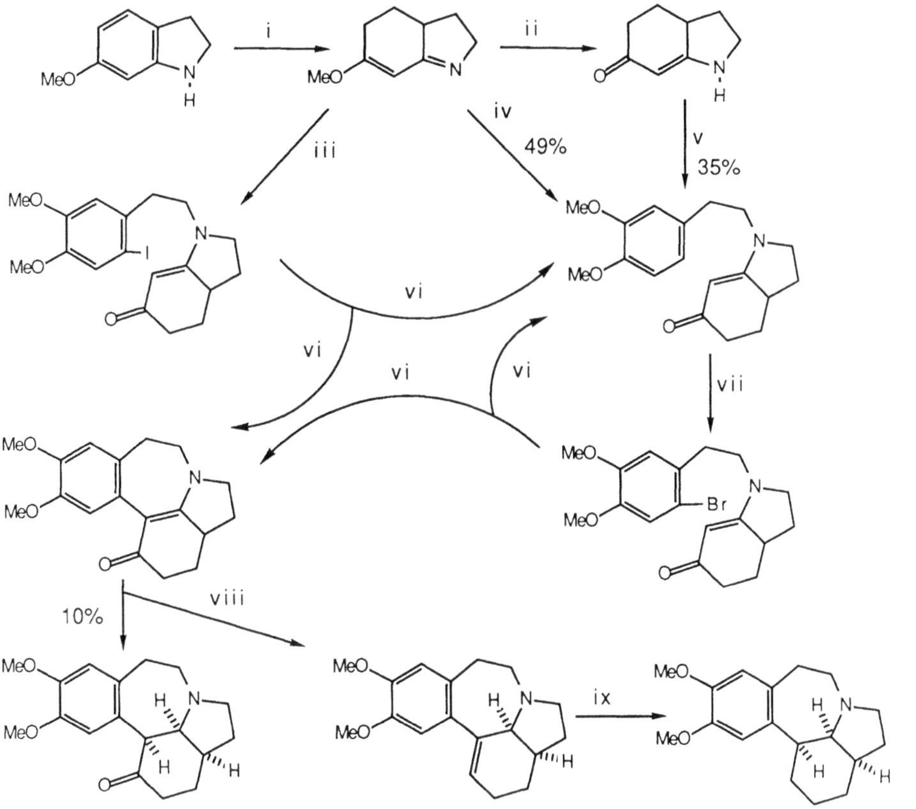

f. Birch reduction

i. Li, lqd NH$_3$, MeOH ii. HCl iii. 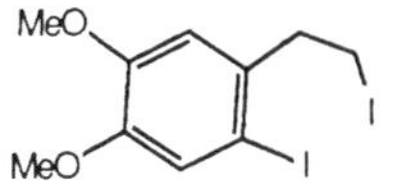 , heat iv. 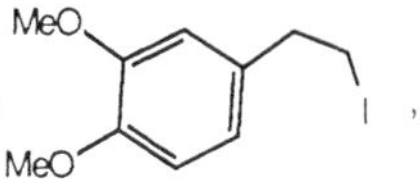 , heat

v. 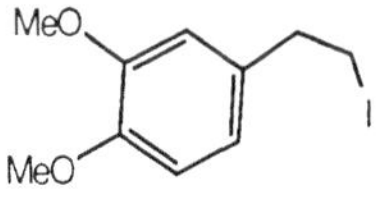 , NaH, DMSO vi. hv vii. Br$_2$, CHCl$_3$ viii. LAH ix. Pt, H$_2$, AcOH

From H. Iida, T. Takarai, and C. Kibayashi, *J. Org. Chem.*, *43*, 975 (1978).

88

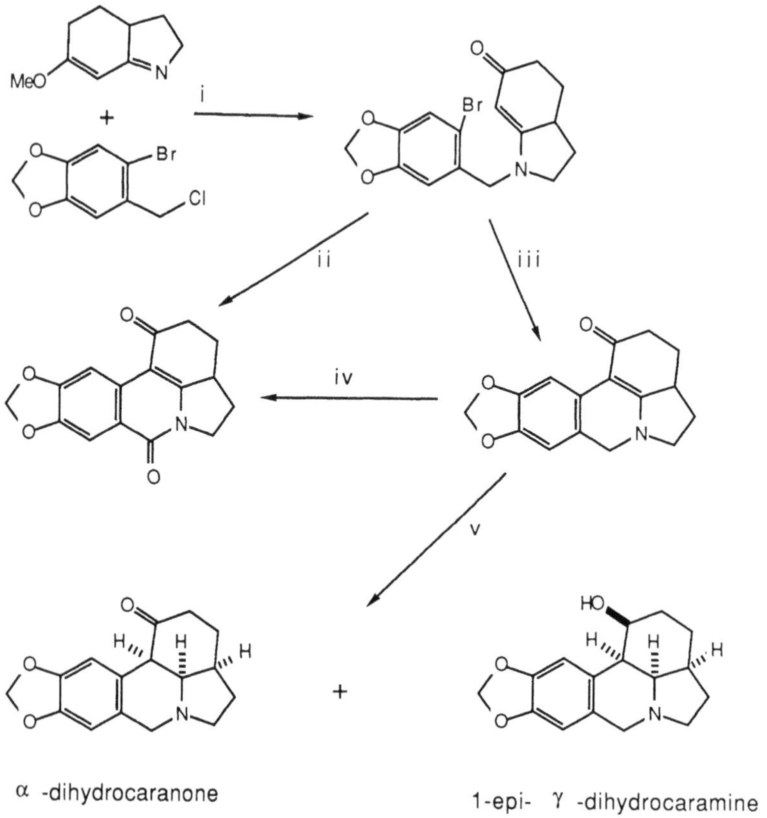
MeO
N
+
Br
Cl
i
Br
O
N
ii
iii
iv
v
α -dihydrocaranone
1-epi- γ -dihydrocaramine

f. imine alkylation

i. PhMe, reflux ii. hν iii. Li-ethylamide iv. O_2, KOH, aq EtOH v. LAH, THF

From H. Iida, Y. Yuasa, and C. Kibayashi, *J. Org. Chem.* *44*, 1074 (1979).

89

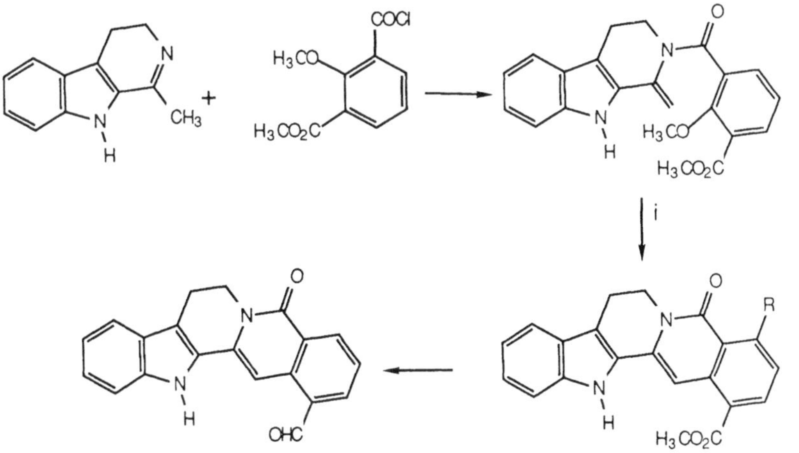

nauclefine

	R=H		R=OCH₃
CH₃CN	53%	:	26%
PhH	58%	:	19%
ether/MeOH 40:1	56%	:	19%

f. condensation

i. hv, pyrex

From T. Naito, E. Doi, O. Miyata, and I. Ninomiya, *Heterocycles*, *24*, 903 (1986).

90

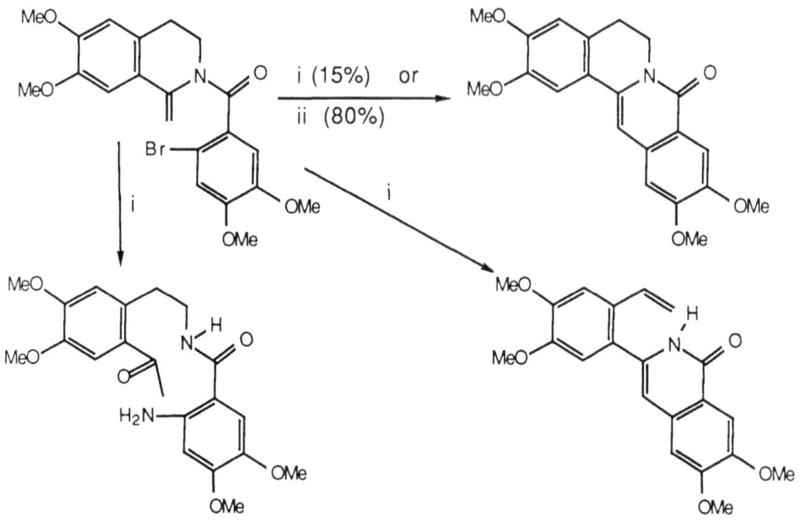

i. NaNH$_2$-liq NH$_3$ ii. hv

From T. Kametani, T. Sugai, T. Honda, F. Satoh, and K. Fukumoto, *J. Chem. Soc., Perkin Trans. I,* 1151 (1977).

91

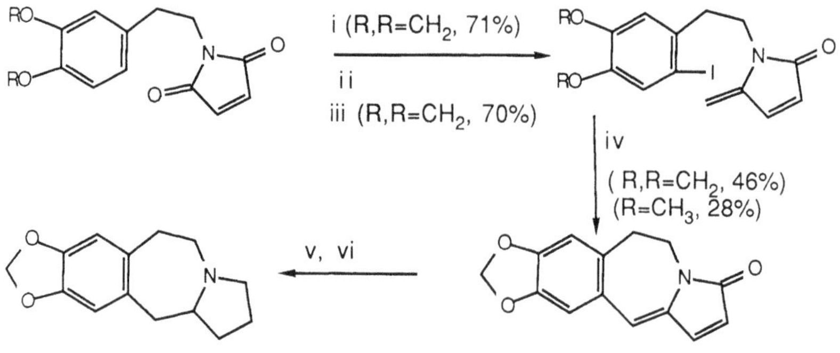

i. I$_2$-CF$_3$CO$_2$Ag, CH$_2$Cl$_2$ ii. MeMgI, ether, PhH   iii. PTSA, PhH
iv. hv (253.7 nm), PhH, Et$_3$N   v. H$_2$-PtO$_2$, MeOH   vi. LAH

From I. Tse and V. Snieckus, *J. Chem. Soc., Chem. Commun.,* 505 (1976).

92

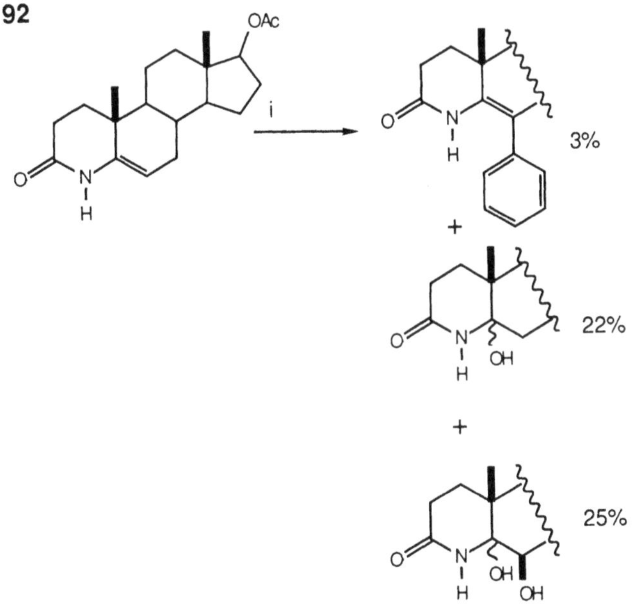

i. hv, (low press. Hg), N₂, 22h, PhH

i. hv, (low press. Hg), N_2, 22h, PhH

From J. Boix, J. Gomez, and J.-J. Bonet, *Helv. Chim. Acta, 58,* 2545 (1975) and F. Abello, J. Boix, J. Gomez, J. Morell, and J.-J. Bonet, *Helv. Chim. Acta, 582549* (1975).

93

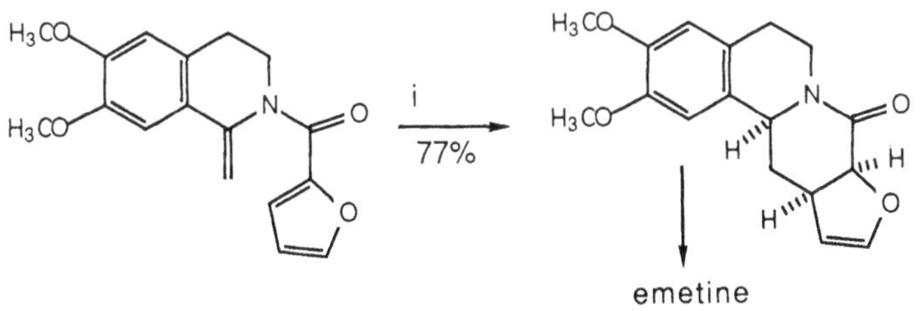

emetine

i. hv, NBH

From T. Naitor, N. Kojima, O. Miyata, and I. Ninomiya, *J. Chem. Soc., Chem. Commun.,* 1611 (1985).

94

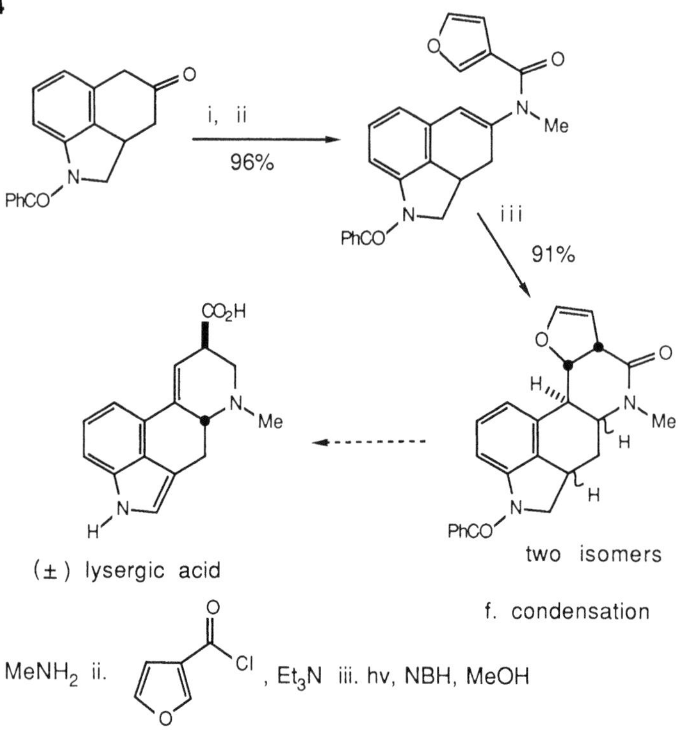
i, ii
96%
PhCO
N
O
Me
iii
91%
CO2H
two isomers
(±) lysergic acid
f. condensation
i. MeNH2 ii.
Cl
, Et3N iii. hv, NBH, MeOH

From T. Kiguchi, C. Hashimoto, T. Naito, and I. Ninomiya, *Heterocycles*, *19*, 2279 (1982).

95

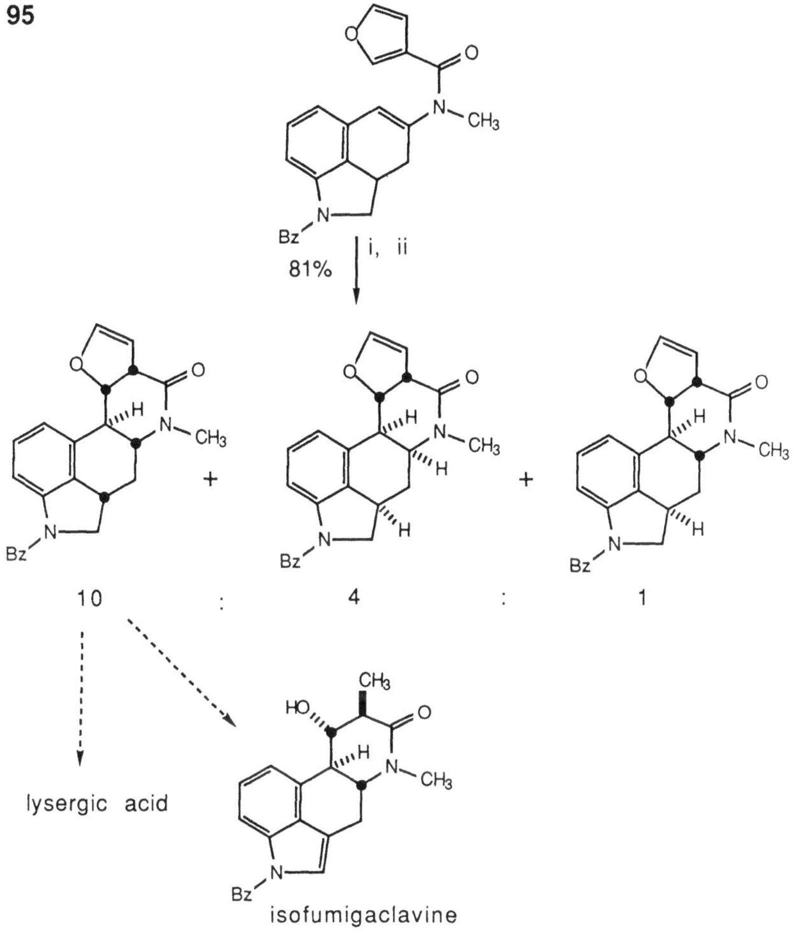

i. hv   ii. NBH, PhH : MeOH (5:1), 4-10°C

From I. Ninomiya, C. Hashimoto, T. Kigushi, and T. Naito, *J. Chem. Soc., Perkin Trans. I,* 941 (1985).

IV. ALKYLATIONS AT THE β'-POSITION: 96—97

96

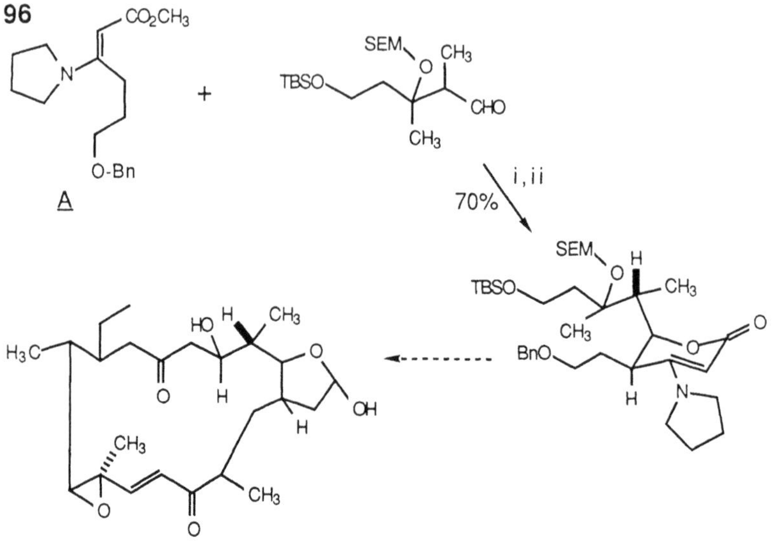

+

i, ii
70%

(+)-rosaramycin aglycone

f. condensation

i. LDA, -78°C   ii. add aldehyde to A in THF, -78°C to 0°C

From R. H. Schlessinger, M. A. Poss, and S. Richardson, *J. Am. Chem. Soc.*, *108*, 3112 (1986).

97

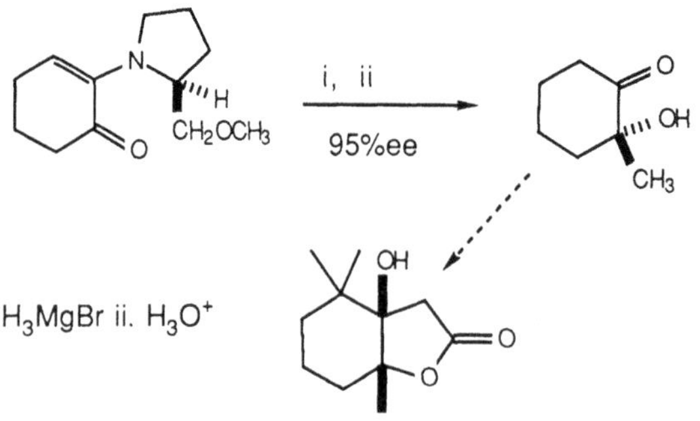

i, ii
95%ee

i. CH₃MgBr ii. H₃O⁺

aeginetolide

From T. Sato, M. Funabora, M. Watanabe, and T. Fujisawa, *Chem. Letters*, 1391 (1985).

V. δ-ALKYLATION REACTIONS OF DIENAMINES: 98—106

98

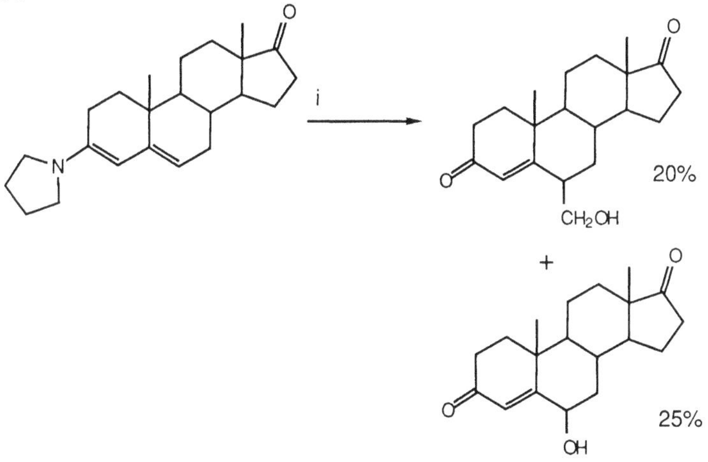

20%

+

25%

i. HCHO, PhH:EtOH (1:2), Ar, 30min

From F. Schneider, A. Boller, M. Müller, P. Müller, and A. Fürst, *Helv. Chim. Acta, 56,* 2396 (1973).

99

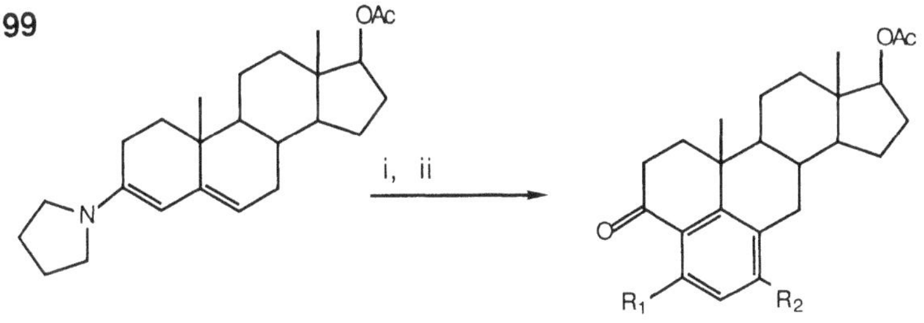

R_1 = H; R_2 =CH_3, 11%
R_1 = CH_3; R_2 = H, 15%

i. 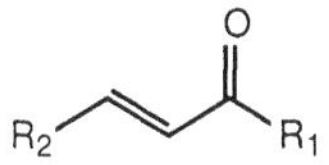 R_2⌃⌃R_1 ,toluene, 135-140°C, 40h ii.H_3O^+

From P. Houdewind, J. C. L. Armande, and U. K. Pandit, *Tet. Letters,* 591 (1974).

100

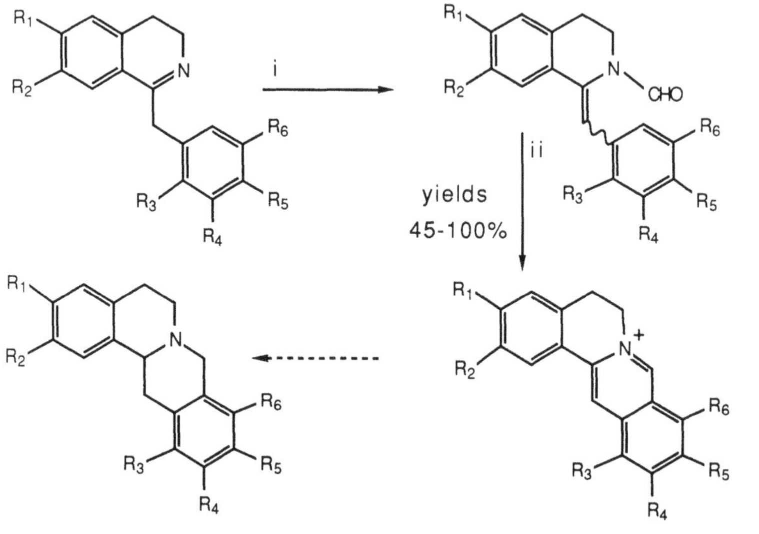

yields: 45%-100%, protoberberine series
66%-100%, berberine series

R_1-R_6 = H or OCH_3

i. Ac_2O-HCHO, NaOAc ii. hv, HI

From G. R. Lenz, *J. Org. Chem.*, *42*, 1117 (1977).

101

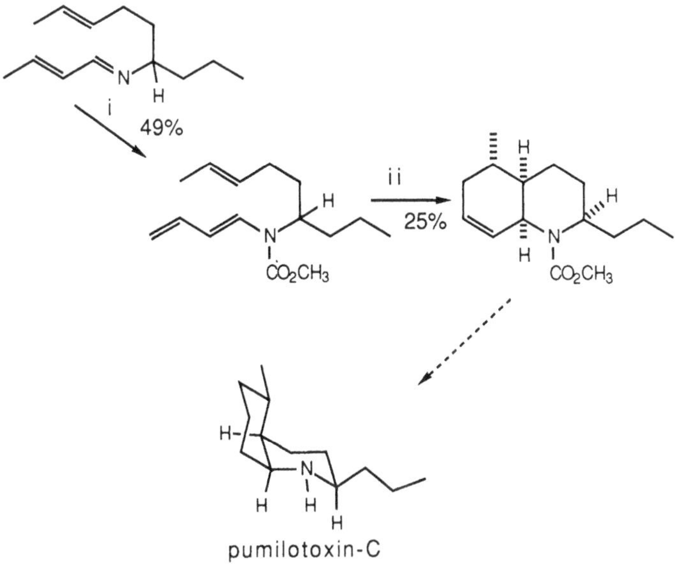
i
49%
ii
25%
CO_2CH_3
CO_2CH_3
H
pumilotoxin-C

f. Diels-Alder

i. $NaN(SiMe_3)_2$, $ClCO_2Me$ ii. Toluene,215°C

From W. Oppolzer, W. Fröstl, and H. P. Weber, *Helv. Chim. Acta*, *58*, 593 (1975).

102

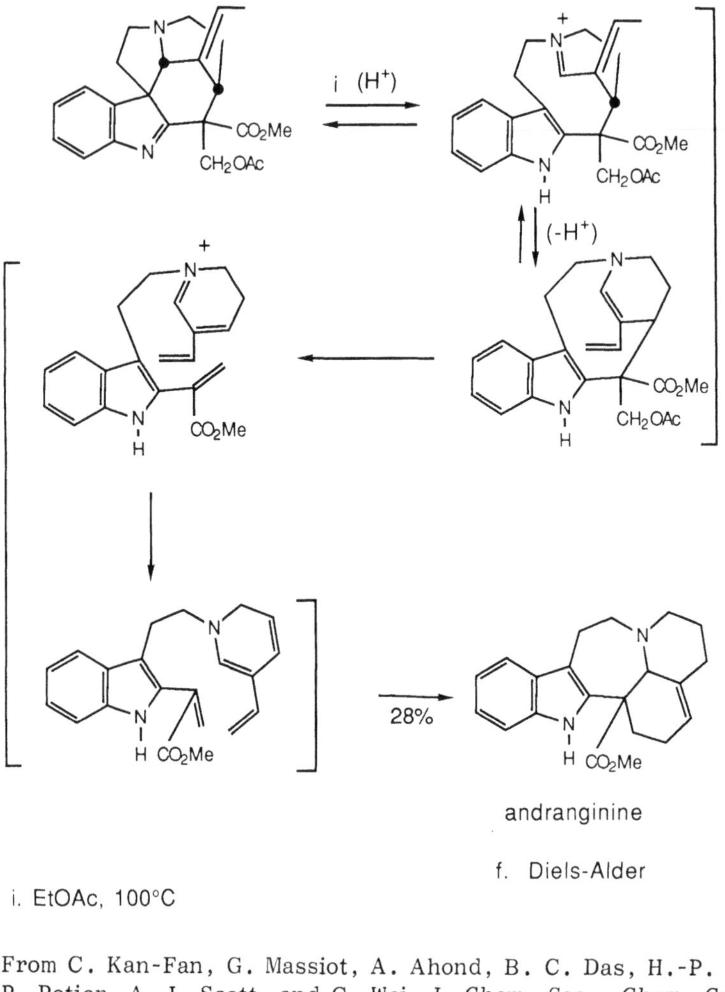

i (H⁺)
(-H⁺)
28%
CO2Me
CH2OAc
andranginine
f. Diels-Alder

i. EtOAc, 100°C

From C. Kan-Fan, G. Massiot, A. Ahond, B. C. Das, H.-P. Husson, P. Potier, A. I. Scott, and C. Wei, *J. Chem. Soc., Chem. Commun.*, 164 (1974).

103

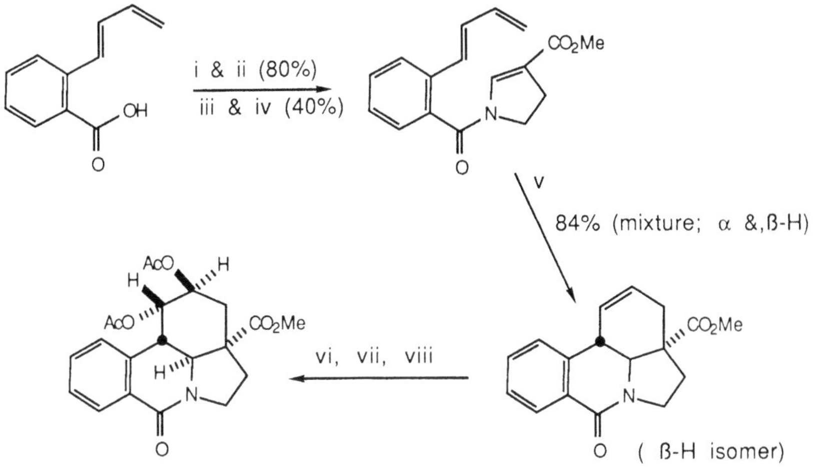
OH
O
i & ii (80%)
iii & iv (40%)
CO2Me
N
O
v
84% (mixture; α &,β-H)
AcO
H
H
AcO
CO2Me
H
N
O
vi, vii, viii
CO2Me
N
O
(β-H isomer)

i. Ph_3P, $BrCCl_3$, THF ii. 3-carbomethoxypyrrolidine, 75°C, 1h iii. Li-tetramethyl piperidine, THF, -78°C iv. I_2, THF, -78°C, DBU
v. o-dichlorobenzene, reflux, 2.5 h
vi. mCPBA, CH_2Cl_2 vii. cat. $HClO_4$, THF, H_2O, 60°C viii. Ac_2O, py

From D. J. Morgans, Jr., and G. Stork, *Tet. Letters*, 1959 (1979).

104

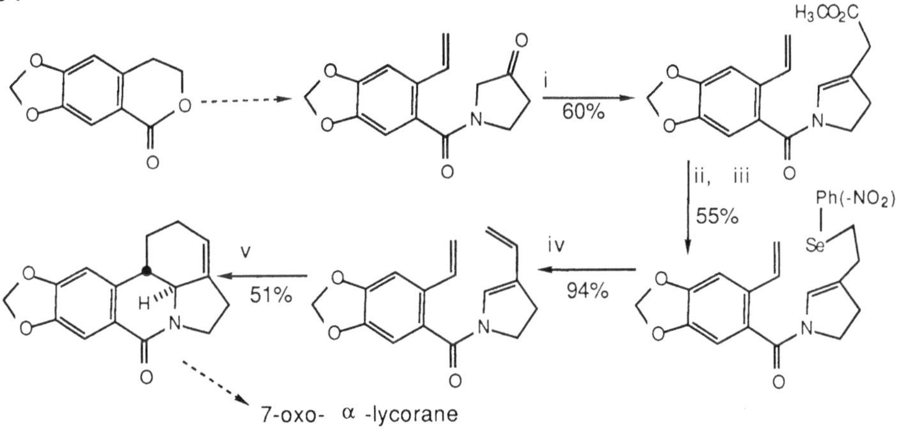
O
O
O
O
O
O
N
O
i
60%
H3CO2C
O
O
N
O
ii, iii
55%
Ph(-NO2)
Se
O
O
N
O
iv
94%
O
O
N
O
v
51%
O
O
H
N
O
7-oxo- α -lycorane

i. $Ph_3P{=}CHCO_2CH_3$, DBU, THF ii.LBH, THF, 22h iii. o-nitro phenyl selenocyanate, Bu_3P
iv. $NaIO_4$, $NaHCO_3$ v. PhCl, 3-Bu^t-4-OH-5-Me-C_6H_2SH (trace),
$MeC(OSiMe_3){=}NHSiMe_3$, 140°C

From G. Stork, D. J. Morgans, Jr., *J. Am. Chem. Soc.*, *101*, 7110 (1979).

105

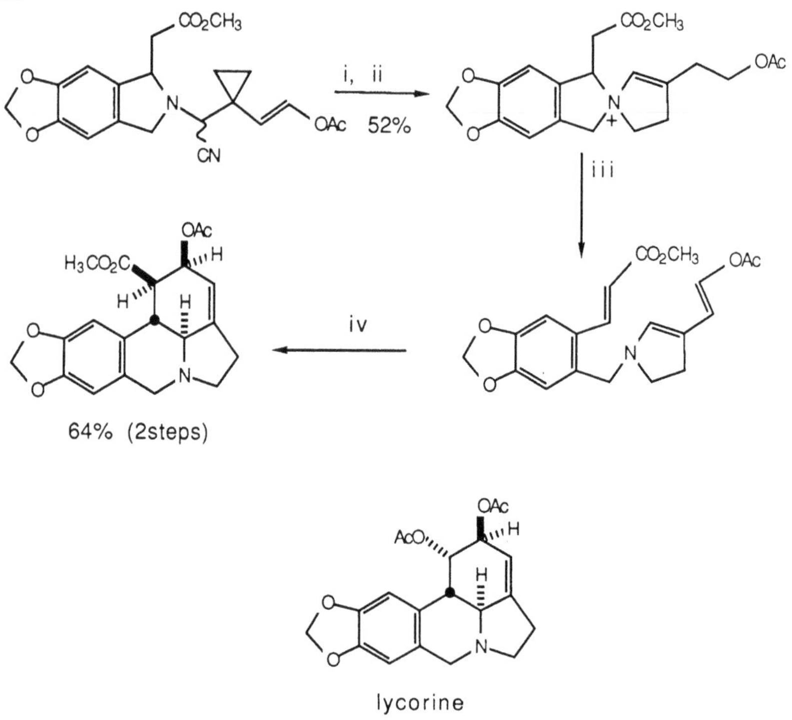

64% (2steps)

lycorine

f. Diels-Alder

i. AgBF$_4$, RT, 30 min, DME ii. LiBr, RT, 1h, CH$_3$CN   iii. DBU, 0°C, 1h, CH$_2$Cl$_2$
iv. 110°C, 4h

From R. K. Boeckman, J. P. Sabatucci, S. W. Goldstein, D. M. Springer, and P. F. Jackson, *J. Org. Chem.*, *51*, 3740 (1986).

106

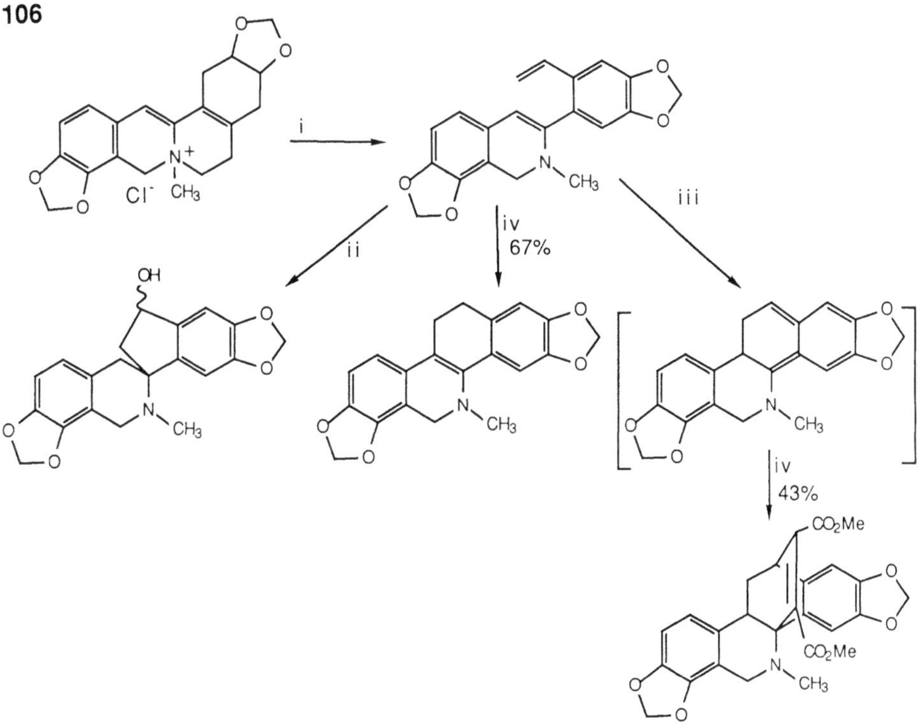

i. KOH-MeOH ii. 10% HCl iii. hv, PhH IV. $H_3CO_2C \equiv CO_2CH_3$

From M. Onda, K. Yonezawa, K. Abe, and H. Toyama, *Chem. and Pharm. Bull.* (Japan), *19*, 31 (1971).

VI. N-ALKYLATIONS: 107–109

In addition to these examples see also example 15.

107

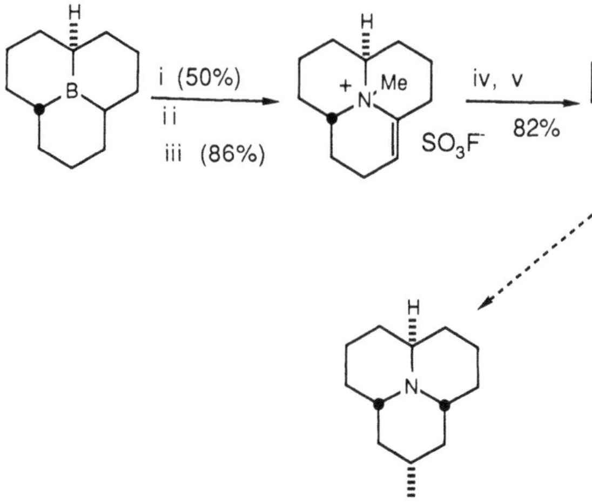

precoccinellin

f. condensation

i. H₂O₂, ClNHODNP, NaOH ii. CrO₃, H₂SO₄ iii. MeOSO₂F, EtOEt, -78°C to 25°C
iv. LDA, THF v. LiSEt, DMF

From R. H. Mueller and M. E. Thompson, *Tet. Letters*, 1991 (1979).

108

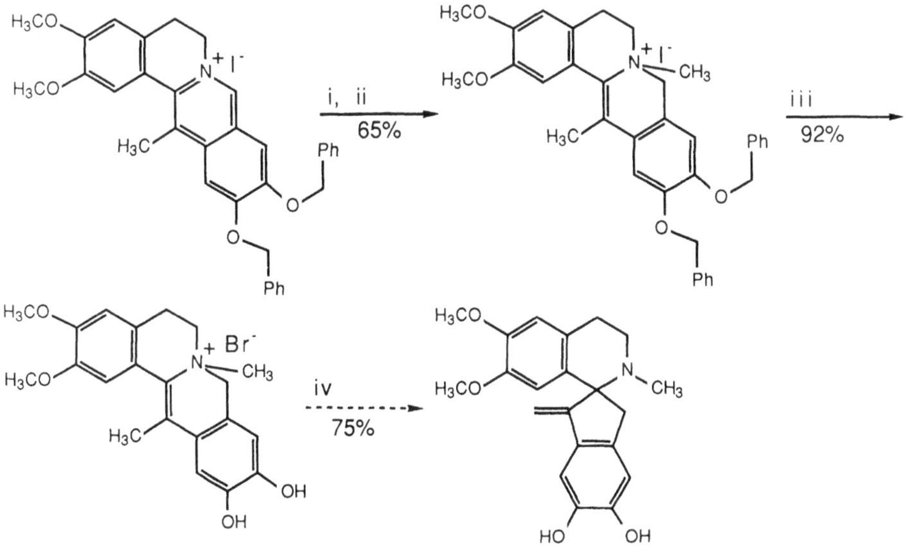

i. Partial redn. with LAH   ii. CH$_3$I   iii. HBr-EtOH, reflux   iv. NaOH-aq. EtOH

From M. Shamma and C. D. Jones, *J. Am. Chem. Soc.*, *91*, 4009 (1969), M. Shamma and C. D. Jones, *J. Am. Chem. Soc.*, *92*, 4943 (1970), M. Shamma and J. F. Nugent, *Tet. Letters*, 2625 (1970), and M. Shamma and J. F. Nugent, *Tetrahedron*, *29*, 1265 (1973).

109

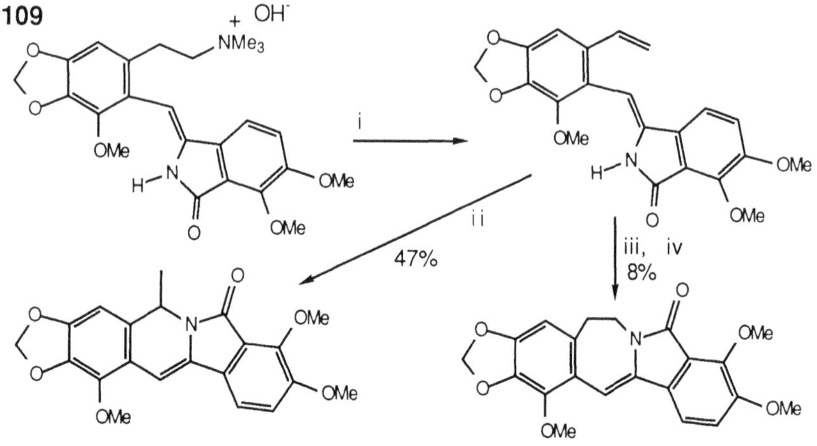

f. N-alkylation

i. heat ii. HCl, aq MeOH iii. 30% aq KOH, 7h, reflux iv. dil. HCl

From J. Trojanek, Z. Koblicova, Z. Uesely, V. Suchan, and J. Holubek, *Coll. Czech. Chem. Commun.*, *40*, 681 (1975).

VII. ADDITION TO THE α-POSITION (OR γ-POSITION FOR DIENAMINES): 110–125

110

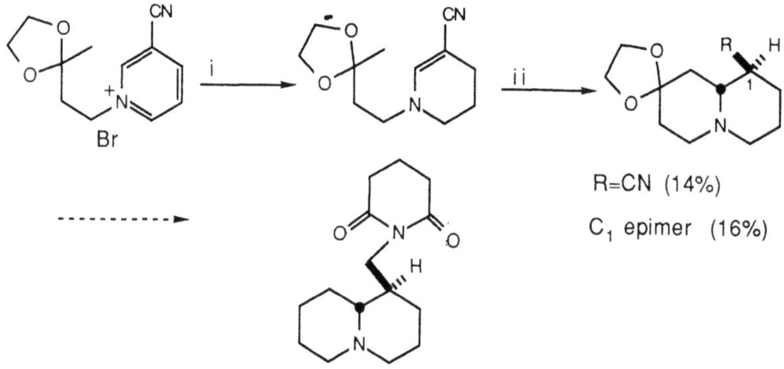

R=CN (14%)

C$_1$ epimer (16%)

i. 10% Pd-C, H$_2$, Et$_3$N ii. PTSA, PhH   f. reduction

From E. Wenkert and A. R. Jeffcoat, *J. Org. Chem.*, *35*, 515 (1970).

111

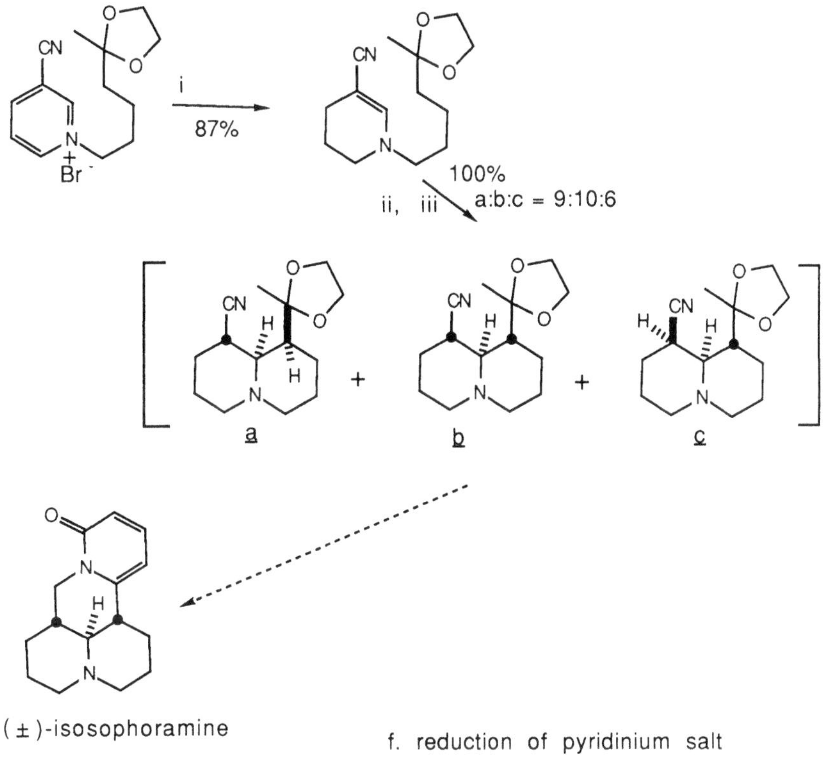

CN
Br⁻
i
87%
ii, iii
100%
a:b:c = 9:10:6
a
b
c
(±)-isosophoramine
f. reduction of pyridinium salt

i. H_2, Pd-cat. ii. Anh. TsOH, PhH, Heat, 24h iii. Et_3N

From E. Wenkert, B. Channcy, K. G. Dave, A. R. Jeffcoat, F. M. Schell, and H. P. Schenk, *J. Am. Chem. Soc.*, *95*, 8427 (1973).

112

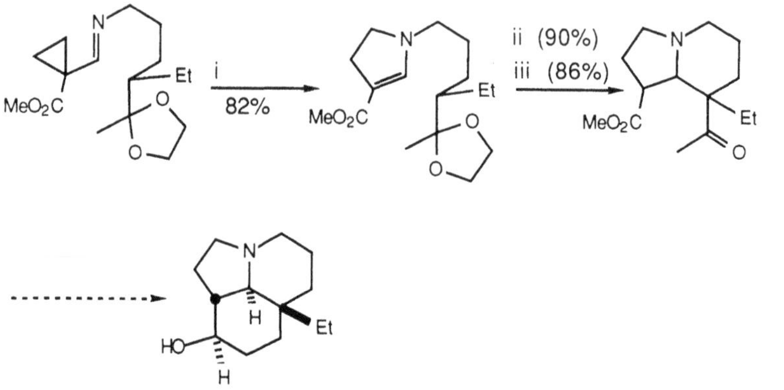
MeO2C
Et
i
82%
ii (90%)
iii (86%)
HO
H

f. cyclopropylimine

i. NH_4Cl-160°C-0.1 mm Hg ii. HCl (gas)-Et_2O iii. Aq. HCl iii. HCl (gas)

From R. V. Stevens, J. M. Fitzpatrick, M. Kaplan, and R. L. Zimmerman, *J. Chem. Soc., Chem. Commun.*, 857 (1971).

113

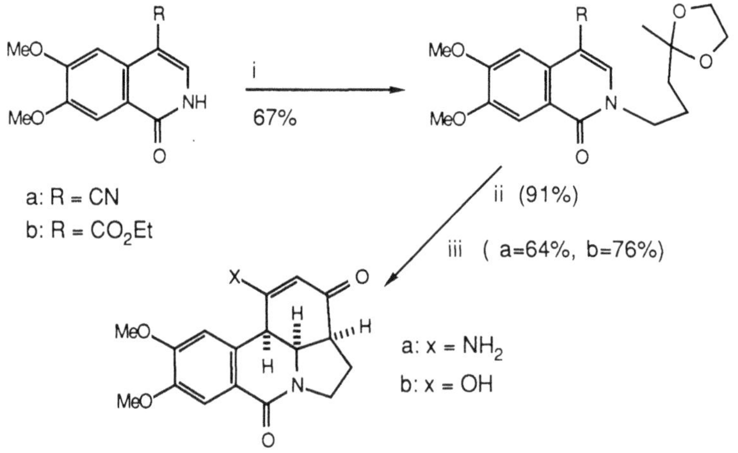
MeO
R
NH
i
67%
a: R = CN
b: R = CO2Et
ii (91%)
iii (a=64%, b=76%)
a: x = NH2
b: x = OH

i. $Br(CH_2)_3C(OCH_2)_2Me$, K_2CO_3, DMF ii. 2% HCl, THF
iii. t-BuOK, t-BuOH

From E. Wenkert, H. P. S. Chawla, and F. M. Schell, *Synth. Commun.*, 3, 381 (1973).

114

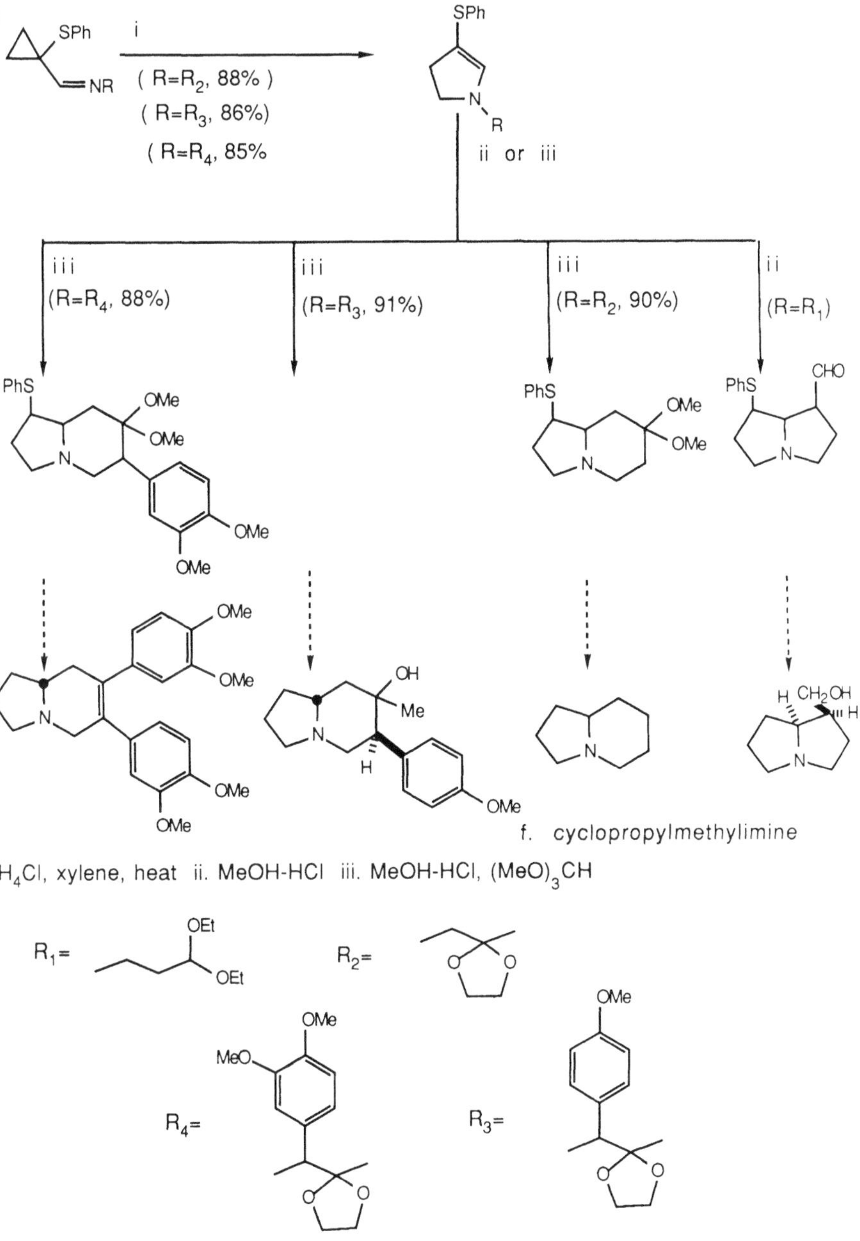

i. NH_4Cl, xylene, heat   ii. MeOH-HCl   iii. MeOH-HCl, $(MeO)_3CH$

f. cyclopropylmethylimine

$R_1 =$

$R_2 =$

$R_3 =$

$R_4 =$

From R. V. Stevens, Y. Luh, and J. Shen, *Tet. Letters*, 3799 (1976), R. V. Stevens, *Acc. Chem. Res.*, *10*, 193 (1977), and R. V. Stevens and Y. Luh, *Tet. Letters*, 979 (1977).

115

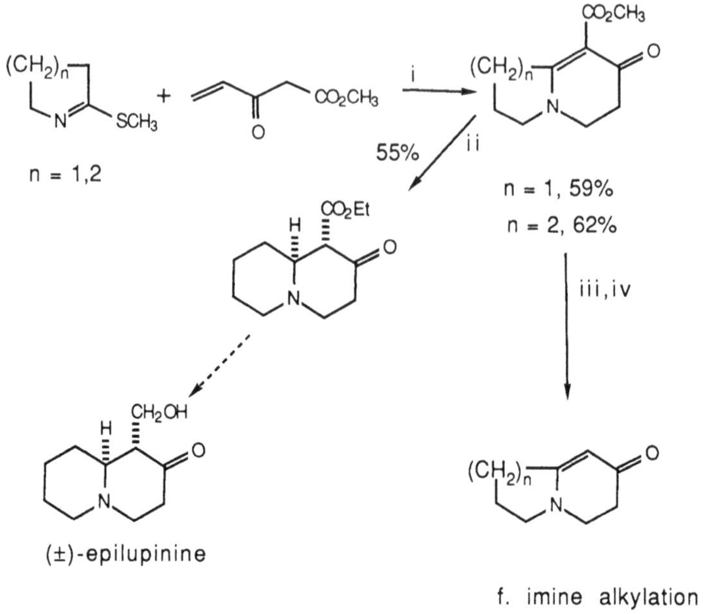

55%

n = 1,2

n = 1, 59%
n = 2, 62%

(±)-epilupinine

f. imine alkylation

i. HgCl$_2$, MeOH   ii. DIBAL, -50°C   iii. 5% KOH   iv. 20% HCl

From H. Takahata, K. Yamabe, T. Suzuki, and T. Yamasaki, *Hetero-cycles*, *24*, 37 (1986).

116

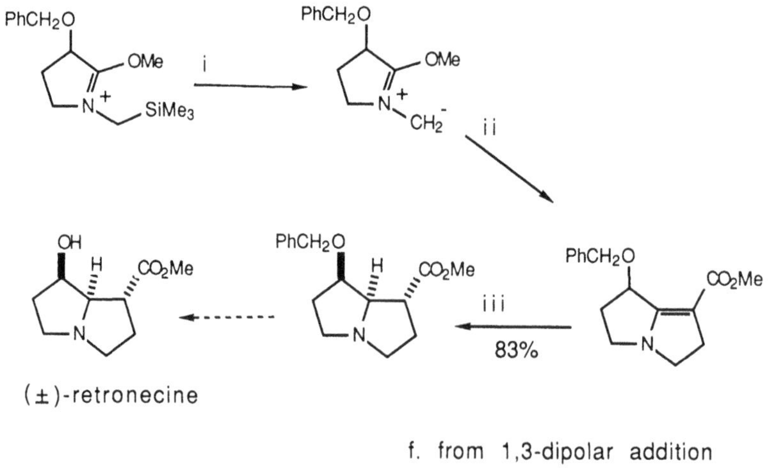

(±)-retronecine

83%

f. from 1,3-dipolar addition

i. CeF   ii. CH$_2$=CHCO$_2$CH$_3$ iii. cat. hydrogenation

From E. Vedejs and G. R. Martinej, *J. Am. Chem. Soc.*, *102*, 7994 (1980).

117

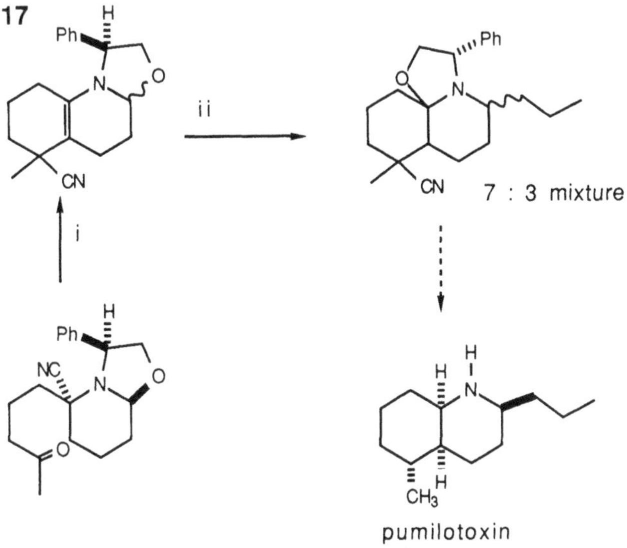

i
ii
7 : 3 mixture
pumilotoxin

f. condensation

i. Al_2O_3, CH_2Cl_2, 1.5h, reflux ii. PrMgBr, 1.5h, 15°C

From M. Bonin, J. Royer, D. S. Grierson, and H. P. Husson, *Tet. Letters*, 1569 (1986) and H. P. Husson, *J. Nat. Prod.*, *48*, 894 (1985).

118

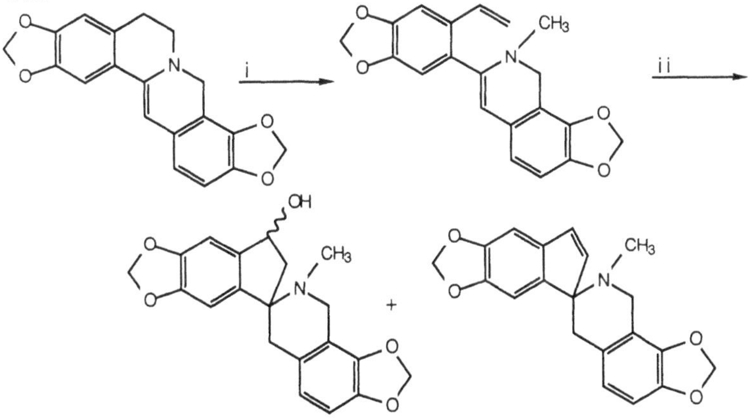

i
ii

2 isomers (64%)

i. KOH-CH_3OH, CH_3I ii. 10% HCl

From M. Onda, K. Abe, and K. Yonezawa, *Chem. Pharm. Bull.*, *16*, 2005 (1968).

119

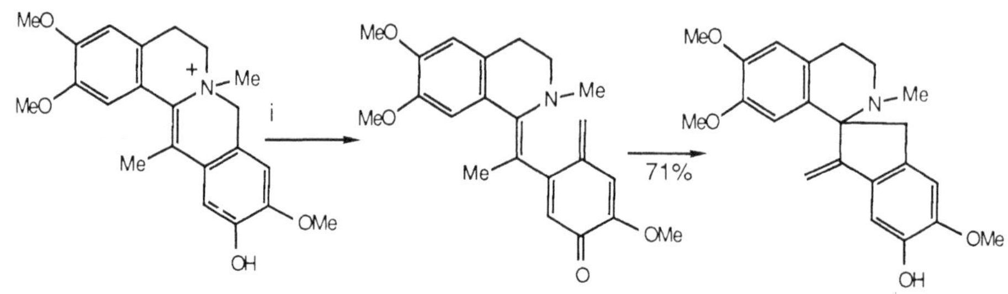

i. ⁻OH, heat

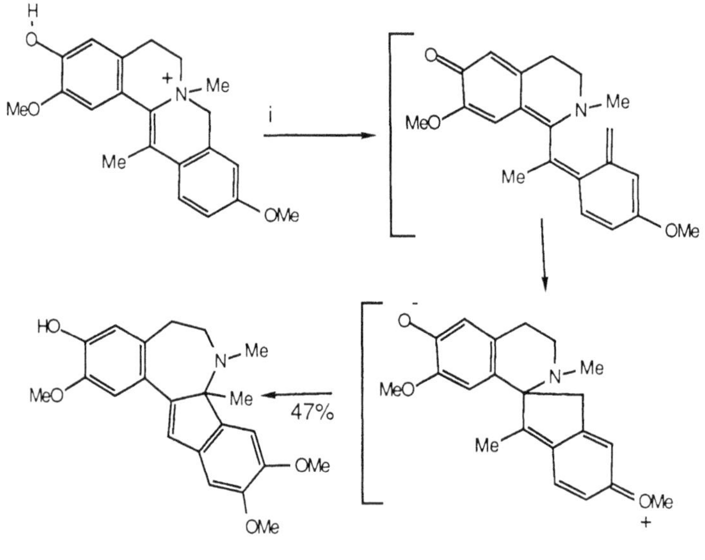

i. ⁻OH, heat

From M. Shamma and J. F. Nugent, *Tetrahedron*, *29*, 1265 (1973).

120

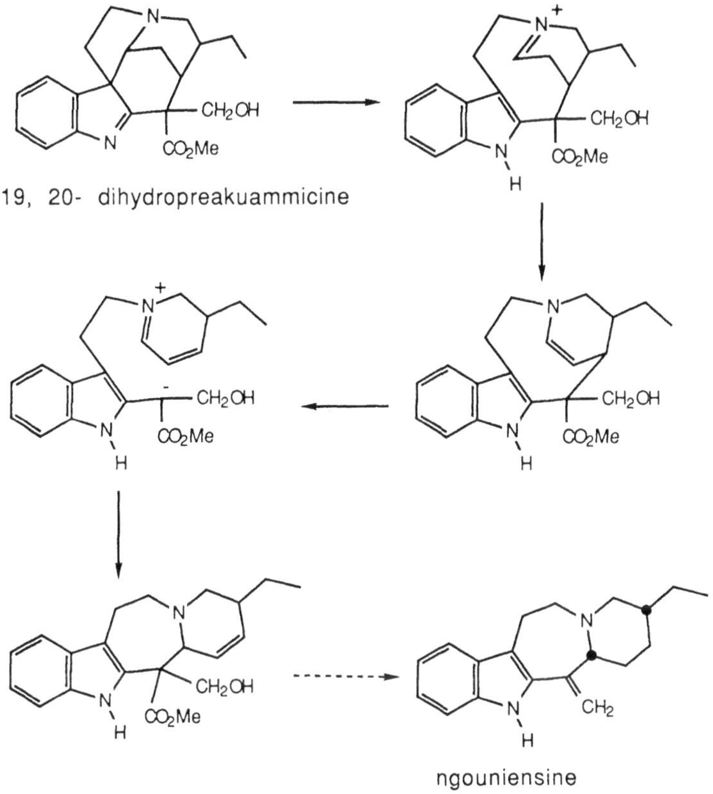
CH2OH
CO2Me
19, 20- dihydropreakuammicine
CH2OH
CO2Me
CH2OH
CO2Me
CH2OH
CO2Me
CH2OH
CO2Me
CH2
ngouniensine

f. fragmentation

From G. Massiot, M. Zeches, P. Thepenier, M. J. Jacquier, L. Le Men-olivier, and C. Delande. *J. Chem. Soc., Chem. Commun.*, 768 (1982).

Conversion of enamines to imonium intermediates by protonation of other electrophilic reactions followed by addition of nuclephiles. In addition to the examples given in this reaction see also examples: 15, 25, 30, 31, 32, 34, 36, 37, 47, 48, 49, 64, 69, 103, 104, 105.

121

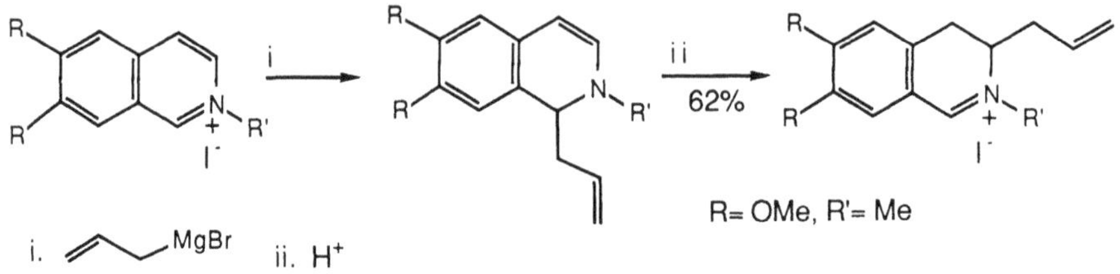
i
ii
62%
R= OMe, R'= Me
i. MgBr
ii. H+

From M. Sainsbury, S. F. Dyke, D. W. Brown, and R. G. Kisman, *Tetrahedron, 26*, 5265 (1970) and J. Knabe and H. D. Holtje, *Tet. Letters, 6*, 433 (1969).

122

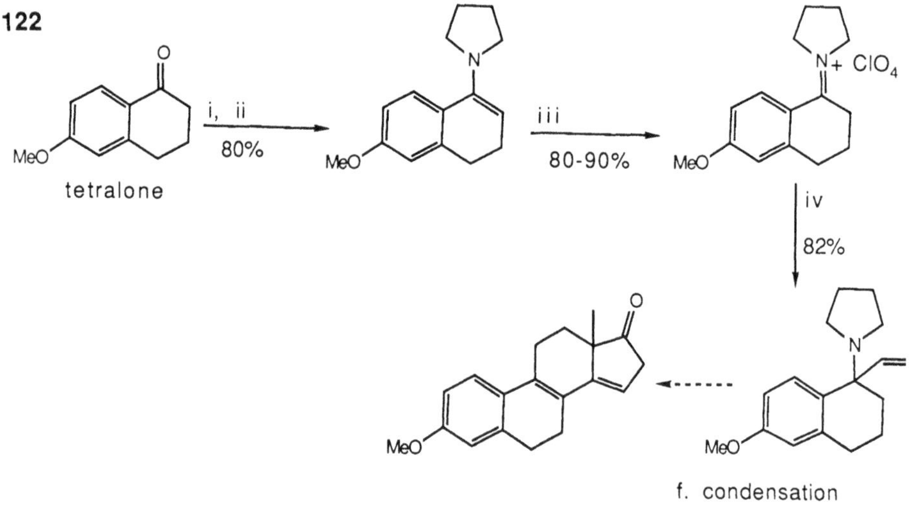
tetralone
i, ii
80%
iii
80-90%
iv
82%
f. condensation

i. pyrrolidine, phH or n-pentane ii. $TiCl_4$, 0-10°C then RT

iii. $HClO_4$, gl. AcOH iv. vinylmagnesium bromide

From U. K. Pandit, F. A. Van der Vlugt, and A. C. Van Dalen, *Tet. Letters*, 3697 (1969).

123

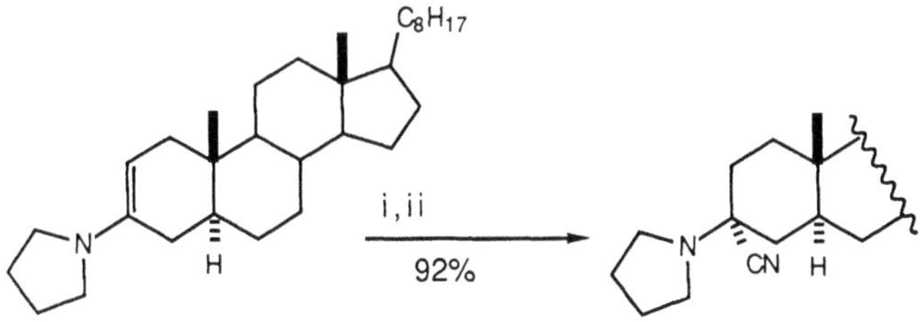

5 α-cholesta-3-one

i. $(C_2H_5O)_2P(O)CN$ (3eq) ii. K_2CO_3 (3eq), THF, reflux, 1h

From S. Harusawa, Y. Hamada, and T. Shiori, *Synthesis*, 716 (1979).

124

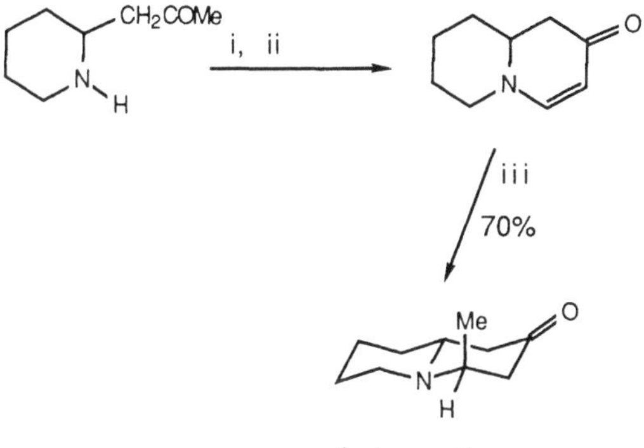

(±)-myrtine

i. AcOCOH, py ii. $Al(OBu^t)_3$, PhMe, reflux iii. MeMgI, PhH

From P. Slosse and G. Hootele, *Tet. Letters*, 4587 (1979).

125

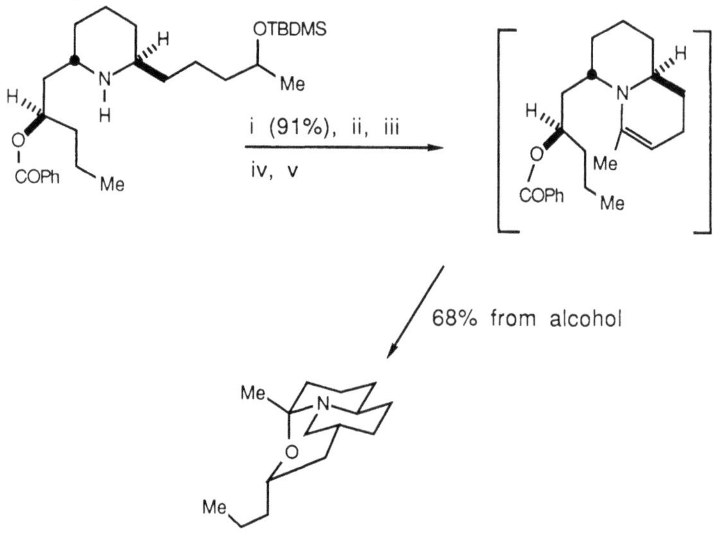

i (91%), ii, iii

iv, v

68% from alcohol

porantheridine

f. condensation

i. 10% aq. HCl, THF, H$_2$ ii. Jones oxidn. iii. aq. Na$_2$CO$_3$, 0°C

iv. KOH, MeOH v. H$_2$O

From E. Gössinger, *Tet. Lett.*, *21*, 2229 (1980) and E. Gössinger, *Monatsh. Chem.*, *111*, 143 (1980).

VIII. α-ARYLATION REACTIONS AND γ-ARYLATIONS FOR DIENAMINES: 126—144

126

19,20-dihydro-20-epi-akuammicine

i. abs. MeOH, sealed tube, 95°C, 50h

From A. I. Scott and C. L. Yeh, *J. Am. Chem. Soc.*, **96**, 2273 (1974).

127

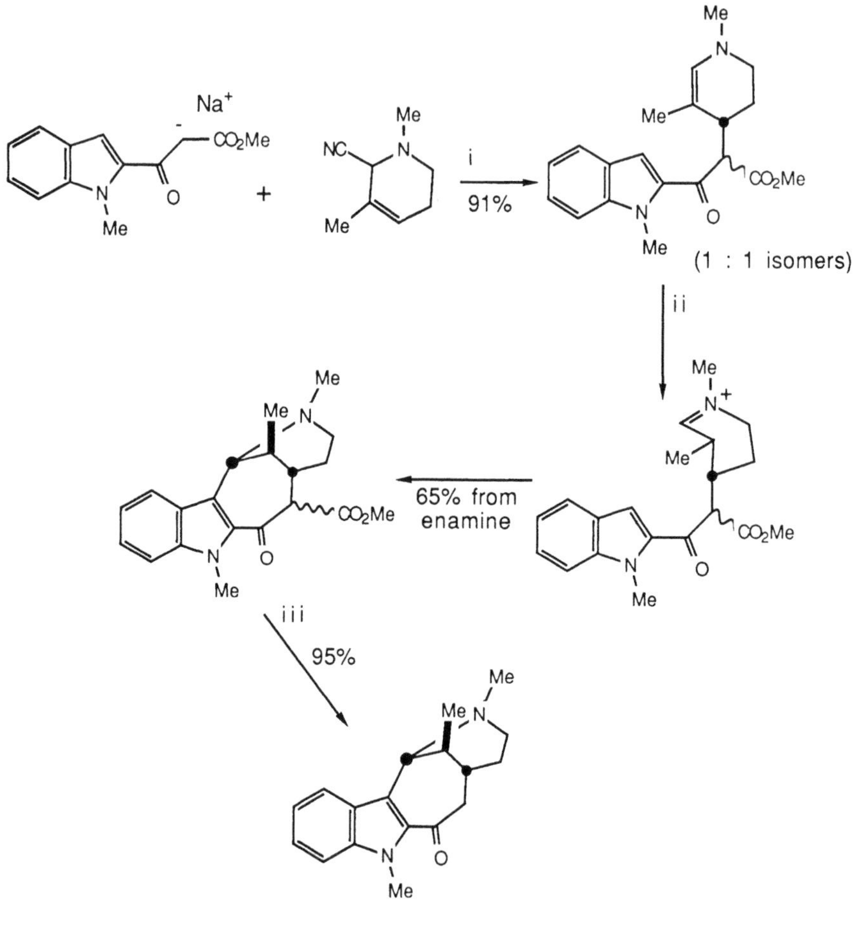

i. AgBF$_4$, THF ii. TsOH, PhMe, heat iii. AcOH, H$_2$O, H$_2$SO$_4$

From M. Harris, D. S. Grierson, C. Riche, and H. P. Husson, *Tet. Letters*. *21*, 1957 (1980).

128

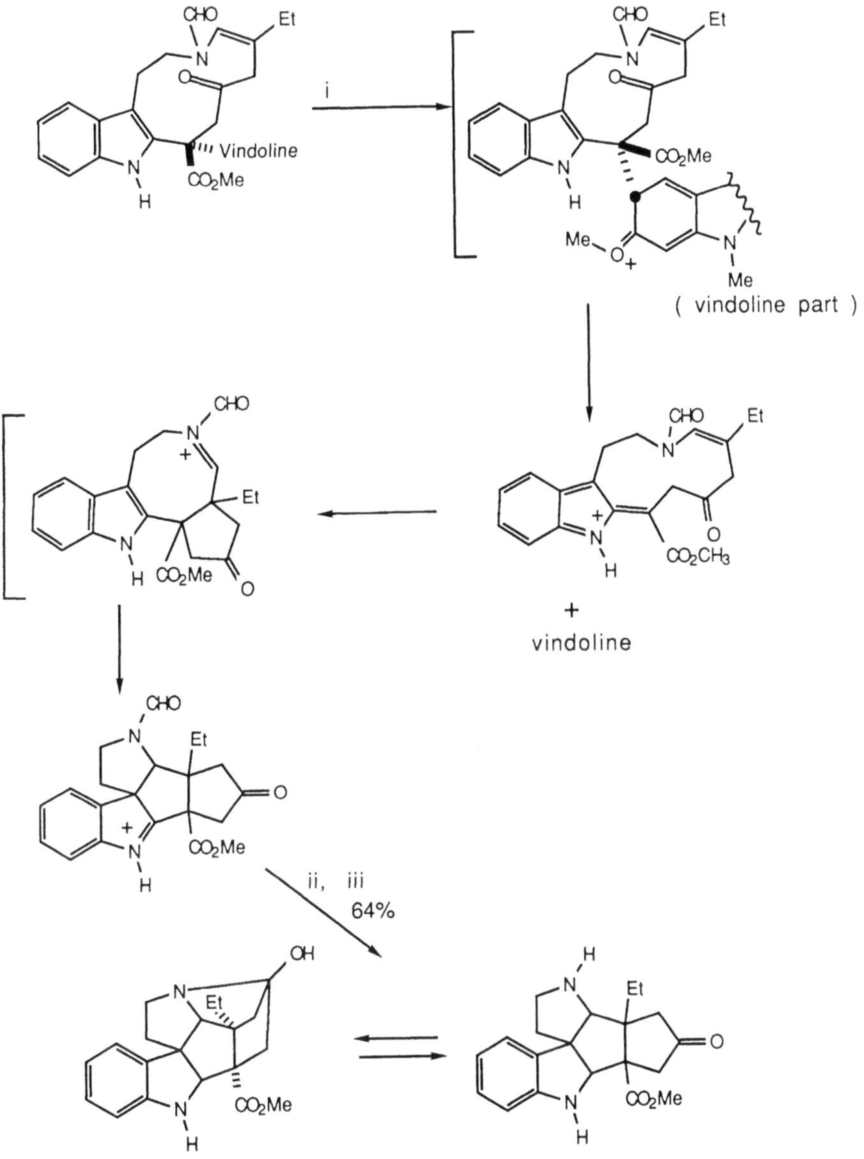

i. TFA   ii. NaBH₃CN   iii. HCl/MeOH

From R. Rasoanaivo, N. Langlois, A. Chiaroni, and C. Riche, *Tetrahedron*, **35**, 641 (1979).

129

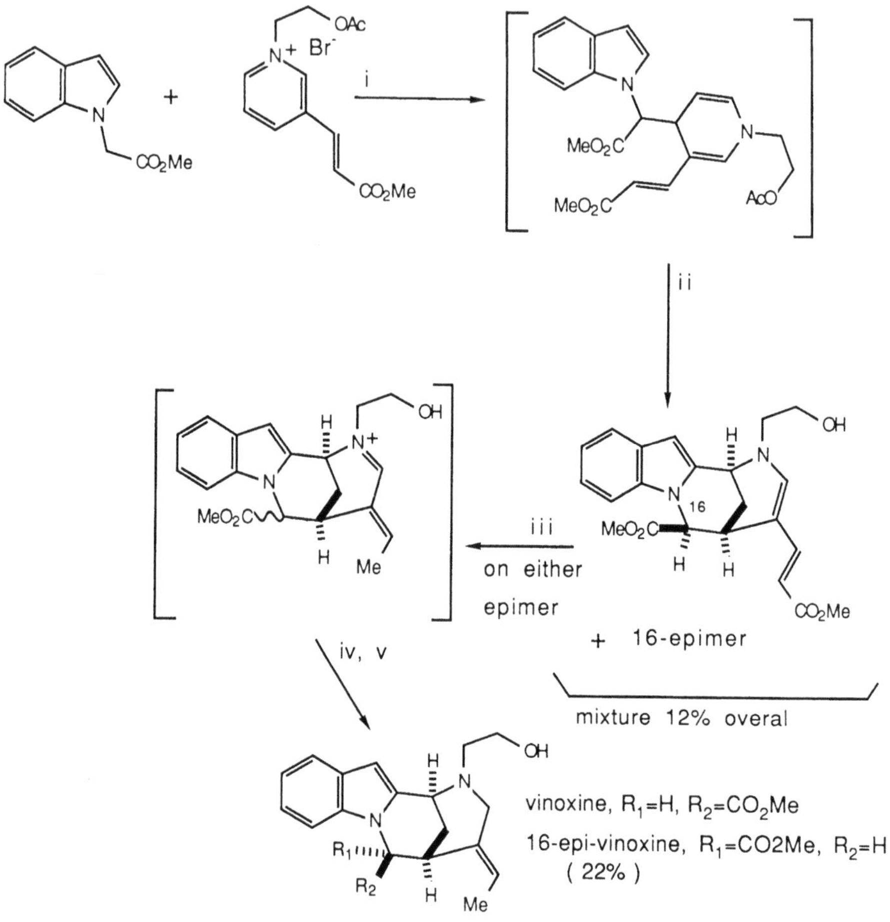
OAc
Br⁻
i
MeO2C
AcO
ii
OH
16
iii
on either epimer
+ 16-epimer
mixture 12% overal
iv, v
vinoxine, R1=H, R2=CO2Me
16-epi-vinoxine, R1=CO2Me, R2=H
(22%)

f. pyridinium salt

i. LDA, THF, -30°C ii. HCl, PhH, pH=3.5-4 iii. 4M HCl
iv. 1.5M HCl, MeOH, RT v. NBH, MeOH, 0°C

From J. Bosch, M. L. Bennasar, E. Zulaica, and M. Feliz, *Tet. Lett.*, *25*, 3119 (1984).

130

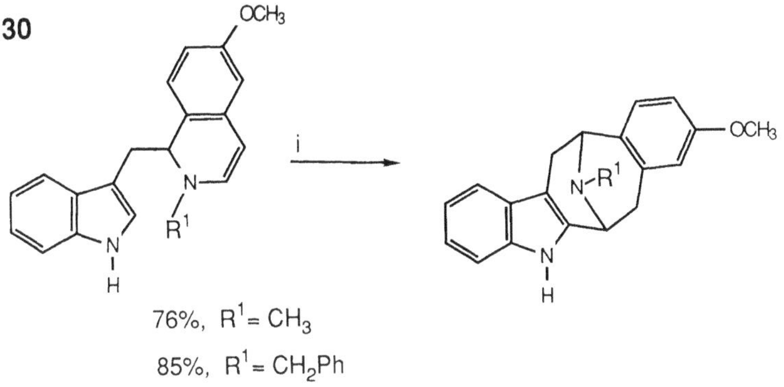

76%, R^1 = CH$_3$

85%, R^1 = CH$_2$Ph

i. excess dil. aq. HCl

From H. Zinnes, F. R. Zuleski and J. Shavel, Jr., *J. Org. Chem.*, *33*, (9), 3605 (1968).

131

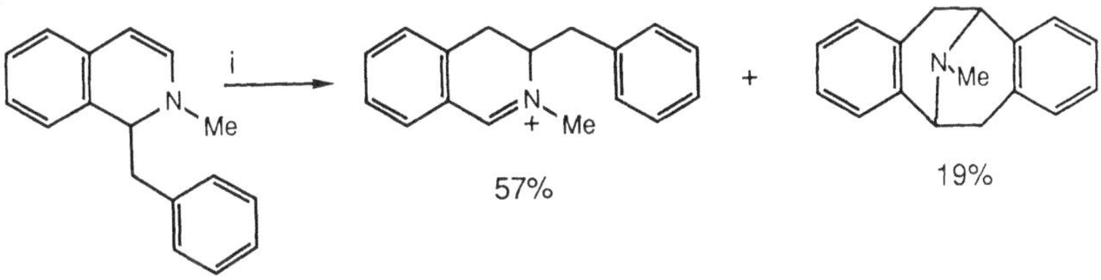

57%

19%

From M. Sainsbury, D. W. Brown, S. F. Dyke, R. G. Kinsman, and B. J. Moon, *Tetrahedron*, *24*, 6695 (1968).

132

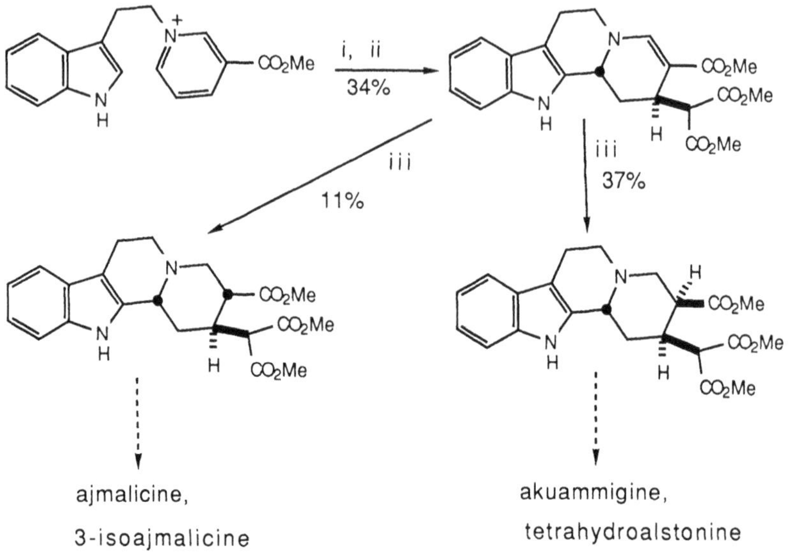

i, ii
34%

iii
11%

iii
37%

ajmalicine,

3-isoajmalicine

akuammigine,

tetrahydroalstonine

f. reduction

i. NaCH(CO$_2$Me)$_2$-DME, N$_2$ ii. PhH, HCl iii. H$_2$, Pt-AcOH

From E. Wenkert, C. Chang, H. Chewla, D. Cochran, E. Hagaman, J. King, and K. Orito, *J. Am. Chem. Soc.*, *98*, 3645 (1976).

133

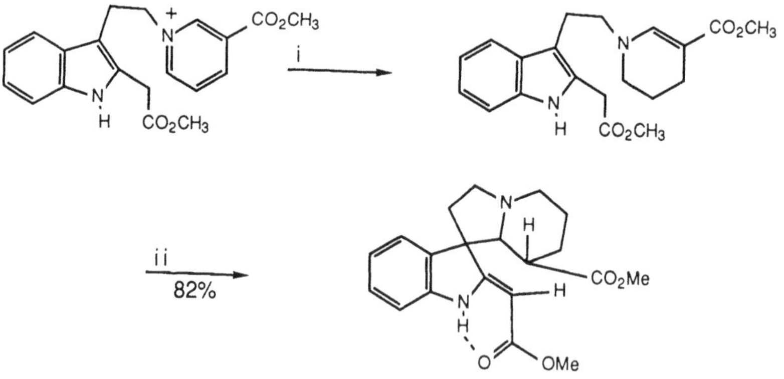

i

ii
82%

i. 10% Pd/C, H$_2$ ii. HCl-MeOH

f. reduction

From E. Wenkert, K. G. Dave, C. T. Gnewuch, and P. W. Sprague, *J. Am. Chem. Soc.*, *90*, 5251 (1968).

134

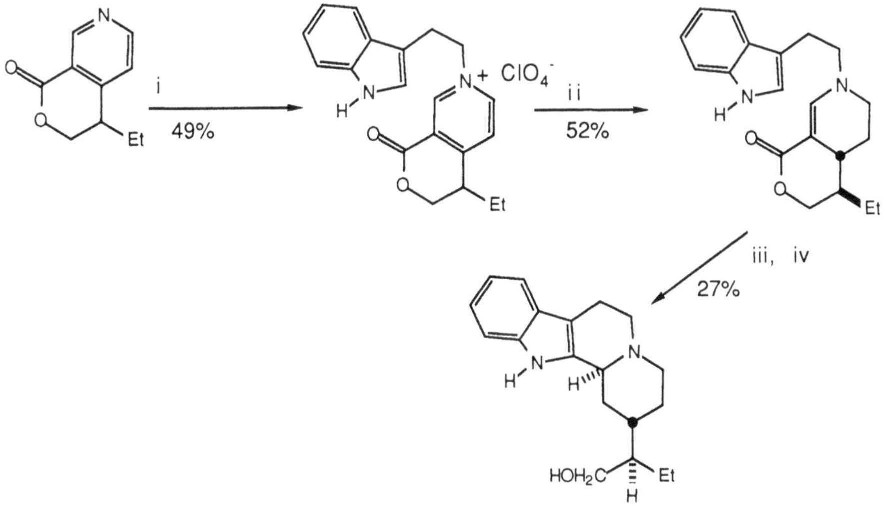

(±)-18,19-dihydroantirhine
f. reduction of pyridinium salt

i. Indol-3-yl-CH$_2$CH$_2$Br, ether, 36h ii. H$_2$, Pd-C, Et$_3$N,

iii. Aq KOH, MeOH, RT, 30h iv. 10% HCl, RT

From E. Wenkert, P. W. Sprague, and R. L. Webb, *J. Org. Chem.*, *38*, 4305 (1973).

135

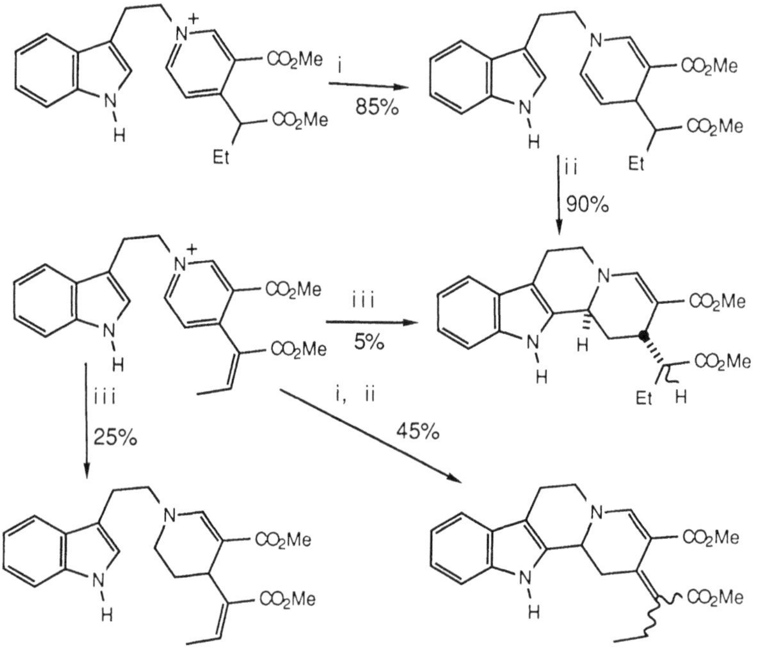

f. redn. of pyridinium salt

i. Na$_2$S$_2$O$_4$, NaHCO$_3$ ii. MeOH, HCl iii. Na$_2$S$_2$O$_4$, H$_2$O, MeOH

From M. Lounasmaa, P. Juu Tinen, and P. Kairisalo, *Tetrahedron, 34*, 2529 (1978), M. Lounasmaa and R. Jokela, *Tetrahedron, 34*, 1841 (1978), M. Lounasmaa and R. Jokela, *Tet. Letters*, 3609 (1978), and M. Lounasmaa, H. Merikallio and M. Puhakka, *Tetrahedron, 34*, 2995 (1978).

136

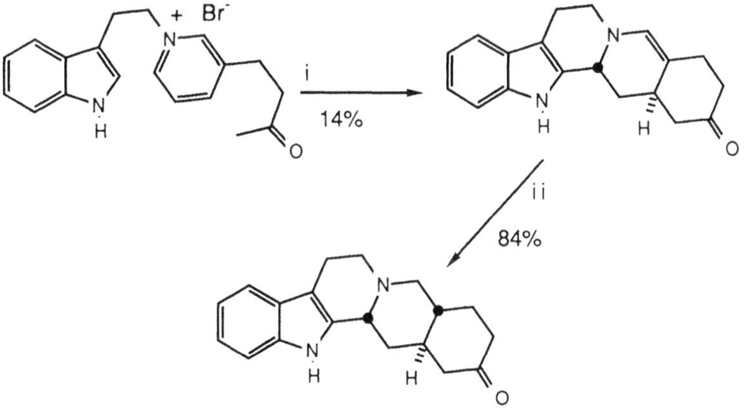

f. reduction

i. KOH-H$_2$O-Et$_2$O, N$_2$ ii. H$_2$-Pd/C, EtOH

From E. Wenkert, C. Chang, H. Chawla, D. Cochran, E. Hagaman, J. King, and K. Orito, *J. Am. Chem. Soc.*, *98*, 3645 (1976).

137

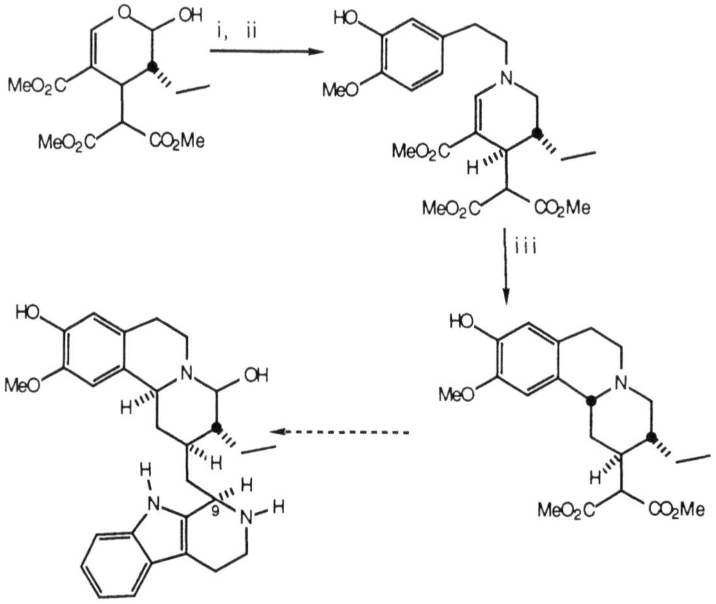

(±)- deoxytubulosine and

f. condensation

i. 3- hydroxy-4- methoxy- phenethylamine   ii. NaBH$_3$CN   iii. H$_3$O$^+$, MeOH, heat

From R. T. Brown and M. F. Jones, *Tet. Lett.*, *25*, 3127 (1984).

138

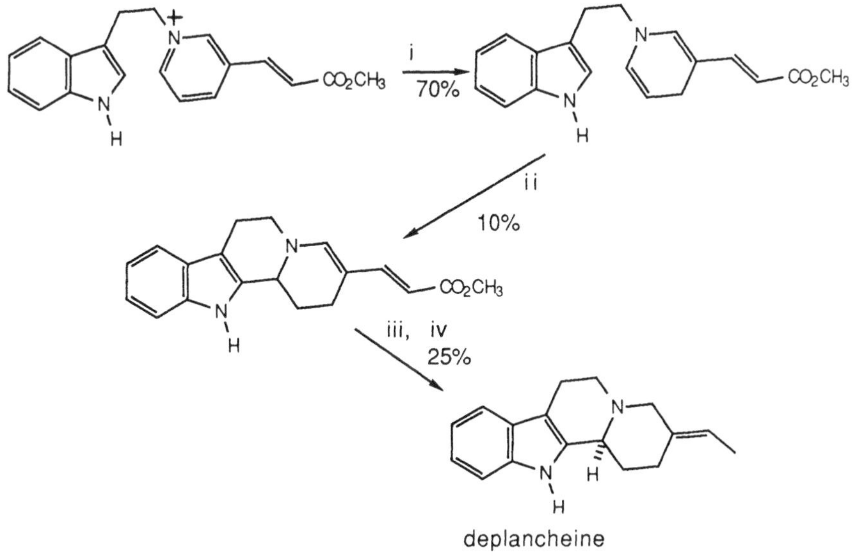

deplancheine

f. pyridinium salt redn.

i. $Na_2S_2O_4$, $KHCO_3$, H_2O, CH_2Cl_2 ii. HCl, MeOH, RT, 5h

iii. 4M-HCl, heat iv. NBH, MeOH

From R. Besselievre, J. P. Cosson, B. C. Das, and H. P. Husson, *Tet. Letters*, *21*, 63 (1980).

139

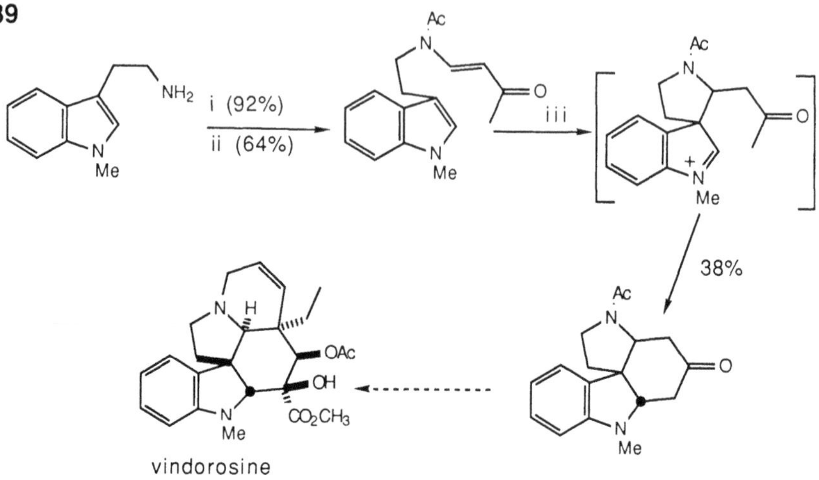

vindorosine

i. ClCH=CHCOCH$_3$, Et$_3$N, EtOH   ii. Ac$_2$O, Et$_3$N, benzene   iii. BF$_3$.Et$_2$O, 90°C, 27 min

From G. Büchi, K. E. Matsumoto, and H. Nishimura., *J. Am. Chem. Soc.*, *93*, 3299 (1971).

140

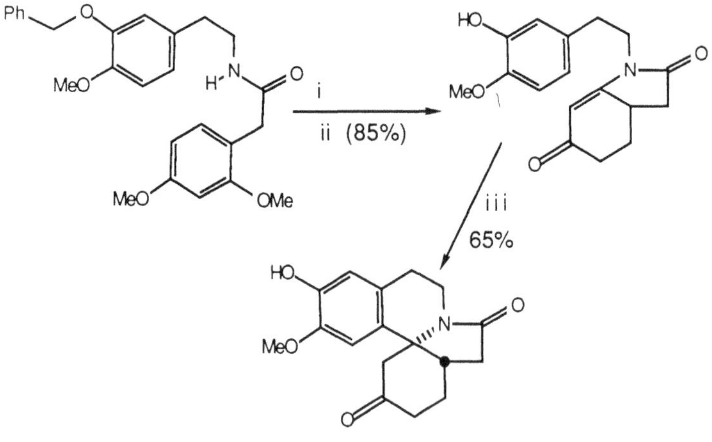

i. liq NH$_3$-Na, NH$_3$-MeOH   ii. 10% H$_2$SO$_4$-DMF   iii. 98% HCO$_2$H

From K. Ito, M. Haruna, and H. Furnkawa, *J. Chem. Soc., Chem. Commun.*, 681 (1975).

141

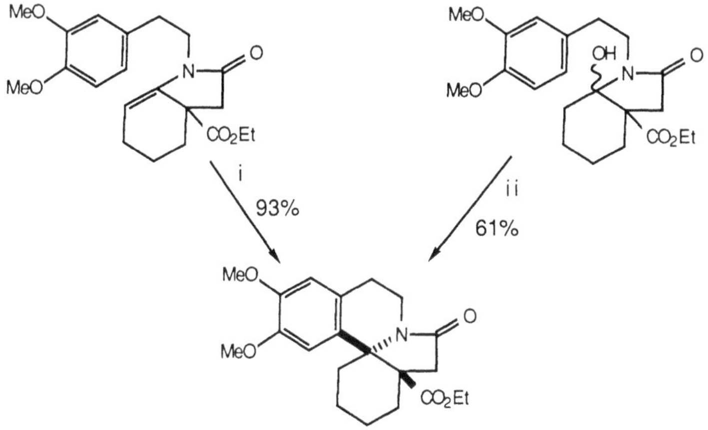

i. H$_3$PO$_4$, heat   ii. HCl-dioxane-H$_2$O, heat

From H. J. Wilkens and F. Troxler, *Helv. Chim. Acta*, 58, 1512 (1975).

142

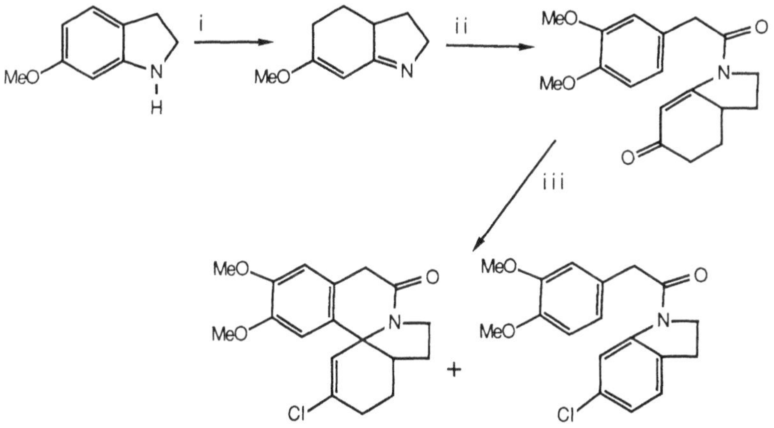

i. Na-liq NH$_3$, MeOH   ii. 3,4-dimethoxyphenylacetylchloride,
Et$_3$N, CHCl$_3$ iii. POCl$_3$, MeCN

From H. Ida, S. Aoyagi, K. Kohno, N. Sasaki, and C. Kibayashi,
Heterocycles, 4, 1771 (1976).

143

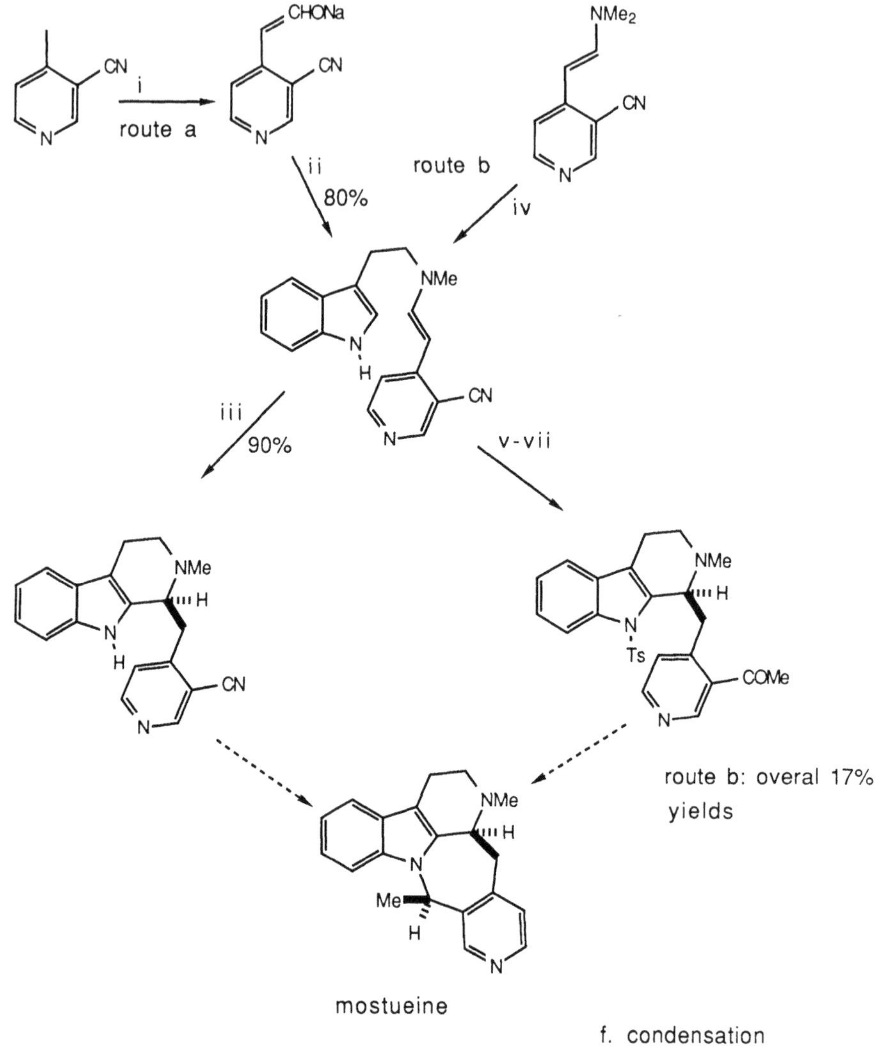
CHONa
CN
NMe2
i
route a
ii
80%
route b
iv
NMe
N
H
iii
90%
v-vii
Ts
COMe
route b: overal 17% yields
Me
mostueine
f. condensation

i. HCO_2Et, NaH ii. N_b-methyltryptamine, AcOH, THF, RT iii. H_2SO_4,THF,0°C, then at RT for 24hrs iv. N_b-methyltryptamine, toluene, TsOH, heat v. HCl, MeOH vi. TsCl, KOH, glyme vii. MeLi, LiBr, THF, -78°C

From M. Onanga and F. Khuong Huu, *Tet. Letters*, *24*, 3627 (1983) and L. R. McGee, G. S. Reddy, and P. N. Confalone, *Tet. Letters*, *25*, 2115 (1984).

144

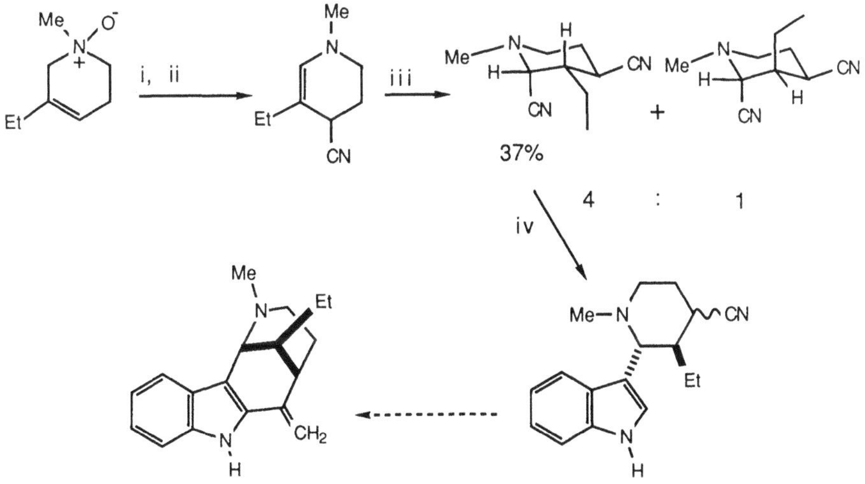

20- epi- uleine

f. Polonovsky reaction

i. TFA anhydride, CH$_2$Cl$_2$ ii. Et$_2$AlCN, PhH iii. KCN, CH$_2$Cl$_2$, H$_2$O
iv. indole, AgNO$_3$, AcOH, H$_2$O, 60°C, 24h

From D. S. Grierson, M. Harris, and H. P. Husson, *Tetrahedron, 39,* 3683 (1983).

IX. OXIDATION REACTIONS, INCLUDING REACTIONS WITH ELECTROPHILIC SULFUR AND HALOGEN: 145–169

145

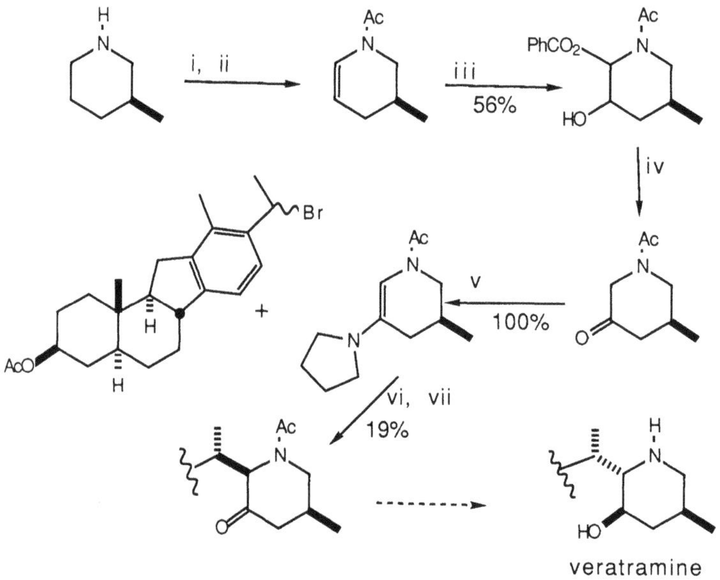

veratramine

i. NaOCl, KOH   ii. Ac$_2$O   iii. PhCO$_3$H   iv. heat in vacuo   v. Pyrrolidine, PhH, reflux
vi. dioxane, sealed tube   vii. H$_2$O

From T. Masamune, M. Takasngi, and A. Murai, *Tetrahedron*, *27*, 3369 (1971).

146

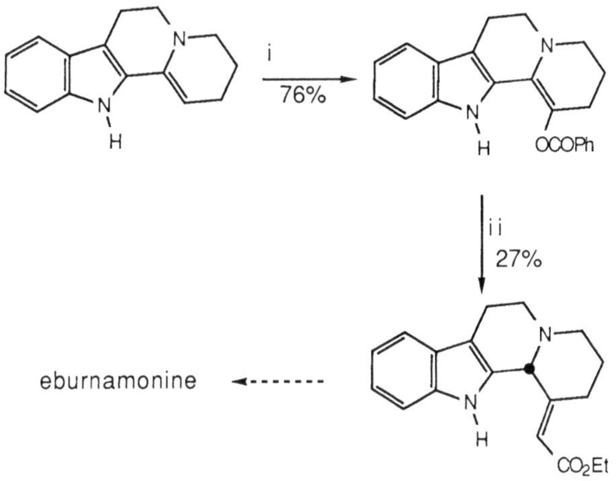

eburnamonine ◄-------

i. (PhCO$_2$)$_2$, hydroquinone, dioxane, 25°C, 30 min then NH$_4$OH

ii. (EtO)$_2$POCH$_2$CO$_2$Et, NaH, DME, N$_2$

From G. Costerousse, J. Buendia, E. Toromanoff, and J. Martel, *Bull. Soc. Chim. Fr.*, Part 2, 355 (1978).

147

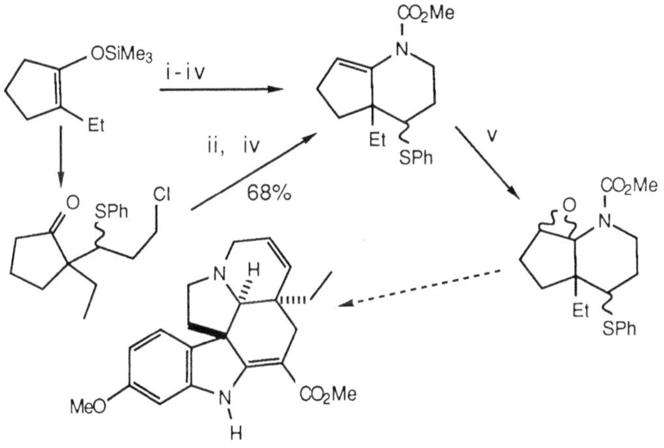

11- methoxy tabersonine

f. condensation

i. PhSCH(Cl)CH$_2$CH$_2$Cl, ZnBr, CH$_2$Cl$_2$, 25°C ii. NaI, MeCOEt, Ar, heat

iii. NH$_3$, CHCl$_3$, RT, heat iv. ClCO$_2$Me, PhNEt$_2$, PhMe v. mCPBA, CHCl$_3$

From L. E. Overman, M. Sworin, and R. M. Burk, *J. Org. Chem.*, 48, 2685 (1983).

148

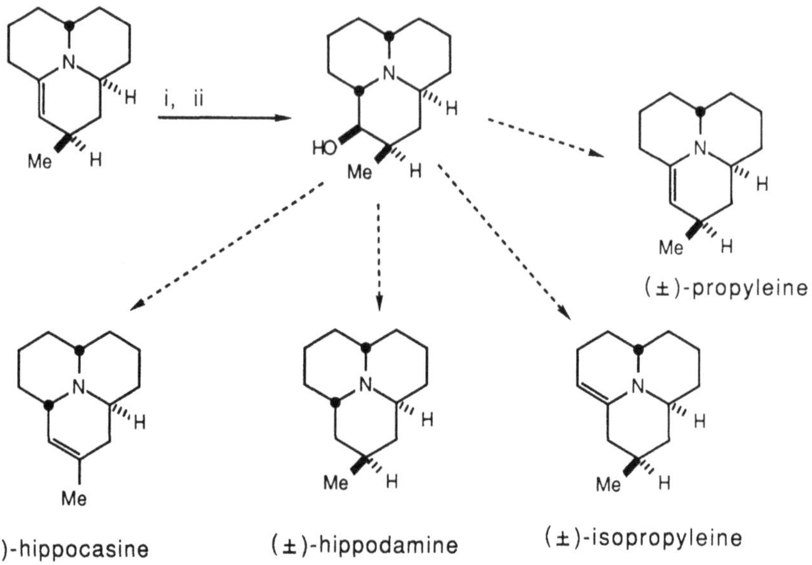

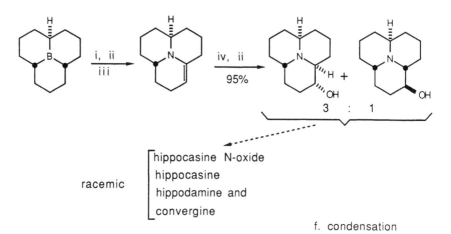

i. BF₃.DMS ii. H₂O₂, NaOH

i. BF$_3$.DMS ii. H$_2$O$_2$, NaOH

(±)-hippocasine   (±)-hippodamine   (±)-isopropyleine

(±)-propyleine

racemic

hippocasine N-oxide
hippocasine
hippodamine and
convergine

f. condensation

95%

3 : 1

i. ClNHDNP, ii. H$_2$O$_2$, NaOH iii. CrO$_3$, H$_2$SO$_4$ iv. BH$_3$.DMS

From R. H. Mueller and M. E. Thompson, *Tet. Letters*, *21*, 1093 (1980).

149

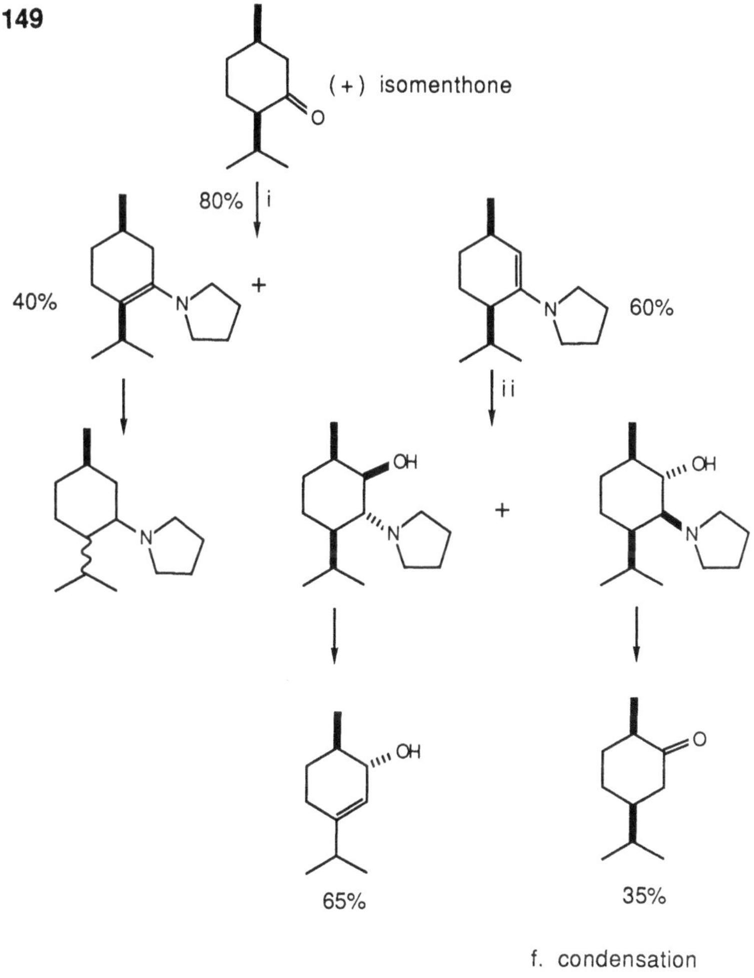

(+) isomenthone

80% | i

40% +

60%

| ii

OH

+

OH

OH

+

O

65%

35%

f. condensation

i. pyrrolidine, $TiCl_4$ ii. B_2H_6, Oxidn.

From J.-J. Barieux and J. Gore, *Bull. Soc. Chim.* France, 3978 (1971).

150

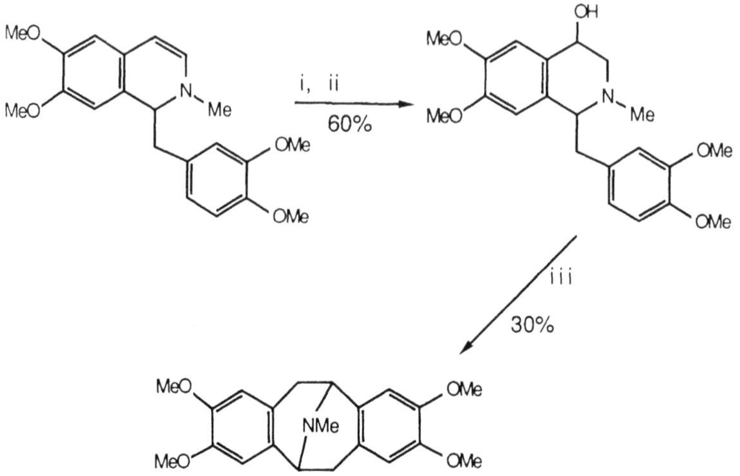

i. B_2H_6 ii. H_2O_2-OH⁻ iii. conc. HCl, 5 days, RT

From S. F. Dyke and A. C. Ellis, *Tetrahedron*, **27**, 3803 (1971).

151

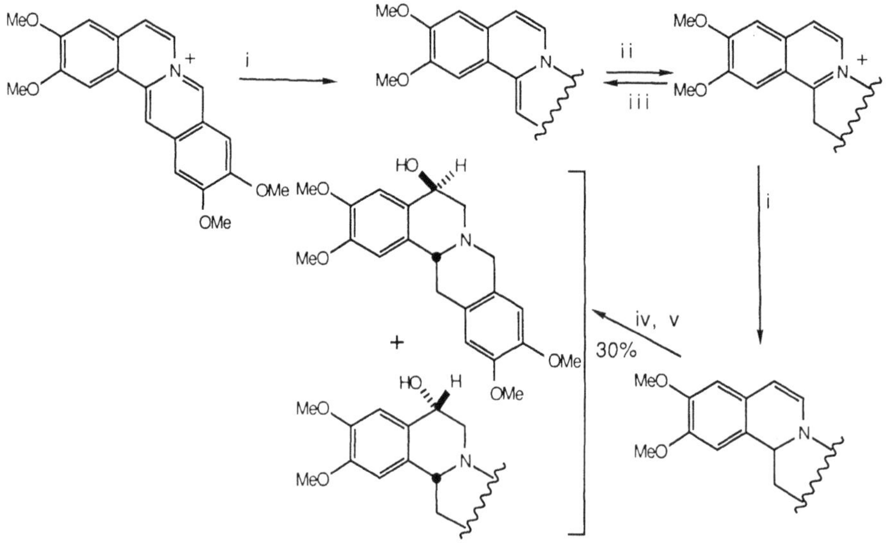

f. reduction of isoquinolinium salt

i. NBH-py ii. H_3^+O iii. ⁻OH iv. hydroboration v. 30% H_2O_2, 20% aq NaOH

From M. Shamma and L. A. Smeltz, *Tet. Letters*, 1415 (1976).

152

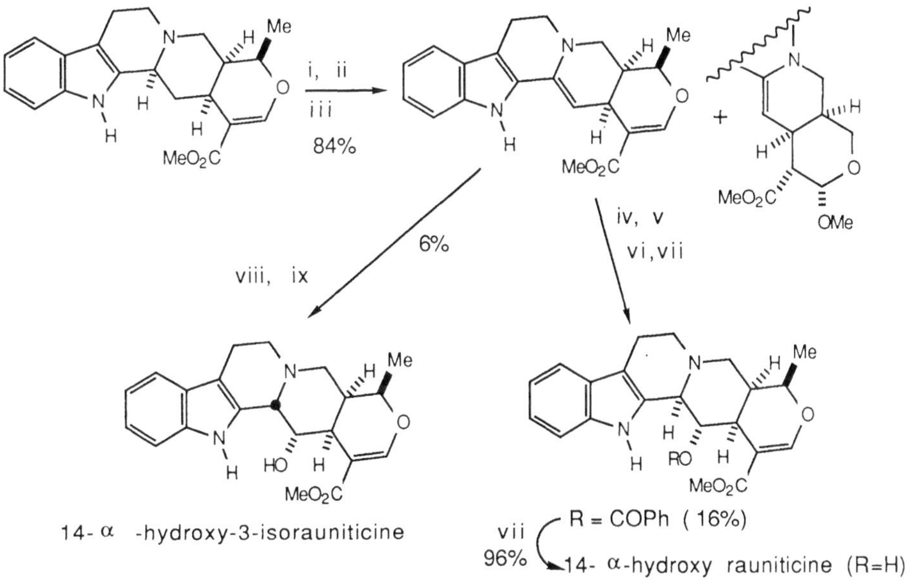

f. oxidation

14-α-hydroxy-3-isorauniticine

vii
96%

R = COPh (16%)
14-α-hydroxy rauniticine (R=H)

i. Bu^{t}OCl ii. HCl, DME iii. KOH, MeOH iv. (PhCO$_2$)$_2$ v. MeOH, HCl
vi. NBH vii. NaOMe viii. BH$_3$, THF ix. 3M-NaOH, 30% H$_2$O$_2$

From E. Yamanaka, E. Maruta, S. Kasamatsu, N. Aimi, and S. Sakai,
Tet. Letters, 24, 3861 (1983).

153

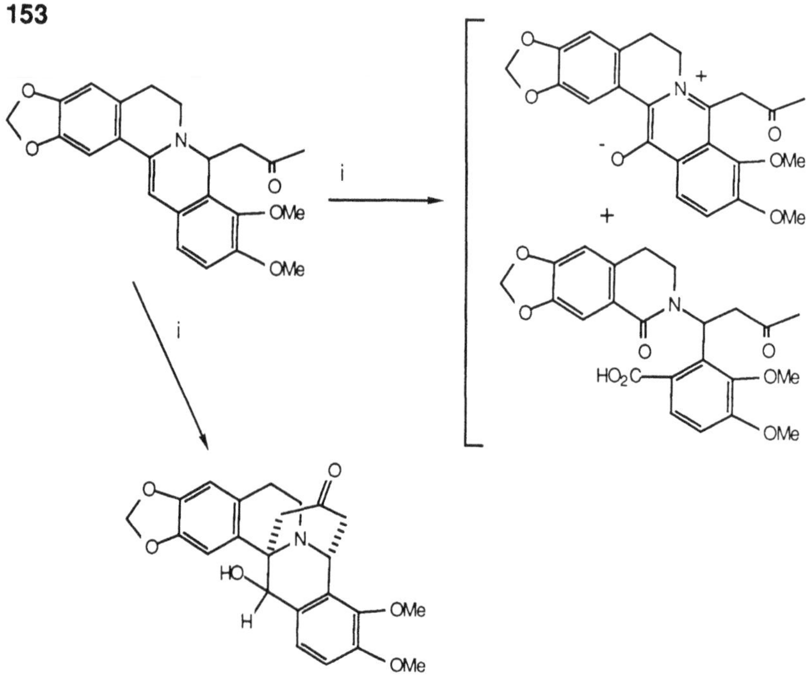

i. KMnO₄

From Y. Kondo and T. Takemoto, *Yakugaku Zasshi*, **95**, 1161 (C. A. 1976, *84*, 17585) (1975) and S. Naruto, H. Nishimura, and H. Kaneko, *Chem. Pharm. Bull.*, Japan, **23**, 1276 (1975).

154

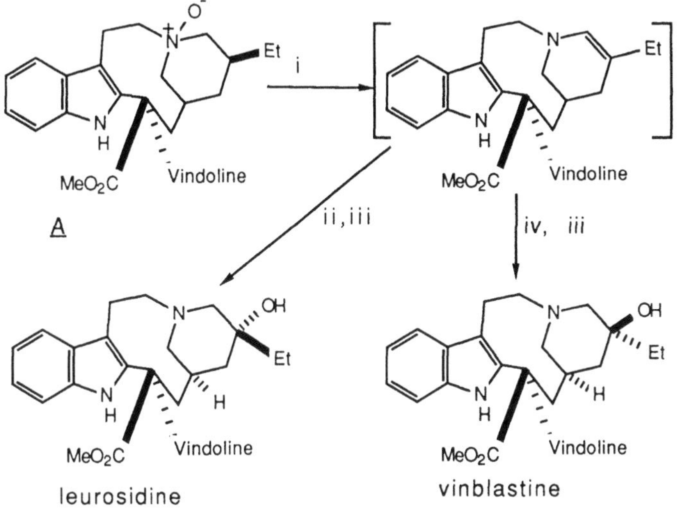

leurosidine

25% from **A**

vinblastine

30% from **A**

f. oxidation

i. Polonovski reaction   ii. OsO₄   iii. NBH   iv. Tl(OAc)₃

From N. Langlois and P. Potier, *Tet. Lett.*, *14*, 1099 (1976), and
P. Mangeney, R. Z. Andriamialisoa, N. Langlois, Y. Langlois, and
P. Potier, *J. Am. Chem. Soc.*, *101*, 2243 (1979).

155

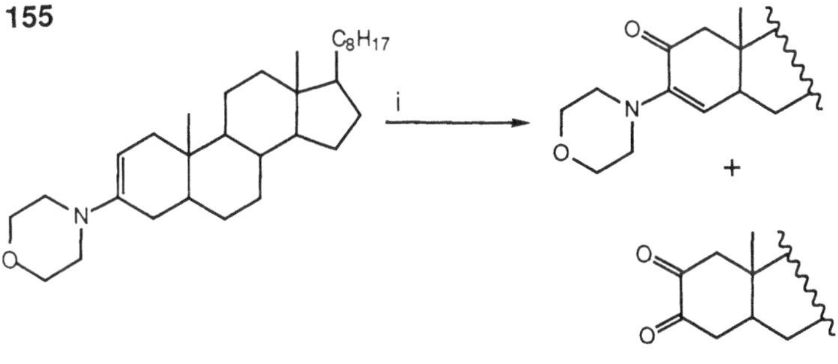

(1:1)   50%

i. singlet O₂, hv , ether, -78°C

From H. H. Wasserman and S. Terao, *Tet. Letters*, 1735 (1975).

156

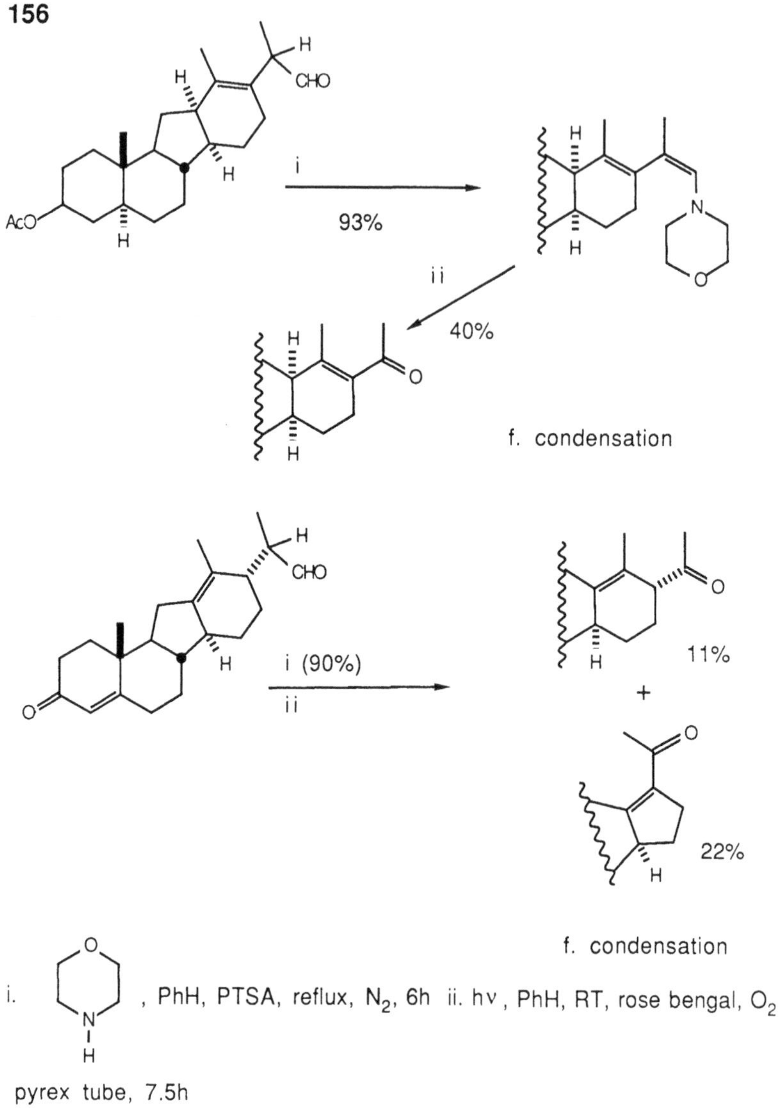
H
CHO
H
H
AcO
H
i
93%
H
N
O
H
ii
40%
H
O
H
f. condensation
H
CHO
H
O
i (90%)
ii
O
H
11%
+
O
H
22%
f. condensation
i.
O
N
H
, PhH, PTSA, reflux, N_2, 6h ii. hν, PhH, RT, rose bengal, O_2
pyrex tube, 7.5h

From A. Murai, C. Sato, H. Sasamori, and T. Masamune, *Bull. Chem. Soc.* (Japan), *49*, 499 (1976).

157

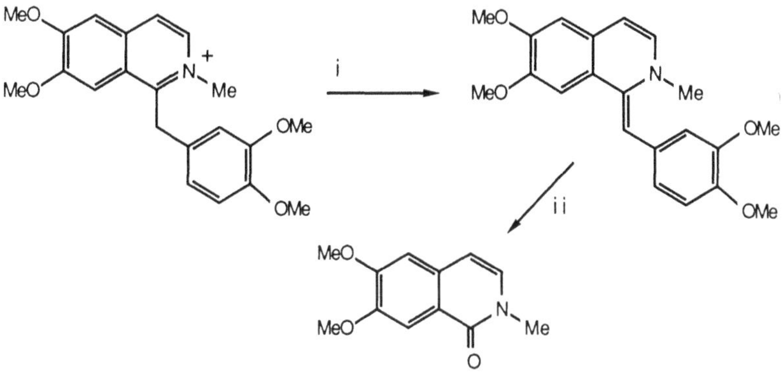

i. NaOH   ii. singlet O₂ or O₂/CuCl

From S. Ruchirawat, *Heterocycles.*, **6**, 1855 (1977).
From S. Ruchirawat, U. Borvornvinvanaut, K. Haintawong, and Y. Thebtaranonth, *Heterocycles.*, **6**, 1119 (1977).

158

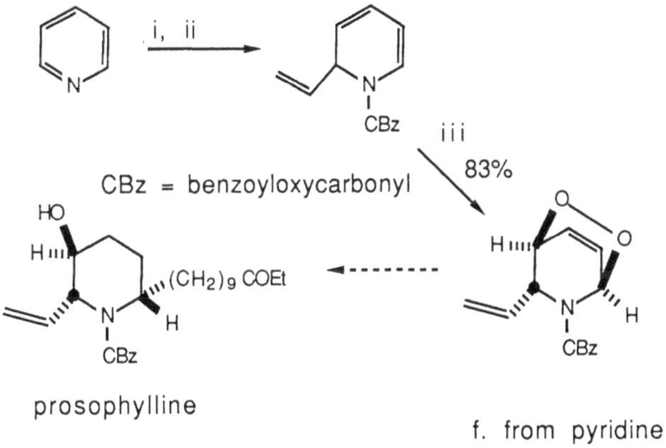

CBz = benzoyloxycarbonyl

prosophylline

f. from pyridine

i. CH₂=CHMgBr, ClCO₂CH₂Ph   iii. photooxidn.

From M. Natsume and M. Ogawa, *Heterocycles*, *16*, 973 (1981).

159

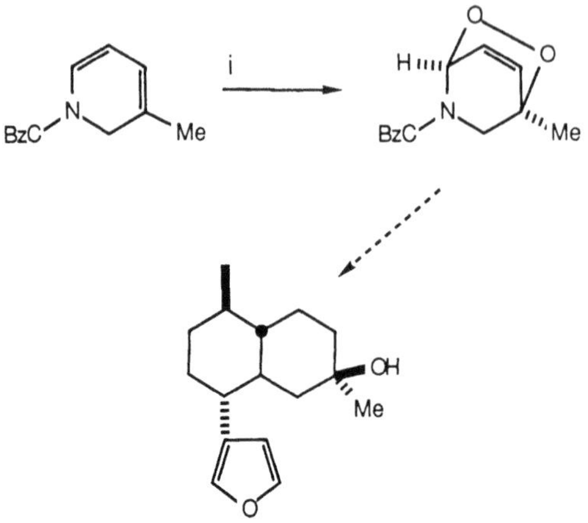

i. O$_2$, hv

From M. Natsume and M. Ogawa, *Heterocycles*, *15*, 237 (1981).

160

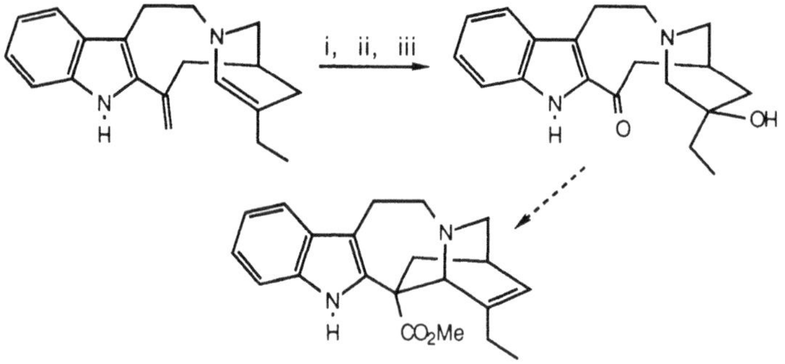

i. OsO$_4$ ii. NBH iii. HIO$_4$ f. oxidation

From J. P. Kutney and F. Bylsma, *J. Am. Chem. Soc.*, *92*, 6090 (1970) and J. P. Kutney, R. T. Brown, E. Piers, and J. R. Hadfield, *J. Am. Chem. Soc.*, *92*, 1708 (1970).

161

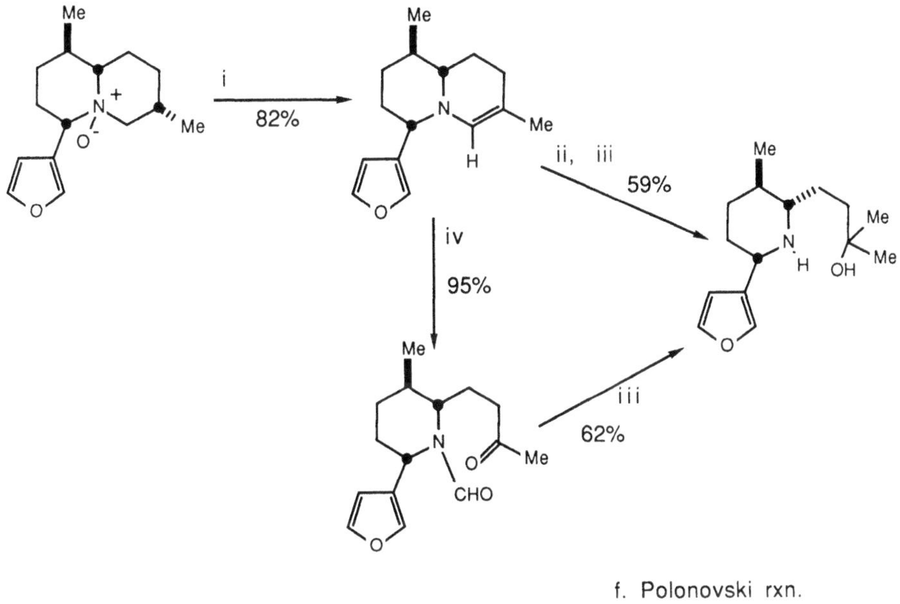

f. Polonovski rxn.

i. Excess Ac$_2$O, 25°C, 75-120hrs   ii. OsO$_4$, HIO$_4$ iii. MeMgI
iv. OsO$_4$, HIO$_4$, Py-H$_2$-dioxane

From R. T. Lalonde, E. Auer, C. F. Wong, and V. P. Muralidharan, *J. Am. Chem. Soc.*, **93**, 2501 (1971).

162

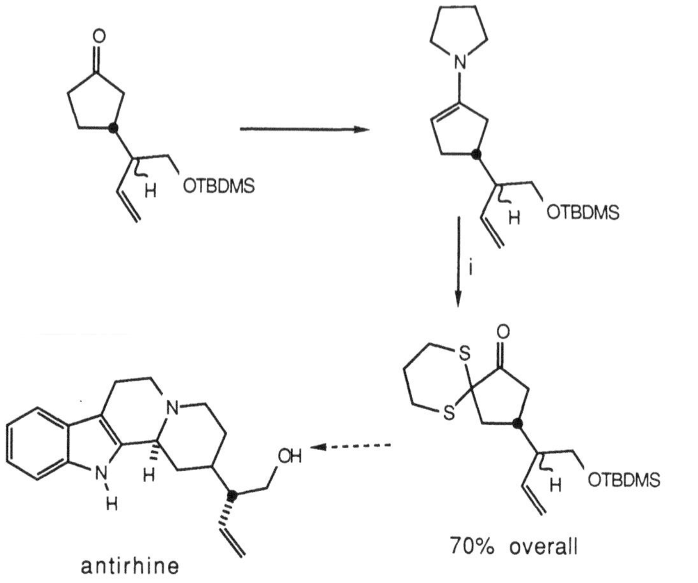

antirhine

70% overall

f. condensation

i. TosS(CH$_2$)$_3$STos

From T. Suzuki, E. Sato, K. Unno, and T. Kametani, *Heterocycles*, 23, 839 (1985).

163

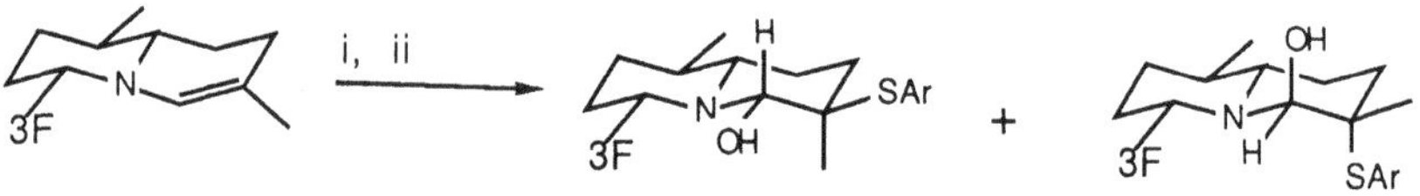

X	Ar	% yield	
SAr	C$_6$H$_5$	0	0
SAr	4-NO$_2$.C$_6$H$_5$	20	20
SO$_2$Ar	C$_6$H$_5$	38	0
Cl	C$_6$H$_5$	75	15

3F =

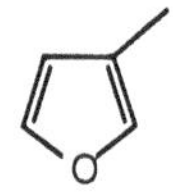

i. CH$_2$Cl$_2$, 25°C, sealed tube, dark   ii. ArSX

From R. T. La Londe and T. S. Eckert, *Can. J. Chem.*, 59, 2298 (1981).

164

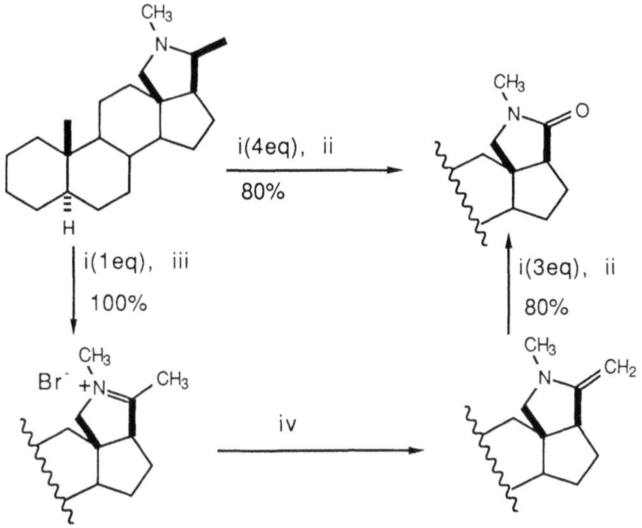

f. oxidation

i. Br$_2$, CH$_2$Cl$_2$ ii. H$_2$O, NaOH iii. Na$_2$CO$_3$ iv OH$^-$

From J. Picot and X. Lusinchi, *Tetrahedron*, *34*, 2747 (1978).

165

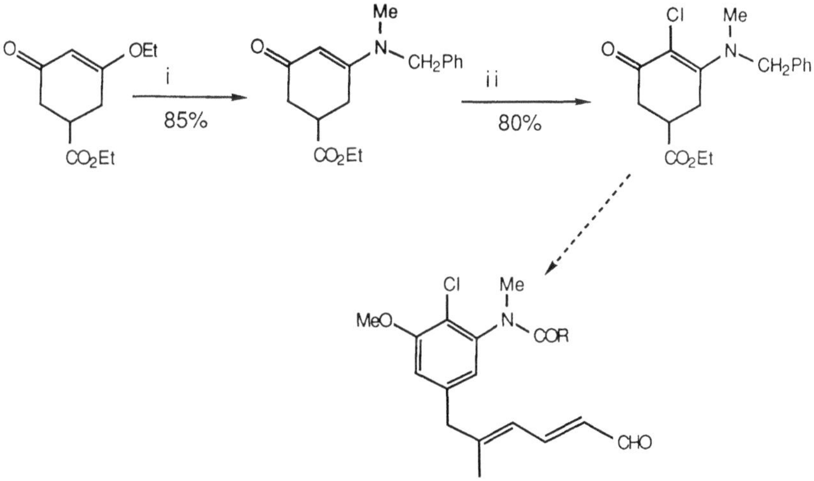

f. condensation

i. N(Me)CH$_2$Ph, 18h ii. t-BuOCl, CHCl$_3$, -50°C

From E. J. Corey, H. F. Wetter, A. P. Kozikowski, and A. V. R. Rao, *Tet. Letters*, 777, (1977).

166

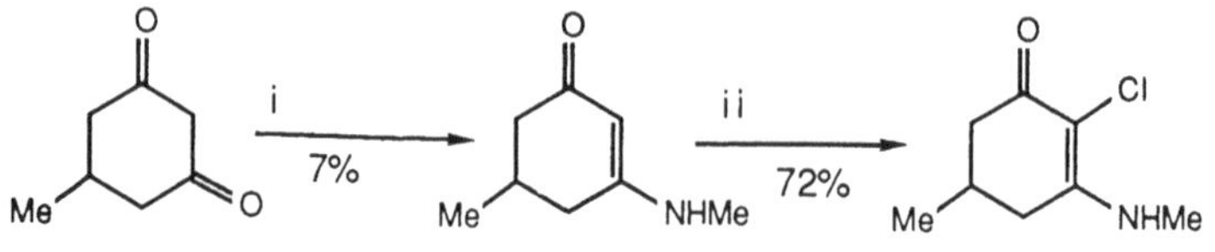

i. 40% MeNH$_2$ ii. NCS　　　　　　　　f. condensation

From J. E. Foy and B. Ganem, *Tet. Letters*, 775, (1977).

167

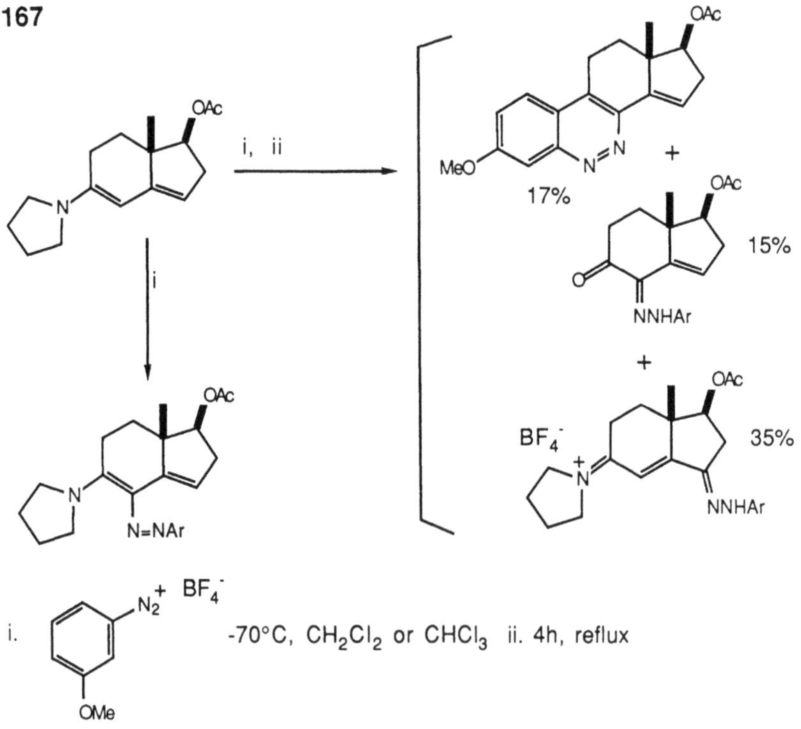

i. (3-methoxyphenyl diazonium N_2^+ BF_4^-, OMe)　-70°C, CH$_2$Cl$_2$ or CHCl$_3$ ii. 4h, reflux

From U. K. Pandit, M. J. M. Pollman, and H. O. Huisman, *J. Chem. Soc., Chem. Commun.*, 527 (1967).

168

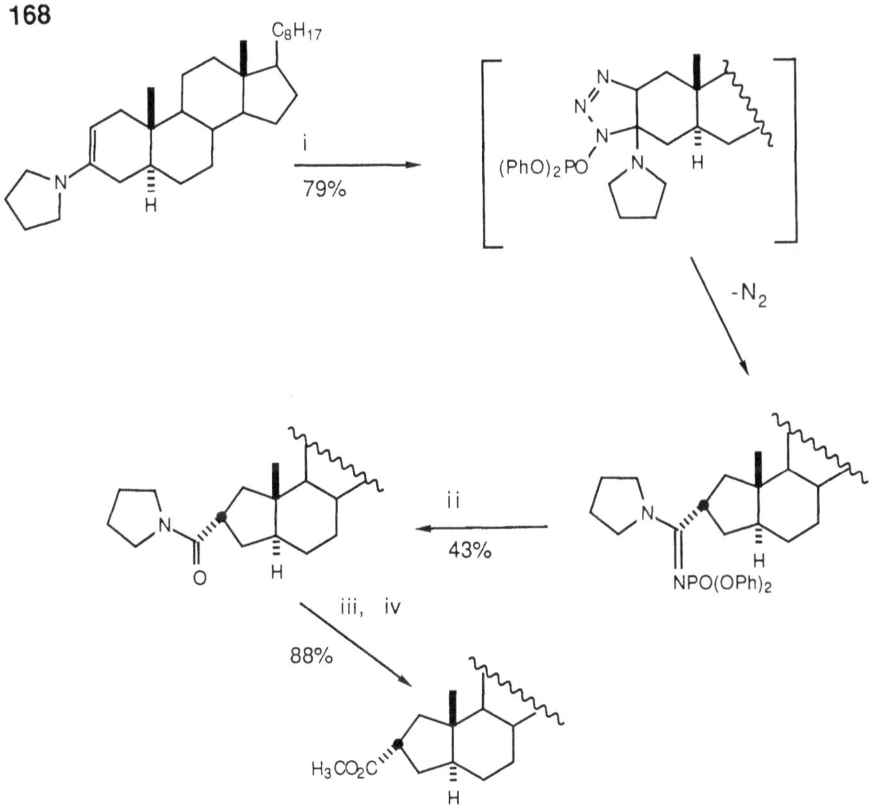

f.1,3-dipolar reaction

i. $N_3PO(OPh)_2$ (2eq) ii. 1N NaOH, dioxane iii. 20% HCl, Dioxane iv. MeOH/HCl

From S. Yamada, Y. Hamada, K. Ninomiya, and T. Shioiri, *Tet. Letters*, 4749 (1976).

169

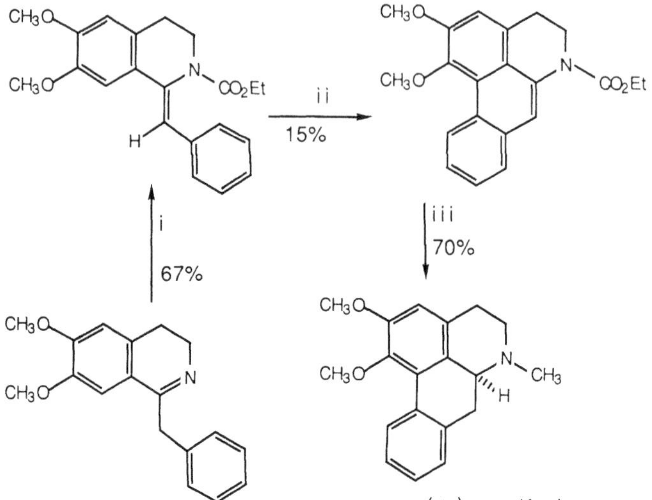

i. ClCO$_2$CH$_2$CH$_3$ ii. hv, I$_2$, EtOH iii. LAH, ether

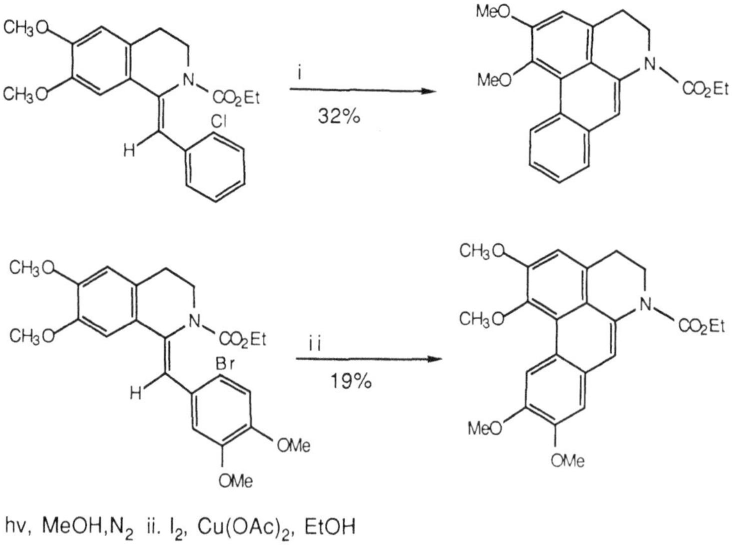

i. hv, MeOH,N$_2$ ii. I$_2$, Cu(OAc)$_2$, EtOH

From M. P. Cava, S. C. Havlicek, A. Lindert, and R. J. Spanglu, *Tet. Letters*, 2937 (1966).

X. REDUCTION REACTIONS: 170—196

In addition to the examples given in this section see also examples 3, 11, 18, 42, 88, 129, 152, 160.

170

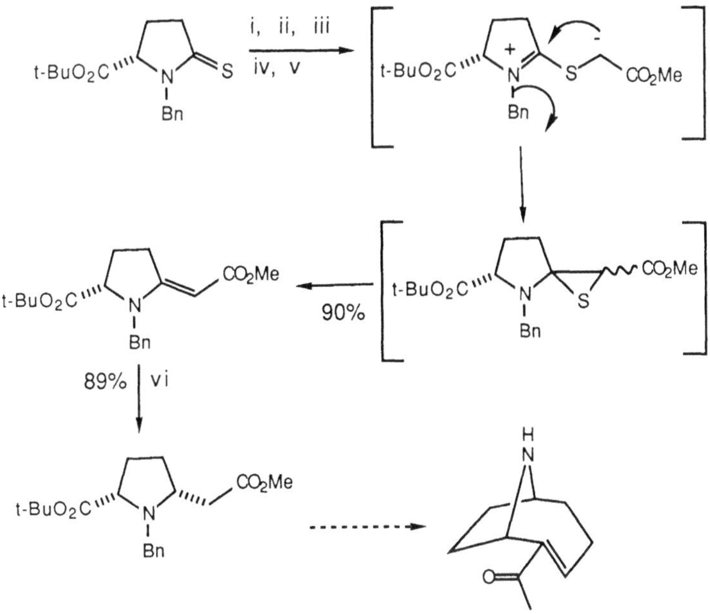

(-)-anatoxin a

f. from amino thioepoxides

i. BrCH$_2$CO$_2$CH$_3$, CH$_3$CN, 40h, rt ii. CH$_2$Cl$_2$, 10 min iii. Ph$_3$P, Et$_3$N, 20h
iv. NaH$_2$PO$_4$ v. CH$_2$Cl$_2$ extrn. vi. 5% Pt/C, EtOAc, 4h

From J. S. Peterson, G. Fels and H. Rapport, *J. Am. Chem. Soc.*, *106*, 4539 (1984).

171

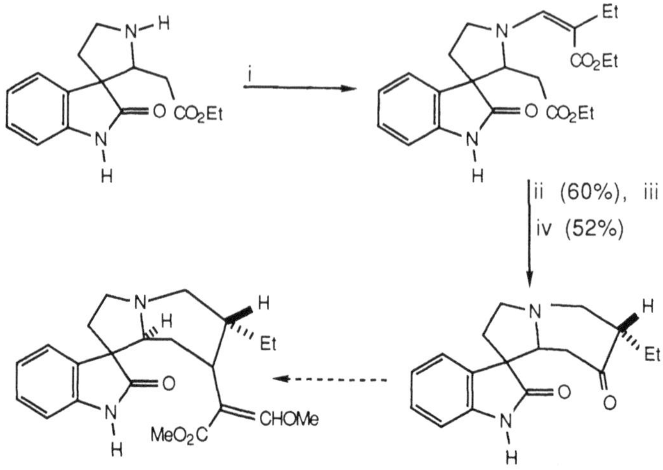

i. EtCH(CHO)-CO₂Et-PhH (using major isomer)   ii. H₂-Pt-AcOH   iii. NaH-PhMe, heat, (using major isomer), iv. aq. HCl, heat.

From Y. Ban, M. Seto, and T. Oishi, *Tet. Letters*, 2113 (1972).

172

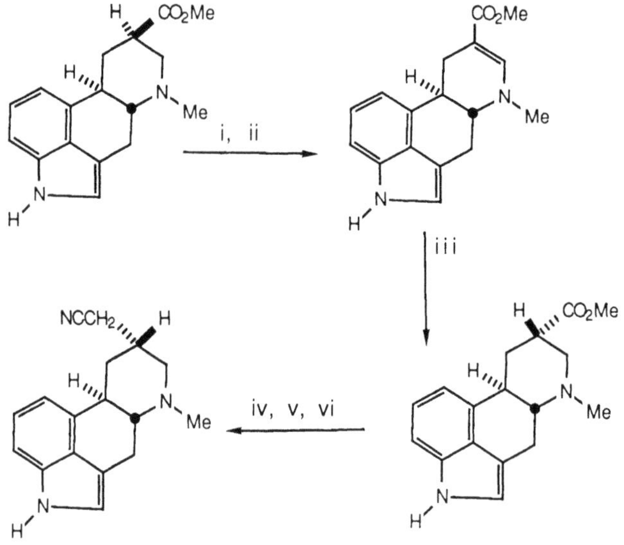

f. amine oxidation

i. mCPBA ii. Ac₂O, TFA iii. H₂, PtO₂, AcOH, DMF iv. LAH v. MeSO₂Cl vi. KCN

From P. L. Stütz, P. A. Stadler, J. M. Vigouret, and A. Jaton, *J. Med. Chem.*, **21**, 754 (1978).

173

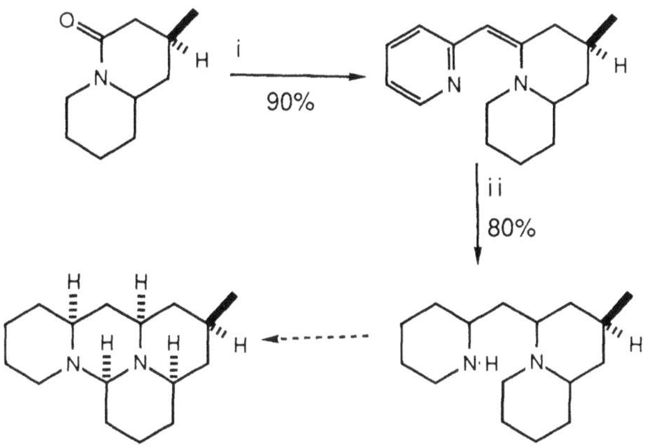

dihydrodeoxyepiallocermine

i. α-picolyl-lithium, THF   ii. H$_2$, Pt-MeCO$_2$H

From Y. Ban, M. Kimura, and T. Oishi, *Chem. Pharm. Bull.*, Japan, *24*, 1490 (1976).

174

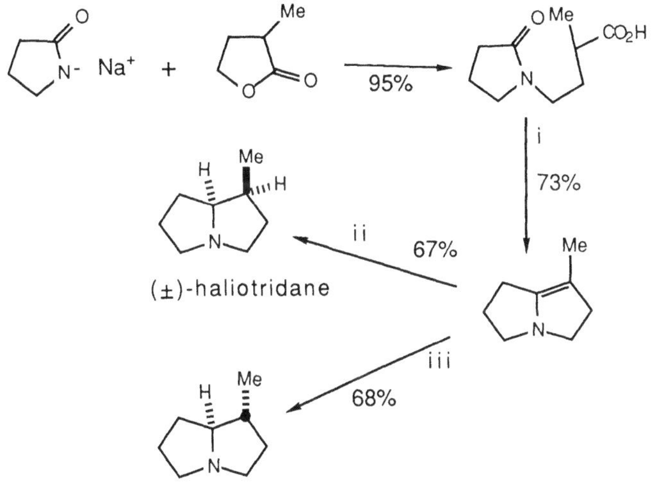

(±)-haliotridane

(±)-psuedohaliotridane

f. from lactam

i. sodalime, heat   ii. cat. hydrogenation   iii. HCO$_2$H, NaOH

From S. Miyano, S. Fujii, O. Yamashita, N. Toraishi, K. Sumoto, F. Satoh, and T. Masuda, *J. Org. Chem. 46*, 1737 (1981).

175

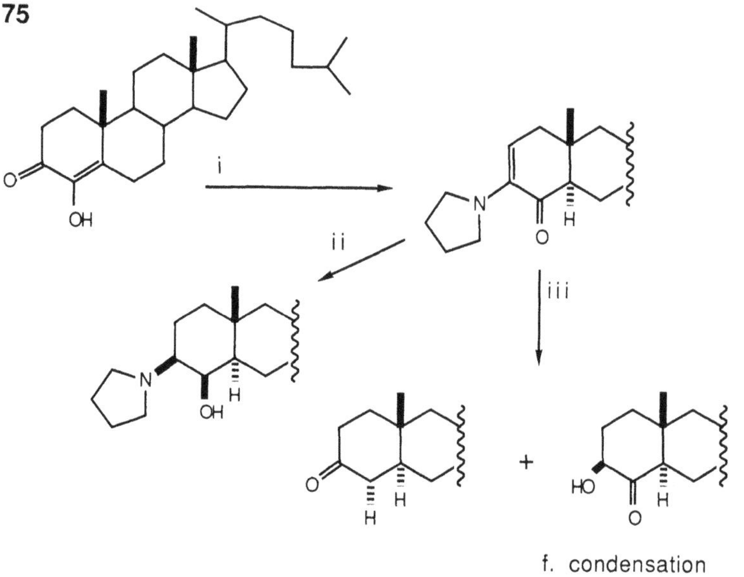
i
ii
iii
OH
N
O
H
OH
HO
+
f. condensation

i. pyrrolidine, MeOH, 30 min, RT ii. cat. redn. iii. LAH, ether, RT, 4h

From C. H. Robinson, L. Milewich, and K. Huber, *J. Org. Chem.*, *36*, 211 (1971).

176

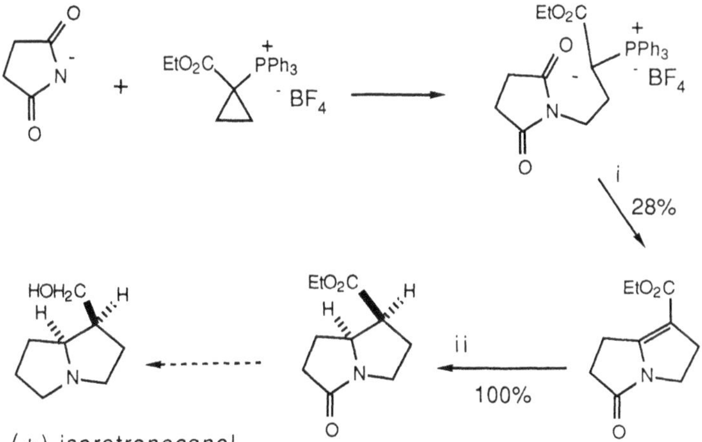

(±)-isoretronecanol

f. from imide

i. heat   ii. 10% Pd/C, H₂, EtOH, 60 psi

From W. Flitsch and P. Wernsmann, *Tet. Letters*, *22*, 719 (1981) and
J. M. Muchowski and P. H. Nelson, *Tet. Letters*, *21* 4585 (1980).

177

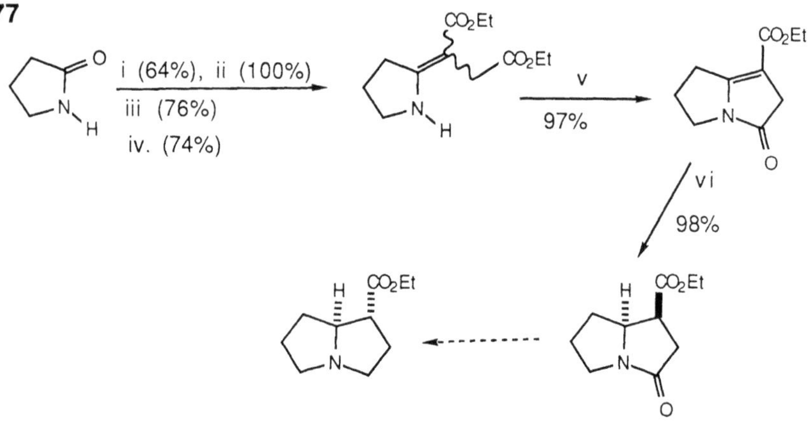

f. from thiolactam
alkylation

i. P_2S_5 ii. $BrCH_2CO_2Et$, $NaHCO_3$ iii. $KOBu^t$, Ph_3P iv. $LiNPr^i_2$,
nBuLi, 0°C, $BrCH_2CO_2Et$ v. KH, 0°C, THF vi. H_2, Pd/C

From H. W. Pinnick and Y. H. Chang, *J. Org. Chem.*, **43**, 4662 (1978).

178

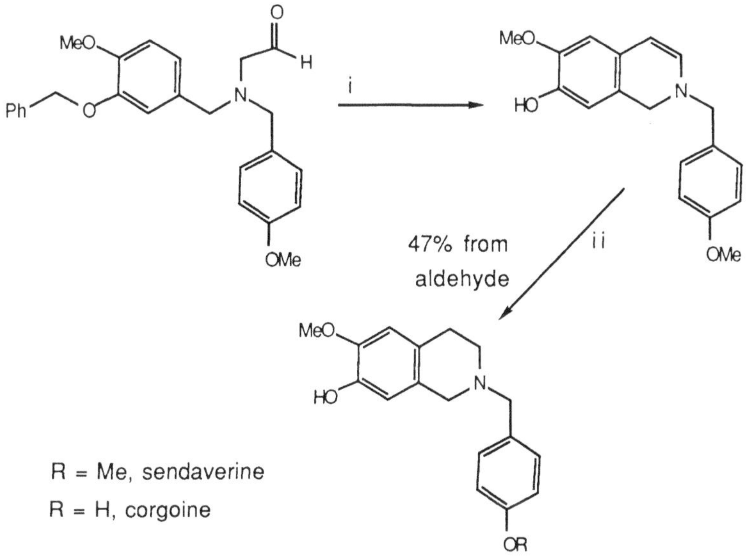

R = Me, sendaverine
R = H, corgoine

f. α -aminoaldehyde arylation

i. 10% HCl in EtOH   ii. NBH

From H. Otomasu, K. Higashiyama, T. Honda, and T. Kametani, *J. Chem. Soc., Perk. Trans I,* 2399 (1982).

179

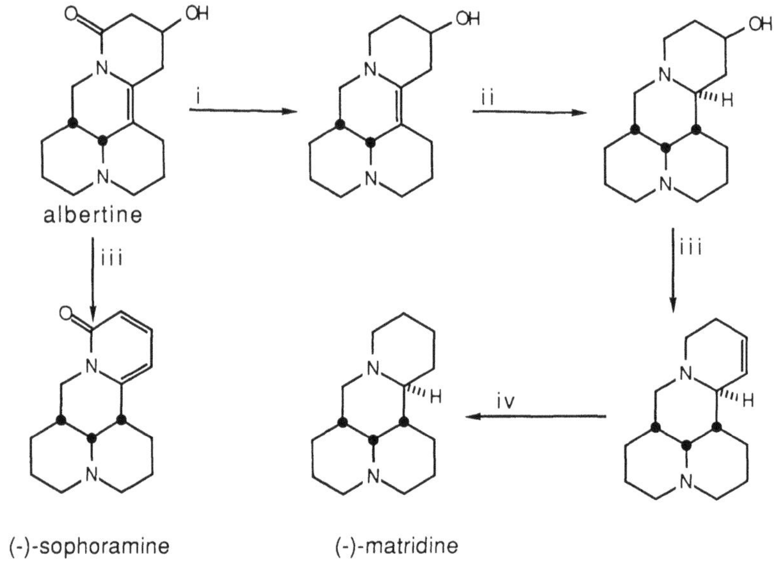
OH
N
N
albertine
i
ii
iii
iv
H
O
(-)-sophoramine
(-)-matridine

i. LAH ii. NBH iii. P_2O_5 iv. H_2-PtO_2

From S. Iskandarov, D. D. Kamalitdinov, and S. Yu. Yunusov. *Khim. Prirod. Soedinenii*, *8*, 628 (C. A. 1973, *78*, 84 616) (1972).

180

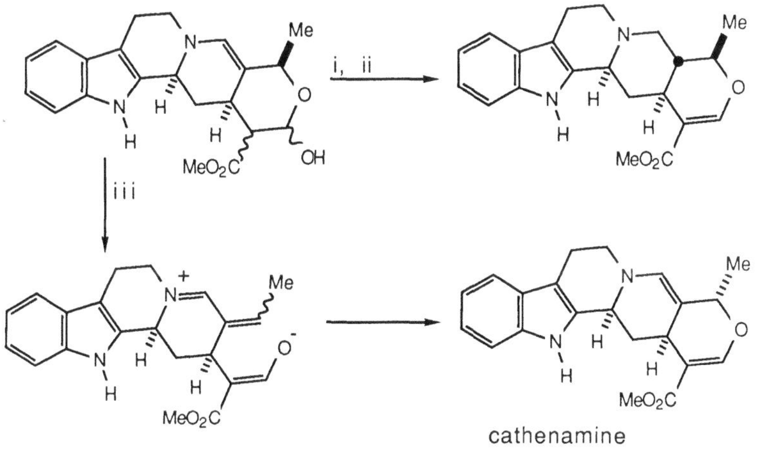

i. NBH   ii. dehydration (TsOH)   iii. CHCl$_3$, silica gel

From C. Kan-Fan and H. P. Husson, *J. Chem. Soc., Chem. Commun.,* 618, (1978).

181

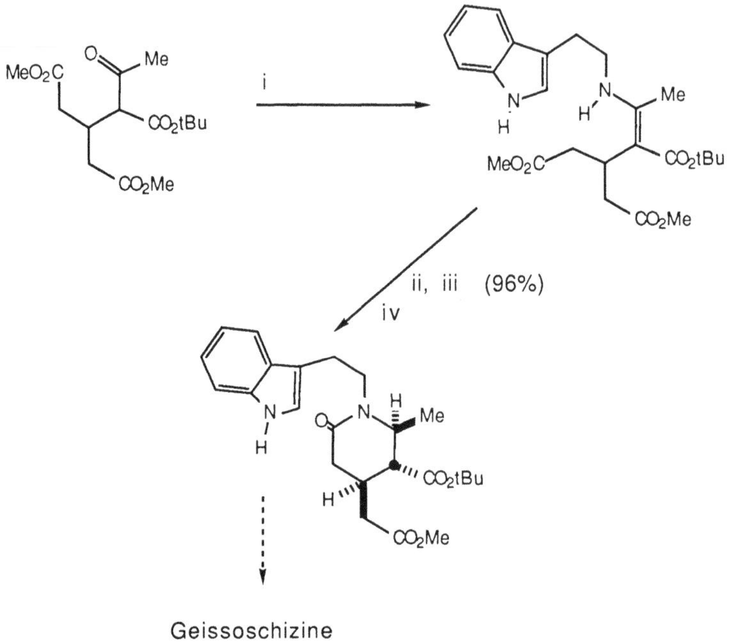

Geissoschizine

i. Tryptamine, dioxane, RT, 2 days   ii. NBH-AcOH   iii. AcOH-py

iv. separate all trans isomer by crystn.

From B. Hachmeister, D. Thielke, and E. Winterfeldt, *Chem. Ber.*, 3825 (1976).

182

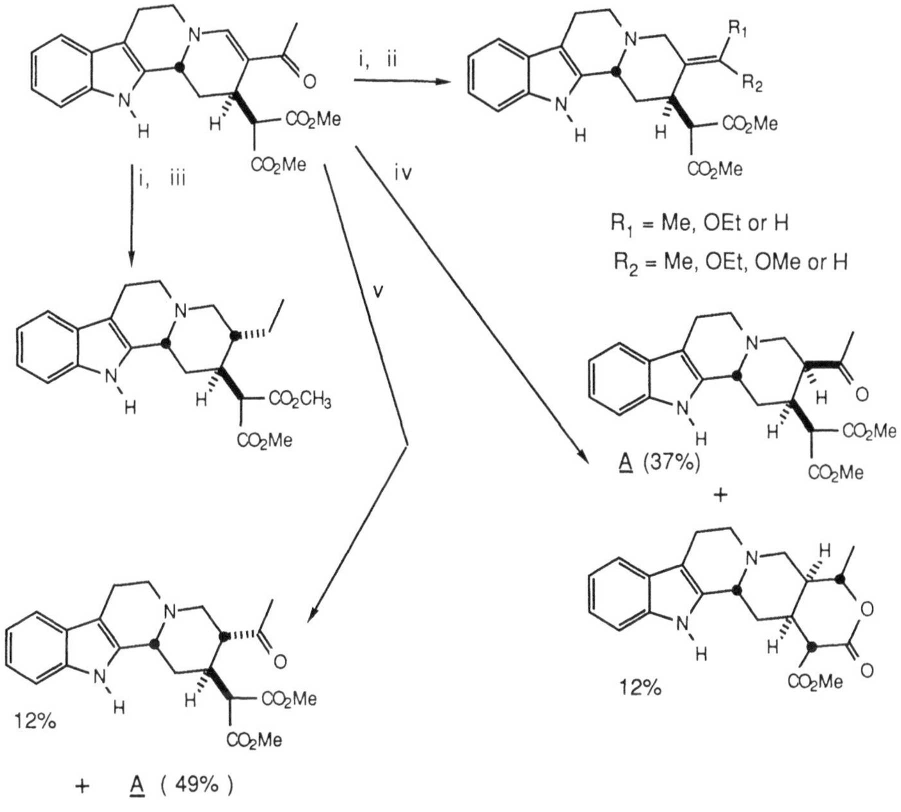

i. EtO⁺.BF₄⁻ ii. NBH, MeOH iii. H₂, Pd/C iv. PtO₂, reduced press.,
AcOH, 15-18°C, 9h v. Pt, H₂, AcOH

From E. Wenkert, C. Chang, H. Chawla, D. Cochran, E. Hagaman,
J. C. King, and K. Orito, *J. Am. Chem. Soc.*, *98*, 3645 (1976).

183

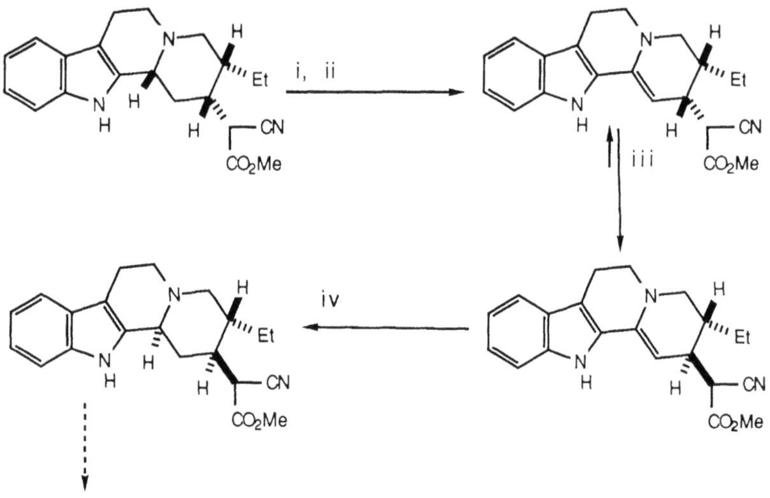

i, ii
iii
iv
H
Et
CN
CO2Me
N
dihydrocorynantheine

f. oxidation

i. HCl,AcOH, $Pb(OAc)_4$, H_2S ii. NH_4OH-H_2O iii. CH_2Cl_2, Argon, 28-48hrs
iv. MeOH-$HClO_4$, NBH-AcOH-MeOH-CH_2Cl_2

From M. Barczai-Beke, G. Dörnyei, G. Toth, J. Tamas, and Cs. Szantay, *Tetrahedron, 32*, 1153 (1976) and M. Barczai-Beke, G. Dörnyei, M. Kajtar, and Cs. Szantay, *Tetrahedron, 32*, 1019 (1976).

184

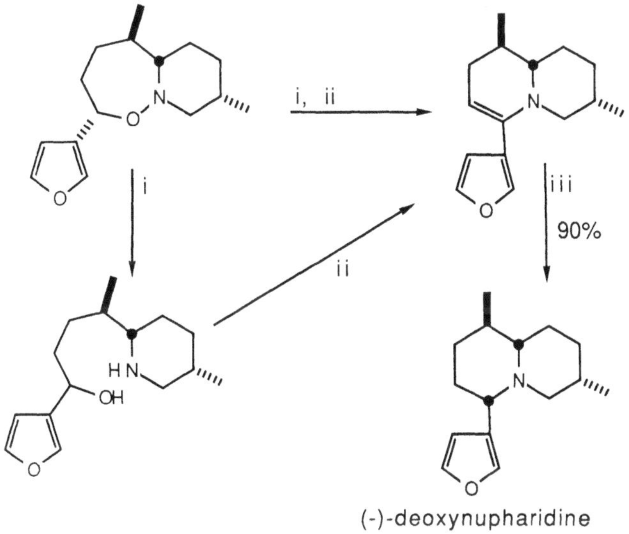

(-)-deoxynupharidine   f. aminoketone

i. Zn, AcOH, H₂O ii. MnO₂   iii. NBH, EtOH

From R. T. La Londe, J. Wooleuer, E. Auer, and C. F. Wong, *Tet. Letters*, 1503 (1972).

185

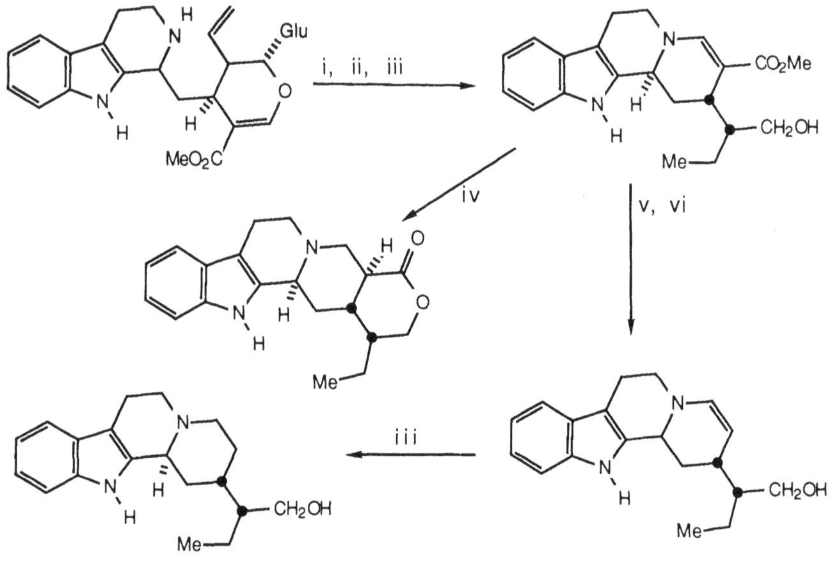

f. vinylogous urethane

i. cat. redn. ii. controlled hydrolysis iii. NBH iv. NBH-AcOH v. mild Basic hydrolysis vi. controlled acid-catalyzed decarboxylation.

From K. T. D. De Sliva, G. N. Smith, and K. E. H. Warren, *J. Chem. Soc., Chem. Commun.*, 905 (1971).

186

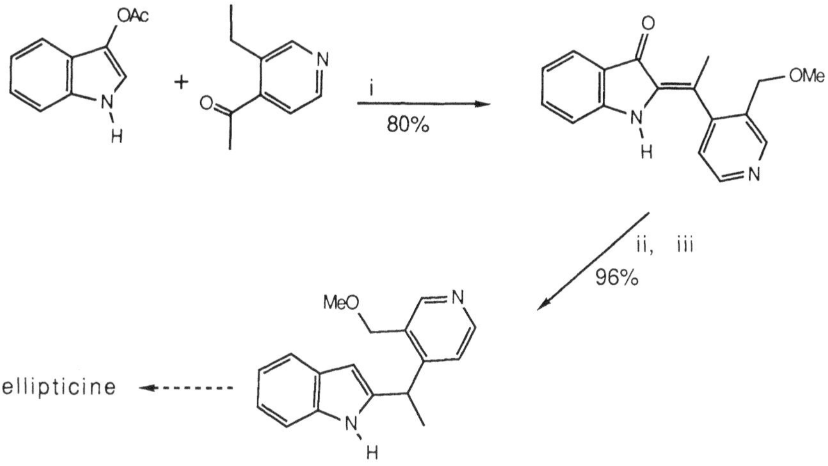

i. KOH, MeOH, 3 days, RT ii. NBH iii. 70% Aq. EtOH, 70°C, 0.5h

From K. N. Kilminstev and M. Sainsbury, *J. Chem. Soc., Perkin I* 2264 (1972).

187

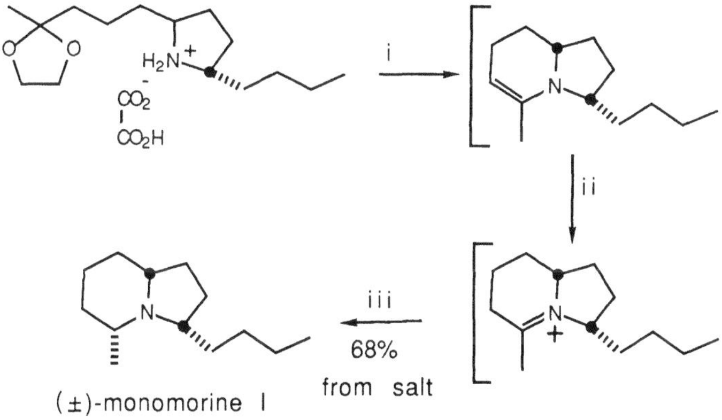

f. condensation

i. H₃O⁺, ⁻OH ii. pH 3.8-5.4 iii. NaCNBH₃

From R. V. Stevens and A. W. M. Lee, *J. Chem. Soc., Chem. Commun.*, 102 (1982).

188

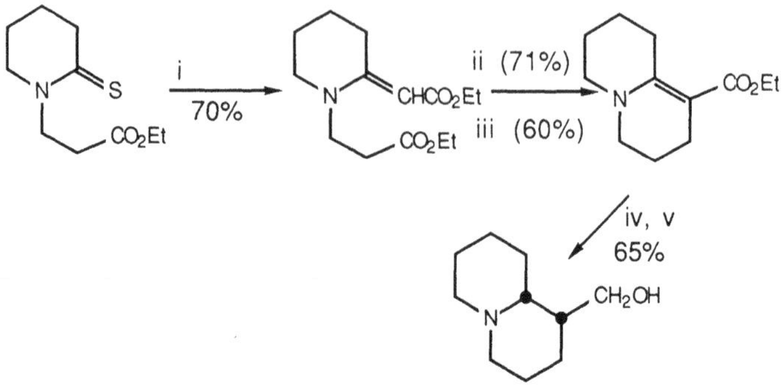

(±)-lupinine

f. thiolactam

i. BrCH₂CO₂Et, Et₃N, Ph₃P ii. LiAlH₄, OEt₂ iii. NaH-TsCl,

MeCN, warm iv. NBH v. LAH

From G. C. Gerrans, A. S. Howard, and B. S. Orlek, *Tet. Letters*,
4171 (1975).

189

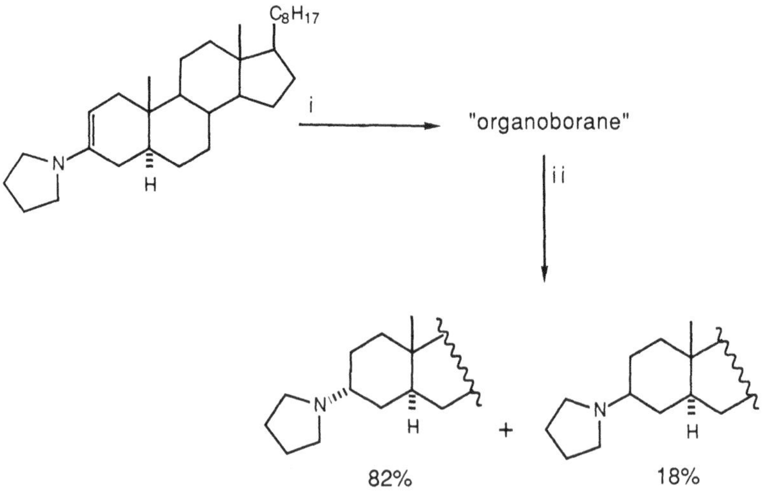

i. B₂H₆, THF ii. CH₃OH, heat

From J. Gore and J. J. Barieux, *Tet. Letters*, 2849 (1970).

190

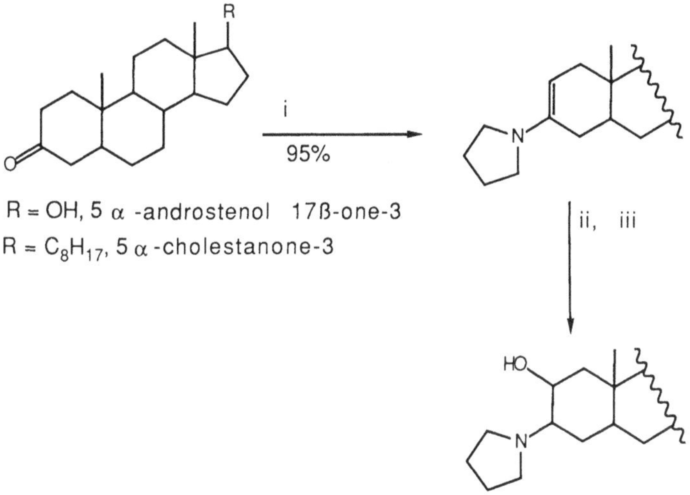

R = OH, 5 α -androstenol   17ß-one-3
R = C$_8$H$_{17}$, 5 α -cholestanone-3

ii, iii

R = OH, (70%)
R = C$_8$H$_{17}$, (80%)

f. condensation

i. pyrrolidine, TiCl$_4$ ii. B$_2$H$_6$ iii. H$_2$O$_2$, ¯OH

From J.-J. Barieux and J. Gore, *Tetrahedron*, *28*, 1537, 1555 (1972).

191

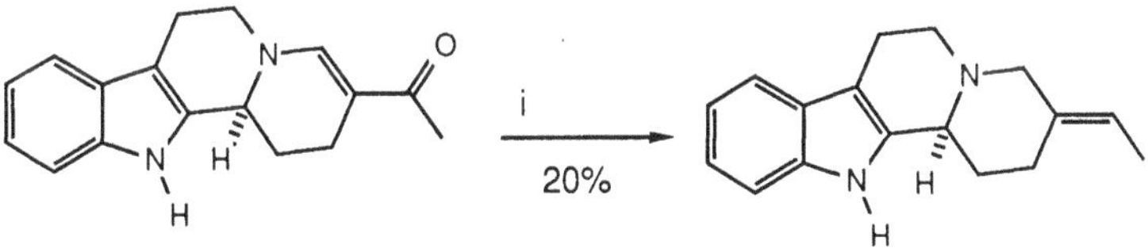

i. LAH

From M. Hämeilä and M. Lounasmaa, *Acta Chem. Scand.*, Sect B, *35*, 217 (1981).

192

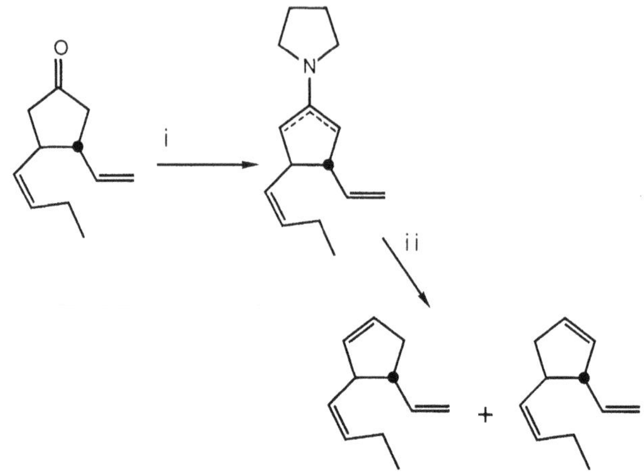

f. condensation

Identical reactions were done with cis compounds
i. pyrrolidine   ii. LAH

From L. Jaenicke and W. Boland, *Just. Lieb. Ann. Chem.*, 1135 (1976).

193

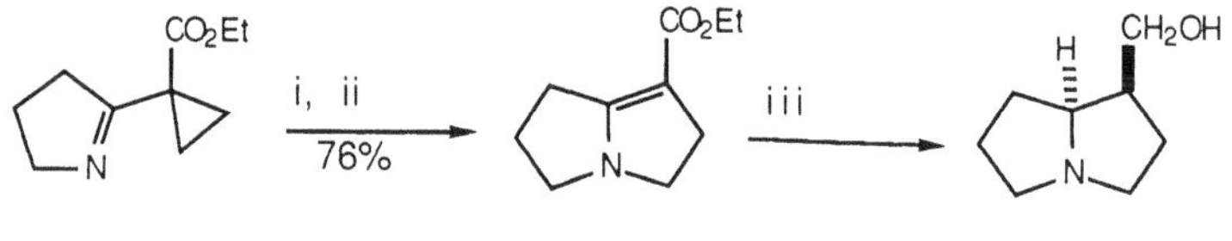

(±)-isoretronecanol

f. cyclopropylimines

i. NH$_4$Cl (cat. amount)   ii. xylene, reflux   iii. LAH

From H. W. Pinnick and Y. H. Chang, *Tet. Letters*, 837 (1979).

194

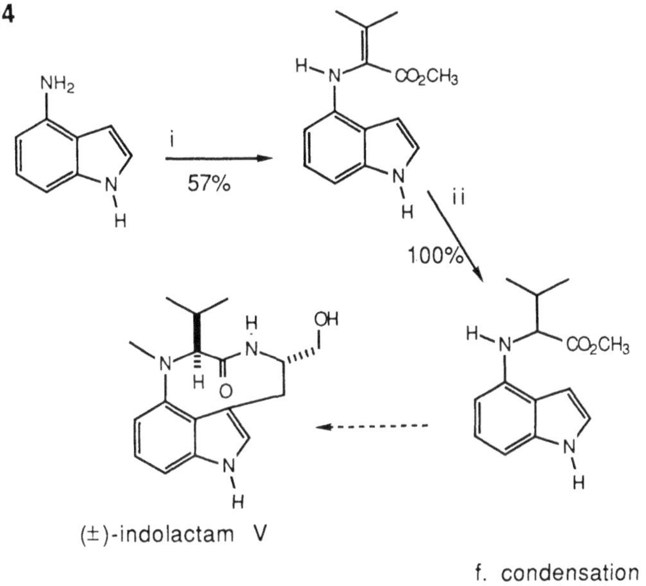

(±)-indolactam V

f. condensation

i. ![structure] —COCO₂CH₃ , TosOH, CHCl₃, heat ii. Mg/MeOH, ultrasound

From S. E. de Laszlo, S. V. Ley, and R. A. Porter, *J. Chem. Soc.,*
Chem. Commun., 344 (1986).

195

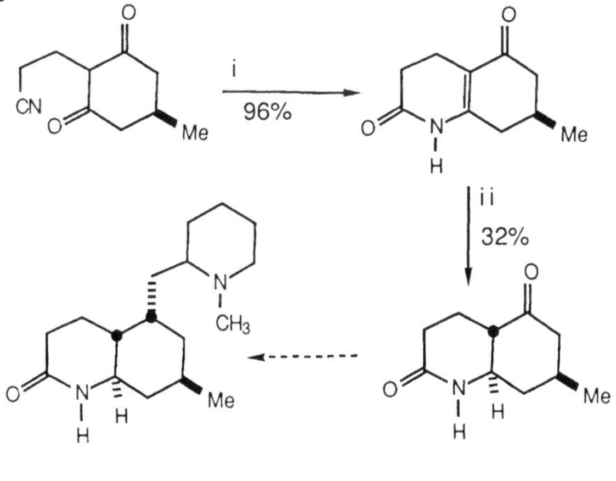

i. AcOH : H₂O (1 : 1) ii. Li/NH₃

From A. Leniewski, J. Szychowski, and D. B. MacLean, *Can. J. Chem.*,
59, 2479 (1981).

196

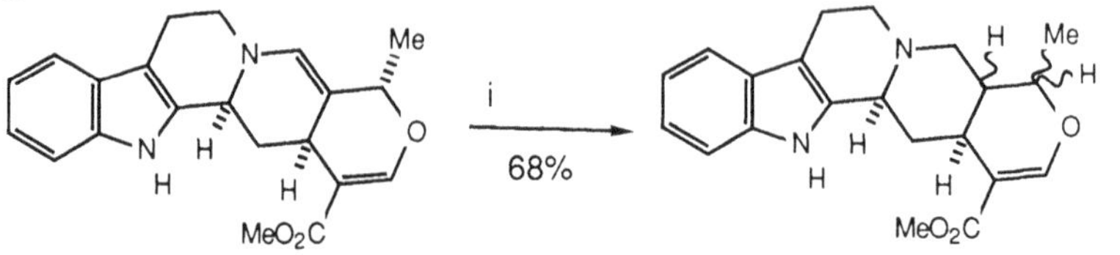

i. NADPH/H⁺

ajmalicine and isomers

From J. Stöckigt, J. Treimer, and M. H. Zenk, *F.E.B.S. Letters*, *70*, 267 (1976) and J. Stöckigt, H. P. Husson, C. Kan-Fan, and M. H. Zenk, *J. Chem. Soc.*, *Chem. Commun.*, 164 (1977).

XI. MISCELLANEOUS EXAMPLES OF ENAMINE GENERATION: 197–211

197

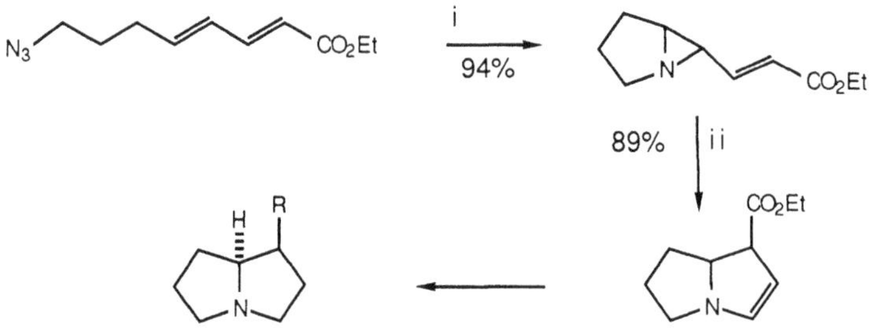

R= α CH₂OH, trachelanthamidine
R= ß CH₂OH, isoretronecanol

i. toluene, reflux, 4h   ii. 450°C, vacuum

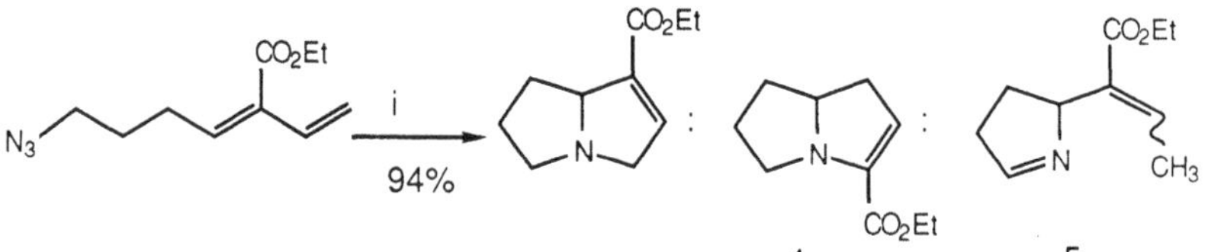

1 : 4 : 5

i. 550°C

From T. Hudlicky, J. O. Frazier, G. A. Seoane, M. Tiedje, A. Seoane, L. D. Kwart, and C. Beal, *J. Am. Chem. Soc.*, *108*, 3755 (1986).

198

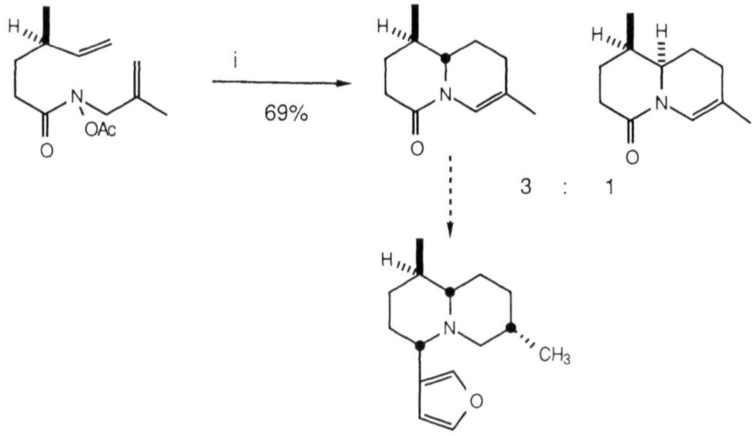

deoxy nupharidine

f. Diels-Alder

i. heat

From Y. C. Hwang and F. V. Fowler, *J. Org. Chem.*, *50*, 2719 (1985).

199

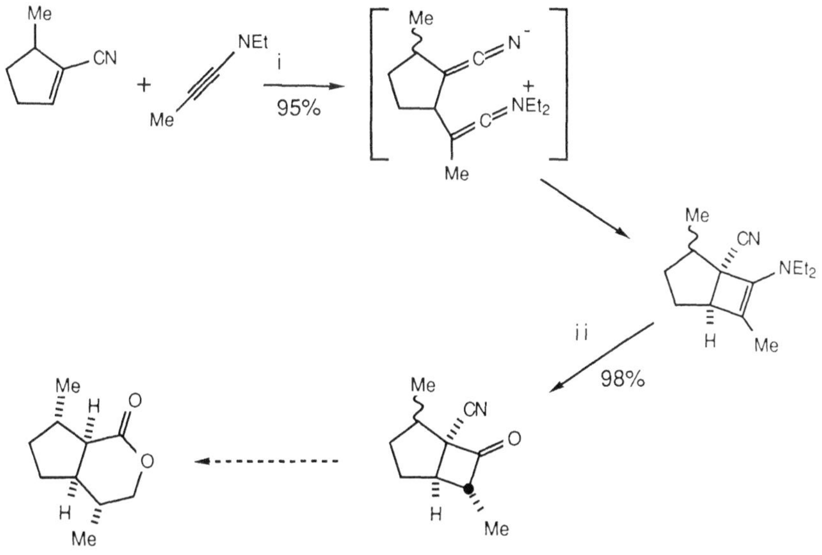

(±)-isodihydronepetalactone

f. from ynamines

i. $MgBr_2$ ii. AcOH (60%)

From J. Ficini and J. d'Angelo, *Tet. Letters*, 687; 679; 683 (1976).

200

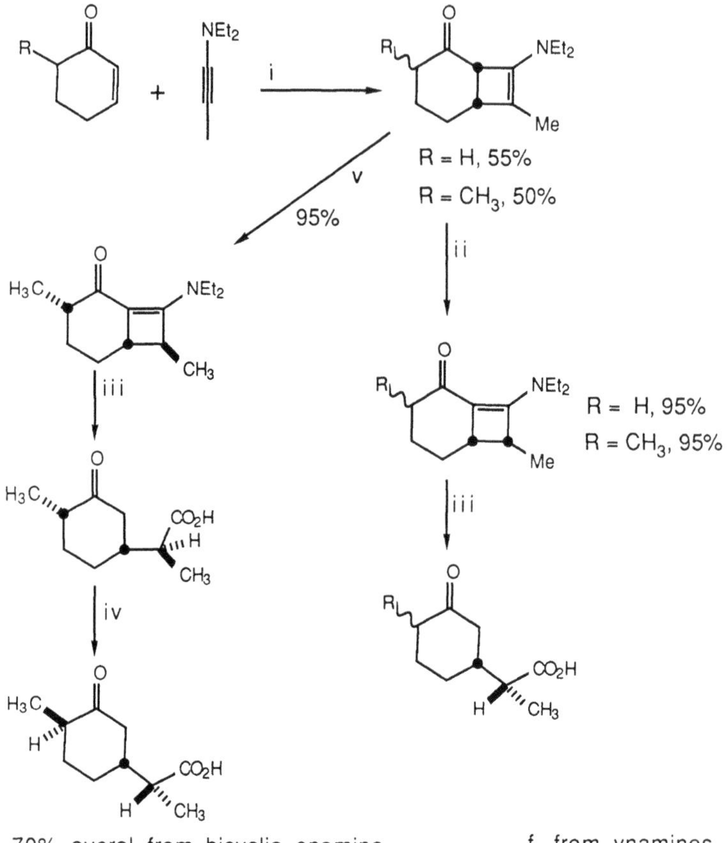

R = H, 55%
R = CH₃, 50%

95%

70% overal from bicyclic enamine   f. from ynamines

i. CH₃CN:THF (5:2), 80°C, 24h ii. THF iii. 60% AcOH iv. 10% HCl v. H₂O

From J. Ficini and A. M. Touzin, *Tet. Letters*, 2093; 2097 (1972).

201

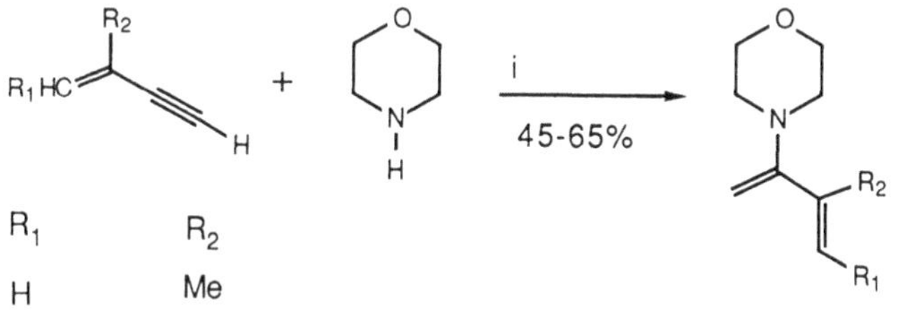

R_1	R_2
H	Me
-(CH$_2$)$_3$-	
-(CH$_2$)$_4$-	
PhCH$_2$	OMe

f. from vinylacetylenes

i. Hg(OAc)$_2$, THF, RT, 6h, Ar

From J. Barluenga, F. Aznar, R. Liz, and M.-P. Cabal, *J. Chem. Soc., Chem. Commun.*, **20**, 1375 (1985).

202

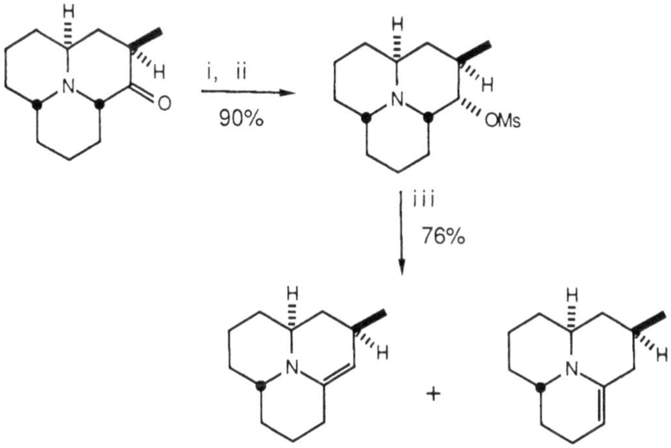

(±)-propyleine

f. ß-elimination of mesilate

i. LAH ii. MsCl, Et$_3$N iii. K$_2$CO$_3$, DMSO, 115°C

From R. H. Mueller and M. E. Thompson, *Tet. Letters*, **21**, 1097 (1980).

203

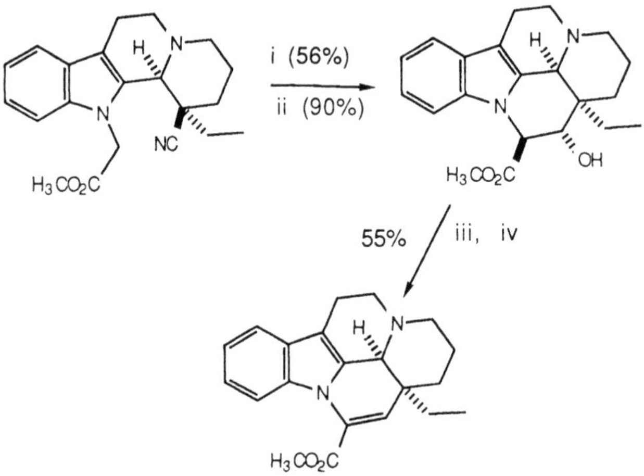

i (56%)
ii (90%)

55% / iii, iv

apovincamine

f. condensation

i. LDA, THF   ii. NBH, MeOH   iii. MsCl, Et₃N   iv. DBU

From H. Rapport and B. D. Christie, *J. Org. Chem.*, *50*, 1239 (1985).

204

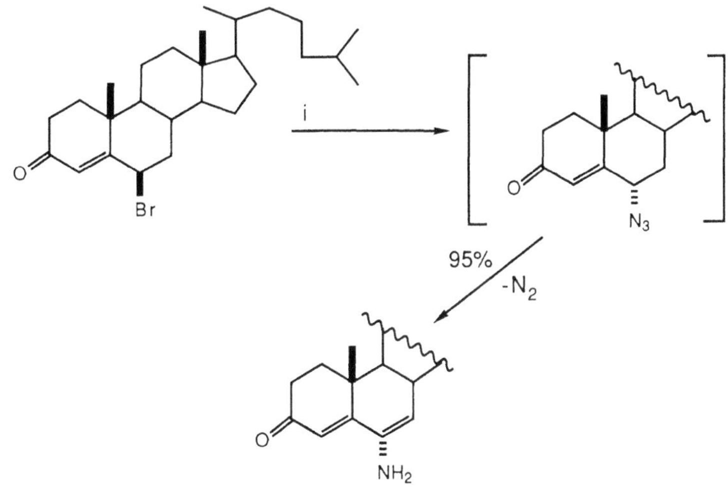

f. from 6 α–aziketones

i. NaN₃, DMSO, heat, 10 min

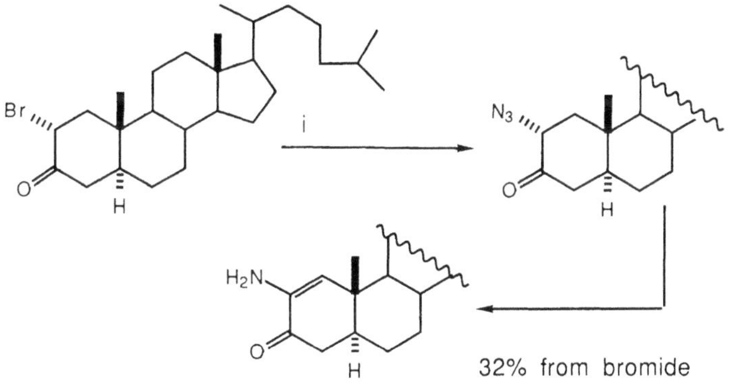

32% from bromide

f. from 2 α–aziketones

i. NaN₃, DMSO, 100°C, 10 min

From J. G. Ll. Jones and B. A. Marples, *J. Chem. Soc.*, 1188 (1970).

205

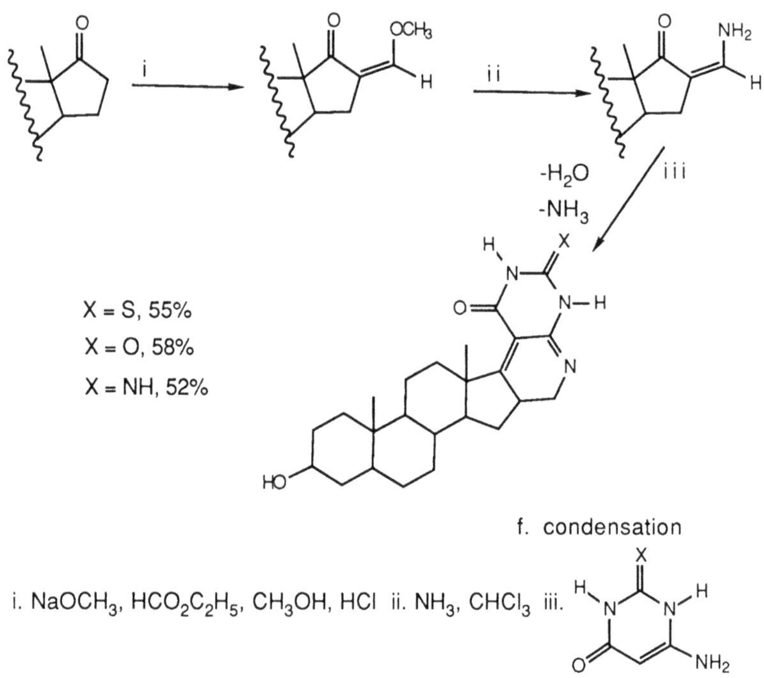
O
OCH3
H
i
ii
NH2
-H2O
-NH3
iii
X
N
N—H
X = S, 55%
X = O, 58%
X = NH, 52%
HO
f. condensation
i. NaOCH3, HCO2C2H5, CH3OH, HCl ii. NH3, CHCl3 iii.
NH2

From G. Bouchon, H. Pech, and E. Breitmaier, *Chimia*, Switzerland, *27*, 212 (1973).

206

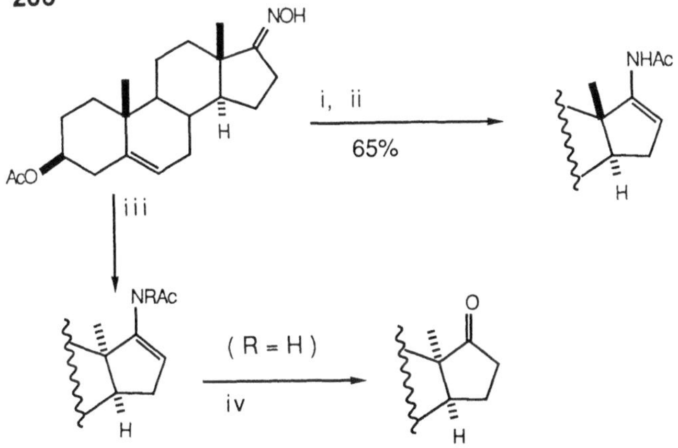

R=H, 13-epi enamide (85%)

R=Ac, 13-epi enimide

f. reductive acylation

i. Ac$_2$O, DMF, N$_2$ ii. Cr(OAc)$_2$, 40°C, overnight   iii. Ac$_2$O, Py, reflux

iv. MeOH/HCl, reflux,1h

From R. B. Boar, F. K. Jetuah, J. F. McChie, M. S. Robinson, and D. H. R. Barton, J. Chem. Soc., Perkin Trans I, 2163 (1977).

207

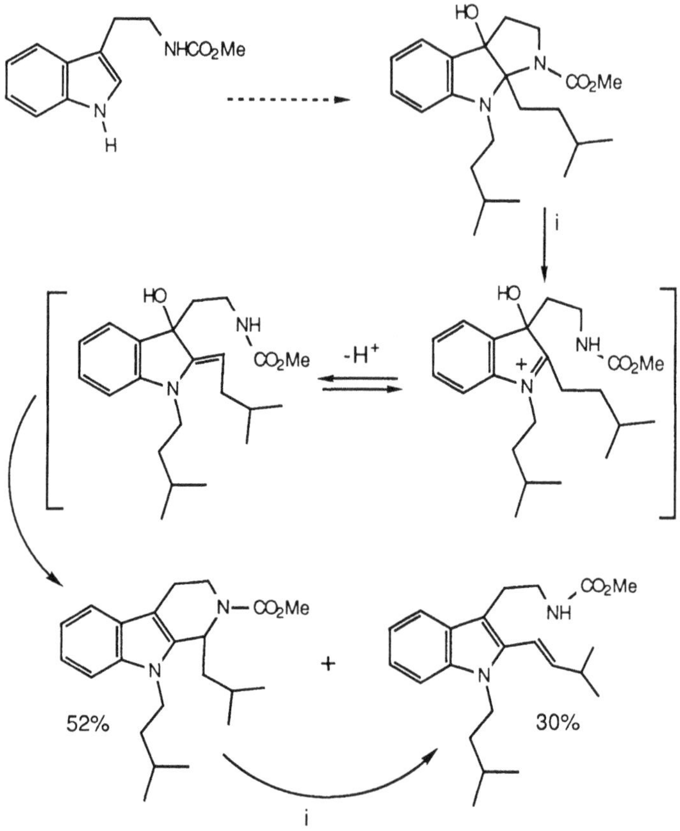

i. CF_3CO_2H

From M. Nakagawa, K. Matsuki, and T. Hino, *Tet. Letters, 24,* 2171 (1983).

208

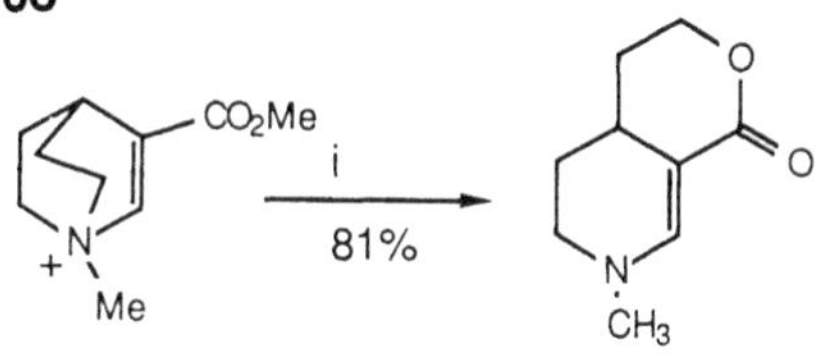

i. 150°C, N₂

f. 4° salt

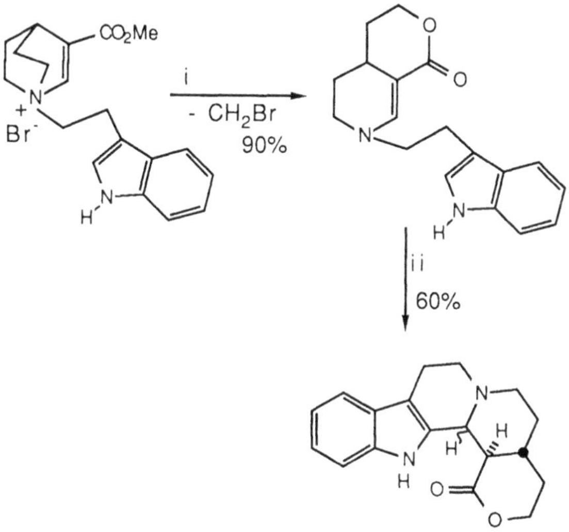

i. heat   ii. H⁺

f. 4° salt

From J. Dolby, R. Doltlbom, K. H. Hasselgren, and J. L. G. Nilsson, *Acta Chem. Scand.*, **25**, 735 (1971).

From J. Dolby, K. H. Hasselgren, S. Castensson, and J. L. G. Nilsson, *Acta Chem. Scand.*, 2469 (1972).

209

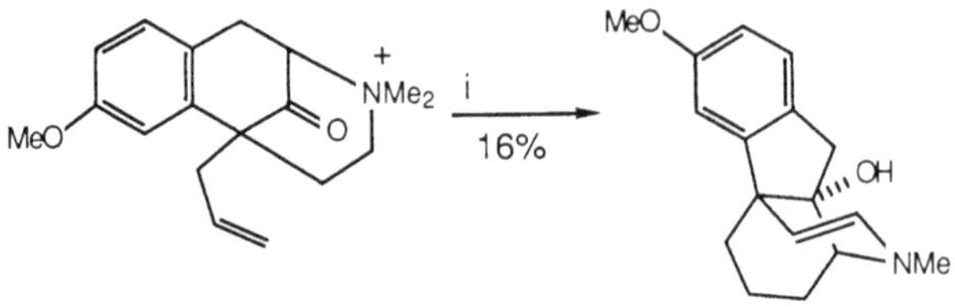

i. heat

From F. R. Ahmed, M. Saucier, and I. Monkovic, *Can. J. Chem.*, 53, 3276 (1975).

210

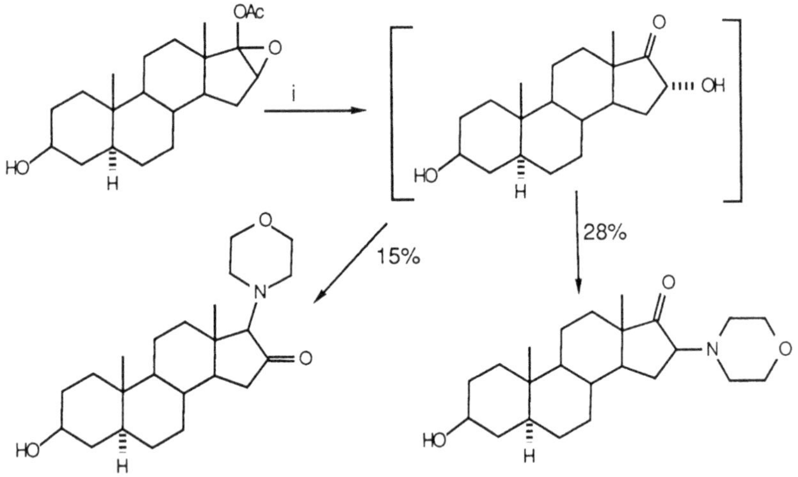

f. from α -hydroxyketones

i. morpholine

From C. L. Hewett and D. S. Savage, *J. Chem. Soc.*, 1180 (1969).

211

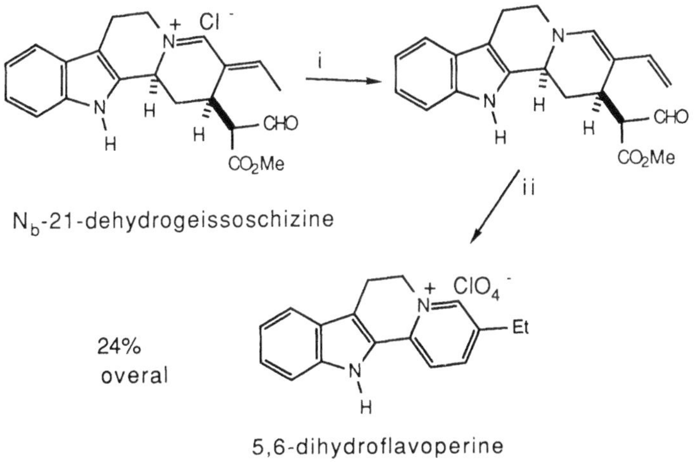

+ Cl⁻
N
H
H
H
CHO
CO2Me
Nb-21-dehydrogeissoschizine
i
CHO
CO2Me
ii
+ ClO4⁻
Et
24%
overal
5,6-dihydroflavoperine

i. KOH, MeOH ii. $NaClO_4$, H_2O, MeOH

From C. Kan Fan and H. P. Husson, *Tet. Lett.*, *21*, 4265 (1980).

ADDENDUM

212

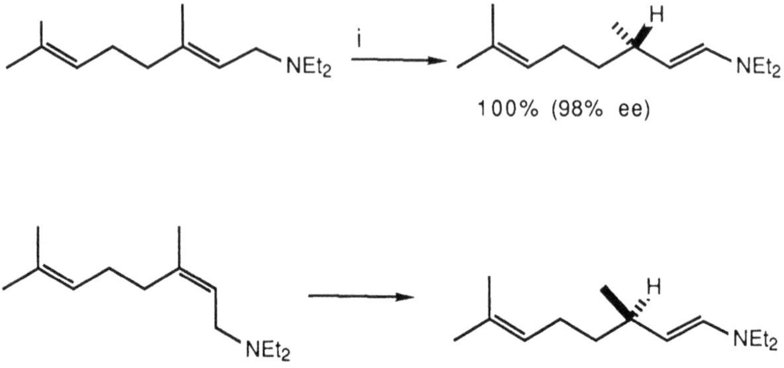

100% (98% ee)

i. Rh(R-binap)$_2$ ClO$_4$, 100°C

binap =

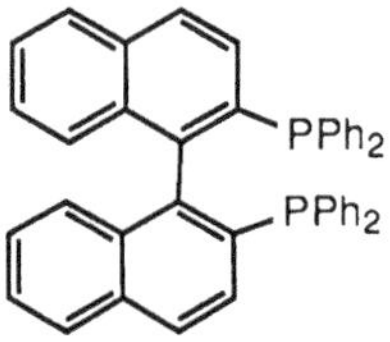

K. Tani, T. Yamagata, Y. Tatsuno, Y. Yamagata, K. Tomita, S. Akutagawa, H. Kumobayashi, S. Otsuka, *Angew Chem. Int. Ed. Engl.*, *24*, 217 (1985).

213

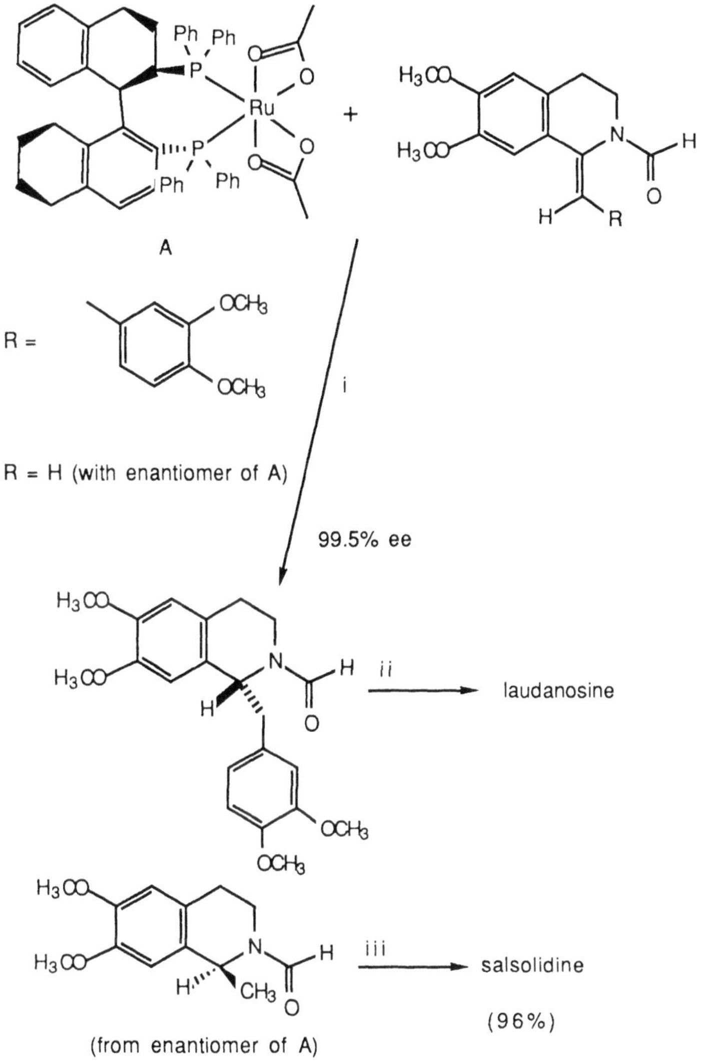

i. H$_2$ ii. LAH iii. hydrol.

R. Noyori, M. Ohta, Y. Hsiao, M. Kitamura, T. Ohta, H. Takaya; *J. Am. Chem. Soc. 108,* 7117 (1986).

REFERENCES CORRESPONDING TO EACH
REACTION EXAMPLE

Number in [ ] corresponds to Reaction Example

F. Abello, J. Boix, J. Gomez, J. Morell and J.-J. Bonet, *Helv. Chim. Acta, 58*, 2549 (1975). [92]

F. R. Ahmed, M. Saucier and I. Monkovic, *Can. J. Chem., 53*, 3276 (1975). [209]

M. Andrierntsiferana, R. Besselievre, C. Riche and H. P. Husson, *Tet. Letters.*, 2587 (1977). [72]

Atta-Ur-Rahman *J. C. S. Perkin I*, 731, 736 (1972). [4]

J. Auerbach and S. M. Weinreb, *J. Am. Chem. Soc.* 94, 7172 (1972). [1]

Y. Ban, M. Seto, and T. Oishi, *Tet. Letters*, 2113 (1972). [171]

Y. Ban, M. Kimura and T. Oishi, *Chem. and Pharm. Bull.* (Japan), *24*, 1490 (1976). [173]

M. Barczai-Beke, G. Dörnyei, G. Toth, J. Tamas and Cs. Szantay, *Tetrahedron, 32*, 1153 (1976). [183]

M. Barczai-Beke, G. Dörnyei, M. Kajtar and Cs. Szantay, *Tetrahedron, 32*, 1019 (1976). [183]

J.-J. Barieux and J. Gore, *Bull. Soc. Chim. France*, 3978 (1971). [149]

J.-J. Barieux and J. Gore, *Tetrahedron, 28*, 1537, 1555 (1972). [190]

J. Barluenga, F. Aznar, R. Liz and M.-P. Cabal, *J. C. S., Chem. Commun., 20*, 1375 (1985). [201]

A. R. Battersby, R. J. Francis, M. Hirst, E. A. Ruveda and J. Stannton, *J. C. S., Perkin I*, 1140 (1975). [55]

A. R. Battersby, J. Stannton, H. R. Wiltshire, R. J. Francis, and R. Sonthgate, *J. C. S., Perkin I*, 1147 (1975). [55]

R. Besselievre, C. Thal, H. P. Husson, and P. Potier, *J. C. S., Chem. Comm.* 90 (1975). [74]

R. Besselievre, J. P. Cosson, B. C. Das, and H. P. Husson, *Tet. Letters, 21*, 63 (1980). [138]

R. B. Boar, F. K. Jetuah, J. F. McChie, M. S. Robinson and D. H. R. Barton, *J. C. S., Perkin Trans I*, 2163 (1977). [206]

R. K. Boeckman, J. P. Sabatucci, S. W. Goldstein, D. M. Springer and P. F. Jackson, *J. Org. Chem., 51*, 3740 (1986). [105]

F. Bohlmann, H. J. Müller, and D. Schumann, *Chem. Ber., 106*, 3026 (1973). [69]

J. Boix, J. Gomez and J.-J. Bonet, *Helv. Chim. Acta, 58*, 2545 (1975). [92]

F. Bondavalli, P. Schenone and A. Ranise, *Synthesis*, 830 (1981). [66]

M. Bonin, J. Royer, D. S. Grierson and H. P. Husson, *Tet. Letters*, 1569 (1986). [117]

J. Bosch, M. L. Bennasar, E. Zulaica and M. Feliz, *Tet. Lett., 25*, 3119 (1984). [129].

G. Bouchon, H. Pech and E. Breitmaier, *Chima (Switz)*, 27, 212 (1973). [205]

J. P. Brennan and J. E. Saxton, *Tetrahedron*, 42, 6719 (1986). [37]

R. T. Brown and M. F. Jones, *Tet. Lett.*, 25, 3127 (1984). [137]

G. Büchi, K. E. Matsumoto and H. Nishimura, *J. Am. Chem. Soc.*, 93, 3299 (1971). [139]

A. Buzas, C. Herisson and C. Lavielle, *Compt. rend.*, 283, C, 763 (1976). [40]

M. P. Cava, S. C. Havlicek, A. Lindert, and R. J. Spanglu, *Tet. Letters*, 2937 (1966). [169]

R. M. Coates, *J. Org. Chem.*, 45, 5430 (1980). [62]

E. J. Corey, Y. Uyeda and R. A. Ruden, *Tet. Letters*, 4347 (1975). [71]

E. J. Corey, H. F. Wetter, A. P. Kozikowski and A. V. R. Rao, *Tet. Letters*, 777 (1977). [165]

G. Costerousse, J. Buendia, E. Toromanoff and J. Martel, *Bull. Soc. Chim. Fr.*, Part 2, 355 (1978). [146]

S. Danishefsky, S. J. Etheredge, R. Volkmann, J. Eggler and J. Quick, *J. Chem. Soc.*, 93, 5575 (1971). [50]

K. T. D. De Silva, G. N. Smith and K. E. H. Warren, *J. C. S., Chem. Comm.*, 905 (1971). [185]

S. E. de Laszlo, S. V. Ley, R. A. Porter, *J. C. S., Chem. Commun.*, 344 (1986). [194]

J. Dolby, R. Doltlbom, K. H. Hasselgren and J. L. G. Nilsson, *Acta Chem. Scand.*, 25, 735 (1971). [208]

J. Dolby, K. H. Hasselgren, S. Castensson and J. L. G. Nilsson, *Acta Chem. Scand.*, 2469 (1972). [208]

L. J. Dolby, S. J. Nelson and D. Senkovich, *J. Org. Chem.*, 37, 3691 (1972). [12]

D. J. Dunham and R. G. Lawton, *J. Am. Chem. Soc.*, 93, 2047 (1971). [44]

D. J. Dunham and R. G. Lawton, *J. Am. Chem. Soc.*, 93, 2075 (1971). [45]

S. F. Dyke and A. C. Ellis, *Tetrahedron*, 27, 3803 (1971). [150]

D. A. Evans, C. A. Bryan and G. M. Wahl, *J. Org. Chem.*, 35, 4122 (1970). [15]

J. Ficini and A. M. Touzin, *Tet. Letters*, 2093, 2097 (1972). [200]

J. Ficini and J. d'Angelo, *Tet. Letters*, 687, 679, 683 (1976). [199]

W. Flitsch and P. Wernsmann, *Tet. Lett.*, 22, 719 (1981). [176]

C. P. Forbes, J. D. Michau, J. Van Ree, A. Wiechers and M. Woudenberg, *Tet. Letters*, 935 (1976). [49]

C. P. Forbes, G. L. Wenteler and A. Wiechers, *Tetrahedron*, 34, 487 (1978). [49]

J. E. Foy and B. Ganem, *Tet. Letters*, 775 (1977). [166]

J. Froborg and G. Magnusson, *J. Am. Chem. Soc.*, 100, 6728 (1978). [51]

T. Gallagher, P. Magnus and J. C. Huffman, *J. Am. Chem. Soc.*, *104*, 1140 (1982). [39]

G. C. Gerrans, A. S. Howard and B. S. Orlek, *Tet. Letters*, 4171 (1975). [188]

S. I. Goldberg and I. Ragade, *J. Org. Chem.*, *32*, 1046 (1967). [3]

S. I. Goldberg and A. H. Lipkins, *J. Org. Chem.*, *35*, 242 (1970). [3]

J. Gore and J. J. Barieux, *Tet. Letters*, 2849 (1970). [189].

E. Gössinger, *Tet. Lett.*, *21*, 2229 (1980). [125]

E. Gössinger, *Monatsh. Chem.*, *111*, 143 (1980). [125]

J.-C. Gramain, H. P. Husson and Y. Troin, *J. Org. Chem.*, *50*, 5517 (1985). [80]

D. S. Grierson, M. Harris, and H. P. Husson, *Tetrahedron*, *39*, 3683 (1983). [144]

G. Habermehl, H. Andres, and B. Witkop, *Naturwiss.*, *62*, 345 (1979). [17]

B. Hachmeister, D. Thielke and E. Winterfeldt, *Chem. Ber.*, 3825 (1976). [181]

M. Hämeilä and M. Lounasmaa, *Acta Chem. Scand.*, *Sect B*, *35*, 217 (1981). [191]

M. Harris, D. S. Grierson, C. Riche and H. P. Husson, *Tet. Letters*, *21*, 1957 (1980). [127]

S. Harusawa, Y. Hamada and T. Shiori, *Synthesis*, 716 (1979). [123]

S. H. Hedges and R. B. Herbert, *J. Chem. Res (M)*, 413 (1979). [58]

J. B. Hendrickson and R. K. Boeckman Jr., *J. Am. Chem. Soc.*, *93*, 1307 (1971). [56]

R. B. Herbert, F. B. Jackson, and I. T. Nicolson, *J. C. S.*, *Chem. Comm.*, 450 (1976). [57]

R. B. Herbert, *J. C. S.*, *Chem. Commun.*, 794 (1978). [59]

C. L. Hewett and D. S. Savage, *J. Chem. Soc.*, (c), 1180 (1969). [210]

P. Houdewind, J. C. L. Armande and U. K. Pandit, *Tet. Letters*, 591 (1974). [99]

T. Hudlicky, J. O. Frazier, G. A. Seoane, M. Tiedje, A. Seoane, L. D. Kwart, C. Beal, *J. Am. Chem. Soc.*, *108*, 3755 (1986). [197]

H. P. Husson, K. Bannal, R. Freire, B. Mompon and F. A. M. Reis, *Tetrahedron*, *34*, 1363 (1978). [73]

H. P. Husson, *J. Nat. Prod.*, *48*, 894 (1985). [53, 117]

Y. C. Hwang and F. V. Fowler, *J. Org. Chem.*, *50*, 2719 (1985). [198]

H. Iida, S. Aoyagi, K. Kohno, N. Sasaki and C. Kibayashi, *Heterocycles*, *4*, 1771 (1976). [142]

H. Iida, S. Aoyagi and C. Kibayashi, *J. C. S.*, *Chem. Commun.*, 499 (1974). [77]

H. Iida, S. Aoyagi and C. Kibayaski, *J. C. S.*, *Perkin I*, 2502 (1975). [77]

H. Iida, T. Takarai and C. Kibayashi, *J. Org. Chem.*, *43*, 975 (1978). [87]

H. Iida, Y. Yuasa and C. Kibayashi, *J. Org. Chem.*, *44*, 1074 (1979). [88]

H. Ishii, K. Harada, T. Ishida, E. Ueda, and K. Nakajama, *Tet. Letters*, 319 (1975). [83]

S. Iskandarov, D. D. Kamalitdinov, and S. Yu, *Yunusov. Khim. Prirod. Soedinenii, 8*, 628 (1972) (C. A. 1973, *78*, 84 616 (1973)). [179]

K. Ito, M. Haruna and H. Furnkawa, *J. C. S., Chem. Commun.*, 681 (1975). [140]

L. Jaenicke and W. Boland, *Just. Lieb. Ann. Chem.*, 1135 (1976). [192]

J. G. Ll. Jones and B. A. Marples, *J. Chem. Soc.*, (c), 1188 (1970). [204]

Gy. Kalaus, M. Kiss, M. Kajtar-Peredy, J. Brlick, L. Szabo and Cs. Szantay, *Heterocycles, 23*, 2783 (1985). [38]

T. Kametani, T. Sugai, T. Honda, F. Satoh and K. Fukumoto, *J. C. S., Perkin I,* 1151 (1977). [90]

T. Kametani, N. Kanaya, H. Hino, S. P. Huang and M. Ihara, *Heterocycles, 14*, 1771 (1980). [43]

C. Kan-Fan and H. P. Husson, *Tet. Lett., 21*, 4265 (1980). [211]

C. Kan-Fan, G. Massiot, A. Ahond, B. C. Das, H.-P. Husson and P. Potier and A. I. Scott and C. Wei, *J. C. S., Chem. Comm.*, 164 (1974). [102]

C. Kan-Fan and H. P. Husson, *J. C. S., Chem. Commun.*, 618 (1978). [180]

S. L. Keeley, Jun., A. J. Martinez and F. C. Tahk, *Tetrahedron, 26*, 4729 (1970). [63]

T. Kiguchi, C. Hashimoto, T. Naito, and I. Ninomiya, *Heterocycles, 19*, 2279 (1982). [94]

K. N. Kilminstev and M. Sainsbury, *J. C. S., Perkin I,* 2264 (1972). [186]

J. Knabe and H. D. Holtje, *Tet. Letters, 6*, 433 (1969). [121]

Y. Kondo and T. Takemoto, Y. Zasshi, *Chem. Pharm. Bull., 95*, 1161 (1975) (C. A. 1976, *84*, 17585). [153]

K. A. Kovar and F. Schielein, *Arch. Pharm.* (Weinheim, Ger.), *311*, 73 (1978). [75]

W. E. Kreighbaum, W. F. Kavanaugh, and W. T. Comes, *J. Heterocyclic Chem., 10*, 317 (1973). [8]

M. E. Kuehne, D. M. Roland and R. Hafter, *J. Org. Chem., 43*, 3705 (1978). [35]

M. E. Kuehne, T. H. Matsko, J. C. Bohnert and C. L. Kirkemo, *J. Org. Chem., 44*, 1063 (1979). [35]

M. E. Kuehne, J. A. Huebner and T. H. Matsko, *J. Org. Chem., 44*, 2477 (1979). [35]

M. E. Kuehne and J. C. Bohnert, *J. Org. Chem.*, *46*, 3443 (1981). [36]

M. E. Kuehne, F. J. Okuniewicz, C. L. Kirkemo, and J. C. Bohnert, *J. Org. Chem.*, *47*, 1335 (1982). [36]

M. E. Kuehne, G. Di Vincenzo, *J. Org. Chem.*, *37*, 1023 (1972). [28]

M. E. Kuehne and J. C. King, *J. Org. Chem.*, *38*, 304 (1973). [27]

J. P. Kutney, and F. Bylsma, *J. Am. Chem. Soc.*, *92*, 6090 (1970). [160]

J. P. Kutney, R. T. Brown, E. Piers and J. R. Hadfield, *J. Am. Chem. Soc.*, *92*, 1708 (1970). [160]

R. T. La Londe, E. Auer, C. F. Wong, and V. P. Muralidharan, *J. Am. Chem. Soc.*, *93*, 2501 (1971). [161]

R. T. La Londe and T. S. Eckert, *Can. J. Chem.*, *59*, 2298 (1981). [163]

R. T. La Londe, J. Wooleuer, E. Auer and C. F. Wong, *Tet. Letters*, 1503 (1972). [184]

N. Langlois and P. Potier, *Tet. Lett.*, *14*, 1099 (1976). [154]

E. Leete, *J. C. S., Chem. Commun.*, 821 (1979). [70]

E. Leete and M. E. Mueller, *J. Am. Chem. Soc.*, *104*, 6440 (1982). [68]

F. Le Goffic, A. Gouyette, and A. Ahond, *Compt. Rend.*, *274* C, 2008 (1972). [21]

A. Leniewski, J. Szychowski and D. B. MacLean, *Can. J. Chem.*, *59*, 2479 (1981). [195]

G. R. Lenz, *J. Org. Chem.*, *42*, 1117 (1971). [100]

M. Lounasmaa, P. Juu Tinen and P. Kairisalo. *Tetrahedron, 34*, 2529 (1978). [135]

M. Lounasmaa and R. Jokela, *Tetrahedron, 34*, 1841 (1978). [135]

M. Lounasmaa and R. Jokela, *Tet. Lett.*, 3609 (1978). [135]

M. Lounasmaa, H. Merikallio and M. Puhakka, *Tetrahedron, 34*, 2995 (1978). [135]

M. Lounasmaa, T. Langenskiöld, and C. Homberg, *Tet. Lett.*, *22*, 5179 (1981). [5]

P. Mangeney, R. Z. Andriamialisoa, N. Langlois, Y. Langlois, P. Potier, *J. Am. Chem. Soc.*, *101*, 2243 (1979). [154]

M. S. Manhas and J. R. McRoy, *J. Chem. Soc.*, (c), 1419 (1969). [65, 67]

S. F. Martin, S. R. Desai, G. W. Phillips and A. C. Miller, *J. Am. Chem. Soc.*, *102*, 3294 (1980). [30]

S. F. Martin and C. Y. Tu, *J. Org. Chem.*, *46*, 3763 (1981). [31]

S. F. Martin and T. S. Chou, *J. Org. Chem.*, *43*, 1027 (1978). [24]

T. Masamune, M. Takasngi and A. Murai, *Tetrahedron, 27*, 3369 (1971). [145]

G. Massiot, M. Zeches, P. Thepenier, M. J. Jacquier, L. Le Men-oliver, and C. Delande, *J. C. S. Chem. Commun.*, 768 (1982). [120]

L. R. McGee, G. S. Reddy and P. N. Confalone, *Tet. Lett.*, *25*, 2115 (1984). [143]

S. Miyano, S. Fujii, O. Yamashita, N. Toraishi, K. Sumoto, F. Satoh and T. Masuda, *J. Org. Chem.*, *46*, 1737 (1981). [174]

O. Miyata, Y. Hirata, T. Naito, and I. Ninomiya, *J. C. S., Chem. Commun.*, 1231 (1983). [81]

D. J. Morgans, Jr. and G. Stork, *Tet. Lett.*, 1959 (1979). [103]

Y. Morita, M. Hesse, U. Renner and H. Schmid., *Helv. Chim. Acta*, *59*, 532 (1976). [25]

J. M. Muchowski and P. H. Nelson, *Tet. Lett.*, *21*, 4585 (1980). [176]

R. H. Mueller and M. E. Thompson, *Tet. Lett.*, 1991 (1979). [107]

R. H. Mueller and M. E. Thompson, *Tet. Lett.*, *21*, 1093 (1980). [148]

R. H. Mueller and M. E. Thompson, *Tet. Lett.*, *21*, 1097 (1980). [202]

A. Murai, C. Sato, H. Sasamori,a nd T. Masamune, *Bull. Chem. Soc. (Japan)*, *49*, 499 (1976). [156]

A. Murai, C. Sato, H. Sasamori and T. Masamune, *Bull. Chem. Soc. (Japan)*, *49*, 499 (1976). [156]

T. Naito, Y. Tada, Y. Nishiguchi and I. Ninomiya, *Heterocycles*, *18*, 213 (1982). [81]

T. Naito, E. Doi, O. Miyata and I. Ninomiya, *Heterocycles*, *24*, 903 (1986). [89]

T. Naito, N. Kojima, O. Miyata and I. Ninomiya, *J. C. S., Chem. Commun.*, 1611 (1985). [93]

M. Nakagawa, K. Matsuki, and T. Hino, *Tet. Lett.*, *24*, 2171 (1983). [207]

S. Naruto, H. Nishimura and H. Kaneko, *Chem. and Pharm. Bull. (Japan)*, *23*, 1276 (1975). [153]

M. Natsume and M. Ogawa, *Heterocycles*, *16*, 973 (1981). [158]

M. Natsume and M. Ogawa, *Heterocycles*, *15*, 237 (1981). [159]

I. Ninomiya, T. Naito and T. Kiguchi, *Tet. Letters*, *51*, 4451 (1970). [77]

I. Ninomiya, T. Naito, T. Kiguichi, *J. C. S., Chem. Comm.*, 1669 (1970). [78]

I. Ninomiya, T. Naito, T. Kiguchi and T. Mori, *J. C. S., Perkin I*, 1696 (1973). [79]

I. Ninomiya, H. Takasugi, and T. Naito, *J. C. S., Chem. Comm.*, 732 (1973). [82]

I. Ninomiya and T. Naito, *Heterocycles*, *2*, 607 (1974). [86]

I. Ninomiya, J. Kiguchi, *J. C. S., Chem. Comm.*, 624 (1976). [42]

I. Ninomiya, J. Yasui and T. Kiguchi, *Heterocycles*, *6*, 1855 (1977). [76]

I. Ninomiya, O. Yamamoto and T. Naito, *Heterocycles*, *4*, 743 (1976). [84]

I. Ninomiya, O. Yamamoto and T. Naito, *J. C. S., Chem. Commun.*, 437 (1976). [84]

I. Ninomiya, Y. Tada, T. Kiguchi, O. Yamamoto and T. Naito, *Heterocycles, 9*, 1527 (1978). [41]

I. Ninomiya, C. Hashimoto, T. Kigushi and T. Naito, *J. C. S., Perkin Trans I*, 941 (1985). [95]

R. Noyori, M. Ohta, Y. Hsiao, M. Kitamura, T. Ohta, *J. Am Chem. Soc., 108*, 7117 (1986). [213]

M. Onanga and F. Khuong Huu, *Tet. Lett., 24*, 3627 (1983). [143]

M. Onda, K. Abe and K. Yonezawa, *Chem. Pharma. Bull. (Tokyo), 16*, 2005 (1968). [118]

M. Onda, K. Yonezawa, K. Abe, H. Toyama, *Chem. and Pharm. Bull. (Japan), 19*, 21 (1971). [106]

M. Onda, K. Yausa, J. Okada, K. Katoaka and K. Abe, *Chem. and Pharm. Bull. (Japan), 21*, 1333 (1973). [22]

W. Oppolzer, W. Fröstl and H. P. Weber, *Helv. Chim. Acta, 58*, 593 (1975). [101]

H. Otomasu, K. Higashiyama, T. Honda and T. Kametani, *J. C. S., Perk. Trans I*, 2399 (1982). [178]

L. E. Overman, M. Sworin and R. M. Burk, *J. Org. Chem., 48*, 2685 (1983). [147]

U. K. Pandit and S. A. G. de Graaf, *J. C. S., Chem. Commun.*, 381 (1970). [26]

U. K. Pandit, F. A. Van der Vlugt and A. C. Van Dalen, *Tet. Letters*, 3697 (1969). [122]

U. K. Pandit, K. de Jonge, K. Erhart and H. O. Huisman, *Tet. Letters*, 1207 (1969). [20]

U. K. Pandit, M. J. M. Pollman and H. O. Huisman, *J. C. S., Chem. Commun.*, 527 (1967). [167]

U. K. Pandit, F. A. Van der Vlugt and A. C. Van Dalen, *Tet. Letters*, 3693 (1969). [7]

G. Pattenden and G. M. Robertson, *Tet. Letters*, 399 (1986). [2]

J. S. Peterson, G. Fels and H. Rapport, *J. Am. Chem. Soc., 106*, 4539 (1984). [170]

J. Picot and X. Lusinchi, *Tetrahedron, 34*, 2747 (1978). [164]

H. W. Pinnick and Y. H. Chang, *Tet. Letters*, 837 (1979). [193]

H. W. Pinnick and Y. H. Chang, *J. Org. Chem., 43*, 4662 (1978). [177]

H. Rapport and B. D. Christie, *J. Org. Chem., 50*, 1239 (1985). [203]

R. Rasoanaivo, N. Langlois, A. Chiaroni and C. Riche, *Tetrahedron, 35*, 641 (1979). [128]

C. H. Robinson, L. Milewich and K. Huber, *J. Org. Chem., 36*, 211 (1971). [175]

G. Rossey, A. Wick and E. Wenkert, *J. Org. Chem., 47*, 4745 (1982). [19]

S. Ruchirawat, *Heterocycles.*, *6*, 1855 (1977). [157]

S. Ruchirawat, U. Borvornvinyanaut, K. Haintawong and Y. Thebtaranonth, *Heterocycles.*, *6*, 1119 (1977). [157]

M. Sainsbury and N. Uttley, *J. C. S., Chem. Comm.*, 319 (1977). [85]

M. Sainbury, S. F. Dyke, D. W. Brown and R. G. Kisman, *Tetrahedron*, *26*, 5265 (1970). [121]

M. Sainsbury, D. W. Brown, S. F. Dyke, R. G. Kinsman and B. J. Moon, *Tetrahedron*, *24*, 6695 (1968). [131]

T. Sano, J. Toda, N. Kashiwaba, Y. Tsuda, and Y. Iitaka, *Heterocycles*, *16*, 1151 (1981). [34]

T. Sato, M. Funabora, M. Watanabe and T. Fujisawa, *Chem. Letters*, 1391 (1985). [97]

R. H. Schlessinger, M. A. Poss and S. Richardson, *J. Am. Chem. Soc.*, *108*, 3112 (1986). [96]

F. Schneider, A. Boller, M. Müller, P. Müller and A. Fürst, *Helv. Chim. Acta*, *56*, 2396 (1973). [98]

A. I. Scott and C. L. Yeh, *J. Am. Chem. Soc.*, *96*, 2273 (1974). [126]

M. Shamma and C. D. Jones, *J. Am. Chem. Soc.*, *91*, 4009 (1969). [108]

M. Shamma and C. D. Jones, *J. Am. Chem. SOc.*, *92*, 4943 (1970). [108]

M. Shamma and J. F. Nugent, *Tet. Letters*, 2625 (1970). [108]

M. Shamma and J. F. Nugent, *Tetrahedron*, *29*, 1265 (1973). [10º, 119]

M. Shamma and L. A. Smeltz, *Tet. Letters*, 1415 (1976). [151]

P. Slosse and G. Hootele, *Tet. Lett.*, 4587 (1979). [124]

E. Stark and E. Breitmaier, *Tetrahedron*, *29*, 2209 (1973). [61]

R. V. Stevens and L. E. Du Pree, jun., *J. Chem. Soc., Chem. Comm.*, 1585 (1970). [48]

R. V. Stevens, R. K. Mehra, and R. L. Zimmerman, *J. C. S., Chem. Comm.*, 877 (1969). [47]

R. V. Stevens, J. M. Fitzpatrick, M. Kaplan, and R. L. Zimmerman, *J. Chem. Soc., Chem. Comm.*, 857 (1971). [112]

R. V. Stevens and J. T. Lai, *J. Org. Chem.*, *37*, 2138 (1972). [64]

R. V. Stevens, Y. Luh and J. Shen, *Tet. Letters*, 3799 (1976). [114]

R. V. Stevens, *Acc. Chem. Res.*, *10*, 193 (1977). [114]

R. V. Stevens and Y. Luh, *Tet. Letters*, 979 (1977). [114]

R. V. Stevens and A. W. M. Lee, *J. C. S., Chem. Commun.*, 102 (1982). [187]

J. Stöckigt, J. Treimer and M. H. Zenk, *F. E. B. S. Letters*, *70*, 267 (1976). [196]

J. Stöckigt, H. P. Husson, C. Kan-Fan and M. H. Zenk, *J. C. S., Chem. Commun.*, 164 (1977). [196]

G. Stork, D. J. Morgans, Jr., *J. Am. Chem. Soc.*, *101*, 7110 (1979). [104]

P. L. Stütz, P. A. Stadler, J. M. Vigouret and A. Jaton, *J. Med. Chem.*, *21*, 754 (1978). [172]

T. Suzuki, E. Sato, K. Unno and T. Kametani, *Heterocycles*, *23*, 839 (1985). [162]

Cs. Szantay, L. Szabo and Gy. Kalaus, *Tet. Letters*, 191 (1973). [16]

Cs. Szantay, L. Szabo and Gy. Kalaus, *Tetrahedron*, *33*, 1803 (1977). [16]

D. F. Taber and B. P. Gunn, *J. Am. Chem. Soc.*, *101*, 3992 (1979). [54]

H. Takahata, K. Yamabe, T. Suzuki and T. Yamasaki, *Heterocycles*, *24*, 37 (1986). [115]

S. Takano, K. Shishido, M. Sato and K. Ogasawara, *Heterocycles*, *6*, 1699 (1977). [11]

S. Takano, Y. Suzuki, and K. Ogasawara, *Heterocycles*, *16*, 1479 (1981). [6]

K. Tani, T. Yamagati, Y. Tatsuno, Y. Yamagata, K. Tomita, S. Akutagawa, H. Kumbayashi, S. Otsuka, *Angew. Chem. Int. Ed. Engl.*, *24*, 217 (1985). [212]

Y. Tamura, H. Maeda, S. Akai, and H. Ishibashi, *Tet. Lett.*, *23*, 2209 (1982). [60]

P. Teisseire, B. Shimizu, M. Plattier, B. Corbier, and P. Rouillier, *Recherches*, *19*, 241 (1974) (CA 1975, 83, 28395v). [46]

C. Thal, T. Sevenet, H. P. Husson and P. Potier, *Compt. rend.*, 275 C, 1295 (1972). [16]

J. Trojanek, Z. Koblicova, Z. Uesely, V. Suchan and J. Holubek, *Coll. Czech. Chem. Comm.*, *40*, 681 (1975). [109]

I. Tse and V. Snieckus, *J. C. S. Chem. Commun.*, 505 (1976). [91]

Y. Tsuda, Y. Sakai, N. Kashiwaaba, T. Sano, J. Toda and K. Isobe, *Heterocycles*, *16*, 189 (1981). [10, 32]

B. Umezawa, O Hoshino, S. Sawaki, S. Sato and N. Numao, *J. Org. Chem.*, *42*, 4272 (1977). [18]

E. Vedejs and G. R. Martinej, *J. Am. Chem. Soc.*, *102*, 7994 (1980). [116]

R. Volkmann, S. Danishefsky, J. Eggler and D. M. Solomon, *J. Am. Chem. Soc.*, *93*, 5576 (1971). [50]

H. H. Wasserman and S. Terao, *Tet. Letters*, 1735 (1975). [155]

B. Weinstein and A. R. Craig, *J. Org. Chem.*, *41*, 875 (1976). [13]

E. Wenkert, C. Chang, H. Chawla, D. Cochran, E. Hagaman, J. C. King and K. Orito, *J. Am. Chem. Soc.*, *98*, 3645 (1976). [136]

E. Wenkert, K. G. Dave, C. T. Gnewuch and P. W. Sprague, *J. Am. Chem. Soc.*, *90*, 5251 (1968). [133]

E. Wenkert, B. Channcy, K. G. Dave, A. R. Jeffcoat, F. M. Schnell and H. P. Schenk, *J. Am. Chem. Soc.*, *95*, 8427 (1973). [111]

E. Wenkert, H. P. S. Chawla and F. M. Schell, *Synth. Comm.*, *3*, 381 (1973). [113]

E. Wenkert, P. W. Sprague and R. L. Webb, *J. Org. Chem.*, *38*, 4305 (1973). [134]

E. Wenkert, C. Chang, H. Chawla, D. Cochran, E. Hagaman, J. King and K. Orito, *J. Am. Chem. Soc.*, *98*, 3645 (1976). [132]

E. Wenkert, T. Hudlicky and H. D. H. Showalter, *J. Am. Chem. Soc.*, *100*, 4893 (1978). [29]

E. Wenkert, B. L. Buckwalter and S. S. Sathe, *Synth. Commun.*, *3*, 61 (1973). [29]

E. Wenkert, C. J. Chang, H. P. S. Chawla, D. Cochran, E. W. Hagaman, J. C. King and K. Orito, *J. Am. Chem. Soc.*, *98*, 3645 (1976). [182]

E. Wenkert and A. R. Jeffcoat, *J. Org. Chem.*, *35*, 515 (1970). [110]

J. D. White, *J. Am. Chem. Soc.*, *103*, 1813 (1981). [52]

H. J. Wilkens and F. Troxler, *Helv. Chim. Acta*, *58*, 1512 (1975). [141]

S. Yamada, M. Shibasaki and S. Terashima, *Tet. Letters*, 381, 377 (1973). [23]

S. Yamada, Y. Hamada, K. Ninomiya and T. Shioiri, *Tet. Lett.*, 4749 (1976). [168]

E. Yamanaka, E. Maruta, S. Kasamatsu, N. Aimi and S. Sakai, *Tet. Lett.*, *24*, 3861 (1983). [152]

K. Yoshida, S. Nomura and Y. Ban, *Tetrahedron*, *41*, 5495 (1985). [9]

F. E. Ziegler, J. A. Kloek and P. A. Zoretic, *J. Am. Chem. Soc.*, *91*, 2342 (1969). [14]

F. E. Ziegler, E. B. Spitzner, *J. Am. Chem. Soc.*, *92*, 3492 (1970). [33]

H. Zinnes, F. R. Zuleski and J. Shavel, Jr., *J. Org. Chem.*, *33*, 3605 (1968). [130]

Index

A

A(1,2) strain, 36
A(1,3) strain, 37
Acorenone B, synthesis of, 574
Acorone, synthesis of, 550
Acrolein, reaction with enamines, 47, 360–362, 481
Acrylamide, reaction with enamines, 408
Acrylonitrile, reaction with enamines, 47, 370
Acryloyl chloride, reaction with enamines, 362
Acyl halides, reaction with enamines, 204–214, 535
Ajmalicine, synthesis of, 133
Alkylation:
  of enamines, 182–203, 474–484, 543–614
    with alkyl halides, 183–189
    with alkynes, 194–195
    with electrophilic olefins, 189–194
    solvent effect, 185, 186

[Alkylation]
  of imines, with alkyl halides, 130, 281
Alkyllithium, reaction with, N-methyl lactams, 137
Alloyohimbine, synthesis of, 594
Aluminum hydrides, reaction with enamines, 228–229, 266
Aminals, 110–114
  iminium salts from, 282–283
Anatabine, synthesis of, 585
Axatoxin a, synthesis of, 657
Andranginine, synthesis of, 608
Angustidine, synthesis of, 594
Antiparallel attack, of an electrophile, 37
Apovincamine, synthesis of, 680
Arylation, of enamines, 195–198
Aspidospermidine, synthesis of, 563
Aspidospermine, synthesis of, 555
Asymmetric induction, in enamine cyclization, 358
Asymmetric photocyclization, 410

Asymmetrical induction, alkylation of enamines, 186
Aza-Claisen rearrangement, 28
Aza-Cope rearrangement, 30, 183–184
1-(N-Azetidino)-3,4-dihydronaphthalene, synthesis of, 47
Azides, reaction with:
  enamines, 412–415
  iminium salts, 324
Aziridinium salts, synthesis of, 313–317
Azomethine ylide, 147–148
  formation of, 330–331
  reaction with enamines, 411
4a-Azonioanthracene ion, reaction with enamines, 380

B

Basicity:
  of enamines, 77–83, 466–468
  solvent effects on, 82–83
  of saturated amines, 77–83
Benzaldehyde, reaction with enamines, 201
Benzene sulfonyl chloride, reaction with enamines, 216
1-Benzenesulfonylallene, reaction with enamiens, 379
Benzonitrile oxide, reaction with enamines, 412
p-Benzoquinone, reaction with enamines, 198
Benzoylperoxide, oxidation of enamines, 252
Benzylideneaniline, reaction with enamines, 406
Benzyne, reaction with enamines, 197–198, 388
Bicyclo[3.1.0]hex-2-ene-6-endo-carboxaldehyde, reaction with pyrrolidine, 117
Birch reduction, 136

Bischler-Napieralski reaction, 123, 133, 511
Bredt's rule, 132
Brevicolline, synthesis of, 587
Bromine, reaction with enamines, 495
Bromodimethylsulfonium bromide, reaction with enamines, 222–223
p-Bromophenyl vinyl sulfone, reaction with enamines, 372
3-Bromopropylamine hydrobromide, reaction with enamines, 408
N-Bromosuccinimide, reaction with enamines, 222
alpha-Broximines, reaction with enamines, 408
3-Buten-2-one, reaction with enamines, 47, 354–361, 482, 507, 582
t-Butylamine borane, reduction of iminium salts, 262, 302

C

Camphor, aminomethylation of, 328
Camptothecin, synthesis of, 572
Carbenoids, reaction with enamines, 389, 553, 554
N-Carbethoxyaziridine, reaction with enamines, 405
Carbon disulfide, reaction with enamines, 395–397
Carboxyhydridoferrates, reduction of iminium salts, 262
Cathenamine, synthesis of, 665
Cephalotaxine, synthesis of, 535
Charge transfer theory, 351
Chloral, reaction with enamines, 199–200, 393
4-Chloro-3-nitropyridine, reaction with enamines, 196

2-Chloro-4,5-dicarbethoxy-pyrimidine, reaction with enamines, 196

2-Chloro-5-nitropyridine, reaction with enamines, 196

α-Chloroacrylonitrile, reaction with:
  dienamine, 368
  enamines, 371

2-Chloronitroethene, reaction with enamines, 372

Cinnamaldehyde, reaction with enamines, 361–362

Citronellal, pyrrolidine enamine of, electrooxidation of, 249

Claisen rearrangement, 28

Clemmensen reduction, 486–487

Colchicine, synthesis of, 107

Concerted cycloadditions, theory of, 348–352

Configuration interaction theory, 351

(+)-Coniine, synthesis of, 575

Convergine, synthesis of, 642

Corgoine, synthesis of, 663

Costaclavine, synthesis of, 566

Cryptopleurine, synthesis of, 579

Curtin-Hammett principle, 56

Cyanide addition to iminium salts, 317, 320, 490–491

3-Cyano-1-(p-toluesulfonyl)-1,3-butadiene, reaction with enamines, 372

Cyanoacetic acid, reaction with, iminium salts, 326

Cyanoallene, reaction with enamines, 378–379

α-Cyanocinnamide, reaction with enamines, 408

Cyanogen azide, reaction with enamines, 415

Cyanogen halides, reaction with enamines, 223–224

Cycloheptene, conformations of, 34

Cyclohexene, conformations of, 35

N-Cyclohexylidinepyrrolidinium perchlorate, reaction with, diazomethane, 313–314

Cyclopentene, conformations of, 34

Cyclopropanes, reaction with enamines, 381–382

Cyclopropenes, reaction with enamines, 382–384

Cyclopropenones, reaction with enamines, 383–384

Cylindrocarine, synthesis of, 561

D

trans-2-Decalone, morpholine enamine of, reaction with DAD, 227–228

Decarbonylation, α-tertiary amino acids, 132

$\Delta^{2,3}$-Dehydro-3-ethyl-3-azonium-bicyclo[3.3.1]nonane, reaction with ethoxide, 322

dl-Dehydrocycloheximide, synthesis of, 46

Dehydronuciferine, reaction with:
  DMAD, 195
  methyl propiolate, 195

1(10)-Dehydroquinolizidine, synthesis of, 118–119

Dehydroquinuclidine, ionization potential, 13

Deoxycorynoline, synthesis of, 595

Deoxynupharidine, synthesis of, 669

Deprotonation of iminum salts, 328–331

1-(N,N-Di-n-butylamino)butene, oxidation of, oxygen, 256

Diazomethane, mechanism of reaction, with iminium salts, 314
  reaction with, iminium salts, 313–319
  stereochemistry of, iminium salt reaction, 314–315
Diazonium salts, aromatic, reaction with enamines, 224–226
Dibenzalacetone, reaction with enamines, 369
Dibenzoylperoxide, reaction with enamines, 485
Diborane, reaction with enamines, 228–229, 266–268, 642–644, 672
  reduction of, iminium salts, 302, 307
Dibromocarbene, reaction with enamines, 389
Dichlorocarbene, reaction with enamines, 388–389, 552
Diethyl azodicarboxylate (DAD), reaction with enamines, 227–228
Diethyl maleate, reaction with enamines, 364–366
Diethyl methylenemalonate, reaction with enamines, 368
Diethyl phosphite, reaction with, iminium salts, 323
Diethyl phosphorocyanidate, reaction with enamines, 321
Diethyl pyrocarbonate, reaction with enamiens, 215
Diethyl vinylphosphate, reaction with enamines, 379
1-(N,N-Diethyl)-1,3-butadiene, reaction with diphenyl-cyclopropenone, 383
1-(N-Diethylamino)-1,3-butadiene, reaction with:
  2,4-pentadienoate, 368
  sulfene, 401–402
1-(N,N-Diethylamino)cyclohexene: disproportionation of, 250–251

[1-(N,N-Diethylamino)cyclohexene] synthesis of, 105, 131
Diethylazodicarboxylate, reaction with enamines, 43
α-Dihydrocaranone, synthesis of, 598
Dihydrocorynantheine, synthesis of, 668
Dihydrocorynantheol, synthesis of, 567
Dihydrodeoxyepiallocermine, synthesis of, 173
5,6-Dihydroflavoperine, synthesis of, 689
Dihydrojoubertiamine, synthesis of, 582
Diisobutyl aluminium hydride, reaction with enamines, 229
Diisobutylaluminum hydride (DIBAL), reduction of amides, 135
Diketene, reaction with enamines, 394
Dimethyl acetylenedicarboxylate (DMAD), reaction with enamines, 194–195, 385–386, 573
Dimethyl bromomesconate, reaction with enamines, 369
Dimethyl fumarate, reaction with enamines, 364–366
Dimethyl maleate, reaction with enamines, 364–366
N,N-Dimethyl(1-isopropyl-2-methylpropenyl)amine, synthesis of, 143
Dimethyl(diazomethyl)phosphonate, reaction with, amine and ketone, 141
1,2-Dimethyl-Δ^2-pyrroline, reaction with, 3-buten-2-one, 360
1,2-Dimethyl-Δ^2-tetrahydropyridine, reaction with, ethyl acrylate, 367
1,2-Dimethyl-3-allyl-Δ^2-tetrahydropyridine, 28

1(N,N-Dimethylamino)1-phenylethene, oxidation of, electrolytic, 248

1-(N,N-Dimethylamino)2,5-dimethylcyclohexene, oxidation of, electrolytic, 248

1-(N,N-Dimethylamino)2-methyl-1-propene:
  alkylation of, 184
  reaction with:
    3-buten-2-one, 356
    p-benzoquinone, 198
    diethyl methylenemalonate, 368
    dimethylketene, 375
    DMAD, 194–195
    N-ethyl maleimide, 369
    ethylenetricarboxylate, 366–367
    fumaronitrile, 371
    ketenes, 206
    methyl acrylate, 363
    methyl propiolate, 194
    paraformaldehyde, 199
    phenyl isocyanate, 218
    phenylacetylene, 194
    phenylisothiocyanate, 397–398
    thiete 1,1-dioxide, 399

2-(N,N-Dimethylamino)-3,3-dimethyl-1-butene, synthesis of, 143

2-(N,N-Dimethylamino)bicyclo-[2.2.1]heptene, reaction with, ethyl propynoate, 387–388
  synthesis of, 142

1-(N,N-Dimethylamino)cyclohexene:
  disproportionation of, 251
  oxidation of, electrolytic, 248
  photoelectron spectrum, 4
  synthesis of, 105

1-(N,N-Dimethylamino)cyclopentene, synthesis of, 105

1-(N,N-Dimethylamino)ethene, reaction with, ethylenetricarboxylate, 366–367

6-(Dimethylamino)fulvene, synthesis of, 147, 288

N,N-Dimethylamino-2-propene, isomerization of, 27

N,N-Dimethylaminoethene:
  geometry of, 15–16
  orbital energies of, 5–6, 8

N,N-Dimethyleniminium ion:
  MNDO calculations, 276
  STO-3G calculations, 276

N,N-Dimethylisopropylideniminium ion, MNDO calculations, 277

N,N-Dimethylisopropylideniminium perchlorate:
  synthesis of, 286
  x-ray diffraction of, 277

1,4-(N,N-Dimorpholino)cyclohexa-1,3-diene, reaction with, cyanoallene, 378–379

2,4-Dinitrochlorobenzene, reaction with enamines, 196

2,4-Dinitrofluorobenzene, reaction with enamines, 590

Diphenylacetylene, photochemical reaction, with enamines, 386

Diphenylcarbene, reaction with enamines, 389

Diphenylcyclopropenone, reaction with enamines, 383–384

Diphenylnitrilimine, reaction with enamines, 412

Diphenylphosphine oxide, reaction with, iminium salts, 324

2,3-Diphenylthiirene 1,1-dioxide, reaction with enamines, 398–399

1,3-Dipolar cycloadditions, 410–415

Dipole moment, of enamines, 17–19, 22

Disproportionation, of enamines, 250–251, 493

1,2-Dithiole-3-thione, reaction with enamines, 397

E

Eburnamonine, synthesis of,
555, 642
Electrocyclic reaction, con-
rotary, 409–410
Electroreduction, of N-methylglutarimide, 452
Ellipticine, synthesis of, 548,
590, 671
Emetine, synthesis of, 601
Enthalpy of formation, of
enamines, 28
Enthalpy of hydrogenation, of
enamines, 28–29
2,7-Epiperhydrohistrionicotoxin,
synthesis of, 587
Ervinceine, synthesis of, 559
Ervistine, synthesis of, 588
Eschenmoser salt, synthesis of,
285
Ethenesulfonyl fluoride, reaction with enamines, 379
Ethoxide, reaction with,
iminium salts, 322
(Ethoxycarbonyl)nitrene,
reaction with enamines,
405
N-Ethoxyquinolinium salts,
reaction with enamines,
380
Ethyl acrylate, reaction with
enamines, 362, 367,
487–484
Ethyl chloroformate, reaction
with enamines, 480, 510
Ethyl cyclopropyl-1,1-cyano-
carboxylate, reaction
with enamines, 381
Ethyl diazoacetate, reaction
with enamines, 415
20-Ethyl eburnamonine, synthesis of, 564
Ethyl p-nitrobenzoate, oxidation of enamines, 252
Ethyl propynoate, reaction with
enamines, 387–388

F

Festaclavine, synthesis of, 566
Flavopereirine perchlorate,
synthesis of, 565
1-Fluoro-2-pyridone, reaction
with enamines, 222
Fluorochlorocarbene, reaction
with enamines, 389
β-Fluoroenamines, synthesis of,
106
Formic acid, reduction of:
enamines, 262, 487–489
iminium salts, 309–313, 487
Frontier molecular orbital theory,
349–351, 414
Fumaronitrile, reaction with
enamines, 371

G

Geissoschizine, synthesis of,
666
Grignard reagents, reaction
with:
iminium salts, 297–302, 489–
490
N-methyl lactams, 136

H

Haliotridane, synthesis of, 659
Halogenation, of enamines,
220–223
Hammett equation:
for enamine hydrolysis, 171,
177
for phenyl azide, addition to
enamines, 414–415
Hantzsch ester, reduction of,
iminium salts, 262–263,
308–309
Hemiaminals, 110–114
3,5-Hexadien-2-one, reaction with
enamines, 360

Hexafluoroacetone azine, reaction with enamines, 408
Hexafluoropropene oxide, reaction with enamines, 214–215
Hexahydrojulolidine, oxidation of, mercuric acetate, 127
2-(N-Hexamethylenimino)-bicyclo[2.2.1]hept-2-ene, synthesis of, 115
Hexamethylphosphoric triamide (HMPT), reaction with ketones, 139
Hippocasine, synthesis of, 642
Hippodamine, synthesis of, 642
Homoenamine, 30
Hydroboration, of enamines, 229, 642–644, 672
Hydrogen peroxide, reaction with enamines, 254–255
Hydrogen sulfide, reaction with enamines, 396
Hydrogenation, of enamines, 268, 659
Hydrogenolysis, of enamines, 228–229
Hydrolysis, of enamines, 165–179
Hydroxylamine-O-sulfonic acid, oxidation of enamines, 262
Hyellazole, synthesis of, 538
Hygrine, synthesis of, 500–501
Hypophosphorous acid, reduction of, iminium salts, 263

I

20-epi-Ibophyllidine, synthesis of, 560
Imine, reaction with, enamines, 405

Iminium salts:
infrared spectra of, 289–293
nmr spectra of, 294–298
preparation of, 277–288
reaction with:
azides, 324
t-butylamine borane, 262, 302
carboxyhydridoferrates, 262
cyanide, 317, 320, 490–491
cyanoacetic acid, 326
diazomethane, 313–319
diborane, 302, 307
diethyl phosphite, 323
diphenylphosphine oxide, 324
ethoxide, 322
formic acid, 309–313, 487
Grignard reagents, 297–302, 489–490
hypophosphorous acid, 263
lithium alum. hydride, 262, 302, 486
NADH, 262
NADPH, 262–263
nitroacetic acid, 326
organometallic reagent, 297–303
phosphorous acid, 263, 302
potassium borohydride, 262, 268, 302
secondary amines, 264–265
L-selectride, 262
sodium cyanoborohydrid, 262, 302, 305
sodium hydrosulfite, 302
trichloroacetate, 324–325
trifluoroacetic acid, 322
triphenylphosphine, 266, 323–324
ynamine, 323
reduction of, 302–313
structure of, 276–277
ultraviolet spectra of, 288–289
Infrared spectra:
of enamines, 60–65, 460–462, 464, 470
of iminium salts, 289–293

Ionization potential:
  of amines, 4–13
  of enamines, 4–13
Ipalbidine, synthesis of, 578
Isoacorone, synthesis of,
  550
Isobellendine, synthesis of,
  537
Isobenzofuroxan, reaction with
  enamines, 407
Isocyanates, reaction with,
  enamines, 218–219
1-Isocyanato-1-cyclohexene,
  reaction with enamines,
  404–405
Isodihydroenpetalactone, synthesis of, 677
Isofumigaclavine, synthesis
  of, 603
Isomerization, of allyl amines,
  27–28, 456–458
Isonitrosomalonitrile tosylate,
  reaction with enamines,
  232
Isopropyleine, synthesis of,
  642
Isoretronecanol, synthesis of,
  661, 674, 676

J

Julolidine, reduction of, 135

K

Ketenes, reaction with enamines,
  204–206, 374–378
Kenteniminium salts:
  cycloaddition with
    alkenes, 331
    alkynes, 332
    imines, 332
  deprotonation of, 331
Kuskhygrine, synthesis of,
  500–501

L

Lead tetraacetate, oxidation of
  enamines, 251–252
Leurosidine, synthesis of, 647
Linear free energy equation, to
  predict E/Z ratios, in
  enamines, 58, 60
Lithium aluminum diethoxyhydride,
  135
Lithium aluminum hydride, reduction of:
  amides, 134–135
  iminium salts, 134, 262, 302–
    308, 486
Lithium tri-sec-butylborohydride
  (L-Selectride), reduction
  of, iminium ions, 262
Lupinine, synthesis of, 672
α-Lycorane, synthesis of, 546
Lysergic acid, synthesis of, 602,
  603

M

MNDO calculations:
  of enamines, 5, 7, 16–19
  of iminium salts, 276–277
Mannich reaction, 326–327
  regioselectivity, 327
  solvent effect, 327
Mass spectra, of enamines, 74–77
Matridine, synthesis of, 664
Mechanism, of enamine formation,
  110–114
Meerwein-Ponndorf reduction, of
  iminium salts, 308
Meisenheimer complex, 391–392
Mercuric acetate:
  oxidation of amines, 118–129,
    453–455, 536, 543
  mechanism of, 123–129
  oxidation of enamines, 252
Metalloenamines, 25
Methoxyphenylcarbene, reaction
  with enamines, 389

Methoxyphenyllead triacetate, reaction with enamines, 196—197

11-Methoxytabersonine, synthesis of, 642

Methuenine, synthesis of, 588

Methyl 4-trimethylsilyl-3-morpholinocrotonate, reaction with enamines, 390—391

Methyl 4-trimethylsilyl-3-pyrrolidinocrotonate, reaction with enamines, 390—391

Methyl acrylate, reaction with enamines, 47, 362—363

Methyl cinnamate, reaction with enamines, 363—364

Methyl crotonate, reaction with enamines, 363—364

Methyl iodide, reaction with enamines, 43—44, 474, 477—478

Methyl methacrylate, reaction with enamines, 363—364

Methyl phenylpropiolate, reaction with enamines, 388

Methyl propiolate, reaction with enamines, 194—195, 386—387

Methyl trans-2,4-pentadienoate, reaction with enamines, 367—368

Methyl vinyl ketone, (see 3-buten-2-one)

Methyl vinyl sulfone, reaction with enamines, 372

N-Methyl-1,2,3,4-tetrahydro-pyridine, ionization potential, 13

N-Methyl-1,2-dihydropyridine, reaction with methyl acrylate, 364

6-Methyl-1-(N-pyrrolidino)-cyclohexene, reaction with acrylonitrile, 370

1-Methyl-2,2,6,6-tetramethyl-piperidine, oxidation of, mercuric acetate, 121

1-Methyl-2-(3-butenyl)-Δ^2-tetrahydropyridine, 28

1-Methyl-2-phenyl-1-azacyclohept-2-ene, synthesis of, 136—137

3-Methyl-3-azabicyclo[3.3.1]-nonane, oxidation of, mercuric acetate, 128

1-(N-Methylanilino)cycloheptene, photolysis of, 409

1-(N-Methylanilino)cyclohexene:
  photolysis of, 409
  reaction with, (ethoxycarbonyl)-nitrene, 405

1-(N-Methylanilino)cyclopentene, photolysis of, 409

N-Methylene-t-butylamine, reaction with enamines, 406

Methylenecyclohexane, geometry of, 35—36

N-Methylisopelletierine, synthesis of, 501

1-Methylpiperidine, oxidation of, mercuric acetate, 121

1-Methylpyrrolidine, oxidation of, mercuric acetate, 122

4-Methylquinolizidine, oxidation of, mercuric acetate, 128

Molecular mechanics calculations, of enamines, 15

Molecular sieves, in synthesis, 143

Monomorine, synthesis of, 671

1-(N-Morpholine)cyclohexene, reaction with, acryloyl chloride, 362

2,5-bis(N-Morpholine)tricyclo-[2.2.1.02,6]heptane, synthesis of, 116

1-Morpholine-2-nitroethene, synthesis of, 149

(E)-1-(N-Morpholino)-1,2-di-phenylethene, oxidation of, electrolytic, 248

1-(N-Morpholino)-1,3-butadiene,
reaction with, α-chloro-
acrylonitrile, 368
2-(N-Morpholino)-1,3-diphenyl-
propene, reaction with,
tosyl azide, 226
2-(N-Morpholino)-1-butene,
reaction with acrolein, 361
1-(N-Morpholino)-1-phenylethene,
reaction with:
  p-quinone, 392
  sulfur, 395
1-(N-Morpholino)-1-propene,
reaction with, 1-nitroso-
2-naphthol, 408
1-(N-Morpholino)-2,6-dimethyl-
cyclohexene, ozonolysis
of, 254
1-(N-Morpholino)-2-methyl-1-
propene, reaction with:
benzylideneaniline, 405—406
chloral, 199
ketene, 376
perfluoroolefins, 392
phenyl seleniocyanate, 231
sulfonyl imides, 399
2-(N-Morpholino)-3,3-dimethyl-
1-butene:
ozonolysis of, 253—254
synthesis of, 143
1-(N-Morpholino)-4-t-butylcyclo-
hexene, reaction with:
benzylideneaniline, 201
diethylmaleate, 365-366
sulfene, 400—401
1-(N-Morpholino)-6-acetylcyclo-
hexene, oxidation of,
hydrogen peroxide, 255
1-(N-Morpholino)butene, reaction with ketene, 376
1-(N-Morpholino)cycloheptene:
alkylation of, 190—191
oxidation of, thallic acetate,
251
1-(N-Morpholino)cyclohexene:
acylation of, 207—211
alkylation of, 192, 193

[1-(N-Morpholino)cyclohexene:]
aromatization of, 261—262
disproportionation of, 250
oxidation of:
  benzoylperoxide, 252
  electrolytic, 249
  nitrosyl chloride, 252
  oxygen, 256—258
  thallic acetate, 251
  trityl ion, 262
reaction with:
  carbon disulfide, 396
  chloral, 393
  cinnamaldehyde, 362
  1-isocyanato-1-cyclohexene,
    404—405
  cyanoallene, 378
  DAD, 227
  dichlorocarbene, 388
  1-fluoro-2-pyridone, 222
  hydrogen sulfide, 396
  2-hydroxy-1-arylaldehyde,
    201
  isobenzofuroxan, 407
  ketene, 377—378
  nitrilium salts, 203
  nitroethylene, 371
  1-nitroso-2-naphthol, 408
  o-quinone, 393
  pentacarbonyltuhgsten, 230
  phenyl isocyanate, 218—219
  salicylaldehyde, 394
  sulfur, 394—395
  tetracyanoethylene, 380
1-(N-Morpholino)cyclononene,
reaction with ketene, 377—
378
1-(N-Morpholino)cyclopentene:
oxidation of:
  benzoylperoxide, 252
  electrolytic, 249
  thallic acetate, 251
reaction with:
  acyl Cl, 535
  chloral, 393
  dichlorocarbene, 388
  ketene, 377—378

[1-(N-Morpholino)cyclopentene:]
  phenyl azide, 413–414
  phenyllead triacetate,
    196–197
  thioacetic acid, 322
2-(N-Morpholino)indene, syn-
    thesis of, 108
exo-5-Morpholinobicyclo[2.2.1]-
    heptan-2-one, synthesis of,
    115–116

N

NADH, reaction with iminium
    salts, 262
NADPH, reaction with iminium
    salts, 262–263
Nauclefine, synthesis of, 599
Nitric acid, oxidation of en-
    amines, 255
Nitrile oxide, reaction with
    enamines, 412
Nitrilimine, reaction with
    enamines, 411–412
Nitrilium salts, reaction with
    enamines, 203
(E)-2-Nitro-2-heptene-1-yl
    pivalate, reaction with
    enamines, 372
6-Nitro-2-quinoxalone, reaction
    with enamines, 232
Nitroacetic acid, reaction with,
    iminium salts, 326
Nitrobenzenesulfenyl chlorides,
    reaction with enamines,
    217
beta-Nitroenamines, 17–18, 21–
    23
  infrared spectrum of, 23
Nitroethylene, reaction with
    enamines, 371
Nitrone, reaction with en-
    amines, 411–412
2-Nitropropene, reaction with
    enamines, 408
5-Nitropyrimidine, reaction with
    enamines, 407

1-Nitroso-2-naphthol, reaction
    with enamines, 408
Nitrosobenzene, reaction with
    enamines, 231–232
β-Nitrostyrene, reaction with
    enamines, 372
Nitrosyl chloride, oxidation of
    enamines, 252
Nitrous acid, oxidation of
    enamines, 253
Nuciferine, synthesis of, 656
Nuclear magnetic resonance s
    spectra:
  of enamines, 66–74, 465
  equation to predict:
    carbon chemical shift, 74
    proton chemical shift, 69

O

Osmium tetroxide, oxidation of
    enamines, 251, 647, 650,
    651
1,3-Oxazolidine, reaction with
    enamines, 407
Oxidation:
  of enamines, 247–262, 640–
    656
    benzoylperoxide, 252
    dibenzoylperoxide, 485
    electrolytic, 248–249
    ethyl p-nitrobenzoate, 252
    hydrogen peroxide, 254–255
    lead tetraacetate, 251–252
    mercuric acetate, 252
    nitric acid, 255
    nitrosyl chloride, 252
    nitrous acid, 252
    osmium tetroxide, 251
    oxygen, 256–261
    oxone, 253–254
    potassium permanganate, 253
    ruthenium tetroxide, 255
    selenium dioxide, 252
    sodium dichromate, 255
    sodium periodate, 255
    trityl ion, 262

[Oxidation]
 of tertiary amines, benzoyl-
      peroxide, 129
   N-bromosuccinimide, 129
   chlorine dioxide, 129
   copper(II) chloride, 129
   diethyl azodiformate, 129
   iodine, 129
   iodine pentafluoride, 129
   manganese dioxide, 129
   mercuric acetate, 118--129,
      453—455
   ozone, 129
   palladium(II)chloride, 129
   permanganate, 129
   photooxidation, 130
   quinone, 129
   trityl ion, 134
Osonolysis, of enamines, 253-
      254

            P

Paraformaldehyde, reaction
      with enamines, 199
Parallel attack, of an
      electrophile, 37
trans-2,4-Pentadienoate, reac-
      tion with dienamines,
      368
tris(2,4-Pentanedionato)colbalt-
      (III), photoreaction with
      enamines, 393
Perchloryl fluoride, reaction
      with enamines, 223
Phenyl azide, reaction with
      enamines, 412—414
Phenyl vinyl sulfone, reaction
      with enamines, 372
Phenyl(trichloromethyl)-
      mercury, reaction with
      triethylamines, 141
Phenylacetylene, reaction with
      enamines, 194
Phenylcarbamoyldiimide, reaction
      with enamines, 408—
      409

Phenylchlorocarbene, reaction
      with enamines, 389
Phenylisocyanate, reaction with
      enamines, 403
Phenylisothiocyanate, reaction
      with enamines, 397—398
Phenylmethane sulfonyl chloride,
      reaction with enamines,
      215—216
4-Phenylquinolizidine, oxidation
      of, mercuric acetate, 128
Phenylseleniocyanate, reaction
      with enamines, 231
Phenylselenium chloride, reac-
      tion with enamines, 231
Phosgene, reaction with enamines,
      215, 408
Phosphorous acid, reduction of,
      iminium salts, 263—302
Photochemistry, 2,6-dimethyl-
      piperidine, 451
   of N-aryl enamines, 409—410
   of dienamines, 392
   of enamines with , dimethyl
      fumarate, 366
   diphenylacetylene, 386
   tris(acetylacetone)Co(III),
      393
   of iminium salts, 332—333
Photoelectron spectra, 4—7
Photooxygenation, of enamines,
      258—261, 394, 485, 650
Piperideines, synthesis of, 443—
      459
1-(N-Piperidino)-1-butene,
      reaction with:
   diazonium salts, 224—225
   methyl acrylate, 363
1-(N-Piperidino)-1-propene,
      reaction with, phenyl
      azide, 412—414,
1-(N-Piperidino)-2-methyl-1-
      propene, reaction with:
   diazonium salts, 225
   paraformaldehyde, 199
1-(N-Piperidino)-4-t-butylcyclo-
      hexene, reaction with sul-
      fene, 400—401

1-(N-Piperidino)cyclohexene, photoreaction, tris(acetylacetone)Co(III), 393
reaction with trichlorosilane, 325
1-(N-Piperidino)cyclopentene, reaction with iminium salts, 323
1-Piperidino-2-nitroethene, synthesis of, 149
1-Piperidinopropene, reaction with aryl sulfonyl chloride, 216
Polonovski reaction, 133
modified, 285
Polyphosphoric acid, reaction with enamines, 541
Potassium borohydride, reduction of, iminium ions, 268
Potassium permanganate oxidation of enamines, 253
Prezizaene, synthesis of, 581
Propyleine, synthesis of, 642, 679
Prosophylline, synthesis of, 649
Proton affinity:
of enamines, 78–82
of saturated amines, 78–82
Protonation, of enamines, 77–83, 166, 182, 277–281
regioselectivity, 280
stereospecificity, 281
Pseudohaliotridane, synthesis of, 659
(+)-Pumiliotoxin, synthesis of, 575
Pumilotoxin-C, synthesis of, 607
Pyramidality of enamines, 15
Pyroclavine, synthesis of, 566
Pyrovellorolactone, synthesis of, 573
1-(N-Pyrrolidino)-1-butene, reaction with, β-nitrostyrene, 372

1-(N-Pyrrolidino)-1-phenylethene, reaction with nitrone, 411–412
1-(N-Pyrrolidine)-2-ethyl-1,3-hexadiene, reaction with acrolein, 361–362
1-(N-Pyrrolidino)-2-methyl-1-propene, oxidation of, oxygen, 257
reaction with:
phosgene, 215
sulfene, 400
3-(N-Pyrrolidino)-2-pentene, reaction with, dihalodimethoxyalkanes, 390
1-(N-Pyrrolidino)-3,4-dihydronaphthalene, 46–47
reaction with, cyanogen halides, 223–224
1-(N-Pyrrolidino)-4-methylcyclohexene, reaction with, enone, 359
1-(N-Pyrrolidino)-4-t-butylcyclohexene:
alkylation of, 109
reaction with sulfene, 400–401
2-(N-Pyrrolidino)bicyclo[2.2.1]-hept-2-ene, reaction with, thiirene, 1,1-dioxides, 398–399
1(N-Pyrrolidino)cycloheptene, reaction with:
acrolein, 361
methyl propiolate, 387
1-(N-Pyrrolidino)cyclohexene:
alkylation of, 190
dipole moment of, 22
disproportionation of, 250–251
oxidation of:
ethyl p-nitrobenzoate, 252
lead tetraacetate, 251–252
oxygen, 256
reaction with:
2,4-pentadienoate, 367–368
3-buten-2-one, 354–355
acrolein, 360–361

[1-(N-Pyrrolidino)cyclohexene]
  acrylonitrile, 370
  benzonitrile oxide, 412
  benzyne, 197–198, 388
  N-carbethoxyaziridine, 405
  cinnamaldehyde, 362
  1-isocyanato-1-cyclohexene,
    404–405
  cyanoacetic acid, 326
  cyanogen chloride, 223
  cyclopropylcyanoester, 381
  diazonium salts, 226
  dibenzalacetone, 369
  diborane, 229
  dimethyl acetylenedi-
    carboxylate, 385
  dimethyl bromomesconate,
    369
  dimethyl maleate, 364–365
  2,4-dinitrochlorobenzene,
    196
  diphenylnitrilimine, 412
  ethyl diazoacetate, 415
  iron carbonyl complex,
    229–230
  phenyl azide, 413–414
  sulfonyl chlorides, 216
  trichloroacetate, 324–325
1-(N-Pyrrolidino)cyclooctene,
    reaction with, methyl
    propiolate, 387
1-(N-Pyrrolidino)cyclopentene:
  dipole moment of, 22
  ionization potential, 12
  reaction with:
    1,2,4-triazine, 408
    1,2-dithiole-3-thione, 397
    1,3,5-triazine, 408
    1,4-diiodobutane, 389
    5-nitropyrimidine, 407
    acrolein, 361
    dimethyl acetylenedica, 386
    methyl propiolate, 386–387
    phenyl azide, 413
1-(N-Pyrrolidino)cycloundecene,
    reaction with methyl
    propiolate, 387

1-(N-Pyrrolidino)dodecene, reac-
    tion with methyl pro-
    piolate, 387
1-(N-Pyrrolidino)indene, 49–51
  synthesis of, 108
1-N-Pyrrolidino-1-cyclohexene,
    alkylation of, 186–187
1-N-Pyrrolidino-1-cyclopentene,
    alkylation of, 187
1-N-Pyrrolidino-2-methyl-1-
    propene, alkylation of,
    184
α-N-Pyrrolidylmethylenecyclo-
    hexane, reaction with,
    dimethyl maleate, 365
Pyrrolines, synthesis of, 443–
    459

Q

Quinolizidine, oxidation of,
    mercuric acetate, 118–119,
    124, 536
Quinone dibenzenesulfonimide,
    reaction with enamines,
    407
p-Quinone, reaction with en-
    amines, 392, 393
Quinuclideine, 466, 472

R

Raman spectra, of nitroenamines,
    23
Rearrangement, of iminium salts,
    333
Red-al, reduction of, iminium
    salts, 302
Reduction:
  of enamines, 115, 123, 247,
    250–251, 262–268, 657–676
    aluminum hydride, 228–229,
    266
    9-BBN, 266

[Reduction]
   diborane, 229, 266-267
   formic acid, 262, 487–
      489
   hydrogenation, 268
   secondary amines, 115
   sodium borohydride, 123,
      536
  of iminium salts, 302–313
   carboxyhydridoferrates,
      262, 302
   diborane, 262, 302
   formic acid, 309–313, 487
   hantzsch ester, 262, 308
   hypophosphorous acid, 263
   lithium alum. hydride,
      262, 302, 486
   NADPH or NADH, 262–263
   phosphorous acid, 263, 302
   potassium borohydride, 262,
      302
   red-al, 262, 302
   secondary amines, 264–266
   L-selectride, 262, 302
   sodium borohydride, 262,
      302
   sodium cyanoborohydride,
      262, 302, 305
   sodium hydrosulfite, 302
   t-butylamine borane, 262,
      302
   triphenylphosphine, 266
   zinc/acetic acid, 262, 302
  of isoquinolinium salt, lithium
      alum. hydride, 134
   sodium borohydride, 134
  of lactams, DIBAL, 135
  of pyridine, 451
  of pyridinium salts, 451–452
Rosaramycin aglycone, synthe-
      sis of, 604
Ruthenium tetroxide, oxidation
      of enamines, 255

S

Salicylaldehyde, reaction with
      enamines, 394

Sanguinarine, synthesis of, 576
Sconlerine, synthesis of, 576
Sedamine, synthesis of, 503
L-Selectride, reduction of,
      iminium salts, 302
Selenium dioxide, oxidation of
      enamines, 252
Sendaverine, synthesis of, 663
Septicine, synthesis of, 577
[3,3]-Sigmatropic rearrangements,
      28, 30, 33, 494–495
  of iminium salts, 333
Sodium bis(2-methoxyethoxy)-
      aluminium hydride, reduc-
      tion of, iminium salts, 262
Sodium borohydride, reduction
      of, iminium salts, 134,
      262, 302–307, 486
Sodium cyanoborohydride, re-
      duction of, iminium salts,
      262, 302, 305
Sodium hydrosulfite, reduction
      of iminium salts, 302
Sodium periodate, oxidation of
      enamines, 255
Sommelet reaction, 263–264
Sophoramine, synthesis of, 664
Sparteine, oxidation of, mercuric
      acetate, 119, 125
  synthesis of, 586
Stereoelectronic effects, electro-
      philic attacks, on enamines,
      37
Styrene oxide, reaction with
      enamines, 393
1-Styryl-2-pyrrolidone, reduc-
      tion, lithium alum. hydride,
      135
Sulfenes, reaction with enamines,
      399–402
Sulfonyl imides, reaction with
      enamines, 399
Sulfonylcarbodiimide, reaction
      with enamines, 408
Sulfur, reaction with enamines,
      394–396
Synchronous cycloadditions,
      theory of, 349

T

Tetracyanoethylene, reaction with enamines, 380

Tetrahydroanabasine, 498

trans-Tetrahydrobenzophen-anthridor, synthesis of, 592

1,2,5,5-Tetramethylpyrrolidine, oxidation of, mercuric acetate, 122

Thallic acetate, oxidation of enamiens, 251

Thermodynamic properties, of enamines, 27−30, 458, 462

1,3,4-Thiadiazolium, reaction with enamines, 196

Thiete 1,1-dioxide, reaction with enamines, 399

Thiocarbene, reaction with enamines, 389

p-Thiocresol potassium salt, reaction with iminium salts, 321

Thiolacetic acid, reaction with enamines, 322

Titanium tetrachloride, in enamine synthesis, 114, 117−118, 544, 547, 643
reaction with, amine & ketone/aldehyde, 143−145

p-Toluenesulfonyl chloride, reaction with enamines, 216

Torreyol, synthesis of, 576

Torsional twist of enamines, 15

Tosyl azide, reaction with enamines, 226

Trachelanthamidine, synthesis of, 676

Transannular reactions, of enamines, 33

1,2,4-Triazine, reaction with enamines, 408

Trichloroacetate, reaction with, iminium salts, 324−325

Trichlorosilane, reaction with enamines, 325

Triethyl orthoformate, reaction with enamines, 215

Trifluoroacetic acid, reaction with, iminium salts, 322

Trifluoroacetic anhydride, reaction with enamines, 209

Trifluoroacetonitrile, reaction with enamines, 408

Trimethyl ethylenetricarboxylate, reaction with enamines, 366−367

1,2,5-Trimethyl-Δ^2-pyrroline, synthesis of, 130

1,3,3-Trimethyl-2-methylenein-doline, acylation of, 210, 479
alkylation of, 188
reaction with:
cyanogen bromide, 224
diazonium salts, 226

Trimethylene dithiotosylate, oxidation of enamines, 253

1,4,4-Trimethylpiperidine, oxidation of, mercuric acetate, 121, 122, 127−128

1,3,4-Trimethylpyrrolidine, oxidation of, mercuric acetate, 122

Triphenylphosphine, reaction with iminium salts, 266 , 323−324

Tripyrrolidinoarsine, reaction with ketones or aldehydes, 142

Tris(dimethylamino)arsine, reaction with ketones or aldehydes, 141−142

Tris(dimethylamino)borane, reaction with, ketone & dimethylamine, 141

Tris(dimethylamino)phosphine, reaction with ketones, 139

Tris(pyrrolidinyl)borane, reaction with, ketone and pyrrolidine, 141
Trityl tetrafluoroborate, oxidation of enamines, 262
3-Tropanone, morpholine enamine of, 537
Two-stage cycloadditions, theory of, 349–352
Two-step cycloadditions, theory of, 348–352

U

Ultraviolet spectra, of enamines, 66, 462, 465, 467

V

Velleral, synthesis of, 573
Vilsmeier salt, reaction with enamines, 203
Vilsmeier-Haack reaction, 147, 202–203, 540
Vinblastine, synthesis of, 647
Vincadifformine, synthesis of, 540, 559, 562
Vincamine, synthesis of, 545, 547
N-Vinylaziridine, 27
2-Vinylpyridine, reaction with enamines, 379
Voaenamine, synthesis of, 551

W

Wiseman's rule (see Bredt's rule)
Wittig reaction, 139–140
  Horner modification of, 140–141
  Wadsworth-Emmons, 140
Woodward-Hoffman rules, 351–352

X

X-ray crystallography:
  of enamines, 14, 20–21
  of iminium salts, 277

Y

Ynamine, reaction with, iminium salts, 323
Yohimbane, synthesis of, 565
Yohimbine, oxidation of, mercuric acetate, 123, 126–127, 453
  synthesis of, 594

Z

Zeise's dimer, reaction with enamines, 231